Tsunami Science:
Ten years after the 2004 Indian Ocean Tsunami
Volume II

Edited by
Alexander B. Rabinovich
Eric L. Geist
Hermann M. Fritz
Jose C. Borrero

Previously published in *Pure and Applied Geophysics* (PAGEOPH), Volume 172, No. 12, 2015

Editors
Alexander B. Rabinovich
Institute of Ocean Sciences
Department of Fisheries and Oceans
9860 West Saanich Road
Sidney, British Columbia
V8L 4B2, Canada
and
P.P. Shirshov Institute of Oceanology
Russian Academy of Sciences
Nakhimovsky Prosp. 36
117997 Moscow
Russia

Eric L. Geist
U.S. Geological Survey
345 Middlefield Road
Menlo Park, California 94025
USA

Hermann M. Fritz
Georgia Institute of Technology
School of Civil and Environmental
Engineering
790 Atlantic Drive
Atlanta, Georgia 30332
USA

Jose C. Borrero
eCoast Marine Consulting and Research
47 Cliff Street
3225 Raglan
New Zealand

ISBN 978-3-0348-0959-7   ISBN 978-3-0348-0960-3 (eBook)
DOI 10.1007/978-3-0348-0960-3

Library of Congress Control Number: 2015934904

Springer Basel Heidelberg New York Dordrecht London
© Springer Basel 2016
This work is subject to copyright. All rights are reserved, whether the whole or part of the material is concerned, specifically the rights of translation, reprinting, re-use of illustrations, recitation, broadcasting, reproduction on microfilms or in other ways, and storage in data banks. For any kind of use, permission of the copyright owner must be obtained.

Cover illustration: © 2005 Space Imaging LLC, ClearView License. Fritz, H.M. and J.C. Borrero (2006). Somalia field survey of the 2004 Indian Ocean, Tsunami. Earthquake Spectra 22(S3):S219-S233.

Cover design: deblik, Berlin.

Printed on acid-free paper

Springer Basel AG is part of Springer Science+Business Media

www.springer.com

Contents

1 Introduction to "Tsunami Science: Ten Years after the 2004 Indian Ocean Tsunami. Volume II."
A.B. Rabinovich · E.L. Geist · H.M. Fritz · J.C. Borrero

7 Evaluation of Intensity of Recent Seismogenic Tsunamis in the World Ocean from 2000 to 2014
V.K. Gusiakov

17 Deep-Ocean Measurements of Tsunami Waves
A.B. Rabinovich · M.C. Eblé

49 A Decade After the 2004 Indian Ocean Tsunami: The Progress in Disaster Preparedness and Future Challenges in Indonesia, Sri Lanka, Thailand and the Maldives
A. Suppasri · K. Goto · A. Muhari · P. Ranasinghe · M. Riyaz · M. Affan · E. Mas · M. Yasuda · F. Imamura

79 Field Survey of the 1945 Makran and 2004 Indian Ocean Tsunamis in Baluchistan, Iran
E.A. Okal · H.M. Fritz · M.A. Hamzeh · J. Ghasemzadeh

93 OSL Dating and GPR Mapping of Palaeotsunami Inundation: A 4000-Year History of Indian Ocean Tsunamis as recorded in Sri Lanka
R. Premasiri · P. Styles · V. Shrira · N. Cassidy · J.-L. Schwenninger

121 Tsunami Impact Computed from Offshore Modeling and Coastal Amplification Laws: Insights from the 2004 Indian Ocean Tsunami
H. Hébert · F. Schindelé

145 Recent Advances in Agent-Based Tsunami Evacuation Simulations: Case Studies in Indonesia, Thailand, Japan and Peru
E. Mas · S. Koshimura · F. Imamura · A. Suppasri · A. Muhari · B. Adriano

161 Interrogation of the Megathrust Zone in the Tohoku-Oki Seismic Region by Waveform Complexity: Intraslab Earthquake Rupture and Reactivation of Subducted Normal Faults
S.K.Y. Lui · D. Helmberger · S. Wei · Y. Huang · R.W. Graves

175 Rapidness and Robustness of Finite-Source Inversion of the 2011 M_w 9.0 Tohoku Earthquake by an Elliptical-Patches Method Using Continuous GPS and Acceleration Data
T. Ulrich · H. Aochi

191 Parallel Implementation of Dispersive Tsunami Wave Modeling with a Nesting Algorithm for the 2011 Tohoku Tsunami
T. Baba · N. Takahashi · Y. Kaneda · K. Ando · D. Matsuoka · T. Kato

209 Investigation of Hydrodynamic Parameters and the Effects of Breakwaters During the 2011 Great East Japan Tsunami in Kamaishi Bay
C. Ozer Sozdinler · A.C. Yalciner · A. Zaytsev · A. Suppasri · F. Imamura

229 On the Leading Negative Phase of Major 2010–2014 Tsunamis
M.C. Eblé · G.T. Mungov · A.B. Rabinovich

245 The Great 2006 and 2007 Kuril Earthquakes, Forearc Segmentation and Seismic Activity of the Central Kuril Islands Region
B.V. Baranov · A.I. Ivashchenko · K.A. Dozorova

273 Estimation of Extreme Sea Levels for the Russian Coasts of the Kuril Islands and the Sea of Okhotsk
G. Shevchenko · T. Ivelskaya

293 Evaluation of the Relationship Between Coral Damage and Tsunami Dynamics; Case Study: 2009 Samoa Tsunami
D.I. Dilmen · V.V. Titov · G.H. Roe

309 Did an underwater landslide trigger the June 22, 1932 tsunami off the Pacific coast of Mexico?
N. Corona · M.-T. Ramírez-Herrera

325 Far-Field Tsunami Impact in the North Atlantic Basin from Large Scale Flank Collapses of the Cumbre Vieja Volcano, La Palma
B. Tehranirad · J.C. Harris · A.R. Grilli · S.T. Grilli · S. Abadie · J.T. Kirby · F. Shi

353 Earthquake Scenario-Based Tsunami Wave Heights in the Eastern Mediterranean and Connected Seas
O. Necmioglu · N.M. Özel

375 Tsunami Squares Approach to Landslide-Generated Waves: Application to Gongjiafang Landslide, Three Gorges Reservoir, China
L. Xiao · S.N. Ward · J. Wang

391 Quantification of Tsunami Bathymetry Effect on Finite Fault Slip Inversion
Q. Bletery · A. Sladen · B. Delouis · L. Mattéo

Introduction to "Tsunami Science: Ten Years after the 2004 Indian Ocean Tsunami. Volume II."

ALEXANDER B. RABINOVICH,[1,2] ERIC L. GEIST,[3] HERMANN M. FRITZ,[4] and JOSE C. BORRERO[5,6]

Abstract—Twenty papers on the study of tsunamis and respective tsunamigenic earthquakes are included in Volume II of the PAGEOPH topical issue "Tsunami Science: Ten Years after the 2004 Indian Ocean Tsunami". The papers presented in this second of two special volumes of Pure and Applied Geophysics reflect the state of tsunami science during this time, including five papers devoted to new findings specifically in the Indian Ocean. Two papers compile results from global observations and eight papers cover Pacific Ocean studies, focusing mainly on the 2011 Tohoku earthquake and tsunami. Remaining papers in this volume describe studies in the Atlantic Ocean and Mediterranean Sea and tsunami source studies. Overall, the volume not only addresses the pivotal 2004 Indian Ocean and 2011 Tohoku tsunamis, but also examines the tsunami hazard posed to other critical coasts in the world.

Key words: 2004 Indian Ocean and 2011 Tohoku earthquakes and tsunamis, source parameters, Pacific and Indian oceans, Mediterranean Sea, DART, tsunami warning system, tsunami records, tsunami modeling, spectral analysis.

1. Introduction

The Indian Ocean tsunami of 26 December 2004 was one of the world's most destructive natural disasters. Spawned by a magnitude (M_w) 9.1 earthquake (third strongest ever instrumentally recorded), the 'Boxing Day' tsunami killed approximately 230,000 people in 14 countries around the Indian Ocean; among the victims were citizens of more than 60 countries. The tsunami propagated as far as the North Pacific and North Atlantic (RABINOVICH *et al.* 2006, 2011) and was probably the most catastrophic and deadliest tsunami in recorded history.

Various countries from around the globe contributed major funding to tsunami research, enabling the installation of hundreds of new high-precision instruments, the development of new technology and the establishment of more modern communication systems. As a result, incredible progress has been achieved in tsunami research and operation during the 10 years after the 2004 Indian Ocean tsunami. Tsunami warning and hazard mitigation systems have dramatically improved; the tsunami observational network of coastal tide gauges has been significantly reconstructed, upgraded, and expanded. Tsunami waves began to be monitored in both the deep ocean and from space. A large number of Deep-ocean Assessment and Reporting of Tsunamis (DART) stations have been emplaced in optimized alignment with the subduction zones encircling the entire Pacific Ocean; DARTs are now also deployed in the Indian and Atlantic oceans (MUNGOV *et al.* 2013). Numerous high-resolution cable bottom stations are installed offshore of the Pacific coast of Japan; similar geophysical stations are also deployed near the coast of Vancouver Island, Canada (RABINOVICH and EBLÉ 2015). These new, precise instruments have yielded thousands of coastal and hundreds of deep-water high-quality tsunami records enabling researchers to refute some previous misconceptions and to significantly improve knowledge about tsunami physics. Modern numerical models, combined

---

[1] Department of Fisheries and Oceans, Institute of Ocean Sciences, 9860 West Saanich Rd., Sidney, BC V8L 4B2, Canada. E-mail: A.B.Rabinovich@gmail.com

[2] P.P. Shirshov Institute of Oceanology, Russian Academy of Sciences, 36 Nakhimovsky Pr., Moscow 117997, Russia.

[3] U.S. Geological Survey, 345 Middlefield Rd., MS 999, Menlo Park, CA 94025, USA. E-mail: egeist@usgs.gov

[4] School of Civil and Environmental Engineering, Georgia Institute of Technology, Atlanta, GA 30332, USA. E-mail: fritz@gatech.edu

[5] eCoast Ltd., Box 151, Raglan 3225, New Zealand. E-mail: jose@ecoast.co.nz; jborrero@usc.edu

[6] Department of Civil and Environmental Engineering, University of Southern California, Los Angeles, CA, USA.

with open-ocean records, make it possible to forecast tsunami waves for coastal sites with reliable accuracy soon after a major earthquake.

However, despite the recent advances, tsunamis remain a major threat to coastal infrastructure and human life. Destructive tsunami events continue to kill people and create enormous damage. Several catastrophic events occurred in the 10 years after the 2004 Indian Ocean (Sumatra) tsunami, including the 2006 Java, 2009 Samoa, 2010 Chile, and 2010 Mentawai tsunamis with hundreds of fatalities per event. The Tohoku (Great East Japan) tsunami of 11 March 2011, which killed almost 20,000 people and destroyed the Fukushima Daiichi nuclear power plant, was a tragic example of a cascading chain of devastating events (SATAKE et al. 2013). We can state with some certainty that the number of victims would have been many times higher without the existing tsunami mitigation programs and effective tsunami warning services in Japan and other countries.

The present topical issue is prepared by the Tsunami Commission that was established within the International Union of Geodesy and Geophysics (IUGG) following the 1960 Chile tsunami generated by the largest (M_w 9.5) instrumentally recorded earthquake. This tsunami propagated throughout the entire Pacific Ocean and highlighted the need for tsunami warning throughout the region (IGARASHI et al. 2011). The topical issue is mainly based on papers presented at the 26th International Tsunami Symposium that was held in September 25–28, 2013 in Göcek, Turkey and Rhodes, Greece. Volume I of this issue (RABINOVICH et al. 2015; PAGEOPH, vol. 172, N 3/4, 2015) comprised 22 papers, which were prepared for publication by December 2014. Volume II of this issue comprises 16 papers, which became ready for publication by July 2015, as well as four additional papers (LUI et al. 2015; ULRICH and AOCHI 2015; BABA et al. 2015; XIAO et al. 2015) originally submitted to a regular issue but transmitted to the current topical issue because their thematic content was closely related to this issue. For convenience, all papers from this volume are separated into five groups according to geography.

2. Global Observations

Various observation systems of tsunamis, including real-time tide gauge stations, DART stations, and field surveys allow for a determination of tsunami characteristics that are global in scope and related to some very general aspects of the problem. GUSIAKOV (2015) considered the important problem of evaluating tsunami intensity for recent events in the world oceans. The tsunami intensity on the Soloviev-Imamura scale is one of the key parameters characterizing the "strength" of a tsunami. Consequently, it is included in global tsunami databases maintained by the National Centers for Environmental Information/World Data Service (NCEI/WDS) and the Novosibirsk Tsunami Laboratory of the Institute of Computational Mathematics and Mathematical Geophysics (NTL/ICMMG), however, only for events before 2003. The same author tries to extend the temporal range of this parameter for events occurring in the twenty-first century and discusses the problems of its determination.

Deep-ocean tsunami measurements play a major role in understanding the physics of tsunami wave generation and propagation, and in improving the effectiveness of tsunami warning systems. RABINOVICH and EBLÉ (2015) provide an overview of the history of such measurements from the earliest days (~50 years ago) to the present time. Modern tsunami monitoring systems such as the self-contained DART and innovative cabled sensing networks, including, but not limited to, the Japanese deep seafloor observatories and the NEPTUNE-Canada geophysical bottom observatory are highlighted. The specific peculiarities of seafloor longwave observations in the deep ocean are discussed and compared with observations recorded in coastal regions.

3. Indian Ocean

As with Volume I of this topical issue, a number of papers in Volume II describe recent advances in understanding the occurrence of tsunamis in the Indian Ocean and progress in the tsunami preparedness and early warning for this region.

SUPPASRI et al. (2015) provide examples of tsunami mitigation measures included in the reconstruction after the 2004 Indian Ocean tsunami from Indonesia, Thailand, Sri Lanka, and the Maldives. The performance of these mitigation measures is discussed based on recent less-devastating tsunami events in the region. Future challenges based on current tsunami countermeasures, such as land use planning, warning systems, evacuation facilities, disaster education, and disaster monuments are explained.

OKAL et al. (2015) provide the results of a 2010 survey of elderly survivors from the 1945 Makran subduction zone tsunami—a key event for hazard assessment in the Arabian Sea and the western Indian Ocean. Results from this survey indicate that there was a significant delay in the arrival of the tsunami relative to that expected from an earthquake source, suggesting that an ancillary event such as a submarine landslide could be responsible for the tsunami.

Paleotsunami studies are a crucial instrument to identify and date past events and to evaluate/mitigate tsunami hazard. Coastal sediments deposited by tsunamis (*tsunamites*) can potentially provide information on tsunami dating and inundation distance. PREMASIRI et al. (2015) suggested a new method to detect tsunami sediments based on the analysis of tsunamite's dielectric properties, which are found to be significantly different from those for surrounding non-tsunami sediments. The method was used to trace tsunami inundation on the coast of Sri Lanka associated with the 2004 Indian Ocean tsunami.

HÉBERT and SCHINDELE (2015) analyze three coastlines impacted by the 2004 tsunami in the Indian Ocean: Thailand-Myanmar, SE India–Sri Lanka, and SE Madagascar. The results of a nonlinear shallow water tsunami model based on coarse and refined grids are compared against the field data collected in the aftermath of the 2004 event. The study includes simple runup approaches based on classic energy conservation laws. A preliminary statistical assessment of tsunami impact is provided for a given earthquake magnitude along the Indonesian subduction based on scenarios.

MAS et al. (2015) describe agent-based modeling of tsunami evacuation in which an "…agent or individual part of a system is modeled as an autonomous decision-making entity." Several case studies of evacuations during recent tsunamis are presented for sites in both the Indian (Thailand and Indonesia) and Pacific (Japan and Peru) Oceans. Their analysis looks at factors such as evacuation times and routes as well as estimates of the demand on shelters relative to the number of evacuees for a hypothetical event. Their case studies and analyses should be useful to hazard-management practitioners looking to devise or revise municipal tsunami evacuation strategies.

4. Pacific Ocean

Tsunami studies in this Volume related to Pacific Ocean events are focused primarily on the 2011 Tohoku earthquake and tsunami and events off the Kuril Islands in 2006 and 2007. The tsunami generated by the 2011 Tohoku earthquake was largely caused by high amount of slip on the megathrust near the trench. Deeper in the subduction zone, however, another component of the 2011 rupture generated energetic, high-frequency seismic waves and numerous intraslab aftershocks investigated by LUI et al. (2015). They conclude that many of these aftershocks reactivated subducted normal faults, which, prior to being subducted, had been a significant tsunami hazard themselves (e.g., 1933 Sanriku event). ULRICH and AOCHI (2015) demonstrate that the elliptical slip patch method for inverting GPS and seismic acceleration data is an efficient and robust method to determine the rupture process of major subduction zone earthquakes, such as the 2011 event. Slip on the western part of the fault (deeper portion) was found to contribute more significantly to low-frequency, on-land displacement signals than shallow slip near the trench.

BABA et al. (2015) applied the nonlinear Boussinesq dispersive equations in spherical coordinates to simulate the 2011 Tohoku tsunami and to estimate the influence of dispersive effects on tsunami propagation and inundation of the Japan coast. The authors used the finest bathymetric grid of approximately 5 m and achieved good agreement between the model results and the tsunami waveforms observed offshore. OZER et al. (2015a, b) examined the effect of breakwaters and critical coastal defense structures on tsunami runup and inundation. The main attention was paid to Kamaishi Bay, the region that was strongly devastated by the 2011 Tohoku tsunami waves.

EBLÉ et al. (2015) investigated the effect of a small negative phase leading the first major positive wave for four major tsunami events in the Pacific Ocean: 2010 Chile, 2011 Tohoku, 2012 Haida Gwaii, and 2014 Chile (Iquique). Importantly, this phase is not reproduced by conventional tsunami models nor is related to earthquake process in generating the tsunami. Nevertheless, the authors found leading negative-phase signatures in examples from the more than 40 deep-ocean bottom pressure and numerous tide gauge records. These findings are in good agreement with the recent theoretical results of WATADA et al. (2014) suggesting that this effect is related to the elasticity of the Earth and seawater compressibility.

BARANOV et al. (2015) examined seismotectonic features of the Central Kuril Islands forearc region, where major tsunamigenic earthquakes took place on November 15, 2006 and January 13, 2007. Based on a segmented forearc structure, the authors suggest three possible scenarios of great earthquake occurrence within the region. If a rupture occupies the entire seismic gap with a total length of about 500 km, then the earthquake magnitude (M_w) might be greater than 8.5, producing tsunami waves significantly higher than the 2006–2007 tsunamis.

One of the most important problems of coastal engineering is estimating the maximum sea levels associated with various natural factors and their probability. SHEVCHENKO and IVELSKAYA (2015) evaluated extreme sea levels arising from the combination of tides, storm surges, seasonal oscillations, and tsunamis for the Russian coast of the Sea of Okhotsk and the Pacific coast of the Kuril Islands. Different factors were found to be dominant in estimating maximum sea level for different regions, but, in general, the importance of tsunamis significantly increases for longer return periods.

In an interesting study that attempts to relate tsunami hydrodynamics to coral reef damage, DILMEN et al. (2015) find that while numerical models adequately hindcast runup and inundation, there seems to be little correlation between runup and the extent of coral damage, suggesting more complex processes of tsunami wave propagation over reefs than that can be currently modeled. CORONA and RAMÍREZ-HERRERA (2015) reanalyze the 1932 event off the Pacific coast of Mexico. The associated earthquake ($M_s = 6.9$) is questioned as source, given the locally observed order of 10-m tsunami runup heights. Historical documents, survivor testimony, tsunami catalogs, a post-tsunami survey report, together with geomorphological interpretation of the continental shelf and slope, and numerical modeling were combined to characterize the tsunami parameters. Their results suggest that tsunami observations may be compatible with a submarine landslide source in the Armería Canyon.

5. Atlantic Ocean and Mediterranean Sea

Two papers in this Volume relate to estimating the tsunami hazard along populous coasts of the Atlantic Ocean and Mediterranean Sea. TEHRANIRAD et al. (2015) reconsider tsunami impacts across the North Atlantic from potential large-scale flank collapse scenarios of the Cumbre Vieja Volcano (CVV) on La Palma (Canary Islands). They modeled tsunami generation using a 3D Navier–Stokes (NS) multifluid VOF model, while the near-field tsunami impact was simulated with a dispersive and fully nonlinear longwave Boussinesq model. The far-field tsunami impact is modeled for two extreme CVV flank collapse scenarios with some detailed inundation models for the US east coast.

NECMIOGLU and ÖZEL (2015) conduct a tsunami hazard assessment generated by earthquakes in the Eastern Mediterranean, Aegean, and Black Seas, where they find that tsunamis can reach 1–3 m along coastlines that border these seas. Maximum wave heights of >3 m may be expected on the coasts of Crete, southern Aegean Sea, and the area between northeast Libya and Alexandria (Egypt).

6. Tsunami Sources

In the final section of Volume II, two papers are presented that describe new research related to landslide and earthquake tsunamis sources. XIAO et al. (2015) introduce a new method to model the combined movement of subaerial landslides falling into water bodies and the waves these landslides generate, using as a case study the 2008 landslide that caused destructive waves within the Three Gorges Reservoir, China.

BLETERY et al. (2015) study the contribution to the tsunami excitation of the so-called bathymetry effect describing the slip in the shallowest portion of the fault interface characterized by steep bathymetry combined with horizontal motion. They find that the bathymetry effect locally exceeds 10 % of the tsunami excitation in all subduction zones. The bathymetry effect improves the consistency of the slip model inverted from tsunami data with seafloor geodesy observations, implying that taking the bathymetry effect into account reduces the epistemic uncertainties on tsunami modeling.

Acknowledgments

We would like to thank Dr. Renata Dmowska, the Editor-in-Chief for Topical Issues of PAGEOPH for arranging and encouraging us to organize these topical volumes. We also thank Ms. Priyanka Ganesh at the Journal's Editorial Office of Springer for her timely editorial assistance. Finally, we would like to thank all the authors and reviewers who contributed to these topical volumes.

REFERENCES

BABA, T., TAKAHASHI, N., KANEDA, Y., ANDO, K., MATSUOKA, D., and KATO, T. (2015), *Parallel implementation of dispersive tsunami wave modeling with a nesting algorithm for the 2011 Tohoku Tsunami*, Pure Appl. Geophys., *172* (12) (this issue); doi 10.1007/s00024-015-1049-2.

BARANOV, B.V., IVASHCHENKO, A.I., and DOZOROVA K.A. (2015), *The Great 2006 and 2007 Kuril earthquakes, forearc segmentation and seismic activity of the Central Kuril Islands region*, Pure Appl. Geophys., *172* (12) (this issue); doi 10.1007/s00024-015-1120-z.

BLETERY, Q., SLADEN, A., DELOUIS, B., and MATTÉO, L. (2015), *Quantification of tsunami bathymetry effect on finite fault slip inversion*, Pure Appl. Geophys., *172* (12) (this issue); doi 10.1007/s00024-015-1113-y.

CORONA, N., and RAMÍREZ-HERRERA, M.-T. (2015), *Did an underwater landslide trigger the June 22, 1932 tsunami off the Pacific coast of Mexico?* Pure Appl. Geophys., *172* (12) (this issue); doi 10.1007/s00024-015-1171-1.

DILMEN, D.I., TITOV, V.V., and ROE, G.H. (2015), *Evaluation of the relationship between coral damage and tsunami dynamics; Case study: 2009 Samoa tsunami*, Pure Appl. Geophys., *172* (12) (this issue); doi 10.1007/s00024-015-1158-y.

EBLÉ, M.C., MUNGOV, G., and RABINOVICH, A.B. (2015), *On the leading negative phase of major 2010–2014 tsunamis*, Pure Appl. Geophys., *172* (12) (this issue); doi 10.1007/s00024-015-1127-5.

GUSIAKOV, V.K. (2015), *Evaluation of intensity of recent seismogenic tsunamis in the World Ocean from 2000 to 2014*, Pure Appl. Geophys., *172* (12) (this issue); doi 10.1007/s00024-015-1101-2.

HÉBERT, H., and SCHINDELE, F. (2015), *Tsunami impact computed from offshore modelings and coastal amplification laws: Insights from the 2004 Indian Ocean tsunami*, Pure Appl. Geophys., *172* (12) (this issue); doi 10.1007/s00024-015-1136-4.

IGARASHI, Y., KONG, L., YAMAMOTO, M., and MCCREERY, C.S. (2011), *Anatomy of historical tsunamis: Lessons learned for tsunami warning*, Pure Appl. Geophys., *168*, 2043–2063; doi 10.1007/s00024-011-0287-1.

LUI, S.K.Y., HELMBERGER, D., WEI, S., HUANG, Y., and GRAVES, R.W. (2015), *Interrogation of the megathrust zone in the Tohoku-Oki seismic region by waveform complexity: Intraslab earthquake rupture and reactivation of subducted normal faults*, Pure Appl. Geophys., *172* (12) (this issue); doi 10.1007/s00024-015-1042-9.

MAS, E., KOSHIMURA, S., IMAMURA, F., SUPPASRI, A., MUHARI, A., and ADRIANO, B (2015), *Recent advances in agent-based tsunami evacuation simulations: Case studies in Indonesia, Thailand, Japan and Peru*, Pure Appl. Geophys., *172* (12) (this issue); doi 10.1007/s00024-015-1105-y.

MUNGOV, G., EBLÉ, M., and BOUCHARD, R. (2013), *DART® tsunameter retrospective and real-time data: A reflection on 10 years of processing in support of tsunami research and operations*, Pure Appl. Geophys., *170*, 1369–1384; doi 10.1007/s00024-012-0477-5.

NECMIOGLU, O., and ÖZEL, N.M. (2015), *Earthquake scenario-based tsunami wave heights in the Eastern Mediterranean and its connected seas*, Pure Appl. Geophys., *172* (12) (this issue); doi 10.1007/s00024-0115-1069-y.

OKAL, E.A., FRITZ, H.M., HAMZEH, M.A., and GHASEMZADEH, J. (2015), *Field survey of the 1945 Makran and 2004 Indian Ocean tsunamis in Baluchistan, Iran*, Pure Appl. Geophys., *172* (12) (this issue); doi 10.1007/s00024-015-1157-z.

OZER C.S., YALÇINER, A.C. and ZAYTSEV, A. (2015a), *Investigation of tsunami hydrodynamic parameters in inundation zone with different structural layout.* Pure Appl. Geophys., *172* (3-4), 931–952; doi 10.1007/s00024-014-0947-z.

OZER, C.S., YALÇINER, A.C., ZAYTSEV, A., SUPPASRI, A., and IMAMURA, F. (2015b), *Investigation of hydrodynamic parameters and the effects of breakwaters during the 2011 Great East Japan tsunami in Kamaishi Bay*, Pure Appl. Geophys., *172* (12) (this issue); doi 10.1007/s00024-015-1051-8.

PREMASIRI, R., STYLES, P., SHRIRA, V., CASSIDY, N., and SCHWENNINGER, J.-L. (2015), *OSL Dating and GPR mapping of palaeotsunami inundation: A 4000 year history of Indian Ocean tsunamis as recorded in Sri Lanka*, Pure Appl. Geophys., *172* (12) (this issue); doi 10.1007/s00024-015-1128-4.

RABINOVICH, A.B., THOMSON, R.E., and STEPHENSON, F.E. (2006), *The Sumatra tsunami of 26 December 2004 as observed in the*

North Pacific and North Atlantic Oceans, Surveys Geophys. *27*, 647–677.

RABINOVICH, A.B., CANDELLA, R.N., and THOMSON, R.E. (2011), *Energy decay of the 2004 Sumatra tsunami in the world ocean*, Pure Appl. Geophys., *168* (11), 1919–1950; doi 10.1007/s00024-01-0279-1.

RABINOVICH, A.B., GEIST, E.L., FRITZ, H.M., and BORRERO, J.C. (2015), *Tsunami Science: Ten Years after the 2004 Indian Ocean Tsunami Volume I*, Pure Appl. Geophys., *172* (3–4), Topical Issue.

RABINOVICH, A.B., and EBLÉ, M C. (2015), *Deep-ocean measurements of tsunami waves*, Pure Appl. Geophys., *172* (12) (this issue); doi 10.1007/s00024-015-1058-1.

SATAKE K., RABINOVICH, A.B., DOMINEY-HOWES, D., and BORRERO, J.C. (2013), *Historical and Recent Catastrophic Tsunamis in the World: Past, Present, and Future. Volume I: The 2011 Tohoku Tsunami.* Pure Appl. Geophys., *170* (6–8), Topical Issue.

SCHINDELÉ, F., GAILLER, A., HÉBERT, H., LOEVENBRUCK, A., GUTIERREZ, E., MONNIER, A., ROUDIL, P., REYMOND, D., and RIVERA, L. (2015), *Implementation and challenges of the Tsunami Warning System in the Western Mediterranean.* Pure Appl. Geophys., *172* (3–4) (this issue); doi 10.1007/s00024-014-0950-4.

SHEVCHENKO, G., and IVELSKAYA, T. (2015), *Estimation of extreme sea levels for the Russian coasts of the Kuril Islands and the Sea of Okhotsk*, Pure Appl. Geophys., *172* (12) (this issue); doi 10.1007/s00024-015-1077-y.

SUPPASRI, A., GOTO, K., MUHARI, A., RANASINGHE, P., RIYAZ, M., MAS, E., YASUDA, M., and IMAMURA, F. (2015), *A decade after the 2004 Indian Ocean tsunami—The progress in disaster preparedness and future challenges in Indonesia, Sri Lanka, Thailand and the Maldives*, Pure Appl. Geophys., *172* (12) (this issue); doi 10.1007/s00024-015-1134-6.

TEHRANIRAD, B., HARRIS, J.C., GRILLI, A.R., GRILLI, S.T., ABADIE, S., KIRBY, J.T., and SHI, F. (2015), *Far-field tsunami hazard in the north Atlantic basin from large scale flank collapses of the Cumbre Vieja Volcano, La Palma*, Pure Appl. Geophys., *172* (12) (this issue); doi 10.1007/s00024-015-1135-5.

ULRICH, T., and AOCHI, H. (2015), *Rapidness and robustness of finite-source inversion of the 2011 Mw 9.0 Tohoku earthquake by an elliptical-patches method using continuous GPS and acceleration data*, Pure Appl. Geophys., *172* (12) (this issue); doi 10.1007/s00024-014-0857-0.

WATADA, S., KSUMOTO, S. and SATAKE, K. (2014), *Traveltime delay and initial phase reversal of distant tsunamis coupled with the self-gravitating elastic Earth*, J. Geophys. Res. Solid Earth, *119*, 4287–4310; doi: 10.1002/2013JB010841.

XIAO, L., WARD, S.N., and WANG J. (2015), *Tsunami squares approach to landslide-generated waves: Application to Gongjiafang landslide, Three Gorges Reservoir, China*, Pure Appl. Geophys., *172* (12) (this issue); doi 10.1007/s00024-015-1045-6.

(Received September 8, 2015, accepted September 9, 2015, Published online October 14, 2015)

Evaluation of Intensity of Recent Seismogenic Tsunamis in the World Ocean from 2000 to 2014

VIACHESLAV K. GUSIAKOV[1]

Abstract—Tsunami intensity on the SOLOVIEV–IMAMURA scale is one of the most important parameters for characterizing the overall size of a tsunami generated by submarine earthquakes. Consequently, this parameter is included in both global tsunami databases maintained by the National Centers for Environmental Information/World Data Service (NCEI/WDS) and the Novosibirsk Tsunami Laboratory of the Institute of Computational Mathematics and Mathematical Geophysics (NTL/ICMMG). S. Soloviev made the initial evaluation of the intensities of a large number of destructive historical tsunamis while compiling his two historical catalogs of tsunamis in the Pacific. The Novosibirsk Tsunami Laboratory under the Expert Tsunami Database Project made further determinations of tsunami intensity for the events after 1975. These intensities have been periodically incorporated into the NCEI/WDS tsunami database under the Global Tsunami Database Joint ICG/ITSU-IUGG/TC Project. In the on-line version of the NCEI/WDS Tsunami Database, the data on tsunami intensity are available only for the events prior to 2003. The main purpose of this paper is to extend the temporal coverage of this important parameter for characterizing tsunamigenic events to the present in order to provide researchers with more data for analyzing the temporal and spatial tsunami occurrence. However, of the 164 tsunamigenic events in the World Ocean from 2000 to the present, we could determine the intensity value for only 44 events that is less than 27 % of the total. For the rest of the events (that is, 73 %), the intensity value cannot be determined due to the lack of data on wave heights from the nearest coast. This shows that despite a great improvement in the tsunami-recording network in the Pacific and other oceanic basins during the last two decades, the data for reliable estimates of tsunami intensity are still problematic.

Key words: Tsunami, submarine earthquakes, earthquake magnitude, tsunami intensity, tsunami warning.

1. Introduction

One of the main problems in cataloging historical tsunamis is to measure the overall "size" or "force" of an event. To compare different tsunamigenic events, we need some scale for their measurement. A number of descriptive and quantitative scales were proposed for quantification of historical tsunamis (their survey can be found in GUSIAKOV 2009), however, only one of them, the SOLOVIEV–IMAMURA intensity scale *I* became *de facto* the most widely used scale for measuring the size of tsunamis. This scale is incorporated into both global historical tsunami databases maintained by NOAA's National Centers for Environmental Information/World Data Service (NCEI/WDS 2015) and the Novosibirsk Tsunami Laboratory (NTL) of the Institute of Computational Mathematics and Mathematical Geophysics of the Siberian Division, Russian Academy of Sciences (ICMMG SD RAS) (HTDB/WLD 2015). What is more important, the intensity *I* is now determined for more than 90 % of all significant historical tsunamis worldwide; thus, allowing us to rank the main tsunamigenic regions by their tsunamigenic potential and different tsunamigenic events by their overall size.

The first quantitative scale to compare the Pacific tsunamis was introduced by (IIDA 1963) who connected the grade number *m* of Imamura's descriptive scale, proposed in 1942 in Japan (IMAMURA 1942) with a maximum observed run-up value at the coast H_{max} by the formula:

$$m = \log_2 H_{max}. \qquad (1)$$

IMAMURA called it a tsunami magnitude scale, but it is a typical example of intensity scale, since it is based on coastal tsunami effects and does not contain

---

[1] Institute of Computational Mathematics and Mathematical Geophysics, Institute of Computational Technologies, Siberian Division, Russian Academy of Sciences, 6 Pr. Lavrentieva, Novosibirsk 630090, Russia. E-mail: gvk@sscc.ru

any correction for distance from source. SOLOVIEV (1972) made an important modification of this scale proposing to use the average height H_{av} over some extended part of the nearest coast for calculation of tsunami intensity I by the formula:

$$I = 1/2 + \log_2 H_{av}. \qquad (2)$$

SOLOVIEV argued that this value is a more reliable characteristic of a tsunami and is most closely related to the total tsunami energy radiated from a source. In his original paper, SOLOVIEV did not quantify the notion of "extended part of the nearest coast," noting only that the stronger the earthquake the longer the part of the coast that should be considered. This uncertainty is one of the main reasons why the procedure of intensity calculation cannot be completely formalized. A simple "rule of thumb" is that the extent of the coast considered should be about double the size of the earthquake source length, that is to be approximately 200 km for sources with $M_w = 7.8$, 400 km for sources with $M_w = 8.4$ and 800 km for mega-events with $M_w = 9.0$.

The intensity I in the SOLOVIEV–IMAMURA scale is based on the coastal tsunami effect (run-up heights), therefore, it is clearly an intensity-type scale. However, in practice it is widely used as a magnitude-type scale allowing a direct comparison of different tsunamigenic events. This is possible because most tsunamis occur in subduction zones where their sources are located at nearly the same distance from the nearest coast (50–100 km).

With this scale, SOLOVIEV evaluated the intensity for a large number of the Pacific tsunamis when compiling his catalogs (SOLOVIEV and GO 1974, 1975; SOLOVIEV 1978). Further determinations of tsunami intensity were made by the Novosibirsk Tsunami Laboratory under the ETDB (Expert Tsunami Database) Project (GUSIAKOV et al. 1997) and periodically have been incorporated into the NCEI/WDS tsunami database under the GTDB (Global Tsunami Database) Joint ICG/ITSU- IUGG/TC Project (GUSIAKOV 2003). In the current on-line version of the NCEI/WDS Tsunami Database, the tsunami intensity data are available only for the events prior to 2003. The main purpose of this paper is to extend the temporal coverage of this important parameter for characterizing historical tsunamigenic events to the present in order to provide researchers with more data for analyzing their temporal and spatial occurrence. Due to a continuous extension and technical improvement of the tsunami observation network in the Pacific and elsewhere, the amount and quality of data on tsunami measurements have been drastically improved for the events during the XXI century as compared with earlier historical events (MOFJELD 2009; RABINOVICH and EBLÉ 2015). It is also interesting to evaluate the recent and older historical tsunamis in terms of the accuracy and reliability of their intensity determination.

2. Data

The data retrieval from the NCEI/WDS database was made for tsunamigenic events from 2000 to 2014 with a validity index from 1 to 4, thus excluding from consideration the events with validity −1 (erroneous entry) and 0 (disturbance in inland rivers). The retrieval returned a list of 164 tsunamigenic events or, on average, almost 11 tsunamis annually occurring in the World Ocean. The geographical distribution of their sources is shown in Fig. 1: 128 events occurred in the Pacific, 21 in the Indian Ocean, 10 in the Atlantic, and 5 in the Mediterranean region. Among the 164 events, 146 had a seismogenic origin, 6 events resulted from volcanic eruptions and associated failures of volcanic slopes, 8 were generated by submarine or coastal landslides, and the remaining 4 events had a non-tectonic origin (meteotsunamis or freak waves, their actual number might be much greater). Only 15 of 164 events (that is less than 10 %) were accompanied by fatalities, all the rest were non-fatal events despite some having great maximum run-ups (up to 50 m, as in the case of the Greenland landslide tsunami of 21 November 2000). The total death toll due to tsunami during the last 15 years is great: 248,085 fatalities, but it is important to note that 99.3 % of them resulted from just two transoceanic mega-tsunamis—the 2004 Sumatra tsunami (227,899 deaths) and the 2011 Tohoku tsunami (18,482 deaths). The remaining 13 fatal events are responsible only for 0.7 % of all tsunami fatalities.

Both global tsunami databases, maintained by the NCEI/WDS and the NTL/ICMMG, consist of two

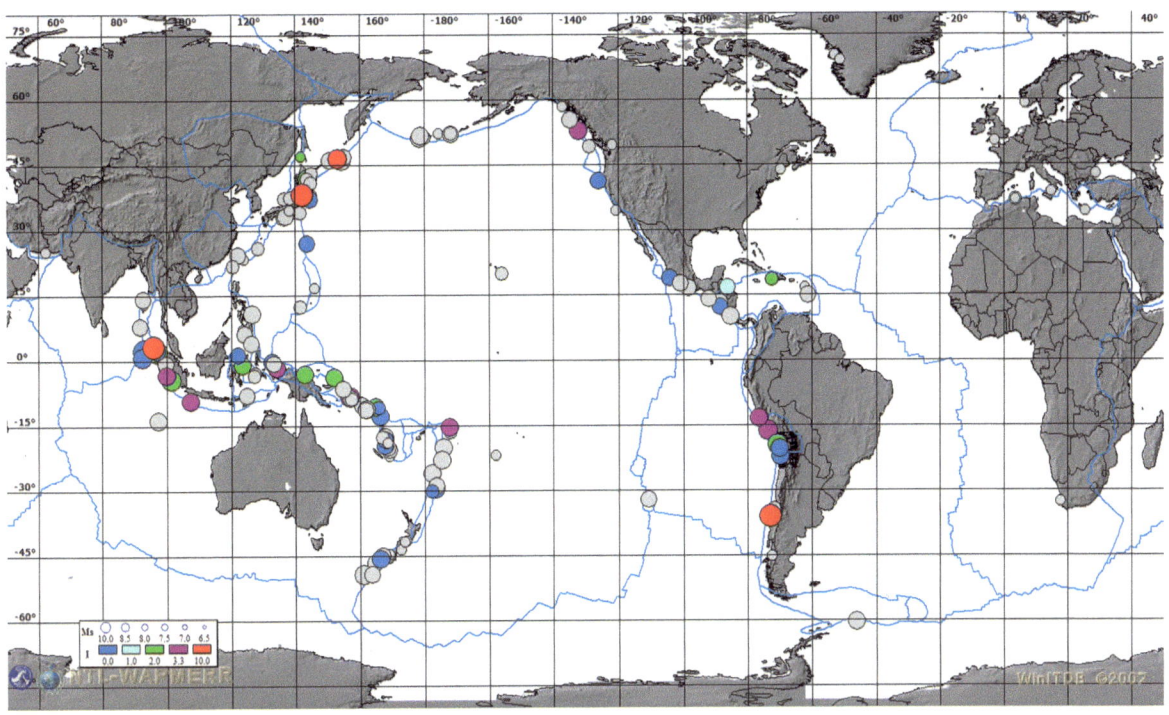

Figure 1
A source map of 164 tsunamigenic events in the World Ocean from 2000 to 2014. The *size of circles* is proportional to the event magnitude; the *color* represents the tsunami intensity on the SOLOVIEV–IMAMURA scale. *Gray color* shows the events for which the tsunami intensity is not determined. *Solid blue lines* are the main plate boundaries

main parts: tsunami source event data containing basic parameters of a source event and tsunami run-up data listing available wave heights for a particular event. For this study, the most important is the second part of the database, containing the data on water height measurements at a particular location: run-up height, inundation depth, maximum amplitudes on tide gage records, etc. (the full list of parameters available in the NCEI/WDS database includes 10 different types of height measurements).

The NCEI/WDS database has 11,761 height measurements for these 164 recent historical events. On average, there are 72 heights per event that in principle allows a reliable determination of the tsunami intensity. However, the actual distribution of height measurements over events is highly inhomogeneous and by far is dominated by two transoceanic tsunamis generated by the M9 class earthquakes on March 11, 2011 in Tohoku, Japan (6054 measurements) and on December 26, 2004 in the Indian Ocean (1507 measurements). These two mega-tsunamis give on the whole nearly 65 % of all the measurements available in the database for this period. Eleven events have more than 100 measurements and 22 events have more than 10 measurements (Fig. 2). Thus, on the whole, we can expect obtaining more or less reliable estimates of the tsunami intensity for only 35 events out of 164, provided that the measurements are mainly represented by the coastal run-up heights. Of the remaining 129 events, having a small number of wave measurements, the vast majority were weak local or regional tsunamis recorded only by coastal and ocean bottom instruments.

In this study, calculation of the intensity I was made with the help of a special built-in procedure available in the WinITDB graphic shell (WINITDB 2009) that is used for visualization and handling of the HTDB/WLD data. The procedure is based on formula (2) and consists of the selection of a historical event from the list, selection of a geographical area around the source, plotting available heights, editing them, if necessary (e.g., by deleting some values closely distributed within a small distance

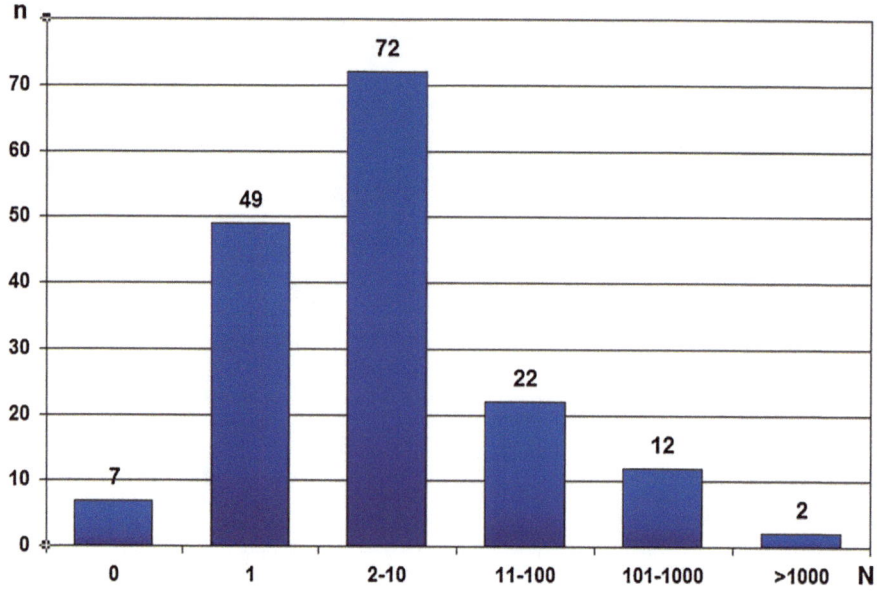

Figure 2
Distribution of number of tsunamigenic events n for the period from 2000 to 2014 over the number of height measurements N available in the NCEI/WDS database

along the coast). Upon obtaining the final list of heights, by pressing just a single button, one can obtain a number of statistical parameters for the selected heights including the I value.

However, practical application of this procedure faces several problems. Certain tsunamigenic earthquakes occur within island archipelagos (like Micronesia or Indonesia), where the coastal geometry is rather complex and the coastline is not straight. For these events, one must be sure to use the heights measured at different azimuths and covering a certain range (10–200 km) of epicentral distances.

The evaluation of intensity for tsunamigenic events in the island arc regions creates another problem. In these areas, the height measurements are available only from the nearest islands with large spatial gaps between them. Examples of these events are the Simushir tsunami of 15 November 2006 and the Haida Gwaii tsunami of 28 October 2012. If we formally apply the above procedure to the available heights for these events, we will obtain an artificially increased I value, as the average will be shifted to greater heights measured on the nearest islands due to the dearth of or the absence of smaller heights in the dataset. For these events we have to introduce some correction to the calculated I value, based on the attenuation of tsunami heights along the coast.

The availability of only tide gage records from remote locations along with the absence of run-up heights from the nearest coast is a typical situation for weak tsunamis. Being non-fatal and non-damaging, these events are rarely followed by a field survey, so the coastal height measurements for them are scarce or absent. However, a correct quantification of them is important for the rational selection of threshold magnitudes for the early tsunami warning. For these events, only expert estimates are possible and they must be based on some correlation between the near-field run-up heights and the far-field wave amplitudes as extrapolated from larger events having both run-up and mareograph data.

All of these factors mean that the process of tsunami intensity determination, despite the availability of data in digital and computer-readable form and the presence of specialized retrieval and visualization software (like WinITDB graphic shell), cannot be completely computerized and therefore requires some expert knowledge.

The final results of intensity determination for the recent tsunamigenic events are presented in Table 1.

Table 1

Tsunamigenic events in the World Ocean from 2000 to 2014 that have a sufficient number of height measurements in the NCEI/WDS tsunami database to calculate tsunami intensity I

No.	Date (YYYY, MM, DD)	Time (UTC)	Source area	M_w	N	H_{max}	N_{nf}	H_{max_NF}	H_{min_NF}	H_{av_NF}	I
1	2000.05.04	04:21:18.0	Sulawesi, Indonesia	7.50	2	6.00	2	6.00	1.00	3.50	2.0
2	2000.11.16	04:54:59.3	Papua New Guinea	8.00	9	3.00	8	3.00	1.00	1.50	1.08
3	2001.06.23	20:33:15.4	Peru	8.39	143	8.77	42	8.77	1.29	3.74	2.40
4	2002.09.08	18:44:25.3	Papua New Guinea	7.59	87	5.50	67	5.50	1.00	1.86	1.39
5	2003.01.22	02:06:36.4	Colima, Mexico	7.48	3	0.61	3	0.61			−0.5
6	2003.05.21	18:44:21.3	Boumerdes, Algeria	6.81	13	3.00	1	1.00			0.5
7	2003.09.25	19:50:08.4	Tokachi-oki, Japan	8.26	262	4.40	210	4.40	1.0	2.44	1.77
8	2004.12.26	00:58:53.7	Sumatra, Indonesia	9.00	1508	50.90	495	50.90	2.00	12.95	4.19
9	2005.03.28	16:09:37.5	Sumatra, Indonesia	8.62	61	4.20	42	4.20	0.50	2.37	1.74
10	2005.06.15	02:50:54.8	Nothern California, USA	7.20	4	0.40	4	0.40	0.07	0.20	−1.8
11	2005.08.16	02:46:28.5	Honsu, Japan	7.20	5	0.13	5	0.13	0.05	0.08	−3
12	2006.07.17	08:19:27.6	Java, Indonesia	7.70	196	20.90	155	20.90	1.00	6.20	2.95
13	2006.11.15	11:14:16.7	Kuril Islands	8.30	254	21.90	114	21.90	1.30	10.24	3.36
14	2007.04.01	20:39:58.0	Solomon Islands	8.10	234	12.10	169	12.10	1.00	3.38	2.26
15	2007.08.02	02:37:43.4	Nevelsk, Sakhalin	6.19	22	3.20	16	3.20	0.70	1.64	1.21
16	2007.08.15	23:40:57.9	Peru	8.00	136	10.05	69	10.05	1.00	3.15	2.16
17	2007.09.12	11:10:27.7	Sumatra, Indonesia	8.49	47	5.00	25	5.00	1.20	2.32	1.71
18	2007.11.14	15:40:50.5	Chile	7.73	6	0.13	3	0.13			−3
19	2008.11.16	17:02:33.7	Sulawesi, Indonesia	7.35	3	0.13	1	0.13			−3
20	2009.01.03	19:43:54.9	Papua, Indonesia	7.67	24	0.39	1	0.39			−2
21	2009.05.28	08:24:46.9	Honduras and Belize	7.35	2	4.00	2	4.00			1
22	2009.07.15	09:22:32.6	NewZealand	7.78	9	0.47	4	0.47	1.12	0.22	−1.7
23	2009.09.29	17:48:10.9	Samoa Islands	8.09	622	23.35	404	23.35	2.00	5.33	2.92
24	2009.10.07	22:03:15.1	Vanuatu	7.62	37	0.31	2	0.31	0.10	0.20	−2
25	2010.01.12	21:53:10.0	Haiti	7.10	7	3.00	3	3.00	1.00	2.33	1.7
26	2010.02.27	06:34:14.0	Chile	8.80	600	29.00	415	29.00	2.00	7.93	3.49
27	2010.04.06	22:15:02.0	Sumatra, Indonesia	7.80	5	0.44	5	0.44	0.07	0.19	−2
28	2010.10.25	14:42:22.0	Mentawai, Indonesia	7.70	89	9.30	69	9.30	1.20	4.56	2.69
29	2010.12.21	17:19:40.0	Bonin Islands	7.40	4	0.13	4	0.13	0.03	0.07	−3
30	2010.12.25	13:16:37.0	Vanuatu	7.30	4	0.15	4	0.15	0.02	0.06	−3.5
31	2011.03.09	02:45:18.0	Honsu, Japan	7.50	11	0.60	11	0.60	0.20	0.33	−1.1
32	2011.03.11	05:46:23.0	Tohoku	9.10	6051	40.57	1476	40.57	2.00	13.61	4.22
33	2011.07.06	19:03:18.2	Kermadec Islands	7.60	23	1.20	4	1.20	0.22	0.66	0
34	2011.08.20	16:55:02.5	Vanuatu Islands	7.10	4	0.18	4	0.18	0.07	0.12	−2.5
35	2011.08.20	18:19:23.5	Vanuatu Islands	7.10	4	0.18	1	0.18			−3
36	2012.04.11	08:38:36.7	Sumatra, Indonesia	8.60	20	1.08	3	1.08	0.24	0.56	−0.5
37	2012.04.11	10:43:10.8	Sumatra, Indonesia	8.20	4	0.22	1	0.22			−2
38	2012.08.27	04:37:19.4	Salvador	7.30	11	0.36	5	0.36	0.03	0.18	−2.0
39	2012.10.28	03:04:10.0	Haida Gwaii, Canada	7.80	176	12.98	62	12.98	2.50	5.83	2.54
40	2013.02.06	01:12:25.8	Solomon Islands	7.90	129	11.0	7	11.00	0.51	2.63	1.9
41	2013.02.08	15:26:38.4	Solomon Islands	7.00	1	0.09	1	0.09			−4
42	2013.10.25	17:10:18.0	Honsu, Japan	7.10	9	0.40	6	0.40	0.20	0.30	−1.3
43	2014.04.01	23:46:46.0	Chile	8.20	164	4.40	8	4.40	0.57	1.78	1.2
44	2014.04.03	02:43:13.0	Chile	7.70	5	0.74	5	0.74	0.19	0.48	−1

N is the total number of height measurements available for a particular event, N_{NF} is the number of near-field heights, H_{max_NF} is the near-field maximum height (m), H_{min_NF} is the near-field minimum height (m), H_{av_NF} is the average height (m) that is used in I calculations. Earthquake parameters (origin time and magnitude M_w) are taken from the ISC-GEM catalog (Storchak et al. 2015)

It contains a list of 44 tsunamigenic events in the World Ocean from 2000 to 2014 that have a sufficient number of height measurements in the NCEI/WDS tsunami database. The events in Table 1 are characterized by their date, source time, source location, moment-magnitude M_w, and the maximum reported wave height H_{max}. In addition to the main calculated value, that is the tsunami intensity I, the table contains several additional parameters, summarizing the data available for the intensity calculation. These parameters are the total number of height measurements (N), available for a particular event, the number of near-field heights (N_{NF}), near-field maximum height H_{max_NF} (in all but one case H_{max_NF} coincides with the H_{max} value), the near-field minimum height H_{min_NF} and the average height H_{av_NF}, that is used in the I calculation.

In Table 1, the values of tsunami intensity I are listed with different accuracy that depends on the value N_{NF} used in the calculation. For the events with $N_{NF} > 10$, the value I is listed as it comes from the calculation, with 2 digits after the decimal point, for the events with $N_{NF} = 6–10$ it is listed with 1 digit after the decimal point, for the events with $N_{NF} = 1–5$ it is rounded to the nearest half of grade (0.5), and to the integer value for the events with no height measurements on the nearest coast. The latter values represent the expert estimates based on the correlation between the near-field run-up heights and the far-field wave amplitudes.

3. Discussion

Table 1 lists 44 recent tsunamigenic events for which we could calculate or somehow estimate the tsunami intensity on the SOLOVIEV–IMAMURA scale. It represents only 27 % of the total number of tsunamigenic events available in the NCEI/WDS database for 2000–2014. For the remaining 120 events even a rough estimation of the tsunami intensity is not possible. Some of these events were non-seismic (volcanic, meteorological, resulting from slope failure, harbor seiches), for which this scale is not applicable. They can only be quantified with a scale based on their overall energy, for instance, ML scale, proposed by (MURTY and LOOMIS 1980). However, nearly 100 events occurring between 2000 and 2014 were generated by submarine earthquakes and their absence in Table 1 results from the dearth of near-field height measurements. Quantification of these events in terms of their intensity is important both for studying the temporal tsunami recurrence in the different tsunamigenic regions and for the rational selection of magnitude thresholds for issuing operational tsunami warnings.

The most important issue about seismogenic tsunamis is how their overall size or energy depends on the source magnitude of the parent seismic event. As was stated in "Introduction," in the absence of practical methods of estimating the total tsunami energy, the intensity I on the SOLOVIEV–IMAMURA scale is a parameter that most closely relates to the overall size of a tsunami. For seismogenic tsunamis, the best parameter for quantification of a source event is, of course, its moment-magnitude M_w that is now routinely being determined for all earthquakes above magnitude threshold 5.5 or so with an accuracy about ±0.1 (STORCHAK et al. 2015).

The relationship between the tsunami intensity on the SOLOVIEV–IMAMURA scale and the source magnitude was first studied by (CHUBAROV and GUSIAKOV 1985) and later, using more complete data by (GUSIAKOV 2011). The main feature of this relationship is a large scattering of I values, exceeding six grades on the intensity scale, for events with a given magnitude M_w in the magnitude range 7.0–8.0, where most tsunamigenic earthquakes occur. This fact makes the operational prediction of tsunami heights at the coast, based solely on seismic data to be a difficult, if at all possible, task.

Figure 3a reproduces Fig. 3 from GUSIAKOV (2011) modified for the period from 1900 to 1999 and represents the $I(M_w)$ dependence for tsunamigenic events of the XX century having estimates for I and M_w. A similar $I(M_w)$ relationship for the last 15 years (2000–2014) is shown in Fig. 3b. As one can see, increasing the accuracy of intensity determination for recent tsunamis (in particular, for those that were followed by post-event field surveys) does not narrow the scattering in $I(M_w)$ dependence. In fact, looking at Fig. 3a, one can conclude that this dependence actually does not exist. There is only a general tendency for an increase in the tsunami intensity with an increase in the source magnitude. This tendency pretty well fits the predicted dependence I on M_w.

$$I = 3.55 M_w - 27.1, \quad (3)$$

obtained in (CHUBAROV and GUSIAKOV 1985) on the basis of numerical simulation of tsunami generation

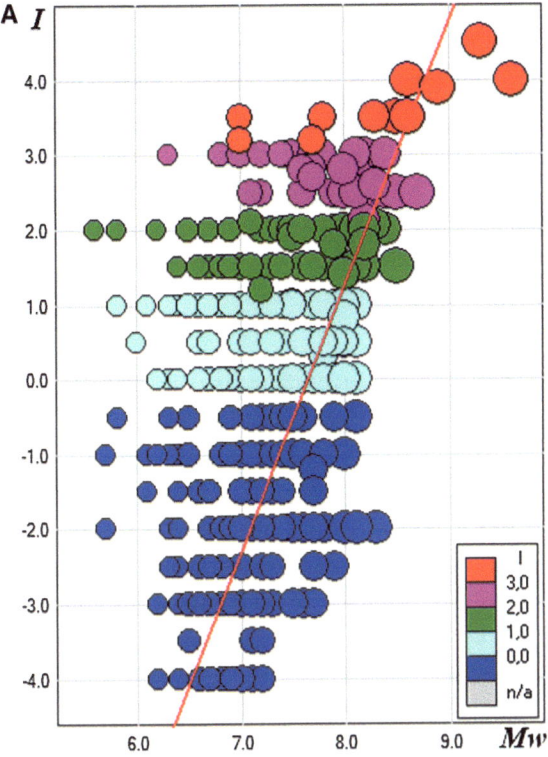

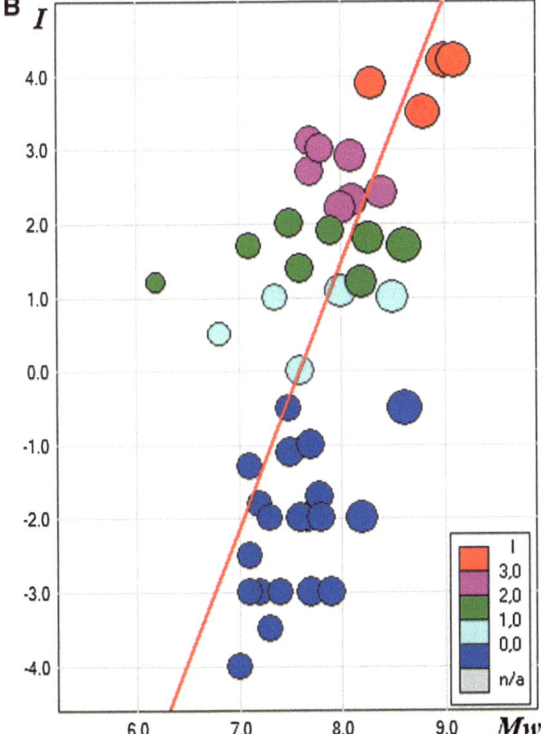

Figure 3
a Tsunami intensity I on the Soloviev–Imamura scale versus magnitude M_w for tsunamigenic earthquakes in the World Ocean from 1900 to 1999. Events are shown as *circles with the color* depending on tsunami intensity and size proportional to the earthquake magnitude. *Solid red line* shows the dependence I on M_w as obtained in (Chubarov and Gusiakov 1985). **b** Tsunami intensity I on the Soloviev–Imamura scale versus magnitude M_w for tsunamigenic earthquakes in the World Ocean from 2000 to 2014. Events are shown as *circles* with the *color* depending on tsunami intensity and size proportional to the earthquake magnitude. *Solid red line* shows the dependence I on M_w as obtained in (Chubarov and Gusiakov 1985)

and propagation on a model relief typical for island arc regions.

The reasons for this large scattering are multifold. First, there are differences in the focus depth and the source mechanisms. Second, there are differences in the source location (marginal seas, subduction zones, deep-water oceanic plate, etc.). Third, and possibly the most important, is the degree of involvement of secondary mechanisms (foremost being submarine slides and slumps) in the tsunami generation process. In greater details, with examples for specific historical tsunamigenic events, including the known tsunami-earthquakes which occurred from 1896 to 2006, this subject is discussed in two earlier publications (Gusiakov 2001, 2011).

In this paper, we have confined our analysis to the 164 tsunamigenic events in the World Ocean from 2000 to 2014. However, the actual number of recent tsunamis may be higher. Table 2 lists submarine earthquakes with magnitude $M_w \geq 7.5$ and a source depth less than 100 km for which the occurrence of tsunami is still unknown. This magnitude is equal to or well above the threshold value for immediate issuing of a tsunami warning for all the main tsunamigenic regions in the Pacific and elsewhere. Normally, such strong submarine earthquakes generate at least a weak tsunami whose signature is clearly visible on coastal or deep-water records. The absence of these events in the tsunami databases means either the absence of recording instruments in the vicinity of the source area or that the routine search for tsunami signals on the instrumental records was not made carefully enough. More progress in both cases will further facilitate the work on improvement of tsunami cataloging.

Table 2

List of strong ($M_w \geq 7.5$) shallow (depth <50 km) submarine earthquakes in 2000–2014 having no entries in the NGDC tsunami database

Date (YYYY, MM, DD)	Time (UTC)	Lat.	Long.	Depth (km)	M_w	Source region
2000.06.04	16:28:28	−4.61	102.06	35	7.9	Sumatra, Indonesia
2000.06.04	16:39:48	−4.64	102.01	35	7.6	Sumatra, Indonesia
2000.11.16	07:42:20	−5.20	153.14	32	7.8	New Ireland region
2000.11.17	21:01:59	−5.54	151.94	37	7.8	New Ireland region
2001.07.07	09:38:43	−17.50	−71.76	10	7.6	Peru
2003.07.15	20:27:53	−2.60	68.35	10	7.5	Carlsberg Ridge
2003.08.04	04:37:24	−60.56	−43.52	22	7.6	Scotia Sea
2007.09.12	23:49:04	−2.57	100.76	35	7.9	Sumatra, Indonesia

Earthquake parameters (origin time in UTC and magnitude M_w) are taken from the ISC-GEM catalog (STORCHAK et al. 2015)

4. Conclusion

1. Based on the available run-up and tide gage measurements, the tsunami intensity on the SOLOVIEV–IMAMURA scale has been determined for 44 seismogenic tsunamis that occurred in the World Ocean during the last 15 years. Among them there were two transoceanic mega-tsunamis (the 2004 Indian Ocean and 2011 Tohoku) and ten destructive tsunamis, which resulted in considerable damage and human fatalities. The rest were non-fatal regional and local events, even though some of them resulted from submarine earthquakes with great source magnitude ($M_w = 8.0$ or higher).

2. Even for the most recent events, the accuracy of intensity values is quite different and varies from ±0.1 up to ±1 depending on the number of available wave heights at the nearest coast. An accurate and reliable intensity determination is possible for the major tsunamis, having more than 10 near-field height measurements. Most of these tsunamis (17 of 18) were followed by post-tsunami field surveys carried out by national and international survey teams.

3. The determination of intensity for non-destructive events is typically based on a small number (1–10) of height measurements, some of them being non-instrumental (witness accounts) or made far away from a source area. The accuracy of these determinations is about ±0.5. The most problematic are intensity estimates for weak tsunamis recorded only by coastal or ocean bottom instruments. For these events, only expert estimates are possible based on an approximation (in the area of small amplitudes) of the rough correlation between the near-field run-up heights and the far-field wave amplitudes obtained for larger events provided both with the run-up and instrumental data. The accuracy of these intensity estimates is of order ±1.

4. Increasing the accuracy of the intensity determination for many recent tsunamis (in particular, for those that were the subject of post-event field surveys) does not narrow the scattering in the $I(M_w)$ dependence. This shows that in addition to the source magnitude, there are some other parameters controlling the tsunami generation process (i.e., source mechanism, source depth, depth of water in the generation area). Another important factor responsible for the absence of direct correlation of the tsunami intensity with the earthquake magnitude is the degree of involvement of secondary generation mechanisms such as submarine slumping. As was shown in (GUSIAKOV 2001), slumping could significantly augment the tsunami potential of nearly 30 % of all tsunamigenic earthquakes that occurred in the Pacific in the XXth century.

5. Despite the great improvement in the tsunami-recording network of coastal tide gages and deep-water instruments during the last two decades, the data for reliable estimates of the tsunami intensity are still problematic. Moreover, for 8 strong ($M_w > 7.5$) submarine earthquakes, which occurred during the last 15 years, there are no entries in the global tsunami databases that would provide evidence for the occurrence of any tsunami.

Acknowledgments

The author wishes to thank Fred Stephenson and Paula Dunbar for carefully reading the manuscript and making numerous suggestions for its improvement as well as Tamara Kalashnikova for assistance in figures and tables preparation. The work was supported by the Russian Science Foundation Grant 14-17-00219.

REFERENCES

CHUBAROV, L.B., and GUSIAKOV, V.K. (1985), *Tsunamis and earthquake mechanism in the island arc region*, Sci. Tsunami Hazards 3(1), 3–21.

GUSIAKOV, V.K. (2001), "Red", "Green" and "Blue" Pacific tsunamigenic earthquakes and their relation with conditions of oceanic sedimentation, In *Tsunamis at the End of a Critical Decade*, (Ed. G. HEBENSTREIT) (Kluwer Academic Publishers, Dordrecht-Boston-London), pp. 17–32.

GUSIAKOV, V.K. (2003), *NGDC/HTDB meeting on the historical tsunami database proposal*, Tsunami Newsletter, XXXV(4), 9–10.

GUSIAKOV, V.K. (2009), Tsunami history—recorded, In *The Sea, Vol.15, Tsunamis*, (Eds. A. Robinson and E. Bernard) (Cambridge, USA, Harvard University Press), pp. 23–53.

GUSIAKOV V.K. (2011), *Relationship of tsunami intensity to source earthquake magnitude as retrieved from historical data*, Pure Appl. Geophys., 168(11), 2033–2041. doi:10.1007/s00024-011-0286-2.

GUSIAKOV, V.K., MARCHUK, AN.G., and OSIPOVA, A.V. (1997). *Expert tsunami database for the Pacific: motivation, design, and proof-of-concept demonstration*. In *Perspectives on Tsunamis Hazard Reduction*. (Ed. G. HEBENSTREIT) (Kluwer Academic Publishers, Dordrecht-Boston-London), 21–34.

HTDB/WLD Historical Tsunami Database for the World Ocean (2015), Web-version is available at http://tsun.sscc.ru/nh/tsunami.php.

IIDA, K., (1963), *Magnitude, energy and generation mechanisms of tsunamis and a catalogue of earthquakes associated with tsunamis*. Proc. Tsunami Meeting, 10th Pacific Sci. Congress, IUGG Monograph, 24, 7–18.

IMAMURA, A., (1942), *History of Japanese tsunamis*. Kayo-No-Kagaku (Oceanography), 2(2), 74–80 (in Japanese).

MOFJELD, H. (2009), Tsunami measurements, In *The Sea, Vol.15, Tsunamis*, (Eds. A. ROBINSON and E. BERNARD) (Cambridge, USA, Harvard University Press), pp. 201–235.

MURTY, T.S., and LOOMIS, H.G. (1980), *A new objective tsunami magnitude scale*, Geod., 4, 267–282.

National Centers for Environmental Information/World Data Service (NCEI/WDS): Global Historical Tsunami Database. National Centers for Environmental Information, NOAA. doi:10.7289/V5PN93H7.

Preliminary Determination of Epicenters (PDE), a weekly and monthly publication, National Earthquake Information Center, U.S. Geological Survey, Golden, Colorado, 1971 to present.

RABINOVICH, A., and EBLÉ, M. (2015), *Deep-ocean measurements of tsunami waves*, Pure Appl. Geoph., doi:10.1007/s00024-015-1058-1.

SOLOVIEV, S.L. (1972), *Recurrence of earthquakes and tsunamis in the Pacific Ocean*. In Volny Tsunami, Trudy SakhCSRI, 29, 7–47 (in Russian).

SOLOVIEV, S.L., and GO, CH.N. (1974), *Catalogue of Tsunamis on the Western Shore of the Pacific Ocean*, Nauka, Moscow, 309 pp. [in Russian; English Translation: Canadian Transl. Fish. Aquatic Sci., No. 5077, Ottawa, 1984, 439 pp.].

SOLOVIEV, S.L., and GO, CH.N. (1975), *Catalogue of Tsunamis on the Eastern Shore of the Pacific Ocean*, Nauka, Moscow, 202 pp. [in Russian; English Translation: Canadian Transl. Fish. Aquatic Sci., No. 5078, Ottawa, 1984, 293 pp.].

SOLOVIEV, S.L. (1978), *Basic data on tsunamis on the Pacific coast of the USSR*, In IZUCHENIE Tsunami v Otkrytom Okeane, Nauka, Moscow, pp. 61–135 (in Russian).

STORCHAK, D.A., GIACOMO, D.DI, ENGDAHL, E.R., HARRIS, J., BONDÁR, I., LEE, W.H.K., BORMANN P., and A. VILLASEÑOR (2015), *The ISC-GEM Global Instrumental Earthquake Catalogue (1900–2009): Introduction*, Phys. Earth Planet. Int., 239, 48–63; doi:10.1016/j.pepi.2014.06.009.

WINITDB (2009) *Window-based graphic shell for the Integrated Tsunami Data Base, Version 5.16 of December 31, 2009, CD-ROM*, Tsunami Laboratory, ICMMG SD RAS, Novosibirsk, Russia.

(Received April 8, 2015, revised May 2, 2015, accepted May 4, 2015, Published online May 31, 2015)

Deep-Ocean Measurements of Tsunami Waves

ALEXANDER B. RABINOVICH[1,2] and MARIE C. EBLÉ[3]

Abstract—Deep-ocean tsunami measurements play a major role in understanding the physics of tsunami wave generation and propagation, and in improving the effectiveness of tsunami warning systems. This paper provides an overview of the history of tsunami recording in the open ocean from the earliest days, approximately 50 years ago, to the present day. Modern tsunami monitoring systems such as the self-contained Deep-ocean Assessment and Reporting of Tsunamis and innovative cabled sensing networks, including, but not limited to, the Japanese bottom cable projects and the NEPTUNE-Canada geophysical bottom observatory, are highlighted. The specific peculiarities of seafloor longwave observations in the deep ocean are discussed and compared with observations recorded in coastal regions. Tsunami detection in bottom pressure observations is exemplified through analysis of distant (22,000 km from the source) records of the 2004 Sumatra tsunami in the northeastern Pacific.

Key words: Tsunami measurements, bottom pressure, bottom cable stations, tsunameter, DART stations, sea level, long waves, tsunami spectra, signal extraction, longwave background noise.

1. Introduction: Historical Overview

Records along coastal margins dominate measurements of sea level changes that have been made by instruments over the recent historical span of approximately 200 years. The main purpose of these measurements was to study tides, storm surges, seasonal oscillations, and long-term sea level variability (PUGH and WOODWORTH 2014). This in large measure

---

In memory of G. R. Miller (1930–1976), R. R. Harvey (1939–1978), and S. L. Soloviev (1930–1994).

[1] Department of Fisheries and Oceans, Institute of Ocean Sciences, 9860 West Saanich Road, Sidney, BC V8L 4B2, Canada. E-mail: a.b.rabinovich@gmail.com

[2] P.P. Shirshov Institute of Oceanology, Russian Academy of Sciences, 36 Nakhimovsky Prosp., Moscow 117997, Russia.

[3] Pacific Marine Environmental Laboratory, National Oceanic and Atmospheric Administration, 7600 Sand Point Way NE, Seattle, WA 98115, USA.

reflects the accessibility of the coastal environment and the need of information for navigation and safety. Embedded in the records, however, were relatively high-frequency oscillations attributed to seiche and tsunami events (with periods from a few minutes to several hours). In particular, the first records of tsunami waves were discovered in marigrams on December 23, 1854 at the ports of San Francisco and San Diego (California, USA). The tsunami waves, induced by a magnitude 8.3–8.4 earthquake off the coast of Japan, crossed the Pacific Ocean and reached the coast of California in 12.5 h (LANDER *et al.* 1993); it is interesting that this particular tsunami strongly damaged the Russian frigate *Diana* in the Port of Shimoda (Japan) (SOLOVIEV and GO 1974), introducing Russians to this phenomenon for the first time. Twenty-five years later, one of the first tsunami records in Europe was obtained at Genoa (Italy) after the catastrophic 1879 Ligurian earthquake ($M_w = 7.9$) (EVA and RABINOVICH 1997). Later still, fundamental examination of seiche and tsunami oscillations in numerous bays and harbors of Japan was provided by HONDA *et al.* (1908); the results of that study are still actively used by specialists (e.g., RABINOVICH 1993).

A huge amount of sea level data has been accumulated over the years. As previously noted, most of these data are observations related to coastal measurements, which only indirectly characterize sea level variations in the open ocean. Deep-ocean recordings of sea level variations have, until recently, remained relatively rare due to the significant technological and engineering challenges. In coastal regimes, nonlinearity, friction, aliasing, instrumental noise, temperature fluctuations, and the influence of wind waves act in concert to complicate tsunami detection and distort the tsunami signal itself (RABINOVICH 1993; PUGH and WOODWORTH 2014). Although many of these same factors provide challenges in the deep ocean, the most specific problems are primarily due to the great depth of instrument deployment (and, hence, the great pressure that bottom sensors, acoustic modems, acoustic releases, and power supplies must be engineered to withstand). In practice, an instrument deployed at a depth of several kilometers is required to provide the measurements necessary to resolve tsunami waves in the open ocean where typical amplitudes of tsunami waves range from a few millimeters to a few centimeters, i.e., ensuring accuracies of the order 10^{-6}–10^{-7}. Deployment and maintenance of deep-ocean systems that are required to transmit data in real time for effective tsunami warning can also be logistically difficult and expensive endeavors. Because of these challenges, direct sea level measurements in the open ocean began only in the second half of the 20th century (see overviews of SNODGRASS 1969; UNESCO 1975; RABINOVICH 1993; MOFJELD 2009; MUNGOV et al. 2013). Ocean tides, as well as investigation of general astrophysical and geophysical problems, marine geodesy, and navigation, were the main factors stimulating those studies (PUGH and WOODWORTH 2014).

Specifically, tsunami was one of the most urgent problems requiring open-ocean sea level measurements. Scientific knowledge of the generation and physical properties of tsunami waves has for centuries been based only on coastal observations. Meanwhile, direct measurements of tsunami waves in the open ocean provide important information on the physical nature of these waves that is otherwise not possible to derive from coastal observations. First, the recording of tsunami waves near the source of generation is of primary significance for examination and further development of the current theories of tsunami generation and propagation. Second, the creation of an automated system of sea level monitoring in the open ocean is the most promising approach for providing reliable and advance tsunami warning. Tsunami forecasting methods based on seismological considerations alone have, until recently, been exclusively used. In these methods, the measure of earthquake energy (momentum magnitude M_w), is the basis for tsunami alarm when the earthquake exceeds a certain threshold (usually $M_w \geq 7$–7.5, depending on the source location). This method has historically resulted in frequent false alarms since the amount of rupture energy imparted to the water column varies so that not every earthquake, even a strong one, induces a tsunami, and not every tsunami is dangerous beyond a local region near the source (SOLOVIEV 1968). The hydrophysical forecast method (SOLOVIEV 1968; JAQUE and SOLOVIEV 1971), which is based on direct deep-ocean measurements of tsunami waves, provides more accurate information about the tsunami wave itself, making more reliable forecasts possible.

Deep-ocean tsunami measurements are also important because they are not distorted by near-coast refraction, nonlinear wave interactions, and scattering by coastal and bottom irregularities. Also, unlike near-coast observations, those in the deep ocean are not affected by wave trapping and shelf resonance, bottom currents in shallow waters, and resonance of local water basins (bays, inlets, and harbors, where tide gauges are normally situated). The study of these phenomena is of undoubted theoretical and applied interest. However, from a practical point of view, they are of limited use to tsunami forecasters and researchers and may impede the extraction of a "pure" (undistorted) tsunami signal. For these constituencies, deep-ocean and shelf measurements of longwave sea level oscillations provide highly valuable information about tsunami physics. In addition, direct measurements promote the significant improvement of numerical tsunami models for all applications. The level of background longwave noise in the open ocean is much smaller than near the coast, while the accuracy of modern bottom pressure sensors is typically substantially higher than that of coastal tide gauges. These coupled together enable researchers to investigate some fine physical features of tsunami waves, unrecognizable in coastal records.

In April 1965, the Lamont Geological Observatory deployed the cable station OBS-II 180 km offshore of the coast of California at a depth of ~ 4 km (NOVROOZI et al. 1966). The station worked for about 4 months. A new station, OBS-III, was deployed in approximately the same region in May 1966; this station successfully worked for more than 7 years (NOVROOZI 1972). Both stations were multipurpose in that they were equipped with varied instrumentation that, in addition to bottom pressure, measured currents, water temperature, sea bottom seismicity, and other geophysical parameters. The stations were connected to shore by heavily armored and quite expensive cable lines, transmitting all data to the Lamont Observatory in real time. Tsunami measurements were not the direct purpose of these stations; besides, the entire period 1965–1972 was seismically "quiet." Nevertheless, the stations provided a significant amount of valuable information about background longwave oscillations in the tsunami frequency band.

Similar instruments based on bottom pressure sensors installed in the shelf zone and connected to shore by a cable were specifically designed to provide tsunami measurements. In the former Soviet Union, the first attempts to provide such measurements were made in the late years of the 1960s at the Sakhalin Complex Scientific Research Institute (SakhCSRI, Yuzhno-Sakhalinsk) under the direction of S. L. Soloviev; respective Vibrotron recorders were elaborated on by V. M. Jaque (JAQUE and SOLOVIEV 1971). Parallel works were initiated by G. R. Miller, leader of the Joint Tsunami Research Effort (JTRE), NOAA (Honolulu, Hawaii, USA) based on instruments constructed by M. Vitousek (VITOUSEK 1965; VITOUSEK and MILLER 1970). Close cooperation between the USA and USSR in the tsunami studies, particularly between Sergey Soloviev and Gaylord Miller, resulted in two joint Soviet–American tsunami expeditions (in 1975 and 1978) (LAPPO and SOLOVIEV 1976; SOLOVIEV et al. 1976; KULIKOV et al. 1979). These expeditions represented milestones in international tsunami cooperation (Fig. 1) and were an important step in recording tsunamis in the open ocean. No strong earthquake occurred during the expeditions in the Pacific Ocean; hence, no deep-ocean tsunami records were obtained. Nevertheless, the expeditions brought priceless materials on long waves in the tsunami frequency band that allowed for detailed examination of the mechanisms of wave trapping and shelf resonance. These processes determine the formation of tsunami waves in the shelf zone and offshore. The materials of those expeditions provided the backbone of the monograph by EFIMOV et al. (1985) and were published in a series of papers (see, e.g., POOLE et al. 1980; KULIKOV et al. 1983). Recognition must here be given to Robert Harvey (Joint Institute for Marine and Atmospheric Research, University of Hawaii, Honolulu) for the key role he played in the preparation and elaboration of deep-ocean tsunami recorders and in the successes of these two expeditions. Robert was tragically lost to the sea in December 1978 during the "Holoholo" accident (KARL 2004).

The first reliable deep-ocean tsunami records were obtained on the shelf of Shikotan Island (DYKHAN et al. 1981, 1983) and in the deep ocean near California (FILLOUX 1982, 1983). An earthquake with magnitude $M_w = 7.1$ occurred near the Lesser Kuril Islands on February 23, 1980; it induced a small tsunami recorded by tide gauges at Shikotan, Kunashir, Iturup, and Hokkaido Islands (Fig. 2a, b). A seafloor station operated on the shelf of Shikotan Island at the time of the event; the station was installed in the summer of 1979 at a depth of 113 m and 8 km offshore. The station recorded tsunami waves with maximum trough-to-crest height of about 15 cm (Fig. 2c).

At approximately the same time, a tsunami record was obtained in the deep ocean. An autonomous bottom pressure station with a sensor developed by FILLOUX (1982) was deployed 150 km off the coast of southern California at a depth of about 3200 m. An earthquake of magnitude $M_w = 7.6$ occurred on March 14, 1979, near the southern coast of Mexico, about 1000 km to the south of the station site (Fig. 3). After retrieving the instrument and analyzing the internally stored data (about 1.5 years after the earthquake), it was found that tsunami waves of about 1 cm in height and having a period of 40 min were produced by the earthquake and recorded (FILLOUX 1982). It is clear from the record (Fig. 3) that the natural longwave background noise (including instrumental noise) observed at the station before tsunami arrival was quite low (<1 mm), and therefore a 1-cm tsunami wave was evident.

Even these two earliest of records enabled specialists to gain a general idea about the character of tsunami waves near the areas of their generation and yielded important scientific results. First, it was found that, despite previous ideas, a tsunami in the open ocean is not a single wave, but a train of waves that can be quite long. Second, these records confirmed the principal possibility and effectiveness of deep-ocean observations for timely tsunami forecast: the tsunami

Figure 1
a Graphical representation of a poster prepared in memory of the Soviet–American tsunami expedition on the r/v "Valerian Urivaev," August–October 1978. **b** Members of the crew of the expedition on Saipan Island, Micronesia (September 1978): (*1*) Boris Bobrovsky (Sakhalin Hydromet), head of the expedition; (*2*) Ted Murphy (UH, USA); (*3*) Igor Panpurin, captain; (*4*) Evgueni Kulikov (SakhCSRI), deputy head of the expedition, at present: head of the Tsunami Laboratory, IORAS, Moscow; (*5*) Robert Harvey (UH, USA); (*6*) Alexander Spirin (SakhCSRI), (*7*) Steve Poole (UH, USA); and (*8*) Alexander Rabinovich (SakhCSRI). Photo by Andrey Kharlamov (SakhCSRI; at present: Tsunami Laboratory, IORAS, Moscow)

signals were recorded at these stations about 1–2 h in advance of wave arrivals at the nearest coastal settlements. A tsunami record obtained by FILLOUX (1982, 1983) demonstrated that even weak tsunami waves with a height of 1–2 cm can be reliably identified based on deep-ocean measurements.

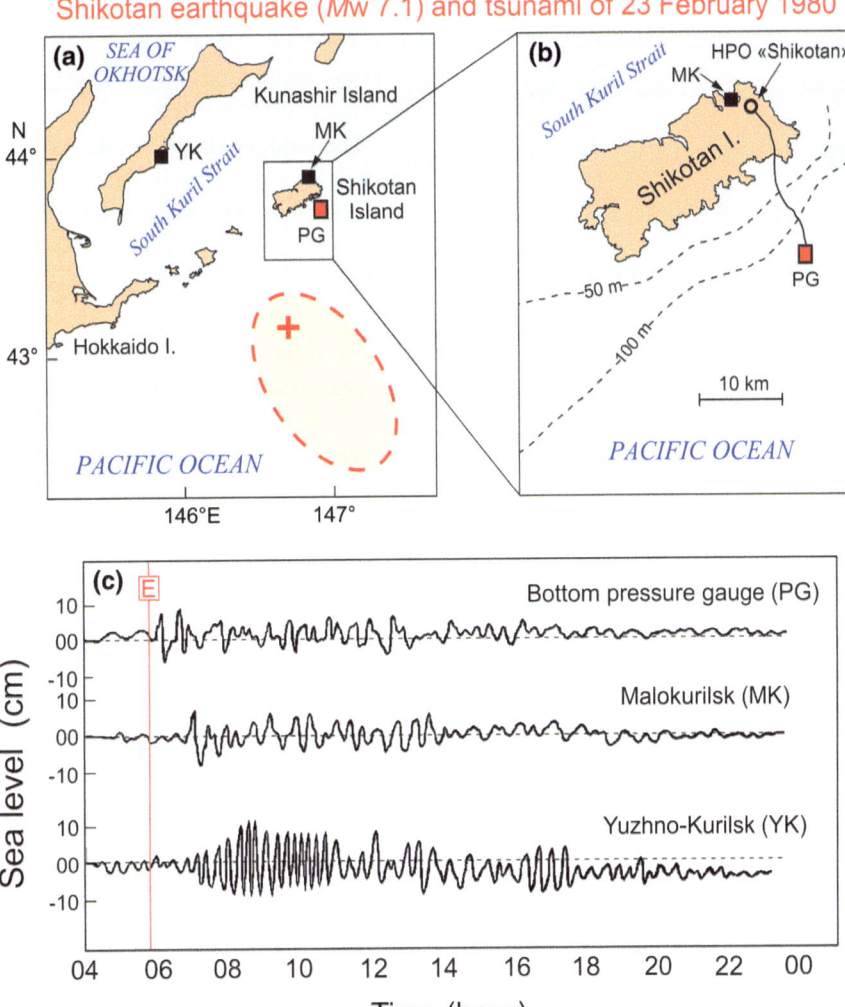

Figure 2
Tsunami of 23 February 1980 recorded off the coast of Shikotan Island, South Kuril Islands. **a** Map of the region showing the location of the M_w 7.1 earthquake epicenter (*red cross*), tsunami source region (*oval area* bounded by the *red dashed line*), position of the cabled bottom pressure gauge (PG), and coastal tide stations at Yuzhno-Kurilsk (YK) and Malokurilsk (MK). **b** The Shikotan Island cable line that runs from the Hydrophysical Observatory (HPO) "Shikotan" to the bottom pressure gauge. **c** Tsunami records from the bottom and coastal gauges; the *vertical solid red line* labeled "E" marks the time of the earthquake

Another important finding of FILLOUX (1982, 1983) was that, after strong earthquakes, bottom pressure sensors record not only tsunami waves, but also seismic waves, and first of all, surface Rayleigh waves (Fig. 3). Actually these earthquake-generated waves were found to be a distinctive property of all deep-ocean bottom pressure records (see comprehensive overview by WEBB 1998).

Typical phase speeds of Rayleigh waves are 3.9–4.0 km/s (FILLOUX 1983), i.e., approximately 20 times faster than the speed of tsunami waves in the open ocean. That is why Rayleigh waves arrive at stations significantly ahead of tsunami waves; For example, Rayleigh waves from the Chile earthquake of 27 February 2010 ($M_w = 8.8$) reached the NEPTUNE-Canada[1] cable geophysical stations installed

---

[1] NEPTUNE-Canada = Canadian North-East Pacific Underwater Networked Experiments (NEPTUNE) component of the Ocean Networks Canada (ONC).

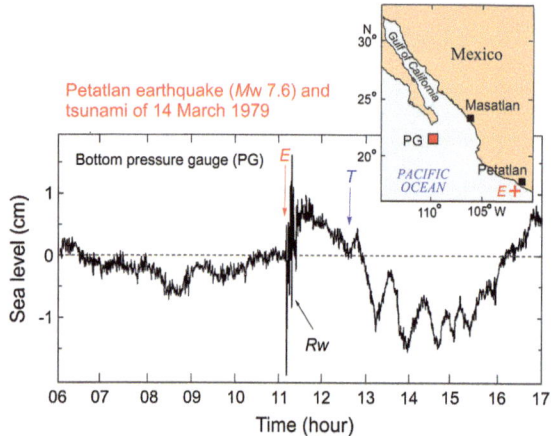

Figure 3
Record of the first tsunami measured in the open ocean. The tsunami was generated by the Petatlán earthquake (M_w 7.6) of 14 March 1979 off the southern coast of Mexico in the Pacific Ocean. The *red* and *blue vertical arrows* mark the time of the earthquake (*E*) and arrival of tsunami waves (*T*), respectively; "*Rw*" are the recorded Rayleigh waves that were observed almost immediately after the earthquake. The *inset* shows a map of the region with the earthquake epicenter indicated by the *red cross* labeled "*E*" and the location of the *bottom* pressure gauge is denoted by the *red square* labeled "*PG*" (modified from FILLOUX 1982)

on the southwestern shelf of Vancouver Island (Canada) within 47 min, while tsunami waves arrived approximately 15 h later (RABINOVICH et al. 2013a).

In fact, when seismic waves are passing by, a bottom sensor registers the pressure changes Δp_s, determined by oscillations of the seafloor vertical accelerations (KULIKOV and GONZÁLEZ 1996):

$$\frac{\Delta p_s}{\rho} = h\frac{d^2 z}{dt^2} = -\alpha h \omega^2, \qquad (1)$$

where ρ is the sea water density, z is the vertical displacement of the seafloor at moment t, a is the amplitude of the seismic wave, and ω is the angular frequency. As follows from (1), the range of seismically caused bottom pressure oscillations is proportional to the water depth of the respective instrument. This is the reason why Rayleigh waves were clearly recorded at the deep-ocean station near California (Fig. 3), but were not recorded near Shikotan (Fig. 2c). Similarly, after the 2010 Chilean earthquake, Rayleigh waves were recorded at three deep NEPTUNE stations near Vancouver Island: 1027-S (2660 m), CORK (2660 m), and 889 (1260 m), but were not observed at another NEPTUNE station, Barkley Canyon (410 m) (RABINOVICH et al. 2013a). The different types of seismic waves recorded by bottom pressure recorders are discussed by WEBB (1998).

The success of the Soviet–American tsunami expeditions and acquisition of the first open-ocean tsunami measurements gave rise to possibilities as to data application. Questions arose about the selection of instrument deployment sites optimal for forecasting, as well as for automatic detection of tsunami waves (SAXENA and ZIELINSKI 1981; ZIELINSKI and SAXENA 1983a, b). These efforts in the Soviet Union were largely stimulated by the 1980 decree of the Council of Ministers on development of the United Automatic System of Tsunami monitoring (UAST). The results were summarized in POPLAVSKY et al. (1988). Unfortunately, for known historical reasons, this program, as envisioned, remained unrealized. In addition to political realities, the theoretical and methodological designs were performed within the program but were significantly hampered by the technological capabilities existing at that time. However, 15–20 years later, during the active deployment of the autonomic and cable station systems throughout the world oceans, such studies turned out to be in demand (see, e.g., MOFJELD 2009).

The appearance of new digital technologies at the end of the 20th century significantly accelerated the development of instruments for measurement of tsunami waves in the deep ocean. Effective deep-ocean real-time reporting autonomous stations were designed in the USA in the early 1990s following a successful research program that began in the early 1980s. Referred to first as simply "RT" (real time) and then Deep-ocean Assessment and Reporting of Tsunamis (DART) systems, these instruments later became collectively known as "tsunameters" (GONZÁLEZ et al. 2005), an instrument known to yield reliable measurements of tsunami waves and provide real-time data transmission via modem and satellite communication technology. Based on the experience and success of the earliest research deployments of seafloor instruments and tsunami measurements (GONZÁLEZ et al. 1991; GONZÁLEZ and KULIKOV 1993), the DART program for continuous tsunami monitoring was developed and established in the USA (TITOV et al. 2005a; GONZÁLEZ et al. 2005). By

December 2004, the original North Pacific research network was replaced with six US operational DART stations. Additional sensing capability was provided by another DART station owned by Chile, which was sited offshore the central South American coast. The catastrophic Sumatra tsunami of 26 December 2004 that claimed an enormous number of victims resulted in an accelerated effort to develop today's global DART network and provide monitoring along those seismically active zones of the World Ocean known to have potential for tsunami generation. The DART network and forecasting strategy are considered in detail in following sections.

Other technologies, such as bottom cabled systems, have been developed to exploit all available resources for specific applications and scientific discovery. Autonomous systems are designed to balance power requirements with sampling frequency within a self-contained system deployed in an environment that is not easily accessible. In contrast, deep-ocean cabled stations take advantage of coastal proximity and unrestricted power. Bottom pressure transducers are installed as a cabled module anchored on the seafloor, while data recorders are mounted on coastal or floating platforms. These components may remain separate in space but are connected to one another by an underwater cable that enables them to function as a single unit to provide continuous monitoring of sea level activity. The continual transmission of data from a cabled pressure transducer to an observer on shore makes deep-ocean cable stations well suited to long-term sea level investigations. Such stations were found to be quite effective for the examination of tsunami waves and can be directly used for operative purposes, in particular, in the Tsunami Warning System (POPLAVSKY et al. 1988). Numerous cable measurements of long waves were secured in the past and are now maintained by the Institute of Marine Geology and Geophysics, Far Eastern Branch, Russian Academy of Sciences (IMGG, formerly SakhCSRI) in the coastal region and on the Far Eastern shelf (see, e.g., KOVALEV et al. 1991; DJUMAGALIEV et al. 1993; DJUMAGALIEV and RABINOVICH 1993; SHEVCHENKO et al. 2013, 2014).

The Japanese deep-ocean cable stations installed on the outer (oceanic) shelf and the continental slope of the Japanese Islands are of special interest for the tsunami problem. The first cable line was laid in August 1978 to the bottom station deployed ~ 100 km southeast of Omaezaki at a depth of 2202 m (the total length of the cable line was 154 km; TAIRA et al. 1985). The total cost of this line, including the instruments, exceeded $10 million dollars (US in 1978 prices). Since that time, the station complex (three bottom seismographs and one hydrophysical station with a bottom pressure sensor) has been operating for >36 years. In the early years of the 1990s, the first Japanese deep-ocean tsunami records were obtained from the cable stations (OKADA 1993, 1995); at present there are several tens of tsunami records (see, e.g., HIRATA et al. 2002; SATAKE et al. 2005; SAITO et al. 2011).

The weakest components of cable systems are the cables themselves. Mechanical damage of a cable inshore, corrosion, and wicking most often result in station mortality. Therefore, increased requirements are imposed to ensure cable quality. The cable armor is thickened, which, in turn, substantially increases the cost of each cable station. The cost to deploy and maintain a cable system compared with an autonomous tsunameter system is relative, depending on project needs and objectives. Cabled sensors have the potential to provide very high-frequency observations for tens of years but have upfront costs in the millions of dollars (US). Autonomous tsunameters, such as DART, have a much lower upfront cost, a few hundred thousand dollars (US), but may experience failures that require more frequent maintenance over their expected 3–5-year life span. With autonomous systems, too, there are data transmission considerations that include power and satellite telemetry costs.

2. Modern Deep-Ocean Systems for Tsunami Measurements

Modern tsunami observation platforms and systems can be broadly classified by power supply, sensor type, and application. In this section, we consider in more detail the autonomous DART system, which has to date proven to be a reliable platform for observing open-ocean tsunamis in support of research and operational forecasting, and cabled systems. Each system provides capabilities that are complementary to one another.

2.1. The DART System

The DART system is maybe the most important among all tsunami sensing technologies in view of global coverage and effective application for the Tsunami Warning System. The progression from a research program to the operational DART sensing network now in place began in the early 1980s with the successful deployment and recovery in 1983 of a bottom pressure recorder (BPR) package in the Equatorial Pacific. By the mid-1980s, a research array of four BPRs was established and maintained south of the Aleutian Islands in the North Pacific. An additional BPR was maintained offshore the US Oregon coast at Axial Seamount along the Juan de Fuca Ridge (MOFJELD et al. 1996). Over the next 10 years, post-recovery analyses of BPR data from nearly 100 deployments led to requirements of the tsunami measurement system through evolution in both technology and understanding of deep-ocean tsunami dynamics. Technology advancements resulted in improvements in everything from data sampling and storage to transducer and BPR platform designs. Data initially sampled at a rate of 56.25 s were later sampled every 15 s in order to minimize aliasing of the tsunami signal while still retaining the required resolution of approximately 0.5 (mm) (EBLÉ and GONZÁLEZ 1991). The high level of instrument noise introduced by reel-to-reel magnetic tape storage was eliminated in the late 1980s with the transition to microchip and, later, flash storage. The Paroscientific, Inc. pressure transducer too underwent one very important change. The early pressure transducers included in the evacuated cavity only a quartz crystal for the measurement of absolute pressure. The oscillation frequency of the crystal is a known function of the pressure exerted by the overlying water column, but because pressure is a function of temperature, a YSI thermistor external to the evacuated transducer chamber provided temperature measurements for correction of pressure. Time lag between thermistor and pressure measurements were later addressed when Paroscientific, Inc. added the measurement of temperature as a function of frequency counts within the evacuated chamber. The conformance equation used to convert frequency counts to engineering units of pressure was modified to replace temperature (°C) with frequency counts associated with changes in temperature. It is noted here that, while the relationship between temperature frequency counts and pressure is accurately determined by the manufacturer for each transducer, the calibration of these frequency counts to temperature (°C) is performed coarsely so that temperature frequency counts should be the basis for studies (GONZÁLEZ et al. 1991).

The BPR underwent a series of modifications during the years leading up to a real-time reporting capability. A circular BPR-A platform typically held, in an upright position, one acoustic release and a cylindrical titanium unit that housed all sensing and recording electronics. In the trial BPR-B design, all sensing and recording electronics were encapsulated in an evacuated dome situated atop encased syntactic foam meant to provide self-floatation. The BPR-C design was similar to that of the BPR-A in that the pressure sensing and recording electronics were again housed in a cylindrical titanium case. In this design though, the BPR lay prone on a square platform while the acoustic release remained upright. A fourth design, the BPR-D, introduced telemetry hardware onto the BPR-C platform for communication between the moored surface buoy and both the anchored BPR and land stations via satellites. Figure 4 shows the three principal BPR designs as they each appeared in the field during deployment or recovery operations.

The first-generation DART (DART I; see Fig. 5a) was prototyped in the early 1990s. With the tsunami detection capability established, the engineering focus was on a proof of concept for real-time data reporting and communications. A seafloor-anchored BPR-D was paired with a separately moored surface communication buoy and deployed for testing as the first real-time station in 1995. The DART I featured an automatic detection and reporting algorithm triggered by a threshold wave-height value selected in consideration of ambient background noise. The DART I performed as designed but proved to be logistically demanding and costly in terms of ship resources to deploy and recover due to the dimensions of the tower. The second-generation DART, the

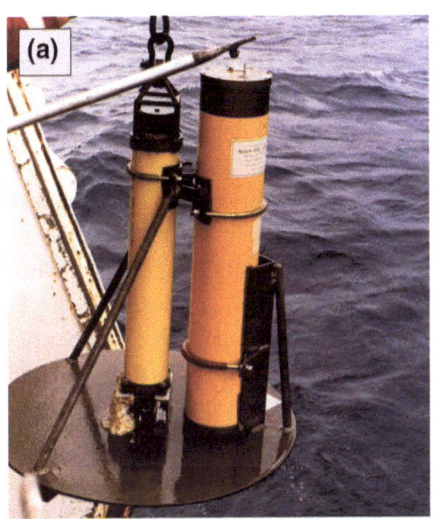

Figure 4
Evolution of bottom pressure recorders (BPRs) developed over the course of the research program that led to development of the DART system initially deployed in the late 20th century. **a** BPR-A. Deployed in the mid- to late 1980s, the BPR-A was designed to store data on magnetic tape at a rate of 56.25 s. **b** BPR-B. A limited number of gauges were deployed over two field seasons before being replaced with (**c**) BPR-C, the last design deployed as part of the research program. Acoustic modems were added to the BPR-C platform for communication with a surface buoy. The resultant BPR-D design is now deployed as part of the DART II system

DART II (Fig. 5b), incorporated two-way communications for tsunami data transmission on demand. Two-way communication allows authorized personnel remote control of the BPR for troubleshooting or for manually forcing "event" mode should data be needed for operations or research. The third-generation DART (Fig. 5c), termed Easy-to-Deploy (ETD), was designed to take full advantage of technology advancements both to address obsolescence and fail point issues and to reduce overall cost. In striking difference to predecessor designs, the ETD incorporates the BPR gauge and communication functions into a single, self-contained unit. The compact unit is typically deployed off a small vessel in areas where current velocities are not extreme enough to threaten instrument function and survival. Unit operational life span is typically greater than for the two-component design of the predecessor DART I and II systems because the ETD is designed to be disposable so is not limited by the life of the acoustic release battery, typically 2 years. A fourth-generation DART, the 4G, retains the ETD design

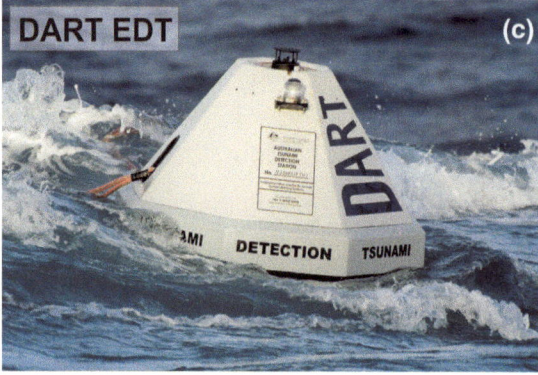

Figure 5
Evolution of the DART communication surface buoys: **a** DART I, **b** DART II, and **c** DART-EDT. The designs evolved from the DART I with its large tower to the low-profile, all-in-one ETD system to reduce deployment costs and take advantage of technological advances

(Fig. 5c) but addresses component obsolescence and further reduces the number of fail points while delivering improved sensing capabilities. The sampling scheme has been improved to burst sample at a rate of 1 Hz in order to separate a seismic signal from that of the tsunami itself. Data from a DART 4G sited in close proximity to a source region are rapidly available, thus reducing the time between tsunami generation and detection and in turn providing more rapid warning.

The US network of tsunameters currently deployed throughout the world oceans comprises almost exclusively DART II systems. The DART II operational system is illustrated in Fig. 6. The US network is augmented by a number of both ETDs and DART IIs that are owned and maintained by other nations and are operating in the Pacific Ocean. During the 2011 Tohoku Japan tsunami, it was a Russian-owned DART II that provided much of the data for forecasting tsunami wave propagation across the Pacific Ocean and inundation at selected communities. Station description and the installation scheme can be found at http://www.ndbc.noaa.gov/dart/dart.shtml (see also MOFJELD 2009; MUNGOV et al. 2013). An autonomous bottom pressure transducer, communication transducers, and a surface buoy with satellite antenna are the main elements of the stations (tsunameters). The seafloor communication transducer transmits to the buoy through an acoustic channel, and data are then relayed to Tsunami Warning Centers and the National Data Buoy Center (NDBC) via an Iridium satellite link.

In October 2014, 60 DART systems were in operation throughout the world oceans, including 47 stations in the Pacific Ocean (USA: 32; Australia: 6; Japan: 3; Chile: 2; Russia: 2; Ecuador: 2), 7 in the Atlantic Ocean (all USA), and 6 in the Indian Ocean (Australia: 3; India: 2; Thailand: 1) (Fig. 7). The system operates in two modes. In standard or normal operating mode, four samples per hour are transmitted to monitor system health. When an event is sensed, the system switches into event mode, in which the station transmits data for a few minutes at full 15-s resolution before transmitting strictly 1-min averages at predefined intervals. Event mode is triggered either (1) manually (by an operator) or (2) automatically with passage of a wave having an amplitude threshold higher than the undisturbed level (MOFJELD 2009).

DART stations are typically installed at depths of 1500–6000 m and represent today's advanced technology for tsunami wave monitoring. Most large

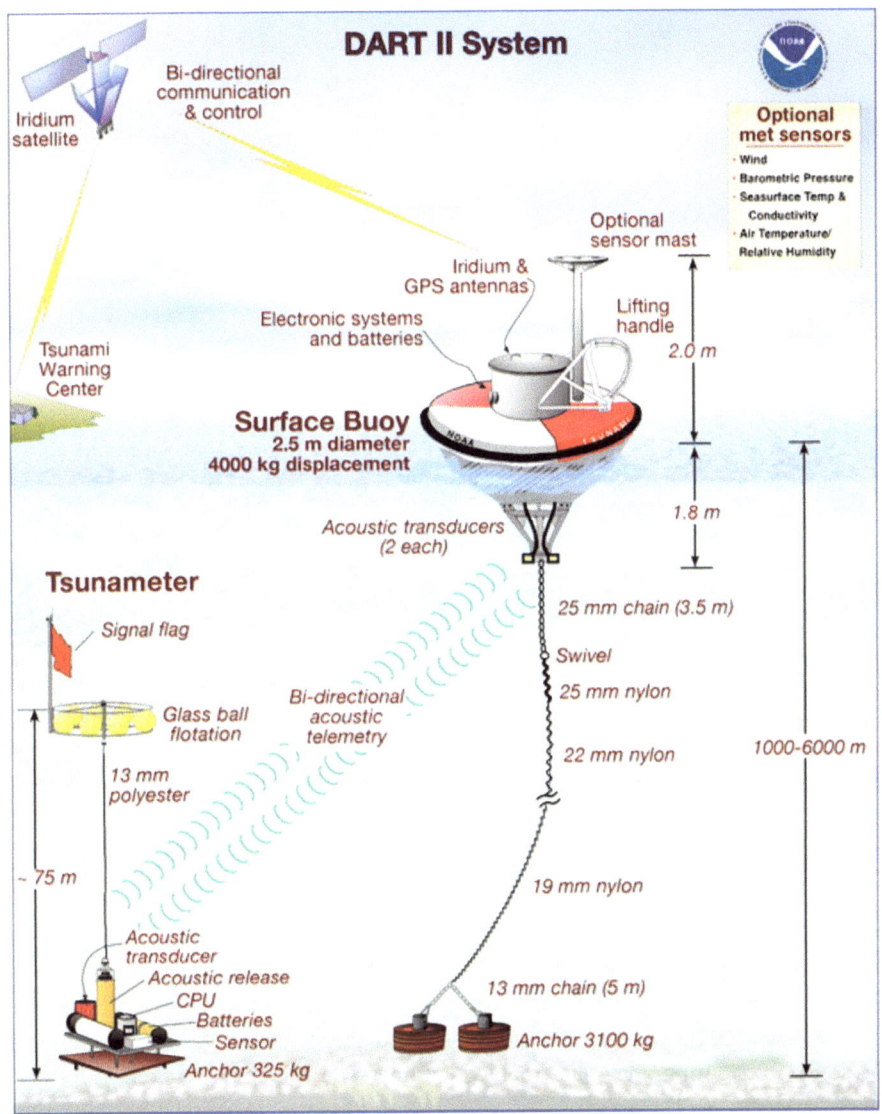

Figure 6
Schematic of the DART II system. Details of the bottom pressure recorder (BPR) and surface buoy design components are identified

tsunamis in the Pacific Ocean were recorded by DART sensors over the past 8 years, resulting in an accumulation of unique scientific material which sharply expanded our knowledge of tsunami wave generation and propagation in the deep ocean (MUN-GOV et al. 2013). In addition, these data have significantly improved real-time tsunami warning, changing from a qualitative forecast (will occur or not) to a quantitative one, i.e., the forecast of expected tsunami flooding at specific coastal sites (TITOV 2009), which is most important. This has made tsunami warning more reliable and precise and has reduced the cost of evacuation. The successful experience with DARTs in the world oceans and their high effectiveness for the Tsunami Warning System (TWS) persuaded the Russian Hydrometeorological Committee (ROSHYDROMET) to purchase a few of these stations: DART 21401 was put into operation on November 8, 2010, with an installation depth of 5264 m and installation site southeast of Iturup Island; DART 21402 was established on September 29, 2012 at depth of 5150 m, 530 km to

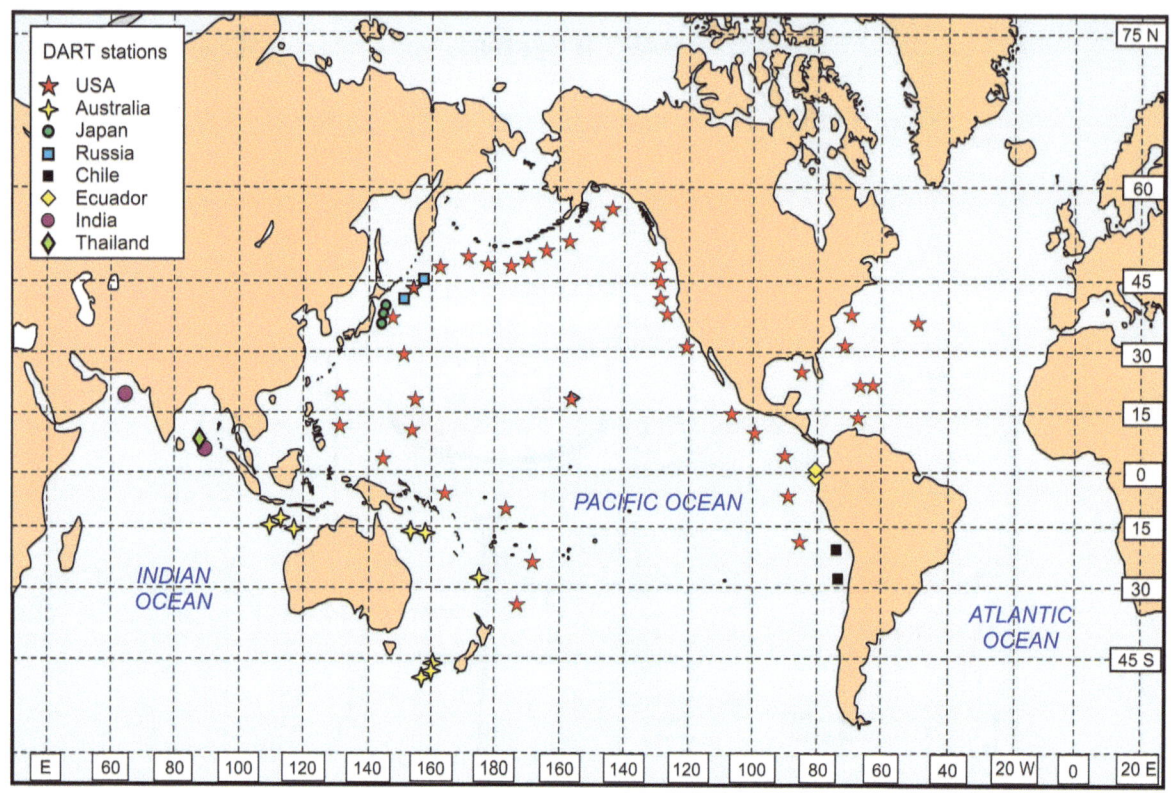

Figure 7
Location of the DART stations deployed throughout the world oceans. The array maintained by the USA is currently composed almost exclusively of DART II systems

the east of Simushir Island. During the first 4 months of operation (November 2010–March 2011), station 21401 recorded five tsunamis with sources off the coast of Japan. This station played an extremely important and useful role during the catastrophic Tohoku tsunami of 11 March 2011. Data provided for the accurate and timely forecast of tsunami wave heights along the Kuril Islands and in more remote regions of the Pacific Ocean (SHEVCHENKO et al. 2014).

2.2. Japanese Network of Cable Stations

To date, 11 cable lines with deep-ocean bottom stations have been installed in Japan. These lines were laid by different institutions (Table 1, Fig. 8a): (1) the Japan Meteorological Agency (JMA), (2) the Earthquake Research Institute (ERI), the University of Tokyo, (3) the National Research Institute for Earth Science and Disaster Prevention (NIED), and (4) the Japan Agency for Marine Earth Science and Technology (JAMSTEC) (MATSUMOTO and KANEDA 2009). The JMA stations are directly incorporated into the Tsunami Warning System of Japan. Other lines and stations are intended for scientific research (see reviews in MOFJELD 2009; LEVIN and NOSOV 2009; JOSEPH 2011). However, according to an agreement, the organizations and owners of these cable lines should transmit real-time data to JMA in the event of dangerous tsunamis. Data from Japanese cable stations have actively been used for the reconstruction of the tsunami source produced by the catastrophic Tohoku earthquake of 11 March 2011, and for examination of parameters of these waves in the near-field zone (see, e.g., SATAKE et al. 2013).

The large-scale DONET-1 and DONET-2 (Dense Oceanfloor Network System for Earthquakes and Tsunamis) networks established by JAMSTEC should be especially noted (Fig. 8b). The DONET-1

Table 1

Deep-ocean cable stations and instruments offshore of Japan [according to (JOSEPH 2011; KANEDA et al. 2011; KAWAGUCHI et al. 2012), and personal communications with Japanese specialists]

No.	Region (Prefecture, Island)	Organization	Year	Cable line length (km)	Instruments (max. depth)
1	Omaezaki (Shizuoka Pref., Honshu I.)	JMA	1978	154	4 OBS, 1 PG (2200 m)
2	Katsuura (Chiba Pref., Honshu I.)	JMA	1985	126	4 OBS, 2 PG (4000 m)
3	Ito (Shizuoka Pref., Honshu I.)	ERI	1993	33	3 OBS (1350 m)
4	Hatsushima Island (Sagami Bay)	JAMSTEC	1993	9	1 OBS, 1 PG (1176 m)
5	Hiratsuka (Kanagawa Pref., Honshu I.)	NIED	1993	125	6 OBS, 3 PG (2340 m)
6	Kamaishi (Iwate Pref., Honshu I.)	ERI	1996	126	3 OBS, 2 PG (2700 m)
7	Muroto (Kochi Pref., Shikoku I.)	JAMSTEC	1997	130	2 OBS, 2 PG (2310 m)
8	Kushiro (Hokkaido Pref., Hokkaido I.)	JAMSTEC	1999	242	3 OBS, 2 PG (3460 m)
9	Omaezaki (Shizuoka Pref., Honshu I.)	JMA	2008	210	5 OBS, 3 PG (2060 m)
10	Owase[a] (Mie Pref., Honshu I.)	JAMSTEC	2010–2011	320 (+20 × 10)	21 + 2 OBS, 21 + 2 PG (4378 m)
11	Kii Suido[b] (Tokushima Pref., Shikoku I.)	JAMSTEC	2014–2016	502 (+29 × 10)	29 OBS, 29 PG (3532 m)

OBS ocean-bottom seismograph, *PG* pressure gauge

[a] Altogether, 21 stations have already been deployed in the frame of the DONET-1 project; two more stations are planned to be deployed in 2015

[b] This system, DONET-2, is under development (Fig. 3b); the first station was deployed in March 2014. All 29 stations are planned to be installed by the end of 2016

project started in 2006, and the first stations began operation in 2010. All five "clusters" (antennas) of four stations each began to function by the end of July 2011; the stations in clusters are connected to the main nodes with 10-km cables. One more station is superposed with an ocean drilling borehole. Each station is equipped with a broadband seismometer, accelerometer, and bottom pressure gauge. The main cable length is 320 km; it comes into the coast in Owase City on the Kii Peninsula (KANEDA 2011; KAWAGUCHI et al. 2012). The Tohoku earthquake of 11 March 2011 occurred before all the DONET-1 stations were installed. Nevertheless, ten stations were in operation working in real time and were enabled to collect highly valuable data on the physical mechanism of the earthquake and tsunami (MATSUMOTO and KANEDA 2013). The DONET-2 project is, in general, similar to DONET-1. It is planned to include seven clusters that include altogether 29 stations (Fig. 8b). The first station went online in March 2014. The entire network is planned for completion by the end of 2016 (KANEDA 2011; KAWAGUCHI et al. 2012). The main purpose of the project is the study of the Nankai seismic zone, which is considered as the most active and potentially dangerous zone off the coast of Japan.

The catastrophic Tohoku earthquake and tsunami of 11 March 2011 showed that the existing monitoring system, based mainly on coastal stations, is unable to ensure reliable Tsunami Warning System operation or support the required level of scientific research in this region. In view of this, an ambitious project has been designed by NIED with the participation of the University of Tokyo. The plan is to cover the entire subduction zone near the Japan and southern Kuril Trenches with a network of 154 bottom stations connected by a heavy cable of 5100 km in total length (Fig. 8c) (UEHIRA et al. 2012). The network, termed "S-net" (Seafloor Observation Network for Earthquakes and Tsunamis along the Japan Trench) (SATAKE 2014), is envisioned to provide Japanese researchers with prompt and precise detection of parameters of an occurring earthquake so as to reveal in advance the "hot zones" of strong expected earthquake motion. The NIED is installing the third of six loops (Fig. 8c) now (in February, 2015) (KENJI SATAKE 2015, Pers. Comm.).

2.3. Ocean Network Canada

The NEPTUNE-Canada, which is a component of the Ocean Network Canada, is a geophysical seafloor

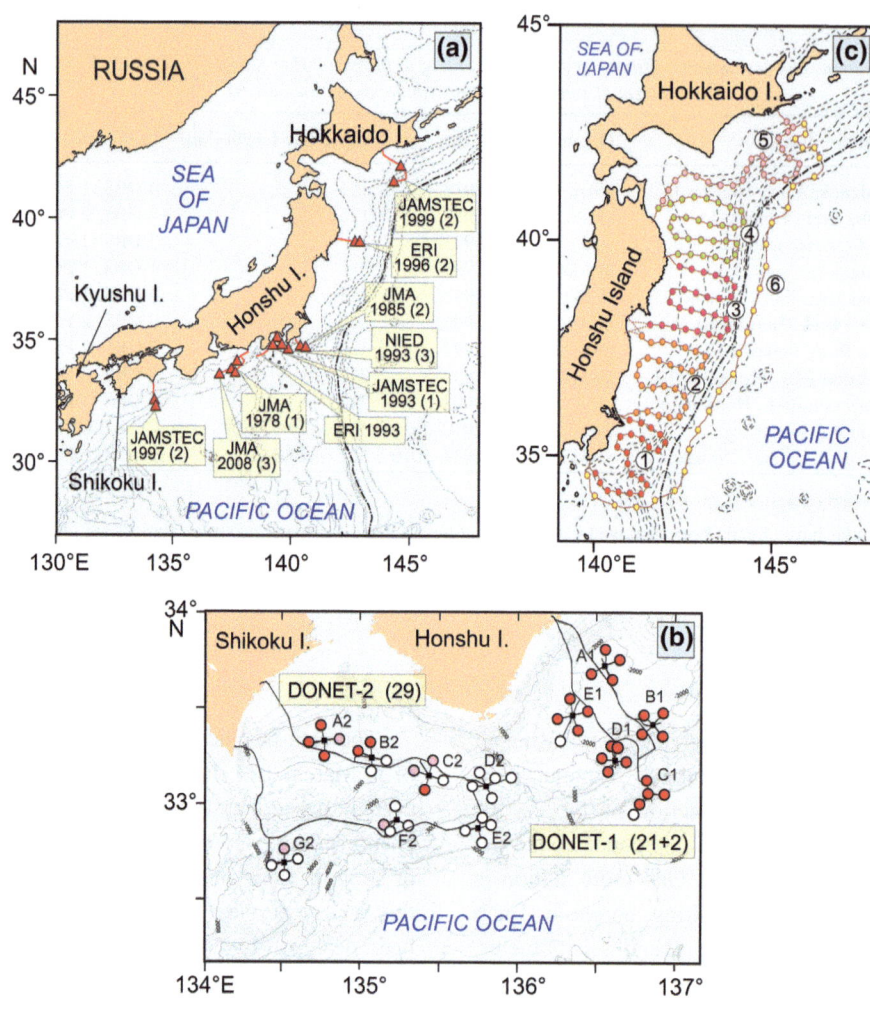

Figure 8
Japanese cable lines and deep-ocean stations. a General scheme of cable lines near the Pacific coast of Japan with indicated institutions–owners of the corresponding lines, years of deployment, and number of bottom pressure stations (*in brackets*); the stations themselves are denoted by *red triangles*. b Scheme of bottom stations within DONET-1 and DONET-2 projects. *Filled red circles* denote stations that are in operation, *pink circles* denote stations installed but not yet connected, and *empty circles* show locations of planned stations. c The NIED project ("S-net") of 154 bottom cable stations to be installed along the Japan Trench and southern Kuril Trench regions. *Numbers* in *circles* indicate six separate loops of the network; each loop is specified by *different color*

network that is another important system of cabled stations. This is a multipurpose observatory installed by the University of Victoria, British Columbia, on the shelf–continental slope of Vancouver Island to research geophysical and oceanographic processes (www.neptunecanada.com) (BARNES *et al.* 2008). Originally, the system included six bottom stations installed at depths of 100–2600 m. The total length of the underwater cable loop that connected the stations with each other and with the coast was 825 km; the cable is linked with landfall at Port Alberni, from where data are transmitted directly to the University of Victoria (Fig. 9). The station network started operating in real time on September 26, 2009, and four days later on September 30, 2009, the first tsunami, induced by a strong earthquake ($M_w = 8.1$) near the Samoa Islands, was detected (THOMSON *et al.* 2011). The tsunami waves crossed the Pacific Ocean in 11.5 h and were recorded by all NEPTUNE transducers and by several coastal tide gauges. These records provided information on the topography-influenced evolution of tsunami waves as they propagated along a route of open ocean–continental slope–shelf–coast. Data too, allowed for construction

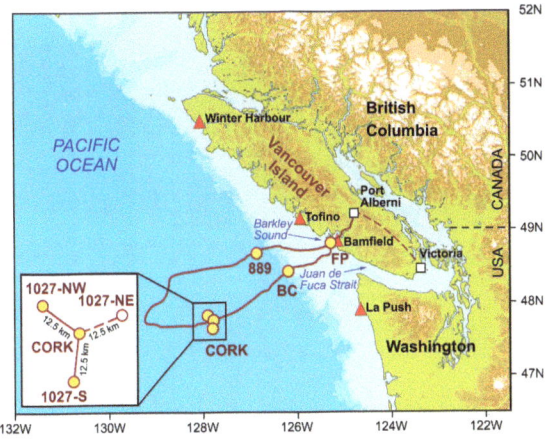

Figure 9
The Ocean Network Canada (NEPTUNE) bottom cable geophysical observatory on the shelf of Vancouver Island. The *red circles* show the location of bottom stations (BPRs), the *white circle* is the planned station, the *solid brown lines* are the bottom cables, the *dashed brown line* is the surface cable line, and the *red triangles* are coastal tide gauges

of a numerical model, one effectively for use as an operative real-time model in a regional TWS (see also RABINOVICH et al. 2013a). The NEPTUNE network has now recorded more than ten tsunamis in little more than 4 years of operation. These include comparatively weak tsunamis from distant sources, e.g., the Bonin tsunami of 21 December 2010, induced by the $M_w = 7.4$ earthquake near Japan.

3. Main Purposes and Problems of Deep-Ocean Tsunami Measurements

More than 120 tsunameter and cable stations are currently operating in the deep ocean. Most of these stations were installed over the past 10 years; after the catastrophic tsunami event of 26 December 2004 in the Indian Ocean. Many more are planned for installation by a number of nations in the near future. All stations transmit data in real time and can be used for a wide range of oceanographic, hydrodynamic, and geophysical studies.

3.1. Open-Ocean Tsunami Monitoring

This is a classic problem as formulated by SOLOVIEV (1968): early detection of a tsunami signal based on open-ocean tsunami monitoring data and, if necessary, the issuance of a tsunami warning. The reverse problem is equally important, i.e., to determine that a tsunami warning should not be issued or, alternatively, that an issued tsunami warning can be cancelled. Open-ocean measurements after any strong earthquake not only allow for detection and confirmation of tsunami presence (or absence), but also provide a means to estimate the danger posed to individual coastal regions (BERNARD et al. 2001).

3.2. Tsunami Flooding Forecast along the Coast

Deep-ocean DART stations provide a high level of real-time tsunami forecast for coastal sites. The forecast technique was developed at PMEL/NOAA (Seattle, USA) in conjunction with their development of the DART system. The approach is based on applying a set of unit prototype earthquake sources with predefined tsunami waveforms. In an operative regime, i.e., in the case of a tsunamigenic earthquake, the Short-term Inundation Forecasting for Tsunamis (SIFT) forecasting system forms a superposition of these unit sources approximating a real tsunami source. A linear combination of precomputed tsunami waveforms produces an initial tsunami forecast within a few minutes. Once a tsunami wave reaches the nearest DART, inversion is performed by a fit of model output to DART observations to refine the source. Further tsunami source refinement may occur during tsunami wave propagation as the wave reaches more DART stations (TITOV 2009; PERCIVAL et al. 2011). This iterative numerical approach allows accurate calculation of a tsunami wave field in the open ocean and the real-time tsunami flooding forecast at particular coastal sites (Fig. 10). Such calculations are currently carried out for 75 sites along the US coast; inclusion of several sites along the coasts of Canada and Russia is under discussion. Details of the forecast methodology employed by NOAA to assess the danger posed by a tsunami after generation is provided in Sect. 4.

3.3. Numerical Model Fitting and Verification

Deep-ocean data are currently widely used for fitting and verification of numerical tsunami models

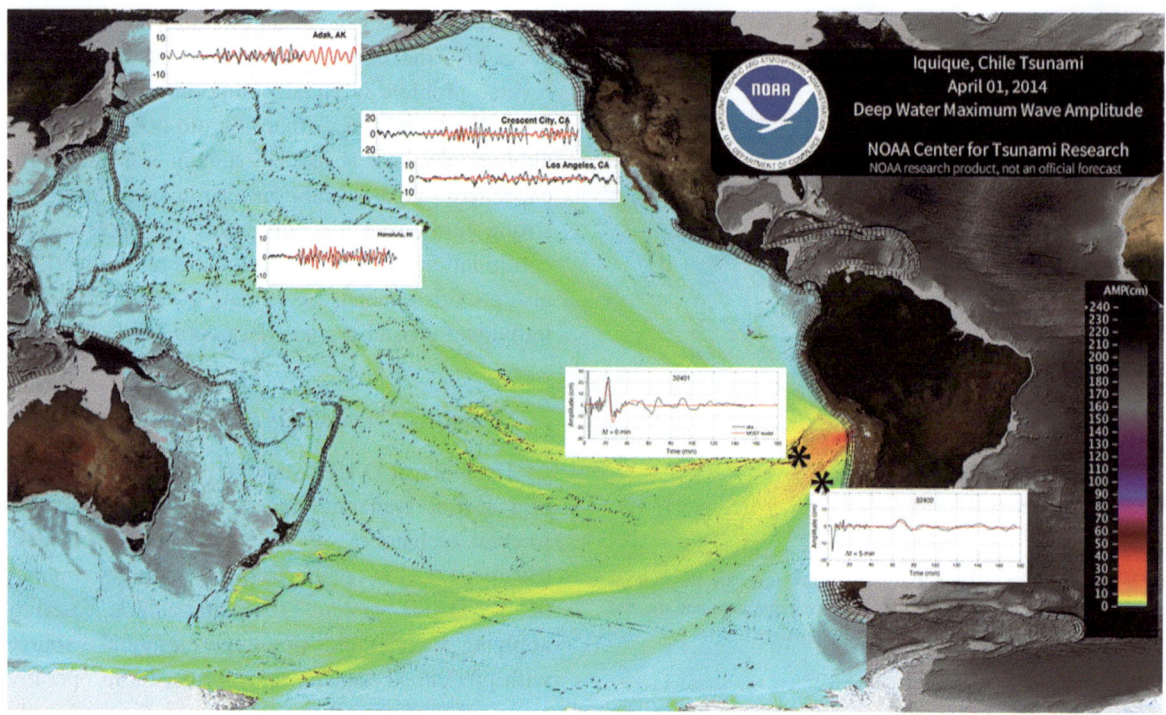

Figure 10
Bathymetric map with modeled maximum tsunami amplitudes during the 2014 Iquique, Chile tsunami. Modeled time series (*red line*) using the source inverted from observations at DART stations 32401 and 32402 (denoted by *asterisk*) are presented in comparison with observation time series (*black line*) at four tide gauge locations

(see, e.g., TITOV 2009; SAITO et al. 2011; THOMSON et al. 2011; RABINOVICH et al. 2011b; SHEVCHENKO et al. 2014, and tens of other papers). In comparison with nearshore measurements, deep-ocean tsunami measurements offer significant advantages: higher accuracy, lower noise, absence of effects of nonlinearity, friction, relief, and associated natural oscillations, as well as other factors that distort the tsunami signal near the coast. The main properties of a tsunami signal in the open ocean are determined by the source; i.e., a "pure" tsunami signal is actually measured. Specifically, open-ocean tsunami measurements promoted considerable progress in the numerical simulation of tsunami waves in recent years.

3.4. Source Reconstruction

The purity of a tsunami signal measured in the open ocean makes these data extremely valuable for reconstruction of the initial source of a given tsunami and for analysis of the source physics. In addition, these measurements, supplemented by coastal tsunami data, enable us to fully characterize the source and, hence, to achieve higher completeness and precision of the calculations. Among various algorithms designed for the tsunami problem, the studies of Fujii and Satake are the best known; the authors have performed the corresponding calculations for all strong tsunamis since 2004 (see, e.g., FUJII and SATAKE 2008; SATAKE et al. 2013).

3.5. Study of Physical Properties of Tsunami Waves

Open-ocean measurements of tsunami waves, away from coast distortion, are invaluable for understanding the nature of the phenomenon and for studying its physical properties, including evolution of the propagating waves and their transformation when approaching the coast. The high precision of the deep-ocean instruments and low noise level permit analysis of various fine effects that cannot be investigated with coastal tide gauges. In particular, the effect of dispersion of tsunami waves was revealed specifically based on deep-ocean measurements (GONZÁLEZ and KULIKOV 1993); this effect was taken into account to reconstruct the initial source signal (KULIKOV and GONZÁLEZ 1996). These measurements are also quite effective for estimating spectral properties of the source (Fig. 11) and tsunami energy decay in the open ocean (RABINOVICH et al. 2013b).

4. Forecast Methodology Based on Deep-Ocean Tsunami Measurements

A very important advantage of deep-ocean bottom pressure measurements is that they can be efficiently incorporated into numerical models to improve the accuracy of tsunami forecast along a coast. The respective methodology was elaborated on by NOAA and successfully incorporated into the US Tsunami Warning System (TITOV 2009).

The general tsunami forecast methodology is to combine real-time, deep-ocean tsunami observations made by the DART array with numerical models to produce estimates of tsunami wave arrival times, amplitudes, and inundation (flooding) at coastal communities pre-identified as being at risk from tsunamis generated in the Pacific, Atlantic, and Caribbean. An operational NOAA forecast system, termed the Short-term Inundation Forecast of Tsunamis (SIFT), integrates four key components: (1) a basin-wide precomputed propagation database, (2) deep-ocean observations, (3) an inversion algorithm, and (4) tsunami forecast models specifically developed to provide 4 h of simulation in approximately 10 min of wall clock time.

A basin-wide propagation database of precomputed water level and flow velocities for unit sources (50 km by 100 km) that form a continuum along continental margins is maintained for the purpose of expediting forecasts (GICA et al. 2008). Within a few minutes of an earthquake, database sources provide an estimate of tsunami energy based solely on seismic parameters, first of all on earthquake moment magnitude. As a tsunami propagates across the ocean and successively reaches DART stations, deep-ocean sea level observations transmitted in response to automatic or manual triggering into event sampling and reporting mode are ingested into SIFT by a process that continuously monitors the DART network. These data are then incorporated into an inversion (fitting) algorithm to produce a refined estimate of the tsunami source that was initially based on seismic parameters alone. A linear combination of the refined sources within the precomputed database produces synthetic boundary conditions of water elevation and flow velocities to initiate the forecast model computation.

Tsunami impact at a coast is forecast based on results provided by numerical models. The Method of Splitting Tsunami (MOST) model is used in the SIFT system to provide real-time tsunami forecasts at selected coastal communities in minutes while employing high-resolution grids constructed by the National Geophysical Data Center. MOST is a suite of numerical simulation codes capable of simulating three processes of tsunami evolution: generation, transoceanic propagation, and inundation of dry land. The overarching goal is to maximize the amount of time that a specific community has to react to a tsunami threat by providing accurate information quickly. To achieve this goal, a set of three nested grids (A, B, C), each of which is successively finer in resolution, are constructed. Near-shore details within the highest-resolution, and smallest, "C" grid are so finely resolved that tide gauge data from historical tsunamis that previously impacted the area can be compared with model results for validation. To generate these nested grids, a large spatial extent merged bathymetric topographic grid is first constructed at the highest possible grid resolution to best reproduce correct wave dynamics during inundation computation. This high-resolution, or reference grid is then iteratively coarsened in resolution and decreased in size until forecast runs complete within specified time constraints. This allows for the significant portion of the modeled tsunami waves, typically 4–10 h, to pass through the model domain without significant model signal degradation.

The forecast approach as described, has been used during tens of tsunami events. Results have shown the value and efficiency of the forecast strategy as implemented in the SIFT system and used to forecast tsunami impact in real time. More detailed information on the technical aspects of tsunami forecast

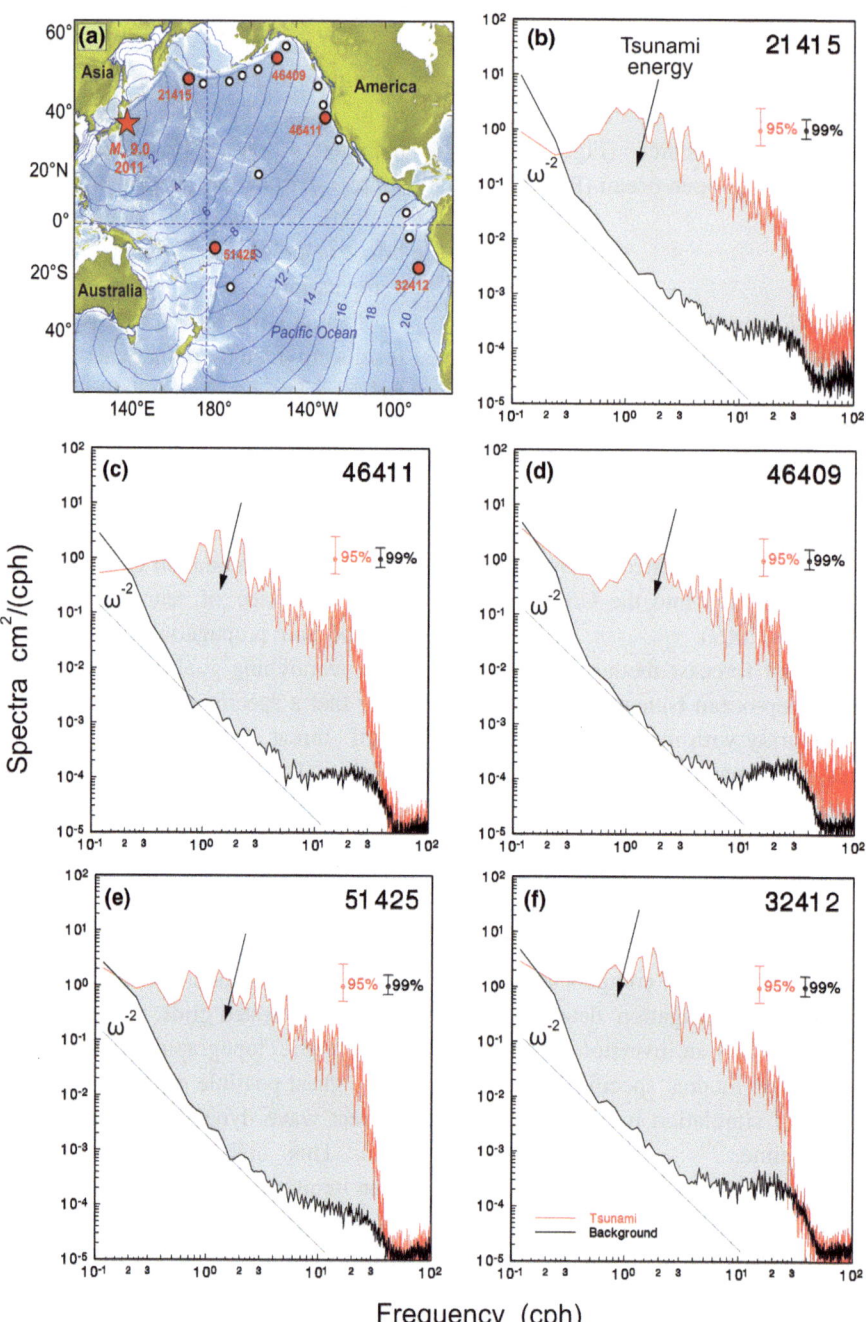

Figure 11

a Map of the Pacific Ocean showing the epicenter of the M_w 9.0 Tohoku earthquake of 11 March 2011 (*red star*) and location of 18 DART stations. **b–f** Spectra of the tsunami and background (pre-tsunami) oscillations recorded at five DART stations (*red circles* and *numbers* in the map show the location of these stations in various regions of the ocean). The locations of other DART stations are indicated by *white circles*. *Solid thin dark-blue lines* on the map are computed hourly isochrones of tsunami travel time from the source area. In the five spectra plots, the spectral "humps" associated with the tsunami energy are shaded; a reference power law ω^{-2} is denoted. Also shown are the 95 % (for tsunami) and 99 % (for background) confidence levels. The spectacular property of all 2011 Tohoku tsunami spectra is their similarity that appears to be related to the spectral properties of the source

5. Some Properties of Deep-Ocean Sea Level Measurements

There is a significant difference between coastal and deep-ocean measurements of sea level oscillations. Ordinary tide gauges, used in hydrographic studies, measure absolute changes in sea level determined by various factors. In contrast, measurement of sea level by instruments in the open ocean are based on recording of variations in hydrostatic bottom pressure, i.e., in the weight of the water column above the instrument (RABINOVICH 1993). The displacements of the free surface caused by the static ocean response to the variations in air pressure do not change the column weight and, therefore, are not recorded. Thus, only the deviations from the equilibrium,

$$\eta = \zeta + \Delta P_a / \rho g, \qquad (2)$$

are recorded, where ζ is the displacement of the free ocean surface, ΔP_a is the change in the air pressure, ρ is the water density, and g is the gravity acceleration. Bottom pressure sensors also do not record sea level oscillations caused by steric effects (variations in the sea water density) due to the same lack of effect the ocean response has on the water column cited above.

One should take into account that the hydrostatic equation is true for long (compared with the ocean depth) waves, i.e., while $\lambda \gg h$; for short (high-frequency) waves, the response of bottom pressure depends on the wavelength λ and the instrument deployment depth h. The attenuation coefficient of surface waves has the form

$$R = 1/\cosh(kh), \qquad (3)$$

where $k = 2\pi/\lambda$ is the wavenumber, which is related to the angular frequency, ω, by the dispersion equation

$$\omega^2 = gk \tanh(kh); \qquad (4)$$

For example, waves with a period of 1 min attenuate by 6 % at a depth of 100 m, by about half at 1000 m, and become undetectable at a depth of 5 km.

To illustrate this feature and describe the general character of long waves in the deep ocean, we consider a concrete example. For this purpose we use three deep-ocean stations in the northeastern part of the Pacific Ocean (Fig. 12): DART 46405 (depth of 3480 m), CORK[2] 1026 (2658 m), and NeMO[3] (1510 m). The stations are located 200–300 km apart and operated in December 2014, recording waves of the 2004 Sumatra tsunami that arrived from the Indian Ocean (RABINOVICH et al. 2011a) (see the next section). The native sampling interval at all three stations was 15 s. Figure 13 shows the bottom pressure spectra, constructed based on observations during the week preceding tsunami arrival, to characterize the spectrum of background oscillations (natural longwave noise). Three features of the spectra are the most important. (1) The similarity of the spectra: in the deep ocean, where topographic effects are of little consequence, the longwave background spectra are of a universal character, indicating the homogeneity of the longwave field [this question is discussed in (RABINOVICH 1993, 1997) in detail]. (2) The longwave spectrum is described well by the power law ω^{-2} in a wide frequency range, from 0.3 to 6 cph (cycles per hour), i.e., at periods from approximately 3.5 h to 10 min. This has been confirmed by numerous measurements in various regions of the Pacific Ocean (KULIKOV et al. 1983; FILLOUX et al. 1991; RABINOVICH 1993, 1997). (3) A pronounced "bulge" at frequencies of 10–50 cph, noted as artificial. A general increase in spectra at frequencies higher than 6 cph is associated with influence of infragravity (IG) waves generated by the nonlinear interaction of wind waves or swell (RABINOVICH 1993). This question has been recently considered in detail in AUDHAN and ARDHUIN (2013) based on DART data. The smaller the depth, the closer the coast, the more actively these waves are generated. For this reason, waves are higher at NeMO than at CORK 1026, and at CORK 1026, higher than at

---

[2] CORK = Circulation Obviation Retrofit Kit; the station has been deployed by the Pacific Geophysical Centre, Canada, within the Ocean Drilling Program.

[3] NeMO = New Millennium Observatory; the station was deployed by PMEL/NOAA (USA) in 2000 within the program of underwater volcano monitoring.

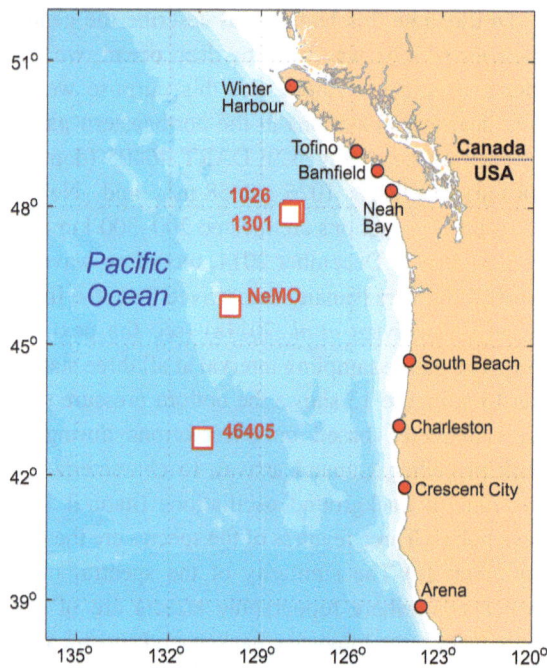

Figure 12
Locations of the deep-ocean bottom pressure stations (*white squares*) that were operational in the northeastern Pacific Ocean in 2004. The *red circles* mark the locations of some of the many coastal tide gauge stations

DART 46405. However, the waves become short at high frequencies (the deeper they are, the lower the frequencies at which this occurs) and rapidly attenuate. In other words, the effect described above "kills" the high-frequency part of the observed spectrum.

The bottom recorded sea level spectrum, $S_{\text{bott}}(\omega)$, may be corrected and the surface wave spectrum, $S_{\text{surf}}(\omega)$, may be reconstructed as

$$S_{\text{surf}}(\omega) = \frac{1}{R^2(\omega)} S_{\text{bott}}(\omega) = \cosh^2(kh) \, S_{\text{bott}}(\omega).$$

(5)

In Fig. 13 this was done and the corrected spectra were reconstructed for all three stations.

The spectrum becomes uniform at high frequencies (white level noise) due to the instrumental noise of the corresponding sensor (it is significantly lower at the DART and NeMO sensors than at the CORK sensor; see Fig. 13).

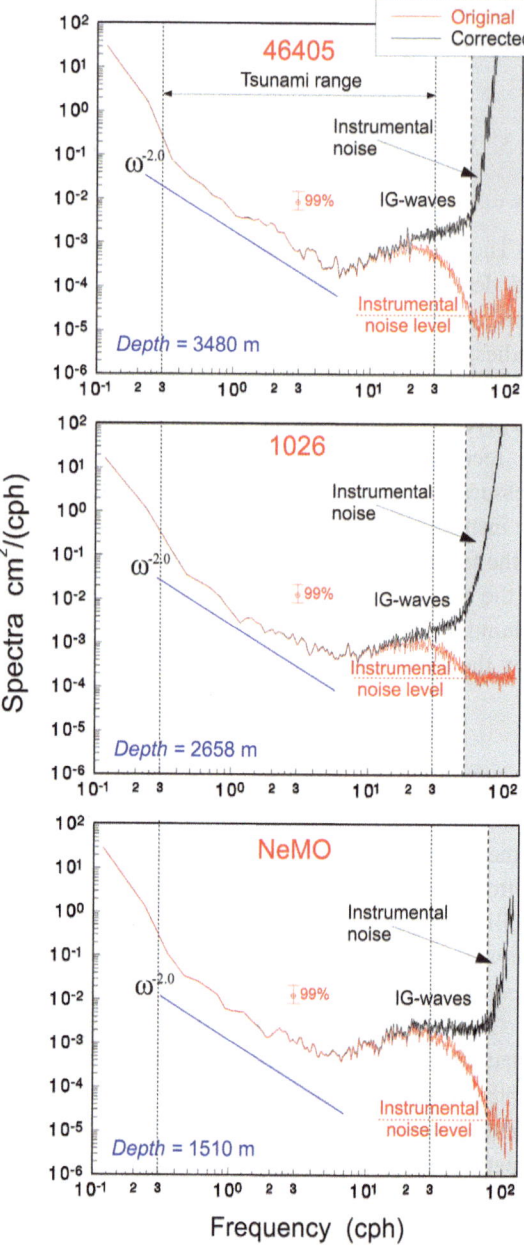

Figure 13
Spectra of background longwave oscillations recorded at three deep-ocean stations in the northeastern Pacific Ocean. Locations of the stations, DART 46405 (3480 m), CORK 1026 (2658 m), and NeMO (1510 m), are shown in Fig. 12. The *red lines* are the spectra of the original bottom records, the *black lines* are corrected (surface) spectra; a reference power law ω^{-2} is denoted. The *shaded areas* show those parts of the spectra controlled by instrumental and environmental noise, i.e., where spectra became unreliable

6. Detection and Analysis of Tsunami Waves in the Open Ocean

The detection and extraction of tsunami waves from original and possibly noisy records is a key problem of tsunami analysis. The problem of isolating a signal embedded in noise is typical for radioelectronics, acoustics, radiolocation, etc. The signal-to-noise (s/n) ratio is the crucial consideration of the problem; it is estimated as the ratio of the maximum signal amplitude to the standard noise deviation (rms). For example, the signal-to-noise ratio for the tsunami recorded during the 2011 Tohoku event was large across all observation locations due to consistently large tsunami amplitudes, making tsunami isolation quite easy. By contrast, the tsunami signal in records during the 2012 Haida Gwaii event was much more difficult to isolate in many locations where observations were available. The purpose of preliminary tsunami data analyses is to diminish the background noise level and to improve the s/n ratio without affecting the tsunami signal itself. In the case of open-ocean tsunami measurements, a pressure recorder installed on the seafloor at a depth of several kilometers is generally required to record variations of 0.1 cm amplitude, thereby making it possible to distinguish tsunami fluctuations from other processes which can have amplitudes 100–1000 times greater.

The problem of tsunami wave detection is challenging since initial signal parameters are unknown and, moreover, it is often unknown if there is a tsunami signal in the particular record at all. In fact, the research objective is the discovery and extraction of a certain anomaly that differs from the background oscillations by its properties (frequency content, amplitude, general character of oscillations) and has physical features typical of tsunami waves. If the s/n ratio is high, i.e., when tsunami waves are strong, their identification and extraction are straightforward (it is only important not to distort the tsunami signal during this extraction). However, if the ratio is low, the problem becomes much more difficult.

In a standard situation, the tsunami wave height, and hence the signal-to-noise ratio, decreases with distance from the source approximately as $1/\sqrt{r}$, where r is the distance between the source and the observation site. The catastrophic 2004 Sumatra tsunami is a typical example. Anomalous sea level oscillations observed in the Indian Ocean were clearly seen directly in the records of coastal tide gauges; the ratio of the tsunami signal to the nontidal background was from 25:1 to 40:1 (RABINOVICH and THOMSON 2007). At the same time, in the North Atlantic and North Pacific Oceans, the tsunami waves were much lower and the s/n ratio ranged from 5:1 to 1:1, making tsunami detection difficult (RABINOVICH et al. 2006).

In general, the identification and analysis of any tsunami event, even a weak one, is undoubtedly interesting. First, each new tsunami record, especially one in the open ocean, yields important scientific material for understanding this phenomenon, enables us to get essential information about the source region of the respective earthquake, and allows for investigation of tsunami wave evolution during propagation. Second, even tsunamis with a wave height of a few centimeters in the open ocean can considerably amplify when approaching the coast and produce dangerous oscillations in the coastal zone.

The process of tsunami detection is based on the consecutive improvement of the signal-to-noise ratio. The main principle to be adhered to during the process is: "do not distort a tsunami signal while extracting." Otherwise, wrong parameters of a specific tsunami and its source may be obtained, and therefore an inadequate model of the event may result. The character and appearance of a tsunami signal is not known a priori, but careful processing techniques allow consecutive suppression of oceanic sea level oscillations in a given record that definitely are not tsunami related.

The entire process can be exemplified considering the extraction of the 2004 Sumatra tsunami from the records of DART 46405, NeMO, and CORK 1026, described in the previous section, and from CORK 1301 (with sampling interval of 1 min) installed near to CORK 1026 (Fig. 12) (RABINOVICH et al. 2011a). The stations were located approximately 22,000 km travel distance from the source (along Australia, New Zealand, and South America); however, the tsunami signal identified in records of coastal tide gauges located in the vicinity (RABINOVICH et al. 2006) provided the reason for looking for the corresponding signal in bottom pressure records. Figure 14 (upper row) shows the initial records from DART 46405 and NeMO (CORK records look similar). There are no visual tsunami traces in the record; the only pronounced oscillations are tidal waves (station NeMO was located closer to the coast than DART 46405—see Fig. 8—and tides at this station were significantly stronger; see Table 2). Figure 14 (second row) shows the same records after the subtraction of tides estimated by the least-squares method from the records themselves. The residual energy level decreased by

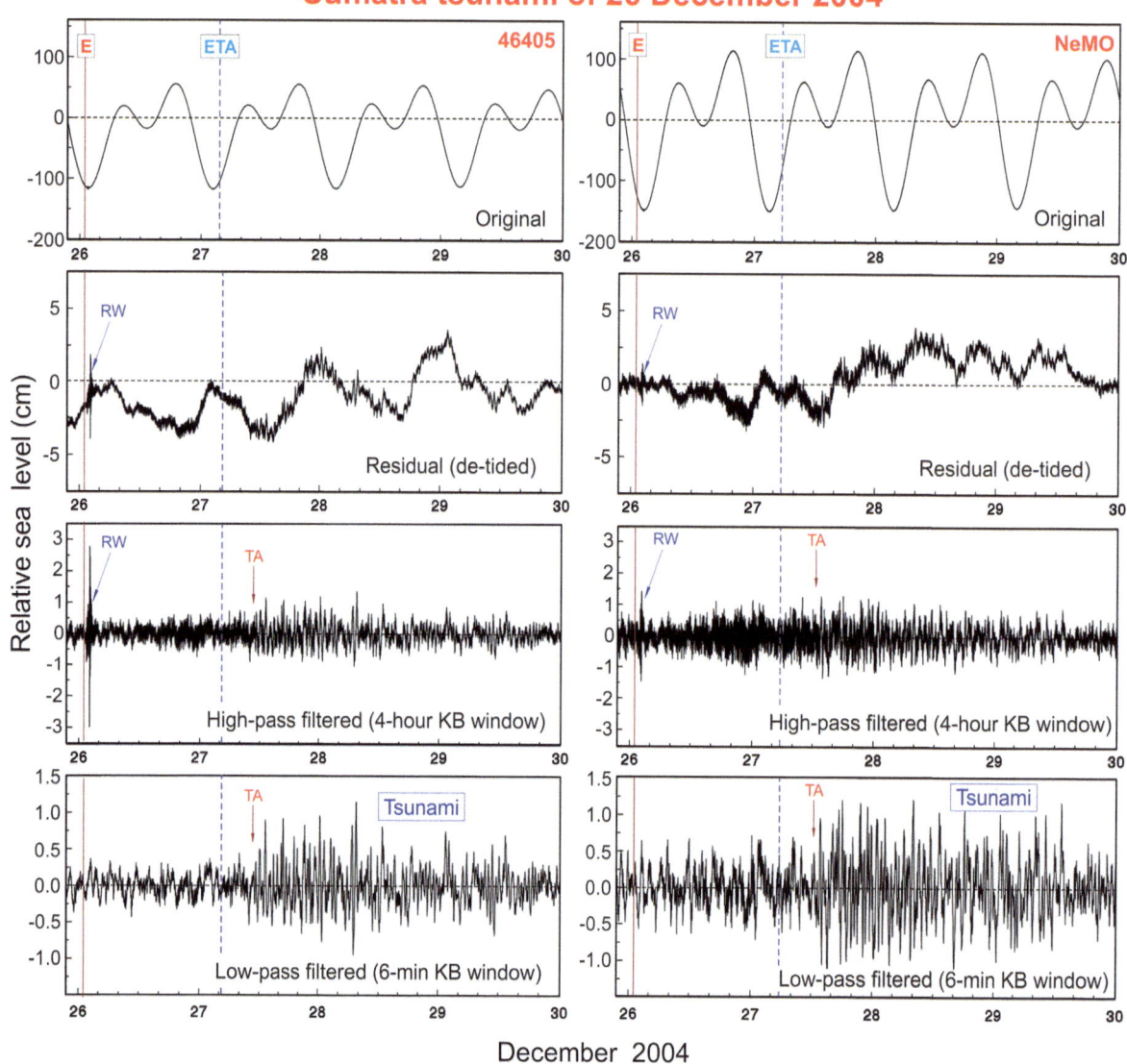

Figure 14
Successive stages of the 2004 Sumatra tsunami detection at deep-ocean stations DART 46405 and NeMO in the northeastern Pacific. Upper frames show the original records strongly dominated by ocean tides. The second panel shows the original residual after subtraction of predicted tides; low-frequency oscillations predominate. The third frame shows the records after high-pass [with a 4-h Kaiser–Bessel (KB) window] filtering of low-frequency oscillations. The Rayleigh waves (RW) are clear, and the tsunami waves became visible (their arrival is denoted by *red arrows* labeled "*TA*"), but high-frequency background noise associated with infragravity (IG) waves still impedes reliable detection of the tsunami. The lower frame shows the records, with the clearest tsunami signal, after additional low-pass (with 6-min KB window) filtering. Parameters of the filters (window lengths) were chosen based on the results of spectral analysis of "tsunami" and "background" (pre-tsunami) records to reduce the distortion of the tsunami signal. The *red solid vertical line* labeled "*E*" denotes the time of the earthquake, and the *dark-blue dashed vertical lines* labeled "*ETA*" indicate the estimated time of tsunami arrival at each station

about three orders of magnitude; estimated tides account for 99.84–99.95 % of the longwave energy in the study region (Table 2). It is important to emphasize that subtraction of the predicted tides does not affect the tsunami signal and still permits continued isolation of tsunami waves with much better s/n ratio. Low-frequency residual oscillations seen in the records are mainly associated with atmospheric processes and with some random near-tidal oscillations that always exist in sea level records[4] (e.g., MUNK and BULLARD 1963; MUNK et al. 1965). Typical residual amplitudes range from 2.5 to 4 cm. Tsunami waves are still not apparent, but a train of intense high-frequency oscillations caused by the seismic Rayleigh waves (RW in Fig. 14), arriving from the source to the observation site about 1 h after the earthquake, is evident.

High-pass filtering of nontidal records with a $\Delta T = 4$ h Kaiser–Bessel window (THOMSON and EMERY 2014) suppressed low-frequency oscillations. The energy of the remaining oscillations was diminished to 0.06–0.12 cm^2 (Table 2) compared with the original variance of 4000–7500 cm^2. The main property of the respective records (third row in Fig. 14) is the pronounced train of Rayleigh waves. However, an increase in the noise level beginning at about 11:00–12:00 UTC on December 27, 2004, i.e., 34–35 h after the earthquake, is also noticeable; a similar increase is seen in the records of the CORK deep-ocean stations. The timing agrees well with the arrival time of the 2004 Sumatra tsunami waves at the western coast of the USA (RABINOVICH et al. 2006). Thus, we can attribute this increase to the tsunami. Extraction and isolation of tsunami waves, however, is hindered by high-frequency IG waves, known to be a typical feature of open-ocean records (e.g., WEBB et al. 1991; RABINOVICH 1993). To suppress these waves, additional 6-min low-pass filtering was applied, making the tsunami waves more evident in the record (bottom row in Fig. 14).

Although the root-mean-square (rms) background noise in the tsunami frequency band is statistically different at the four stations (due to differences in instrument responses and depths), the tsunami rms is similar, ~ 3.1–3.4 mm, for all instruments, i.e., about 2–2.5 times higher than the noise level in the tsunami-wave frequency range (see the two rightmost columns in Table 2). The mean tsunami amplitudes identified in the study region were approximately 5 mm, compared with 1 m for tidal waves and 4 cm for low-frequency oscillations. Nevertheless, even such a weak tsunami signal, obtained at a distance of 22,000 km from the source, was successfully extracted and identified by thorough analysis, and characteristics of isolated tsunami waves at various stations look alike (Fig. 15). The observed tsunami travel time agrees well with the global tsunami propagation model by TITOV et al. (2005b), which showed that mid-ocean ridges served as waveguides for the 2004 tsunami, efficiently transmitting the tsunami energy from the source area in the Indian Ocean to far-field regions in the Pacific Ocean (RABINOVICH et al. 2011a). The tsunami phase speed on this route was approximately 180 m/s (~ 650 km/h).

The noteworthy property of the 2004 tsunami waves, evident in the isolated tsunami records at these stations, is very long ringing. The recorded tsunami is not a solitary wave, not a train of waves, not even several trains of waves, but rather a prolonged polychromatic signal propagating northwestward along the coast of North America without a noticeable decay (Fig. 15). An efficient way to estimate the evolution of these waves and their frequency content is frequency–time (f–t) diagrams (RABINOVICH et al. 2006; RABINOVICH and THOMSON 2007), which are similar to wavelet plots (e.g., THOMSON and EMERY 2014). These diagrams (Fig. 16) show that there is good agreement of the observed waves at various sites and also that the 2004 tsunami energy was concentrated in a broad frequency range with peak energy at frequencies of 0.8–2.0 cph (i.e., at periods of ~ 30–80 min). In general, f–t diagrams help to detect the exact arrival of tsunami waves at specific sites and to examine particular features of these waves, including the effect of tsunami wave dispersion (e.g., GONZÁLEZ and KULIKOV 1993; KULIKOV and GONZÁLEZ 1996; RABINOVICH et al. 2013a, SHEVCHENKO et al. 2014) and from this point of view are quite useful for tsunami analysis.

---

[4] An attempt to eliminate these near-tidal oscillations by increasing the number of computed tidal constituents may create a negative effect and distort the tsunami signal.

Table 2

Sea level variance at different stages of the 2004 Sumatra tsunami signal extraction for four deep-ocean stations in the northeastern part of the Pacific Ocean (from RABINOVICH et al. 2011a)

Station (depth, sampling)	Variance (cm^2); period of 24–30 December, 2014				One-day rms (mm)	
	Initial	Nontidal (residual)	After HF (4-h) filtering	After LF (6-min) filtering	Background	Tsunami
DART 46405 (3480 m, 15 s)	4336.4	6.92	0.060	0.043	1.37	3.11
NeMO (1510 m, 15 s)	5720.9	5.70	0.097	0.053	1.79	3.38
CORK 1026 (2658 m, 15 s)	7487.6	4.32	0.122	0.101	1.80	3.24
CORK 1301 (2658 m, 1 min)	7094.8	3.69	0.115	0.066	2.48	3.18

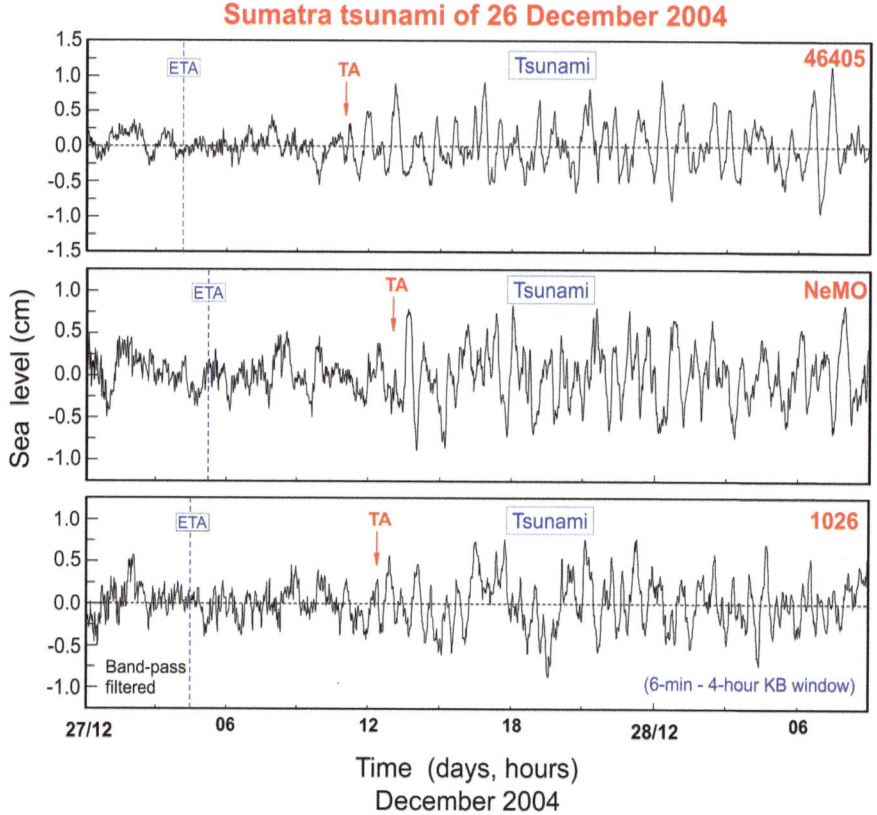

Figure 15
Records isolating the 2004 Sumatra tsunami in the northeastern Pacific at deep-ocean stations DART 46405, NeMO, and CORK 1026 after suppression of non-tsunami oscillations. The *dark-blue dashed vertical lines* labeled "*ETA*" indicate the estimated time of tsunami arrival at each station. The *red arrows* denote the actual tsunami arrival time

Tsunami records shown in Fig. 12 are band-pass filtered records (6 min–4 h). It is important to take into account that any filter introduces certain distortion into the analyzed tsunami signal. This especially affects peak values of tsunami waves and the first wave arrival. Filtering is a simple and effective way to isolate tsunami waves and visualize them (as Figs. 14, 15 demonstrate). However, filtering can significantly corrupt the statistical characteristics of the observed waves, so a filter should be designed quite carefully. The choice of optimum parameters for the filter is critical. Too wide a filter band will pass through too much non-tsunami energy, impeding the isolation of tsunami waves. Conversely, too narrow a band will

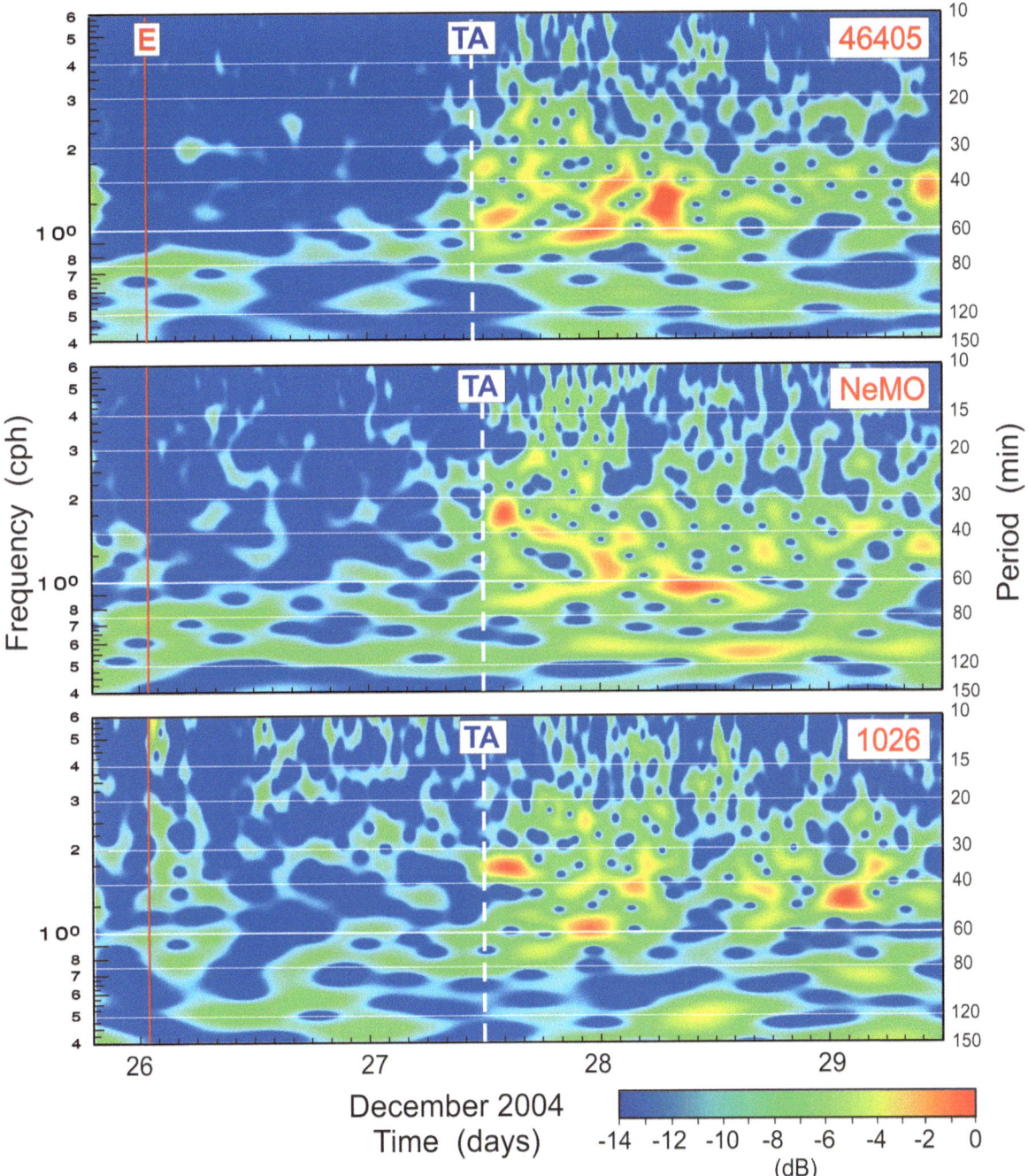

Figure 16
Frequency–time (*f–t*) diagrams of the 2004 Sumatra tsunami records at deep-ocean stations DART 46405, NeMO, and CORK 1026. The *red solid vertical line* labeled "*E*" denotes the time of the earthquake, and the *dashed white vertical line* with label "*TA*" indicates the actual tsunami arrival time at each station

distort the signal. Given the choice, we should take into consideration that normally strong earthquakes with extensive source area have more energy at low frequencies, while earthquakes with lesser source areas typically produce high-frequency tsunamis (see, for example, RABINOVICH et al. 2013b).

For spectral analysis of tsunami waves it is always better to use unfiltered nontidal records. As an example, Fig. 17a shows the results of spectral analysis of the 2004 Sumatra tsunami records from stations DART 46405 and NeMO. Comparison of these spectra with the spectra of background oscillations at the same stations enables us to specify the tsunami energy (e.g., RABINOVICH 1997). Background spectra have been estimated using 4-day pre-tsunami data segments, while tsunami spectra have been estimated from 1-day records immediately following the first tsunami wave arrival. The results of spectral analysis for the two stations are generally similar; the "tsunami spectra," $S_{\text{tsu}}(\omega)$, match the "background spectra," $S_{\text{bg}}(\omega)$, at low and high frequencies but, at intermediate frequencies, have a pronounced "bulge" associated with the arriving tsunami energy that occupies the frequency range of 0.4–7.0 cph (i.e., the range of periods from 2.5 h to 8 min) with more than 90 % of this energy in frequencies of 0.9–2 cph (periods 70 to 30 min) and a peak value near 1.2 cph (period of 50 min).

Following RABINOVICH (1997), we calculated spectral ratios

$$R_s(\omega) = S_{\text{tsu}}(\omega)/S_{\text{bg}}(\omega), \qquad (6)$$

which give the amplification of the longwave spectrum during the tsunami event relative to the background conditions and characterize the spectral properties of the source (which is why it can be called the "source function") (e.g., RABINOVICH et al. 2013b). These functions are presented in Fig. 17b. It is likely that the dominant long periods of the source functions (which are very similar for the two stations) are associated with the extensive ($\sim$1300 km) initial source area of the Sumatra earthquake of 26 December 2004 (STEIN and OKAL 2005).

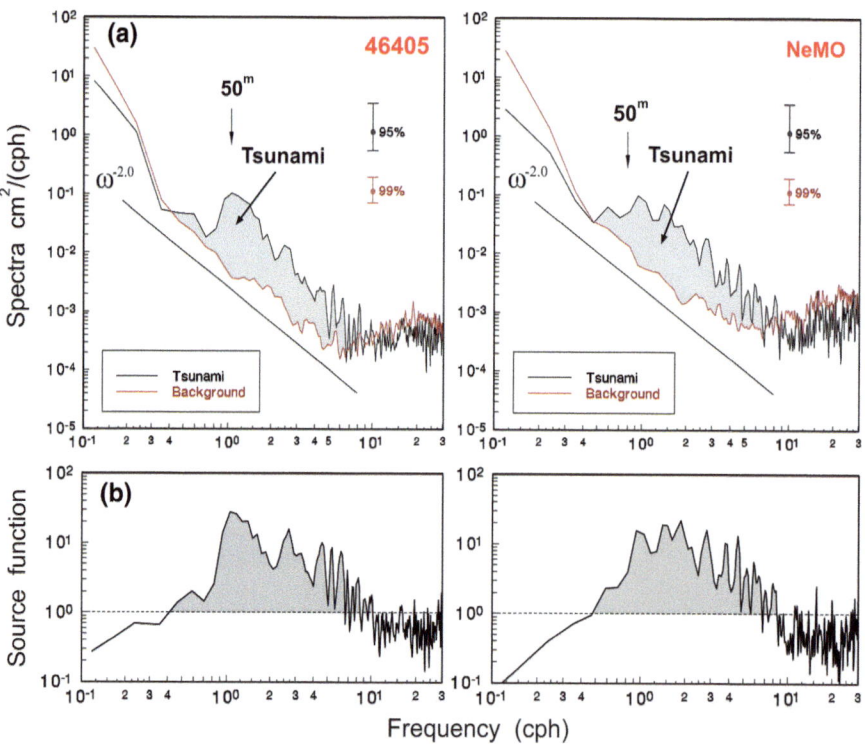

Figure 17

a Spectra of the background (pre-tsunami) and the 2004 Sumatra tsunami oscillations recorded in the northeastern Pacific at deep-ocean stations DART 46405 and NeMO. The spectral "humps" associated with the tsunami are shaded; a reference power law ω^{-2} is denoted. Also shown are the 95 % (for tsunami) and 99 % (for background) confidence levels. **b** "Source functions" (spectral ratios tsunami/background) for the respective records

7. Noise Level

In the considered example, the tsunami waves were extracted and identified for research purposes. The detection and isolation of a probable tsunami wave within the TWS has some specificity (see, for example, POPLAVSKY et al. 1988; MOFJELD 2009). The real-time detection of a signal that is potentially dangerous for a coast is of principal importance. Specific algorithms for the extraction of anomalies against the background noise are used for this purpose (e.g., KULIKOV 1990; MOFJELD 1997; TOLKOVA 2010; BRESSAN and TINTI 2012). The establishment of tsunami early warning systems in various regions of the world oceans makes this problem very important (BELTRAMI 2011; BELTRAMI and DI RISIO 2011; DI RISIO and BELTRAMI 2014).

As indicated above, the detection and extraction of tsunami waves from bottom pressure or coastal tide gauge records strongly depends on the signal-to-noise (s/n) ratio. The limitation of some methods, at least as they are described in the respective papers (e.g., TOLKOVA 2010; BELTRAMI and DI RISIO 2011; DI RISIO and BELTRAMI 2014), is that they are oriented towards analysis of the original records, i.e., records with tides. The energy of tides in the open ocean is typically two to four orders higher than the energy of recorded tsunami waves (see Table 2). For this reason, the corresponding algorithms struggle to balance suppression of tides with retention of the tsunami; either leaving a few percent of tidal energy (e.g., DI RISIO and BELTRAMI 2014), or distorting the tsunami signal.

It is important to emphasize that, for each DART or cable station (as well as for coastal tide gauges), tides are normally well known and so can be predicted and subtracted from the original records without distortion of tsunamis or any other nontidal processes.[5] The methods and algorithms that target detection and isolation of tsunami waves would be much more effective in that case, making it more straightforward to recognize much weaker events (see previous section and KULIKOV 1990).

In fact, the tsunami signal identification threshold is determined by the noise parameters. The nontidal (residual) noise is produced by two major factors: (1) instrumental or (2) environmental. Instrumental noise is associated with sensor properties, recorder resolution (number of digits retained when analyzing records), sampling interval, and influence of the water media. Environmental factors are due to natural longwave background oscillations mainly associated with atmospheric processes.

Instrumental noise affects all frequencies; the respective variance (σ^2_{inst}) may be estimated as

$$\sigma^2_{inst} = \int_0^{f_N} S_{inst}(\omega)\, df, \qquad (7)$$

where $S_{inst}(\omega)$ is the spectrum of the instrumental noise and $f_N = 1/2\Delta t$ is the Nyquist frequency for sampling interval Δt. The instrumental noise level is clearly seen in the bottom pressure spectra: it is the level of the uniform ("white noise") spectrum observed at high frequencies (Fig. 13). Assuming $S_{inst}(\omega) = \hat{S}_{inst} = $ const., we can easily evaluate the rms of the instrumental noise as

$$\sigma_{inst} = \sqrt{f_N \hat{S}_{inst}}. \qquad (8)$$

As an example, we present the results of spectral analysis of 15-s records from 18 DART stations for the period 5–10 March 2011, i.e., just before the Tohoku tsunami of 11 March 2011. The location of the stations is shown on the map in Fig. 11a (see also RABINOVICH et al. 2013b); five selected spectra are shown in Figs. 11b–f. The level of white noise is evident in the spectra, thus permitting an estimate of σ_{inst} for all 18 instruments; the results are presented in Table 3. The σ_{inst} values vary from 0.35 mm (DARTs 32411 and 43413) to 0.79 mm (DART 46408) with a mean value of σ_{inst} (averaged over 18 instruments) of 0.50 mm. From these estimates, it is apparent that the accuracy of DARTs is high enough that the random instrumental noise does not create any difficulties for tsunami measurements. It is, however, important to take into account that records at deep-ocean stations

---

[5] It would be useful if de-tided DART records were given on the DART website (http://www.ndbc.noaa.gov/dart.shtml), similarly as is done on the Center for Operational Oceanographic Products and Services (CO-OPS) NOAA website for coastal tide gauge records.

Table 3

Background noise at various DART stations in the Pacific Ocean estimated for the frequency band of 0.46–30 cph (periods from 2.2 h to 2 min) and the period of 5–10 March 2011, and the instrumental noise estimated for the same DARTs

Station (DART)	Region	Coordinates		Depth (m)	Noise level (rms, mm)	
		Latitude	Longitude		Background	Instrumental
21415	Aleutian Is.	50.1762°N	171.8486°E	4710	1.42	0.61
21414	Aleutian Is.	48.9367°N	178.2514°E	5465	1.06	0.60
46408	Aleutian Is.	49.6244°N	169.3056°W	5379	0.96	0.79
46402	Aleutian Is.	51.0683°N	164.0053°W	4718	0.97	0.72
46403	Alaska	52.6486°N	156.9208°W	4514	0.96	0.66
46409	Alaska	55.3000°N	148.4923°W	4189	0.97	0.44
46410	Alaska	57.6317°N	143.8017°W	3728	1.01	0.38
46419	S. Vancouver I.	48.7596°N	129.6104°W	2775	1.21	0.49
46407	US West Coast	42.5890°N	128.9002°W	3266	1.54	0.50
46411	US West Coast	39.3238°N	126.9910°W	4260	1.05	0.39
46412	North Mexico	32.4553°N	120.5569°W	3718	1.18	0.44
43413	Mexico	11.0650°N	99.8567°W	3404	1.14	0.35
32411	Central America	4.9417°N	90.6522°W	3166	1.19	0.35
32413	North Peru	7.4003°S	93.4989°W	3894	1.10	0.38
32412	South Peru	17.9865°S	86.3887°W	4325	1.19	0.44
51407	Hawaii Is.	19.6169°N	156.5106°W	4682	1.44	0.53
51425	Samoa Is.	9.5044°S	176.2297°W	4978	1.11	0.44
51426	Tonga Is.	22.9911°S	168.1031°W	5686	1.00	0.54

Comment: All DART coordinates and depths relate to March 2011 and are slightly different from the DART positions/depths at the present time

(as well as at coastal tide gauges) may include spikes, shifts, gaps, etc., that require remediation. The reasons for these are many and include transmission errors, power jumps, instrument movements, and others. Some of these disturbances are sporadic, whereas others are regular. Thus, certain DART records include single or triple spikes of periodicity associated with the modem schedule for data transmission. Typical heights of such spikes or shifts are on the order of 5–10 cm. In the original records (with tides of ~0.5–1.0 m amplitudes) these spikes and shifts are hardly recognizable, but in de-tided records, they are clearly evident. Spikes, shifts, gaps, and other types of instrumental errors can significantly contaminate a record, so identification of them by subtraction of the predicted tides is critical. Moreover, these instrumental artifacts can erroneously be identified as a tsunami signal (see OKADA 1993, 1995 as an example). That is why it is so important to detect and eliminate various types of instrumental errors. GORING (2008) suggested some efficient algorithms to "clean" the records; however, some visual control is always useful.

Of the 18 DARTs listed in Table 3 (see also RABINOVICH et al. 2013b), 8 (specifically, DARTs 32411, 32413, 46403, 46407, 46409, 46410, 46411, and 46412) included numerous spikes, shifts or other types of instrumental errors in the original records. All errors were identified and thoroughly eliminated before providing spectral analysis. The noise values presented in Table 3 are for the corrected records.

Thus, except for episodic instrumental errors, instrument noise does not impact tsunami detection, and our ability to identify and isolate tsunami waves is determined by the longwave background noise level in the tsunami frequency band. We estimated the respective variance (σ_{bg}^2) as

$$\sigma_{bg}^2 = \int_{f_l}^{f_h} S(\omega)\, df, \quad (9)$$

where $f_l = 0.46$ cph (period of ~2.2 h) and $f_h = 30$ cph (2 min) roughly bracket the tsunami frequency band. The background rms values (σ_{bg}) for 18 DART stations are given in Table 3.

In general, the background noise cumulative energy depends on the depth, offshore distance, and atmospheric activity. In turn, the latter factor depends on the concrete region and season (for example, in the northeastern Pacific, the atmospheric activity and associated sea level activity is substantially higher in winter). Nevertheless, the σ_{bg} estimates evaluated for different parts of the Pacific Ocean are consistent with overall variation between approximately 1.0 and 1.5 mm. These values are in good agreement with those presented in Table 2 for the background noise in December 2004 at DART 46405, NeMO, and CORK 1026. Slightly higher values of σ_{bg} at the two latter stations are the apparent result of being closer to the shore and having lesser depths than that of the DART stations.

The tsunami detection algorithm incorporated into onboard DART software and therefore used by NOAA to identify and report tsunamis is based on a typical threshold of 3 cm relative to the undisturbed sea level (Mofjeld 1997). The threshold is set higher in DART instruments deployed in regions with greater background noise. Analysis of the background noise leads to the conclusion that even a tsunami of 0.5 cm amplitude can be reliably detected by the DART technology, a conclusion supported by the numerous tsunamis with sub-0.5-cm amplitude identified in the many event records.

8. Discussion and Conclusions

Deep-ocean bottom pressure recorders deployed along almost all seismically active zones of the world oceans have provided over recent years a huge amount of observational data, much of which have not yet been completely exploited. Observations at several tens of deep-ocean stations in the Pacific Ocean have been recorded during several great recent tsunami events (2009 Samoa, 2010 Chile, 2011 Tohoku, and 2012 Haida Gwaii). In the 1980s, theoretical tsunami research efforts significantly exceeded investigations based on actual observations of tsunami waves. Now, it seems, the situation has reversed as the large quantity of recently available data significantly exceeds the possibilities of processing and interpretation. In addition to seismically generated tsunami investigations, the high value of these data has been shown for other areas of research including global tides, climate, and in the identification of other anomalous waves, in particular, meteorological tsunamis. This latter consideration is especially important for the northwestern region of the Atlantic Ocean, where such events occur quite frequently. Atlantic DART stations clearly recorded an intense meteotsunami induced by the derecho (a rapidly propagating system of destructive thunderstorm fronts and squalls) of 13 June 2013. Therefore, deep-ocean pressure records can play a key role in understanding the physical mechanism of extreme seiche oscillations recorded onshore.

The experience of JMA, other owners of Japanese cable lines, and the Ocean Networks Canada with the bottom geophysical observatory provides a good indication of the particular features of cable stations in comparison with autonomous tsunameter stations. In general, tsunameter and cable stations complement one another. There are, however, considerations specific to each technology that, in a sense, are related to efficient operation and maintenance of a station rather than to the specific station itself. Bottom cable stations provide enduring effectiveness as demonstrated by high reliability and longevity up to 35 years. A suite of configurable data in addition to bottom pressure (seismicity, ocean currents, water temperature, and conductivity) can be transmitted from these stations in real time without loss of sampling resolutions. The main shortcomings of bottom cable stations are the very high cost associated with cable lines and station deployments, power requirements, and the need for coastal infrastructure connectivity. By contrast, tsunameter stations are less costly to establish but have a shorter operational life span, have power and data transmission limitations, and must be routinely serviced in a hostile environment.

Relative advantages and disadvantages of sensing technologies aside, the overarching consideration is the continued and increased collection of high-quality data for furthering our understanding of tsunami properties.

The main purpose of the instruments deployed in recent years in the open ocean is to provide reliable and precise early tsunami warning, i.e., to improve the existing Tsunami Warning System. From this point of view, these instruments have already proven their high efficiency and probably have saved innumerable lives. At the same time, deep-ocean tsunameters, in particular DART stations, were found to be invaluable for research purposes and have significantly widened our knowledge and understanding of the tsunami physics and generation mechanism. At the present time, tsunameters provide coverage along most of the seismically active zones in the Pacific Ocean as well as some other regions of the world oceans. However, there are still recognized gaps (for example, the region of Haida Gwaii in the northeastern Pacific) where such instruments will be highly important.

Acknowledgments

We gratefully acknowledge the Japan Agency for Marine Earth Science and Technology (JAMSTEC) and specific agency researchers H. Matsumoto and K. Kawaguchi for the information provided on Japanese cable stations. We thank Drs. Eddie Bernard (NOAA/PMEL, Seattle, WA) and Derek Goring (Mulgor Consulting, Christchurch, NZ) for their valuable comments and suggestions. We especially thank Dr. George Mungov (National Geophysical Data Center, Boulder, CO) for his advice and assistance with DART data processing. Russian Science Foundation, Grant 14-50-00095, provided partial support for the contribution of A.B.R. This is NOAA Pacific Marine Environmental Laboratory contribution number 4297.

REFERENCES

AUCAN, J., and ARDHUIN, F. (2013), *Infragravity waves in the deep ocean: An upward revision*, Geophys. Res. Lett., 40, 3435–3439; doi:10.1002/grl.50321.

BARNES, C., BEST, M., JOHNSON, F., PHIBBS, P., and PIRENNE, B. (2008), *Transforming the ocean sciences through cabled observatories*, Marine Technology Reporter, October 2008, 30–36.

BERNARD, E.N., GONZÁLEZ, F.I., MEINIG, C., and MILBURN, H.B. (2001), *Early detection and real-time reporting of deep-ocean tsunamis*, Proc. Int. Tsunami Symp. 2001, Seattle, WA, pp. 97–108.

BELTRAMI, G.M. (2011), *Automatic, real-time detection and characterization of tsunamis in deep-sea level measurements*. Ocean Eng. 38(14–15), 1677–1685; doi:10.1016/j.oceaneng.2011.07.016.

BELTRAMI, G. M. and DI RISIO, M. (2011), *Algorithms for automatic, real-time tsunami detection in wind-wave measurements. Part I: implementation strategies and basic tests*. Coastal Eng. 58(11), 10621071; doi:10.1016/j.coastaleng.2011.06.004.

BRESSAN, L., and TINTI, S. (2012), *Detecting the 11 March 2011 Tohoku tsunami arrival on sea-level records in the Pacific Ocean: application and performance of the Tsunami Early Detection Algorithm (TEDA)*, Nat. Hazards Earth Syst. Sci. 12, 1583–1606; doi:10.5194/nhess-12-1583-2012.

DI RISIO, M. and BELTRAMI, G.M. (2014), *Algorithms for automatic, real-time tsunami detection in wind-wave measurements: using strategies and practical aspects*, Procedia Eng. 70, 545–554.

DJUMAGALIEV, V.A., and RABINOVICH, A.B. (1993), *Long wave investigation at the shelf and in the bays of the South Kuril Islands*, J. Korean Soc. Coast. Ocean Eng., 5(4), 318–328.

DJUMAGALIEV, V.A., KULIKOV, E.A., and SOLOVIEV, S.L. (1993), *Analyses of ocean level oscillations in Malokurilskaya Bay caused by tsunami on February 16, 1991*, Sci. Tsunami Hazards, 11(1), 47–55, 1993.

DYKHAN, B.D., JAQUE, V.M., KULIKOV, E.A., et al. (1981), *The first registration of tsunamis in the open ocean*, Dokl. Akad. Nauk USSR, 257(5), 1088–1092. [in Russian].

DYKHAN, B.D., JAQUE, V.M., KULIKOV, E.A., et al. (1983), *Registration of tsunamis in the open ocean*, Mar. Geodesy, 6. 303–310.

EBLÉ, M.C., and GONZÁLEZ, F.I. (1991), *Deep-ocean bottom pressure measurements in the Northeast Pacific*, J. Atmos. Oceanic Technol., 8, 221–233.

EFIMOV, V.V., KULIKOV, E.A., RABINOVICH, A.B. and FINE, I.V. (1985), *Ocean Waves in Boundary Regions*, Gidrometeoizdat, Leningrad, pp 280. [in Russian].

EVA, C. and RABINOVICH, A.B. (1997), *The February 23, 1887 tsunami recorded on the Ligurian coast, Western Mediterranean*, Geophys. Res. Lett., 24(17), 2211–2214.

FILLOUX, J.H. (1982), *Tsunami recorded on the open ocean floor*, Geophys. Res. Lett. 9(1), 25–28.

FILLOUX, J.H. (1983), Pressure fluctuations on the open ocean floor off the Gulf of California: Tides, earthquakes, tsunamis, J. Phys. Oceanogr. 13(5), 783–796.

FILLOUX, J.H., LUTHER, D.S., and CHAVE, A.D. (1991), *Update on seafloor pressure and electric field observations from the north-central and northeastern Pacific: Tides, infratidal fluctuations, and barotropic flow*, In: *Tidal Hydrodynamics* (Ed. B.B. Parker), Wiley, New York, pp. 617–639.

FUJII, Y., and SATAKE, K. (2008), *Tsunami sources of November 2006 and January 2007 Great Kuril earthquakes*, Bull. Seismol. Soc. Am., 98, 1559–1571.

GICA, E., SPILLANE, M., TITOV, V.V., CHAMBERLIN, C., and NEWMAN, J.C. (2008), *Development of the forecast propagation database for NOAA's Short-term Inundation Forecast for Tsunamis (SIFT)*. NOAA Tech. Memo. OAR PMEL-139, 89 pp.

GONZÁLEZ, F.I., and KULIKOV, E.A. (1993), *Tsunami dispersion observed in the deep ocean*. In: *Tsunamis in the World*, (Ed. S. Tinti), Kluwer Academic, Dordrecht, pp. 7–16.

GONZÁLEZ, F.I., MADER, C.L., EBLÉ, M.C., and BERNARD, E.N. (1991), *The 1987–88 Alaskan Bight tsunamis: Deep ocean data and model comparisons*, Nat. Hazards, 4(1/2), 119–139.

GONZÁLEZ, F.I., BERNARD, E.N., MEINIG, C. et al. (2005), *The NTHMP tsunameter network*, Nat. Hazards, 35(1), 25–39.

GORING, D.G. (2008), *Extracting long waves from tide-gauge records*, J. Waterw. Port Coast. Ocean Eng. 134(5), 306–312.

HIRATA, K., AOYAGI, M., MIKADA, H. et al. (2002), *Realtime geophysical measurements on the deep seafloor using submarine* cable in the Southern Kurile Subduction Zone, IEEE J. Oceanic Eng., 27, 170–181.

HONDA, K., TERADA, T., YOSHIDA, Y., and ISITANI, D. (1908), *An investigation on the secondary undulations of oceanic tides*, J. College Sci., Imper. Univ. Tokyo, 108 pp.

JAQUE, V.M., and SOLOVIEV, S.L. (1971), *Remote registration of tsunami type weak waves on the shelf of the Kuril Islands*, Dokl. Akad. Nauk USSR, 198(4), 816–817. [in Russian].

JOSEPH, A. (2011), *Tsunamis: Detection, Monitoring and Early-Warning Technologies*, Elsevier, Amsterdam, 436 pp.

KANEDA, Y. (2011), *Advanced ocean floor network system for mega thrust earthquakes and tsunamis*, Proc. Underwater Techn. 2011 IEEE Symp. doi:10.1109/UT.2011.5774149.

KARL, D. M. (2004), *UH and the Sea: The Emergence of Marine Expeditionary Research and Oceanography as a Field of Study at the University of Hawaii at Manoa*, University of Hawaii, Honolulu, SOEST Report 04-01.

KAWAGUCHI, K., KANEDA, Y., ARAKI, E. et al. (2012), *Reinforcement of seafloor surveillance infrastructure for earthquake and tsunami monitoring in western Japan*, Proc. Oceans2012 Mts/Ieee Yeosu, May 21–24, 2012. Yeosu, Republic of Korea.

KOVALEV, P.D., RABINOVICH, A.B., and SHEVCHENKO, G.V. (1991), *Investigation of long waves in the tsunami frequency band on the southwestern shelf of Kamchatka*, Nat. Hazards, 4(2/3), 141–159.

KULIKOV, E.A. (1990), *Sea level measurement and tsunami forecasting*, Sov. Meteorol. Hydrol., 6, 61–68.

KULIKOV, E.A., and GONZÁLEZ, F.I. (1996), *Recovery of the shape of a tsunami signal at the source from measurements of oscillations in the ocean level by a remote hydrostatic pressure sensor*, Trans. (Doklady) Russian Acad. Sci., Earth Sci. Sect. 345A, 585–591.

KULIKOV, E.A., PAVLENKO, V.G., LAPPO, S.S., and RABINOVICH, A.B. (1979), *The Second Soviet–American Expedition to Study Tsunamis in the Open Ocean*, Oceanology, 19(2), 235–236.

KULIKOV, E.A.; RABINOVICH, A.B., SPIRIN, A.I., POOLE, S.L., and SOLOVIEV, S.L. (1983), *Measurement of tsunamis in the open ocean*, Mar. Geodesy 6(3–4), 311–329.

LANDER, J.F., LOCKRIDGE, P.A., and KOZUCH, M.J. (1993), *Tsunamis affecting the West Coast of the United States, 1806–1992.* Boulder, Colorado, National Geophysical Data Center, 242 pp.

LAPPO, S.S., and SOLOVIEV, S.L. (1976), *The First Soviet American Open-Ocean Tsunami Expedition*, Oceanology, 16(4), 412–413.

LEVIN, B., and NOSOV, M. (2009), *Physics of Tsunamis*, Springer, Dordrecht, 327 pp.

MATSUMOTO, H., and KANEDA, Y. (2009), *Review of recent tsunami observation by offshore cabled observatory*, J. Disaster Res. 4(6), 1–9.

MATSUMOTO, H., and KANEDA, Y. (2013), *Some features of bottom pressure records at the 2011 Tohoku earthquake—Interpretation of the far-field DONET data.* Proc. 11th SEGJ Intern. Symp.

MOFJELD, H.O. (1997), *Tsunami detection algorithm*, NOAA/PMEL, Seattle, WA. Only available online at http://nctr.pmel.noaa.gov/tda_documentation.html.

MOFJELD, H.O. (2009), *Tsunami measurements*, In: *The Sea*, Vol. 15, Tsunamis, (Eds. A. Robinson and E. Bernard), Harvard University Press, Cambridge, USA, pp. 201–235.

MOFJELD, H.O., WHITMORE, P.M., EBLE, M.C., GONZÁLEZ, F.I., and NEWMAN, J.C. (2001), *Seismic-wave contributions to bottom pressure fluctuations in the North Pacific—Implications for the DART tsunami array*, Proc. Int. Tsunami Symp. 2001, Seattle, WA, CD, pp.97–108.

MOFJELD, H.O., GONZÁLEZ, F.I., and EBLE, M.C. (1996), *Subtidal bottom pressure observed at Axial Seamount in the northeastern continental margin of the Pacific Ocean*, J. Geophys. Res. 101(C7), 16381–16390.

MUNGOV, G., EBLE, M., and BOUCHARD, R. (2013), *DART® tsunameter retrospective and real-time data: A reflection on 10 years of processing in support of tsunami research and operations*, Pure Appl. Geophys., 170, 1369–1384; doi:10.1007/s00024-012-0477-5.

MUNK, W.H., and BULLARD, E.C. (1963), *Patching the long-wave spectrum across the tides*, J. Geophys. Res. 68(12), 3627–3634.

MUNK, W.H., ZETLER, B., and GROVES, G.W. (1965), *Tidal cusps*, Geophys. J. R. Astron. Soc. 10(2), 211–219.

NOWROOZI, A.A. (1972), *Long-term measurements of pelagic tidal height off the coast of northern California*, J. Geophys. Res., 77(3), 434–443.

NOWROOZI, A.A., SUTTON, G., and AULD, B. (1966), *Oceanic tides recorded on the sea floor*, Ann. Geophys., 22(3), 512–517.

OKADA, M. (1993), *Tsunami observation by ocean bottom pressure gauge*, In: Proc. IUGG/IOC Int. Tsunami Symp., Wakayama, Japan, pp. 385–396.

OKADA, M. (1995), *Tsunami observation by ocean bottom pressure gauge*. In: *Tsunami: Progress in Prediction, Disaster Prevention and Warning* (Eds. Y. Tsuchiya and N. Shuto), Kluwer, Dordrecht, pp. 287–303.

PERCIVAL, D.B., DENBO, D.W., EBLE, M.C., GICA, E., MOFJELD, H.O., SPILLANE, M.C., TANG, L., and TITOV, V.V. (2011), *Extraction of tsunami source coefficients via inversion of DART® buoy data*, Nat. Hazards, 58(1), 567–590; doi:10.1007/s11069-010-9688-1.

POOLE, S.L., RABINOVICH, A.B., SPIELVOGEL, L.Q., and HARVEY R.R. (1980), *Study of ocean tides in the region of the Kuril-Kamchatka and Japan Trenches*, Oceanology 20(6), 655–659.

POPLAVSKY, A.A., KULIKOV, E.A., and POPLAVSKAYA, L.N. (1988), *Methods and Algorithms of Automatic Tsunami Warning*, Moscow, Nauka, 128 pp. [in Russian].

PUGH, D., and WOODWORTH, P. (2014), *Sea-Level Science: Understanding Tides, Surges, Tsunamis and Mean Sea-Level Changes*, Cambridge University Press, 395 pp.

RABINOVICH, A.B. (1993), *Long Ocean Gravity Waves: Trapping, Resonance, and Leaking*, Gidrometeoizdat, Leningrad, 325 pp. [in Russian].

RABINOVICH, A.B. (1997), *Spectral analysis of tsunami waves: Separation of source and topography effects*, J. Geophys. Res. 102(C6), 12,663–12,676.

RABINOVICH, A.B., and THOMSON, R.E. (2007), *The 26 December 2004 Sumatra tsunami: Analysis of tide gauge data from the World Ocean Part 1. Indian Ocean and South Africa.* Pure Appl. Geophys. 164(2/3), 261–308.

RABINOVICH, A.B., THOMSON, R.E., and STEPHENSON, F.E. (2006), *The Sumatra Tsunami of 26 December 2004 as observed in the North Pacific and North Atlantic Oceans*, Surveys Geophys. 27, 647–677.

RABINOVICH, A.B., STROKER, K., THOMSON, R., and DAVIS, E. (2011a), *DARTs and CORK in Cascadia Basin: High-resolution observations of the 2004 Sumatra tsunami in the northeast Pacific*, Geophys. Res. Lett. 38, L08607; doi:10.1029/2011GL047026.

RABINOVICH, A.B., WOODWORTH, P.L., and TITOV, V.V. (2011b), *Deep-sea observations and modeling of the 2004 Sumatra tsunami in Drake Passage*, Geophys. Res. Lett., 38. L16604; doi:10.1029/2011GL048305.

Rabinovich, A.B., Thomson, R.E., and Fine, I.V. (2013a), *The 2010 Chilean tsunami off the west coast of Canada and the northwest coast of the United States*, Pure Appl. Geophys., *170*, 1529–1565; doi:10.1007/s00024-012-0541-1.

Rabinovich, A.B., Candella, R.N., and Thomson, R.E. (2013b), *The open ocean energy decay of three recent trans-Pacific tsunamis*, Geophys. Res. Lett., *40*, doi:10.1002/grl.50625.

Saito, T., Ito, Y., Inazu, D., and Hino, R. (2011), *Tsunami source of the 2011 Tohoku-Oki earthquake, Japan: Inversion analysis based on dispersive tsunami simulations*, Geophys. Res. Lett., *38*; doi:10.1029/2011GL049089.

Satake, K. (2014), *Advances in earthquake and tsunami sciences and disaster risk reduction since the 2004 Indian Ocean tsunami*, Geosci. Lett. *1*(15), 1–13.

Satake, K., Baba, T., Hirata, K. et al. (2005), *Tsunami source of the 2004 off the Kii Peninsula earthquakes inferred from offshore tsunami and coastal tide gauges*, Earth Planets Space, *57*, 173–178.

Satake, K., Fujii, Y., Harada, T., and Namegaya, Y. (2013), *Time and space distribution of coseismic slip of the 2011 Tohoku earthquake as inferred from tsunami waveform data*, Bull. Seismol. Soc. Am., *103*, 1473–1492.

Saxena, N., and Zielinski, A. (1981), *Deep-ocean system to measure tsunami wave height*, Mar. Geodesy, *5*(1), 55–62.

Shevchenko, G., Ivelskaya, T., Loskutov, A., and Shishkin, A. (2013), *The 2009 Samoan and 2010 Chilean tsunamis recorded on the Pacific coast of Russia*, Pure Appl. Geophys., *170*, 1511–1527.

Shevchenko, G., Ivelskaya, T., and Loskutov, A. (2014), *Characteristics of the 2011 Great Tohoku tsunami on the Russian Far East coast: Deep-water and coastal observations*, Pure Appl. Geophys., *171*(12), 3329–3350; doi:10.1007/s00024-014-0727-1.

Snodgrass, F.E. (1969), *Study of ocean waves, 10-5 to 1 Hz*, Inst. Geophys. Planet. Phys., University of California, San Diego, Surv. Paper No. 8, 34 pp.

Soloviev, S.L. (1968), *The tsunami problem and its significance for Kamchatka and the Kuril Islands*, In: The Tsunami Problem, Nauka, Moscow, pp. 7–50 [in Russian].

Soloviev, S.L., and Go, Ch. N. (1974), *Catalogue of Tsunamis on the Western Shore of the Pacific Ocean*, Nauka, Moscow, 309 pp. [in Russian; English Traslation: Canadian Transl. Fish. Aquatic Sci., No. 5078, Ottawa, 1984, 439 pp.].

Soloviev, S.L., Popov, V.M., Miller, G.R., et al. (1976), *Preliminary Results of the First Soviet-American Tsunami Expedition*. Hawaii Institute of Geophysics, NOAA-JTRE-162, HIG-76-8, 74 pp.

Stein, S., and Okal E.A. (2005), *Speed and size of the Sumatra earthquake*, Nature, *434*, 581–582, doi:10.1038/434581a.

Taira, K., Teramoto, T., and Kitagawa, S. (1985), *Measurements of ocean bottom pressure with quartz sensor*, J. Oceanogr. Soc. Jpn., *41*(3), 181–192.

Tang, L., Titov, V.V., Wei, Y., Mofjeld, H.O., Spillane, M., Arcas, D., Bernard, E.N., Chamberlin, C., Gica, E., and Newman, J. (2008), *Tsunami forecast analysis for the May 2006 Tonga tsunami*, J. Geophys. Res. *113*(C12015); doi:10.1029/2008JC004922.

Thomson, R.E., and Emery, W.J. (2014), *Data Analysis Methods in Physical Oceanography*, Third and revised edition, Elsevier, New York, 716 pp.

Thomson, R.E., Fine, I.V., Rabinovich, A.B., Mihaly, S.F., Davis, E.E., Heesemann, M., and Krassovski, M.V. (2011), *Observations of the 2009 Samoa tsunami by the NEPTUNE-Canada cabled observatory: Test data for an operational regional tsunami forecast model*, Geophys. Res. Lett. *38*, L11701; doi:10.1029/2011GL046728.

Titov, V.V. (2009), *Tsunami forecasting*, In: The Sea, Vol. 15, Tsunamis, (Eds. A. Robinson and E. Bernard), Harvard University Press, Cambridge, USA, pp. 371–400.

Titov, V.V., González, F.I., Bernard, E.N.Mofjeld, H.O., Newman, J.C., and Venturato, A.J. (2005a), *Real-time tsunami forecasting: Challenges and solutions*, Nat. Hazards, *35*(1), 41–58.

Titov, V.V., Rabinovich, A.B., Mofjeld, H., Thomson, R.E., and González, F.I. (2005b), *The global reach of the 26 December 2004 Sumatra tsunami*, Science 309, 2045–2048.

Tolkova, E. (2010), *EOF analysis of a time series with application to tsunami detection*, Dyn. Atmos. Oceans *50*(1), 35–54.

Uehira, K., Kanazawa, T., Noguchi, S.I. et al. (2012), Ocean bottom seismic and tsunami network along the Japan Trench, AGU 2012 Fall Meeting, OS41C-1736. http://www.fallmeeting.agu.org/2012/eposters/eposter/os41c-1736/.

UNESCO (1975), *An Intercomparison of Open Sea Tidal Pressure Sensors*. Techn. Papers in Marine Sciences, No. 21, 67 pp.

Vitousek, M.J. (1965), *An evolution of the vibrotron pressure transducer as a mid-ocean tsunami gage*, Hawaii Institute of Geophysics, HIG-65-13, University of Hawaii, Honolulu, 12 pp.

Vitousek, M.J., and Miller, G.R. (1970), *An instrumentation system for measuring tsunamis in the deep ocean*, Honolulu: University Press., 239–252.

Webb, S.C. (1998), *Broadband seismology and noise under the ocean*, Rev. Geophys., *36*(1), 105–142.

Webb, S.C., Zhang, X., and Crawford, W. (1991), *Infragravity waves in the deep ocean*. J. Geophys. Res., *96*(C2), 141–144.

Zielinski A., and Saxena N. (1983a), *Rationale for measurement of midocean tsunami signature*, Mar. Geodesy, *6*(3–4), 331–337.

Zielinski A., and Saxena N. (1983b), *Tsunami detectability using open-ocean bottom pressure fluctuations*, IEEE J. Oceanic Eng, *OE-8*(4), 272–280.

(Received December 2, 2014, revised February 5, 2015, accepted February 17, 2015, Published online March 11, 2015)

A Decade After the 2004 Indian Ocean Tsunami: The Progress in Disaster Preparedness and Future Challenges in Indonesia, Sri Lanka, Thailand and the Maldives

Anawat Suppasri,[1] Kazuhisa Goto,[1] Abdul Muhari,[2] Prasanthi Ranasinghe,[3] Mahmood Riyaz,[4] Muzailin Affan,[5] Erick Mas,[1] Mari Yasuda,[1] and Fumihiko Imamura[1]

Abstract—The 2004 Indian Ocean tsunami was one of the most devastating tsunamis in world history. The tsunami caused damage to most of the Asian and other countries bordering the Indian Ocean. After a decade, reconstruction has been completed with different levels of tsunami countermeasures in most areas; however, some land use planning using probabilistic tsunami hazard maps and vulnerabilities should be addressed to prepare for future tsunamis. Examples of early-stage reconstruction are herein provided alongside a summary of some of the major tsunamis that have occurred since 2004, revealing the tsunami countermeasures established during the reconstruction period. Our primary objective is to report on and discuss the vulnerabilities found during our field visits to the tsunami-affected countries—namely, Indonesia, Sri Lanka, Thailand and the Maldives. For each country, future challenges based on current tsunami countermeasures, such as land use planning, warning systems, evacuation facilities, disaster education and disaster monuments are explained. The problem of traffic jams during tsunami evacuations, especially in well-known tourist areas, was found to be the most common problem faced by all of the countries. The readiness of tsunami warning systems differed across the countries studied. These systems are generally sufficient on a national level, but local hazards require greater study. Disaster reduction education that would help to maintain high tsunami awareness is well established in most countries. Some geological evidence is well preserved even after a decade. Conversely, the maintenance of monuments to the 2004 tsunami appears to be a serious problem. Finally, the reconstruction progress was evaluated based on the experiences of disaster reconstruction in Japan. All vulnerabilities discussed here should be addressed to create long-term, disaster-resilient communities.

Key words: 2004 Indian Ocean tsunami, reconstruction, disaster preparedness, disaster risk reduction.

---

[1] International Research Institute of Disaster Science, Tohoku University, Sendai, Japan. E-mail: suppasri@irides.tohoku.ac.jp
[2] The Ministry of Marine Affairs and Fisheries, Jakarta, Indonesia.
[3] Lanka Hydraulic Institute Ltd, Moratuwa, Sri Lanka.
[4] Maldives Energy and Environmental Company, Male, Maldives.
[5] Faculty of Mathematics and Natural Sciences, Syiah Kuala University, Banda Aceh, Indonesia.

1. Introduction

The 2004 Indian Ocean tsunami remains the deadliest tsunami in recorded history. The earthquake, which had a magnitude M_w 9.3 and a rupture length of approximately 1200 km (Stein and Okal 2005, 2007), triggered a tsunami that reached 30 m in height (Synolakis and Kong 2006) and caused at least 230,000 fatalities in 15 African (Fritz and Borrero 2006; Weiss and Bahlburg 2006) and Asian countries, such as Indonesia, Sri Lanka, Thailand, the Maldives (Borrero et al. 2006; Jaffe et al. 2006; Goff et al. 2006; Ruangrassamee et al. 2006; Fritz et al. 2006; Okal et al. 2006a), and other island countries in the Indian Ocean (Okal et al. 2006b, c). Lessons were learned and good practices were developed as a result of this event. Far-field tsunami hazards from other possible sources were also studied after the 2004 event, focusing on various countries (Løvholt et al. 2006; Burbidge et al. 2008; Latief et al. 2008; Okal and Synolakis 2008; Suppasri et al. 2012a, b). In addition, disaster risk-reduction elements, such as tsunami early warning systems, evacuation buildings, tsunami memorials, tsunami museums and disaster education programs, were developed in this region. This paper presents a summary of the progress made with regard to disaster preparedness over the last 10 years in Indonesia, the Maldives, Sri Lanka and Thailand based on studies, observations and our collaborative activities with government and international counterparts in Japan. This is somehow representative of efforts made by other nations, including the US, Germany, France, Australia, New Zealand and Chile. The main objectives are as follows: (1) to report on the reconstruction and vulnerabilities after 10 years, (2) to evaluate the present tsunami warning systems and emergency response measures and (3) to discuss disaster awareness in terms of education, memorials and the geological evidence that remains and conveys the risk of tsunami impacts. In considering the first objective, we present a summary of the reconstruction of housing and vital infrastructure and the restoration of businesses, land use and education in the affected countries during the initial years after the disaster. To address the second objective, large earthquakes that have generated ocean-wide tsunami warnings and have impacted the region are evaluated. Finally, to address the third objective, we assess the progress made with regard to disaster preparedness and the future challenges faced in these areas.

2. Early Stage Reconstruction After the 2004 Indian Ocean Tsunami

This section describes several examples of the early-stage reconstruction conducted during the first few years after the 2004 tsunami, including housing, lifeline and business restoration, land use management, disaster reduction education and the internal conflicts.

2.1. Housing Reconstruction

Sri Lanka is used as an example here. The tsunami hit the Eastern, Southern and Western coasts of Sri Lanka 2–3 h after the earthquake, with maximum measured run-up heights from several meters to 10 m or more (GOFF et al. 2006). The tsunami killed more than 31,000 people and destroyed 80,000 houses, displacing more than 500,000 people (KHAZAI et al. 2006). Housing reconstruction was studied by MURAO and NAKAZATO (2010), who used data on more than 56,000 transitional houses and 28,000 permanent houses in the tsunami-damaged areas of Sri Lanka. They computed the recovery ratio—the ratio of the number of buildings constructed per month—to the total number of completed buildings as of February 2006, 15 months after the tsunami. Approximately 50 % of the reconstruction was completed after 5 months for the transitional houses and after 8 months for the permanent houses. The transitional houses were constructed mainly during the first period after the tsunami, whereas the construction of permanent houses began after careful evaluation by the government and stakeholders. In Sri Lanka, no-construction zones existed in areas that underwent the rehabilitation process. The reconstruction and the resettlement has been performed considering the initial no-construction zones (FRANCO et al. 2013) declared as 100 m in the South and West, 200 m in the East, and 500 m in the North. However, these initial no-construction zones obtained opposition from the local community as they created severe constraints on their daily lives and livelihoods, and as a result the buffer zones were revised to 25–50 m in the Southern Districts and a minimum of 50 m on the Eastern coast. Although the 200, 500 and 25–50 m no-construction zones are not arbitrary, broad no-build zones still exist and affect people's livelihoods.

2.2. Lifeline Infrastructure Reconstruction

A comparative study of lifeline reconstruction between the cases of Okushiri Island, Japan, after the 1993 tsunami and Nam Khem village in Phang Nga province of Thailand was performed by TAKADA et al. (2010). They reported that the water and power supply reconstruction in Nam Khem village took approximately 1 and 3 months, respectively, whereas the full recovery of lifelines took 5 years on Okushiri Island. This difference was due to the preference of the local residents of Nam Khem village to rebuild in the same place where they had lived before the tsunami. Therefore, the local residents were able to move into their new permanent houses directly from the evacuation shelter. Conversely, on Okushiri Island, the local residents worked for many years together with the local government to design a new, resilient town to mitigate future disasters.

2.3. Business Restoration

A survey of business restoration was carried out in Galle, Sri Lanka, in 2005 (KUWATA and TAKADA 2010). Galle, the capital city of Southern Sri Lanka and known as the location where the tsunami swept away a train with almost 2000 passengers, is located on the south tip of Sri Lanka where the tsunami inundation depth measured more than 5 m (GOFF *et al.* 2006). In this study, business restoration was defined by the amount of sales of a product after the tsunami compared to the amount of sales of the same product before the tsunami. The data were obtained through interviews with shop owners. Financial services and lifelines were the two types of businesses that were restored most quickly. These businesses were fully restored after 2 months because both were considered important for the overall reconstruction. Moreover, agricultural restoration took approximately 6 months. Although the farmers may be situated far away from the sea, the agricultural sector took longer time to completely restore as it was necessary to first remove contamination of the soil due to salinity (KUWATA 2011). Tourism, manufacturing and both wholesale and retail businesses were slowly restored; approximately 50 % were restored 6 months after the tsunami. One of the reasons for this was the reduction in visiting international tourists. Fisheries suffered the most; less than 40 % were restored 1 year after the event. Reasons for this include the damage incurred by fishing boats and the rumors of incurring health risks from eating seafood from the area.

2.4. Land Use Management

An example of land use management as an important adaptive strategy for tsunami resilience in the Maldives was presented by RIYAZ and PARK (2010). The Maldives is a country that is extremely vulnerable to many types of hazard, i.e., storm surges, torrential monsoon rain, sea level rise, tidal waves and tsunamis. Unlike other countries in the Indian Ocean, the average height of Maldivian islands is 1.5 m above Mean Sea Level which exacerbates the challenges associated with the tsunami evacuation process (Ministry of Environment, Energy and Water 2007). Therefore, the "Safer Island Concept" (FRITZ *et al.* 2006) was proposed as follows: (1) establish environmental protection zones (EPZs): high-level sand bunds or embankments to protect the islands from a high rise in sea level, (2) create lower drainage areas on the landward side between the ring road and the EPZ for proper water drainage and (3) build elevated ground or high-rise buildings for vertical evacuation. Vertical evacuation sites can act as a good solution to the problem, but these sites must be built to withstand the great impact, which includes floating debris. The government of the Maldives decided to pursue this concept on ten islands as a long-term goal. However, only two islands have presently been developed as "safer islands". In addition, the two islands developed as "safer island" do not strictly follow the concept outlined above.

2.5. Disaster Reduction Education

Education is important in addition to reconstruction. Examples of new efforts in disaster reduction education in Thailand, at the government level, and Indonesia, at the local level, are discussed in this section. SIRIPONG (2010) reported that, in Thailand, three government agencies provide disaster reduction education: (1) the Ministry of Education, for schools and universities, (2) the Department of Disaster Prevention and Mitigation, for operational and rehabilitative academies, and (3) the National Disaster Warning Center. The role of each agency and their collaboration with international organizations are explained in the study. GOTO *et al.* (2010) introduced an example of educational materials and methods at the local level: a collaborative workshop between Japanese and Indonesian universities. Visual educational aids, such as computer graphic of tsunami, and exercises, such as evacuation drills and the preparation of evacuation maps, were judged to be effective educational tools. YASUDA *et al.* (2014) present similar activities in the 2004 tsunami-affected countries (Indonesia and Thailand), the 2011 tsunami-affected area in Japan, and the 2013 Typhoon Haiyan-affected area in the Philippines and Hawaii. They found that after the activity, for example, 20 % of the students changed their mind from agreeing to totally agreeing to tell their parents what they have learned in class.

2.6. Internal Conflicts

There are some examples of internal conflicts associated with the early-stage reconstruction of this tsunami disaster as both Indonesia and Sri Lanka had been affected by civil war for several decades before the tsunami (BAUMAN et al; 2007; ENIA et al. 2008; HYNDMAN 2009; KUHN 2009). Two of the main issues raised during the early-stage reconstruction are (1) public safety of post-tsunami buffer zone/empowerment issues among the government and rebels, and (2) distribution of aid and goods from international institutions (HYNDMAN 2009). However, the result of the conflicts after the tsunami is different in both countries. Banda Aceh achieved peace after an agreement in 2005 and greater access of natural resources such as oil and gas even before the tsunami. On the other hand, civil war was renewed in Sri Lanka with the loss of government power. These examples suggest the need to pay more attention to post disaster conflict zones given their potential of both positive and negative ends (ENIA 2008).

3. Review of the Indian Ocean Tsunami Events After 2004 and Their Impacts

This section describes the three major tsunamis triggered by earthquakes with magnitudes of approximately M 8.0 or greater and their impacts on countries bordering the Indian Ocean. These three events occurred after the 2004 Indian Ocean tsunami: the 2005 Nias earthquake, the 2007 Sumatra earthquake, the 2010 Mentawai earthquake and the 2012 Indian Ocean earthquakes, as shown in Fig. 1.

3.1. The 2005 Nias Earthquake and Tsunami

This earthquake occurred on 28 March 2005, 3 months after that of December 2004, with a magnitude of M 8.7 (USGS 2005). The earthquake was felt on the entire west coast of Malaysia and in high-rise buildings in Singapore and Thailand, among other countries. The ground motion reached as far as the Andaman Islands, but it was not felt on the Indian mainland (ASC 2005). The epicenter was located in a shallow water region, producing a high-intensity earthquake but a small tsunami. Based on field survey data (BORRERO et al. 2011), the tsunami runup height was approximately 4 m or less at areas near the epicenter. These values are comparable to those of the 2004 tsunami at the same location. Smaller waves were detected at tide gauges in the Indian Ocean in comparison to the 2004 event shown in brackets: for example, 0.10 m (1.08 m) in Male, the Maldives; 0.13 m (0.30 m) in the Cocos Islands, Australia; and 0.21 m (1.50 m) in Colombo, Sri Lanka (NOAA/NGDC 2014). Tsunami warnings were issued by the Pacific Tsunami Warning Center to many countries in the region but were later cancelled. People in Indonesia, India, Malaysia, Sri Lanka and Thailand evacuated to safer places by moving to higher ground, whereas traffic jams occurred on the roads leading out of Banda Aceh, Indonesia, and the coastal areas of Chennai, India (ASC 2005). Fatalities were due primarily to the strong ground motion in areas near the epicenter; however, the USGS (2005) reported that at least ten people died from panic during the evacuation from the Sri Lankan coast.

3.2. The 2007 Sumatra Earthquake and Tsunami

Another large earthquake of magnitude M 8.5 (USGS 2007) occurred in this region on 12 September 2007, 3 years after the 2004 event. Similar to the event in 2005, the earthquake was felt in countries as far as Malaysia, Singapore and Thailand, and high-rise buildings were used for evacuation (BBC 2007). In addition, tsunami warnings were issued to several countries in and around the Indian Ocean, including small islands and Indonesia, India, Malaysia, Sri Lanka and even Kenya (BBC 2007). A moderate tsunami with a runup height of approximately 4 m was measured in the areas near the epicenter (BORRERO et al. 2009). Relatively smaller tsunami waves were observed at tide gauges in the Indian Ocean, including 0.40 m in Phuket, Thailand; 0.11 m in Male, the Maldives; 0.12 m in the Cocos Islands, Australia; and 0.30 m in Colombo, Sri Lanka (NOAA/NGDC 2014). Although the tsunami caused destructive damage to buildings and facilities, no one was killed due to the successful tsunami evacuation following warning messages that reached residents via TV or radio (IMAMURA 2008).

3.3. The 2010 Mentawai Earthquake and Tsunami

Although the magnitude of this earthquake (M 7.8) was smaller than the others discussed in this

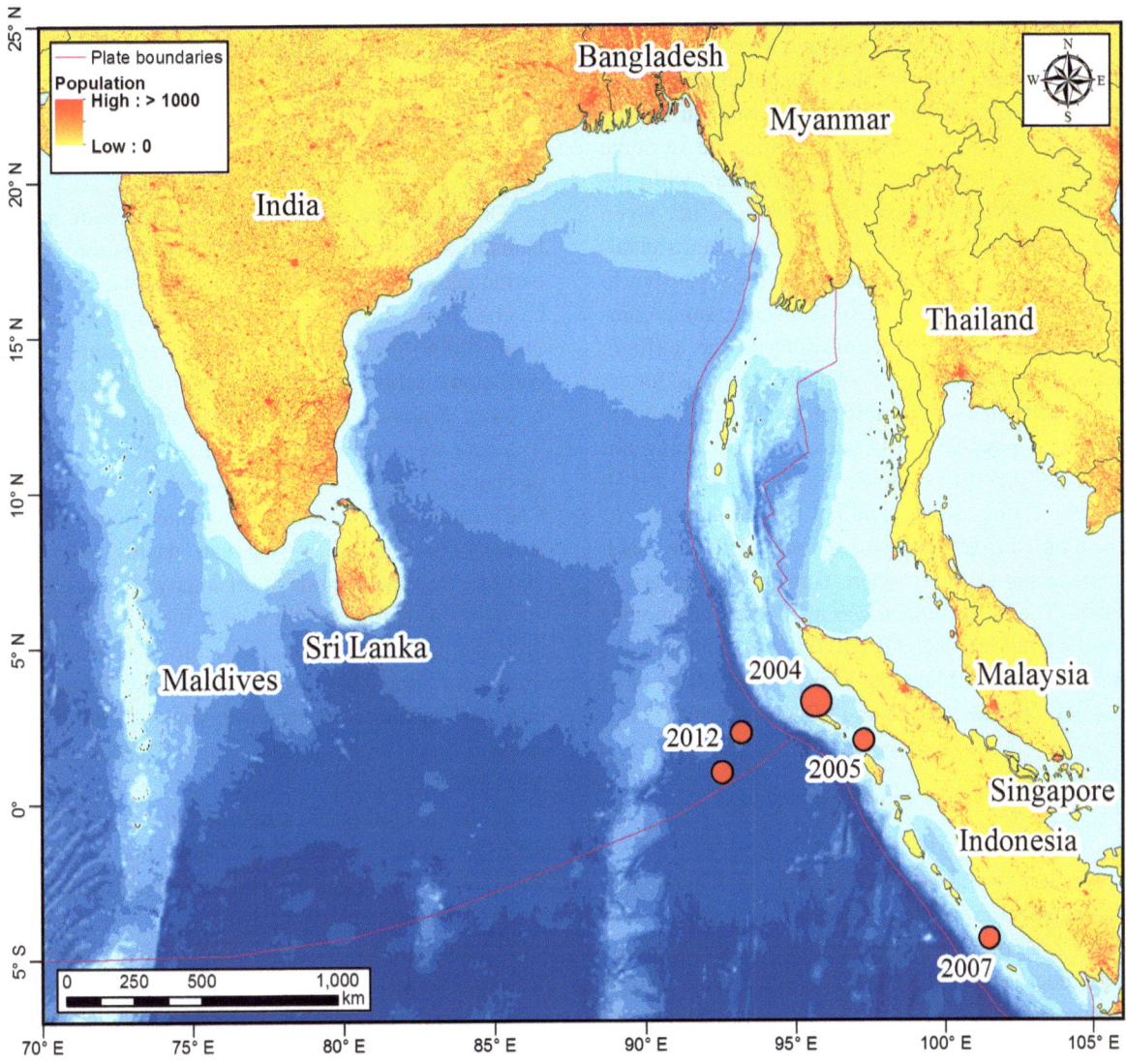

Figure 1
The locations of the major earthquakes in the Indian Ocean since 2004 and the population distribution, revealing the vulnerability of this region. (Original data from Oak Ridge National Laboratory 2006)

section, the number of casualties was the largest (as high as 509) (NOAA/NGDC 2014). In fact, this earthquake was considered a "tsunami earthquake" that generated a much larger tsunami than expected from the seismic magnitude, insofar as the observed tsunami was 4–7 m high (TOMITA et al. 2011; SATAKE et al. 2013) and the maximum runup was measured as high as 16.8 m (HILL et al. 2012). Issues that might have magnified the damage and impact include the following: (1) risk bias: the previous large earthquakes generated strong ground motions but did not generate significant tsunami; the slow ground motion during the 2010 earthquake thus brought a false sense of safety, and most people thought that no tsunami would follow (MUHARI and IMAMURA 2014); (2) the earthquake occurred at night; (3) the earthquake occurred in a location where there was less awareness; and (4) there was no adequate tsunami warning infrastructure to warn the local people. There was no damage reported in the mainland areas where the earthquake was felt, but the tsunami damaged the remote Mentawai Islands, where the tsunami arrived as soon as 5–15 min after the earthquake in some areas. Here, there was also a problem with the tsunami warning systems. Villagers were not given early warning because some buoys had been vandalized, and the equipment had been too expensive to replace (The Telegraph 2010). However, some communities, particularly those residing in the northern part of the affected areas—which is famous for its surfing activities—were able to evacuate; the evacuation was possible not because of the warning but because of the surfers' guidance (MIKAMI et al. 2014). Therefore, this event signifies the importance of education for self-evacuation, especially in such remote areas or areas where tsunamis could arrive swiftly. The 2010 tsunami was also observed in other countries, reaching heights of 0.09 m in Colombo, Sri Lanka; 0.11 m in Male, the Maldives; and 0.16 m in the Cocos Islands, Australia (NOAA/NGDC 2014).

3.4. The 2012 Indian Ocean Earthquakes and Tsunami

The last two sizable events were both outer-rise earthquakes (an unusual type of earthquake that occurs near oceanic trenches; GEIST et al. 2009) of magnitudes over M 8.0 that occurred in North Sumatra on 11 April 2012. One of the earthquakes had a magnitude of M 8.6, whereas the second earthquake presented with a magnitude of M 8.2; they were located 100 and 200 km southwest of the major subduction zone (USGS 2012), respectively. They were felt in areas as far away as Malaysia, Thailand, India and the Maldives, and tsunami warnings were issued for all countries bordering the Indian Ocean (Aljazeera 2012). Residents of Indonesia, Thailand, Sri Lanka and India were advised to move to high ground or to stay far away from the sea. Nevertheless, due to the strike-slip type of these earthquakes, no significant damage was reported in the countries surrounding the Indian Ocean, and the tsunami height was recorded at approximately 0.05 m in Phuket, Thailand; 0.08 m in the Cocos Islands, Australia; 0.21 m in Male, the Maldives; and 1.08 m in the port of Meulaboh, Aceh, Indonesia (NOAA/NGDC 2014).

3.5. Return Period of the 2004 Event-Like Earthquake

One important issue is the recurrence of major tsunami such as the 2004 event. Information of the earthquake return period in the Indian Ocean basin can be obtained from the following studies. BURBIDGE et al. (2008) derived hazard curves representing the earthquake return period as a function of magnitude in Java, Sumatra, Nankai, Seram and South Chile. The data were mainly based on the earthquake focal mechanisms of all the earthquakes in the Global CMT catalogue for the eastern Indian Ocean region since 1976 with $M_w \geq 7.0$ and depths less than 100 km. Because records of seismicity are rarely long enough, they proposed their own method developed to avoid underestimating the return period due to infrequent earthquake events in the area. From their proposed hazard curve, it was estimated that a 2004 tsunami-class earthquake might occur every 1000 years.

A similar earthquake return period was shown in a study by LATIEF et al. (2008). They considered four main subduction segments as tsunamigenic sources in Aceh-Seumelue-Andaman and Nias. The analysis was conducted using the EZ-FRISK program to provide a hazard curve correlating earthquake return periods and the potential moment magnitudes while considering the seismic parameters adopted in the recurrence models. Their results indicate that the recurrence of an event comparable to the 2004 tsunami is approximately 520 years.

Another method for calculating the earthquake return period was used in an investigation of sand sheets in Phang Nga province, Thailand (JANKAEW et al. 2008). From the results of this study, it was concluded that the full-sized predecessor to the 2004 tsunami occurred about 550–700 years ago. The same survey method was conducted in Aceh province, Indonesia (MONECKE et al. 2008). They concluded that the recurrence of the 2004 tsunami is about 600 years which agrees with the estimation by LATIEF et al. (2008) and JANKAEW et al. (2008). Therefore, approximately 600 years may be considered a suitable number for the return period of the 2004 event such as earthquake. This information is very important for disaster planning and policy-making.

4. The Situation in Indonesia

4.1. Overview

In contrast to Japan, which has a long history of tsunami disasters and has implemented countermeasures, the word "tsunami" was not commonly heard and its impacts were not well understood in Sumatra prior to the 2004 Indian Ocean tsunami, even though there were some major tsunami events prior to the 2004 event such as the 1992 Flores and the 1994 Java tsunamis. The tsunami of 2004 destroyed 120,000 homes and heavily damaged 70,000 more, destroying 3000 government buildings (including hospitals and schools) and 14 seaports. Nearly 3000 km of roads in this region were damaged, and 150,000 people lost their lives. Overall, the tsunami affected 800 km of Indonesia's coastline, with a total affected area of 413 km^2 (SAMEK et al. 2004).

Because of the breadth and scale of the disaster, the Indonesian government established a special agency, called the Agency for Reconstruction and Rehabilitation in Aceh and Nias (BRR Aceh-Nias), to conduct reconstruction activities in Banda Aceh and the surrounding areas. This agency began to operate in mid-2005. With only a 4-year tenure, reconstruction activities focused on the development of housing (2005 to mid-2007), the development of public infrastructure (2005–2008), institutional and social development (2005 to mid-2009) and economic development (2005–2009). The agency's overall reconstruction expenses were 6.7 billion USD, which came from the Indonesian government (37 %), bilateral and multilateral donors (36 %), and national and international non-governmental organizations (27 %).

The redevelopment of the Aceh region was conducted with a community participatory method at the village level. For instance, a community would decide whether its area would be rebuilt using land consolidation or if it would be left as it was.

Along with these village-level reconstruction efforts, facilities and infrastructure for future tsunami mitigation were also designed at a higher level. A field visit was organized in November 2014 to observe these facilities as shown in Fig. 2. Buildings for vertical evacuation are now ready for use in Banda Aceh (Fig. 3a), and evacuation routes have been tested in national and international tsunami drills, in which all countries affected by the 2004 tsunami performed the drills. Recognizing that experiencing the enormity of a tsunami is important, the media has been employed to disseminate the tsunami-related experiences from 2004 to future generations in various forms. A tsunami museum (Fig. 3b) was established in Banda Aceh along with 85 tsunami poles that indicate the height of the tsunami that hit the area. Other forms of memorials, such as ships (Fig. 3c), including a large diesel power plant vessel that was carried approximately 3 km inland, were left in place as a memorial for future generations (Fig. 3d).

4.2. Development of Tsunami Early Warning Systems

The development of a tsunami early warning system has been one of the major initiatives in Indonesia since the 2004 Indian Ocean tsunami. It was first developed under cooperation between the Indonesian and German governments in 2005 (MÜNCH et al. 2011; PARIATMONO 2012). It was launched in 2009 and the present status is reviewed by LAUTERJUNG et al. (2015). Since its development, the system has been tested by both near- and far-field actual tsunamis (MUHARI and IMAMURA 2014). The primary aim of an early warning system is to disseminate the information that a tsunami might occur after the occurrence of an earthquake; the Indonesian tsunami early warning system (Ina-TEWS) is able to issue a tsunami warning within 5 min after an earthquake (PARIATMONO 2012).

Tsunami buoys are not yet integrated in the overall warning system. The buoys are managed by an institution, which is separate from BMKG (Badan Meteorology, Klimatologi dan Geofisika, or the Meteorology, Climatology and Geophysics Agency). The challenge this system faces is not only when the warning is issued but also when the warning should

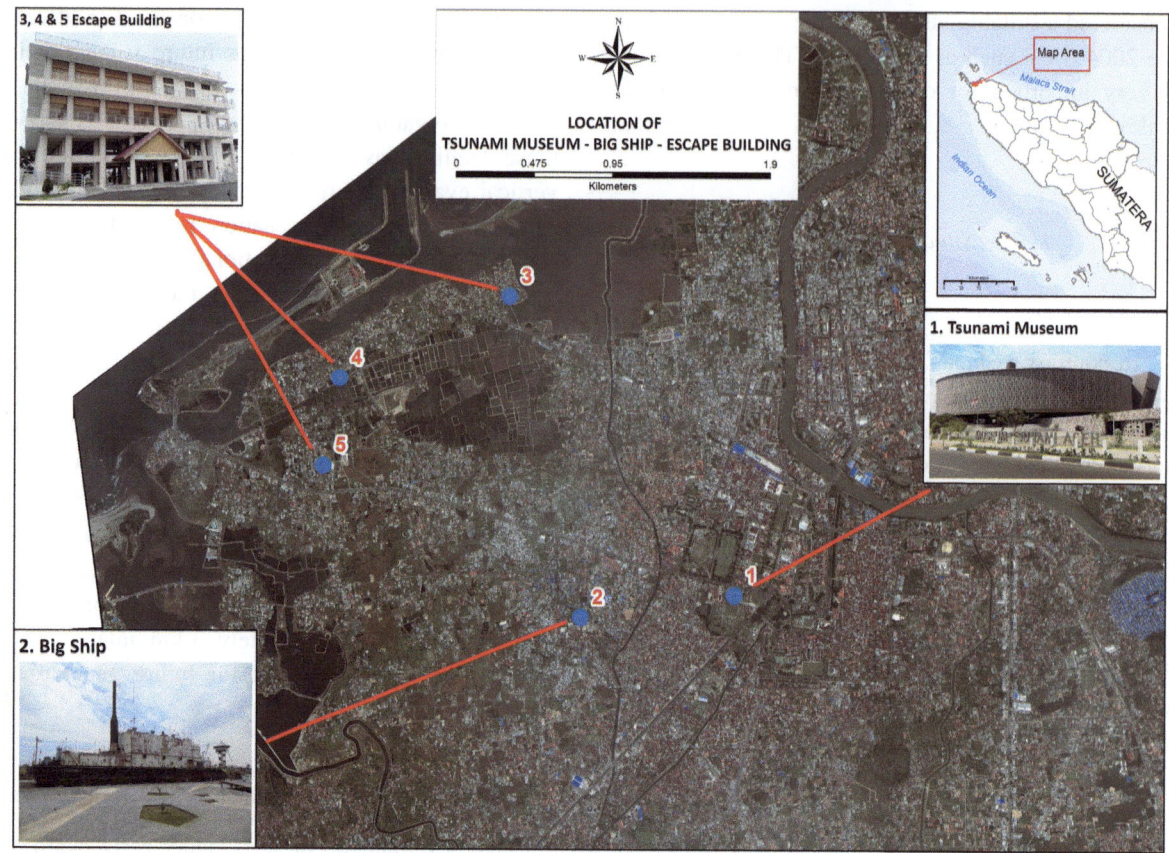

Figure 2
Locations in Banda Aceh, Indonesia mentioned in this study

be cancelled. During the 2010 Mentawai earthquake, the warning was terminated 51 min after the earthquake, although the tsunami was still occurring up to 2 h later, particularly in the small bays in the remote Mentawai Islands (MUHARI and IMAMURA 2014). The decision to terminate the warning was made after considering real-time data from the nearest tide gauges, which indicated the tsunami height was only as high as 0.22 m (in Enggano) prior to the warning's termination (BMKG 2010). However, those tide gauges were located approximately 300 km away from the source (Fig. 5). Near the source, in the southern part of the Mentawai Islands, located only 50 km from the rupture area, the tsunami height was much higher, ranging from 2 m to a maximum of 17 m, as observed on the small Sibigau Island located near South Pagai Island (HILL et al. 2012; SATAKE et al. 2013).

Moreover, during the 2011 Japan tsunami, the tsunami warning in the east Indonesian region was terminated just after the first wave arrived. However, the largest wave came 1 h later and caused the death of a local who had returned to the lowland after receiving notice that the warning had been terminated (DIPOSAPTONO et al. 2013). The lack of real-time measurement data for continuously updating the warning level or for indicating whether a tsunami has been generated is one of the issues that needs to be resolved. In addition, disseminating warnings in remote areas, such as small islands, is another challenge for future improvements to the system.

From the perspective of warning dissemination, even though the National Tsunami Warning Center (BMKG) had successfully issued a warning within 5 min after the earthquake, the local sirens did not sound because they relied on the affected city's

Figure 3
a One of the four vertical evacuation buildings in Banda Aceh, **b** the tsunami museum in Banda Aceh, with an evacuation sign indicating the entrance, **c** the memorial site of a stranded ship and **d** the 2600-ton floating power plant that was swept inland by the tsunami

electricity, which was usually automatically shut off during the earthquake. Manual activation of the sirens was difficult because the officer in charge of them also evacuated after the earthquake. Such a practical problem should be addressed to ensure safe evacuation in the future.

4.3. Tsunami Evacuation Condition

The other issue is the human response to warnings. The 2012 Indian Ocean earthquake was likely the best opportunity to test the preparedness of the people of Banda Aceh against tsunamis. The magnitude M 8.6 earthquake triggered a massive evacuation even though the tsunami sirens in Banda Aceh did not sound until 70 min after the initial shock. Most of the people evacuated using motorcycles to quickly escape the lowland area. Although vertical evacuation buildings were available in the lowland, most of the residents did not believe that the structures could withstand the earthquake and tsunami. The majority of people in Banda Aceh thought that horizontal evacuation was the best choice. As a result, massive traffic jams were observed along the main road of Banda Aceh City during the evacuation. According to a survey conducted after the event (Goto et al. 2012c) among 161 people who evacuated (of the total 220 respondents), 153 were trapped in the traffic jam during the evacuation. Respondents did not believe that the vertical evacuation structures could increase their chances of successfully evacuating given the limited time before the tsunami would arrive. The lack of community participation in determining the location and design of the vertical evacuation structures is believed to be one of the reasons why, even now, people do not trust that vertical evacuation structures can save their lives. In addition, the lack of interest in these structures has led to poor maintenance of the facilities (Fig. 4), which are rarely used, even for social activities, by the residents who live near the buildings.

Banda Aceh city has made several plans to reduce the problem of traffic jams during emergency situations to facilitate evacuation of future disaster. Tsunami evacuation simulation was conducted (GOTO et al. 2012b) and some plans have been realized such as widening existing roads which are to act as escape roads, and also making emergency crossings on two-way main roads. This emergency crossing will be used in emergency situations to shorten U-turns or to cut the road block so that traffic jams can be avoided or reduced. Another plan to be implemented in 2015 is to make a flyover on roads with heavy traffic and to also make an underpass on the designated road that has been identified by the city government. With support from the National Agency for Disaster Management (BNPB) Banda Aceh city has made a tsunami evacuation road map. This tsunami evacuation road map should be publicized to Banda Aceh city residents to facilitate proper evacuation for future disaster mitigation.

4.4. Design of Evacuation Buildings

An evacuation building in Banda Aceh has been designed with four stories and an overall height of 18 m, and incorporates 54 columns each having a diameter of 70 cm. The roof includes a helipad for helicopter landings. In a large-scale tsunami evacuation drill which was held on 2 November 2008, a helicopter was landed there smoothly. The second floor has a height of approximately 10 m, as indicated by the 26 December 2004 tsunami wave height at the location of the building. The first floor is left open with no partitions or hollow structures, following the concept of the mosque. The aim is to avoid the wave force of future tsunamis. The building can accommodate the evacuation of 500 people and is designed to withstand earthquakes with a moment magnitude of 10 on the Richter scale. The stairs leading to the upper floor is made of two parts. One main staircase has a width of approximately 2 m and another one has a width of 1 m with the slope designed to accommodate the use of wheel chairs in emergency situations. The building is also equipped with facilities for emergency situations. The building serves as a community center that is surrounded by villagers who are alert and ready to mitigate the effects of disasters.

5. The Situation in Sri Lanka

The 2004 Indian Ocean tsunami is remembered as the most devastating catastrophe in Sri Lanka, having caused 35,000 fatalities on two-thirds of the coastal belt, across 13 coastal districts (ADB 2005). The waves penetrated approximately 500 m inland, on average, to a maximum of 2 km on the east coast and

Figure 4
a A wooden bar blocking the entrance for disabled persons to the evacuation building and **b** the first floor of the evacuation building

with a maximum tsunami runup recorded on the southern coast over 10 m in Yala (LIU *et al.* 2005) and Hambantota (WIJETUNGE 2009). The damage in Sri Lanka was particularly severe along the coast from east to south. The damage was dramatically more significant because no tsunami early warning system existed in Sri Lanka at the time due to the country being located very far from active faults.

Since the 2004 event, several individuals and national and international organizations have supported the recovery of the damaged areas on a short-term and long-term basis. Providing food, clothing and temporary shelter to displaced populations comprised some of the immediate actions taken by authorities. Resettlement and reconstruction activities supervised by the Ministry of Relief, Rehabilitation and Reconstruction commenced after the completion of the relief activities (ADB 2005). Reconstruction activities were further enhanced by short-term and long-term plans that led to the establishment of a powerful body called the Disaster Management Center (DMC) in 2005. However, there have been many challenges regarding the policy to create urban development strong enough to withstand future disasters. The 2004 tsunami itself created the unforgettable example of a lack of awareness of tsunami events leading to devastating damage in Sri Lanka even though it was a far-field tsunami. This occurred, in part, because there are still very few tsunami experts in Sri Lanka; consequently, hazard and risk evaluations have been insufficient. Our group monitored the recovery process in Sri Lanka from 2010 to 2013 with the cooperation of the Japan International Cooperation Agency and the Miyagi Prefecture in Japan. Here, we briefly summarize our work and discuss the current problems related to the recovery process and to future tsunami countermeasures in Sri Lanka.

5.1. The Current Conditions of Severely Damaged Areas

Galle City was severely damaged by the 2004 Indian Ocean tsunami (Fig. 5). The tsunami runup height in this city was estimated at 4–5 m (LIU *et al.* 2005; WIJETUNGE 2009). The placement of tsunami signs, the designation of hazard zones and disaster prevention education activities have been conducted

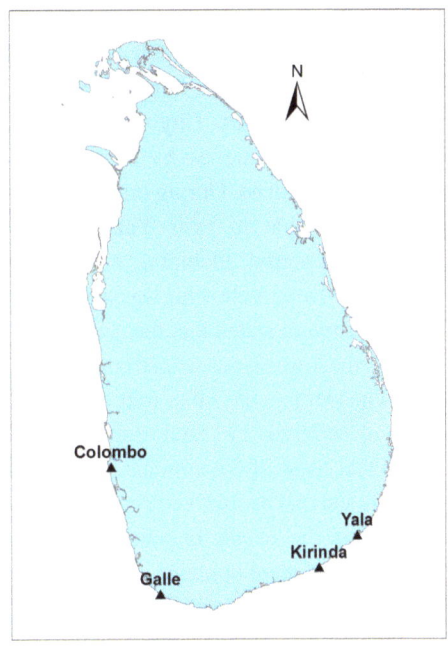

Figure 5
Locations of coastal cities in Sri Lanka mentioned in this study

by city authorities, universities and schools. A group from the Faculty of Engineering at the University of Ruhuna in Sri Lanka prepared a tsunami hazard map and evacuation plan for Galle City following the guidelines that were being used in Japan (Coastal Development Institute of Technology 2004), and this was one of the initial activities that led to increasing people's tsunami awareness (WIJAYARATNA *et al.* 2006). However, it is difficult to continue disaster-prevention activities with a limited budget and changing levels of awareness of tsunamis in the local population. This problem is likely to be widespread among tsunami-affected countries. Consequently, it is important to determine how to continue disaster-prevention activities over the long term. For example, there is a Tsunami Photo Museum in Telwatta containing thousands of photographs taken during and after the tsunami as well as paintings by children. However, the museum is operated by the private sector, so it is important to determine how to continue to support such a museum. According to the DMC, the establishment of a disaster information system has been completed. However, the hazard map for various types of natural disasters and the action agenda for disaster prevention require improvement.

Hambantota and Yala are other locations on the southern coast that were severely affected by the Indian Ocean tsunami (LIU et al. 2005; GOFF et al. 2006). Being mostly flatland, Hambantota City was affected by the tsunami, and a lack of awareness by its people increased the damage it experienced. During our 2012 visit to Sri Lanka, we were able to visit Yala, where many casualties were reported, including 15 foreigners. A safari tour guide from Yala who was a witness to the Indian Ocean tsunami stated that the tsunami flow that came through the lower lands was strong enough to kill the people who were on the safari tour even though the Yala area is covered by sand dunes. He further emphasized that most of the animals in the Yala wild park were safe and that people died because of their lack of awareness of the event. In addition to localized reconstructions and developments, the DMC was able to create a well-planned early warning system for disasters, including 74 early warning towers that cover the entire coastal belt of Sri Lanka; the aim of these towers is to issue tsunami and storm surge warnings.

5.2. The Recovery of Harbor Functions

The damage the Indian Ocean tsunami caused to coastal infrastructure, ecosystems and the fishery sector was significant. Nearly 80 % of fishing boats and ten out of the twelve fishery harbors were badly damaged (CH2MHill 2006). Hikkaduwa, Mirissa, Galle, Puranawella, Hambantota and Kirinda are harbors that were considered highly damaged due to the tsunami. The Kirinda fishery harbor is a good example of engineering projects that mitigate natural phenomena. After its initial construction, the harbor faced a serious problem in 1992: sand accumulation blocked the harbor entrance. Some civil engineering projects, such as dredging and the construction of new breakwaters, were introduced to solve the problem, but the problem continued until before the 2004 tsunami occurred (Fig. 6). The Kirinda fishery harbor was severely affected by the 2004 Indian Ocean tsunami in terms of both coastal structures and coastal morphology (e.g., GOTO et al. 2011; RANASINGHE et al. 2013). During the tsunami, an approximately 8-m-high tsunami flood was recorded, and a large dredging ship called the *Weligowwa* was cast ashore. Because of the sand blockage, bathymetric change was being measured in detail at regular intervals. The last measurement before the tsunami was taken in November 2004. Post-tsunami bathymetry was measured from February to March 2005. According to GOTO et al. (2011) and RANASINGHE et al. (2013), the first runup tsunami wave transported large amounts of offshore, sea bottom sediment and deposited it in a layer up to 4 m thick along the shoreface slope. Bathymetry values returned to normal by November 2005, approximately 1 year after the tsunami. After the tsunami, sand accumulation began again, and sand blockage had re-occurred by 2006. Because the harbor was an important facility for fishermen, dredging work has been conducted such that, as of 2013, the harbor is in use.

In the case of Kirinda, the tsunami's impact on bathymetry was limited, and it recovered very quickly. However, several studies have reported that coastal geomorphology has not fully recovered from the 2011 Tohoku-oki tsunami (UDO et al. 2013). It is very important that we understand what determines whether coastal geomorphology recovers.

5.3. Scientific and Technological Support from Japan

In this project, the Japan International Cooperation Agency (JICA) led a collaboration with the Miyagi prefectural government and Tohoku University. The project was implemented for 3 years in 2009, 2010 and 2012. To expand the understanding of city planning in preparation for future disasters in Sri Lanka, Japanese knowledge of disaster-prevention-specific activities in the region was taught to trainees from Sri Lanka. The goals were the comprehensive support and practice of science and technology, the practice of disaster prevention, enlightenment and education. During their training in Japan, trainees learned basic disaster prevention, disaster history and the primary countermeasures and current challenges in both countries. The trainees are expected to contribute to disaster prevention and mitigation activities in Sri Lanka. In addition, education on earthquake and tsunami disaster prevention assistance in Sri Lanka was provided through seminars and workshops, the presentation and sharing of information, the exchange of ideas and the presentation of challenges. A tsunami deposit site visit to assess the risk of low frequency hazards will also be required in the future. In

Figure 6
Bathymetric changes in the Kirinda fishery harbor in 1995, 2004, 2009 and 2013. [Image source: Lanka Hydraulic Institute Ltd (LHI) and Google earth]

addition, disaster prevention education, awareness-raising activities and community-based disaster management activities in schools and the education community, based on the Participatory Technology Assessment (PTA), were organized with local professionals. Some examples of our activities, such as a field survey of reconstruction conditions and disaster risk reduction activities are shown in Fig. 7.

5.4. Problems with Preserving Monuments

Unlike Indonesia and Thailand, Sri Lanka does not have a damaged structure that has been preserved as a tsunami monument. One possible monument is a train named "the queen of the sea" that was hit by the 2004 tsunami with more than 1700 passengers aboard; they became casualties due to their belief that the train would be strong enough to withstand the tsunami. This became the deadliest train disaster in recorded history. Nevertheless, this train might be too emotionally difficult for the local people to preserve as a monument. This example demonstrates the importance of disaster risk education for local people. Among the tsunami memorials that were constructed, the memorial at Peraliya in the Galle district is still in good condition and has become an attraction for tourists as well. This memorial was constructed in remembrance of the people who died in the train accident. The remaining cars of the damaged train had been kept in Peraliya but were then moved to Hikkaduwa and kept there for some time. The train was repaired and put into service, and it ran the same trip on 26 December 2008 as a memorial to the people who lost their lives. In Yala, a resort hotel and most of the vegetation were severely damaged by the tsunami (GOFF et al. 2006). After this disaster, the remnants of the bungalow were preserved, and a tsunami monument was constructed in 2005.

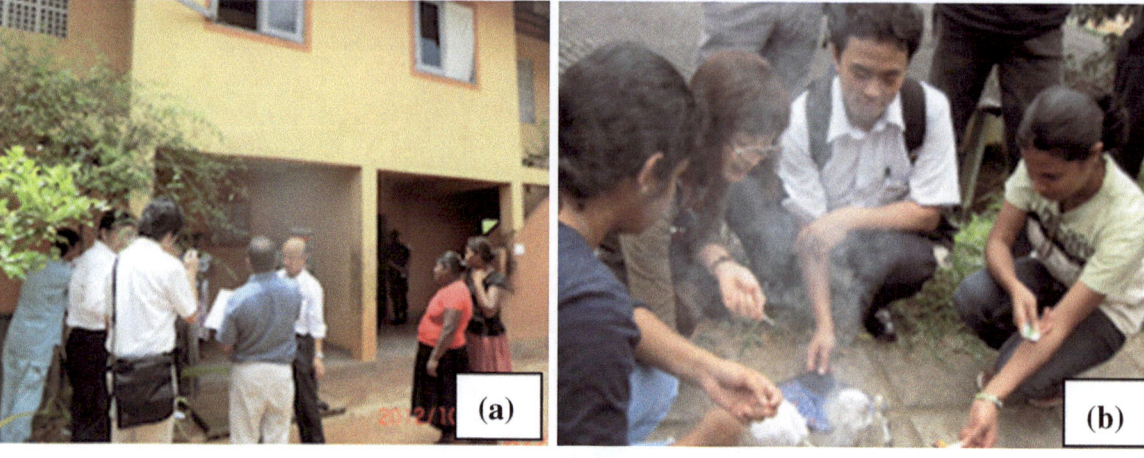

Figure 7
Interviews on disaster response and recovery conditions in Galle City and practicing survival (making rice using cans)

Figure 8
a A steel monument and **b** a tsunami memorial in 2005 at the Yala safari bungalow

However, as seen in Fig. 8, part of the painted writing on the monument disappeared in 2012, and it is difficult to read the text. Although it is very important to memorialize the tsunami disaster by visiting the monument, its preservation is even more important.

6. The Situation in Thailand

6.1. General Observations and Future Challenges in Each Location

A significant change in Thailand after the 2004 tsunami was the establishment of the National Disaster Warning Center (NDWC). The NDWC obtains disaster observation data from the Thai Meteorological Department (TMD) and other international organizations and acts as the center for disaster warning in Thailand. Using this system, a tsunami warning can be issued in Thailand within 5 min after an earthquake occurs. In the six provinces that were damaged by the 2004 tsunami, there are also many examples of change, such as the preparation of tsunami hazard maps and tsunami signs and the construction of tsunami evacuation buildings. Field visits were organized in 2009 and 2013 in areas shown in Fig. 9 in collaboration with the TMD, the NDWC, the Department of Disaster Prevention and Mitigation

Figure 9
Locations in Phang Nga and Phuket, Thailand mentioned in this study

(DDPM) and the Regional Integrated Multi-Hazard Early Warning System for Africa and Asia (RIMES).

6.1.1 Nam Khem Village, Phang Nga Province

Nam Khem was one of the worst-hit areas in Thailand, where more than a thousand lives were lost as a result of flooding by a 10-m-high tsunami wave. Most villagers were fishermen or worked in aquaculture and sightseeing. A tsunami museum and memorial were constructed near the coast, containing pictures, damage information and a list of the victims. Another tsunami memorial was made out of two large fishery boats that were transported inland by the tsunami, killing people and destroying houses. The population in Nam Khem village has since decreased, and the tsunami evacuation plan has been very much improved. A number of activities such as tsunami evacuation drills, reconstruction forums, and disaster-related lectures are being conducted here, and residents are well-prepared and highly aware compared to other areas. They have their own tsunami evacuation drills and emergency plans. A sea observation team will check on the receding wave when they receive a warning. This procedure is risky and dangerous because it is not necessarily the case that a tsunami will be preceded by drawdown (SUPPASRI 2010). Education regarding to the correct basic understanding of tsunamis is still needed. A patrol team will monitor residents' property. A traffic team will implement their own rules during evacuation, having pedestrians keep left and vehicles keep right. A rescue team will check every house for remaining and possibly disabled residents. A nursing team will provide first aid. The school was moved further from the sea and to higher ground as a result of previous experience. The evacuation route for vehicles is the main road (a 2-lane road). Two types of evacuation drills are conducted, and the residents understand the problems that occur when they have to evacuate in the rain or during nighttime. As a result, their evacuation process in response to the tsunami event on 11 April 2012 was a complete success, without any problems or traffic jams. Nam Khem serves as a good example of a tsunami disaster-resilient community. In 2013, the tsunami memorial park looks similar to what was observed in 2009; however, the two fishing boat memorials (Fig. 10a) have been seriously deteriorated by corrosion due to a lack of maintenance. In addition, tsunami-related signs, such as hazard maps, tsunami height indicators and evacuation direction signs, have faded due to sunlight and other weathering processes (Fig. 10b).

6.1.2 Pakarang Cape, Phang Nga Province

The tsunami evacuation building in Pakarang Cape (Fig. 11a) is adjacent to the beach, in compliance with the safe zone distance of more than 1 km, and there is no greenbelt to dissipate tsunami energy. Construction began for the evacuation shelter in March 2009 and was completed in November 2009. There is a sign there that indicates the 2004 tsunami level of 5 m, but the actual height of the sign is approximately 1 m. In this area, local residents find a height of 5 m difficult to imagine (SUPPASRI 2010). This may lead to an underestimation of the tsunami hazard. Therefore, the height of the sign should be set equal to the height of the tsunami. A detailed study of tsunami evacuation simulation in this area, including the mentioned evacuation building, can be seen from a study by MAS et al. (2015-the same issue).

6.1.3 Khao Lak Area, Phang Nga Province

Khao Lak is the most popular tourist area in the Phang Nga province, where a large number of non-Thai tourists were killed by the tsunami. Many hotels and resorts were also destroyed by the tsunami but had been fully reconstructed a few years after the tsunami. When the tsunami occurred, the water police patrol boat T.813 was on its way to secure one of the royal families and was swept inland across the main road, which is located approximately 2 km from the sea. Some of the boat's crew, including the captain, survived. The boat is now preserved as a tsunami memorial similar to those in Nam Khem village. Tsunami evacuation signs in both Thai and English were erected in many places in Khao Lak to maintain tsunami awareness. Figure 12a shows the undeveloped area containing the boat memorial with a wooden bridge crossing the river in 2009. By 2013 (Fig. 12b), the bridge structure had become concrete, and the area near the boat had been developed and now contains a memorial park, exhibition hall, travel center and other activity spaces.

6.1.4 Patong Beach, Phuket Province

Patong Beach is one of the most popular tourist areas in Phuket and contains many hotels, shops and restaurants. The tsunami in 2004 was not as high there as in Phang Nga. Therefore, only the first floors of buildings were damaged, and the tsunami only reached approximately 300 m inland. There are also a large number of tsunami signs in Patong. Large

Figure 10
a Corrosion of a memorial fishing boat and **b** poor maintenance of tsunami-related signs in Nam Khem village

Figure 11
Tsunami evacuation building in Pakarang Cape (**a**) during construction in 2009 and (**b**) in 2013

numbers of tourists stay in the tsunami hazard zone or spend most of their time there. Although evacuation signs are located every 300, 200 and 100 m from the beach to the safe zone, some of them are not well maintained. Moreover, the evacuation route planned by the local government goes through private property and is full of shops and stalls that could lead to a difficult evacuation. The traffic jam in Patong was the most serious among the 2004 tsunami-affected areas. Traffic jams occur daily during rush hour in the morning and evening. One reason for these traffic jams is the large number of motorcycles that are rented by tourists. The three evacuation shelters comprise a school, a temple and a market. All of these shelters are located more than 1 km from the sea. The school appears to be the most suitable shelter because of its space and accessibility. The tsunami warning in 2012 (without a damaging tsunami) confirmed that this situation is very hazardous. The warning occurred during the peak traffic period. As a result, all traffic was at a standstill from the top of the mountain to the road along the beach, inside the 2004 tsunami inundation zone. Figures 13 and 14 show the one-way traffic routes in Patong beach, which extend

Figure 12
The T.813 patrol boat as a tsunami memorial (a) in 2009 and (b) in 2013

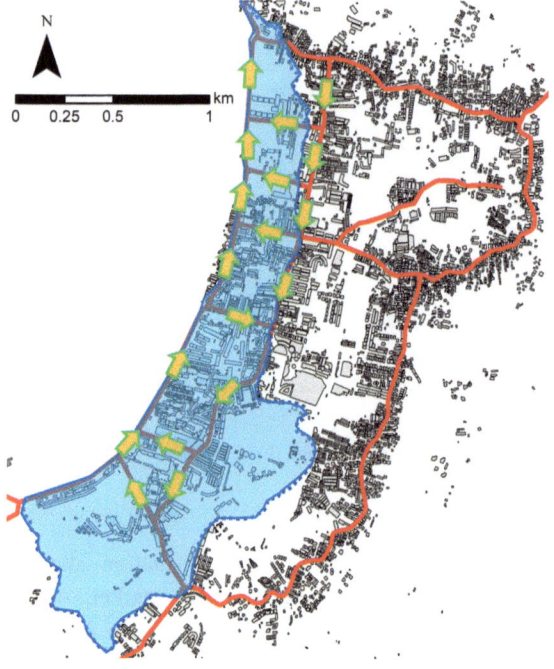

Figure 13
Map of Patong beach; the 2004 tsunami inundation depth (*blue*), buildings (*gray*), roads (*red*) and one-way traffic routes (*arrows*)

in a clockwise direction, going north along the shoreline and south on the inner road. Most of the small roads that link these two roads extend toward the sea except for one road that extends inland. The direction of these small roads is one of the reasons for the traffic jam, and the traffic rules during tsunami evacuations in Patong must be improved in preparation for future tsunamis.

6.2. Geological and Environmental Changes

6.2.1 Tsunami boulders at Pakarang Cape

The tsunami in 2004 transported thousands of meters-long boulders shoreward at Pakarang Cape, Thailand. Goto et al. (2007) investigated the size, position and long-axis orientation of 467 boulders on the cape. Most of the boulders found on the cape were well rounded, ellipsoid and without sharp, broken edges. They were fragments of reef rocks, and their sizes were estimated to be an average of 14 m^3 (22.7 t). Goto et al. (2007) calculated the tsunami inundation at Pakarang Cape. They suggested that the tsunami waves, which were directed eastward, struck both reef rocks and coral colonies that were originally located on the shallow sea bottom near the reef edge; these detached, and the boulders were transported shoreward. During our visit in February 2013 (Fig. 15a), the boulders were still there. Conversely, sand accumulation was remarkable on the tidal bench, and the beach berm had quickly recovered.

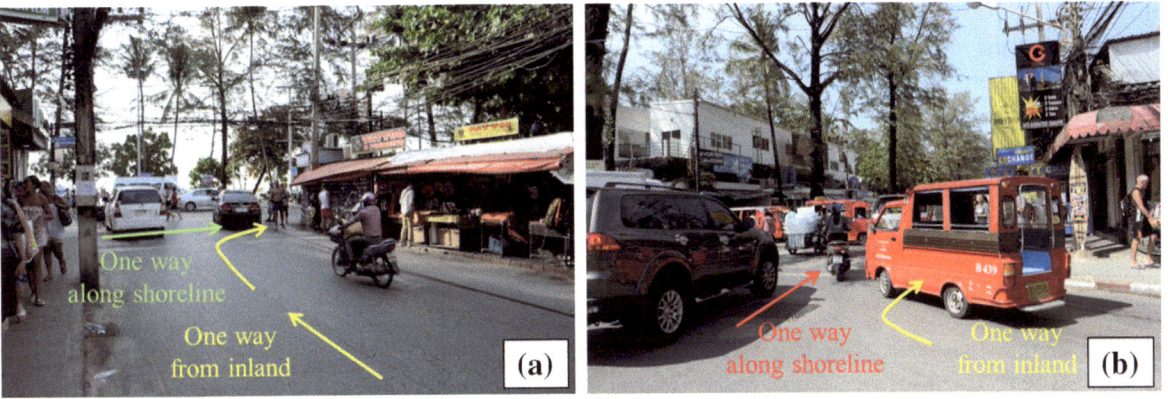

Figure 14
Traffic direction in Patong beach (**a**) in the south and (**b**) in the north

Figure 15
Remaining geological evidence at Pakarang Cape, **a** tsunami boulders and **b** tsunami sand deposits

6.2.2 Tsunami Deposits in Bang Sak Beach

JANKAEW et al. (2008) was the first group who discovered paleo-tsunami deposits prior to the 2004 event in Thailand. Although after a decade, there are well-preserved 2004 tsunami deposits on Bang Sak Beach. This area includes a beach and shallow marine sediments with abundant marine plankton. GOTO et al. (2012a) conducted field surveys in March 2005 and December 2008 in southwestern Thailand to investigate local variation in the thickness and the preservation potential of onshore deposits formed by the 2004 tsunami. The 2008 survey results revealed that the thickness of deposits varied by a few centimeters in pits located less than 10 m apart because of the local undulation of the topography and possible bioturbation. At 13 of the 24 sites, the difference in thickness between the 2005 and 2008 surveys was within the range of local variation. In fact, very thin tsunami deposits (1 cm thick) in the 2005 survey were well preserved during the 2008 survey. Furthermore, tsunami deposits near the maximum inundation limit were found in the 2008 survey, with thicknesses that were consistent with those reported in the 2005 survey. At no site was a tsunami deposit eliminated completely. Based on these observations, GOTO et al. (2012a) inferred that tsunami deposits tend to be well preserved, even in a tropical climate with heavy rains, such as that of Thailand. Our visit in 2013 further supports the observation by GOTO et al. (2012a), insofar as we observed that the deposits were well preserved in cases where there were no human disturbances (Fig. 12b).

6.2.3 Mangrove Forests at Pakarang Cape

YANAGISAWA et al. (2009) investigated the damage to mangroves caused by the 2004 tsunami at Pakarang Cape. Comparing pre- and post-tsunami satellite imagery of the study area, they found that approximately 70 % of the mangrove forest was destroyed by the tsunami. Based on field observations, they determined that the survival rate of mangroves increased with increasing stem diameter. Specifically, they found that 72 % of Rhizophora trees with a 25–30 cm stem diameter survived the tsunami impact, whereas only 19 % of the trees with a 15–20 cm stem diameter survived. After the tsunami, the mangrove forest was not replanted, and parts of the forest became resort areas.

7. The Situation in the Maldives

In comparison to other countries, such as Indonesia, Sri Lanka and Thailand, the Maldives experienced less damage not only because of their geological setting but mainly because of their geological features: the coral atolls with small islands separated by deep channels (FRITZ et al. 2006). Even though the tsunami arrived 4 h after the Maldives, a 9 m runup resulted in 300 fatalities in Somalia (FRITZ and BORRERO 2006). However, if we compare the displaced population with the total population of the Maldives, the damage was very significant. Although the elevation of the entire island chain is lower than 2 m, the 26 December tsunami had a limited impact on the Maldives due to its unique geological features as mentioned earlier. In addition, because of the country's low elevation, the tsunami could completely overtop or submerge the island. Therefore, there was no runup process amplifying the tsunami as high as 8 m in Sri Lanka (LIU et al. 2005) and the maximum tsunami height observed in the Maldives was only up to 4.1 m (FRITZ et al. 2006). A field visit to discuss the present countermeasures prepared by the central and local governments and to observe the present condition of the Maldives was held from 23 to 29 December 2013 with support from the Asian Disaster Reduction Center (ADRC).

7.1. General Observations

Male, the capital city (Fig. 16), with an area of 1×2 km^2, was the main target area for our official visits. The city was protected during the 1990s by seawalls and breakwaters (Fig. 17a) that were built before the 2004 tsunami; however, measures for supporting evacuation during future tsunamis were not observed. Unlike in Indonesia, Sri Lanka and Thailand, there were no tsunami-related signs to provide information on the tsunami hazard zone, the tsunami height, evacuation routes or evacuation buildings. On the 9-year anniversary of the tsunami (26 December 2013), there were no major events in Male except several news reports, but there were some small events on other islands. There was no sign indicating the tsunami memorial monument shown in Fig. 17b. This picture was taken on the anniversary, and the monument was no busier than on other days. The main road along the shoreline, which links the entire city, is a one-way street, similar to the roads in Patong beach in Phuket, Thailand. The two-lane road is always crowded with parked cars and motorcycles that could obstruct evacuation (Fig. 18a, b). The small streets that link the main road to the inland are narrow and obstructed by many objects on the ground. In addition, although most buildings along the shoreline have at least two stories, there are no signs indicating tsunami evacuation shelters. The coast guard building and other government buildings near the shoreline could be used as evacuation buildings.

7.2. Central Government Offices

Meetings with several offices of the central governmental were conducted to investigate the ongoing disaster risk reduction efforts of each organization—namely, the Maldives National Defense Force (MNDF), the Ministry of Education (ME), the Ministry of Tourism (MT), the Ministry of Housing and Infrastructure (MHI) and the Maldives Meteorological Service (MMS).

Figure 16
Locations of atolls and islands in the Maldives mentioned in this study

Figure 17
a Seawall and breakwater in the south of Male and **b** memorial statue of the 2004 Indian Ocean tsunami in Male

Figure 18
a Crowded traffic during rush hour in the road along the coast and **b** an evacuation route that is blocked by parked motorcycles

The MNDF is the most important of these agencies because it has the necessary manpower and facilities that can be used during an emergency. It would be one of the first teams to reach the affected areas to solve physical problems and support further activities. However, some of its assets that can be used in case of emergency, such as speed boats, seaplanes and helicopters, will encounter logistical problems because the country contains numerous islands and a limited budget for new assets. Response time is likely the most important factor in an emergency. The shorter the response time of the MNDF it, the more time that is available for other activities, such as the medical team response, the receipt of supporting materials and volunteer work.

The ME is now attempting to implement a disaster-related curriculum in schools starting at the kindergarten level because children may have to lead adults to evacuate in future disasters. For example, during the 2004 tsunami in Thailand, a young British girl helped more than a hundred people evacuate and survive. The ministry is also adapting the teaching materials so that they will be interesting to children.

At present, good practices have been integrated into tourism—i.e., a brief introduction to disaster-related information is provided at check-in at each resort. The MT considers the total exposed population (the local population plus the number of tourists) to be sustainable. Because tourism is the main economic sector in the country, the MT is now addressing any type of disaster that might interrupt tourism activities. Information related to disasters should also be well disseminated because tourists are more sensitive to the media, which mainly focus on bad news.

The MHI is the key ministry for long-term reconstruction insofar as its work is related to daily life. Some problems identified during the interview were (1) the lack of logistics for importing construction resources from nearby countries (e.g., India or the Middle East), (2) the lack of skilled manpower, (3) the necessity of group relocation to reduce construction costs for basic infrastructure and (4) the necessity of providing education to people who receive compensatory money but who spend it to meet different objectives. To solve one of these problems, the MHI is now encouraging children to relocate to bigger islands for a better education. If this occurs, parents must follow their children, and infrastructure construction costs can be reduced.

Although the MMS does not have its own warning system and must rely on other international services, the present warning and observation systems appear to be sufficient. In terms of tsunami warning, the MMS has a network linked to all four major institutions that take measurements in the Indian Ocean. Warning information from these institutions, such as the expected tsunami arrival time and amplitude, is sufficient for such short-term measures; however, for the long term, more detailed information, such as hazard maps for predicted future events, may be necessary. The warning dissemination method still appears to pose problems. A communication network and a commitment to use it during a disaster should be established to disseminate vital information, which should be made more easily understandable for less educated people.

7.3. Relocated Residents and Local Stakeholders

A field site visit was organized to observe and interview local residents and stakeholders. Thulusdhoo Island, the capital of the Kaafu Atoll, was selected as the target area because of its large proportion of relocated residents from other islands affected by the 2004 tsunami—340 out of the total population of 1361 on the island. The island is approximately 25 km northeast of Male. Typical disasters in Thulusdhoo Island are storm surges, floods and coastal erosion; however, the island was hit by waves 1–2 m high during the 2004 tsunami. After the tsunami, the government requested that some companies move off the island, and the factories were used as temporary shelters for 4 years. Figure 19a shows that the deserted areas remained the same for 9 years. New houses were constructed with donations from Saudi Arabia to those who used to own land before the tsunami, as shown in Fig. 19b.

The island is small, and the sea is a 5–10 min walk from the center of the island. There is only one reinforced concrete high-rise building (Fig. 20a) for tsunami evacuation on the island, which belongs to the MNDF. This building has speakers on the roof to issue warnings. During the day of our visit, we attended a disaster resilience workshop held by the MNDF in this building. Participants were asked to create an exposure map, a hazard map and an evacuation map for each type of hazard (Fig. 20b). Residents on this island appear to have a high awareness of disasters as a result of the 2004 tsunami.

Interviews were organized with the local residents who lost their houses in the 2004 Indian Ocean tsunami and relocated to Thulusdhoo Island. The government gave them a choice of three possible islands for their relocation. All of the residents are satisfied with their new homes, and they have better economic opportunities. Although 9 years have passed since the 2004 tsunami, residents still have a

Figure 19
The present conditions of **a** deserted temporary shelters and **b** reconstructed houses in Thulusdhoo Island

Figure 20
a The MNDF building and **b** an example of a tsunami evacuation route map made by residents in Thulusdhoo Island

high awareness and a strong will to participate in disaster-related activities. However, there has been a problem regarding the amount of space in residents' new houses. Because extended families commonly live together in this country, often one house with three rooms cannot accommodate the size of a household.

A meeting was organized with representatives from schools, health centers, the police force and the atoll council. Although the local people are satisfied with the present conditions, the stakeholders who attended the meeting are not because the facilities and infrastructure do not meet the standards to which the government had committed itself. The population size has been increasing, but the capacity of institutions such as schools and health centers has remained the same. In particular, the stakeholders indicated the need to upgrade the health center to a hospital. At the moment, there is still a lack of necessary medical equipment. These problems resemble those experienced in the areas mentioned in the previous section, where the country's geography makes it difficult for the central government to reach remote areas and provide support.

8. Conclusion

Based on the findings obtained during our visits to the target countries, we identified the three key issues that reflect the preparedness of these areas and the future challenges that exist 10 years after the 2004 Indian Ocean tsunami. For this long interval event, it is necessary to prepare resilient communities for future generations monitoring such progress. In addition, the progress achieved in disaster preparedness after 3 years of reconstruction was compared to that achieved after 10 years and evaluated based on experience in disaster reconstruction in Japan.

8.1. Town Reconstruction and Land Use Planning

The countries in our study area were similar in terms of the installation of tsunami warning signs and the construction of tsunami evacuation shelters, which were constructed while town reconstruction was occurring. In general, the town reconstructions in most of the countries in our study were completed within a few years after the 2004 tsunami, except in the Maldives, where the logistics of obtaining building materials appeared to be the most important factor causing the delay. Improving the logistics system for a country that contains many islands is a challenge. In addition, the country contains a large number of small islands on coral atolls without building materials apart from the corals. In terms of land use planning, there have been no major changes in the popular sightseeing areas, such as those in Thailand and the Maldives. However, future disaster-related property loss is expected to be larger than that from the 2004 event along the Andaman coast of Thailand because that area is more exposed due to the larger number of hotels and resorts. In addition, serious traffic jams might occur similar to those experienced in 2012. For tourism-designated islands in the Maldives, where the number of tourists is significant, taking into account the expected number of tourists appears to be one of the most important measures for emergency response.

8.2. Warning Systems and Emergency Response

Tsunami warning services can be classified into three categories: (1) a regional service (Indonesia), (2) a national service (Thailand) and (3) the need for information from other regional service countries (the Maldives and Sri Lanka). After the 2004 tsunami, the Pacific Tsunami Warning Center (PTWC) acted as the center for tsunami warnings in the Indian Ocean for several years, but currently, this service is being provided by regional tsunami service providers in Australia, India and Indonesia and by other regional services, such as RIMES. Thailand has a national service provided by the NDWC. Sri Lanka and the Maldives do not have their own services, but they receive warning messages from the aforementioned regional service countries. This approach appears to be sufficient in terms of tsunami evacuation. A tsunami evacuation drill is performed yearly and simultaneously in all study areas to test their tsunami warning systems. Nevertheless, further research and development of tsunami-related topics, such as land use planning in preparation for future tsunamis using detailed probabilistic tsunami hazard maps and vulnerability functions, are needed.

8.3. Disaster Reduction Education and Other Disaster-Awareness-Related Topics

Disaster reduction education is an issue to which all countries in our study area gave considerable attention during the beginning stages of reconstruction. Basic knowledge of disasters was added to school curricula, and workshops and lectures were organized in cooperation with both national and international organizations. These efforts will help local people understand the meaning of warnings and the proper action to take during an evacuation. Furthermore, tsunami museums and tsunami memorials have been built in most locations. There are some differences among the countries; for example, Indonesia and Thailand succeeded in maintaining some marine vessels that were carried ashore as memorials, but this was not the case for the damaged train in Sri Lanka. Nevertheless, these attempts will help maintain awareness of the 2004 disaster. Geological evidence will also maintain people's awareness. The number of boulders that remains in Thailand is a good example for telling the story of the 2004 tsunami.

8.4. Evaluation of the 2004 Affected Countries Using the Four Main Components of Disaster Reconstruction

We further evaluated reconstruction after the 2004 Indian Ocean tsunami based on observations from our field visits using the four main components of disaster reconstruction [Disaster Prevention Research Institute (DPRI) 2004]. This type of long-term reconstruction plan is based on experience with historical disasters in Japan, including the 1611 Keicho-Sanriku tsunami and the 1854 Ansei-Nankai tsunami. The four components are described below.

1. Town reconstruction and planning (Living): town and regional planning conducted alongside housing reconstruction.
2. Self-reliance and linkage (Communication): communication during normal and emergency periods among local residents, their self-reliance, the linkages among them and evacuation drills.
3. Community and local industry formation (Occupation): recovery of local industries and related facilities.
4. Disaster education and culture (Experience transfer): Maintenance of disaster awareness and research on disaster observations and assessments.

Our evaluations of the early-stage reconstruction period (within three years after the 2004 tsunami, as introduced in Sect. 2, and reconstruction after a decade, as discussed in Sects. 4, 5, 6, 7, are shown in Tables 1 and 2, respectively.

Using three symbols and their accompanying definitions—(–) ongoing, (▲) existing with some limitations, and (●) well-prepared—and Thailand as an example, explanations are presented below.

1. Town reconstruction and planning: (▲) → (▲)
   The reconstruction of houses and hotels was completed in most areas during the first stage, but problems with evacuation routes and disaster zoning remain after 10 years.

Table 1

Evaluation of the 2004 tsunami-affected countries using the four main components of disaster reconstruction [after 3 years (i.e., the early-stage reconstruction period): Sect. 2]

Four components	The 2004 Indian Ocean tsunami-affected countries			
	Indonesia	Sri Lanka	Thailand	The Maldives
Town reconstruction and planning	▲	▲	▲	–
Self-reliance and linkage	–	▲	▲	▲
Community and local industry formation	▲	▲	▲	–
Disaster education and culture	–	–	–	–

(–) Ongoing, (▲) exists with some limitations, (●) well prepared

Table 2

Evaluation of the 2004 tsunami-affected countries using the four main components of disaster reconstruction (after a decade: Sects. 4, 5, 6, 7)

Four components	The 2004 Indian Ocean tsunami-affected countries			
	Indonesia	Sri Lanka	Thailand	The Maldives
Town reconstruction and planning	▲	▲	▲	▲
Self-reliance and linkage	▲	●	●	▲
Community and local industry formation	●	●	●	▲
Disaster education and culture	▲	▲	▲	▲

(▲) → (▲) Ongoing, (▲) → (●) exists with some limitations, (–) → (▲) well prepared

2. Self-reliance and linkage: (▲) → (•)

   The performance of tsunami warnings has been improved from 20 min after earthquake occurrence during the early stage of NDWC reconstruction to less than 5 min after a decade. Evacuation drills are organized every year and receive positive feedback and cooperation from local residents.

3. Community and local industry formation: (▲) → (•)

   The recovery of hotels and other tourism-related businesses was completed in most areas a few years after the tsunami, and they were fully operational after 10 years.

4. Disaster education and culture: (—) → (▲)

   During the early stage, a framework for disaster education was applied in some university-level curricula, but not at the lower pre-university levels. Some damaged structures were preserved as tsunami monuments, but most of these have been poorly maintained.

We hope that our evaluation results will be used as indicators for a regional comparative study of disaster reconstruction and that they can be applied to evaluate progress in disaster preparedness for other disaster events in the future.

Acknowledgments

Dr. Anawat Suppasri would like to express sincere gratitude to ADRC for supporting his trip to the Maldives and to TMD and RIMES for their local support. Dr. Prasanthi Ranasinghe would like to convey her gratitude to the Lanka Hydraulic Institute Ltd (LHI) for providing literature that contributed to the Sri Lankan section. This research was funded by the Reconstruction Agency of the Government of Japan and Tokio Marine & Nichido Fire Insurance Co., Ltd. through IRIDeS, Tohoku University. The authors greatly appreciate Dr. Hermann M. Fritz, guest editor, two anonymous reviewers and Joshua Macabuag from University College London for valuable comments on the entire paper.

Open Access This article is distributed under the terms of the Creative Commons Attribution 4.0 International License (http://creativecommons.org/licenses/by/4.0/), which permits unrestricted use, distribution, and reproduction in any medium, provided you give appropriate credit to the original author(s) and the source, provide a link to the Creative Commons license, and indicate if changes were made.

REFERENCES

ALJAZEERA NEWS (2012) Indian Ocean on tsunami alert after quakes http://www.aljazeera.com/news/asia-pacific/2012/04/201241185126462944.html (Accessed 25 March 2014)

AMATEUR SEISMIC CENTER (ASC) (2005) M8.7 Nias-Simeulue Earthquake, 2005 http://www.asc-india.org (Accessed 25 March 2014)

ASIAN DEVELOPMENT BANK (ADB) (2005), Japan Bank for International Cooperation and World Bank, Preliminary Damage and Needs Assessment-2005 Post Tsunami Recovery Program, Sri Lanka.

BAUMAN, P., AYALEW, M. and PAUL, G. (2007) *Beyond disaster: Comparative analysis of tsunami interventions in Sri Lanka and Indonesia/Aceh*, Journal of Peacebuilding and Development, 3 (3), 6–21.

BBC News (2007) Powerful quake shakes Indonesia http://news.bbc.co.uk/2/hi/asia-pacific/6991134.stm (Accessed 25 March 2014)

BORRERO, J. C., MCADOO, B., JAFFE, B., DENGLER, L., GELFENBAUM, G., HIGMAN, B., HIDAYAT, R., MOORE, A., KONGKO, W., LUKIJANTO, PETERS, R., PRASETYA, G., TITOV, V. YULIANTO, E. (2011) *Field survey of the March 28, 2005 Nias-Simeulue earthquake and tsunami*, Pure and Applied Geophysics, 168, 6–7, 1075–1088.

BORRERO, J. C., WEISS, R., OKAL, E. A., HIDAYAT, R., SURANTO, ARCAS, D. and TITOV, V. (2009) *The tsunami of 2007 September 12, Bengkulu province, Sumatra, Indonesia: post-tsunami field survey and numerical modeling*, Geophysical Journal International, 178 (1), 180–194.

BORRERO, J. C., SYNOLAKIS, C. E. and FRITZ, H. (2006) *Northen Sumatra field survey after the December 2004Great Sumatra earthquake and Indian Ocean tsunami*, Earthquake Spectra, 22 (S3), 93-104.

BURBIDGE, D., CUMMINS, P. R., MLECZKO, R. and THIO, H. K. (2008) *Probabilistic tsunami hazard assessment for Western Australia*, Pure and Applied Geophysics, 165, 2059–2088.

CH2MHILL and LANKA HYDRAULIC INSTITUTE LTD (2006) Main report of Fish Harbor Repairs and Improvements-Preliminary Assessment Report-February 2006, Sri Lanka Tsunami Reconstruction Program,1–8.

DIPOSAPTONO, S., A. MUHARI, F. IMAMURA, S. KOSHIMURA and H. YANAGISAWA (2013). *Impacts of the 2011 East Japan tsunami in the Papua region, Indonesia: field observation data and numerical analyses*. Geophysical Journal International, Volume 194, Issue 3, p.1625–1639

Disaster Prevention Research Institute (DPRI) (2004) Handbook of disaster prevention science, Edited by DPRI, Kyoto University, 2nd Edition, Asakura publishing, 724 pp. ISBN 4-254-26012-1 C3051

ENIA, J. S. (2008) *Peach in its wake? The 2004 tsunami and internal conflict in Indonesia and Sri Lanka*, Journal of Public & International Affairs 19(1): 7–27.

FRANCO, G., SHETH, A. and MEYER M. (2013) Observations on the Recovery and Reconstruction in Sri Lanka Following the December 26, 2004 Tsunami, An Earthquake Engineering Research Institute (EERI) Field Report, 44 pp.

FRITZ, H.M., J.C. BORRERO (2006). *Somalia field survey of the 2004 Indian Ocean Tsunami*. Earthquake Spectra 22(S3):S219–S233.

FRITZ, H. M., SYNOLAKIS, C. E. and MCADOO, B. G. (2006) *Maldives field survey after the December 2004 Indian Ocean tsunami*, Earthquake Spectra, 22, S3, S137–S154.

GEIST E., KIRBY, S., ROSS, S. and DARTNELL, P. (2009) Samoa disaster highlights danger of tsunamis generated from outer-rise earthquakes, Sound Waves, Coastal and Marine Research News from Across the USGS, December 2009. Available at: http://soundwaves.usgs.gov/2009/12/research.html (Accessed date: 30 May 2015)

GOFF, J., LIU, P.L.-F., HIGMAN, B., MORTON, R., JAFFE, B.E., FERNANDO, H., LYNETT, P., FRITZ, H., SYNOLAKIS, C. (2006). *The December 26th 2004 Indian Ocean tsunami in Sri Lanka*. Earthquake Spectra 22(S3):S155–S172.

GOTO, K., CHAVANICH, S. A., IMAMURA, F., KUNTHASAP, P., MATSUI, T., MINOURA, K., SUGAWARA, D., YANAGISAWA, H. (2007) *Distribution, origin and transport process of boulders transport by the 2004 Indian Ocean tsunami at Pakarang Cape, Thailand*. Sedimentary Geology, Vol. 202, 821–837

GOTO, Y., OGAWA, Y. and KOMURA, T. (2010) *Tsunami disaster reduction education using town watching and moving tsunami evacuation animation—Trial in Banda Aceh*, Journal of Earthquake and Tsunami, 4 (2), 115–126.

GOTO, K., TAKAHASHI, J., OIE, T., IMAMURA, F. (2011) *Remarkable bathymetric change in the nearshore zone by the 2004 Indian Ocean tsunami: Kirinda Harbor, Sri Lanka*. Geomorphology 127 (1–2), 107–116.

GOTO, K., TAKAHASHI, J., FUJINO, S. (2012a) *Variations in the 2004 Indian Ocean tsunami deposits thickness and their preservation potential, southwestern Thailand*. Earth, Planets and Space, Vol. 64, 923–930.

GOTO, Y., AFFAN, M., AGUSSABTI, NURDIN, Y., YULIANA, D. and ARDIANSYAH (2012b) *Tsunami evacuation simulation for disaster education*, Journal of Disaster Research, 7 (1), 92–101.

GOTO, Y., AFFAN, M. and FADLI (2012c) Response of the People in Banda Aceh just after the 2012 April 11 Off-Sumatra Earthquake (M8.5), Quick report, Syiah Kuala University, 16 pp.

HILL, E.M., J.C. BORRERO, Z. HUANG, Q. QIU, P. BANERJEE, D.H. NATAWIDJAJA, P. ELOSEGUI, H.M. FRITZ, B.W. SUWARGADI, I.R. PRANANTYO, L. LIN, K.A. MACPHERSON, V. SKANAVIS, C.E. SYNOLAKIS, and K. SIEH (2012). *The 2010 Mw 7.8 Mentawai earthquake: Very shallow source of a rare tsunami earthquake determined from tsunami field survey and near-field GPS*, J. Geophys. Res. Solid Earth, 117, B06402.

HYNDMAN J. (2009) *Siting Conflict and Peace in Post-Tsunami Sri Lanka and Aceh, Indonesia*, Norwegian Journal of Geography 63.1 (2009): 89–96.

IMAMURA, F. (2008) *Dissemination of Information and Evacuation Procedures in the 2004-2007 tsunamis, including the 2004 Indian Ocean*, Journal of Earthquake and Tsunami, 3(2), 59–65.

JAFFE, B., BORRERO, J. C., PRASETYA, G. S., PETERS, R., MCADOO, B., GELFENBAUM, F., MORTON, R., RUGGIERO, P., HIGMAN, B., DENGLER, L., HIDAYAT, R., KINGSLEY, E., KONGKO, W., MOORE, A., TITOV, V. and YULIANTO, E. (2006) *Northwest Sumatra and offshore islands field survey after the December 2004 Indian Ocean tsunami*, Earthquake Spectra, 22 (S3), 105–135.

JANKAEW, K., ATWATER, B. F., SAWAI, Y., CHOOWONG, M., CHAROENTITIRAT, T., MARTIN, M. E. and PRENDERGAST, A. (2008) *Medieval forewarning of the 2004 Indian Ocean tsunami in Thailand*, Nature, 455, 1228–1231.

KUHN, R. (2009) Tsunami and conflict in Sri Lanka, The World Bank-UN Project report on the economics of disaster risk reduction, 41 pp.

KUWATA, Y. and TAKADA, S. (2010) *Business restoration related to lifeline after tsunami disaster*, Journal of Earthquake and Tsunami, 4 (2), 73–82.

KUWATA, Y. (2011), Post-Tsunami Lifeline Restoration and Reconstruction, Tsunami—A Growing Disaster, Prof. Mohammad Mokhtari (Ed.), ISBN: 978-953-307-431-3, InTech

KHAZAI, B., FRANCO, G., INGRAM, J. C., RUMBAITIS DEL RIO, C., DIAS, P., DISSANAYAKE, R., CHANDRATILAKE, R., and KANNA, S. J., (2006). *Post-December 2004 tsunami reconstruction in Sri Lanka and its potential impacts onfuture vulnerability*, Earthquake Spectra, 22 (S3), 829–844.

LATIEF, H., SENGARA, I. W. and KUSUMA, S. B. (2008) Probabilistic seismic and tsunami hazard analysis model for input to tsunami warning and disaster mitigation strategies, Int. Conf. Tsunami Warning (ICTW) Bali, Indonesia.

LIU, P.L.-F., LYNETT, P., FERNANDO, J., JAFFE, B.E., FRITZ, H.M., HIGMAN, B., MORTON, R., GOFF, J., SYNOLAKIS, C.E. (2005). *Observations by the International Tsunami Survey Team in Sri Lanka*, Science 308(5728):1595.

LØVHOLT, F., BUNGUM, H., HARBITZ, C. B., GLIMSDAL, S. LINDHOLM, C. D. and PEDERSEN, G. (2006) *Earthquake related tsunami hazard along the western coast of Thailand*, Natural Hazards and Earth System Sciences, 6, 979–997.

LAUTERJUNG, J., MÜNCH, U., and RUDLOFF, A. (2010). *The challenge of installing a tsunami early warning system in the vicinity of the Sunda Arc, Indonesia*, Nat. Hazards Earth Syst. Sci., 10, 641–646.

MAS, E., KOSHIMURA, S., IMAMURA, F., SUPPASRI, A., MUHARI, A. and ADRIANO, B. (2015) *Recent advances on agent based tsunami evacuation simulation. Case studies at Indonesia, Thailand, Japan and Peru*, Pure and Applied Geophysics, (in this issue)

MINISTRY OF ENVIRONMENT, ENERGY and WATER (2007) National adaptation program of action—Republic of Maldives, 114 pp. Available online at: http://unfccc.int/resource/docs/napa/mdv01.pdf

MONECKE, K., FINGER, W., KLARER, D., KONGKO, W., MCADOO, B. G., MOORE, A. L. and SUDRAJAT, S. U. (2008) *A 1,000–year sediment record of tsunami recurrence in northern Sumatra*, Nature 455, 1232–1234.

MÜNCH, U., RUDLOFF, A. and LAUTERJUNG, A. (2011). *Postface "The GITEWS Projec—results, summary and outlook"*, Nat. Hazards Earth Syst. Sci., 11, 765–769

MURAO, O. and NAKAZATO, H. (2010) *Recovery curves for housing reconstruction in Sri Lanka after the 2004 Indian Ocean tsunami*, Journal of Earthquake and Tsunami, 4 (2), 51–60.

MIKAMI, T., SHIBAYAMA, T., ESTEBAN, M., OHIRA, K., J. SASAKI,T., SUZUKI, H., ACHIARI, H., WIDODO, T. (2014). *Tsunami vulnerability evaluation in the Mentawai Islands based on the field survey of the 2010 tsunami*, Nat. Hazards, 71, 851–870.

MUHARI, A., and IMAMURA, F. (2014). *When to cancel tsunami warning? a comprehensive review from the recent tsunamis in Indian Ocean and Pacific Ocean*, Input paper for the Global Assessment Report (GAR) on Disaster Risk Reduction 2015. Available on http://www.preventionweb.net/english/hyogo/gar/2015/en/home/documents.html

NOAA National Geophysical Data Center (NGDC) (2014) Tsunami Event Database http://www.ngdc.noaa.gov/nndc/struts/form?t=101650&s=70&d=7 (accessed 25 March 2014)

OAK RIDGE NATIONAL LABORATORY (2006) LandScan 2006, Available at http://www.ornl.gov/sci/gist/landscan/landscan2006/.

OKAL, E.A., H.M. FRITZ, P.E. RAAD, C.E. SYNOLAKIS, Y. AL-SHIJBI, and M. AL-SAIFI (2006a) *Oman field survey after the December 2004 Indian Ocean tsunami*, Earthquake Spectra, 22, S203–S218.

OKAL, E.A., H.M. FRITZ, R. RAVELOSON, G. JOELSON, P. PANCOSKOVA, and G. RAMBOLAMANANA (2006b) *Madagascar field survey after the December 2004 Indian Ocean tsunami*, Earthquake Spectra, 22, S263–S283, 2006.

OKAL, E.A., A. SLADEN, and E.A.-S. OKAL (2006c) *Rodrigues, Mauritius and Réunion Islands, field survey after the December 2004 Indian Ocean tsunami*, Earthquake Spectra, 22, S241–S261.

OKAL, E.A., and C.E. SYNOLAKIS (2008) *Far-field tsunami hazard from mega-thrust earthquakes in the Indian Ocean*, Geophys. J. Intl., 172, 995–1015.

PARIATMONO (2012). *The Influence of Mentawai Tsunami to Public Policy on Tsunami Warning in Indonesia*, Journal of Disaster Research, 7(1), 102–106.

RANASINGHE, D. P. L., GOTO, K.., TAKAHASHI, T., TAKAHASHI, J., WIJETHUNGE, J. J., NISHIHATA, T. and IMAMURA, F. (2013) *Numerical assessment of bathymetric changes caused by the 2004 Indian Ocean tsunami at Kirinda Fishery Harbor, Sri Lanka*, Coastal Engineering 81(2013), 67–81.

RIYAZ, M. and PARK, K. H. (2010) *"Safer Island Concept" developed after the 2004 Indian Ocean tsunami: A case study of Maldives*, Journal of Earthquake and Tsunami, 4 (2), 135–143.

RUANGRASSAMEE, A., YANAGISAWA, H., FOYTONG, P., LUKKUNAPRASIT, P., KOSHIMURA, S., and IMAMURA, F., (2006). *Investigation of tsunami-induced damage and fragility of buildings in Thailand after the December 2004 Indian Ocean tsunami*, Earthquake Spectra, 22(S3), 377–401.

SAMEK, J.H., SKOLE, D.L., CHOMENTOWSKI, W (2004). Assessment of impact of the December 26, 2004 tsunami in Aceh Province Indonesia. Center for global change and earth observations. Available at http://www.landsat.org/trfic/tsunami2004/Assessment_of_Tsunami04.pdf

SATAKE, K., NISHIMURA, Y., PUTRA, P. S., GUSMAN, A. R., SUNENDAR, H., FUJII, Y., TANIOKA, Y., LATIEF, H. and YULIANTO, E. (2013) *Tsunami source of the 2010 Mentawai, Indonesia earthquake inferred from tsunami field survey and waveform modeling*, Pure and Applied Geophysics, 170 (9–10), 1567–1582.

SIRIPONG, A. (2010) *Education for disaster risk reduction in Thailand*, Journal of Earthquake and Tsunami, 4 (2), 61–72.

STEIN, S., and OKAL, E. A., (2005). *Size and speed of the Sumatra earthquake*, Nature, 434, 581-582.

STEIN, S., and E. A., OKAL, *Ultra-long period seismic study of the December 2004 Indian Ocean earthquake and implications for regional tectonics and the subduction process* (2007) Bull. Seismol. Soc. Amer., 97, S279–S295, 2007.

SUPPASRI, A. (2010). Tsunami Risk Assessment to Coastal Population and building in Thailand. PhD thesis. Tohoku University.

SUPPASRI, A., IMAMURA, F. and KOSHIMURA, S. (2012a) *Tsunami hazard and casualty estimation in a coastal area that neighbors the Indian Ocean and South China Sea*, Journal of Earthquake and Tsunami, 6(2), 1250010.

SUPPASRI, A., MUHARI, A., RANASINGHE, P., MAS, E., SHUTO, N., IMAMURA, F. and KOSHIMURA, S. (2012b) *Damage and reconstruction after the 2004 Indian Ocean tsunami and the 2011 Great East Japan tsunami*, Journal of Natural Disaster Science, 34 (1), 19–39.

SYNOLAKIS, C. E., and KONG, L., (2006) *Runup measurements of the December 2004 Indian Ocean tsunami*, Earthquake Spectra, 22 (S3), 67-91.

TAKADA, S., KUWATA, Y. and PINTA, A. (2010) *Damage and reconstruction of lifelines in Phang Nga province, Thailand after the 2004 Indian Ocean earthquake and tsunami*, Journal of Earthquake and Tsunami, 4 (2), 83–94.

TOMITA, T., ARIKAWA, T., KUMAGAI, K., TATSUMI, D. and YEOM, G. S. (2011) Field survey on the 2010 Mentawai tsunami disaster, technical note of the port and airport research institute, Japan, No. 1235, 26 p.

THE TELEGRAPH (2010) Indonesian tsunami warning system 'was not working'http://www.telegraph.co.uk/news/worldnews/asia/indonesia/8090377/Indonesian-tsunami-warning-system-was-not-working.html (accessed 27 October 2010)

UDO, K., TANAKA, H., MANO, A., TAKEDA, Y., 2013. Beach morphology change of southern Sendai coast due to 2011 Tohoku Earthquake Tsunami. Journal of Japan Society of Civil Engineers Series B2 69, I1391–I1395. (In Japanese with English abstract)

US GEOLOGICAL SURVEY (USGS) (2005) M8.7 Northern Sumatra Earthquake of 28 March 2005 ftp://hazards.cr.usgs.gov/maps/sigeqs/20050328/20050328.pdf (accessed 25 March 2014)

US GEOLOGICAL SURVEY (USGS) (2007) Magnitude 8.5 - SOUTHERN SUMATRA, INDONESIA http://earthquake.usgs.gov/earthquakes/eqinthenews/2007/us2007hear/ (accessed 25 March 2014)

US GEOLOGICAL SURVEY (USGS) (2012) M8.6 and M8.2 Northern Sumatra, Indonesia Earthquakes of 11 April 2012 ftp://hazards.cr.usgs.gov/maps/sigeqs/20120411/20120411.pdf (accessed 25 March 2014)

WIJETUNGE, J.J., (2009). *Fieldmeasurements and numerical simulations of the 2004 tsunami impact on the south coast of Sri Lanka*. Ocean Engineering 36, 960–973.

WIJAYARATNA, N., RANASINGHE, D.P.L., JAYANTHIRAN, A., WEERAKKODI, C.M. and WIJESENA, W.M.N. (2006) Hazard Map as a Means of Tsunami Disaster Mitigation in Galle, Sri Lanka, Proceedings of the International Conference on Tsunami Storm Surge & other Coastal Disasters in Colombo, Sri Lanka.

WEISS, R. and BAHLBURG, H., (2006). *The coast of Kenya field survey after the December 2004 Indian Ocean tsunami*, Earthquake Spectra, 22(S3), 235-240.

YANAGISAWA, H., KOSHIMURA, S., GOTO, K., MIYAGI, T., IMAMURA, F., RUANGRASSAMEE, A., and TANAVUD, C., 2009, *The reduction effects of mangrove forest on a tsunami based on field surveys at Pakarang Cape, Thailand and numerical analysis*. Estuarine, Coastal and Shelf Science, Vol. 81, 27–37.

YASUDA, M., IMAMURA, F. and SUPPASRI, A. (2014) Practical application of disaster education for children in Japan, in Proceedings of the international seminar of the 10 years commemoration of the 2004 Indian Ocean tsunami, Jakarta, Indonesia 24–27 November 2014.

(Received April 28, 2014, revised June 2, 2015, accepted June 29, 2015, Published online July 18, 2015)

Field Survey of the 1945 Makran and 2004 Indian Ocean Tsunamis in Baluchistan, Iran

EMILE A. OKAL,[1] HERMANN M. FRITZ,[2] MOHAMMAD ALI HAMZEH,[3,4] and JAVAD GHASEMZADEH[5]

Abstract—We report the result of a 2010 survey of the effects on the Iranian coastline of the tsunami which followed the earthquake of 27 November 1945 ($M_0 = 2.8 \times 10^{28}$ dyn cm; $M_w = 8.2$), the only large event recorded along the Makran subduction zone since the onset of instrumental seismology. Based on the interview of elderly survivors of the event, we obtained a database of nine values of run-up or splash amplitudes on a segment of shore extending 280 km from Souraf in the West to Pasabandar near the Pakistani border, and ranging in vertical amplitude from 2.3 to 13.7 m. Witness reports are consistent with a significant delay (estimated at ~2.5 h) of the tsunami waves, suggesting that they were generated by an ancillary phenomenon, such as a landslide triggered by the earthquake. None of our witnesses bore ancestral memory of comparable events in the past, suggesting that reported predecessors to the 1945 earthquake may have been smaller in size. The survey also allowed the compilation of previously unreported data concerning the effects of the 2004 Sumatra–Andaman tsunami.

1. Introduction and Background

This paper presents the results of a field survey of the 1945 Makran tsunami conducted in October 2010 in the provinces of Sistan-e-Baluchistan and Hormozgan, Southeastern Iran. This survey was organized under the auspices of UNESCAP and the Intergovernmental Commission of UNESCO in the aftermath of a workshop held in Tehran in May 2010, with the principal goal of establishing of a homogeneous scientific database of field measurements of the 1945 tsunami, based on the interview of elderly survivors.

1.1. The 1945 Makran Earthquake

The Makran earthquake of 27 November 1945 (local date 28 November) stands alone as the only interplate thrust subduction event having inflicted a devastating tsunami in the Indian Ocean West of the Indian subcontinent since the dawn of instrumental seismology. It was assigned a "Pasadena" magnitude of 8.3 by GUTENBERG and RICHTER (1954), and studied in detail by BYRNE et al. (1992).

As shown on Fig. 1, the convergence between the stable blocks of the Arabian and Eurasian plates takes place at a local rate of ~2.7 cm/year in the azimuth N10°E (SELLA et al. 2002), with a sizable fraction taken up by deformation in the mountains of Southeastern Iran (e.g., VERNANT et al. 2004). In this context, the 1945 earthquake is one of the rare events expressing active subduction at the plate interface.

We relocated the 1945 earthquake using arrival times listed in the International Seismological Summary (ISS) and the technique of WYSESSION et al. (1991), which estimates a confidence ellipse based on a Monte Carlo procedure consisting of injecting Gaussian noise (with standard deviation $\sigma_G = 4$ s) into the data. The resulting epicenter (24.88°N; 63.53°E) is shown in Fig. 1, and compared with other available locations. The origin time 21:56:51.9 GMT corresponds to a local time of 01:27 in Iran and 03:27 Indian Standard Time, both on the 28 November. Our solution is essentially identical to the epicenters obtained by both the ISS and ENGDAHL and VILLASEÑOR (2002) as part of their Centennial catalog, and also in excellent agreement with QUITTMEYER and JACOB's (1979). Only GUTENBERG

---

[1] Department of Earth and Planetary Sciences, Northwestern University, Evanston, IL 60208, USA. E-mail: emile@earth.northwestern.edu

[2] School of Civil and Environmental Engineering, Georgia Institute of Technology, Atlanta, GA 30332, USA. E-mail: fritz@gatech.edu

[3] Iranian National Institute of Oceanography and Atmospheric Sciences, Chabahar, Iran.

[4] *Present Address*: Department of Geology, Ferdowsi University of Mashhad, Mashhad, Iran.

[5] Department of Fisheries and Marine Sciences, Chabahar Maritime University, Chabahar, Iran.

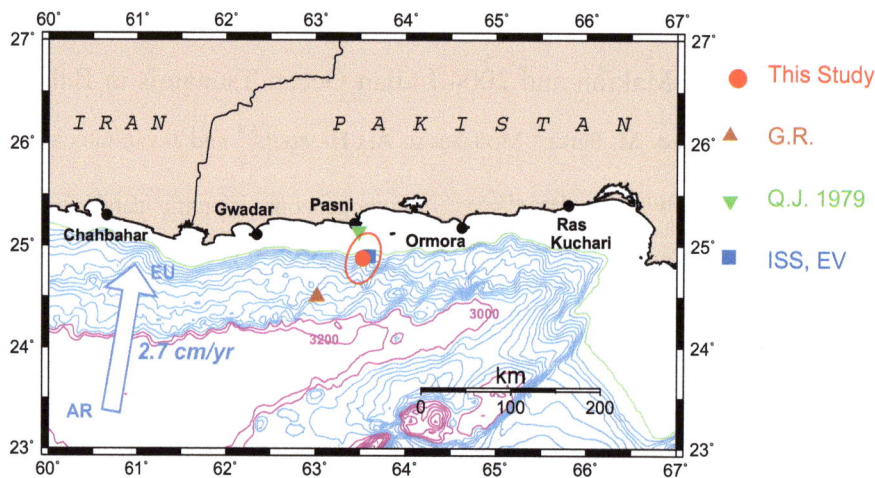

Figure 1
Epicentral estimates for the 1945 Makran earthquake. The *red dot* shows the epicenter relocated in this study, with associated Monte Carlo ellipse, the *blue square* the ISS location, which on this scale, is undistinguishable from ENGDAHL and VILLASEÑOR's (2002) (EV), the *green inverted triangle* QUITTMEYER and JACOB's (1979) source. Only GUTENBERG and RICHTER's (1954) location (*brown triangle*) is significantly displaced. The *large arrow* at *left* shows the convergence vector of the Arabian and Eurasian plates (SELLA et al. 2002). Isobaths are at 200-m interval with depths greater than 3000 m (labeled) in *magenta*

and RICHTER's (1954) original solution, 68 km from our preferred epicenter, lies outside our confidence ellipse. Note that depth could not be resolved from the available dataset of arrival times.

The focal mechanism of the earthquake was studied by BYRNE et al. (1992), who advocated a low-angle thrust mechanism ($\phi = 246°$; $\delta = 7°$; $\lambda = 89°$), with a seismic moment of 1.8×10^{28} dyn cm. Note that the azimuth of their slip vector is 34° away from that of the relative plate motion, and also ~25° from that of the normal to the plate boundary as expressed by the coastline. As part of the present study, we inverted a moment tensor from spectral amplitudes of mantle surface waves at five stations, using the preliminary determination of focal mechanism (PDFM) method (REYMOND and OKAL 2000), which is particularly well suited to historical events (OKAL and REYMOND 2003). The resulting mechanism, shown in Fig. 2, confirms a shallow-angle thrusting geometry ($\phi = 261°$; $\delta = 9°$; $\lambda = 89°$), rotated 15° from BYRNE et al.'s

(1992) in the formalism of KAGAN (1991), and in better agreement with local plate tectonics geometry. Our seismic moment, $M_0 = 2.8 \times 10^{28}$ dyn cm ($M_w = 8.2$), is significantly larger than proposed by BYRNE et al. (1992), but identical to the value obtained by HEIDARZADEH and SATAKE (2015) from the inversion of maregraph and geodetic data.

In more recent times, the catalog of GlobalCMT solutions in the region is extremely scarce, featuring only two small thrust fault events in the vicinity of the 1945 epicenter, on 07 December 1991 and 30 January 1992, rotated 8° and 5°, respectively, from our PDFM mechanism. QUITTMEYER and JACOB (1979) and BYRNE et al. (1992) also compiled WWSSN-era seismicity along the Eastern Makran, featuring a wide variety of mechanisms, including several thrust events of a geometry comparable to that of the 1945 earthquake. Farther West, we note one significant GlobalCMT thrust faulting solution on the Iranian coast, albeit with much steeper dip ($\phi = 305°$;

MAKRAN EARTHQUAKE 27 November 1945

PDFM Inversion

Stations used:

COL, HON, PAS, SJG, WES

$M = 2.8 \times 10^{28}$ dyn*cm

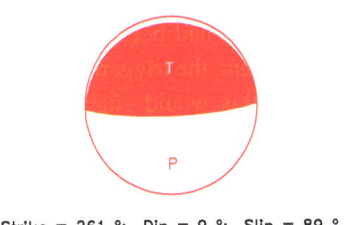

Strike = 261°; Dip = 9°; Slip = 89°.

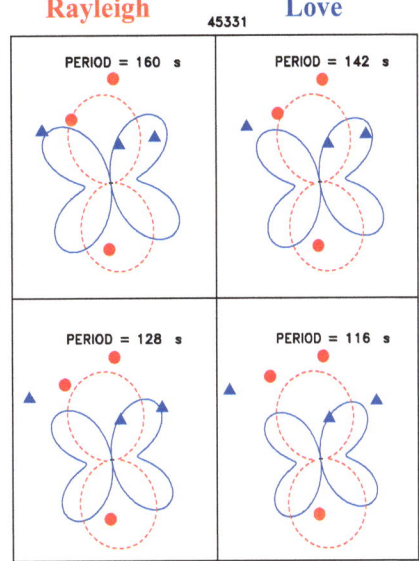

Figure 2

Results of PDFM inversion of mantle waves for the 1945 Makran event: *Left*: inverted focal mechanism and scalar moment. *Right*: inverted spectral amplitudes at representative periods. The theoretical azimuthal radiation patterns are shown as the *blue solid* (Love) and *red dashed* (Rayleigh) *lines*, the observed ones as individual symbols (*red circles*: Rayleigh; *blue triangles*: Love). See REYMOND and OKAL (2000) for methodological details

$\delta = 54°$; $\lambda = 80°$) at 25.9°N, 59.0°E (07 December 1989), and a small one (22 July 2009) inside the Gulf of Oman at 26.8°N, 55.8°E.

1.2. The 1945 Tsunami

The 1945 Makran earthquake generated a tsunami which can be described as devastating along the coast of the Eastern Makran (present-day Pakistan), despite the scarcity of documentation from what was then a very remote area, in the immediate aftermath of the second World War. Run-up was reported to have reached 12–15 m at Pasni (PENDSE 1946), with "serious loss of life and property" at both Pasni and Ormara. Waves were reported at Karachi (1.5 m) and Bombay (2 m with fatalities). The total number of casualties due to the tsunami is difficult to assess given the lack of precise reports and the impossibility to separate victims of the earthquake and of the waves. It could be as high as several thousand, but is given conservatively as 300 by AMBRASEYS and MELVILLE (1982).

A most remarkable aspect of the 1945 tsunami is that its main waves were significantly delayed (on the average 2.5 h) with respect to the earthquake. This is described in some detail along the Eastern Makran coast, at Karachi and Bombay in PENDSE's (1946) report, and at Mahé, Seychelles in a short note by BEER and STAGG (1946); as detailed below, it is also clear from the testimony of several of our witnesses from the Iranian coast. To explain this delay in the maximum wave, several models were proposed, notably the development of an edge wave along the continental shelf of the Indian subcontinent (NEETU et al. 2011). More generally, HEIDARZADEH and SATAKE (2015) were able to model the tidal gauge records at Karachi and Bombay (including their arrival times) using heterogeneous coseismic slip on a fault plane consistent with BYRNE et al.'s (1992) focal geometry, but their model can be reconciled neither with the extreme amplitudes reported at Pasni, nor with the delays observed in Iran and the Seychelles. BEER and STAGG (1946) noted that interpreting the earthquake as the source of the tsunami would result in an average ocean depth between its source and the Seychelles of only 1.3 km (assuming an undispersed celerity $c = \sqrt{gh}$), and went on to suggest that the first wave recorded at Mahé, arriving approximately ~170 min late, had to be the eighth one in the wavetrain, without providing a rationale for this singular sequencing of tsunami energy among subsequent waves. A much more probable model would invoke generation by a landslide, itself triggered by the earthquake with a delay of ~2.8 h, in a scenario generally similar to the case of the 1998 Papua New Guinea or 1990 Rudbar events (SYNOLAKIS et al. 2002; SALAREE and OKAL 2015). A critical observation in this respect is the fact that the Karachi–Muscat section of the Bombay–London telegraphic cable was severed on that day, although the precise timing is not available (TIMES OF INDIA 1945a; AMBRASEYS and MELVILLE 1982). A long history of such cable breaks and of their subsequent repairs after the earthquakes in Grand Banks (1929), Luzon (1934), Suva (1953), El Asnam (*ex*-Orléansville, 1954; 1980), Boumerdes (2003), and even possibly Rukwa (1910), has shown that cables were always ruptured during a major underwater mass movement, rather than by the earthquake shaking, and thus cable breaks have become a proxy for the generation of a tsunami by underwater landslides (REPETTI 1934; HEEZEN and EWING 1952, 1955; HOUTZ 1962; EL-ROBRINI et al. 1985; BOUHADAD et al. 2004). In addition, the TIMES OF INDIA (1945b) reports the presence of subaerial landslides on the coast, suggesting their probable triggering underwater as well. Note finally that HEIDARZADEH and SATAKE (2014) have suggested an underwater landslide in essentially the same area as the source of the tsunami generated by the 2013 Pakistani earthquake whose fault zone was located more than 200 km inland.

1.3. Other Historical Events

There exist a number of historical reports of severe earthquakes in the immediate vicinity of the 1945 event. An earthquake is mentioned at Ras Kuchari, Pakistan (~65.7°E) around 1765, which allegedly caused an entire hill to slump into the sea, as compiled by AMBRASEYS and MELVILLE (1982). Another shock is described at Gwadar (present-day Pakistan) in 1851 with a possible aftershock in 1864 (OLDHAM 1893; WILSON 1930; QUITTMEYER and JACOB 1979), but these earthquakes are not compiled as a separate entry in AMBRASEYS and MELVILLE's (1982) catalog. These authors simply mention the 1864 event in a footnote to their description of the 1945 earthquake; remarkably, they cite a report by WALTON (1865) of a break in the telegraphic cable between Karachi and Telegraph Island near the Strait of Hormoz, which had been laid a few months earlier, again suggesting the triggering of an underwater mass movement; this would suggest that the 1864 event may have been the greater of the two. However, the cable rupture is not mentioned in the original reference (WALTON 1865); it may have been included in a document originally appended to WALTON's correspondence, but presently unavailable. In addition, a large earthquake was reported widely felt in Western Makran in 1483 and "about the same time" in Oman (AMBRASEYS and MELVILLE 1982; BYRNE et al. 1992), which would suggest a source in the Western part of the Gulf of Oman.

Finally, it is worth recalling that, under the command of Nearchus of Crete in 325–324 B.C., the fleet of Alexander the Great was impacted by a phenomenon often interpreted as a tsunami similar to the 1945 event. However, a careful compilation of the chronicle by ARRIAN (1983, pp. 367–415) indicates that at that point, Nearchus had sailed an estimated 12,500 stadia from the Indus River delta. This can be converted to a distance of 1950 km, using Eratosthenes' estimate of 40,800 stadia for the Earth' radius. The distance between the Indus delta and present-day Chabahar would be only 830 km hugging the coast, while the Strait of Hormoz is around 1450 km from the Indus. Cabotage along the coast could increase distances by 35 %, but probably not by a factor of 3. Furthermore, ARRIAN (1983; p. 417) indicates that only 600 stadia (~100 km) after the wave impact, Nearchus left the coast of "Carmania" for that of "Persia", which if Carmania is taken to be equivalent to present-day Kerman province in Iran, would correspond grossly to the boundary between the present-day Sea of Oman and Persian Gulf. It is, therefore, probable that the wave impact took place in the vicinity of the Strait

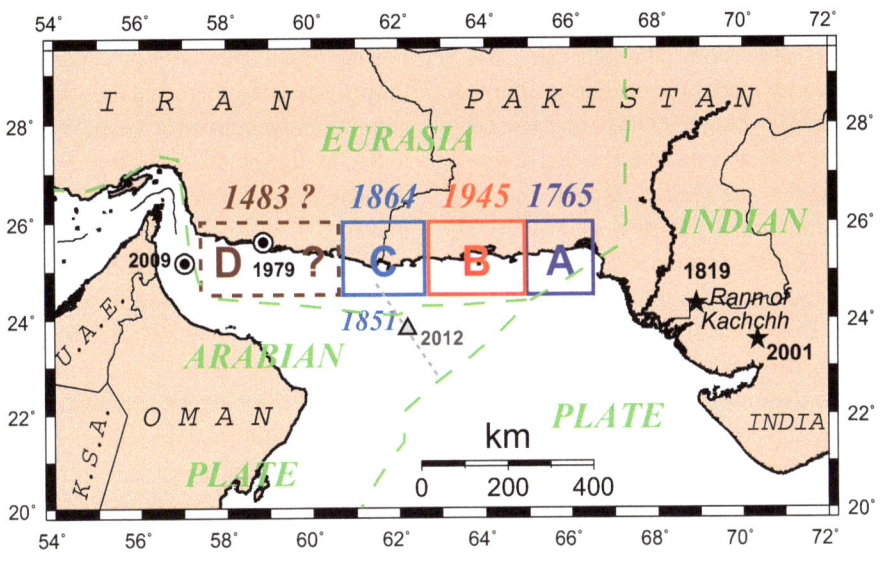

Figure 3
Schematic map of the possible layout of historical earthquakes along the Makran subduction zone. Inspired by ANDO (1975) and adapted from BYRNE et al. (1992) and OKAL and SYNOLAKIS (2008). *Green dashed lines* are plate boundaries (BIRD 2003). The historical events in 1765, 1945 and 1864 may correspond to the rupturing of individual fragments (*A, B, C*) along the Arabian–Eurasian boundary. The hypothetical 1483 event may involve a fourth (*D*) block, shown as the *dashed box* to the West. The intriguing GlobalCMT solutions of 1979 and 2009 are shown as bull's eye symbols. The Sonne fault is sketched as the *gray dotted line*, with the 2012 strike-slip event shown as a *gray triangle*

of Hormoz, and thus would have been more likely similar to the putative earthquake in 1483, than to the recent one farther East in 1945.

Figure 3, adapted from Fig. 6 of BYRNE et al. (1992), summarizes the distribution of these large earthquakes along the Makran coast. It is tempting to interpret them in the context of ANDO's (1975) now classical study of historical earthquakes along the Nankai trough, where he documented a fragmentation of the subduction zone into several segments (A, B, C,…), with subsequent earthquakes rupturing one or several segments in an apparently random way. This model was later confirmed by tsunami modeling and paleotsunami studies in the Kuriles, Cascadia, Peru and Southern Chile (NANAYAMA et al. 2003; CISTERNAS et al. 2005; KELSEY et al. 2005; OKAL et al. 2006a). Under this model, OKAL and SYNOLAKIS (2008) have envisioned as worst-case scenarios for tsunami hazard in the Makran, a simultaneous rupture of blocks A, B, C on Fig. 3 (their scenario 5), corresponding to an earthquake of moment $M_0 = 2.7 \times 10^{29}$ dyn cm ($M_w = 8.9$).

An unresolved problem in regional tectonics is whether convergence in Westernmost Makran is (1) entirely accommodated by onland orogeny; or (2) possibly by creep at the plate contact; or (3) partially taken up during infrequent large subduction earthquakes. In terms of tsunami potential, only (3) carries significant hazard. Under models (1) and (2), the difference in the modes of convergence between Western and Eastern Makran (in simple terms at the Iranian and Pakistani coasts) could be attributed to a fragmentation of the Arabian plate into two blocks separated by the left-lateral strike-slip Sonne fault (KUKOWSKI et al. 2000), where one compatible strike-slip GlobalCMT solution has recently been documented (01 June 2012; triangle on Fig. 3). These models would also be supported by the low level of background seismicity in Western Makran, and by an inconclusive search for ancient earthquakes using paleoseismicity (RAJENDRAN et al. 2013). MUSSON (2009) also used historical documents to argue against a common source to the earthquake reports in Hormoz and Oman at the end of the XVth century.

On the other hand, several regional characteristics could favor seismic subduction through rare, but great earthquakes, under model (3), including significant GPS residual motion between Oman and coastal Iran, the presence of a deep oceanic basin and active sediment deformation in the Gulf of Oman, and of terraces on the Iranian coast (MASSON et al. 2007; MOKHTARI et al. 2008). Such patterns, as well as the absence of background seismicity during a short OBS deployment (NIAZI et al. 1980), are indeed reminiscent of their counterparts off Cascadia (WHITE and KLITGORD 1976), an area then regarded as featuring aseismic subduction (2), but now known to entertain occasional megathrust earthquakes (3) (SATAKE et al. 1996; KELSEY et al. 2005). This legitimizes, at least in principle, the concept of rare megathrust events in the Western Makran (although not necessarily in 1483), leading to a tentative "D" block at the Western end of the region (Fig. 3), and hence to the possibility of an extreme worst-case scenario for tsunami risk in the Makran, involving all four blocks (Model 6 in OKAL and SYNOLAKIS 2008; $M_0 = 4.7 \times 10^{29}$ dyn cm; $M_w = 9.0$).

It is in this general context for tsunami hazard in the Makran region that we conducted a field survey of the 1945 tsunami in Iran in October 2010. A similar program across the border in Pakistan has resulted in the identification and interview of scores of witnesses, and the quantification of the relevant dataset is presently under way (KAKAR et al. 2015; ANONYMOUS 2015).

2. Survey Methodology

The methodology of the present project follows the now classical techniques initially developed by OKAL et al. (2002) for the survey of the 1946 Aleutian tsunami in the Pacific, and since then extended to several other events, e.g., Amorgos (Greece), 1956 (OKAL et al. 2009a). They consist of identifying elderly witnesses of the tsunami, conducting their interview (if possible recorded), and traveling with the witness to the location inundated during the tsunami, to document and survey the extent of penetration of the wave. We recall that *Inundation* is defined as the horizontal extent of penetration of the tsunami on initially dry land; *Flow level*, the sum $(z + h)$ of the depth h of the water column flowing past a reference point at an altitude z ($z = 0$ at the original coast line); and *Run-up*, the altitude z, with respect to unperturbed sea level at the time of the tsunami, of the point of maximum penetration where inundation is measured (or flow level at the limit $h = 0$).

A particularly challenging aspect of the present survey was the very advanced age of any reliable witnesses, who would have had to be at least 10 years old in 1945, and thus 75 at the time of the interview in 2010. In this context, and while we tried very hard to recover information about our witnesses, the question of their exact age often remained unresolved. Several witnesses claiming to be in their early sixties had unquestionably vivid memories of the tsunami which should then have predated their birth. By contrast, younger family members had a general tendency to extrapolate the age of their elderly relatives, one adolescent even describing his grandfather as being 100 years old while offering him a sporty ride on his moped; in our opinion, this witness was probably in his seventies, as suggested by his status as a bachelor at the time of the 1945 tsunami. It is with such reservations that we offer estimates of the age of our witnesses. Figure 4 shows a sample of interviews of our witnesses.

Similarly, absolute times given by witnesses for the observation of natural phenomena are generally approximate. Whenever possible, we attempted to ascertain them using reference activities (e.g., morning prayer). We regard as generally more reliable relative timing differences between various aspects of the phenomenon, most notably between the feeling of earthquake tremors and the arrival of the waves.

3. Tsunami Survey, 10–15 October 2010

The team assembled in Tehran on 09 October 2010, and flew to the field on 10 October 2010. Authors EAO and MAH covered the Eastern part of the province, operating out of the INIO office at Chabahar, where they were met by JG. Over the next 5 days, this Eastern team visited the coastal areas between Bir Daf (Longitude 59.8°E) and Pasabandar (Longitude 61.4°E). Because of security constraints,

Figure 4
Examples of interviews of witnesses of the 1945 tsunami. **a** Chabahar, 11 October. *Left* to *right*: JG, EAO, MAH, INIO driver, Witness. **b** On the beach at Lipar, 12 October. *Left* to *right*: MAH, Witness Mr. Dadshahpour, EAO. **c** Beris, 13 October. *Clockwise* from *left*: JG, EAO, Witness Mr. Khaled Baluch, INIO driver (with back to camera). **d** Pasabandar, 13 October. *Left* to *right*: MAH, INIO driver, Witness Mr. Adam Baluch and grandsons, EAO

it was not possible to pursue the survey East of Pasabandar to the village of Gavater at the Pakistani border, but it is doubtful that any elderly witnesses could have been found in Gavater given that the area has seen development only recently, in the form of an extensive shrimp farm. In the meantime, traveling from Bandar Abbas, Author HMF covered the Western part of the Makran coast, from the Strait of Hormoz, including both Qeshm and Hormoz Islands, to Abkouhi, with an additional emphasis on damage inflicted by recent meteorological events such as Cyclone Gonu, the strongest tropical cyclone on record in the Arabian Sea (FRITZ *et al.* 2010). This aspect of the survey, being unrelated to the 1945 tsunami, is not reported in the present work.

In addition to identifying and interviewing elderly witnesses of the 1945 tsunami, we also gathered information on the effects of the 2004 Sumatra–Andaman tsunami, which apparently had not been surveyed along the Iranian coast. All data are compiled in Tables 1 and 2.

Flow level and run-up values were measured using conventional techniques (leveling rod and hand-held eye level; Eastern team) and laser ranging (Western team). The precision of these measurements is estimated at ~5 cm. Inundation was derived from GPS measurements with a precision of 1 m.

- *Tiss*, Sunday, 10 October 2010

On Sunday afternoon, 10 October, we visited the jetty in the port area of Tiss, a suburb of Chabahar. We met a fisherman in his sixties, who related the story of his late father who had observed the water running up in the drainage channel of the village of Tiss, across the bay, up to the vicinity of the village mosque. This transect was surveyed on Monday, 11 October, with a run-up of 2.00 m and an inundation of 1289 m (Site Number 01).

An additional testimony, not directly relevant to the Iranian survey, was obtained from a 62-year-old man who had heard from a sailor who had been in Jiwani (in present-day Pakistan) during the 1945 tsunami, and witnessed the emergence of a new

Table 1

Dataset for 1945 Makran tsunami

Number	Latitude (°N)	Longitude (°E)	Vertical amplitude (z; m)		Nature[a]	Inundation (m)	Date and time surveyed		Notes
			Raw	Corrected				(UTC)	
01	25.35663	60.60562	2.00	2.26	R	1289	11-Oct-2010	5:50	Tiss
02	25.35383	60.40205	3.05	3.24	R	52	12-Oct-2010	7:25	Konarak
03	25.29643	60.62480	3.65	2.76	R	367	13-Oct-2010	12:08	Chabahar
04	25.26528	60.75295	7.15	7.14	S		13-Oct-2010	7:59	Ramin
05	25.25090	60.83202	2.80	2.64	R	540	13-Oct-2010	6:26	Lipar
06	25.15668	61.17877	5.60	5.02	R	212	14-Oct-2010	11:54	Beris
07	25.08618	61.40825	10.00	9.71	R	660	14-Oct-2010	10:10	Pasabandar
08	25.06988	61.41115	13.90	13.66	S	61	14-Oct-2010	9:43	Pasabandar
09	25.57596	58.69028	7.00	6.99	R	406	13-Oct-2010	7:58	Souraf

[a] Codes to nature of vertical measurements: *R*: run-up, *S*: splash, z = terrain elevation

Table 2

Dataset for 2004 Sumatra tsunami

Number	Latitude (°N)	Longitude (°E)	Vertical amplitude (m)		Nature[a]	Inundation (m)	Date and time surveyed		Notes
			Raw	Corrected				(UTC)	
21	25.35523	60.60022	3.35	2.73	F		10-Oct-2010	11:05	Tiss Harbor
22	25.35383	60.40205	2.20	3.38	R	26	12-Oct-2010	6:52	Konarak
23	25.36018	60.31382	1.10	2.22	R		12-Oct-2010	8:23	Pozm
24	25.35823	60.31350	1.95	3.04	R		12-Oct-2010	8:30	Pozm
25	25.06838	61.41687	1.00	1.75	F		14-Oct-2010	9:55	Pasabandar
26	25.35720	59.89457	1.10	1.69	R		15-Oct-2010	8:47	Tang
27	25.40330	59.81385	3.80	4.43	R	35	15-Oct-2010	10:02	Bir Daf
28	25.56027	58.81188	2.30	3.63	R	38	12-Oct-2010	6:48	Gugsar
29	25.56922	58.70397	1.80	2.59	R	62	13-Oct-2010	5:31	Geshmi
30	25.56746	58.20059	1.60	2.18	R	21	15-Oct-2010	6:00	Surgalm
31	25.74628	57.73546	3.10	3.56	R	55	16-Oct-2010	7:27	Lafi

[a] Codes to nature of vertical measurements: R: run-up, F: flow level $(z + h)$ with z = terrain elevation, h = flow depth above terrain

island, to a reported height of 15 m above sea level; this island is said to be gradually coming down now.

Regarding the 2004 Indonesian tsunami, we obtained the testimony of Mr. Abdalghani Siahuee, who described an anomalous sound around 11 a.m. (07:30 GMT), followed an hour later by flooding of the jetty up to a height of 3.35 m (uncorrected for tides; Site Number 21), resulting in several boats breaking their moorings.

- *Konarak*, Tuesday 12 October 2010

On Tuesday, 12 October 2010, we traveled to the western shores of the Bay of Chabahar, home of the major fishing community of Konarak, and later to the port of Pozm.

A first witness gave us a precise report of the inundation of the 1945 tsunami reaching to the middle of the block past the mosque, and related that the water penetrated into a house adjacent to the mosque, disrupting a week-long wedding ceremony. The witness remembers being about 12–13 years old at the time, but claims to be only 67 in 2010.

A second witness confirmed inundation of the mosque, but was clearly confused in terms of dates, claiming to have been in his twenties at the time, but only 60 years old in 2010.

A third, and in our opinion more reliable, witness confirmed that the water penetrated the mosque, disrupted morning prayers (estimated around 4 a.m. local time), broke windows and left a few people injured with broken limbs, although no deaths were known to him in Konarak. One month later, he traveled to Pasni and Ormara (in Pakistan), where the city had been totally destroyed. This witness claimed to have been 20–25 years old at the time and to be in his eighties now.

The combination of these testimonies is translated into a surveyed run-up of 3.05 m and an inundation of 52 m (Site Number 02).

Regarding the 2004 tsunami, witnesses reported flooding to the front of the coastal road, for a run-up of 2.20 m and an inundation of 26 m (Site Number 22).

- *Pozm*, Tuesday 12 October 2010

We could not find witnesses of the 1945 tsunami in Pozm, where the 2004 tsunami was remembered by a number of witnesses on the port jetty. The latter, measured at 1.80 m above sea level, was not flooded to its top. In the village, a witness indicated that his boat, which he was cleaning, was beached during the ebbing phase, and later floated back. At the estuary of the river, the sea ran up to 1.10 m on the local beach (Site Number 23), and 1.95 m on a bluff near the town schoolyard (Site Number 24).

- *Chabahar*, Monday 11 October, and Wednesday, 13 October 2010

We were able to interview several reliable witnesses in the city of Chabahar itself on Monday, 11 October (Fig. 4a). Among them, we had a long interview in his house with Mr. Rahmat Khodadian, who claimed to be 75 years old, clearly a very educated person, who indicated an arrival of the waves around 03:30 a.m., and an inundation reaching to the present location of the Bank Melli, near the downtown beach. This location (Site Number 03) was surveyed on 13 October for a run-up of 3.65 m and an inundation of 367 m. The witness mentioned as his only known ancestral memory (i.e., transmitted by previous generations of family members) of major inundation following an earthquake a case "about 200 years ago in India", which is most probably the 1819 earthquake in Gujarat, during which the Rann of Kuchchh was flooded, but without generation of a genuine tsunami by this continental earthquake. This observation reinforces the absence of ancestral memory from true tsunamigenic events other than the 1945 one along the Makran coast. Mr. Khodadian also mentioned that the telegraph cable (presumably from Europe) was not severed in Chabahar, by contrast with the situation farther East, as reported in the Indian press, suggesting that the Indian cable was cut by a local submarine landslide.

We also interviewed a female witness, aged about 13 at the time of the tsunami, who recalled the destruction, by the waves, of straw houses in the low lands neighboring the beach, resulting in several fatalities, a description which would generally agree with the run-up surveyed at Site Number 03.

During the 2004 Indonesian tsunami, the main cargo harbor of Chabahar saw some 500-ton ships break their moorings and wander inside the harbor, reportedly around 5 p.m. local time, i.e., 13:30 GMT or 5–6 h after the arrival of the first waves. These episodes are reminiscent of similar effects in Toamasina, Madagascar; Le Port, Réunion; Dar-es-Salaam, Tanzania, and Salalah, Oman (OKAL et al. 2006b, c, d, 2009b). They are attributable to the arrival of higher-frequency components of the tsunami, dispersed outside the shallow-water approximation, and capable of setting ports in resonance.

- *Ramin*, Tuesday 12 October 2010

Ramin is a village located 12 km East of Chabahar on the coastal highway. It is built at an altitude of about 15–20 m, behind dunes topping at an estimated 40 m. It features an improved harbor, next to a small beach which drains a river bed (dry at the time of our visit). To the East, the coast transforms into a series of cliffs, on which lie a large number of imbricated boulders (Fig. 5). These boulders were described by SHAH-HOSSEINI et al. (2011) who have suggested that maximum-scenario storm waves are insufficient to have deposited them on the cliff, and that they must, therefore, have been deposited by a tsunami. We identified a witness who reported to us that the boulders were reached by the 1945 tsunami, but that they had been present before, which would support SHAH-HOSSEINI et al.'s (2011) much greater radiocarbon ages (1100–1600 years) for bivalves collected from the boulders. We quantify our witness' report as a splash on the cliff at a height of 7.15 m (Site Number 04).

Figure 5
Imbricated boulders on the cliff near Ramin, 12 October. These boulders were described by SHAH-HOSSEINI et al. (2011); they were reached by the 1945 tsunami, but their deposition predates this event

- *Lipar*, Wednesday 13 October 2010

Lipar is a small community located at the mouth of a major river (also named Lipar) which drains a vast hinterland plain, capable of retaining some water in the form of intermittent lakes and marshes after heavy precipitations, following the building of a levee damming the Lipar. The coastal road crosses the valley on an artificial causeway. Lipar is presently inhabited by about a dozen people, but in 1945, a community existed at the foot of the cliffs, some 3 km inland. We identified as a witness Mr. Ghalamhossein Dadshahpour, who used to live in Lipar but now living in Ramin, who remembered being in his early teens at the time of the tsunami (and claimed to be 73 years old in 2010). We drove him in our field vehicle to the beach in Lipar, where he guided our survey (Fig. 4b). He described three big waves coming in "the early morning", and which carried boulders from the sea to the berm at the estuary (lately covered with sediments). Based on his recollections, we measured (Site Number 05) a run-up of 2.80 m for an inundation of 540 m, to the present huts (which did not exist in 1945); the tsunami penetrated some 3–4 km inside the river bed (now beyond the damming levee).

- *Beris*, Thursday 14 October 2010

On Thursday 14 October, we traveled to the easternmost part of the shoreline, and visited the communities of Beris and Pasabandar.

Beris is a fishing port established in a natural bay sheltered by a prominent headland. The main village is, at least at present, established behind the headland, at an altitude of about 50 m, but in 1945, some people lived in huts on the low lands in the vicinity of the beach. Our witness, Mr. Khaled Baluch, remembers being about 10 years old at the time of the tsunami, but claimed to be 80–90 years old in 2010; a more probable figure being about 75 (Fig. 4c). He recalled that the earthquake was indeed felt, but weakly, around 1 a.m. (a remarkable estimate, the correct time being 1:30 a.m. GMT + 3:30), and that three waves came 3–4 h later. Ships in the harbor were smashed against each other, breaking anchorages. Mr. Baluch had no knowledge of any ancestral memory of similar events going back at least three generations. His descriptions were corroborated by a second witness, Mr. Khalil Taherat, said to be 82 years old in 2010, who confirmed three waves reaching to the front of the beachside mosque. This point (Site Number 06) was surveyed for a run-up of 5.60 m and an inundation of 212 m.

In 2004, currents in the port resulted in the break-up of a vessel which had remained docked; the other boats were at sea at the time.

- *Pasabandar*, Thursday 14 October 2010

This is the last important fishing village before the border with Pakistan. The village is built on an uplifted promontory reaching an estimated altitude of 25–30 m above sea level. We obtained a testimony from Mr. Adam Baluch (Fig. 4d), obviously an educated man, fluent in Arabic, and who recalled being about 12 years old at the time of the tsunami (which essentially agrees with his claimed age of 76 years in 2010). He described three waves coming at dawn (around 4 a.m.), splashing about midway up the promontory (Site Number 08), at an altitude surveyed at 13.9, and 61 m from the sea. He further described the water as reaching to the base of the hills along the road to Gavater. This point (Site Number 07) was estimated at an altitude of 10.0 m, for an inundation of 660 m, measured to the nearest point of coastline. Mr. Baluch further commented that the waves reached up to 10 km inland along the river bed.

Regarding the 2004 tsunami, the wharf at Pasabandar was not inundated, corresponding to a flow depth of 1.0 m, based on a description by a number of witnesses (Site Number 25).

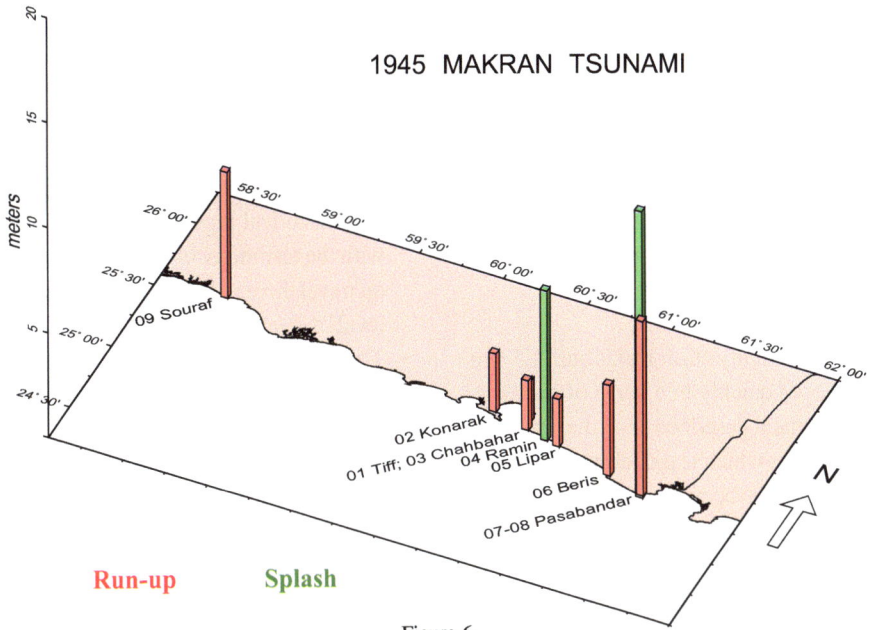

Figure 6
Map of survey results for the 1945 Makran tsunami. *Vertical bars* are scaled with the amplitudes of run-up (in *red*) or splash (in *green*). Site numbers refer to Table 1

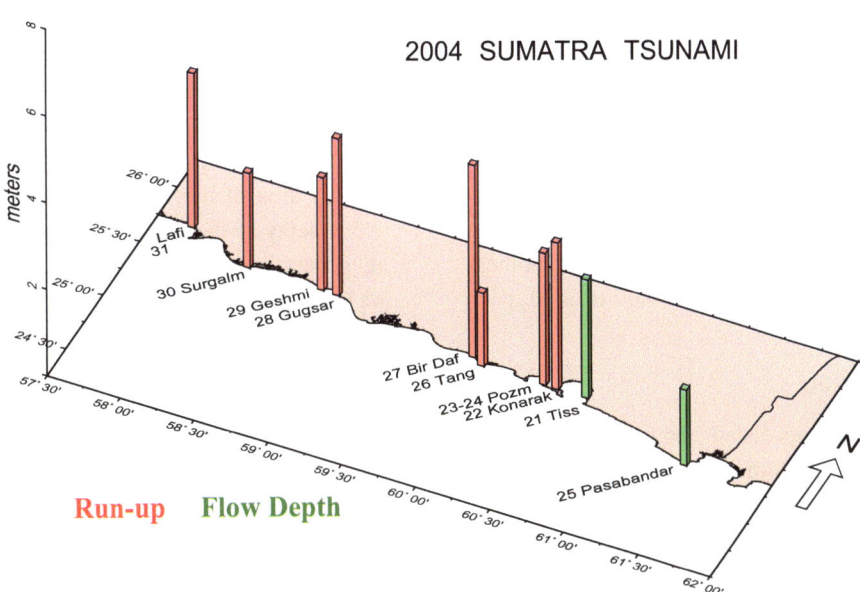

Figure 7
Map of survey results for the 2004 Sumatra tsunami. *Vertical bars* are scaled with the amplitudes of run-up (in *red*) or flow level (in *green*). Site numbers refer to Table 2

- *Tang*, Friday 15 October 2010

Finally, on Friday 15 October, we explored the Western part of the province in the communities of Tang and Bir Daf. In neither of them, could we find witnesses of the 1945 tsunami. In Tang, the only elderly man in town, Mr. Shahul Neshat, claiming to be 75 years old, could not remember any such event. By contrast, in 2004 several witnesses described 3 waves arriving in late morning which deposited a boat at the maximum storm surge water mark, and sank another one during an ebbing phase. Based on these witnesses' testimony, we surveyed a 2004 run-up of 1.1 m for an inundation of 229 m at this location (Site Number 26), protected by a barrier island.

- *Bir Daf*, Friday 15 October 2010

To the West, the community of Bir Daf is built 1.5 km inland. The beach slopes quickly to a berm of dunes. A witness of the 2004 tsunami described to us the arrival of two waves, around 8 a.m. (which is probably erroneous), and a run-up half-way up the berm, to a height of 3.8 m, for an inundation of 35 m (Site Number 27).

- *Souraf*, Wednesday 13 October 2010

This location was visited on 13 October 2010 by the Western team led by Author HMF; they were able to obtain one witness report of the 1945 tsunami at Souraf (Site Number 09), from a resident aged over 90 years, Mrs. Khairi, who described how she "saw the water coming around 6 a.m.", resulting in the destruction of the coastline settlement, which was eventually rebuilt farther inland. This translates into a run-up of 7.0 m, for an inundation of 406 m.

Furthermore, four reports were obtained by the Western team regarding the 2004 Indonesian tsunami along a 115-km stretch of coastline. (Site Numbers 28–31), with run-up ranging from 1.60 to 3.10 m.

3.1. Data Processing

All the data points listed above were surveyed using GPS devices and the exact time at which the survey took place was recorded to allow correction of the raw values listed above for tide level at the time of tsunami arrival. This correction used the OSU tidal prediction software (EGBERT and EROFEEVA 2002). The measured data of the western and eastern survey teams were corrected using tide predictions for Jask and Chabahar, respectively.

The entire dataset for 1945, including raw and corrected data, is listed in Table 1, that for 2004 as Table 2. The datasets are presented graphically on Figs. 6 and 7.

4. Conclusions

- We were able to build a set of 9 data points for the 1945 Makran tsunami. While most run-up values are in the 3–7 m range, they ramp up to significantly higher values near the Pakistani border, in general agreement with historical reports from present-day Pakistan, and with the testimony of some of our witnesses who were on travel there in 1945, and who confirm a much higher level of inundation and destruction.

- The witnesses generally described a series of three waves, arriving between 4:00 and 5:00 a.m., local time, in the Eastern part of the survey, and possibly 6 a.m. at Souraf. In this respect, the phenomenon is clearly delayed with respect to direct propagation from the source area, which should not take more than 30 min (HEIDARZADEH and SATAKE 2015), resulting in an expected arrival time of $\sim$02:00 local time (GMT + 3:30; 30 min later at Souraf). This delay of approximately 2.5 h is comparable to, if slightly smaller than, that inferred from BEER and STAGG's (1946) observation at Mahé, and therefore does not lend itself to interpretation as a dispersive effect, which should be a function of distance, or as propagation as an edge wave on the continental shelf, the latter being poorly developed along the Makran coastline, as compared to the coastal area of the Indian subcontinent. Rather, a 2.5-h delay in the arrival of the tsunami advocates its generation by an ancillary phenomenon, such as an underwater landslide, which is also the preferred model to explain the rupture of the Bombay–London telegraphic cable. A crucial datum in this respect would be the exact timing of that rupture.

- The witnesses we interviewed did not possess any ancestral memory of comparable events prior to the occurrence of the 1945 tsunami. By contrast, one of our witnesses expressed ancestral memory for the distant earthquake of 1819 in the Rann of Kuchchh. This would suggest that the Makran earthquakes of 1851 and 1864, originally reported by OLDHAM (1893) to have occurred to the West of the 1945 shock, were of comparatively lower magnitude.

- The 2004 Sumatra tsunami had significant run-up, in the 3–4 m range, along the whole surveyed segment of coastline. This is comparable to observations in Oman (OKAL et al. 2006c), while significantly less than at the southern boundary of the Arabian Sea along the Horn of Africa in Somalia (FRITZ and BORRERO 2006) and Socotra Island, Yemen (FRITZ and OKAL 2008). The 2004 tsunami involved delayed response effects in the port of Chabahar, due to the dispersed arrival of the higher-frequency components of the tsunami.

Acknowledgments

This project was made possible by a grant from UNESCAP. In addition, we gratefully acknowledge the support of the Iranian National Institute for Oceanography, particularly its Director, Dr. Vahid Chegini, as well as Mr. Majid Naderi, Mr. Nima Kiani and Ms. Fahimeh Foroughi in Tehran. In Chabahar, we are grateful to the Staff of Research Station of INIO, in particular to our driver, who took a personal interest in the project, and provided translation from the Balochi language during our interviews. Finally, we thank all our witnesses for their eagerness to share with us the precious memories of their unique experience during the 1945 tsunami. The paper was improved by the comments of reviewers.

REFERENCES

AMBRASEYS, N.N., and MELVILLE, C.P. (1982), *A history of Persian earthquakes*, Cambridge Univ. Press, Cambridge, 219 p.

ANDO, M. (1975), *Source mechanism and tectonic significance of historical earthquakes along the Nankai Trough, Japan*, Tectonophysics, 27, 119–140.

ANONYMOUS (2015), *Remembering the 1945 Makran tsunami: Interviews of survivors beside the Arabian Sea*, UNESCO — IOC Brochure 2015-1, 79 pp., Paris.

ARRIAN OF NICOMEDIA (1983), *Anabasis Alexandri, Books V–VII; Indica*, ed. and transl. by P.A. Brunt, Harvard Univ. Press, Cambridge, Mass., 589 p.

BEER, A., and STAGG, J.M. (1946), *Seismic sea-wave of November 27, 1945*, Nature, 158, 63.

BIRD, G.P. (2003), *An updated digital model of plate boundaries*, Geochem., Geophys. Geosyst., 4, (3), 1027, 52 pp.

BOUHADAD, Y., NOUR, A., SLIMANI, A., LAOUAMI, N., and BELHAI, D. (2004), *The Boumerdes (Algeria) earthquake of May 21, 2003 (M_w = 6. 8): Ground deformation and intensity*, J. Seismol., 8, 497–506.

BYRNE, D.E., SYKES, L.R., and DAVIS, D.M. (1992), *Great thrust earthquakes and aseismic slip along the plate boundary of the Makran subduction zone*, J. Geophys. Res., 97, 449–478.

CISTERNAS, M., ATWATER, B.F., TORREJÓN, F., SAWAI, Y., MACHUCA, G., LAGOS, M., EIPERT, A., YOULTON, C., SALGADO, I., KAMATABI, T., SHISHIKURA, M., RAJENDRAN, C.P., MALIK, J.K., RIZAL, Y., and HUSNI, M. (2005), *Predecessors of the giant 1960 Chile earthquake*, Nature, 437, 404–407.

EGBERT, G.D., and EROFEEVA, S.Y. (2002), *Efficient inverse modeling of barotropic ocean tides*, J. Atmos. Ocean. Tech., 19, 183–204.

EL-ROBRINI, M., GENNESSEAUX, M., and MAUFFRET, A. (1985), *Consequences of the El-Asnam earthquakes: Turbidity currents and slumps on the Algerian margin (Western Mediterranean)*, Geo-Marine Letts., 5, 171–176.

ENGDAHL, E.R., and VILLASEÑOR, A. (2002), Global Seismicity: 1900–1999, **In:** *International Handbook of Earthquake and Engineering Seismology, Part A*, ed. by W.H.K. Lee, H. Kanamori, P.C. Jennings, and C. Kisslinger, Chapter 41, pp. 665–690, Academic Press, London.

FRITZ, H.M., and BORRERO, J.C. (2006), *Somalia field survey of the 2004 Indian Ocean tsunami*, Earthq. Spectra, 22, S219–S233.

FRITZ, H.M., and OKAL, E.A. (2008), *Socotra Island, Yemen: Field survey of the 2004 Indian Ocean tsunami*, Natural Haz., 46, 107–117.

FRITZ, H.M., BLOUNT, C.D., ALBUSAIDI, F.B., and AL-HARTHY, A.H.M. (2010), *Cyclone Gonu storm surge in Oman*, Estuar. Coastal Shelf Sci., 86, 102–106.

GUTENBERG, B., and RICHTER, C.F. (1954), *Seismicity of the Earth and Associated Phenomena*, Princeton Univ. Press, Princeton, 310 p.

HEEZEN, B.C., and EWING, W.M. (1952), *Turbidity currents and submarine slumps, and the 1929 Grand Banks earthquake*, Amer. J. Sci., 250, 849–878.

HEEZEN, B.C., and EWING, W.M. (1955), *Orléansville earthquake and turbidity currents*, Bull. Amer. Soc. Petrol. Geol., 39, 2505–2514.

HEIDARZADEH, M., and SATAKE, K. (2014), *Possible sources of the tsunami observed in the Northwestern Indian Ocean following the 2013 September 24 M_w = 7.7 Pakistan inland earthquake*, Geophys. J. Intl., 199, 752–766.

HEIDARZADEH, M., and SATAKE, K. (2015), *New insights into the source of the Makran tsunami of 27 November 1945 from tsunami waveforms and coastal deformation data*, Pure Appl. Geophys., 172, 621–640.

HOUTZ, R.E. (1962), *The 1953 Suva earthquake and tsunami*, Bull. Seismol. Soc. Amer., 52, 1–12.

KAGAN, Y.Y. (1991), *3-D rotation of double-couple earthquake sources*, Geophys. J. Intl., 106, 709–716.

KAKAR, D.M., NAEEM, G., USMAN, A., and ATWATER, B.F. (2015), *The Balochistan and Sindh coast survivors revealed facts of the 1945 Makran tsunami*, Proc. Intl. Conf. Reducing Tsunami Risk in Western Indian Oc., Muscat, Oman, UNESCO-IOC, Muscat (abstract).

KELSEY, H.M., NELSON, A.R., HEMPHILL-HALEY, E., and WITTER, R.C. (2005), *Tsunami history of an Oregon coastal lake reveals a 4600-yr record for great earthquakes on the Cascadia subduction zone*, Geol. Soc. Amer. Bull., 117, 1009–1032.

KUKOWSKI, N., SCHILLHORN, T., FLUEH, E.R., and HUHN, K. (2000), *Newly identified strike-slip plate boundary in the Northeastern Arabian Sea*, Geology, 28, 355–358.

MASSON, F., ANVARI, M., DJAMOUR, Y., WALPERSDORF, A., TAVAKOLI, F., DAIGNIÈRES, M., NANKALI, H., and VAN GORP, S. (2007), *Large-scale velocity field and strain tensor in Iran inferred from*

GPS measurements; new insight for the present-day deformation pattern within NE Iran, Geophys. J. Intl., 170, 436–440.

MOKHTARI, M., FARD, I.A., and HESSAMI, K. (2008), Structural elements of the Makran region, Oman Sea, and their potential relevance to tsunamigenesis, Natural Hazards, 47, 185–199.

MUSSON, R.M.W. (2009), Subduction in the Western Makran: the historian's contribution, J. Geol. Soc. London, 166, 387–391.

NANAYAMA, F., SATAKE, K., FURUKAWA, R., SHIMOKAWA, K., ATWATER, B.F., SHIGENO, K., and YAMAKI, S. (2003), Unusually large earthquakes inferred from tsunami deposits along the Kuril trench, Nature, 424, 660–663.

NEETU, S., SURESH, I., SHANKAR, R., NAGARAJAN, R., SHARMA, R., SHENOI, S.S.C., UNNIKRISHNAN, A.S., and SUNDAR, D. (2011), Trapped waves of the 27 November 1945 Makran tsunami: observations and numerical modeling, Natural Hazards, 59, 1609–1618.

NIAZI, M., SHIMAMURA, H., and MATSU'URA, M. (1980), Microearthquakes and crustal structure off the Makran coast of Iran, Geophys. Res. Letts., 7, 297–300.

OKAL, E.A., and REYMOND, D. (2003), The mechanism of the great Banda Sea earthquake of 01 February 1938: Applying the method of Preliminary Determination of Focal Mechanism to a historical event, Earth Planet. Sci. Letts., 216, 1–15.

OKAL, E.A., and SYNOLAKIS, C.E. (2008), Far-field tsunami hazard from mega-thrust earthquakes in the Indian Ocean, Geophys. J. Intl., 172, 995–1015.

OKAL, E.A., SYNOLAKIS, C.E., FRYER, G.J., HEINRICH, P., BORRERO, J.C., RUSCHER, C., ARCAS, D., GUILLE, G., and ROUSSEAU, D. (2002), A field survey of the 1946 Aleutian tsunami in the far field, Seismol. Res. Letts., 73, 490–503, 2002.

OKAL, E.A., BORRERO, J.C., and SYNOLAKIS, C.E. (2006a), Evaluation of tsunami risk from regional earthquakes at Pisco, Peru, Bull. Seismol. Soc. Amer., 96, 1634–1648.

OKAL, E.A., FRITZ, H.M., RAVELOSON, R., JOELSON, G., PANČOŠKOVÁ, P., and RAMBOLAMANANA, G. (2006b), Madagascar field survey after the December 2004 Indian Ocean tsunami, Earthquake Spectra, 22, S263–S283.

OKAL, E.A., FRITZ, H.M., RAAD, P.E., SYNOLAKIS, C.E., AL-SHIJBI, Y., and AL-SAIFI, M. (2006c), Oman field survey after the December 2004 Indian Ocean tsunami, Earthquake Spectra, 22, S203–S218.

OKAL, E.A., SLADEN, A., and OKAL, E.A.-S. (2006d), Rodrigues, Mauritius and Réunion Islands field survey after the December 2004 Indian Ocean tsunami, Earthquake Spectra, 22, S241–S261.

OKAL, E.A., SYNOLAKIS, C.E., USLU, B., KALLIGERIS, N., and VOUKOUVALAS, E. (2009a), The 1956 earthquake and tsunami in Amorgos, Greece, Geophys. J. Intl., 178, 1533–1554.

OKAL, E.A., FRITZ, H.M. and SLADEN, A. (2009b), 2004 Sumatra tsunami surveys in the Comoro Islands and Tanzania and regional tsunami hazard from future Sumatra events, South Afr. J. Geol., 112, 343–358.

OLDHAM, R.D. (1893), A manual of the geology of India: stratigraphical and structural geology, Off. Superint. Gov. Printer, Calcutta, 543 p., 1893.

PENDSE, C.G. (1946), The Mekran earthquake of the 28th November 1945, India Meteorol. Dept. Sci. Notes, 10, 141–146.

QUITTMEYER, R.C., and JACOB, K.H. (1979), Historical and modern seismicity of Pakistan, Afghanistan, Northwestern India, and Southeastern Iran, Bull. Seismol. Soc. Amer., 69, 773–823.

RAJENDRAN, C.P., RAJENDRAN, K., SHAH-HOSSEINI, M., NADERI BENI, A., NAUTIYAL, C.M., and ANDREWS, R. (2013), The hazard potential of the Western segment of the Makran subduction zone, Northern Arabian Sea, Natural Hazards, 65, 219–239.

REPETTI, W.C. (1934), The China Sea earthquake of February 14th, 1934, in: Seismological Bulletin for 1934 January-June, Dept. Agriculture and Commerce, Govt. of the Philippine Is., pp. 22–29, Manila.

REYMOND, D., and OKAL, E.A. (2000), Preliminary determination of focal mechanisms from the inversion of spectral amplitudes of mantle waves, Phys. Earth Planet. Inter., 121, 249–271.

SALAREE, A., and OKAL, E.A. (2015), Field survey and modeling of the Caspian Sea tsunami of 20 June 1990, Geophys. J. Intl., 201, 621–639.

SATAKE, K., SHIMAZAKI, K., TSUJI Y., and UEDA, K. (1996) Time and size of a giant earthquake in Cascadia inferred from Japanese tsunami records of January 1700, Nature, 379, 246–249.

SELLA, G.F., DIXON, T.H., and MAO, A. (2002), REVEL: A model for Recent plate velocities from space geodesy, J. Geophys. Res., 107, (B4), ETG_11, 32 p.

SHAH-HOSSEINI, M., MORHANGE, C., NADERI BENI, A., MARRINER, N., LAHIJANI, H., HAMZEH, M.A., and SABATIER, F. (2011), Coastal boulders as evidence for high-energy waves on the Iranian coast of Makran, Mar. Geol., 290, 17–28.

SYNOLAKIS, C.E., BARDET, J.-P., BORRERO, J.C., DAVIES, H.L., OKAL, E.A., SILVER, E.A., SWEET, S., and TAPPIN, D.R. (2002), The slump origin of the 1998 Papua New Guinea tsunami, Proc. Roy. Soc. (London), Ser. A, 458, 763–789.

THE TIMES OF INDIA (1945a), Bombay, 29 and 30 November 1945a.

THE TIMES OF INDIA (1945b), Bombay, 06 December 1945b.

VERNANT, P., NILFOROUSHHAN, F., HATZFELD, D., ABBASSI, M.R., VIGNY, C., MASSON, F., NANKALI, H., MARTINOD, J., ASHTIANI, A., BAYER, R., TAVAKOLI, F., and CHÉRY, J. (2004), Present-day crustal deformation and plate kinematics in the Middle East constrained by GPS measurements in Iran and Northern Oman, Geophys. J. Intl., 157, 381–398.

WALTON, H.I. (1865), Correspondence, Trans. Bombay Geogr. Soc., 17, ccxxv–ccxxvi.

WHITE, R.S., and KLITGORD, K. (1976), Sediment deformation and plate tectonics in the Gulf of Oman, Earth Plan. Sci. Letts., 32, 199–209.

WILSON, A.T. (1930), Earthquakes in Persia, Bull. School Orient. Afr. Stud., 6, 103–131.

WYSESSION, M.E., OKAL, E.A., and MILLER, K.L. (1991) Intraplate seismicity of the Pacific Basin, 1913–1988, Pure Appl. Geophys., 135, 261–359.

(Received May 4, 2015, revised July 21, 2015, accepted July 23, 2015, Published online August 9, 2015)

OSL Dating and GPR Mapping of Palaeotsunami Inundation: A 4000-Year History of Indian Ocean Tsunamis as recorded in Sri Lanka

RANJITH PREMASIRI,[1,3] PETER STYLES,[1] VICTOR SHRIRA,[1] NIGEL CASSIDY,[1] and JEAN-LUC SCHWENNINGER[2]

Abstract—To evaluate and mitigate tsunami hazard, as long as possible records of inundations and dates of past events are needed. Coastal sediments deposited by tsunamis (tsunamites) can potentially provide this information. However, of the three key elements needed for reconstruction of palaeotsunamis (identification of sediments, dating and finding the inundation distance) the latter remains the most difficult. The existing methods for estimating the extent of a palaeotsunami inundation rely on extensive excavation, which is not always possible. Here, by analysing tsunamites from Sri Lanka identified using sedimentological and paleontological characteristics, we show that their internal dielectric properties differ significantly from surrounding sediments. The significant difference in the value of dielectric constant of the otherwise almost indistinguishable sediments is due to higher water content of tsunamites. The contrasts were found to be sharp and not to erode over thousands of years; they cause sizeable electromagnetic wave reflections from tsunamite sediments, which permit the use of ground-penetrating radar (GPR) to trace their extent and morphology. In this study of the 2004 Boxing Day Indian Ocean tsunami, we use GPR in two locations in Sri Lanka to trace four identified major palaeotsunami deposits for at least 400 m inland (investigation inland was constrained by inaccessible security zones). The subsurface extent of tsunamites (not available without extensive excavation) provides a good proxy for inundation. The deposits were dated using the established method of optically stimulated luminescence (OSL). This dating, partly corroborated by available historical records and independent studies, contributes to the global picture of tsunami hazard in the Indian Ocean. The proposed method of combined GPR/OSL-based reconstruction of palaeotsunami deposits enables estimates of inundation, recurrence and, therefore, tsunami hazard for any sandy coast with identifiable tsunamite deposits. The method could be also used for anchoring and synchronizing chronologies of ancient civilisations adjacent to the ocean shores.

Key words: Tsunami hazard, Reconstruction of palaeotsunami deposits, Tsunami sediments, Ground-penetrating radar (GPR), Optically Stimulated Luminescence (OSL), Tsunami in the Indian Ocean.

---

[1] Institute for the Environment, Physical Sciences and Applied Mathematics (EPSAM), Keele University, Keele ST5 5BG, UK. E-mail: p.styles@keele.ac.uk
[2] School of Archaeology, University of Oxford, South Parks Road, Oxford OX1 3PS, UK.
[3] Department of Earth Resources Engineering, University of Moratuwa, Moratuwa, Sri Lanka.

1. Introduction

One of the most promising approaches to evaluate tsunami hazard is through reconstruction of inundations and dates of past events. Coastal sediments deposited by tsunamis (tsunamites) can potentially provide this information, and, therefore, attract intense current interest (JANKAEW et al. 2008; MONECKE et al. 2008; MORTON et al. 2008; RUIZ et al. 2008; YAN and TANG 2008). Finding the evidence of inundation remains the most challenging element of palaeotsunami reconstruction. The difficulty is aggravated by the inevitable long-shore variability of tsunamis which has a significant random component (CHOI et al. 2012). In view of this variability even if the values of inundations or wave height were somehow accurately found in one or two locations this would provide only a relatively weak constraint on tsunami parameters (run-up, number of waves, etc.). Hence, the mapping of inundation or wave height is needed for better constrained estimates. Since the extensive excavation, which is often required, is rarely if ever possible, the use of non-invasive methods is needed. These might include electrical imaging methods and shallow seismic and acoustic methods but these do not have the resolution required (STYLES 2011). The non-invasive technique based upon ground-penetrating radars (GPR) utilizes reflections of emitted microwave electromagnetic pulses from subsurface non-uniformities in dielectric properties. The first GPR investigation of a tsunami deposit (washover sand) was carried out by SWITZER et al. (2006) on the Australian coast, they identified erosional downcutting at the base of the deposit but they did not trace palaeotsunamites as reported here. Very recently GPR has been applied for defining sediment/rock interfaces and the presence of debris with boulders brought in by palaeotsunamis (KOSTER et al. 2013, 2014), i.e. GPR has been used and found effective in situations with a priori large dielectric contrasts between the tsunami deposit and the surrounding topsoil and sand. The big open question is whether the GPR-based approach can work in the generic situations of sandy beaches where tsunamis bring in sand and deposit it upon a sandy beach and there are no noticeable contrasts between the primary physical properties of tsunamite and non-tsunamite sands.

Here, by analysing tsunamites from Sri Lanka identified using sedimentological and paleontological characteristics and confirming and extending our preliminary reports (STYLES et al. 2007; PREMASIRI 2012), we show that their dielectric properties differ significantly from surrounding sediments even when the differences in the primary physical characteristics are very subtle. These contrasts cause sizeable electromagnetic wave reflections from tsunamite sediments, which permit the use of ground-penetrating radar (GPR) to trace their cross-shore extent long-shore variations and three-dimensional morphology. In this pilot study we use GPR in two locations to trace four identified major palaeotsunami deposits upon sandy beaches for at least 400 m inland (constrained by inaccessible military security zones). The subsurface extent of tsunamites (not available without extensive excavation) provides a good proxy for inundation (GOTO et al. 2011; ABE et al. 2012). The deposits were dated using the established method of optically stimulated luminescence (OSL), (HUNTLEY and CLAGUE 1996). This dating, partly corroborated by other authors (RANASINGHAGE 2010, JACKSON et al. 2014) and available historical records, contributes to a new picture of tsunami hazard in the Indian Ocean.

Quantifying tsunami hazard requires extensive records of magnitudes and dates of past events. Historical records are often insufficient or absent altogether, while for the pre-historical time period, the only available sources are sediments deposited by palaeotsunamis. The first key problem is to confidently differentiate tsunami deposits (tsunamites) from storm surge sediments (tempestites). Micropalaeontological and sedimentological analyses allow such a differentiation as described later (BRYANT and NOTT 2001; WITTER et al. 2001; MORTON et al. 2007; PRATT and BORDONARO 2007; DAHANAYAKE and KULASENA 2008a, b). Dating of the identified deposits can be carried out using several techniques (carbon dating, stratigraphy, thermoluminescence and in appropriate circumstances, tephrochronology) with varying accuracies. If deposits contain quartz sands (very often true, although in Japan these are volcanically derived) OSL is the most appropriate technique (DULLER 2008, BLUSZCZ 2006), enabling dating of events from zero to 150,000 years with accuracies of about 5–10 %. The most difficult and yet unresolved part of the reconstruction is finding the amplitude or/and inundation of the tsunami waves. Several studies have sought clues from the hydrodynamics of tsunamis in geological records with success in determining the number of incoming waves using sedimentological textural analysis (SMITH et al. 2004; PETERS et al. 2007; WEISS 2008). However, estimating flow rate or tsunami run-up elevation has proved much more difficult (MORTON et al. 2007; PETERS et al. 2007; BONDEVIK et al. 1997; HINDSON and ANDRADE 1999; MOORE 2000; LE ROUX and VARGAS 2005, 2007; FUJINO et al. 2006; KORYCANSKY and LYNETT 2007; ; MOORE et al. 2007). Recent attempts to link flow velocity and wave heights to grain size distributions are based on a sound fluid mechanical footing (JAFFE et al. 2003; JAFFE and GELFENBAUM 2007, SOULSBY et al. 2007), but permit only rough estimates. Since even solutions of the direct problem have large errors, good quantitative results for the inverse problem are unlikely.

Here, we report work carried out during the PhD thesis of PREMASIRI (2012) on the potential of palaeotsunami reconstruction based on the previously undiscovered sharp dielectric constant contrasts between tsunamites and normal beach sediments. The contrasts between the dielectric constant of Sri Lankan tsunami deposits and surrounding sediments of other origin measured in the laboratory with very subtle differences in the primary physical characteristics were found to be always large enough for confidently employing GPR for 'direct' probing of morphology of palaeotsunami deposits. Subsequent use of OSL provided dating of the palaeotsunami deposits. The paper reports the testing of this approach carried out on Sri Lanka.

The paper is organized as follows. In Sect. 2, we briefly describe the specific locations on the Sri Lanka coast where we identified palaeotsunami deposits. Detailed studies have been confined to two sites: Hambantota and Yala. In Sect. 3, we describe four different independent approaches (standard grain size statistical analysis, micro-petrography, micro-fossil and mineralogical analyses) we employed for identifying the palaeotsunami deposits in the sedimentological sequence using the 2004-tsunami sediments as a reference. In Sect. 4, we describe our laboratory study of dielectric properties of the sampled sediment showing that the values of tsunamite dielectric constant differ significantly from those of surrounding sediments. The resulting contrasts at the boundaries of tsunami sediment layers which, crucially, were found to remain sharp over thousands of years cause sizeable reflections of electromagnetic waves, which enables us to apply GPR to trace the palaeotsunami deposits. Whereas the previous works simply identified the base of tsunami deposits we show that internal multi-layered structure can be derived using GPR. The successful three-dimensional mapping of the palaeotsunami deposits with GPR in two locations is described in Sect. 5. Section 6 describes the procedure and results of the OSL dating of the collected samples of identified palaeotsunami deposits. Three palaeotsunami events have been OSL dated in the interval between the present day and 3.5 thousand years ago. An additional deeper fourth layer identified as a palaeotsunami deposit could not be OSL dated as the samples were accidentally exposed to light; its date was estimated at ~4000BP by extrapolating the estimated average rate of non-tsunamite deposition. OSL is preferable to radiocarbon dating for these tsunami sediments, as it does not require any knowledge concerning the provenance of organic material.

In Sect. 7, we discuss the available historical evidence and folklore on palaeotsunamis and put our finding into the regional historical perspective. Concluding in Sect. 8, we discuss the potential of this efficient, non-destructive high-resolution technique for tracing palaeotsunami deposits and mapping inundation. The specific implications for Indian Ocean hazard are also discussed.

2. Locations of the Study and Sedimentological Descriptions

The research was carried out in an area of the Sri Lankan coast where the 26 December 2004 (Boxing Day) tsunami caused severe impact. Sri Lanka is one of the islands where catastrophic damage was caused even though it is located over 1000 km away from the origin of the 2004 Sumatra Tsunami (Fig. 1). About 75 % of the coast, nearly 1000 km, was affected. The present study mainly concentrates on the south-eastern, southern and western parts of the Sri Lankan coast but detailed analysis is confined to two coastal sites in the southeast of Sri Lanka at the sites shown in Fig. 2 where Quaternary beach deposits (mainly fine-grained garnet-rich, quartz and feldspar mineral sands) overlie Proterozoic metamorphic bedrock, mostly quartz and feldspar-rich granitoid gneisses.

To study coastal stratigraphic sequences and provide ground truth for interpretation of GPR data we dug trenches approximately 1.5 m deep by hand. The 2004 tsunamites are mainly sands, mud, silts and gravel, with occasional organic materials and plant debris and are up to 30 cm in thickness.

The extracted sediment sequences comprise well-stratified beach and fluvial sands, peaty and lagoonal sand or soil interlayered with tsunami deposits. These

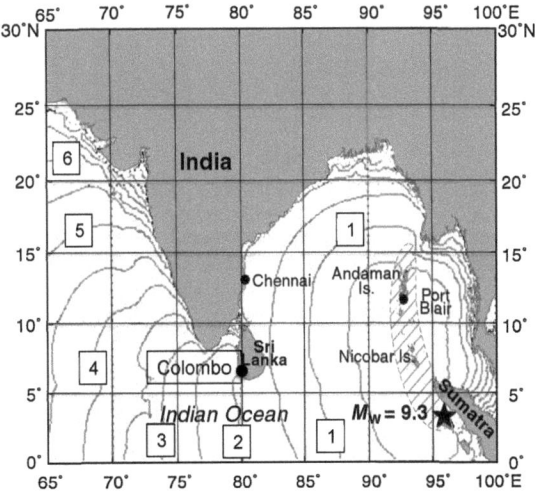

Figure 1
Map of the 2004 Sumatra earthquake epicentre and the surrounding countries affected by the associated tsunami. Isochrones of tsunami travel time at 30 min contours with hourly labels (FINE et al. 2005) are also shown together with the inferred rupture zone

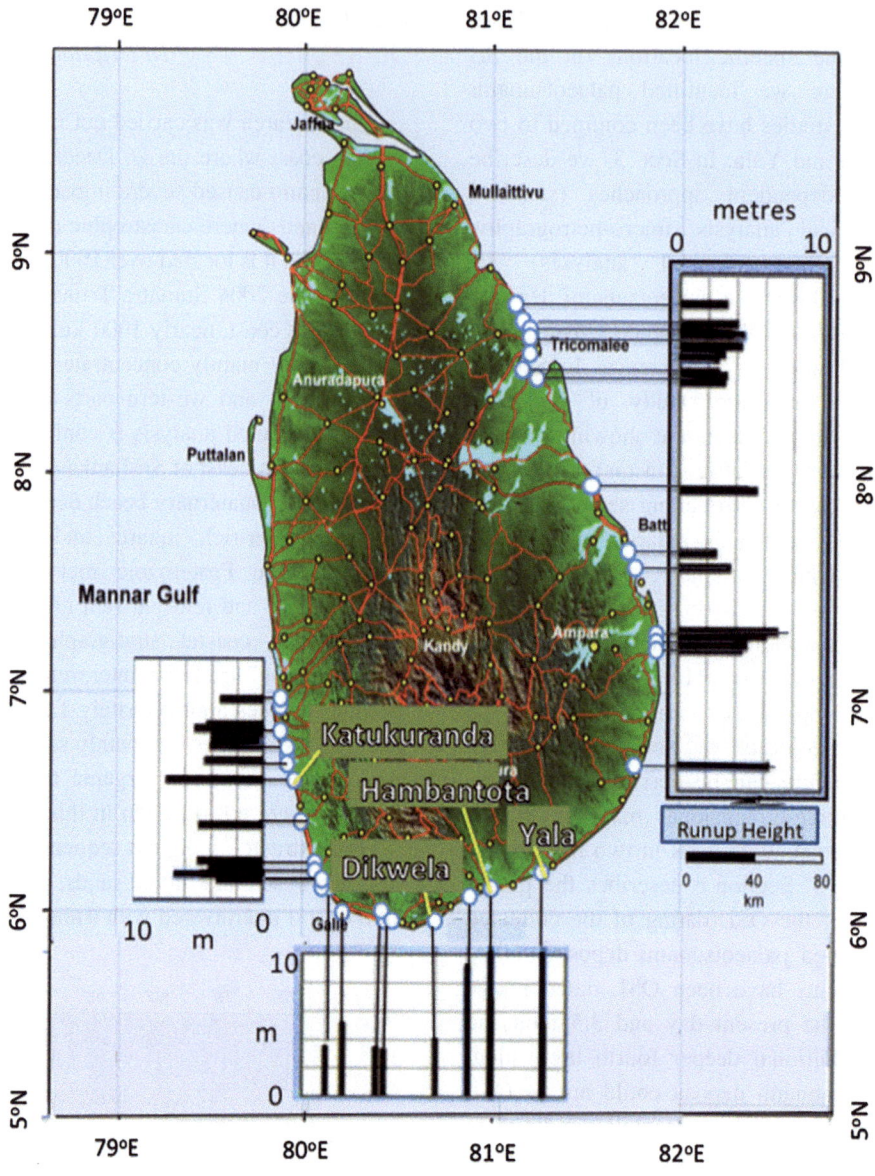

Figure 2
Location map of tsunami sediment studies; Location 1 (Hambantota) and Location 2 (Yala). Dikwela and Katukurunda where GPR profiles were obtained are also shown. *Black bars* the run-up (in metres) observed around the coast of Sri Lanka (pers. comm. Professor A Gunathilake, based on Sri Lanka Geological Survey data, 2005 unpublished information)

different beds are characterized by differing textures, sorting, colour, mineral composition and the presence of shells and fossils. Certain horizons within the sequence are characterized by light colour, poor sorting and comprise gravels or pebbles with fossils and were tentatively inferred to be older palaeotsunami deposits, which has been later confirmed by laboratory analysis on the basis of mineralogical, paleontological, optical microscopic properties. These are discussed fully in Sect. 3.

We used the 2004 tsunamites as a stratigraphic marker, and reference. Both Yala and Hambantota are located on the east–southeast of the island and are likely to have experienced the same general direction of approach (as most great seismic events are located to the east–northeast of the Island) for palaeotsunami deposits as the 2004 Boxing Day tsunami and, therefore, similar pattern of tsunami deposits for the 2004 and palaeo events. In 2004, both sites suffered major inundations of between 500 and 1100 m giving access to sufficiently long profiles to test tracing of the morphology of the deposits using ground-penetrating radar. Figure 2 also shows the run-up and direction of approach based on the study of HETTIARACHCHI and SAMARAWICKRAMA (2005) and the run-up observed around the coast of Sri Lanka based on unpublished data from the Sri Lanka Geological Survey (pers. comm. Professor A. Gunathilake).

3. Sedimentological Analysis and Identification Of 2004 Tsunami Sediments and Palaeotsunami Sediments in Sri Lanka

Descriptions of sediment package morphology and sedimentary grain size analysis for the 2004 tsunami in Sri Lanka are described in STYLES *et al.* (2007) and DAHANAYAKE and KULASENA (2008a, b). The main features of these deposits can be summarized as follows:

- Sand deposits of varying thickness (massive sandy deposits in some places) from a few centimetres to over 30 cm thick) were observed all around the island
- Sediment deposition was higher on the southeast and eastern coasts than on the western coast. The tsunami deposits in Sri Lanka are mainly confined to low relief topographical features such as swells or marshlands. Sediment deposition patterns were more often controlled by local topographical features than by the tsunami wave characteristics.
- The extent of deposition of tsunami sediment inland is strongly dependent on the inundation distance. Both were measured independently. Sediment deposition was observed over at a maximum of 700 m inland at these localities (although much greater distances were observed elsewhere around Sri Lanka) as continuous beds, discontinuous sheets, lenses, and pocket-like deposits.
- Layered deposits were often observed with the number of sediment depositional packages, observed at any location corresponding to the number of waves.

- The tsunamite sediment deposits on the coastal zone from the tsunami were easily distinguishable from normal beach sediments because of their contact morphology (erosive and with the lower contact downcutting peaty or normal dark coloured top soils) and their textural characteristics (often coarse-grained, poorly sorted and light coloured).
- Isolated large boulder deposits (which are called tsunami-ishi by KATO and KIMURA 1983) reflect the high-energy hydrodynamic regimes, and the sandy deposits contain rip-up clasts showing a very energetic environment during the deposition and in fact some regions of the coastline were sometimes subject to erosion rather than deposition.

Following and extending Styles *et al.* (2007) and PREMASIRI (2012) we identify the features listed below as distinctive characteristics of the 2004 tsunami sediments on the Sri Lankan coast.

1. Tsunami sediment depositional characteristics are strongly dependent on local topographical variation and the nature of back-wash flow. The sediment mainly consists of poorly sorted, medium-to-coarse sand.
2. While tsunami sands are not identical to normal beach sand in composition, texture structure and most other physical characteristics they often show compositional heterogeneity and it is difficult to identify them on sedimentological characteristics alone and such identifications should be treated cautiously.
3. The first 50 to 75 m distance from the coast is mainly subjected to erosion rather than deposition.
4. No strong relationship exists between sediment thickness and wave characteristics such as wave direction or run-up height.
5. Mineralogically, tsunamite sands are rich in quartz and feldspar minerals with other heavy or opaque minerals and Fe/Mg silicate minerals present in significant amounts as compared to normal beach sand (HERATH 1985, 1988).
6. The tsunami sands from Sri Lanka show mechanical fracturing at microscopic scales. The specific mechanism of fracturing is not identified; this goes beyond the scopes of the present study. The observed fracturing may be an indicator of provenance rather than process.

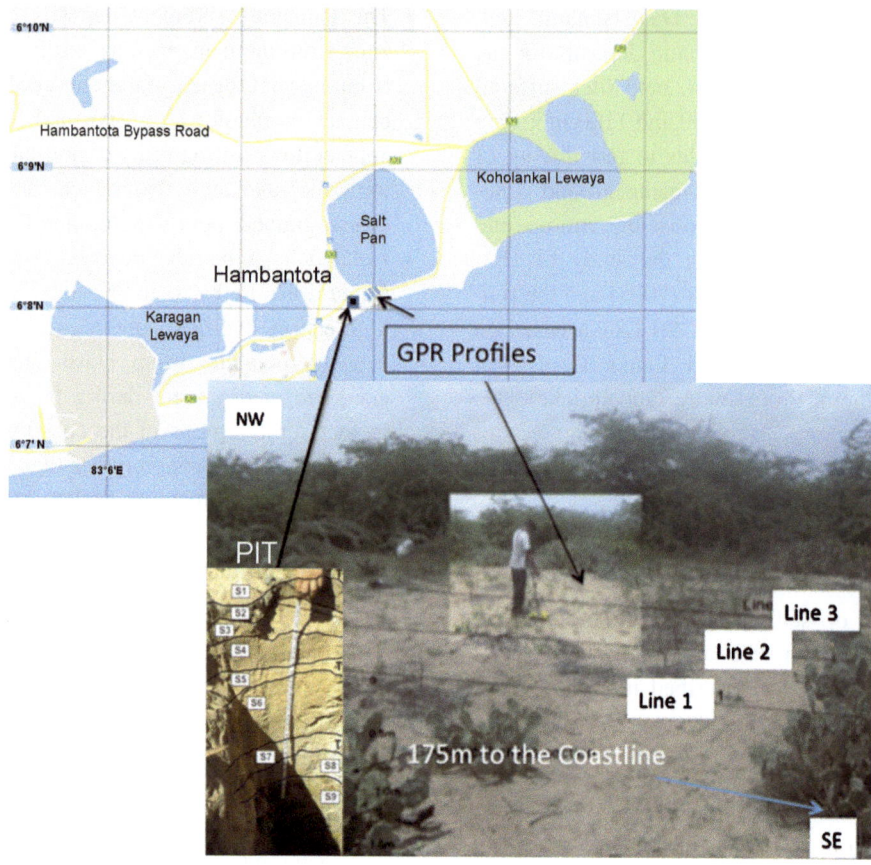

Figure 3
Location of the pit dug for palaeotsunami studies at Location 1 (Hambantota 6.13°N, 81.13°E). The location of the excavated pit and the GPR *Survey lines* are shown

7. With time, the uppermost layer of the tsunami sediments can be altered by natural environmental causes; these changes may depend on sediment type, climatic conditions and height of deposition (SPISKE et al. 2013). However, the effect decreases inland and the impact on thicker deposits is less pronounced. We resampled tsunami sediments after a period of 3 years and found no discernible difference.

3.1. Grain Size Statistical Analysis

3.1.1 Location 1: Hambantota, Southeast Coast of Sri Lanka

Location 1 is a flat beach ridge adjacent to Hambantota-Mahalewaya lagoon where sediment deposition displayed thicknesses exceeding 30 cm for the 2004 tsunami. In July 2008, a 1.5-m-deep trench, located about 175 m inland at an elevation of 0.3 m was dug by hand (Fig. 3), until it reached the lagoon clay bed. Samples were taken for paleontological and sedimentological analysis and OSL dating This area was severely affected by the 2004 tsunami and an associated sediment layer a few centimetres thick could be observed. The stratigraphic section (Fig. 4a) displays conspicuous layering in the sandy sediments. The 2004 tsunami sediment (S1) was clearly identified lying on top of the original peaty ground.

It was also possible to identify a further two layers (S5 and S7) comprising light-coloured, coarse-grained sediments with properties distinct from the background sediments which are inferred from field observation to be palaeotsunami sediments. Grain size analyses and statistical analysis were performed for each tsunami sediment sample collected from

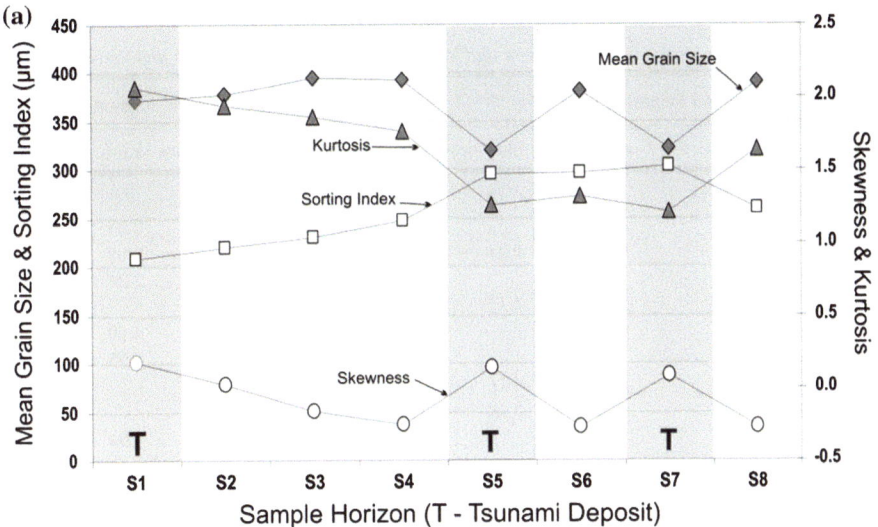

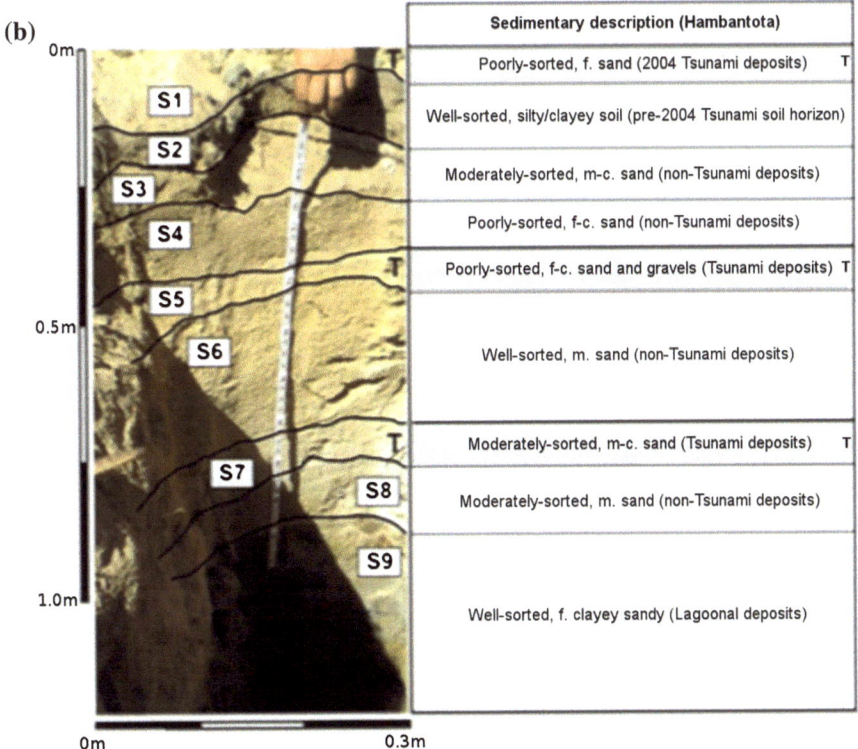

Figure 4

a Palaeotsunami sand layers and non-tsunami deposits (beach sand, fluvial sands and lagoonal clayey sand) at Location 1, Hambantota (layers are labelled as S1, S2 up to S9). Notation: *f* fine, *f–c* fine-to-coarse, *m* medium, *m–c* medium-to-coarse, *T* tsunami sediments. b Statistical characteristics of grain size distribution at Hambantota

both the 2004 tsunami and presumed palaeotsunami deposits (distinguished because of their distinctive colour and texture from the lagoonal sediments) as follows. 400–500 g samples were collected from each layer and sub-sampled by the cone and quartering method and then 100 g of sediment were used for the

Table 1

Sedimentological parameters for the Hambantota and Yala tsunami (italicized numbers) and non-tsunami deposits

Site	Sample	Tsunami Deposits				Fluvial, Marine, Lagoon Deposits			
		Mean Grain Size	Kurtosis	Skewness	Sorting Index	Mean Grain Size	Kurtosis	Skewness	Sorting Index
Site 2—Yala	S1	*377.9*	*2.05*	*0.44*	*190.3*				
	S2					349.9	2.25	0.32	201.2
	S3/4	*214.6*	*2.46*	*0.93*	*242.3*				
	S5					451.5	2.24	−0.56	222.4
	S6	*282.5*	*4.42*	*1.12*	*142.4*				
	S7					320.5	4.10	0.92	141.3
	S9					358.9	2.55	0.19	181.2
	S10	*346.5*	*2.86*	*0.71*	*165.6*				
Site 1—Hambantota	S1	*371.7*	*2.06*	*0.18*	*208.6*				
	S2					377.2	1.94	0.02	219.9
	S3					394.6	1.86	−0.16	230.0
	S4					393.0	1.77	−0.26	247.4
	S5	*320.0*	*1.26*	*0.13*	*295.8*				
	S6					380.9	1.32	−0.27	296.9
	S7	*322.4*	*1.21*	*0.08*	*304.0*				
	S8					390.4	1.64	−0.27	259.7
	SD	56.6	1.10	0.41	62.3	36.2	0.81	0.44	45.2
	Median	322.4	2.06	0.44	208.6	380.9	1.94	−0.16	222.4
	Mean	319.4	2.33	0.51	221.3	379.7	2.19	−0.01	222.2

sieve analysis. The samples were fractionated into grain size using standard sieve sizes, described by SYVITSKI (1991). Some sites had extremely fine-grained layers; the 2004 tsunami sediment deposit at Katukurunda underwent grain size analysis using a Malvern Mastersizer X Ver. 1.2b. Statistical parameters of the grain size distributions such as mean grain size, standard deviation (sorting index), skewness (deviation from a symmetrical Gaussian curve either in a positive or negative direction, i.e. towards coarser or finer grains), and kurtosis (a measure of 'peakiness' where the distribution is more/less sharply peaked than a standard Gaussian bell curve) were determined using the Gradisat software (BLOTT and PYE 2001) calculated on the basis of the percentile statistics of FOLK and WARD (1957).

To provide the context we mention that sorting values may be used to determine the wave period. Sediment layers with higher sorting values may suggest that the sediment has been created by a shorter wave period (WAGNER and SRISUTAM 2011). High values of skewness (positively skewed) correspond to a grain size distribution with a modal peak corresponding to medium-grained sands with the tail corresponding to coarser grained sediments. Low skewness values (negatively skewed) correspond to grain size distributions in which there is a significantly finer tail element. Low kurtosis values usually correspond to grain size distributions which are narrower than Gaussian unimodal distributions. High kurtosis values correspond to broader and flatter, often poly-modal, grain size distributions (DAWSON et al. 1996a, WAGNER and CHANCHAI 2011). In the adopted normalization the Gaussian distribution has kurtosis equal to one.

From the grain size statistical analyses (Table 1; Fig. 4b), we observe that layers S5 and S7 (inferred palaeotsunami) as well as the 2004 tsunami sediment S1 exhibit small positive skewness, while all other layers have negative skewness. However, these differences in the grain size distributions we consider to be too subtle to serve for robust identification of the tsunami sediments.

Table 1 and Fig. 5 both show that tsunami and non-tsunamite deposits are strongly bi-modal in grain size with no immediately obvious differences between them; it would be difficult to make a firm identification simply on sedimentological grounds alone without associated field observation as discussed by (ENGEL and BRÜCKNER 2011).

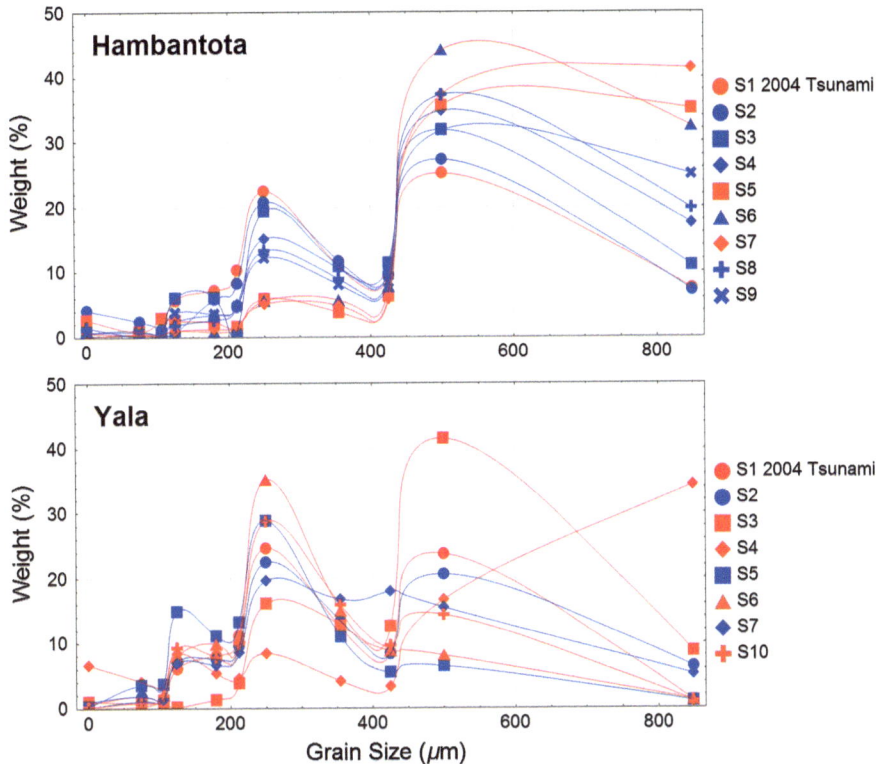

Figure 5
Grain size distributions of tsunami and palaeotsunami (*red*) and beach sediments (*blue*) at Location 1 (Hambantota) and Location 2 (Yala)
(color figure online)

3.1.2 Yala, Southeast Coast of Sri Lanka

Location two was on flat ground at the boundary of the Yala National Wild Life Park. The location was severely affected by the 2004 tsunami, which caused massive damage. In July 2008, a 1.5-m-deep trench, located about 250 m inland at an elevation of 0.3 m above mean level was dug by hand until it reached the lagoon clay bed (Fig. 6). Samples were taken for paleontological and sedimentological laboratory analysis and OSL dating.

The sediment sequence at the location shows well-marked sediment layers (Fig. 7) with distinctly different textural/structural properties and colour. The topmost layer of the sequence (S1) is the 2004 tsunami sand deposit, which is recognized from its erosive lower contact with the original peaty ground surface.

Three other light-coloured, coarse-grained sediment layers (S3/4, S6 and S10) have been identified as tsunamite (palaeotsunami) sediments and are clearly visually distinguishable from other coastal beach sediments on the profile, although the definitive identification is possible only in the laboratory. The grain size distribution patterns of tsunamite sediment do not differ greatly from other sediments, showing similar poly-modal distributions and increased coarse fraction. Although these tsunamite sediment layers show positive skewness, which is not typical of the non-tsunami sediment, these signatures although consistent are considered to be too weak to be a basis for robust distinguishing of tsunami deposits.

3.2. Micro-petrographic Studies

Petrographic studies were carried out on sediment samples from the two locations Hambantota and Yala, using plane-polarized and cross-polarized illumination. A few samples from Hambantota are shown at two different magnifications in Fig. 8a–e. The tsunamite sediments from layers S1 and S5 are

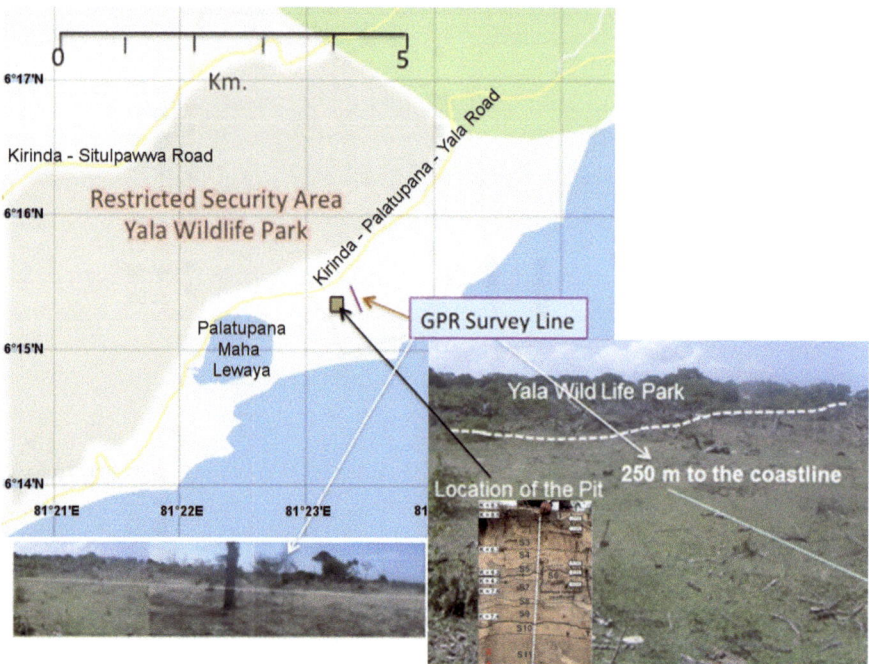

Figure 6
Map of the palaeotsunami studies carried out at Location 2 (Yala 6.25°N, 81.39°E). The position of the excavated pit and the GPR Survey Line are shown

poorly sorted comprising a wide range of grain sizes from fine to very coarse. They mainly comprise well-rounded to angular grains of quartz and feldspars with trace amounts of opaque and Fe–Mg Silicate minerals. Most of the quartz and feldspar grains exhibit micro-fracturing (Fig. 8a, e). The precise mechanism of this fracture process is yet to be established, here we speculate that it is the much more energetic processes involved in tsunami entrainment and deposition as compared to normal beach processes; the issue goes beyond the scope of the present work.

Non-tsunami sediment layers of S2, S3 and S4 from Hambantota comprise mainly fine-to-medium-sized grains, that are moderately to well sorted, angular to sub-rounded. Their mineral composition was mainly quartz and feldspar with a characteristic feature being that they are more opaque and with a larger component of Fe–Mg silicate minerals than are seen in the tsunamite sediments of S1 and S5.

3.2.1 Petrographic Studies at Hambantota

Petrographic studies were also carried out using plane-polarized and cross-polarized illumination. Some samples are shown at two different magnifications in Fig. 9a–e. Left-hand pictures in 9a, 9b and 9d are at 6.3× magnification while the right-hand pictures are at 2.5× magnification, under cross-polarized light. Figure 9c shows the fossil assemblages present.

3.2.2 Petrographic Studies at Yala

Samples were also collected at Yala. The Yala tsunamite sediments show similar micro-petrographic characteristics to those of location 1 (the Hambantota site). They exhibit poor sorting from fine-to-coarse sand, the presence of micro-fracturing, large quartz and feldspar grains, angular to well-rounded mixtures of grains and very low Fe–Mg silicate and opaque mineral content (Fig. 9a, c). The sections of layers S5 and S7, which are identified as non-tsunami sediments show relatively smaller grain size and better sorting together with a higher composition of Fe–Mg opaque minerals (Fig. 9b, d).

The normal beach sediments contain higher compositions of Fe–Mg minerals (Fig. 9b, d) because they are principally of continental

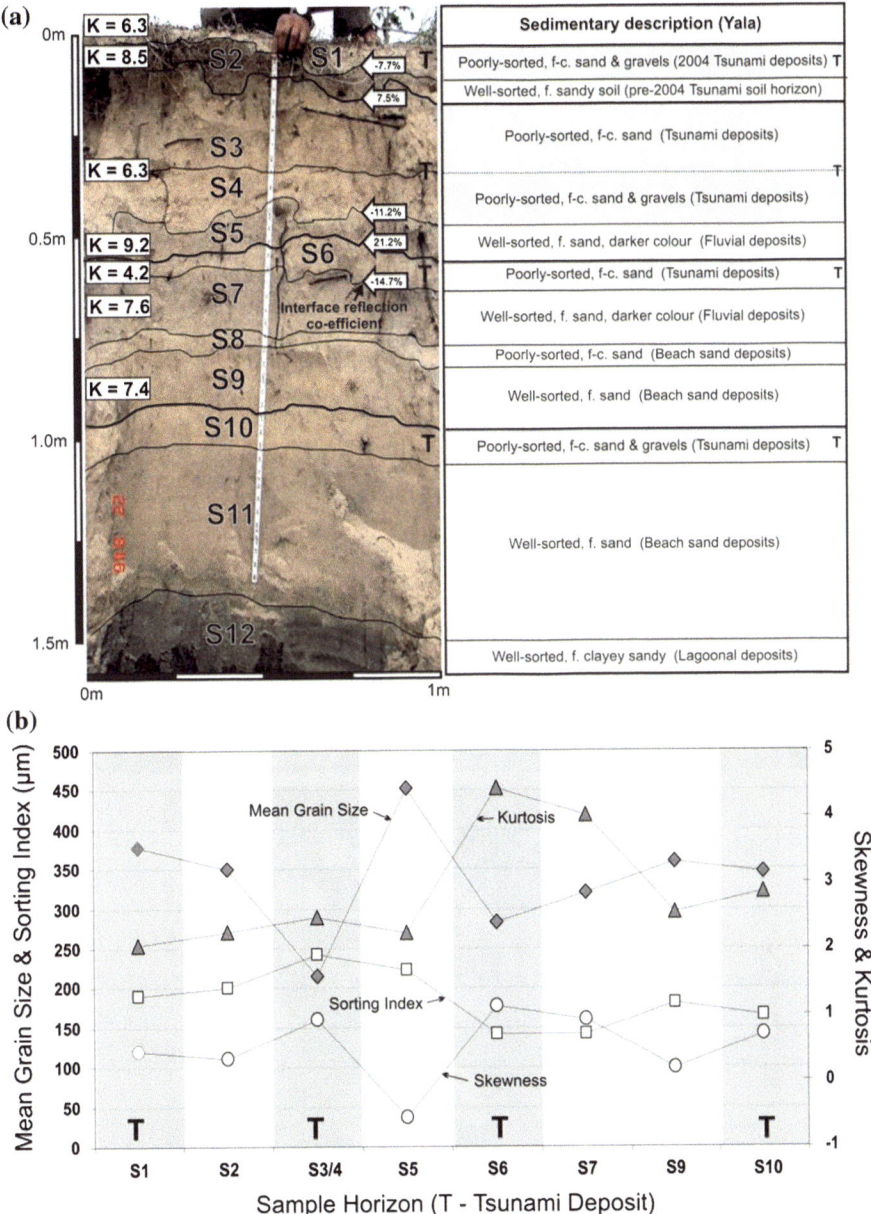

Figure 7
a Tsunami sand layers and non-tsunami deposits (beach sand, fluvial sands overlying the lagoonal clayey sand) at Location 2, Yala (layers are labelled as S1, S2 up to S12). Notation: *f* fine, *f–c* fine to coarse, *m* medium, *m–c* medium-to-coarse, *T* tsunami sediments, *K* electromagnetic impedance contrast. **b** Statistical characteristics of grain size distribution at Yala

provenance and derived locally; the tsunami by contrast mobilizes much more distal sediment which is impoverished in Fe–Mg minerals. This is an additional specific robust discriminating feature between Sri Lankan tsunamite sediments and normal beach deposits.

3.3. Micro-fossil Analysis of Palaeotsunami Sediments

Microfossils from both 2004 tsunami and older palaeotsunami sediments have been identified and studied in detail (DAHANAYAKE and KULASENA 2008b; PREMASIRI 2012). The 2004 tsunami sediment samples

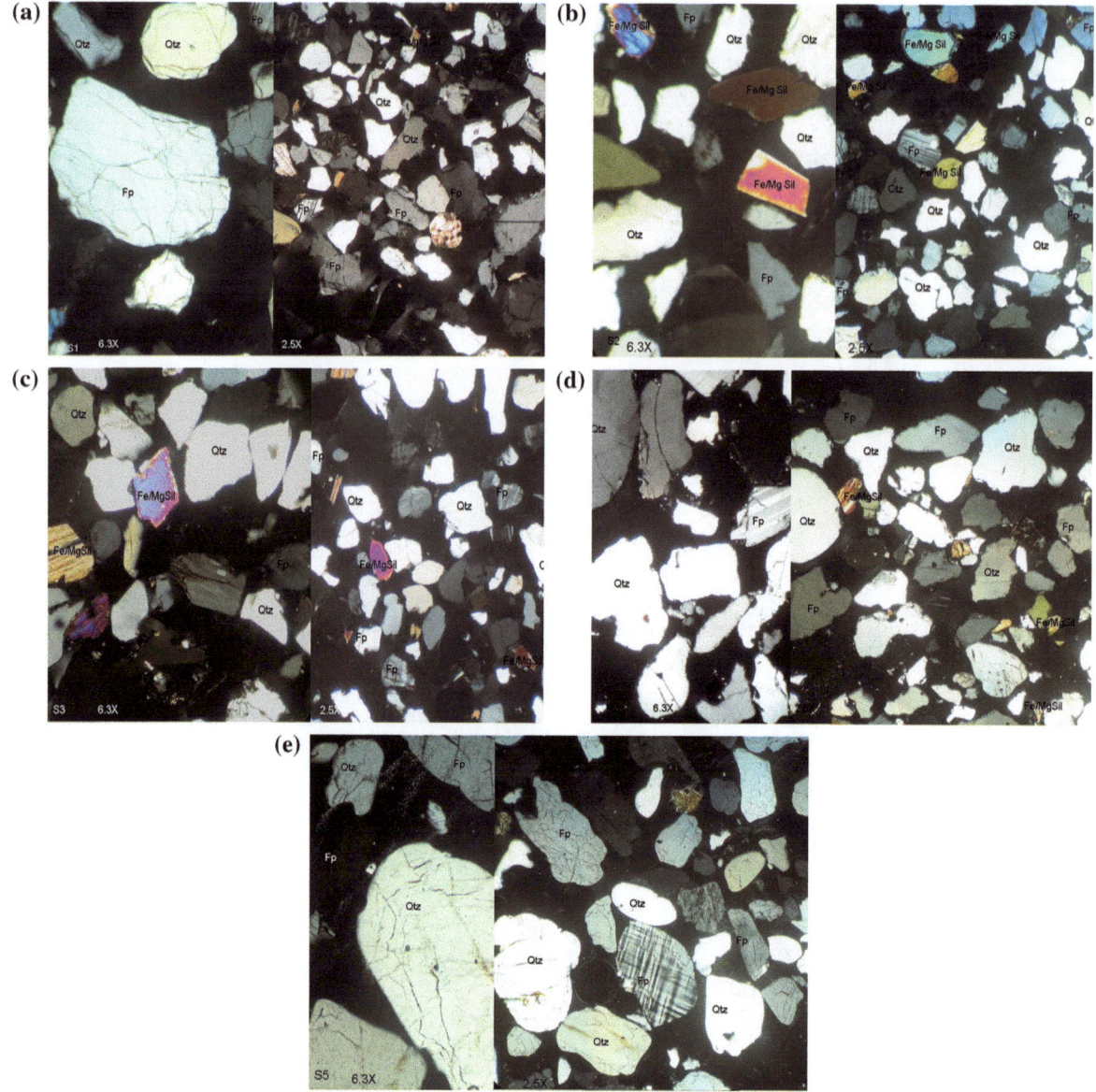

Figure 8
Micro-petrographic images of different sediment layers of coastal sediment sequences at Hambantota (Location 1). *Left-hand side panels* are at ×6.3 magnification and *right-hand side panels* are at ×2.5 magnification, under polarized light. *FeMgSi* Feldspar, *Qtz* Quartz. **a** Layer S1: 2004 Tsunami sediments. Note the strong fracturing present in the quartz grains. **b** Layer S2: non-tsunami sediments (Soil Horizon). The micro-fracturing is not displayed here. Quartz grains can be clearly seen in the *left-hand panel*. **c** Layer S3: non-tsunami; beach sediments. **d** Layer S4: non-tsunami sediment; fluvial. **e** Layer S5: palaeotsunami sediments. Note very strong fracture fabric in the quartz grains on the *left*

were studied at many locations (Fig. 11): Katukurunda (Kaluthara), west coast, Karaganlewaya (Hambantota), southeast coast, Palatupana (Yala), southeast coast and microfossils were analysed for palaeotsunami sediments from Hambantota (Pit at Location1) and Yala (Pit at location 2). One of the most distinctive features of the palaeotsunami sediment deposits we have studied on the Sri Lankan coast which is shared by the 2004 tsunami sediments is the presence of different species of microfossils in all the tsunamite layers (identified on the basis of totality of the features) and their absence in all other

Figure 9
Micro-petrographic images of different sediment layers of coastal sediment sequences at Yala (Location 2). *Left-hand side pictures* are at ×6.3 magnification and *right-hand side pictures* are at ×2.5 magnification, under polarized light. *FeMgSi* Feldspar, *Qtz* Quartz. **a** Layer S3: palaeotsunami sediments, no fractured Quartz. **b** Layer S5: non-tsunamigenic sediments, no fractured Quartz. **c** Layer S6: palaeotsunami sediments showing fossil assemblages in plane-polarized light. A number of foraminiferal species are present (Premasiri 2012). **d** Layer S7: non-tsunamigenic sediment, no fractured quartz

layers. Most of the microfossils have been identified as extinct benthic and large benthic species of (Dawson et al. 1996b) which belong to a wide age range. Taxa identified after both reflection microscopy and scanning electron microscope (SEM) studies by (Dahanayake and Kulasena 2008a, b) and by (Premasiri 2012) are described in detail in (Premasiri 2012).

3.4. Identification of Palaeotsunami Sediments

Although the palaeotsunami sediments are characterized by positive skewness and exhibit strong negative correlations between grain size and skewness ($r = -0.8$ to -0.9) and a less pronounced relationship between mean grain size and kurtosis ($r = -0.6$ to -0.9) (Table 1) we do not employ these features for discriminating tsunamites, relying instead upon more robust methods of identification.

Micropalaeontological analysis of the 2004 and palaeotsunami deposits shows that, in contrast to sediment layers of other origin, they contain diverse foraminifera. Since three different, totally independent tsunami diagnostic criteria (micro-petrography, microfossil and mineralogical analyses) are satisfied for five distinct layers in the sedimentological sequence with the 2004 Boxing Day sediments serving as a reference, we identify the lower four of them as palaeotsunamites.

4. Laboratory Analysis of the Dielectric Properties of Tsunami Sediments

Ground-penetrating radar uses electromagnetic waves in the frequency range from c 50 MHz up to

1 GHz, emitted from specialized tuned antennae placed directly on the ground surface. The electromagnetic waves of this range are modified by propagation through geological materials, being attenuated strongly by conductive layers including significant clay content and saline water but propagating well through quartz-rich sediments such as beach sands (Figs. 4, 7). GPR-based techniques have been used extensively for mapping the internal configuration of sandy deposits such as dunes (JOL 2008). Where there are sharp changes in electromagnetic (dielectric) properties (electrical resistivity and magnetic susceptibility) strong reflections take place with the reflected waves received on complementary tuned antenna enabling these interfaces to be traced for large distances (EZZY et al. 2003, 2011). The signals attenuate with depth due to energy loss and absorption but in favourable circumstances penetrations of some tens of metres and reflections from multiple layers and their geometries and spatial relationships can be clearly seen. The use of shallow GPR for other sedimentological purposes has been widely reported (e.g. MEYERS et al. 1996; TAMURA et al. 2008) and its use for environmental applications is described in detail in STYLES et al. (2011). GPR is an established, non-invasive, high-resolution geophysical technique with wide application for imaging shallow sedimentary sequences for sedimentological and stratigraphic studies and has been successfully utilized for mapping several differing types of Quaternary quartz-rich clastic deposits but only recently (KOSTER et al. 2013, 2014) in the tsunami context. Our laboratory measurements of samples taken from our sites show that tsunamite dielectric properties differ significantly from ambient soil and normal beach sediments, causing distinct reflections from the boundaries of the tsunamite layers, with reflection coefficients shown in Table 2 for the site at Yala. The dielectric properties of saturated sediments from the Yala site show that very high reflection coefficients of between 7 and 21 % exist at interfaces between the tsunami-deposited sedimentary layers and the beach/lagoonal layers. Measurements were also made on a subset of dry samples where the reflection coefficients are all about 1 %. On this basis we suggest that the water content residing within the complex internal pore geometry of the immature, poorly sorted sediment of the tsunami deposits plays a critical role in the differential increase of the dielectric constant of tsunamites, which results in the generation of the observed continuous and very distinct reflections from the tsunami deposit boundaries. Pinpointing the specific properties of tsunamite sand grains which lead to a higher porosity of the tsunamites and hence result in such strong contrasts in water content and dielectric properties remains an outstanding fundamental problem outside the scope of this work. Here, we just note that since normal beach sand grains are more rounded than the sand grains deposited by tsunami, we can hypothesize that more rounded grains allow tighter packing and hence less voids providing space for water.

This dielectric study work was carried out at Keele University, and not all sediment samples were available for analysis but the dielectric properties (relative permittivity) of the identified tsunami and non-tsunami sediments from the sediment profile at Yala, Sri Lanka (S1, S3/4, and S6 are tsunami

Table 2

Dielectric properties of saturated sediments from the Yala site

Sample (Yala site)	S1	S2	S3/S4	S5	S6	S7	S9
Dielectric constant K	6.27	8.53	6.32	9.92	4.19	7.58	7.43
K lower error limit	5.23	7.30	5.03	8.71	3.72	6.17	6.78
K upper error limit	7.32	9.77	7.61	11.13	4.66	8.99	8.07
Loss tangent	0.11	0.11	0.11	0.09	0.11	0.12	0.13
Attenuation (dB/m)	15.46	20.11	15.89	18.45	12.41	18.76	21.64
Velocity (m/ns)	0.120	0.103	0.119	0.095	0.146	0.109	0.110
Reflection coefficient at basal interface (%)	−7.7	7.5	−11.2	21.2	−14.7		

Measurements were also made on a subset of dry sample and the reflection coefficients are all less than 0.011 showing that the water content within the internal geometry of the samples is critical sediment layers) for both 'dry' and 'wet' conditions have been measured using the standard method following the procedure described in JOL and BRISTOW (2003). In situ, the sediments are almost always wet and so the samples were rewetted with distilled water thus restoring their ambient saturation levels (~25 % by volume moisture content). The samples were then packed into the measurement cell to equal density/volume ratios and the relative permittivity (dielectric constant, K) measured at the central GPR frequency of 500 MHz. Results from these wet sediments are shown in Table 2. The frequency dependence of dielectric properties of sand in this range is known to be small (MÄTZLER and MURK 2010) and so these measurements are also representative for the GPR antennae employing frequencies in the range 450–900 MHz.

We are primarily interested in the reflection coefficients (R) at a sedimentary layer boundary

$$R = (\sqrt{K2} - \sqrt{K1})/(\sqrt{K2} + \sqrt{K1}),$$

where $K1$ and $K2$ are the dielectric constants in the adjacent layers. The coefficients characterize the normalized amplitudes of reflections between the adjacent layers in the sequence. The table demonstrates a noticeable change in dielectric constant between the tsunamites and the adjacent non-tsunamite layers, which, therefore, leads to strong negative reflection coefficients at the bases of the tsunami layers. The subtle differences in the primary physical characteristics of sediment grains (prevalence of angular shaped grains, grain size, sorting, etc.) affect the microscopic volumetric moisture content variation across the interfaces that, in turn, lead to significant macroscopically averaged dielectric property contrasts.

Where sedimentary layer interfaces are sharp, laterally coherent and have metre-scale planar geometries, observable GPR reflections can be obtained from relatively low values of reflection coefficient (<3 %) and here we have reflection coefficients, which far exceed this. Such characteristics are common in both aeolian and fluvial sedimentary environments where GPR-based studies have been highly successful (BRISTOW 2009).

The distinct dielectric signatures of the tsunamite sediments provide the rationale for the use of ground-penetrating radar in the field. Indeed, at the two locations described above the existence of strong reflected signals from each tsunami sediment layer has been confirmed and allowed us to discriminate the very clear layered pattern of tsunamite sediments even in cases exhibiting only subtle differences in the primary physical characteristics.

5. Subsurface Mapping of Palaeotsunami Sediments Using the GPR Technique

To reconstruct the 3D distribution of the tsunamite deposits a non-invasive remote sensing technique is needed. Since our trench excavations indicated shallow, thin layering and thanks to our discovery of a dielectric constant anomaly for the tsunamites, GPR is an ideal technique for tracing the subsurface configuration of the tsunami deposits on the Sri Lankan coast. In this study, two pilot GPR surveys aimed primarily at testing the feasibility of the approach were carried out at two previously described sites (Fig. 2), Yala, Hambantota. GPR traverses normal to the coast at 2.5–5 cm intervals were collected with a PulseEkko PE1000 system using 450 and 900 MHz antennae, and processed with standard methods (JOL and BRISTOW 2003).

More specifically, the GPR profiles were acquired with a Pulse Ekko GPR system with 450 MHz antennae as a sum of short (12.43 m) segments with spacing between acquisition points of 25 cm and with bistatic, parallel antenna geometry with an antenna spacing of 20 cm. The data were sampled at 0.1 ns giving 632 time samples per trace.

The data were processed using the Open Source MATGPR processing package in Matlab: (http://users.uoa.gr/~atzanis/matgpr/matgpr.html) and the processing sequence is described in detail in "Appendix".

Figures 10, 11, 12, and 13, show the raw (unprocessed) and processed GPR data. The detailed interpretation has been overlain on the Yala profile (Fig. 10) but the interpretation of the others is shown in Figs. 14 and 15.

While some features can be seen in the shallowest portions of the raw data they are obscured by pulse shape as the signal has not been adjusted to remove

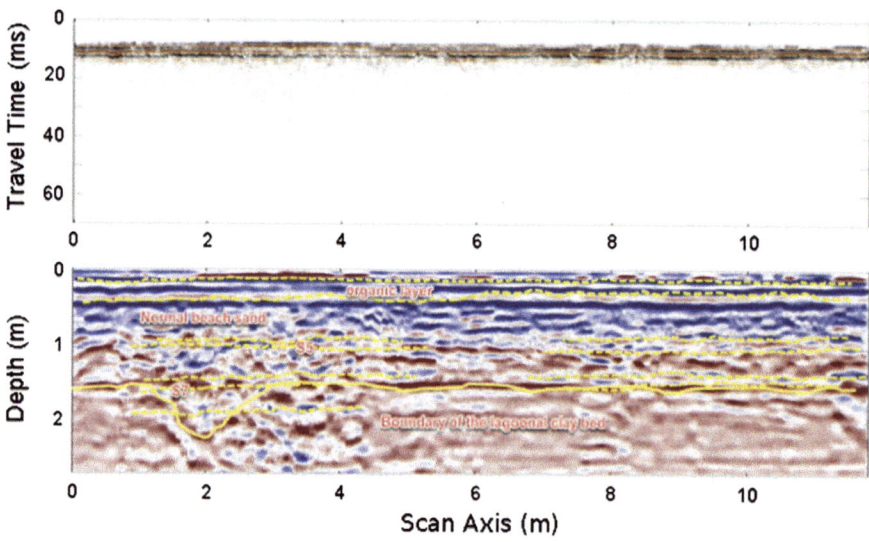

Figure 10
Yala. A 19m segment of the GPR survey line shown on Fig.6. *Upper panel* raw GPR data. *Lower panel* reprocessed and depth converted data. The interpretation of this profile has been overlaid. Coastline is to the *left*

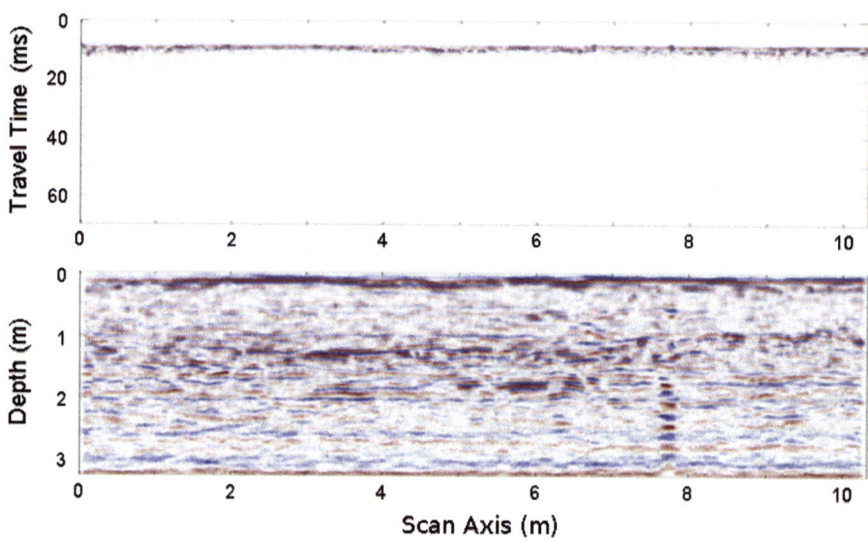

Figure 11
Line 1 Hambantota. *Upper panel* raw GPR data. *Lower panel* reprocessed and depth converted data. Coastline is to the *right*

this by deconvolutions or for geometric spreading ("Appendix") and, therefore, we cannot resolve the deeper layers until after processing when we can clearly make out strong continuous reflections extending across the whole section and down to depths of a few metres. Although there is strong reflection from the water table since it is not a perfect reflector some penetration does occur, which enables us to see a layered structure beneath. The improvement in the clarity and interpretability of these processed sections can be clearly seen. Separate bespoke processing of the very shallowest layers may be able to reveal more detail about the depositional characteristics of the 2004 tsunami and further acquisition is planned to explore this. While dune forests were recognized at other sites (Dikwela) they

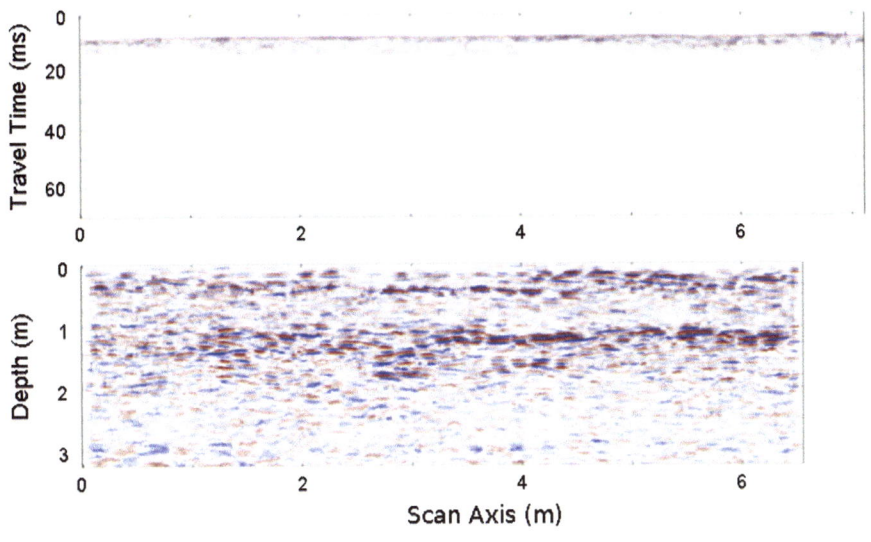

Figure 12
Line 2 Hambantota (as in Fig. 11)

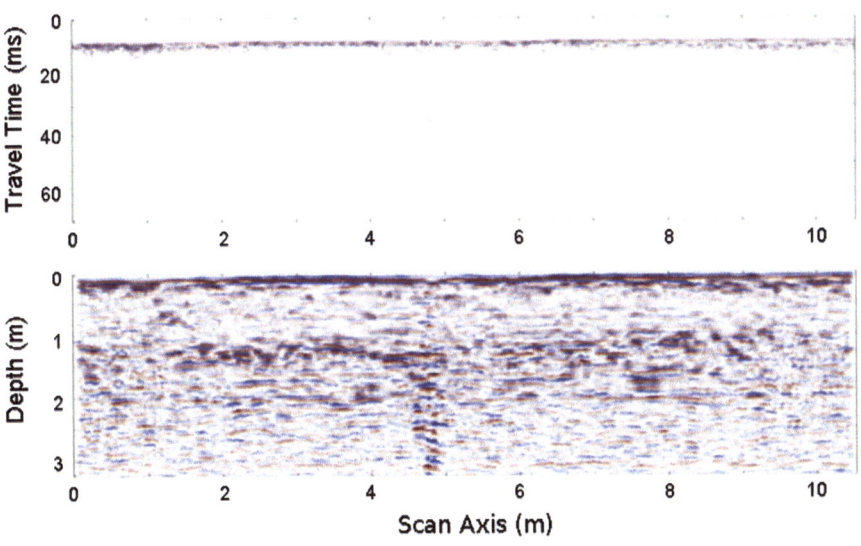

Figure 13
Line 3 Hambantota (as in Fig. 11)

were not present on the GPR profiles here which is indicative of differing flow regimes.

Figure 14 shows an interpretation of the Yala traverse (Fig. 10), where we have identified tsunamite sediment layers based on trenches/pits and geological studies and it was possible to unequivocally trace tsunamite sediment layers with GPR. The validation of the reconstruction was provided by the ground truth data. We also collected samples for OSL dating from these sites. From the GPR survey at Hambantota, two tsunami layers could be traced from all three GPR sections, while at Yala three sediment layers from palaeotsunami events could be traced up to 250 m inland (Fig. 16). The surveys could not be continued further inland because of the security restrictions due to civil war at the time of the field study. Two additional GPR surveys were carried out at Dikwela and Katukuranda (see map on Fig. 2), where there were no indications for tsunamite sediment from trenches or geological field studies to

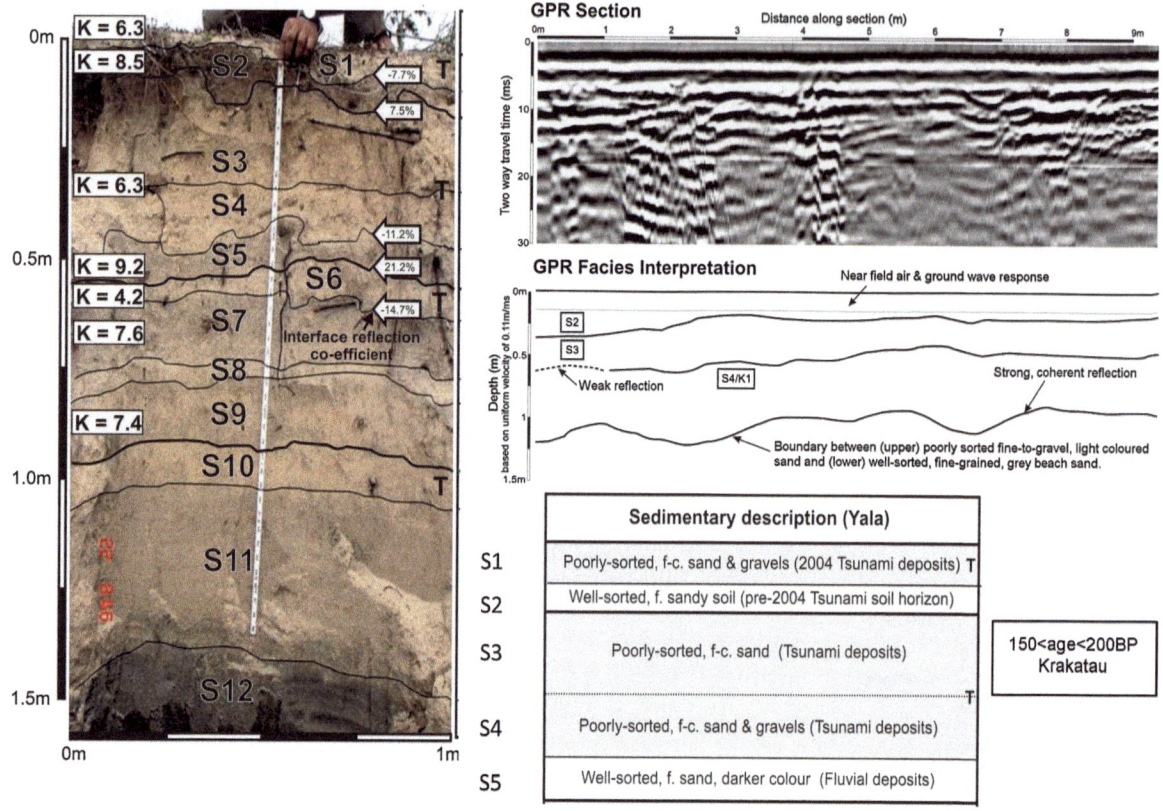

Figure 14
Field identification of palaeotsunamites and correlation with the GPR profile and its interpretation at Yala. K is the electromagnetic impedance which is the contrast between layer boundaries; it determines the reflection coefficient and the layering observed on the GPR section

correlate with the GPR sections. GPR profiles from these sites do not show any positive indications of palaeotsunamite sediments as were recognized at sites Yala and Hambantota (PREMASIRI 2012). The GPR section at Dikwella gave no indication of the presence of coarse-grained sand deposits interlayered with the marshy clay beds and GPR interpretation of site at Katukurunda showed no evidence for the occurrence of palaeotsunami sediment layers in the beach sand profile. They present as normal beach sand deposits exhibiting cross bedding; ripple marks and other poorly differentiated sand layers with textural and mineralogical variations.

These extra GPR surveys (PREMASIRI 2012) illustrate the point that palaeotsunami sediments are not preserved everywhere in the areas inundated by the 2004 tsunami as sometimes the environment is erosive rather than depositional, and obviously we should expect that tsunami deposits might be present in some areas and not in others. However, where the tsunamite deposits do occur in the coastal deposits they manifest as very distinctive geological layers compared with the normal beach deposits and the GPR technique can be invaluable for mapping them.

The results of the laboratory electromagnetic measurements show a clear contrast between the dielectric properties of the tsunamite sediments and normal beach sands and lagoonal sediments (Table 2), which is the reason for the clearly defined reflection signatures observed on the GPR profiles. Inspection of the reflection coefficients shows that for all of the tsunami deposits examined here there is a strong negative reflection coefficient, which defines the tsunami event. The lower values of dielectric constant in the tsunami deposits are consistent with a decrease in pore space and, therefore, water content associated with the less rounded sand grains observed in these sediments.

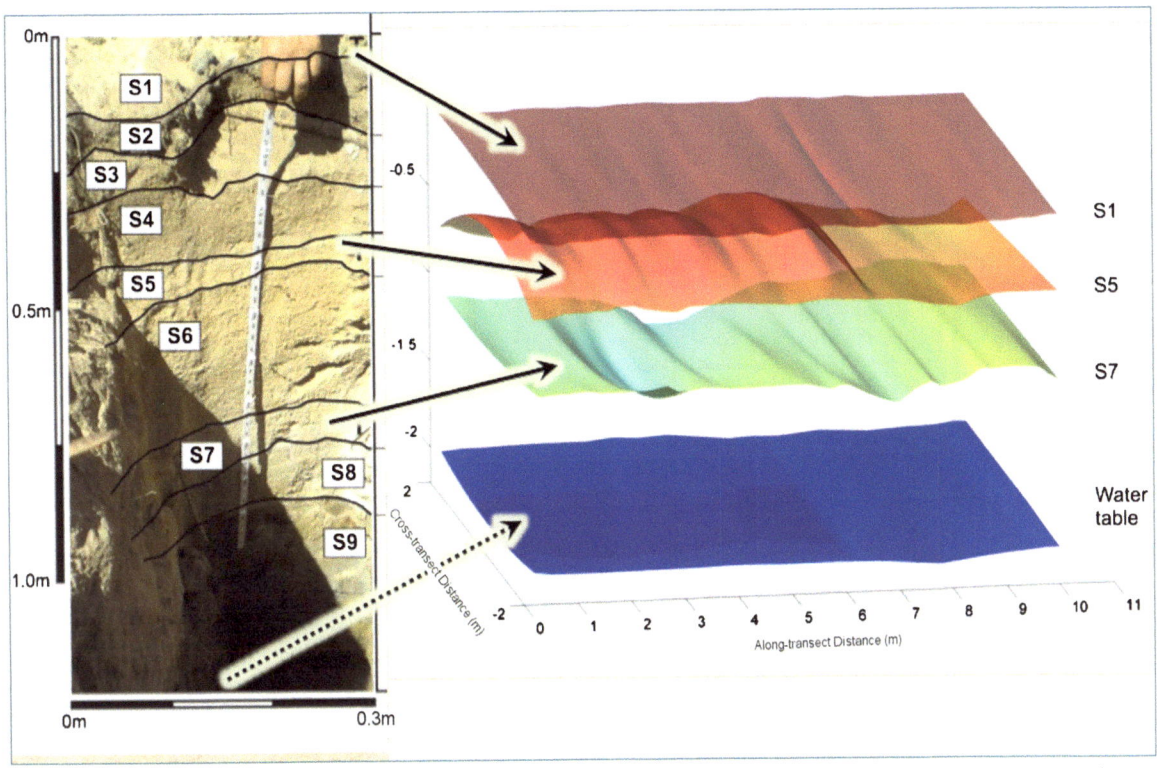

Figure 15
Three-dimensional view of the lower surfaces corresponding to tsunami and palaeotsunamite layers identified by GPR (S1, S5 and S7) in this study at Hambantota showing that even with these limited data sets the three-dimensional topography of the interfaces can be retrieved. The vertical variations are exaggerated

The dating of the sediments discussed below allows us to conclude that the dielectric contrasts remain sharp over several 1000 years (at the very least!) which allows this technique to be used for Palaeotsunami sediment mapping.

At Hambantota we acquired three parallel closely spaced radar sections (lines 6, 9 and 17 in Figs. 11, 12, 13) and we were able to trace at least three palaeotsunami sediment layers identified visually in the excavated trenches. These were confirmed to be tsunamites by laboratory analysis and were traced from 150 to 400 m from the coastline. Three-dimensional projections of the tsunami depositional surfaces derived from these sections are shown in Fig. 15. Tsunami deposits with varying thicknesses and continuity continue further, but restricted security access at that time (although access may be available in the future) precluded tracing them beyond 400 m from the coast. The 2004, benchmark tsunamites are visually observed to extend at least 600 m inland at this location (Hambantota). We are able to produce the internal configuration of the palaeotsunami depositional surfaces. Thus, we identify four major palaeotsunami deposits, all almost certainly associated with major geological catastrophes. Similar results were obtained at the Yala location (Fig. 2), but could not be traced as far inland due to marshland and security restrictions.

Geological and sedimentological studies reveal that tsunamite deposits are confined to certain specific locations on the coastal zone and they have varying thicknesses within continuous or discontinuous sheets or layered deposits. These deposits are preserved within coastal sediment sequences or in low-lying marshy clay beds. The dielectric constants of these sediments were found to differ sufficiently from the surrounding normal beach and lagoonal sediments to provide significantly large electromagnetic reflections at the layer boundaries to enable them to be traced using GPR. The cross-shore extent

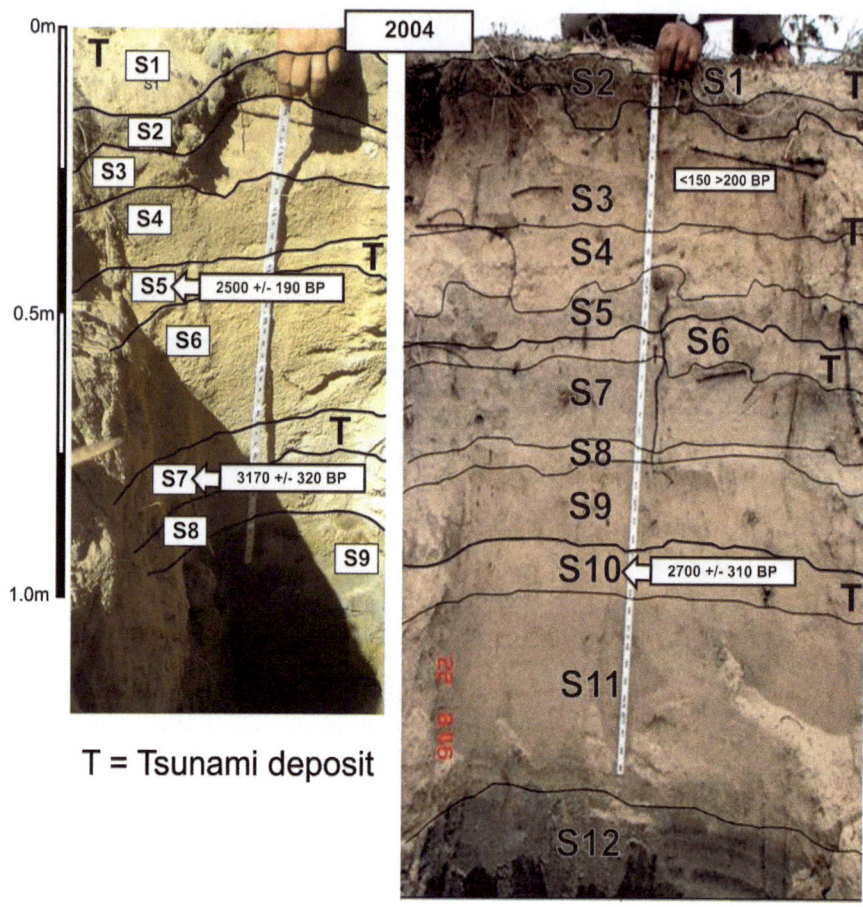

Figure 16
Optically stimulated luminescence dating of palaeotsunami sediment layers at Hambantota and Yala

of the tsunamite deposit supported by mapping of its lateral dependence is a key input parameter for determining the tsunami amplitude which deposited these palaeotsunamites. Recent successful applications of GPR for mapping Quaternary deposits and several types of coastal sand deposits are described in (BOTHA et al. 2003; HAVHOLM et al. 2003; HEINZ and AIGNER 2003; JOL and BRISTOW (2003); WOODWARD et al. 2003). However, KOSTER et al. (2013, 2014) and now this work is the first applications of GPR for mapping tsunamite deposits and this is certainly the first application to Sri Lanka which is not itself tsunamigenic and so experiences only mega-tsunamis generated at great distances.

6. OSL Dating of Palaeotsunami Sediments on the Sri Lankan Coast

Luminescence dating is based on the time-dependent dosimetric properties of silicate minerals, predominately feldspar and quartz (BOS and WALLINGA 2009). The technique has been used to date sediments, usually <200,000 years old, that received sunlight exposure prior to deposition (HASHIMOTO et al. 1986). Exposing sediment to sunlight for hours or heating to above 300 °C eliminates most of the previously acquired luminescence from mineral grains. After the sediment is buried and shielded from further light exposure ionizing radiation from the decay of naturally occurring radioisotopes of U, Th, and K produces free electrons which are subsequently trapped in crystallographic charge defects in silicate minerals. Excitation of minerals by heat or light causes recombination of stored charge, which results in luminescence emissions. The intensity of the luminescence is calibrated in the laboratory to yield an equivalent dose [De, measured in grays (Gy); 100 rads = 1 gray], which is divided by an estimate of the radioactivity that the sample received during burial (dose rate, Dr) to give a luminescence age (WALLINGA 2002). When sediment is buried, the effects of the incoming solar radiation are removed. With this bleaching effect removed, a signal accumulates within individual mineral grains (most commonly quartz and feldspars). It is this signal that is the key to luminescence dating techniques. Given an estimate of the rate of received ionizing radiation and knowing the total accumulated dose (the palaeodose) it is possible to derive an age since burial (MURRAY and WINTLE 2000).

Samples were collected from Yala and Hambantota where the detailed GPR studies were carried out, from each layer identified as a tsunami deposit. To collect samples for OSL dating without exposing them to sunlight, opaque polyvinyl tubes (12 cm length and 5 cm diameter) were used as sample containers and the sample was collected directly into the tube while inserting the tube horizontally into the geological profile and both ends of the tube were shielded immediately. The samples were analysed at the Oxford University Luminescence Dating Laboratory using single-aliquot procedures (MURRAY and WINTLE 2000). Three identified palaeotsunami events have been OSL dated at 150 to 200 BP, 2550 ± 190 BP, 2710 ± 310 BP (probably the same event within the error bounds) and 3170 ± 320 BP years by combining the data from the two profiles (Fig. 16).

In addition to these three OSL-dated palaeotsunami sediments layers, we did identify a fourth layer as a palaeotsunami but samples from the oldest fourth layer were accidentally exposed to light and could not be OSL dated; we plan to resample this layer. Meanwhile, we crudely estimate its date at ~4000 BP by extrapolating the estimated average rate of non-tsunami sedimentation; since three older tsunami deposits with ages of 4200, 4500, and 5000 B.P. have been found on the same stretch of the coast in Kirinda (6.2277°N, 81.3349°E) and Okanda (6.6554°N, 81.7695°E) Lagoons (RANASINGHAGE 2010) this is a credible postulate.

7. Palaeotsunami Records from the Indian Ocean and Region and Their Record on the Sri Lankan Coast

In this study, we confidently identify and can trace for some considerable distance palaeotsunamites with ages: 150–200 BP, 2550 ± 190 BP, 2710 ± 310 BP (probably the same event within the error bounds) and 3170 ± 320 BP. Previous studies (based on carbon dating) on the coast of Thailand and Sumatra have shown a recurrence interval of mega-tsunamis in the Indian Ocean of nearly 600 years. They report four palaeotsunami events at 100–130 BP, 540–600 BP, 966–1170 BP, 2200 BP and at 3050 BP (BONDEVIK 2008; JANKAEW et al. 2008; MONECKE et al. 2008; BRILL et al. 2011, 2012). The first palaeotsunami event in this sequence is the 1883 Krakatau eruption tsunami. This tsunami as well as the 2004 Sumatra one originate from the Sumatra region, which is considered to be the most significant potential tsunami-generating source in the Indian Ocean. (Table 3).

Three older tsunami deposits with ages of 4200, 4500, and 5000 B.P. have been found in Kirinda (midway between Hambantota and Yala) and Okanda Lagoons (inland from Yala, 6.6554°N, 81.7695°E), (RANASINGHAGE 2010) and ABEYRATNE et al. (2007) dated with radiocarbon a sand layer at 4829 ± 362 B.P. in Kirinda Lagoon. A further deposit from Panama Lagoon (6°45′–6°46′N, 81°48–81°49′E) has a reported age of 6817 ± 132 B.P. (RANASINGHAGE 2010).

In a very recent paper, Jackson et al. (2014) report the results of coring in the coastal lagoons from Karagan Lagoon in south-eastern Sri Lanka that provide evidence for palaeotsunami deposits at 2417 ± 152 B.P. to 2925 ± 98 B.P., which we report here. PREMASIRI (2012), traced the same deposits with GPR for a significant distance although as we have no OSL dates we do not discuss it further here. Jackson et al. (2014) also identified six older tsunamis

Table 3

Palaeotsunami records for the Sri Lankan coast and the Indian Ocean from the present study and the literature

Event	Date	References	This study	Historical records and legend
1	2004	Sumatra Tsunami	2004 (Sumatra)	2004 Sumatra Mega Tsunami
2	1883	Newspapers, colonial archives Choi et al. (2003), Pelinovsky et al. (2005)	150–200 BP	Krakatau eruption
3	100–130 BP	Brill et al. (2012)	Not recognized	1881 Rupture of the Sunda Arc
4	380 + 130 BP 540–600 BP 500–700 BP	Jankaew et al. (2008) Prendergast et al. (2012) Brill et al. (2011)	Not recognized	
5	990–1400 BP 966–1170 BP	Prendergast et al. (2012) Monecke et al. (2008)	Not recognized	
6	2200 BP 2100 ± 260	Monecke et al. (2008) Prendergast et al. (2012)	2500 ± 190 BP	Viharamahadevi Event (Sri Lanka)
7	3050 BP	Brill et al. (2012) identified an event in Phra Thongs as marine sand but it may be a palaeotsunamite)	3170 ± 320 BP	Mentioned in the Ramayana (c3000 BP) Thailand

between 4064 ± 128 B.P. and 6665 ± 110 B.P. They do not seem to have identified the event at 3179 ± 320 BP which we can trace at Hambantota, as we have noted above, not all events are preserved everywhere depending on coastal orientation and littoral topography. This event appears to have historical corroboration as we describe below.

7.1. Krakatau Event

The 1883 tsunami caused by the Krakatau eruption was reported by numerous eyewitnesses and was widely described in newspaper articles of the day (e.g. Sunday Observer of 27th August 1883). However, there are no quantitative data and, in particular, the run-ups were not recorded, eyewitnesses reported heights of c. 1 m (Choi et al. 2003). All available information was summarized in (Choi et al. 2003, Pelinovsky et al. 2005) where the Krakatau tsunami was modelled retrospectively utilizing all the data worldwide. The model yields the maximum wave amplitude at Sri Lanka coast of c.3.6 m, the calculated run-up height at Colombo was 0.5 m.

7.2. Princess Viharamahadevi's Tsunami

There are no contemporary historical records confirming the precise date corresponding to the tsunami event, which deposited the layer S5 for which the OSL dating yields 2500 ± 190 BP. However, the event left a lasting historical impression, which has survived as a legend, first as an oral tradition and then in written form, to the present day. The legend links the event to King Kelanitissa who ruled over a part of Ceylon from the second century BC and his daughter, Princess Viharamahadevi. The king sacrificed Viharamahadevi by sailing her out to sea in a golden vessel to protect the populace from giant sea waves, which inundated most of the coast. This story has been reported in classical historic Sri Lankan chronicles, Mahawamsa and Rajawaliya (Suraweera 2000; Sumangala 1996; Codrington, 1994; Hardy 1866).

We provide long quotes from the books to identify key actors, some of which are known contemporary historical figures and to give an idea of the scale of devastation brought by that tsunami.

The queen of this king (king of Kelaniya) Tissa had carried on an intrigue with her brother-in-law, who on being detected fled and corresponded with her by a messenger disguised as a priest. The man attached himself to the attendants of the chief priest who was visiting the palace, and catching the eye of the queen dropped his master's letter. Unfortunately the palm-leaf missive made a noise in falling; the correspondence was detected, and the king in his fury slew not only the messenger but also the chief priest, whose complicity he suspected. Thereupon the sea, which according to the Rājāvaliya was then about seven gaus (some fifteen miles) from Kelaniya, overwhelmed the land, submerging many towns and villages. To put an end to this the king placed his daughter Dēvī in a golden vessel and launched it into the sea: she was carried southwards and cast ashore near the temple (vihara), when she became the queen consort of Kākavanna Tissa under the name of Vihāra Dēvī. Their sons were Gāmani Abhaya, the future hero, and Tissa.

This quote from (Hardy 1866) identifies the key actors, which at least partially are known historical figures and suggests the inundation distance at 15 miles at Kelaniya, whose location has not been definitely established. The second quote from (Codrington 1994) adds some colourful, but difficult to interpret, details on the scale of devastation:

...to punish the king for this act of impiety towards an innocent priest, the déwas who protect Ceylon caused the sea to encroach on the land, and much damage was done to the country. To appease their wrath, as it was supposed that the country could be saved in no other way, the king resolved to sacrifice his virgin daughter. Placing her in a golden vessel, on which was inscribed the word "rájathítáti", which signified that she was a royal maiden, the vessel was committed to the waves of the sea. But the flood still raged, until 100,000 towns, 970 fisher villages, and 470 villages inhabited by divers for pearls, had been submerged. As the king, from the back of his elephant, was watching the progress of the devastation, the earth opened, and flame burst forth from beneath, and he was no more seen by his people. By this time twenty miles of the coast, extending inland, had been washed away, and the distance from Kalyáni to the sea was reduced to four miles. The royal virgin drifted towards the dominions of the king of Mágam, when some fishermen, who brought it to land, saw the vessel. The monarch Káwantissa, having heard of the wonderful capture made by the fishermen, went to examine it; and when he had read the inscription upon it, he released the princess from her confinement, and she became his queen.

King Dtugamunu, the son of the princess is a prominent historic figure; the period of his reign is considered to be well established. According to SURAWEERA (2000) and (SUMANGALA 1996), he ruled during the interval 161-137 BC, which is close to the OSL dating of 2550 ± 190 BP. The accuracy of the OSL dating could be improved by analysing more samples and by calibrating the results using the known date of the 1883 Krakatau event; however, this goes beyond the scope of the present work.

It should also be mentioned that in November 325 BC (c. 2330 BP), the fleet of Alexander the Great was seriously damaged by a tsunami-like event in the mouth of the Indus (ARRIAN et al. 2004). This tsunami was observed about 1500 km to the North-West of Sri Lanka but this date lies within the error bars of our OSL dating of the layer at 2550 ± 190 BP. While this correlation is tentative at best it is interesting to speculate that the two might be related. However, it is also possible that this was more likely to be of local origin rather than basin-scale, as it may have been caused by an earthquake located within the Makran Subduction Zone, which is a region that has produced numerous earthquakes in recent times (e.g. BYRNE et al. 1992). However, clarification would require significant extra work and identification and dating of tsunamite sediments from sites around the Gulf of Aden, East African and Indian coasts.

7.3. Earlier Events

Layer 7 at Hambantota, which has been identified as a tsunamite, has an OSL age of 3170 ± 320 BP and there is at least a mentioning in historical records of a tsunami which may have been around that time. WATTEGAMA (2005) describes seven significant tsunamis to hit Sri Lanka, the second has already been correlated with the event at c. 2500 BP but he refers to an earlier event, which he calls 'Tsunami One' and is recorded in the epic poem the Ramayana.

The Ramayana is one of the two great Indian epics, the other being the Mahabharata, and describes life in India around 1000 BCE, i.e. 3000 BP. There are many translations/interpretations and their provenance is of variable reliability but all of these share a common belief that there was a great tsunami which devastated the region and so this must have occurred c. 3000 BP.

Sita is projected as the daughter of the Earth. Symbolically, Sita is a small landform or island which emerges out of a big Earthquake or

Volcanic Eruption in the sea. Rama is the symbolic representation of the people from India who invade or conquer the new land form, Sita, with bow and arrows. Lanka is an island and Raven represents the people who settle there. Raven conquers Sita by making the setters run away. Hanuman is the son of the God of Wind or Storm and he is capable of burning Lanka and putting a mountain in the sea so that India and Lanka can be connected. It means out of a Volcanic Eruption, the city of Lanka is burned out and new mountain in the sea emerges connecting India and Lanka. Crossing the sea, through the newly emerged land, Rama conquers Lanka. By a Tsunami, Sita disappears and Rama immerses in the water. Since the sons of Rama and Sita were on the mainland, they escaped. (http://drrajumathew.wordpress.com/2011/03/22/ramayana-the-story-of-a-great-tsunami-in-india-told-symbolically/).

Earlier tsunami events in Sri Lanka seem to have left no traces in collective memory, although at the times of the tsunami which deposited the fourth postulated tsunamite in the sediment sequence at Hambantota which on stratigraphic arguments may possibly be older than 4000 ±/−500 BP, the island had been populated and there was a developed civilisation. There are clear layered sequences on the GPR records which lie beneath those which we have identified and dated in our pits and JACKSON et al. (2014) have demonstrated that at least 6 further tsunamis lie in this age range. We believe that these can also be identified and traced to determine inundation distance and hence through modelling estimate tsunami run-up height and potentially by comparison between different localities the epicentre and magnitude of the causative seismic event.

Thus, the dates of two of the three recent palaeotsunami events, estimated as between 150 and 200 years BP, 2550 ± 190 and 3170 ± 320 years BP, are corroborated by historical records. The youngest layer clearly corresponds to the recorded 1883 Krakatau event, while the second youngest might correspond either to the major tsunami reported in Sri Lankan chronicles during the reign of King Kawantissa at about 2200BP or a tsunami encountered by Alexander the Great's fleet in 325 BC on the shores of the North Arabian Sea (WIJETUNGA 2008). Carbon dating studies on the coasts of Thailand and Sumatra (JANKAEW et al. 2008; MONECKE et al. 2008) report three events: 1300–1450 AD, 780–900 AD and 2200 BP, suggesting a recurrence interval of circa 600 years for this seismically active part of the Indian Ocean, The Sri Lankan sites, far from any seismically active zones, reflect only the basin-scale events with a correspondingly longer recurrence interval (poorly constrained) at circa 850 years. Our potential circa 4000 BP event appears to be confirmed by the results of JACKSON et al. (2014).

Reconstruction of palaeotsunami magnitudes based upon estimates of inundation is sensitive to sea-level variations. The only recorded Holocene sea-level rises which are reported as gradual around Sri Lanka of about one metre each are well established from coral studies and radiocarbon dating (KATUPOTHA and FUJIWARA 1988; WOODROFFE and HORTON 2005) at 5500 ± 500 BP and 2750 ± 400 BP. Therefore, interpretation of the two youngest of the three recent palaeotsunami events examined here, the Krakatau and Princess Viharamahadevi tsunamis, can rely upon the level and position of the present coastline. For the two older events one metre correction should be factored in.

8. Conclusions

We have demonstrated that ground-penetrating radar is an efficient non-destructive high-resolution technique for tracing palaeotsunami deposits and mapping inundation on the shores of Sri Lanka. The discovered significant contrasts in dielectric properties between deposits and surrounding soil even with only subtle differences in their primary physical characteristics are expected to be generic and a potential physical mechanism linked to the higher water content has been identified. The recent papers by KOSTER et al. (2013, 2014) from other regions (Spain, Greece, Arabian Sea) have exploited the robustness of the contrasts and applicability of GPR for a variety of geological and geomorphological contexts. The contrasts were found to remain sharp over thousands of years. Complemented by OSL dating, this allows evaluation of local tsunami hazard through an estimation of tsunami inundation and frequency distribution.

The specific implications for Indian Ocean hazard follow from the remoteness of Sri Lanka from seismically active areas; the four palaeotsunami deposits identified here must have left traces on most Indian Ocean shores aiding discrimination between 'local' and 'basin scale' events. Our estimate of the recurrence interval of 'basin-wide' events as circa 850 years is based on dating of just four palaeotsunami deposits with the oldest tentatively dated at c. 4000 BP. It is noticeably larger than the c. 600 years recurrence interval estimated for Sumatra and Thailand (BONDEVIK 2008; JANKAEW et al. 2008; MONECKE et al. 2008; BRILL et al. 2011, 2012). All the above estimates are weakly constrained and to improve them one needs going to older tsunamites. In our pilot project, GPR clearly shows layered structures well below the c. 4000 BP event; presumably these correspond to the older events identified in cores by JACKSON et al. (2014). We cannot confirm these, as the dug pits cannot extend as deep as sediment coring. However, radiocarbon dating by JACKSON et al. (2014) indicates that the recurrence interval may range from as short as 181–517 years to as long as 1045 ± 334 years, in this upper part of the sequence and our value is within this range.

While coring of sediments can clearly give a longer time range than we have currently obtained using GPR, they only give a snapshot at individual localities of the sediment distribution especially for palaeotsunamites and it is not practicable to excavate the whole area. The ability to correlate and trace these tsunamites using GPR permits a robust estimate of the most landward inundation and hence tsunami run-up.

Since the time horizon for OSL dating is much greater than the ages of palaeotsunami we have recorded here, the combination of the pits, coring and GPR suggests that further reliable age estimates could be obtained when security restrictions are lifted. Increasing the number of sampled locations will further improve the accuracy of dating, which could be also used for synchronizing of chronologies of ancient civilisations adjacent to the ocean shores.

Acknowledgments

The authors are grateful to the anonymous referees for the thorough reading of the manuscript and helpful comments.

Appendix: GPR Processing Sequence

The following processing sequence was applied to the Sri Lankan GPR data:

1. Zero adjust (static shift)—during a GPR survey, the first waveform to arrive at the receiver is the air wave., which is delayed due to propagation in the connecting cables. We, therefore, need associate zero-time with zero-depth, so that any time offset due to instrument recording is removed before interpretation of the radar image. This is done individually for each radar section but in this survey was typically about 7 ns.

2. Background removal—background noise is present due to ringing in the antennae, producing a coherent banding effect across the radar section. We sum all the amplitudes of simultaneous reflections along a profile and divide by the number of traces to give a composite signal, which is an average of all background noise which is then subtracted from the data.

3. Gain—gain compensates for amplitude variations in the GPR image as early signal arrival times have greater amplitude than later arrival times because of geometric spreading and intrinsic attenuation. Time-variable gain functions are to equalize the amplitudes of the recorded signals constant along the trace. We have used an inverse amplitude decay gain, which compensates for both geometric spreading and attenuation simultaneously in conjunction with automatic gain control, which balances the signal amplitudes and emphasizes more subtle features.

4. Frequency filtering—GPR data are collected with source and receiver antennae of specified dominant frequency, however, the recorded signals include a band of frequencies wider than this specific frequency. Frequency filtering removes undesirable higher and/or lower frequencies to produce a more interpretable GPR image. This is

called band-pass filtering and in this survey a pass band was selected interactively for each section but typically this covered a range of 0.1–3.0 GHz.

5. Deconvolution—the recorded radar signal is the interaction between the layered earth and its reflecting horizons which depend on electromagnetic contrast (which is what we require) and the radar source function, which is generated by reflection at each interface and adds to the final recorded signal. Deconvolution is the inverse filtering operation that attempts to remove the effects of the source wavelet to better interpret GPR profiles as images of the earth structure. There many different deconvolution methodologies but the one, which we have used, is known as sparse deconvolution (i.e. the Earth has mostly zero reflectivity with a few 'sparse' strong reflectors) with a Blackman–Harris (a sum of weighted shifted sinc functions) window.

6. Median filter—this enhances the continuity of reflector signals by replacing each value by the 5×5 median of the values around it.

7. Depth conversion—the radar sections are converted from time to depth using the velocities determined experimentally in the laboratory and the depth correlation established in the field.

8. Colour scale—finally, an appropriate colour scale and set of brightness values is chosen to produce a good interpretable radar section. This is a subjective process but can make significant improvements in clarity.

References

ABE, T., GOTO, K., SUGAWARA, D. (2012), *Relationship between the maximum extent of tsunamis and the inundation limit of the 2011 Tohoku-oki tsunami on the Sendai Plain, Japan*, Sedimentary Geology, 282, 142–150.

ABEYRATNE, M., JAYASINGHA, P., HEWAMANNE, R., MAHAWATTA, P., and PUSHPARANI, M.D.S. (2007), Thermo-luminescence dating of palaeo-tsunamis and/or large storm-laid sand deposits of small estuary in Kirinda, southern Sri Lanka. A preliminary study. In Proceedings of the 23rd Annual Session, Geological Society of Sri Lanka, p. 6.

ARRIAN, ANABASIS ALEXANDRI: Book VIII (Indica), (Kessinger Publishing Co, ISBN-10: 1419106783; ISBN-13: 978-1419106781, Whitefish, Montana, USA, 2004).

BLOTT, S.J. and PYE, K. (2001), *Gradisat: A grain size distribution and statistics package for the analysis of unconsolidated sediments*, Earth Surface Processes and Landforms, 26, 1237–1248.

BLUSZCZ A., OSL Dating in Archaeology, in Impact of the Environment on Human Migration in Eurasia (Eds. Scott E. M., Alekseev A. Y. and Zaitseva G), (Springer Netherlands 2006) pp. 137–149.

BONDEVIK, S., SVENDSEN, J.I., JOHNSEN, G., MANGERUD, J., and KALAND, P.E. (1997), The Storegga tsunami along the Norwegian coast, its age and runup, Boreas, 26, 29–53.

BONDEVIK, S. (2008), *Earth science: The sands of tsunami time*, Nature, 455, 1183–1184.

BOS, A.J.J. and WALLINGA, J. (2009), *Analysis of the quartz OSL decay curves by differentiation*, Radiation Measurements 44, 588–593.

BOTHA, G.A., BRISTOW, C.S., PORAT, N., DULLER, G.A.T., ARMITAGE, S.J.,ROBERTS, H.M., CLARKE, B.M., KOTA, M.W., SCHOEMAN, P. Evidence for dune reactivation from GPR profiles on the Maputaland coastal plain, South Africa, In: Ground Penetrating Radar in Sediments, (eds. Bristow, C.S. and Jol, H.M.), (Geological Society, London, Special Publication, 2003), vol. *211*, pp. 29–46.

BRILL D., BRÜCKNER H., JANKAEW K., KELLETAT D., SCHEFFERS, A and SCHEFFERS S. (2011), *Potential predecessors of the 2004 Indian Ocean Tsunami Sedimentary evidence of extreme wave events at Ban Bang Sak, SW Thailand*, Sedimentary Geology, 239, 146–161.

BRILL D., KLASEN N., BRÜCKNER H., JANKAEW K., SCHEFFERS, A., KELLETAT D. and SCHEFFERS S., (2012), *OSL dating of tsunami deposits from Phra Thong Island, Thailand*, Quaternary Geochronology, 10, 224–229.

BRISTOW, C.S., Ground penetrating radar in Aeolian dune sands. In., Ed. Ground Penetrating Radar: theory and applications. (ed. Jol, H.M) (Elsevier Science, 2009) pp. 273–297.

BRYANT, E.A. and NOTT, J. (2001), *Geological indicators of large tsunami in Australia*, Natural Hazards, 24, 231–249.

BYRNE, D. E., L. R. SYKES, and D. M. DAVIS (1992), *Great thrust earthquakes and aseismic slip along the plate boundary of the Makran Subduction Zone*, J. Geophys. Res., 97(B1), 449–478.

CHOI, B.H, PELINOVSKY, E., KIM, K. O. and LEE, J.C. (2003), *Simulation of the trans-oceanic tsunami propagation due to the 1883 Krakatau volcanic eruption*. Natural Hazards and Earth System Sciences, 3, 321–332.

CHOI, B.H, MIN, B. I., PELINOVSKY, E., TSUJI, Y. and KIM, K. O. (2012), *Comparable analysis of the distribution functions of runup heights of the 1896, 1933 and 2011 Japanese Tsunamis in the Sanriku area* Nat. Hazards Earth Syst. Sci., 12, 1463–1467.

CODRINGTON, H.W., A Short History of Ceylon (Asian Educational Services, New Delhi Madras, 1994).

DAHANAYAKE, K. and KULASENA N. (2008a), *Geological evidence for Palaeo-Tsunami in Sri Lanka*, Sci. Tsunami Hazards., 27, 54.

DAHANAYAKE, K and KULASENA, N. (2008b), *Recognition of diagnostic criteria for recent- and palaeo-tsunami sediments from Sri Lanka*, Marine Geology, 254, 180–186.

DAWSON, A.G., SHI, S., DAWSON, S., TAKAHASHI, T. and SHUTO, N. (1996a), *Coastal sedimentation associated with the June 2nd and 3rd, 1994 Tsunami in Rajegwesi, Java*. Quaternary Science Reviews, 15, 901–912.

DAWSON, S., SMITH, D. E., RUFFMAN, A., and SHI, S. (1996b), *The diatom biostratigraphy of tsunami sediments: Examples from recent and middle Holocene events*, Phys. Chem. Earth, 21, 87–92.

DULLER, G. A. T., Luminescence Dating: Guidelines on using luminescence dating in archaeology (English Heritage, Swindon 2008).

ENGEL M. AND BRÜCKNER H., (2011), *The identification of palaeo-tsunami deposits: a major challenge in coastal sedimentary research*, Coastline Reports 17, 65–80.

representative for the recent tsunami history. For instance, the Mediterranean Sea is well known for key historical events, especially in its eastern region (large tsunamis due to rare, large earthquakes in Crete, or cataclysmic volcanic explosion in Santorini) (SOLOVIEV et al. 2000), as is the NE Atlantic Ocean (1755 tsunami off Lisbon) (BAPTISTA et al. 2003). Recent events, often more moderate (BAPTISTA and MIRANDA 2009), tend to underestimate the actual tsunami hazard in most of those areas. Similarly in the Indian Ocean, destructive events were not to be discarded (ORTIZ and BILHAM 2003), even though the extent of the 2004 earthquake and tsunami was honestly not foreseen by the geophysicists. The same remarks also apply to the Tohoku area where the expected magnitude was much lower than the event actually triggered in 2011, although historical and geological evidences could have favored such scenarios (SAWAI et al. 2012).

The 2004 tsunami belongs to the category of events related to subduction processes, whose return periods are quite long (above 500 year), and which give rise to large tsunamis every 500–700 year at least (JANKAEW et al. 2008). The historical seismic activity of the Sunda Arc (ZACHARIASEN et al. 1999; SUBARYA et al. 2006) indicates that important tsunamis may have repeatedly occurred there. The relative regional quiescence during the 20th century together with the poor amount of earlier observations among the Indian Ocean residents was not in favor of a proper awareness of tsunami hazard along the coasts facing this ocean.

In addition to historical catalogs, tsunami hazard studies can rely on paleotsunami investigation on well-preserved paleoshoreline environments, and on numerical modeling whose recent advances allowed refining the quantitative results in terms of tsunami heights, flooding areas characteristics, and also toward the assessment of damage on buildings and facilities (SYNOLAKIS 2002). Tsunami numerical modeling, as for every natural phenomenon, can illustrate every sequence of the process, including all uncertainties therein (SYNOLAKIS and BERNARD 2006).

Three major steps can be defined: the initiation, the deep sea propagation, and the coastal amplification (where the tsunami is actually observed and causes damage). While the offshore propagation can be easily modeled through the simple approximation of the shallow water equations, the source and coastal processes often remain delicate to be numerically handled. As for the initiation, usual elastic dislocation models (e.g. OKADA 1985) provide a good approximation for the initial static coseismic triggering. The limitation here mostly consists of uncertainties on the seismological parameters used for the earthquake source. On the coasts, the limitations arise not only on the parameters to be used, namely the resolution of the bathymetric grids and the extent of the topographic data available, but also on the models used. A non-linear shallow water model is usually used, including non-linear terms to properly take into account the amplifying shoaling effect. However, the validity of such models appears more questionable for very high, short tsunami waves that interact with strong discontinuities (cliffs, buildings) and/or with various ground morphologies that can also be expressed in terms of various bottom friction coefficients. Indeed for instance, including variable friction coefficients in the shallowest bathymetry, following a Chezy approach, significantly modifies amplitudes of synthetic tide gauges (example in the Pacific Ocean, e.g. HÉBERT et al. 2009).

In the following study of the 2004 tsunami, we test a couple of amplification laws among those known in the literature. We avoid intensive modeling in the coastal areas, but provide instead a first order approximation of the tsunami heights that can be rapidly computed. Since these amplification laws implicitly contain the specific characteristics of the coastal response, and also depend on the coarse models performed, they have to be fit for each region studied. Then we attempt to propose a statistical approach to the coastal tsunami hazard, for several earthquake scenarios of a given magnitude along the Java–Sumatra trench. This allows to underline some specific regions as more prone to tsunami amplifications, for a given selection of scenarios.

3. Methods and Data

3.1. Tsunami Numerical Modeling

As explained above, the tsunami source is usually treated as an instantaneous perturbation of the sea bottom in response to the static displacement due to the earthquake. The initial deformation is computed here through a model of elastic dislocation (OKADA 1985) constrained with seismological parameters of the rupture that satisfy the expression of moment magnitude $M_0 = \mu ULW$, where μ is the rigidity, U the average slip amount and $L(W)$ the length (width) of the fault plane (e.g. KANAMORI and ANDERSON 1975). Such seismological parameters describing the fault geometry are mostly obtained from the inversion studies available when large earthquakes occur, or they are adapted from regional realistic seismotectonics, when various scenarios must be defined.

The modeling of tsunami propagation is carried out using algorithms solving the hydrodynamic equations under the non-linear shallow water theory. Equations of continuity (1) and motion (2) are here solved in spherical coordinates by means of a finite-difference method, centered in time, and using an upwind scheme in space (HEINRICH et al. 1998; HÉBERT et al. 2001). Open free boundary conditions are ascribed to the grid boundaries.

$$\frac{\partial (\eta + d)}{\partial t} + \nabla \times [\mathbf{v}(\eta + d)] = 0 \quad (1)$$

$$\frac{\partial (\mathbf{v})}{\partial t} + (\mathbf{v} \times \nabla)\mathbf{v} = -\mathbf{g}\nabla \eta \quad (2)$$

where d is the sea depth, η is the water elevation above mean sea level, $\mathbf{v}$ is the depth-averaged horizontal velocity vector, and $\mathbf{g}$ is the gravitational acceleration. The time increment Δt is constrained by the Courant–Friedrichs–Lewy criterion imposing $\Delta t \ll \Delta x / \sqrt{gh}$ where Δx is the space increment. To properly handle the shoaling effect near the coasts, refined grids, or nested bathymetric grids with increasing resolution, should be used to describe the refinement of bathymetric features when the water depth h decreases, hence when the tsunami wavelengths shorten, as well as the coastal structures. The method also allows to compute the inundation onshore provided topographic high resolution data are available. In the following, where only a coarse approach is tested, neither grid refinement and nesting, nor onshore run-up computations are performed.

3.2. Bathymetric Data

The computational domain relies on bathymetric data describing the sea depth d at each grid point. Usually coarse modeling is performed using bathymetric data from public datasets, such as the one derived from satellite altimetry (ETOPO2 bathymetric, 2′ grid cell size) (SMITH and SANDWELL 1997), or the oceanographic synthesis GEBCO (1′ grid cell) (GEBCO 2006). For this study, a bathymetric grid derived from ETOPO2 has been built in order to encompass the source area as well as the targets to be studied (Fig. 1). In order to reduce the computational time and increase the number of tests, the grid has been resampled to a 4′ cell size. In any case, such a grid should allow relative comparisons and also account for the amplification due to significant bathymetric discontinuities such as submarine ridges.

3.3. The Earthquake Sources

The amplification laws can be tested and compared with observations gathered after the 2004 tsunami. To this end, we used two different sources for the Sumatra–Andaman earthquake (Fig. 2). First, a simplified heterogeneous seismological source (model A) describing the 2004 earthquake was set up (Table 1) and successfully tested in detailed modeling for the far field, namely to discuss the impact in La Réunion island (HÉBERT et al. 2007). Second, a homogeneous source accounting for the same seismic moment has been defined, with a uniform slip all along the fault, as it may be defined in the minutes following the earthquake (model B).

For both used sources, the fault geometry consists of 6 segments characterized by fault slips ranging from 3 to 16 m, yielding a maximum positive seafloor displacement of 7 m, in the heterogeneous case, and with a uniform slip of 8.5 m for all subfaults in the homogeneous case (maximum seafloor deformation is then 4 m). The strike follows the mean trench azimuth, while the dip angle is between 12° and 15° from South to North. The fault width is 130 km, and the depth of the center is 20 km. The seismic moment is 6.75×10^{22} Nm, equivalent to $M_w = 9.2$.

3.4. Coastal Amplification Laws

The amplification of tsunamis close to the shores can be described using the law proposed by GREEN (1837) which predicts that the quantity $\eta d^{1/4}$ remains constant (see definitions for η and d in Sect. 3.1), due to conservation of the wave energy. This implies that the run-up, if equivalent to η_{max} taken at the water depth d_0, obeys $\eta_{max} \times d_0^{1/4} = \eta \times d^{1/4}$, or $\eta_{max} = \eta \ (d/d_0)^{1/4}$. Such an approach obviously provides an approximation of the wave height as close as possible to the shore, but not of the actual run-up which is theoretically measured at the maximum inundation line.

This kind of law was also extended using a description of the run-up of linear solitary waves (SYNOLAKIS 1991; SYNOLAKIS and SKJELBREIA 1993; SYNOLAKIS and BERNARD 2006) which interestingly accounts for the mean bathymetric slope β:

$$\frac{R}{d} = 2.831 \ \sqrt{\cot \beta} \ \left(\frac{\eta}{d}\right)^{5/4}$$

where R is the maximum run-up. In that case, the value of the mean slope β of the last emerged meters must be estimated, and may much influence the values of the run-up. This law was actually further revised and adapted to work on different waveforms, for instance using N waves, allowing distinguishing leading elevation or depression waves (TADEPALLI and SYNOLAKIS 1996). Other laws also take into account periodic waves (PELINOVSKY and MAZOVA 1992) and would also deserve to be used against the 2004 database.

In the following, we apply only two laws to the results of coarse tsunami modeling. While the Green's law is widely used as a simple formulation of conservation of energy, the Synolakis law used here should apply only to non-breaking solitary waves. However, although designed for non-breaking solitary waves, a theoretical law such as the one by TADEPALLI and SYNOLAKIS (1996) actually helped to better understand how, in 2004, the leading depression (toward the eastern Indian Ocean, hence Thailand) implied an increased run-up on average, compared to the leading elevation (toward the western Indian Ocean) (SYNOLAKIS and BERNARD 2006). The use of the original Synolakis law should be only considered here as a comparison with the Green's law, bearing in mind that many limitations remain when applied to actual tsunami data.

In this study, the basic idea is to retrieve the tsunami heights computed on coarse grids, taken along given bathymetric contours (typically within the last 200–50 m water depths), and then to find the proper shallow water depth to which the tsunami height can be extrapolated, depending on the studied site.

Such an extrapolation approach is used for first order tsunami amplitude assessment. The Japanese Meteorological Agency applies an extrapolation to the water depth of 1 m to assess the expected tsunami heights. It was also applied for tsunami hazard assessment based on near field sources, for instance in South Italy, where the coastal tsunami height can then be easily estimated with a simplified Synolakis law, expressing the mean submarine slope $\cot(\beta)$ by the ratio between the horizontal slope L and the water depth d at the source (TINTI et al. 2005). It is worth noting that more recently a probabilistic study involving amplification laws (Green's law; PELINOVSKY and MAZOVA 1992) was conducted to assess the tsunami hazard for the Pacific coastline in central America (BRIZUELA et al. 2014).

4. Application to Several 2004 Tsunami Observation Databases

4.1. Synthesis of Available Data for the 2004 Tsunami for Various Coastal Sites

The 2004 tsunami in the Indian Ocean provided detailed and comprehensive databases of tsunami observations in various coastal contexts, in the near field as in the far field, along linear shorelines as around islands. The tsunami databases usually contain the descriptions of the tsunami arrival, of the water heights, of the run-up heights and the flow depths (IOC 2013), and of the horizontal inundation lengths. Flow depths (within the inundated area) and run-up values (measured inland at the farthest inundation point) should be cautiously distinguished since they are not measured at the same point: a run-up can, for instance, be lower than flow depths at other points on the inundated area, especially when the impacted surface is very flat. This is the case in many 2004 databases, where very low coastal areas produced run-up values smaller than the flow depths over the inundated area (LAVIGNE et al. 2009; BORRERO et al. 2006).

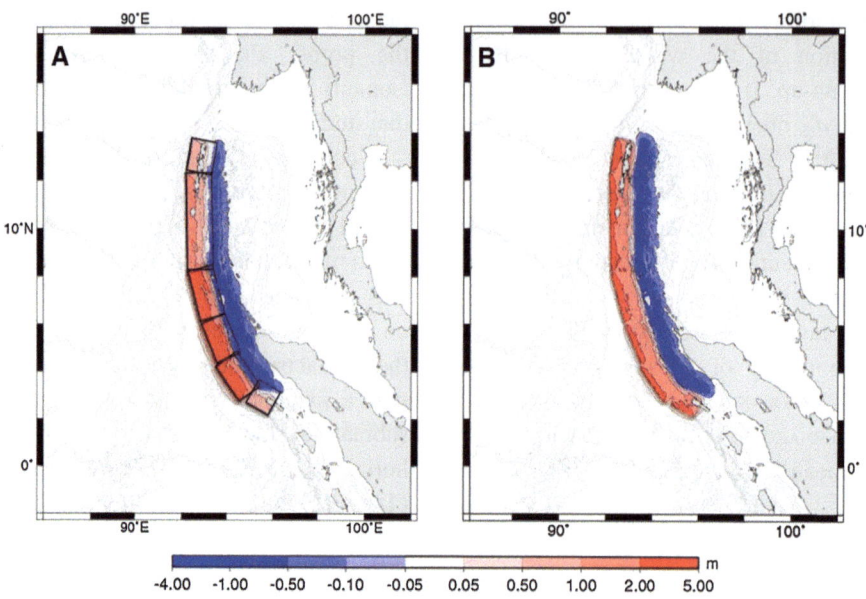

Figure 2
Initial seafloor vertical displacement due to the 2004 earthquake, heterogeneous (**a**), and homogeneous (**b**). Parameters are described in Table 1

Table 1

Parameters of the subfaults used to model the elastic dislocation for the Sumatra earthquake for models A (heterogeneous slip) and B (homogeneous slip)

Subfault	Longitude °E	Latitude °N	Slip (m) Model A	Slip (m) Model B	Strike, dip, rake (°)	Length (km)
1	92.9	13.0	3	8.5	10, 15, 100	150
2	92.8	10.3	5	8.5	359, 14, 98	450
3	93.1	7.3	12	8.5	345, 13, 95	240
4	93.75	5.4	16	8.5	337, 12, 92	200
5	94.5	3.8	12	8.5	325, 12, 92	200
6	95.6	2.9	3	8.5	300, 12, 92	120

The location (longitude and latitude) refers to the central point of the subfault. The central depth is 20 km, the fault width is 130 km throughout the rupture, and the rigidity is fixed to 45 GPa. This model is consistent with a seismic moment of 6.76×10^{22} Nm, or a magnitude M_w 9.16

4.2. Application to Thailand–Myanmar

4.2.1 Observations of the 2004 Tsunami

Thailand coastlines have been heavily affected by the 2004 tsunami, 2 h after the mainshock, and leading to about 8000 casualties among local residents, and on very touristic places. A huge amount of direct observations was besides taken by tourists in pictures, videos, where the catastrophic tsunami impact and energy are well shown. Post-event surveys have been organized to measure the tsunami heights, as early as January 2005, and have confirmed extreme values amounting to 20 m near Khao Lak especially (TSUJI et al. 2005, 2006) (Fig. 3). Only a few run-up values are available, most of the measurements consist of tsunami flow depths (tsunami heights inland close to the shore).

By contrast, Myanmar, although located to the close northern vicinity of Thailand, has been much less affected, although the tide level was high (SATAKE et al. 2006), and tsunami heights did not exceed 3 m here. Using an earthquake source not extending northwards to latitude 9°N, as in the first source models following the 26 December mainshock, was a possible explanation for these

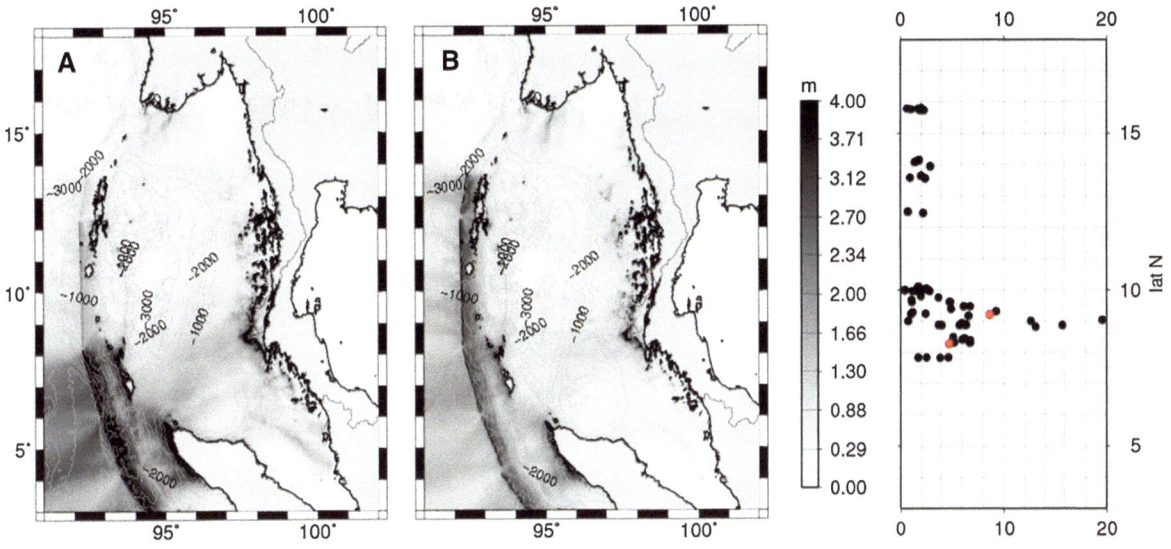

Figure 3
Maximum sea elevations from the heterogeneous (**a**) and homogeneous (**b**) models, after 4 h of propagation, also displaying several annotated bathymetric contours. *Right* tsunami heights measured along the Thailand (south to latitude 10°N) and Myanmar (north to 10°N) coastlines. The *red dots* stand for measured run-up values compared to measured tsunami heights (*black dots*)

contrasting amplitudes (SATAKE *et al.* 2006), but also the protection from offshore islands, and the different bathymetric slopes off Thailand and Myanmar, may be put forward.

4.2.2 Estimation of Tsunami Heights Using the Green's Law

Using both source models described above, we apply the Green's law to the Thailand and Myanmar coastlines (Fig. 4a, b). We modeled water heights along 3 bathymetric contours offshore (−200, −100 and −50 m), using the coarse bathymetric grid presented above (results plotted along the different blue colors on Fig. 4). Then the values extrapolated to a given water depth d_0 close to the shore (−10, −1 and −0.1 m) are computed following the dependence on $h^{-1/4}$, for each original deep bathymetry. For a given extrapolation depth d_0 (−10, −1 and −0.1 m), the obtained values for the 3 bathymetric contours are very close, their mean deviation being much smaller than the variations arising from the various extrapolation depths d_0. The values obtained for the 3 deep bathymetric contours are thus averaged to yield a mean profile for a given depth of extrapolation d_0 (various gray lines on Fig. 4).

In practice, for the heterogeneous 2004 source (Fig. 4b), the values computed along the deep bathymetric profiles (in blue) amount to 2–6 m at most, with a maximum clearly evidenced near the value of 360 km (x-axis on Fig. 4), or close to the 9°N latitude (as seen in Fig. 3). When extrapolated to the various shallow water depths d_0 (plotted along the different gray lines) the obtained profiles still exhibit a maximum near this 9°N latitude, where the maximum observations were indeed reported, in the region of Khao Lak. For a depth of extrapolation of −0.1 m, the maximum values reach about 20–24 m, at the location of the maximum observation of 20 m (Khao Lak). An interesting result is that the same match is obtained when considering a homogeneous source (Fig. 4a), although the fault slip is lowered near the latitudes from 5°N to 10°N, and increased northwards.

The absolute values obtained for both models differ, since in the worst case extrapolation to −0.1 m, with the heterogeneous source, the maximum values reach 24 m (Fig. 4b), while in the other case, they do not exceed 20 m (Fig. 4a). For the heterogeneous source, a depth of extrapolation of −1 m allows a better match to the mean values observed in Thailand near the distance of 200 km (x-axis on Fig. 4).

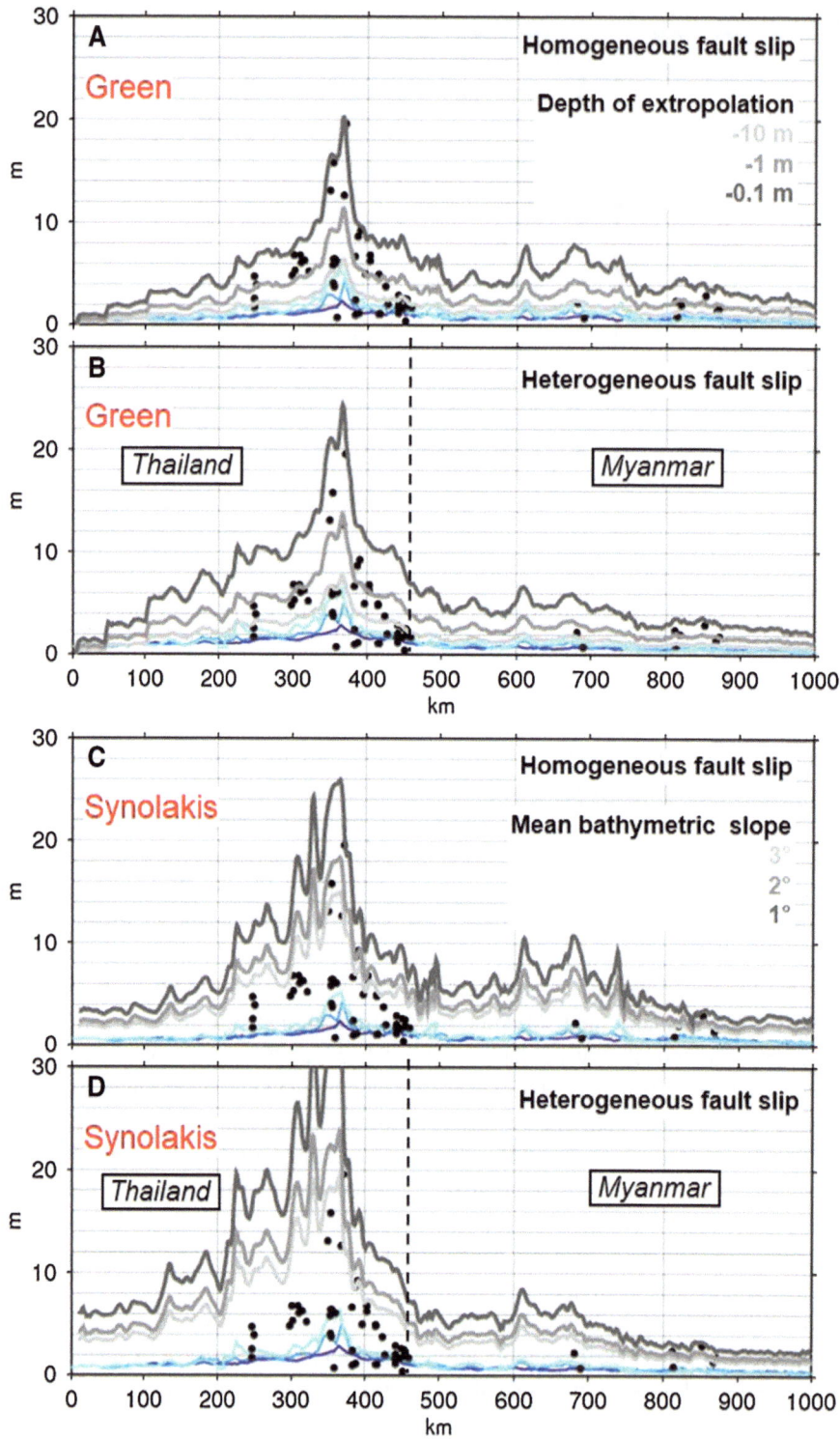

Figure 4
For Thailand and Myanmar, comparison between the measures and the modeling results extrapolated using the Green's law, for the homogeneous (**a**) heterogeneous (**b**) sources, and for the Synolakis law, for the homogeneous (**c**) heterogeneous (**d**) sources. *Colored blue lines* are maximum computed water heights along bathymetric contours of −200 (*dark blue*), −100 (*blue*) and −50 m (*light blue*). Extrapolated values following the Green's law are displayed for depths of extrapolation of −10 m (*light gray*), −1 m (*intermediate gray*) and −0.1 m (*dark gray*). Extrapolated values following the Synolakis law are displayed for the mean bathymetric slopes of 3° (*light gray*), 2° (*intermediate gray*) and 1° (*dark gray*)

In addition, both homogeneous and heterogeneous models display a very interesting trend, underlining the area most exposed for this given source region (Fig. 5). This area, including the Khao Lak and Phuket heavily damaged coastlines in 2004, is clearly identified as a very exposed area for tsunamis coming from this part of the Andaman subduction. This amplification seems to be related to bathymetric features favoring focusing of the tsunami energy at these latitudes, as shown by the maximum tsunami heights computed offshore that appear focused above a submarine ridge directed toward the latitude of Khao Lak (Fig. 5).

4.2.3 Estimation of Tsunami Heights Using the Synolakis Law

The main difficulty to apply the law proposed by Synolakis is to estimate the mean submarine slope β. The analysis of the bathymetry used here in the shallow area (<200 m) reveals that the slopes are extremely low for the area off Thailand, not exceeding 1° or 2°.

Contrary to the results using the Green's law, the extrapolations for the various original profiles (taken at $z = -200, -100$ or -50 m) exhibit a significant standard deviation, however, smaller than the one obtained depending on the slope. We computed the average of the 3 results obtained with the 3 deep contours, to concentrate on the variability upon the bathymetric slope. The results (Fig. 4c, d) show that the overall trend is satisfactory using a mean bathymetric slope from 2° to 3°, thus probably exceeding the actual bathymetric slope (as estimated from the grid used coming from ETOPO2). Here again, the heterogeneous model yields higher values than the homogeneous model, due to the higher slip concentrated near latitude 8°N. But as above, the general trend obtained reveals the maximum impacted area, near latitude 9°N.

4.2.4 Conclusions for the Thailand–Myanmar Case

We analyzed the 2004 database along the Thailand–Myanmar coastline, and compared it with the first order modeling results performed on a coarse grid. Some remarks can already be made. Even though the parameters estimated for both Green or Synolakis laws may depend on the grid resolution (here 4'), for the Green's law, extrapolations to water depths of 1 m satisfactorily fit the mean observed run-up values, while extrapolations to 0.1 m yield a reasonable distribution for the observed flow depths. The bathymetric slope to consider in the Synolakis law is close to the actual one, actually rather steeper. As for the tsunami

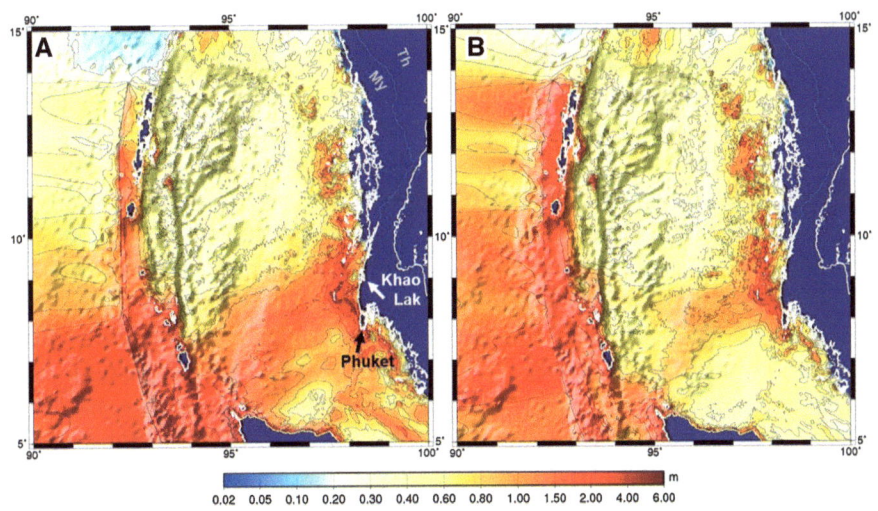

Figure 5
Maximum tsunami heights computed after 4 h of propagation toward Thailand and Myanmar, with the heterogeneous source model (**a**) (*left*) and the homogeneous source model (**b**) (*right*). Note the location of Khao Lak in front of a focusing bathymetric ridge

impact in these areas, the submarine bathymetry off Thailand certainly played a key role in focusing energy toward Phuket and Khao Lak, as was already observed in previous studies (SATAKE et al. 2006).

In the following, we will further check whether these parameters obtained for Thailand–Myanmar can also apply to other coastlines, especially at farther distances, using the same earthquake sources and the same coarse modeling. Since numerical dispersion in the numerical modeling may increase the attenuation of computed amplitudes, these parameters may be consequently revised for other coastlines.

4.3. Application to India and Sri Lanka

4.3.1 Observations of the 2004 Tsunami

Observations gathered after the 2004 tsunami, and used here (Fig. 7), consist of run-up values measured along the southeastern coast of India (JAYAKUMAR et al. 2005), and of water heights including run-up, around Sri Lanka (LIU et al. 2005; GOFF et al. 2006). In India, the measures are reported as real run-up values (taken at the inundation limit), and range from 2 to 6 m approximately. In Sri Lanka, only a few run-up measures are available, reaching the maximum values of 9–13 m, while most of the flow depths range from 2 to 6–8 m.

The largest tsunami heights are observed in the southern part of Sri Lanka where a run-up of about 13 m has been reported (Yala region). On the western coastline, the observed tsunami heights are about 4–6 m and can be explained through detailed numerical modeling (POISSON et al. 2009).

The maximum water heights computed off India and Sri Lanka (Fig. 6), for the homogeneous source, display two areas of higher impact, the first near latitude 11–12°N, off Pondicherry (Tamil Nadu), and the second off the south of Sri Lanka, near latitude 5°N. These two latitudes are also the places where the impact of the tsunami was the highest. Using the heterogeneous source slightly changes that pattern, lowering the impact near the latitude 11°–12°N, but strengthening the impact to the south of Sri Lanka, corresponding to the reinforced slip in the central part of the earthquake rupture.

4.3.2 Estimation of Tsunami Heights Using the Green's Law

Using the same tsunami modeling results as for the Thailand case, the tsunami heights extrapolated to various sea depths have been computed, for the two earthquake sources considered. When the heterogeneous source model is used, the extrapolated values (Fig. 7b) are higher for the southern Sri Lanka area, in good agreement with the observed trend of the tsunami heights, whereas when the homogeneous model is used (Fig. 7a), the highly impacted area is located to the latitude 12°N, moreover quite in disagreement with the observations. This may also confirm that the slip in the northern extremity of the rupture was not that high as the value chosen for this homogeneous model.

Regarding the absolute values obtained, the depths of extrapolation of -10 and -1 m are sufficient to properly adjust the mean observed trend. This is quite in agreement with the study off Thailand, but rather more pronounced here, since the depth of extrapolation of -0.1 m is clearly overestimating the observed values for Sri Lanka and India. This is even worse for India data, where the extrapolation yields too large values for depths of extrapolation shallower than -10 m. In India, the available data are run-up values only, thus flow depths may have been locally much higher, but no information was available to give more precision. In addition, the coastal feature near latitude 11°N–12°N, which corresponds to a less pronounced and milder slope offshore, with a more concave pattern for the submarine bathymetry, seems to favor modeled tsunami amplification for those latitudes: this is, however, not confirmed by the data available here, which are thus more consistent with a moderate source fault slip near 10°E.

In Sri Lanka, both available observations and our models are consistent with a maximum value in the southern extremity of the island, a rather different pattern than the modeling results obtained in another study where this place appears as less impacted compared to the western and eastern coastlines (POISSON et al. 2009). For the western shoreline, it seems that the extrapolation value of 0.1 m better compares well with the modeling performed in the data and detailed results obtained for the southwestern coast of Sri Lanka (POISSON et al. 2009).

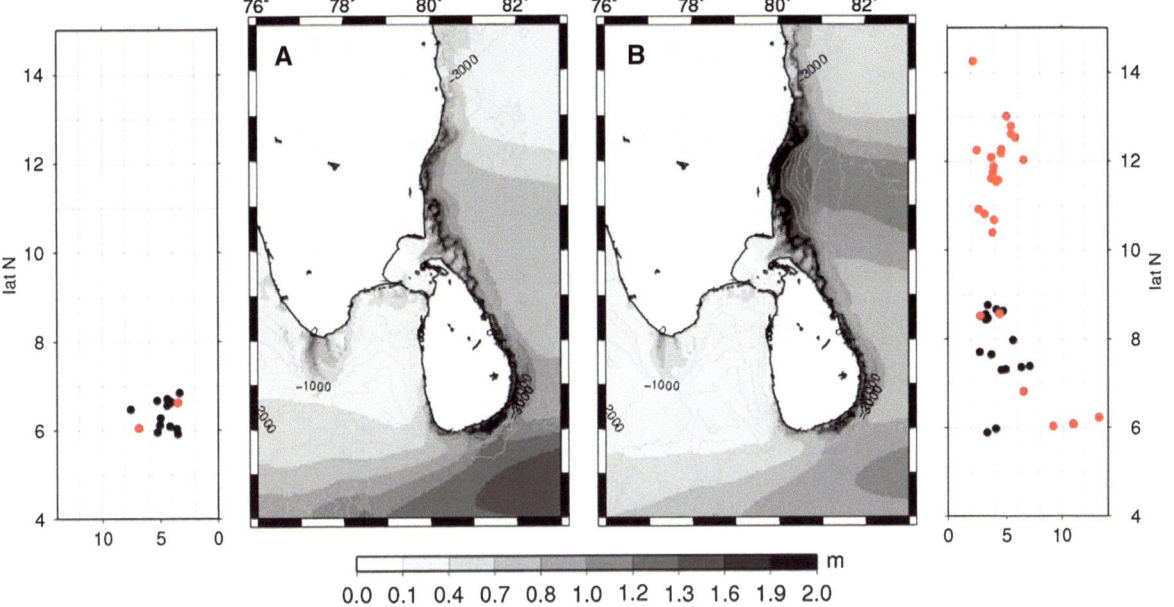

Figure 6
Central panels maximum sea elevations from the heterogeneous (**a**) and homogeneous (**b**) models, after 4 h of propagation, also displaying several annotated bathymetric contours. *Left and right* synthesis of the observations gathered in eastern India (JAYAKUMAR *et al.* 2005) and Sri Lanka (LIU *et al.* 2005) following the 2004 tsunami

4.3.3 Estimation of Tsunami Heights Using the Synolakis Law

Using the same mean bathymetric slopes as for the Thailand case study, the extrapolated values (Fig. 7c, d) exhibit a trend very close to the one obtained for the Green's law, again underlining the importance of the source heterogeneity, and stressing the maximum impact in the southern part of Sri Lanka, and off Pondicherry (Tamil Nadu). As in the Thailand case study, the bathymetric slopes requested to better fit the observations should be close to 3° at least.

A more detailed analysis of the bathymetry off Sri Lanka and India reveals that the continental shelf is narrower than off Thailand, and that submarine slopes within the shelf are very low, and very similarly close to 1°–2°. But the border of the submarine shelf is rather steeper and reaches more than 5°–7°, over distances of a few tenths of kilometers. This may explain why a steeper mean slope must be taken into account to better fit the data with the Synolakis law.

4.3.4 Conclusions for the India–Sri Lanka Case

The results show that the parameters used to fit data for India and Sri Lanka slightly differ from the ones used for Thailand, since extrapolation depths lower than 1 m are not requested here. As for the Synolakis law, taking into account mean slopes larger than 3° are satisfactory.

The source heterogeneity also plays a more important role than in the Thailand case. This is understandable since the tsunami spreading at distance implies a high dependence on the initial slip latitudinal contrasts, for the kind of distant impacted points studied in Sri Lanka and India. In Thailand, the same source heterogeneities produce less contrasted tsunami heights, all the more as the observed highest values in Thailand are confined within less than 200 km. In Sri Lanka and India, these source heterogeneities imply distinct peaks of amplitude at least 500 km apart, and the heterogeneous model allows to better fit the large observations in south Sri Lanka and the more moderate observations off India.

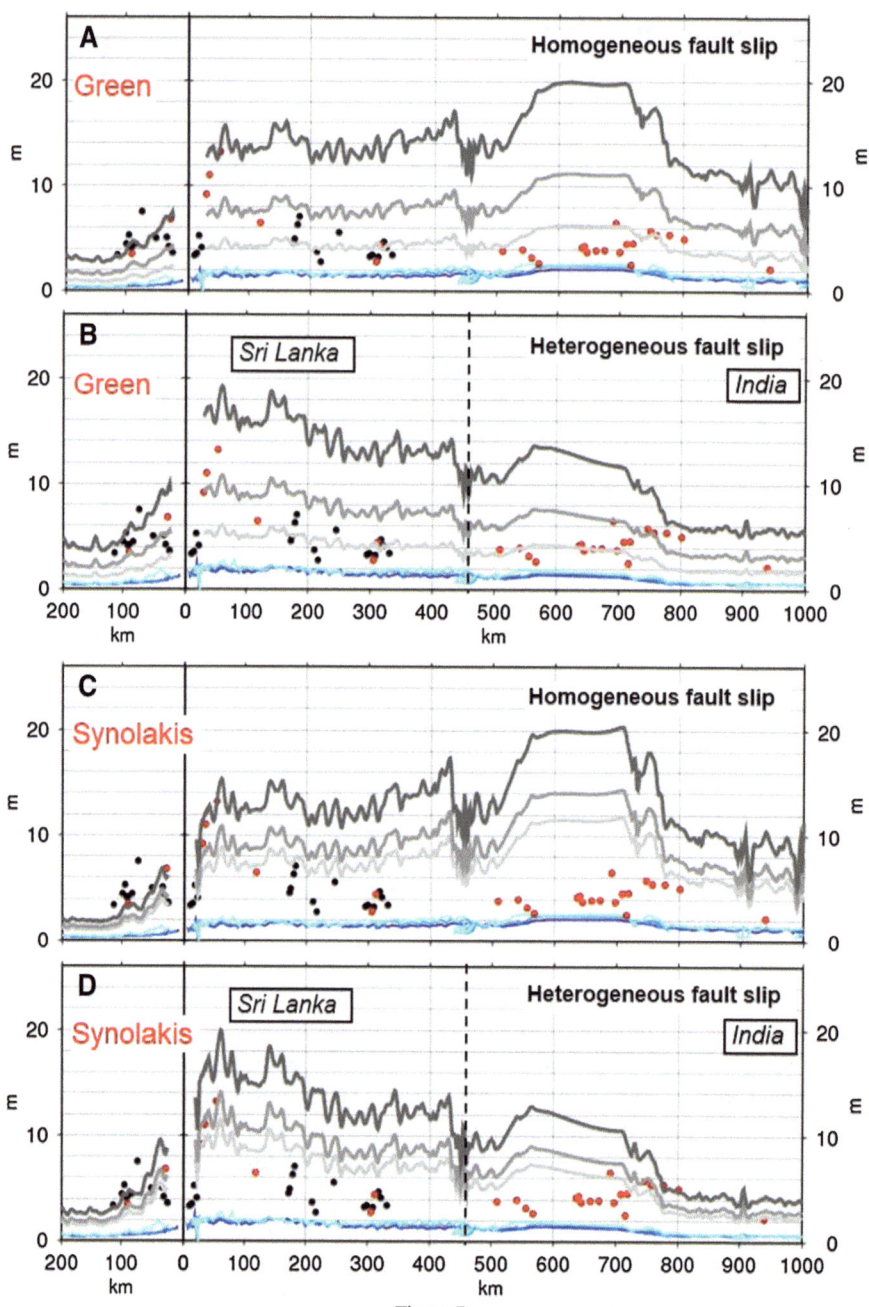

Figure 7
For Sri Lanka and SE India, comparison between the measures and the modeling results extrapolated using the Green's law, for the homogeneous (**a**) heterogeneous (**b**) sources, and for the Synolakis law, for the homogeneous (**c**) heterogeneous (**d**) sources. *Blue* and *gray color scales* as in Fig. 4

4.4. Application to Madagascar

4.4.1 Observations of the 2004 Tsunami

The eastern Madagascar shoreline has been significantly impacted by the 2004 tsunami, with run-up values from 1 to 5 m very locally, but generally comprised between 2 and 3 m (Fig. 8), and with a child drowning to be deplored. Local resonance phenomena led to strong disturbances in the harbor of Toamasina, near latitude 18°S (OKAL *et al.* 2006a). While the western shoreline

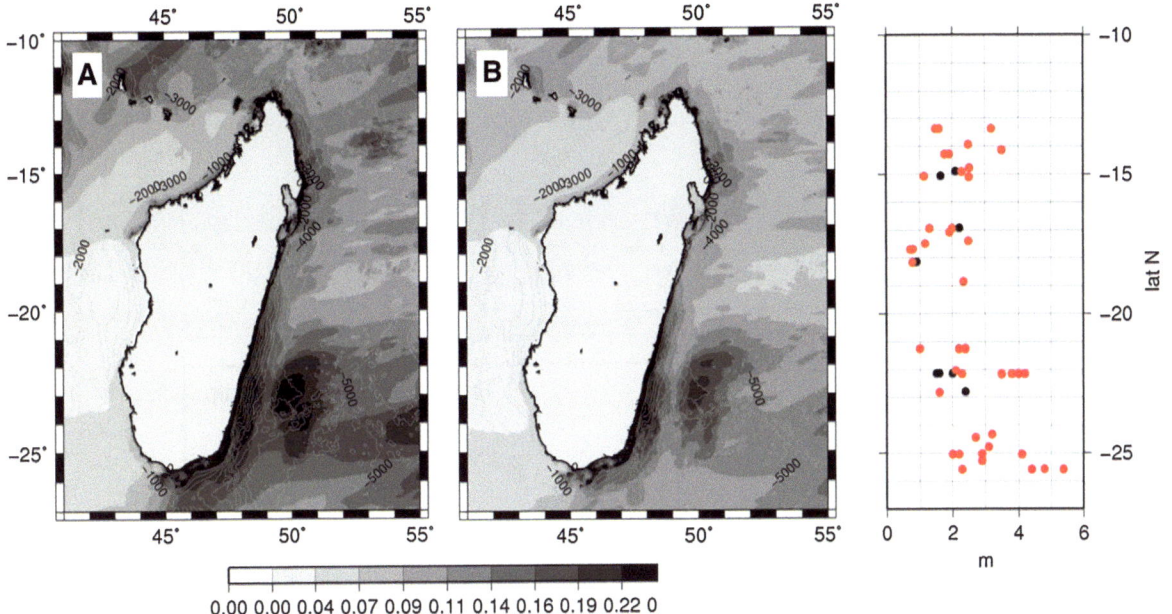

Figure 8
Synthesis of the observations gathered in Madagascar (OKAL *et al.* 2006a) following the 2004 tsunami. The *left panel* displays the maximum water heights computed after 10 h of propagation, using the heterogeneous (**a**) and homogeneous (**b**) source models. Run-up values are in *red*, flow depths in *black*

was poorly impacted, some doubts remain on the southernmost shore where local edge waves may have produced damage (Okal *et al.* 2006a).

4.4.2 Estimation of Tsunami Heights Using the Green's Law

The homogeneous source model (Fig. 9a) underestimates the maximum tsunami height observations, even for the shallowest depths of extrapolation, by a factor of almost 2. However, the overall trend is well reproduced, underlining the larger tsunami impact along the southeasternmost shoreline. This discrepancy could be discussed with respect to the fact that most of the observations used here are run-up values, contrary to the previously studied sites where they mostly consisted in tsunami flow depths. But above all, the observed absolute values considered in Madagascar (from 2 to 4 m on average) are much lower than the two previously studied sites, thus the obtained results are quite satisfactory as far as the relative ratios are concerned.

Using the heterogeneous model (Fig. 9b) improves the fit by enhancing the computed impact to the South. Only the extreme values to the South cannot be satisfactorily fit. As for La Réunion Island (location on Fig. 1) (HÉBERT *et al.* 2007), this source is most probably more realistic for these remote targets, because the highest slip off North Sumatra, in the southern central region of the rupture, contributes to more tsunami energy toward this region of the Indian Ocean.

4.4.3 Estimation of Tsunami Heights Using the Synolakis Law

Again the same parameters have been applied to study how the Synolakis law can reproduce the overall trend of the 2004 observations in Madagascar. As for the Green's law, the heterogeneous model (Fig. 9d) better fits the observation distribution than the homogeneous source (Fig. 9c), but still the extreme values to the South are not well explained.

Using a very low slope of 1° allows fitting well the observed trend. In this case, the extrapolation must therefore be reinforced. And it is worth noting that the submarine slopes off Madagascar are rather steeper than in the two previous studied areas: south to 20°S, they reach 3°–4°, but north to 20°S a narrow shelf break displays slopes from 5° to 7°. These

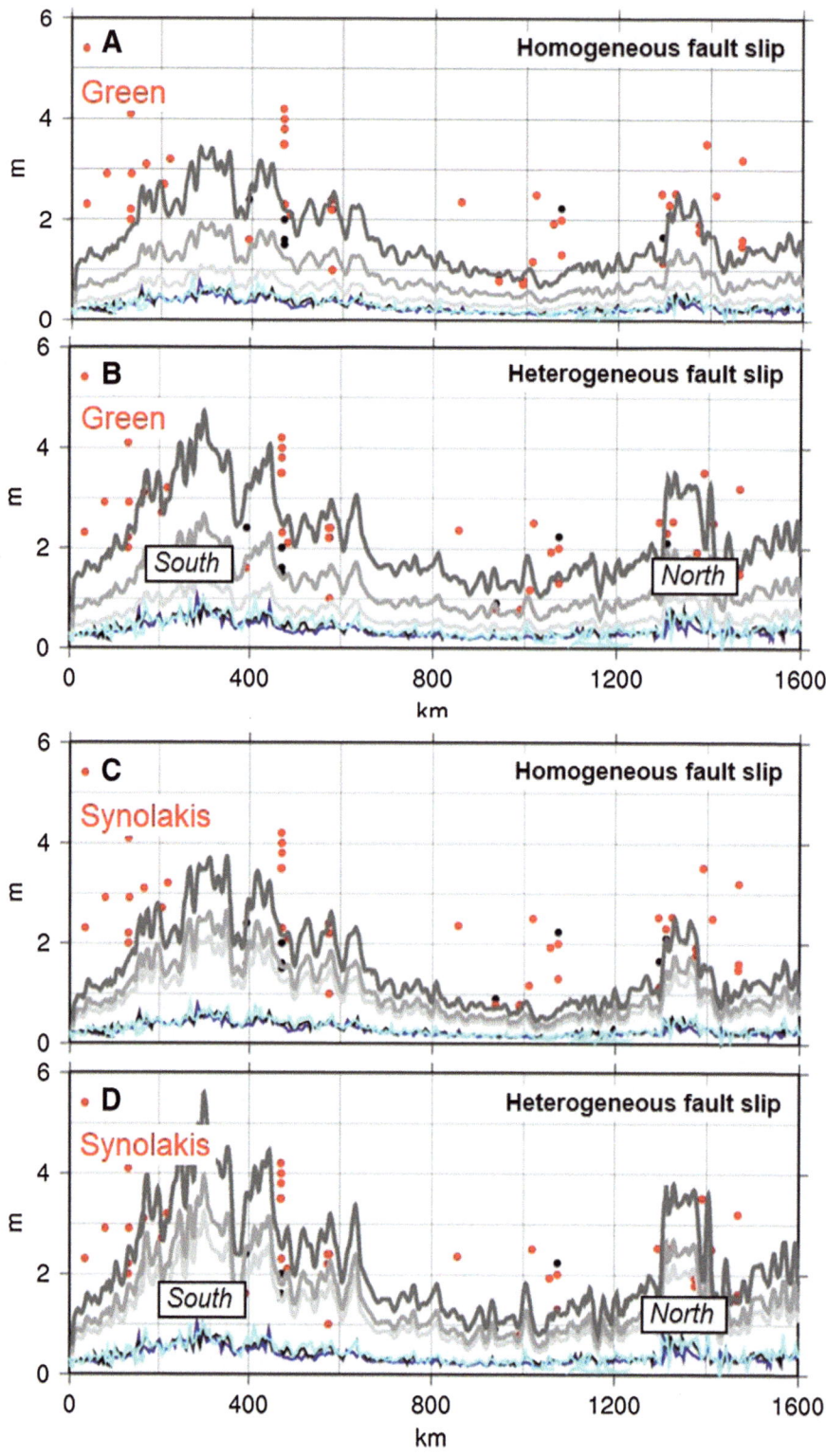

Figure 9
For Madagascar, comparison between the measures and the modeling results extrapolated using the Green's law, for the homogeneous (**a**) heterogeneous (**b**) sources, and for the Synolakis law, for the homogeneous (**c**) heterogeneous (**d**) sources. *Blue and gray color scales* as in Fig. 4

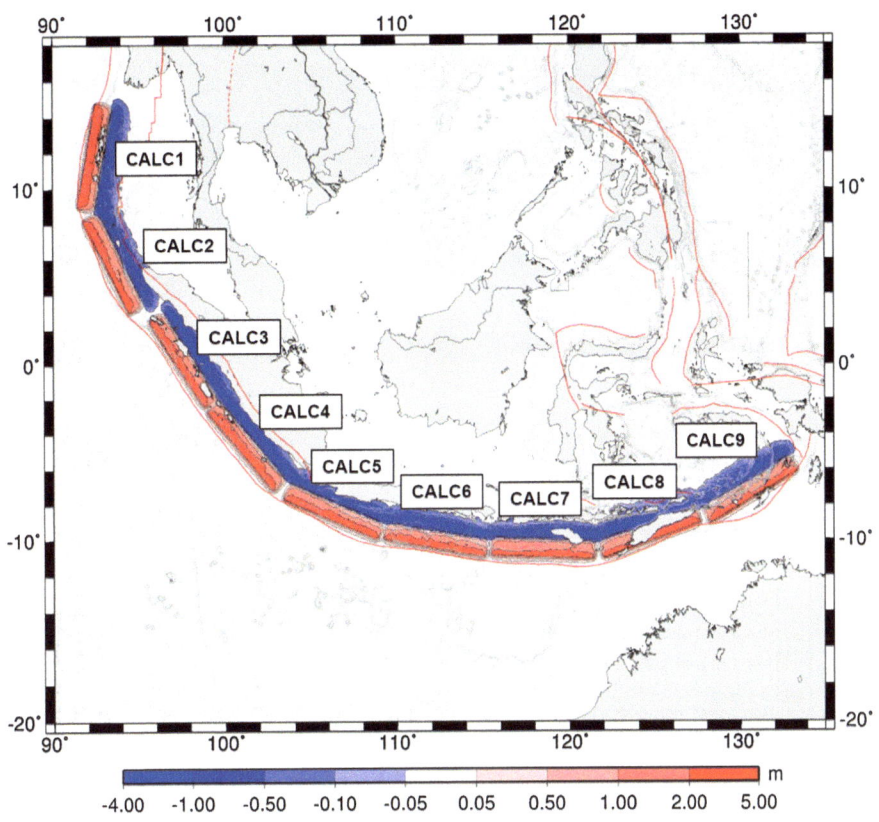

Figure 10
Initial coseismic deformations for the M_w 9.0 earthquakes for the statistical test

contrasting slopes are in good agreement with the contrasting observations in north and south Madagascar, but they also confirm that the extrapolation method is not able to fit the data for such realistic slopes.

4.4.4 Conclusions for the Madagascar Case

For both employed methods, the extrapolated models tend to underestimate the observations, especially when the fault slip model is simply homogeneous. Using a heterogeneous model improves the fit, except for the extreme values observed in the southern part of the island.

A reason for the general discrepancies between the extrapolated models and observations may also come from the tsunami model used in this study: at such propagation distances, the numerical dispersion may appear due to the use of coarse grids, and this may have further reduced the computed amplitudes, compared to Thailand and Sri Lanka. Such a bias could also be taken into account. Nevertheless the method reproduces well the relative, overall trend of the data, with larger values to the south than to the north, and with locally higher tsunami heights near latitude 14°S.

5. Proposition for a Statistical Analysis

The first aim of this study, discussed in the previous section, was to test amplification laws against actual observations gathered during a specific, well documented event, the 2004 tsunami. A second objective is to study a series of earthquake scenarios to assess the variability of their impact for a given coastline. To this end, we conducted a study based on several scenarios in the Andaman Sea and along the Sumatra–Java trench, using magnitude M_w 9.0 events, larger than the March 2005 earthquake which generated a very weak tsunami (GEIST et al. 2006).

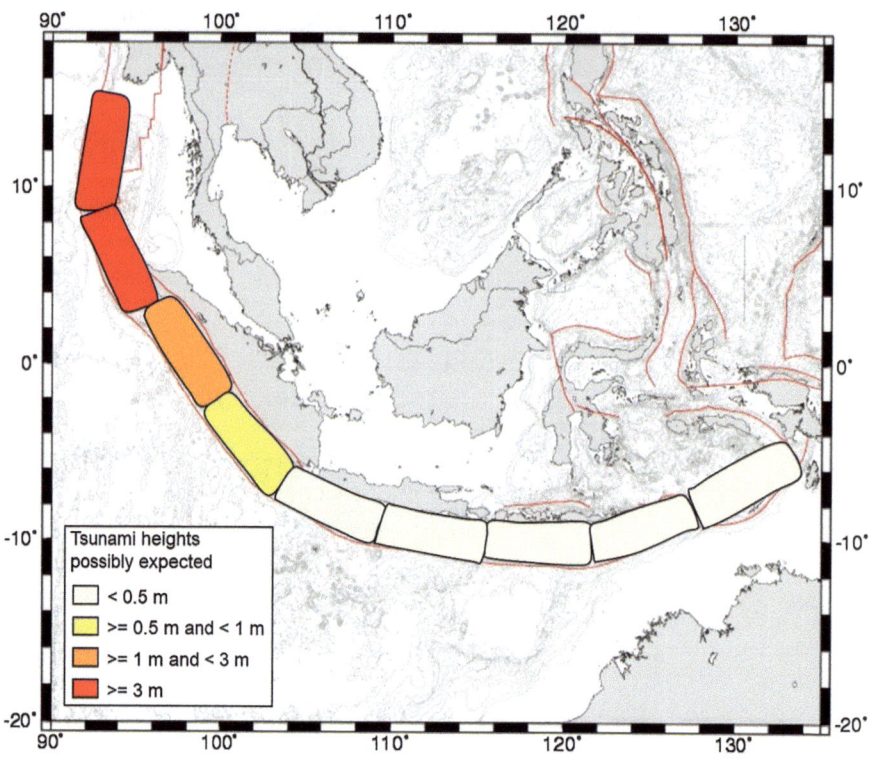

Figure 11
Mapping of the possible tsunami heights to be expected along the Thailand and Myanmar coastline, for the different scenarios

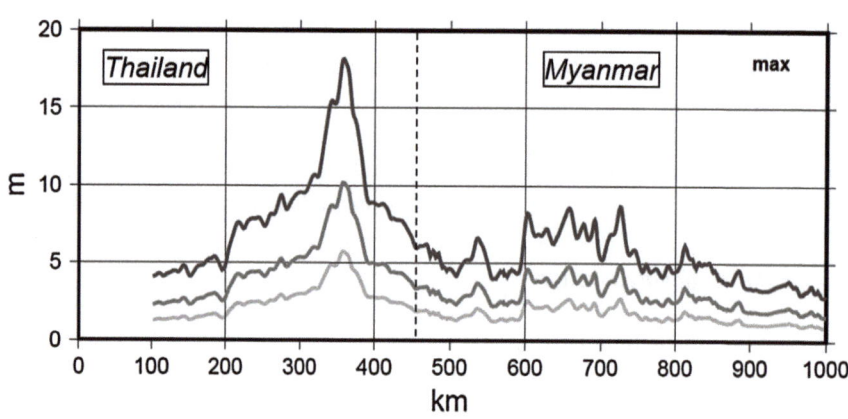

Figure 12
Maximum impact along Myanmar and Thailand, for 4 different M_w 9.0 earthquakes in the Andaman–Sumatra subduction zones, amplified using the Green's law

This level of magnitude is not corresponding to the worst case scenario which could amount to $M_w = 9.3$. Nevertheless, the second largest earthquake that occurred in the last decade is the Tokoku-Oki event on March 11 2011 (KOKETSU et al. 2011) and it reminds us of the possibility of magnitudes above 9.0 in places where they were not foreseen at all. In addition ZACHARIASEN et al. (1999) have suggested that the giant Sumatran subduction earthquake that occurred on November 24 1833 had an 8.8–9.2

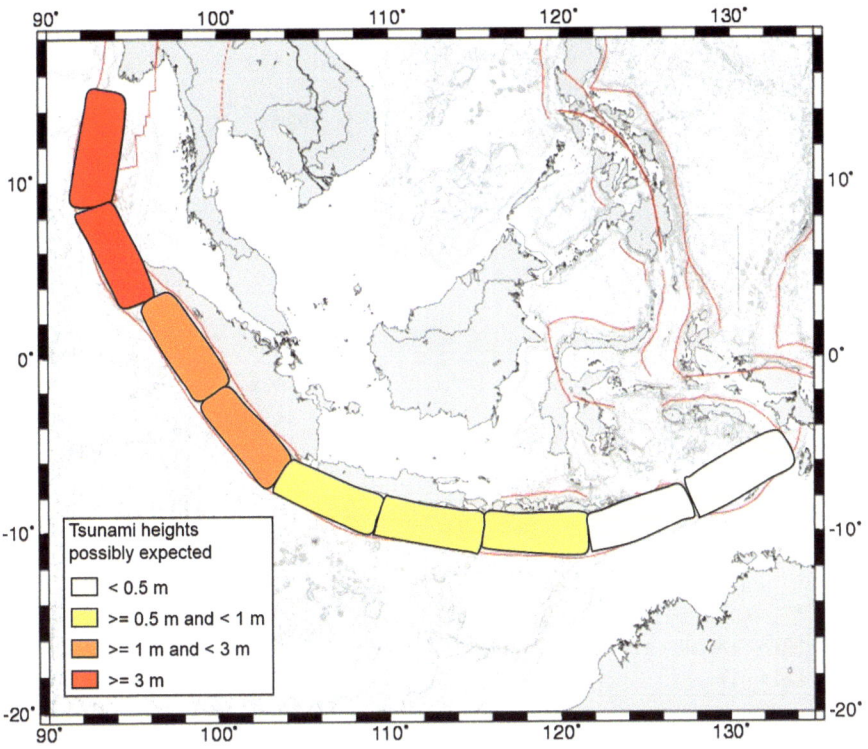

Figure 13
Mapping of the possible tsunami heights to be expected along the SE Sri Lanka and SE India coastlines, for the different scenarios

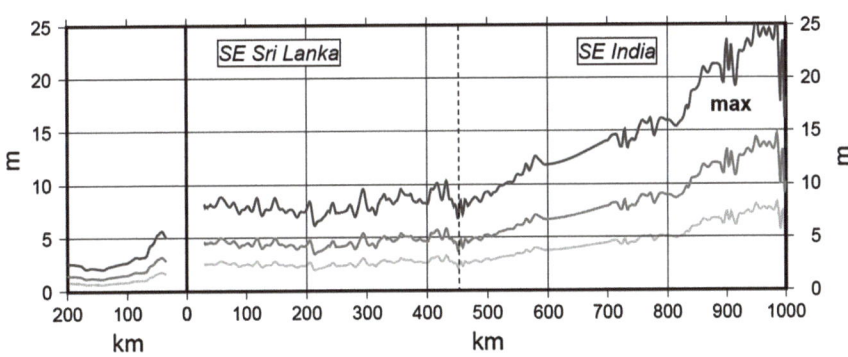

Figure 14
Maximum impact along SE India and Sri Lanka, for 4 different M_w 9.0 earthquakes in the Andaman–Sumatra subduction zones, amplified using the Green's law

M_w magnitude. We considered in this study that a magnitude M_w 9.0 earthquake should be therefore expected all along the Sumatra and Java trenches.

The rectangular sources used are 650 km long and 130 km wide, with a uniform fault slip amounting to 10 m, and an 11° dip angle (Fig. 10). Then the extrapolations have been carried out using the Green's law only.

5.1. The Thailand–Myanmar Case

The estimated impacts can be mapped onto the source areas, in order to represent the variability of the source with respect to the tsunami heights generated, for a given exposed shoreline (Fig. 11). We applied the same extrapolation depths as those previously validated against the 2004 data, i.e. using 0.1–1 m. Obviously this

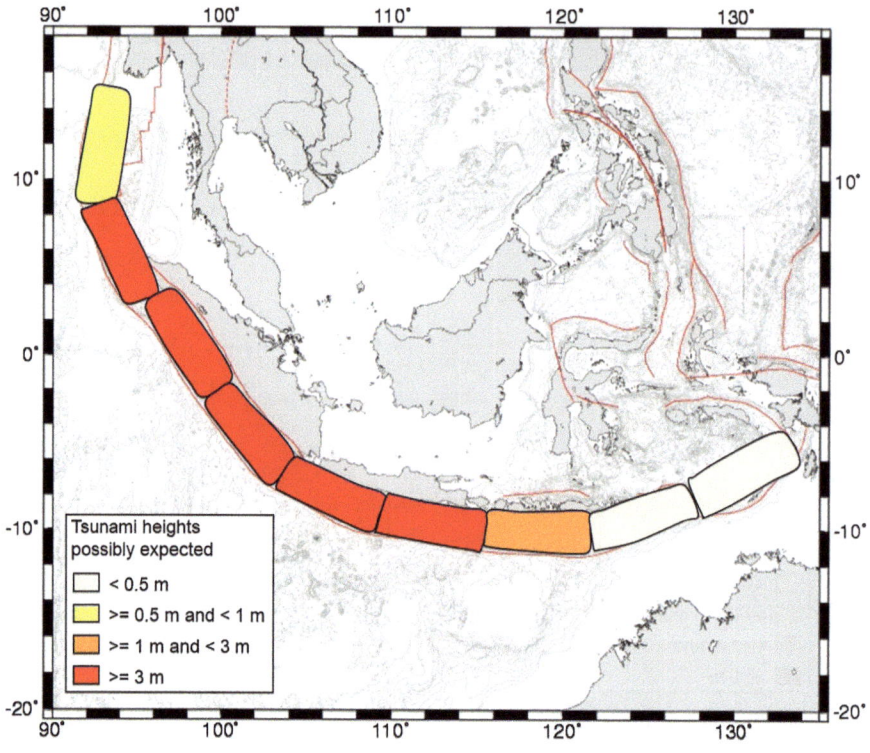

Figure 15
Mapping of the possible tsunami heights to be expected along the Madagascar coastline, for 8 different scenarios

mapping does not provide any information on the variability of the impact along the studied shoreline, but it allows identifying the threatening source areas on average. As expected, none of the sources located southeast to Sumatra produces any impact to Thailand and Myanmar coastlines.

The maximum value of the 4 scenarios exhibits a maximum of 15–18 m near Khao Lak in Thailand (Fig. 12), underlining the important exposure of the area to tsunami hazard, as already observed during the 2004 event (Figs. 4, 5). To a lesser extent the Myanmar coast from the distance 600 to 700 km (x-axis) also offers an important exposure of 7–8 m (near latitude 12°N).

The maximum value along the shoreline stresses again the different behavior of Thailand and Myanmar, mostly due to the bathymetric pattern off these coasts.

5.2. The India–Sri Lanka Coastline

Regarding SE India and Sri Lanka, representing the tsunami impacts onto the source map stresses that the most threatening M_w 9.0 sources are located off Andaman and Sumatra (Fig. 13). The sources located south of Java have a very moderate impact (<1 m) and those located further eastwards have no impact.

The maximum values of the scenarios reach 15–25 m and are much higher in SE India, north to latitude 12°N (Fig. 14). In comparison with the 2004 event, the source which contributes most to this amplification is located near the Andaman Islands.

5.3. The Madagascar Coastline

This last example (Fig. 15) is more comprehensive because seven different scenarios produce an impact along the studied shoreline. Looking at the distribution along the shoreline (Fig. 16), the most influencing sources are CALC3, CALC4 and CALC5, and yield, for the worst case (CALC5), maximum tsunami heights exceeding 5 m in Madagascar. Both sources CALC4 and CALC5 are located in a similar direction, as seen from Madagascar, and

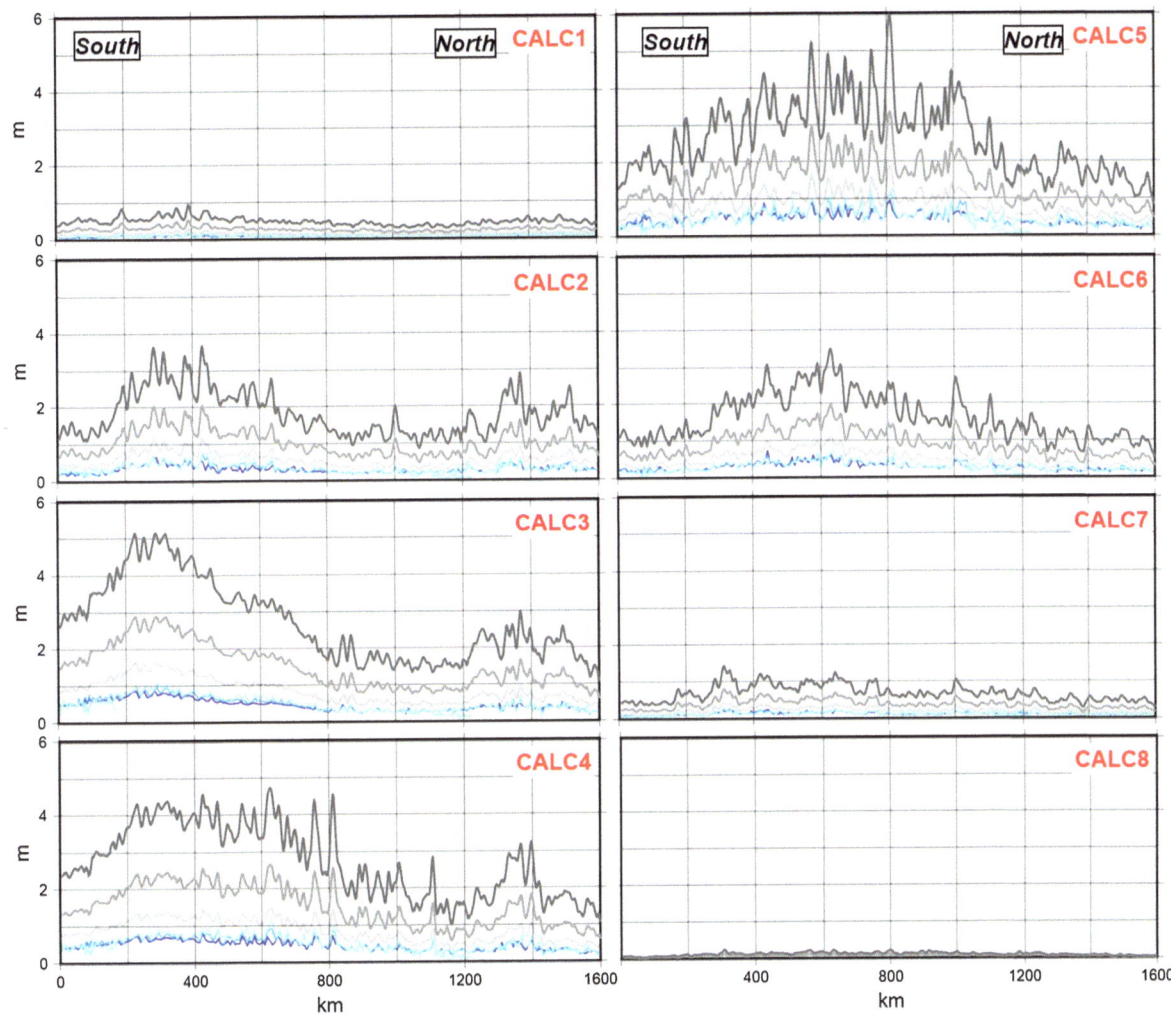

Figure 16
Results of coarse modeling (in *blue*) and extrapolation to various water depths using the Green's law, for the 8 sources studied, and for the Madagascar coastline. The *color codes* are as in Fig. 4 and subsequent

no attenuation of tsunami waves occurs due to archipelagos. The CALC8 and CALC9 sources located to the east do not produce significant heights.

Finally, a preliminary statistical treatment has been applied to all the series obtained from the extrapolations performed with the Green's law (Fig. 17). The idea is to be able to discuss this variability of the impact along a given shoreline. The eight M_w 9.0 scenarios have been used to estimate the 25-th quantile, the median value, the 75-th quantile and the maximum values of the modeled impacts reached along the shoreline. This underlines the more pronounced hazard of South Madagascar with respect to North Madagascar for tsunami sources coming from Sumatra. The median value is lower than 3 m throughout the shoreline, but as the fit using an extrapolation depth of 0.1 m was still underestimating the observations for south Madagascar, this value may be a minimum. 75 % of the scenarios are lower than 4 m, while 25 % of the scenarios are lower than 1.5–2.0 m. The maximum values remind us that the maximum tsunami heights probably range from 5 to 6 m, thus higher than in 2004.

The northern part of Madagascar seems to be more protected from these tsunamis, as in 2004 where the Mascarene Plateau may have played a protective role, and in addition the contrasting mean bathymetric slopes also protect north Madagascar more than

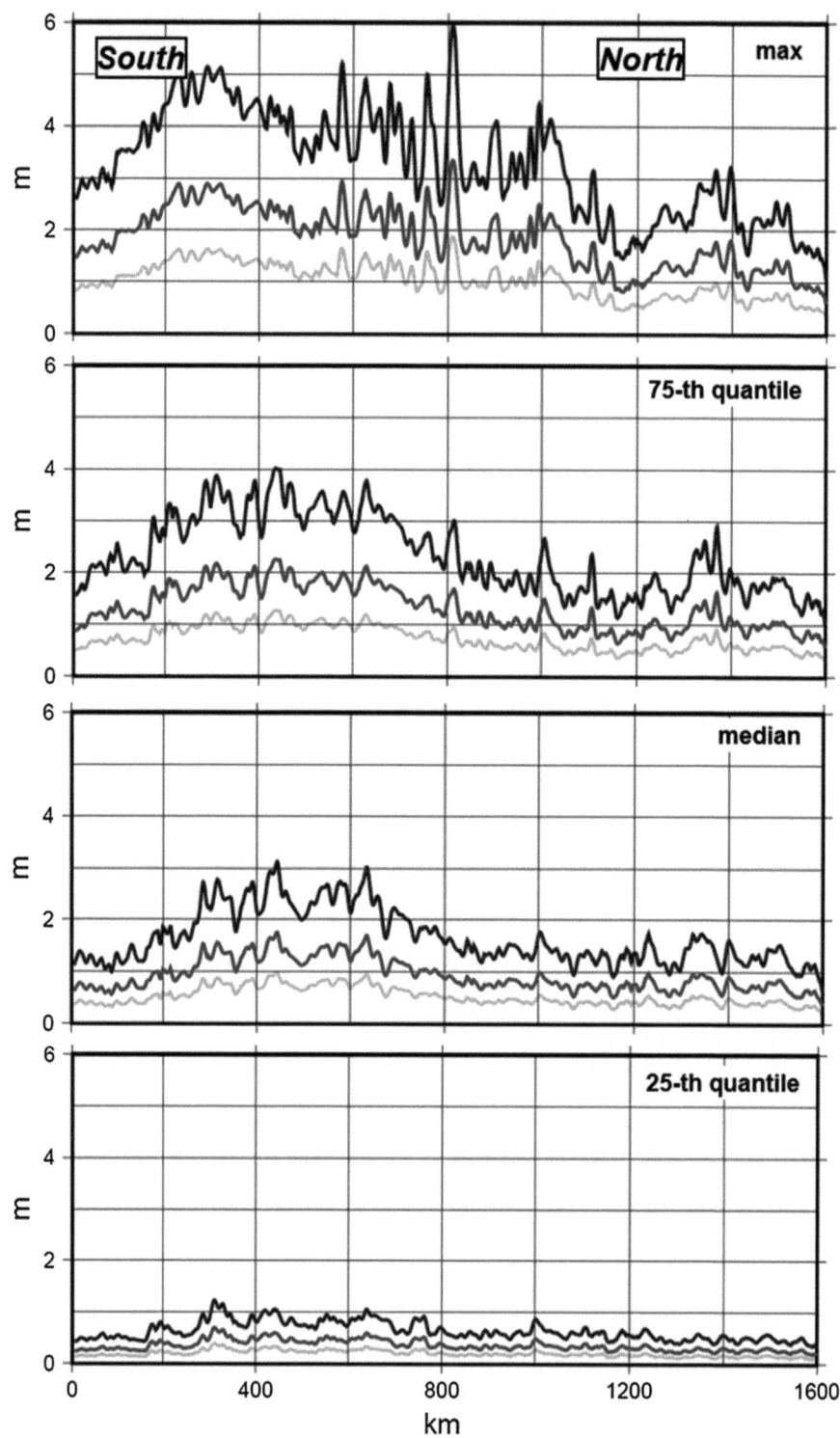

Figure 17
Statistical assessment of the tsunami variability along Madagascar, for 8 different M_w 9.0 earthquakes along the Andaman–Sumatra subduction zone

south Madagascar. However, it is worth noting from the observations in 2004 near latitude 18°S that tsunamis heights from 1 to 2 m can lead to resonance and severe harbor disturbances.

It is also to be mentioned that the magnitude 8.7 March 2005 tsunami (close to, but smaller than the CALC4 used here) was almost unnoticed in Madagascar, possibly because it arrived at night, but also because it was a very weak tsunami (OKAL et al. 2006a) due a coseismic displacement mainly under islands (Geist et al. 2006).

6. Discussion and Conclusions

The analysis of the 2004 tsunami observation databases for 3 selected areas (Thailand–Myanmar, India–Sri Lanka, and Madagascar) has been completed. The distinction between run-up and flow depths must be stressed since the run-up values are not always the maximum values. Using these databases, and modeling results obtained using an offshore low resolution bathymetric grid, from which values are taken along constant depth bathymetric contours, we attempted to fit the observations using either the Green's law or the Synolakis law. The fit is satisfactory in terms of observation variability along the shoreline, and, in many cases, the obtained factor of amplification can be related to particular focusing or defocusing bathymetric features.

Relative higher impacts are well reproduced for some specific sites (Khao Lak, South Madagascar) with respect to other less impacted areas. However, the absolute values are best fit for a set of parameters, mostly similar for all cases in the near field (less than 2 h of propagation), yet slightly different in the far field (Madagascar). This may also be accounted for by the increase of numerical dispersion along with the propagation distance, and thus by a more pronounced spreading for the far field.

When high resolution tsunami models are available, this method can be compared to accurately computed tsunami heights. Most of the amplitudes we obtain with our approach fall well within the range of these highly resolved models, although some specific minimum results in south Sri Lanka are not reproduced here (POISSON et al. 2009). Since a similar trend is also observed in other studies (LØVHOLT et al. 2006), probably our model is not accurate enough for the southernmost extremity of Sri Lanka.

Our previous study on La Réunion island could unfortunately not be compared here (HÉBERT et al. 2007) since the 4′ bathymetric grid used was not able to account for the detailed bathymetric features around the island (whose spatial extent does not exceed 40 km). However, we were able to apply the same method to another 300-m grid, and in that case, an extrapolation to 10 to 1-m contours was sufficient to account for the overall distribution of the few observations available (OKAL et al. 2006b).

Using several scenarios toward the same target, a statistical assessment of the tsunami variability can be proposed, and this was tested more in detail in Madagascar where 8 scenarios can be studied. In this preliminary study, the method allows to map the expected tsunami heights for a given source, or, conversely, to map the overall trend of the expected tsunami heights along the studied shoreline. In brief, the results obtained here underline that:

- Thailand is more exposed to tsunamis than Myanmar. Sources involving M_w 9.0 earthquakes in the Andaman Sea can generally produce tsunami heights much greater than 10–15 m, especially in Thailand, and even more pronounced for the region near Khao Lak, where the submarine morphology seems to contribute to a significant amplification,
- SE India is more exposed than Sri Lanka, with heights greater than 10 m, mostly because of the M_w 9.0 scenario located in the Andaman Sea, but moderate threat from southwards scenarios also exists,
- Madagascar is essentially exposed for its southernmost coasts, where the M_w 9.0 scenarios may produce maximum tsunami heights of 4–6 m.

A statistical assessment has been tried here for Madagascar, which relies on 8 scenarios only, confirming, however, the maximum hazard for South Madagascar. Such an approach could be similarly applied for more numerous, smaller earthquakes to assess tsunami hazard along a given shoreline, for a specific level of magnitude.

Our study also contributes to show some of the uncertainties inherent in the use of amplification laws in the operational context. Indeed, the laws to be applied should depend on the various parameters influencing the results; hence they have to be defined according to each coastline. Some places are well known to amplify more the tsunamis (for instance, depending on the submarine morphology), and the operational approach should account for these regional conditions (REYMOND et al. 2012).

Acknowledgments

This work originally benefited from the support of the "Délégation Interministérielle Pour le Tsunami" set up in 2005 by the French Ministry of Foreign Affairs after the catastrophic 2004 Indian Ocean tsunami. It was then supported by the European FP6 TRANSFER project (2006–2009) under the contract 037058. We thank Alberto Armigliato and an anonymous reviewer, as well as the Guest Editor Hermann Fritz, for their valuable comments allowing to improve the quality of the manuscript.

REFERENCES

BAPTISTA, M.A. and J.M. MIRANDA, *Revision of the Portuguese catalog of tsunamis* (2009), Nat. Hazards Earth Syst. Sci., 9, 25–42, www.nat-hazards-earth-syst-sci.net/9/25/2009/.

BAPTISTA, M.A., J.M. MIRANDA, F. CHIERICI, and N. ZITELLINI (2003), *New study of the 1755 earthquake source based on multi-channel seismic survey data and tsunami modeling*, Nat. Hazards Earth Syst. Sci., 3, 333–340.

BORRERO, J.C., SYNOLAKIS, C.E., FRITZ, H. (2006). *Northern Sumatra field survey after the December 2004 great Sumatra earthquake and Indian Ocean Tsunami*. Earthquake Spectra 22(S3):S93–S104. doi:10.1193/1.2206793.

BRIZUELA, B., A. ARMIGLIATO, and S. TINTI (2014), *Assessment of tsunami hazards for the Central American Pacific coast from southern Mexico to northern Peru*, Nat. Haz. Earth Syst. Sci., 14, 1889–1903.

FUJII, Y., and K. SATAKE (2007), *Tsunami source of 2004 Sumatra-Andaman earthquake inferred from tide-gauges and satellite data*, Bull. Seismol. Soc. Am., 97, S192–S207. doi:10.1785/0120050613.

GEIST, E.L., S.L. BILEK, D. ARCAS and V.V. TITOV (2006), *Differences in tsunami generation between the December 26, 2004 and March 28, 2005 Sumatra earthquakes*, Earth Planets Space, 58, 185–193.

GEIST, E. L., V. V. TITOV, D. ARCAS, F. P. POLLITZ, and S. L. BILEK (2007), *Implications of the December 26, 2004 Sumatra-Andaman earthquake on tsunami forecast and assessment models for great subduction zone earthquakes*, Bull. Seismol. Soc. Am., 97, S249–S270. doi:10.1785/0120050619.

GEBCO: The General Bathymetric Chart of the Oceans (2006), http://www.ngdc.noaa.gov/mgg/gebco/gebco.html.

GOFF, J., LIU, P.L.-F., HIGMAN, B., MORTON, R., JAFFE, B.E., FERNANDO, H., LYNETT, P., FRITZ, H., SYNOLAKIS, C. (2006). *The December 26th 2004 Indian Ocean tsunami in Sri Lanka*, Earthquake Spectra 22(S3):S155–S172.

GREEN G (1837), *On the motion of waves in a variable canal of small depth and width*, Trans. Cambridge Phil. Soc., 6, 457–462.

GUSIAKOV, V.K. (2005), *Tsunami generation potential of different tsunamigenic regions in the Pacific*, Marine Geology, 215, 3–9. doi:10.1016/j.margeo.2004.05.033.

HÉBERT H., P. HEINRICH, F. SCHINDELÉ, and A. PIATANESI (2001), *Far-field simulation of tsunami propagation in the Pacific Ocean: impact on the Marquesas Islands (French Polynesia)*, Journal of Geophysical Research, 106, C5, 9161–9177.

HÉBERT, H., D. REYMOND, Y. KRIEN, J. VERGOZ, F. SCHINDELÉ, J. ROGER, and A. LOEVENBRUCK (2009), *The 15 August 2007 Peru earthquake and tsunami: influence of the source characteristics on the tsunami heights*, Pure and Applied Geophysics, 166, 1–2, 211–232.

HÉBERT, H., A. SLADEN, and F. SCHINDELÉ (2007), *Numerical modeling of the great 2004 Indian Ocean tsunami: focus on the Mascarene Islands*, Bull. Seismol. Soc. Am., 97, 1A, S208–S222.

HEINRICH, P., F. SCHINDELÉ, S. GUIBOURG and P. IHMLÉ (1998), *Modeling of the February 1996 Peruvian tsunami*, Geophys. Res. Lett., 25, 2687–2690.

IOC (Intergovenmental oceanographic commission) (2013), Tsunami Glossary, Technical Series no. 85.

IOUALALEN, M., J. ASAVANANT, N. KAEWBANJAK, S.T. GRILLI, J.T. KIRBY, and P. WATTS (2007), *Modeling the 26 December 2004 Indian Ocean tsunami; case study of impact in Thailand*, J. Geophys. Res., 112, C07024.

JANKAEW, K., B.F. ATWATER, Y. SAWAI, M. CHOOWONG, T. CHAROENTITIRAT, M.E. MARTIN, A. PRENDERGAST (2008), *Medieval forewarning of the 2004 Indian Ocean tsunami in Thailand*, Nature, 455, 1228–1231. doi:10.1038/nature07373.

JAYAKUMAR, S. D. ILANGOVAN, K.A. NAIK, R. GOWTHAMAN, G. TIRODKAR, G.N. NAIK, P. GANESHAN, R.M. MURALI, G.S. MICHAEL, M. V. RAMANA, G. C. BHATTACHARYA (2005), *Run-up and inundation limits along southeast coast of India during the 26 December 2004 Indian Ocean tsunami*, Curr. Sci. India, 88, 11, 1741–1743.

KANAMORI, H., and D.L. ANDERSON (1975), *Theoretical basis of some empirical relations in seismology*, Bull. Seismol. Soc. Am., 65, 1073–1095.

KOKETSU, K., Y. YOKOTA, N. NISHIMURA, Y. YAGI, S. MIYAZAKI, K. SATAKE, Y. FUJII, H. MIYAKE, S. SAKAI, Y. YAMANAKA, T. OKADA (2011), *A unified source model for the 2011 Tohoku earthquake*, Earth Planet. Sci. Lett. 310, 480–487.

LAVIGNE, F., R. PARIS, D. GRANCHER, P. WASSMER, D. BRUNSTEIN, F. VAUTIER, F. LEONE, F. FLOHIC, B. DE COSTER, T. GUNAWAN, C. GOMEZ, A. SETIAWAN, R. CAHYADI and FACHRIZAL (2009), *Reconstruction of Tsunami Inland Propagation on December 26, 2004 in Banda Aceh, Indonesia, through Field Investigations*, Pure and Applied Geophysics, 166, 1–2, 259–281.

LORITO, S., A. PIATANESI, V. CANNELLI, F. ROMANO, and D. MELINI. *Kinematics and source zone properties of the 2004 Sumatra-Andaman earthquake and tsunami: Nonlinear joint inversion of tide gauge, satellite altimetry, and GPS data* (2010), J. Geophys. Res., 115, B02304. doi:10.1029/2008JB005974.

LøVHOLT, F., H. BUNGUM, C.B. HARBITZ, S. GLIMSDAL, C.D. LINDHOLM, and G. PEDERSEN (2006), *Earthquake related tsunami hazard along the western coast of Thailand*, Nat. Hazards Earth Syst. Sci., *6*, 979–997. www.nat-hazards-earth-syst-sci.net/6/979/2006/.

LIU, P.L.-F., P. LYNETT, H. FERNANDO, B.E. JAFFE, H. FRITZ, B. HIGMAN, R. Morton, J. GOFF, and C. SYNOLAKIS (2005), *Observations by the International Tsunami Survey Team in Sri Lanka*, Science, *308*, 1595.

OKADA, Y. (1985), *Surface deformation due to shear and tensile faults in a halfspace*, Bull. Seismol. Soc. Am., *75*, 1135–1154.

OKAL, E.A., H.M. FRITZ, R. RAVELOSON, G. JOELSON, P. PANČOŠKOVÁ and G. RAMBOLAMANANA (2006), *Madagascar Field Survey after the December 2004 Indian Ocean Tsunami*, Earthq. Spectra, *22*, S3, S263–S283.

OKAL, E.A., A. SLADEN and E.A-S OKAL (2006b), *Rodrigues, Mauritius, and Réunion Islands Field Survey after the December 2004 Indian Ocean Tsunami*, Earthq. Spectra, *22*, S3, S241–S261.

OKAL, E.A. and C.E. SYNOLAKIS (2008), *Far-field tsunami hazard from mega-thrust earthquakes in the Indian Ocean*, Geophys. J. Int, *172*, 995–1015.

ORTIZ, M., and R. BILHAM (2003), *Source area and rupture parameters of the 31 December 1881 Mw = 7.9 Car Nicobar earthquake estimated from tsunamis recorded in the Bay of Bengal*, J. Geophys. Res., *108*, B4, 2215. doi:10.1029/2002JB001941.

PELINOVSKY, E.N., and R. Kh. MAZOVA (1992), *Exact analytical solutions of nonlinear problems of tsunami wave run-up on slopes with different profiles*, Natural Hazards, *6*, 227–249.

POISSON, B., M. GARCIN and R. PEDREROS (2009), *The 2004 December 26 Indian Ocean tsunami impact on Sri Lanka: cascade modelling from ocean to city scales*, Geophys. J. Int., *177*, 1080–1090. doi:10.1111/j.1365-246X.2009.04106.x.

REYMOND, D., E. A. OKAL, H. HÉBERT, and M. BOURDET (2012), *Rapid forecast of tsunami wave heights from a database of precomputed simulations, and application during the 2011 Tohoku tsunami in French Polynesia*, Geophys. Res. Lett., *39*, L11603. doi:10.1029/2012GL051640.

RUDLOFF, A., J. LAUTERJUNG, U. MÜNCH, and S. TINTI (2009), *The GITEWS Project (German-Indonesian Tsunami Early Warning System)*, Nat. Hazards Earth Syst. Sci., *9*, 1381–1382, 2009.

SATAKE, K., T.T. AUNG, Y. SAWAI, Y. OKAMURA, K. S. WIN, W. SWE, C. SWE, T.L. SWE, S.T. TUN, M.M. SOE, T.Z. OO, and S.H. ZAW (2006), *Tsunami heights and damage along the Myanmar coast from the December 2004 Sumatra-Andaman earthquake*, Earth Planets Space, *58*, 2, 243–252.

SAWAI, Y., Y. NAMEGAYA, Y. OKAMURA, K. SATAKE, and M. SHISHIKURA (2012), *Challenges of anticipating the 2011 Tohoku earthquake and tsunami using coastal geology*, Geophys. Res. Lett., *39*, L21309. doi:10.1029/2012GL053692.

SCHINDELÉ, F., A. GAILLER, H. HÉBERT, A. LOEVENBRUCK, E. GUTIERREZ, A. MONNIER, P. ROUDIL, D. REYMOND, and L. RIVERA (2015), *Implementation and Challenges of the Tsunami Warning System in the Western Mediterranean*, Pure and Applied Geophysics, *172*, 821–833.

SLADEN, A. and H. HÉBERT (2008), *On the use of satellite altimetry to infer the earthquake rupture characteristics: application to the 2004 Sumatra event*, Geophys. J. Int., *172*, 707–714 doi:10.1111/j.1365-246X.2007.03669.x.

SMITH, W.H.F., and D.T. SANDWELL (1997), *Global seafloor topography from satellite altimetry and ship depth soundings*, Science, *277*, 1956–1962.

SOLOVIEV, S.L., O.N. SOLOVIEVA, C.N. GO, K.S. KIM and N.A. SHCHETNIKOV (2000), *Tsunamis in the Mediterranean Sea 2000 BC–2000 AD*, Kluwer Academic Publishers, Dordrecht, Netherlands, 243 pp.

SUBARYA, C., M. CHLIEH, L. PRAWIRODIRDJO, J.-P. AVOUAC, Y. BOCK, K. SIEH, A.J. MELTZNER, D. H. NATAWIDJAJA, and R. MCCAFFREY (2006), *Plate-boundary deformation associated with the great Sumatra–Andaman earthquake*, Nature, *440*, 46–51.

SYNOLAKIS, C.E. (1991), *Tsunami runup on steep slopes: how good linear theory really is*, Nat Haz., *4*, 221–234.

SYNOLAKIS, C. E. (2002), Tsunami and seiche. In Earthquake engineering handbook (ed. W.-F. CHEN & C. SCAWTHORN), pp. 9-1–9-90. New York, NY: CRC Press.

SYNOLAKIS, C.E., and E. J. SKJELBREIA (1993), *Evolution of maximum amplitude of solitary waves on plane beaches*, J. Water. Harbor Coast. Ocean Eng., *119*, 3, 323–342.

SYNOLAKIS, C.E., and BERNARD, E.D. (2006), *Tsunami science before and beyond Boxing Day 2004*, Phil. Trans. R. Soc. A, *364*, 2231–2265.

TADEPALLI, S, and C.E. SYNOLAKIS (1996), *Model for the leading wave of tsunamis*, Phys. Rev. Lett., 77, 10, 2141–2144.

TINTI, S., A. ARMIGLIATO, R. TONINI, A. MARAMAI and L. GRAZIANI (2005), *Assessing the hazard related to tsunamis of tectonic origin: a hybrid statistical-deterministic method applied to southern Italy coasts*, ISET Journal of Earthquake Technology 42, 4, 189–201.

TINTI, S., L. GRAZIANI, B. BRIZUELA, A. MARAMAI, and S. GALAZZI (2012), *Applicability of the Decision Matrix of North Eastern Atlantic, Mediterranean and connected seas Tsunami Warning System to the Italian tsunamis*, Nat. Hazards Earth Syst. Sci., *12*, 843–857.

TSUJI, Y., H. MATSUMOTO, Y. NAMEGAYA, W. KANBUA, M. SRIWICHAI, and V. MEESUK (2005), Thailand Tsunami Field Investigation Team, Survey 24 Feb. 4 March 2005. http://www.drs.dpri.kyoto-u.ac.jp/sumatra/thailand2/.

TSUJI, Y., Y. NAMEGAYA, H. MATSUMATA, S.I. IWASAKI, W. KANBUA, M. SRIWICHAI, and V. MEESUK (2006), *The 2004 Indian tsunami in Thailand: surveyed run up heights and tide gauge records*, Earth Planets and Space, *58*, 2, 223–232.

YANASIGAWA, H., S. KOSHIMURA, K. GOTO, T. MIYAGI, F. IMAMURA, A. RUANGRASSAMEE, C. TANAVUD (2009), *The reduction effects of mangrove forest on a tsunami based on field surveys at Pakarang Cape, Thailand and numerical analysis*, Estuarine, Coastal and Shelf Science *81*, 27–37. doi:10.1016/j.ecss.2008.10.001.

ZACHARIASEN, J., K. SIEH, F.W. TAYLOR, R.L. EDWARDS and W.S HANTORO (1999), *Submergence and uplift associated with the giant 1833 Sumatran subduction earthquake: Evidence from coral microatolls*, J. Geophys. Res., *104*, 895–919.

(Received November 30, 2014, revised June 20, 2015, accepted June 29, 2015, Published online July 24, 2015)

Pure and Applied Geophysics

Recent Advances in Agent-Based Tsunami Evacuation Simulations: Case Studies in Indonesia, Thailand, Japan and Peru

ERICK MAS,[1] SHUNICHI KOSHIMURA,[1] FUMIHIKO IMAMURA,[1] ANAWAT SUPPASRI,[1] ABDUL MUHARI,[2] and BRUNO ADRIANO[3]

Abstract—As confirmed by the extreme tsunami events over the last decade (the 2004 Indian Ocean, 2010 Chile and 2011 Japan tsunami events), mitigation measures and effective evacuation planning are needed to reduce disaster risks. Modeling tsunami evacuations is an alternative means to analyze evacuation plans and possible scenarios of evacuees' behaviors. In this paper, practical applications of an agent-based tsunami evacuation model are presented to demonstrate the contributions that agent-based modeling has added to tsunami evacuation simulations and tsunami mitigation efforts. A brief review of previous agent-based evacuation models in the literature is given to highlight recent progress in agent-based methods. Finally, challenges are noted for bridging gaps between geoscience and social science within the agent-based approach for modeling tsunami evacuations.

1. Introduction

The 2004 Indian Ocean tsunami (IOT) was one of the deadliest disasters in recent history due to the extreme tsunami heights combined with insufficient warning, awareness, and early evacuation responses. Since its occurrence, the event has been used as a learning tool. This event likely marked the first time that people watched videos or pictures of a devastating tsunami event. Among those pictures, SYNOLAKIS and BERNARD (2006) noted the surprising images of tourists in Phuket, Thailand, watching the approaching tsunami without taking any protective action. Unfortunately, the Indian Ocean did not have a tsunami warning system at that time, so no warnings could reach the people in the affected areas.

However, would these people have reacted to warning information? This is an important question because tsunami risk involves not only a hazard assessment but also a social component of human behavior because evacuation is the best option for saving lives during a tsunami (SHUTO 2009).

Similarly, the 2011 Great East Japan tsunami (GEJT) was one of the most destructive tsunami events in modern history, although it provided lessons for preparing for future events. The large inundation and tsunami heights (SUPPASRI et al. 2012a) destroyed several towns and villages along the Japanese coast. However, the survival rate of the people living in the inundated areas was 96 % (SUPPASRI et al. 2012b; FRASER et al. 2012).

Fatality rates were lower for the 2011 GEJT event, compared with the 2004 IOT in Indonesia (SUPPASRI et al. 2014; MAS et al. 2013b), despite the numerous towns struck by larger tsunami heights. What caused these different outcomes? A plausible answer could be that not only structural countermeasures but also rapid dissemination of warning information, disaster education, tsunami awareness, and, in particular, evacuation responses contributed to strongly reducing the casualties in the 2011 GEJT. Both tsunami events confirmed the importance of early evacuation and tsunami awareness and the need to develop much more resilient communities with effective evacuation plans.

Therefore, following these lessons provided by the most destructive tsunami events in recent times, we aim to highlight the importance of tsunami evacuation planning and to present a tool to support such an activity. Four international case studies were chosen to describe practical applications of casualty and bottleneck estimation and analyses of vehicles in evacuation, human behavior and shelter demand, all

---

[1] International Research Institute of Disaster Science (IRIDeS), Tohoku University, Sendai, Miyagi, Japan. E-mail: mas@irides.tohoku.ac.jp

[2] Ministry of Marine Affairs and Fisheries, Jakarta, Indonesia.

[3] Graduate School of Engineering, Tohoku University, Sendai, Miyagi, Japan.

of which contribute to tsunami mitigation and evacuation planning. The use of tsunami evacuation simulations attempts to support reconstruction activities in Japan and efforts to develop resilient communities in at-risk areas in Indonesia, Thailand and Peru.

2. Tsunami Evacuation Modeling

The second part of this section presents a comprehensive summary of agent-based models for tsunami evacuation, although the section does not provide an extensive review of all the various models available. We wish to clarify our motivation for focusing on the agent-based approach instead of other methods. First, what are agent-based models?

Agent-based model (ABM): it is a bottom–up approach in which each agent or individual part of a system is modeled as an autonomous decision-making entity. Each agent follows particular rules according for their role in the system; thus, they are able to execute various behaviors. The interaction of these parts and their behaviors develops into a macro description of the system based on an emergent phenomenon.

ABMs are flexible and capture the emergent phenomena from a natural description of a system such as a community and its individual members. Therefore, an ABM is ideal for simulating disaster emergency evacuations (MUNADI et al. 2012) because it provides valuable insight into the mechanisms and behavior that result in jamming or casualties.

There are several other approaches used in tsunami evacuation models that can be found in the literature, e.g., genetic algorithms (PARK et al. 2012), Geographic Information System (GIS) (SUGIMOTO et al. 2003; CLERVEAUX et al. 2008; WOOD and SCHMIDTLEIN 2012; DEWI 2012; FREIRE et al. 2012; GONZALEZ-RIANCHO et al. 2013) distinct or discrete element methods (DEMs) (ABUSTAN et al. 2012), and system dynamic approaches (SIMONOVIC and AHMAD 2005; KIETPAWPAN 2008). However, GIS approaches are traditionally top–down methods that use aggregate descriptions of a system. In evacuation, the complexity and diversity of behaviors that are interrelated produce dynamic changes that GIS models are not well suited to tracking (CASTLE and CROOKS 2006) unless they incorporate micro-scale components provided by ABMs (JOHNSTON 2013). Similarly, system dynamic approaches might be able to track system changes throughout a simulation; however, they lack spatial complexity and require many assumptions about the system if applied to simulating evacuation procedures. In addition, DEMs apply physical laws, such as fluid dynamics, to the evacuees. Such a representation may provide a good description of large crowd behavior during an evacuation (HELBING et al. 2005); however, not all phenomena follow Newtonian motion because psychological forces or sudden changes in motion are likely to occur.

In summary, we choose to focus on ABM due to its benefits over other model techniques for capturing emergent phenomena and providing a natural description of a system. ABM is also flexible for scaling, tuning agent complexity and behavior, and has the capability to use modern data with a higher level of detail.

2.1. Agent-Based Tsunami Evacuation Modeling

One of the first tsunami evacuation models published (USUZAWA et al. 1997) can be found in the Japanese literature. The model simulates the evacuation of Aonae on Okushiri Island, which was affected by the 1993 Hokkaido earthquake. In this initial model, a network modeling approach, which is commonly used to simulate evacuations from hurricanes, floods, nuclear disasters and fires in buildings (WATTS 1987), was used as the modeling method. Following the research of USUZAWA et al. (1997), IMAMURA et al. (2001) described another network model for the same area, but this study included a different start time for each agent in the evacuation and control parameters to distinguish pedestrians from vehicles in the calculation. The computational method uses sequential programming, so all of the evacuees move at the same time and decisions are scheduled only at intersections or the nodes of the network. In addition, in this model, agents jump from node to node in the step that corresponds to the relationship between their speed and road length, limiting the analysis of crowding or the dynamics of pedestrians during the simulation. Later, FUJIOKA et al. (2002) proposed a much more complex representation of evacuees by formally introducing multi-agent systems to tsunami evacuation simulations. The model represented evacuees and guides as agents with different objectives and communication capabilities. This study was one of the most advanced representations of human behavior in the tsunami evacuation field at the time. Although the model uses a combination of network- and grid-based roads, the agent collision avoidance dynamics were limited because the speed was fixed for all of the agents throughout the simulation. In addition, the model returned to the total compliance approach for the evacuation start time, which is less realistic than the different timing approach used by IMAMURA et al. (2001). SAITO and KAGAMI (2004) presented an agent-based model similar to the model of IMAMURA et al. (2001); they modeled the movement of agents using results from a questionnaire about residents' preferences for the start time of evacuations. However, preference surveys alone are not sufficient to describe evacuees' behavior because post-event surveys have demonstrated that actual behavior may differ from the expected behavior. The importance of human behavior in tsunami evacuation simulations gained the attention of researchers in the field (SUZUKI and IMAMURA 2005). Nonetheless, from the point of view of pedestrian dynamics, tsunami evacuation models lacked a clear approach for path finding because the main rule given to agents was to proceed to the next highest node in the network instead of searching for the shortest path or a specific goal. The model developed by KATADA et al. (2004) incorporated a routing method to find the shortest path in the network. Moreover, while previous studies tried to answer specific research questions, Katada's model moved tsunami evacuation simulations from scientific research to practical applications for tsunami mitigation, particularly for disaster education and outreach. The Tsunami Scenario Simulator (KATADA et al. 2000) was developed as a GIS model to investigate information dissemination during disasters; it was later modified (KATADA and KUWASAWA 2006) into the Tsunami Dynamic Hazard Map for disaster education purposes. This tool allowed citizens to dynamically observe the consequences of many of their potential evacuation decisions. Since this modification, the literature on tsunami evacuation models using agent-based techniques has increased considerably (MEGURO and ODA 2005; NOZAWA et al. 2006; WATANABE and KONDO 2009; GOTO et al. 2012).

Increases in computational power have enabled the analysis of large amounts of data and have made it possible to shift the modeling approach from network-based (IMAMURA et al. 2001; LÄMMEL et al. 2010) to grid-based (MAS et al. 2012), potential fields (MEGURO and ODA 2005), or hybrid modeling approaches to improve realism in pedestrian dynamics and collision avoidance behaviors (FUJIOKA et al. 2002; KATO et al. 2009; NGUYEN et al. 2012a). Gradually, the research methodology has moved toward using much more data with finer levels of detail through agent-based modeling and high-performance computing (WIJERATHNE et al. 2013). In addition, the importance of human behavior in evacuations is increasingly being considered in models (SUZUKI and IMAMURA 2005; MAS et al. 2012; FUJIOKA et al. 2002). Following the 2004 IOT and 2011 GEJT, tsunami evacuation modelers have focused on providing practical applications of simulations to solve the particular problems that were observed in these events, such as evacuation timing, bottlenecks and traffic congestion from vehicle evacuations, shelter locations, evacuee behavior, and risk communication, among other factors. In addition, reconstruction in tsunami-affected areas requires new evacuation plans that follow new urban layouts. Effective evacuation plans to be executed under new urban spatial conditions can initially be analyzed and evaluated using evacuation models. In the next section, we describe applications of an ABM tsunami evacuation model that integrate tsunami inundation features and human behavior during evacuations.

3. Case Studies of the Practical Applications of Tsunami Evacuation Simulations

The simulation of tsunami evacuations is becoming important to investigate potential responses to warnings, estimate potential casualties, evaluate evacuation plans and explore options for tsunami mitigation. These experiments are guiding the development of more effective educational and mitigation programs in many countries (BERNARD et al. 2006). Here, we demonstrate several examples of case studies of ABM tsunami evacuation modeling applied to verify, analyze and evaluate actual or predicted tsunami scenarios for evacuations. A large body of literature on agent-based tsunami evacuation models is available for several areas (WIJERATHNE et al. 2013; GOTO et al. 2012; NGUYEN et al. 2012b; ABUSTAN et al. 2012; LÄMMEL et al. 2010; MEGURO and ODA 2005; KATADA et al. 2004); however, for brevity and consistency, we will introduce case studies and mitigation-related results from one agent-based model.

3.1. Casualty and Bottleneck Estimation: Cases of Arahama, Japan and Padang, Indonesia

The agent-based model described in this paper incorporates tsunami inundation modeling outputs and pedestrian and vehicle agent simulation (MAS et al. 2012). The model was verified using the 2011 GEJT data from the evacuation in Arahama, Sendai, Japan (Fig. 1).

One thousand iterations of a stochastic simulation were conducted; each simulation provided the number of evacuees at a shelter, the number of evacuees who passed one of the exits ("safe") and the number of evacuees trapped by the tsunami (the probability of becoming a casualty exceeded 50 %) (Fig. 2). This information at such a high level of detail is one of the greatest advantages of agent-based modeling. In addition to emergent behavior, the behavioral details of each agent and local issues (e.g., traffic and crowd congestion due to the number of agents or the presence of slow-speed agents in front of fast agents) can be identified. The exact values from a real situation are difficult to obtain with stochastic simulations; however, the average number of estimated survivors at the tsunami evacuation building (TEB) showed that the model realistically represents some evacuation decisions and outcomes in the area.

Previous models that did not employ an agent-based approach have also been able to provide casualty estimations and sometimes identify bottleneck areas; however, because those methods (i.e., GIS and DEM) are static or aggregated, information on possible timing and details of interactions, e.g., vehicle–pedestrian interactions, cannot be determined.

The tsunami evacuation simulation demonstrated the capability of the model to identify bottlenecks and to verify the evacuation process with several behavioral conditions within a dynamic framework. The stochastic simulation and the individual level of representation in the model provide the modeler with a reasonable amount of data to analyze and identify

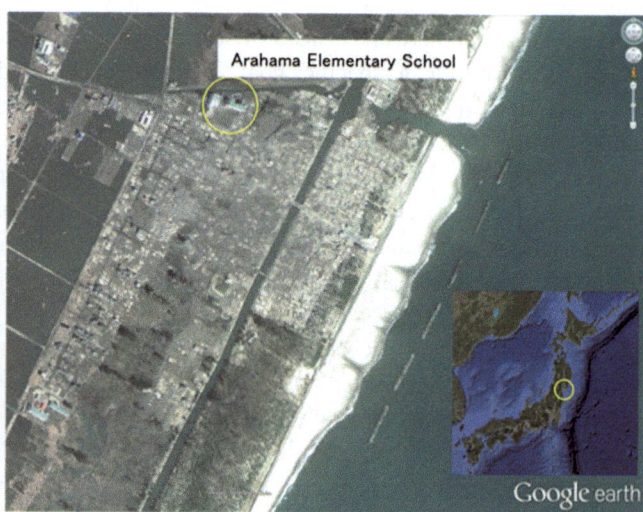

Figure 1
Google Earth images of the Arahama area before (April 4, 2010) and after (March 14, 2011) the tsunami. The *yellow circle* shows the location of Arahama Elementary School

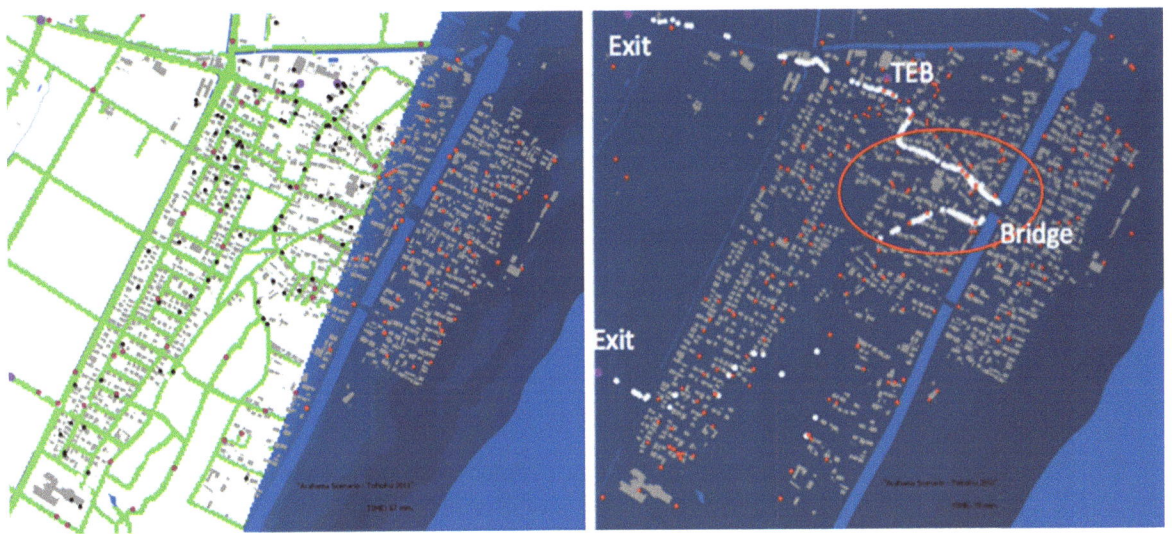

Figure 2
Left A snapshot of the model applied to the Arahama evacuation when the tsunami inundates the study area. The *black dots* are active pedestrian evacuees; the *purple dots* are the evacuees in vehicles, and the *red dots* are fatalities caused by the tsunami. *Right* The last snapshot of the simulation shows bottleneck areas during the evacuation as *white dots*. These bottlenecks mainly occurred on the bridge and in front of Arahama Elementary School, which is the evacuation building

issues at both large and small scales, where agent behavior might not contribute to the safe evacuation of other individuals. For example, agents that are slow to begin evacuating increase their own risk of fatality and reduce the flow of evacuation in the network by creating bottlenecks, especially when their speed is also comparatively low.

In addition to verification of past events, ABM models can be used to assess hypothetical events such as a large tsunami affecting Padang City, Indonesia. IMAMURA *et al.* (2012) described the tsunami hazard in this area based on a mega-thrust earthquake scenario for tsunami simulation using high-resolution bathymetry and topography data. The resulting inundated area was approximately 25 km^2, threatening at least 235,000 people with tsunami depths from 3 m to approximately 8 m (MUHARI *et al.* 2011). Padang city lacks vertical evacuation facilities within the predicted inundation area and based on reports from tsunami evacuations during the 2007, 2009 and 2010 events (HOPPE and MARHADIKO 2009), residents mainly use vehicles and motorcycles for evacuation, despite the experiences of traffic congestion. Consequently, estimating the time needed by evacuees to leave the tsunami inundation area and determining possible congested routes during evacuation are necessary.

A total of 104,352 agents were modeled in a 15 km^2 area in southern Padang using agent-based modeling (MAS *et al.* 2012). Based on the size of the modeled population and the evacuation behavior that considers individual departure times, the model calculated casualty estimates, the time needed for evacuation, and bottleneck points (Fig. 3). In the simulation, the tsunami resulted in the fatalities of approximately 37.7 % of the population. The authors identified several congested streets in the northern study area due to the popular use of the exit points. In addition, the shopping center and traditional market center areas were highly congested due to the high local population density. In the south of the evacuation simulation domain, the agents evacuated to high ground by crossing the river; however, based on the tsunami simulation results, those areas are expected to be inundated due to overtopped river embankments.

The application of the tsunami evacuation simulation in these case studies clarifies the importance of evacuation start times, which cannot be explored with aggregated modeling approaches, the need for high evacuation areas, and the sensitivity of streets to congestion in highly populated urban cities, such as Padang. The same area was evaluated as a case study by LÄMMEL *et al.* (2010) a multi-agent traffic immediate crowding. These characteristics make this scenario suitable for modeling using the agent-based approach. However, areas with wide roads, where crowding might not be expected, may use other approaches, such as the least-cost distance (LCD) in GIS (WOOD and SCHMIDTLEIN 2012), to obtain evacuation timings. In summary, correctly applying tsunami evacuation models depends on the purpose of the simulation and the assumed agent behavior.

3.2. Vehicles in Evacuation: Case of Pakarang Cape, Thailand

Another concern for evacuation planning is the use of vehicles during evacuation. This option must be evaluated according to the characteristics of the environment and the population involved in the evacuation. MAS et al. (2013b) conducted an evacuation simulation for the Pakarang Cape in Thailand, located in the Khao Lak beach resort area of Phang Nga Province on the Andaman Sea, an area devastated by the 2004 IOT (Fig. 4).

A total of 2649 residents were modeled based on a nighttime population scenario. The objective in this case study was to explore the influence of vehicles on the evacuation, combined with different reaction times from the residents. A set of percentages of evacuees in vehicles (passengers and drivers) was assumed for developing several scenarios for the simulation. The evacuation rate as a function of time followed the results of a questionnaire survey (SUPPASRI 2010). In addition to this distribution, three other scenarios were considered: a late evacuation and two intermediate scenarios between the former two distributions. The results of the 20 simulated cases are shown in Fig. 5. Due to differences in reaction time and the long evacuation distance, the use or non-use of vehicles was found to contribute to fatality rates, which ranged from 6 to 34 % of the total population at risk.

The application of the tsunami evacuation simulation in this case study showed the capability of ABM to evaluate the feasibility of evacuations and on-road vehicle–pedestrian interactions. In this case, 20 scenarios with different evacuation starting times and percentages of vehicles used in the evacuation were compared. The results suggest that because of

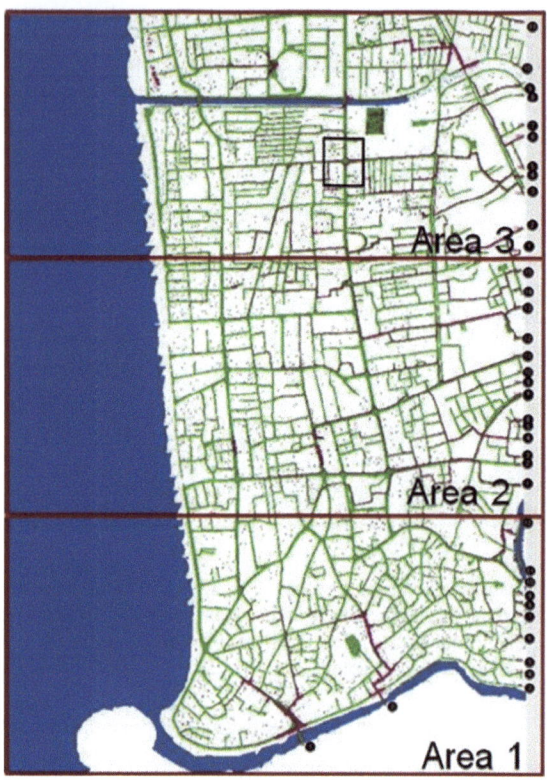

Figure 3
Simulation of evacuation areas, with the bottleneck results shown in *purple*. Details of other results are reported by IMAMURA et al. (2012)

simulator (MATSim) to represent pedestrian agents instead of vehicles. While the approach used in the MATSim simulation differs from the model presented here, similar results were obtained. The models exhibited similar results because the agent behavior in path-finding or route-planning for evacuation was primarily set to use the shortest path. LÄMMEL et al. (2010) noted that a risk–cost function should be added to the route-planning algorithm in MATSim; similarly, in our model, a risk–cost value for the cells near the tsunami flooding areas might be included. However, before such modifications, it is necessary to investigate the residents' preferences in an evacuation to adequately represent the evacuees. It is possible that some people prefer short routes with earlier starting times, while others prefer longer routes with lower risk. It is recommended to pursue an early starting time and low-risk routes during tsunami evacuations. Padang city is a highly populated area with narrow streets that contribute to

Figure 4
Phang Nga devastation from the 2004 Indian Ocean tsunami. *Left* January 13, 2003. *Right* December 29, 2004. Source: Space Imaging/CRISP-Singapore. The *yellow inset* shows the simulated area

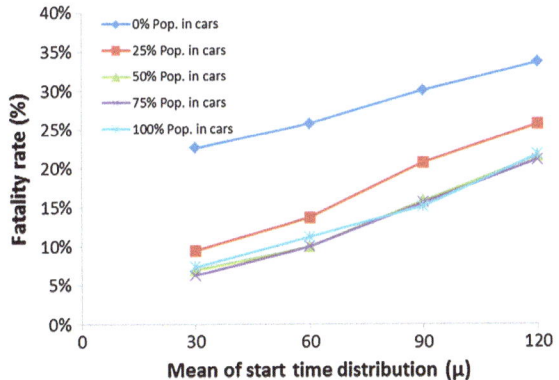

Figure 5
Fatality ratios for different evacuation timing scenarios and the percentages of the population using vehicles. Note the advantage of a fast evacuation decision and the advantage of using cars in this specific case

the distance to the shelters, vehicle evacuation might be necessary. It is possible that the use of vehicles in this area might not result in significant traffic congestion due to the small population and sufficient road capacity. Note that with a larger population than evaluated here, traffic congestion is possible, as shown in the Arahama case study. This suggests that conclusions from tsunami evacuation simulations in one area should not be arbitrarily applied to another area, particularly with regard to restrictions on vehicles for evacuation.

3.3. Evacuation Behavior: Case of Natori, Japan

This is the second application of a tsunami evacuation model to the 2011 GEJT (Fig. 6). TAKAGI *et al.* (2014) simulated the evacuation behavior reported in Yuriage, Natori, to replicate the evacuation process and investigate the reasons for the large number of fatalities in the area. Yuriage is a small town near the Natori River located on the plains of the Miyagi Prefecture. Before the earthquake on March 11, 2011, approximately 5612 residents were living in the area. After the earthquake, 752 people were killed by the resulting tsunami, and 41 are still missing; this event resulted in one of the highest fatality rates in the plains area of Miyagi.

Reports indicate that the residents in the area evacuated to nearby shelters; however, before the tsunami arrived, the tsunami warning was elevated to reflect a larger estimated tsunami height (JMA 2013).

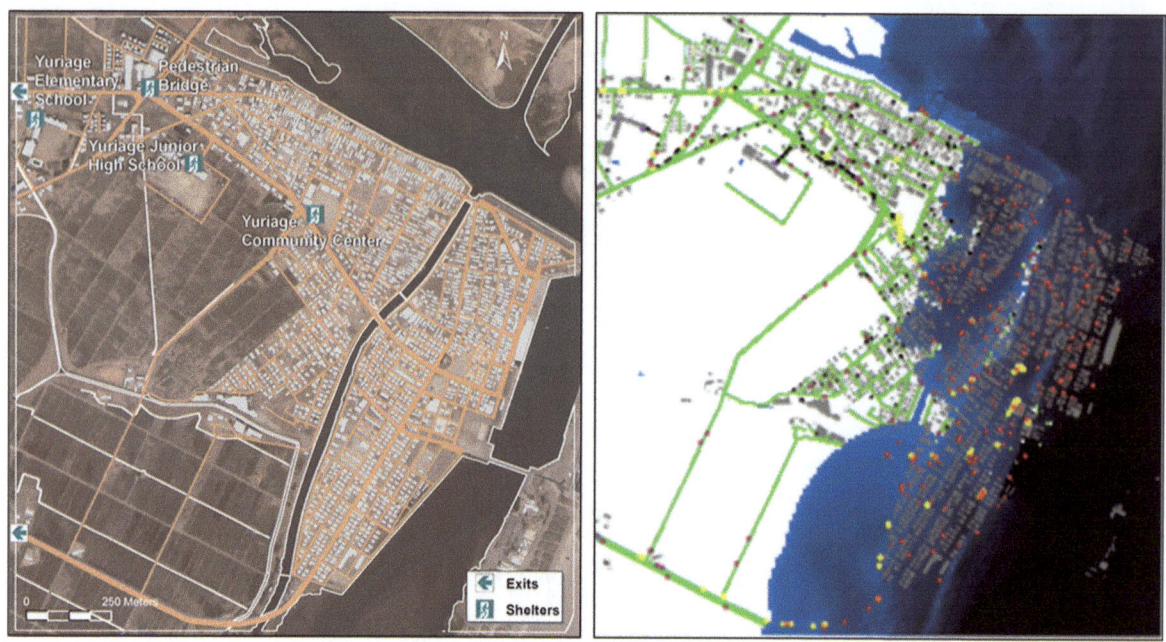

Figure 6
Left The study area of Yuriage in Natori, Miyagi Prefecture, Japan. From *right* to *left* (landwards): the *green signs* indicate the locations of the Yuriage Community Center, Yuriage Junior High School, pedestrian bridge (not official) and Yuriage Elementary School, respectively. The *green arrows* are the exit routes for vehicle evacuations. *Right* A snapshot of the model applied in Yuriage, Natori. The study area is inundated by the tsunami based on the numerical simulation results. The *black dots* are the active pedestrian evacuees; the *purple dots* are evacuees in vehicles; the *red dots* are fatalities due to the tsunami; and the *yellow dots* show the bottleneck points in the simulation

Therefore, over the next few minutes, some of the evacuees decided to conduct a secondary evacuation to a more inland shelter. Evacuees in the community center, which is a 2-story building, moved to Yuriage Junior High School, which is a 3-story building located approximately 500 m inland (Fig. 6). During this second evacuation, the arrival of the tsunami (MUHARI et al. 2012) resulted in pedestrian fatalities. In this case study, the model was applied to two scenarios: (1) Case A, a scenario as close as possible to the real evacuation based on the data reported by local authorities and survivors and (2) Case B, a what-if scenario in which the second evacuation was not performed. The actual reported number of fatalities during the event and the results from the simulation are shown in Table 1 and Fig. 7. From the survey reports, the community center safely sheltered 43 residents (MURAKAMI et al. 2012).

Figure 7 shows the evacuation sequence at each shelter in Case A; note that the community center was filled to its capacity—300 people (Table 1)—approximately 25 min after the earthquake (~15:10 JST). At 15:14 JST, the Japan Meteorological Agency (JMA) issued its first tsunami warning upgrade for the Miyagi coast from 6 m to over 10 m (JMA 2013), which might explain why some people decided to relocate to a more inland shelter; the community center is only a two-story building (KAHOKU SHIMPO 2011), while Yuriage Junior High School is a three-floor building. In addition, information on the damage to the areas in the north where tsunami waves had already struck might also have contributed to their decision. Based on the information provided by survivors, evacuees started moving from the community center to Yuriage Junior High School at approximately 15:30 JST, which agrees with the second tsunami warning information issued by JMA (2013). Therefore, by setting 15:30 JST as the time for the second evacuation in the model, the results showed that a total of 257 evacuees were able to leave the community center before the tsunami arrived. Of the 257 people, only 82 were able to reach

Table 1

Shelter capacity near Yuriage, the outcome of the 2011 tsunami and the results from Case A and Case B

Shelter	Capacity	2011 GEJT	Case A	Case B	B − A
Community center	300	43[a]	43	300	+257
Yuriage Junior High School	2000	∼1000[b]	1050	1067	+17
Yuriage Elementary School	2300	∼870[b]	699	759	+60
Fatalities	–	752[b]	774	436	−338

The last column shows the reduction in the fatalities when no secondary evacuation behavior was exhibited

[a] Murakami et al. (2012)

[b] Natori city in Takagi et al. (2014)

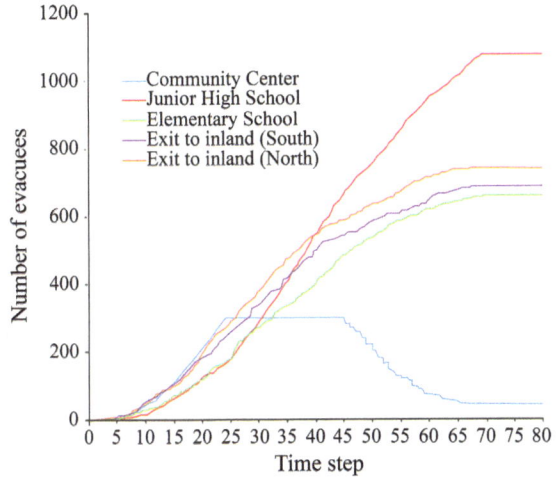

Figure 7
The simulated timeline of the evacuees in the shelters. Note the arrival at the community center (*blue line*). Approximately 25 min was necessary to fill the shelter to its full capacity; no additional evacuees arrived or left until 45 min after the event. Over the next 20 min, people relocated to the junior high school (*red line*), and the tsunami's arrival resulted in the deaths of evacuees en route

the junior high school in time. The reasons for the fatalities and for the evacuees reaching the high school or not are further explored in the simulations. We summarize some of the reasons as follows. (1) The timing of the secondary evacuation: each evacuee conducted a second evacuation between 15:30 JST and approximately 15:50 JST, when the tsunami arrived at the community center. During this 20-min period, the first people to evacuate might have arrived on time, depending on their means of transportation. (2) The means of transportation: based on the survivor accounts and the simulation results, traffic was congested on the road in front of the community center, and people who attempted to evacuate by car might have been delayed because of this situation. However, in Case B, in which agents did not conduct a secondary evacuation and remained at the community center, the total number of fatalities was approximately 44 % less than in Case A, provided that the community center was filled to its capacity.

This case study shows the advantage of agent-based models. Unlike aggregate and static approaches, modeling low-level component behavior and event scheduling is possible in agent-based models. The immense amount of available data related to the GEJT makes it possible to utilize agent-based approaches to examine an evacuation process defined by more than the sum of its parts instead of a global picture with several assumptions. In addition, agent-based models are powerful tools that can be applied to verify and analyze the effects of evacuees' decisions on the outcomes of the evacuation process. Future evacuation plans and activities for the reconstruction process and urban planning can be supported by the results from tsunami evacuation agent-based models. For example, in the Natori area, a new urban layout has been proposed in which new structural countermeasures and resident relocations would be considered. With a more efficient population distribution, improved road networks, shelter availability, and shorter distances to high ground, a lower tsunami risk is expected. However, a tsunami warning can still trigger a massive evacuation in this area, and the characteristics of such a potential evacuation need to be investigated to avoid accidents or fatalities. Plans for a layout suitable for tsunami protection and a massive evacuation can be explored using agent-based tsunami evacuation simulations.

3.4. Shelter Demand: Case of La Punta, Peru

La Punta is a peninsula in the western part of Callao Province. The area is entirely surrounded by the Pacific Ocean, except on its northeastern side where it is bordered by downtown Callao. This district is one of the smallest in Peru, with 4370 inhabitants and a total land area of 0.75 km^2. Historically, earthquakes and tsunamis have struck the area of La Punta in 1586, 1687 and 1746. More than 250 years of seismic inactivity in this region suggests the seismic gap is large enough to trigger earthquakes with an 8.9 magnitude (PULIDO et al. 2013) (Fig. 8).

Tsunami mitigation and preparedness activities have been conducted in La Punta; however, difficulties in conducting frequent evacuation drills with wide population participation suggested it was prudent to apply tsunami evacuation simulations to evaluate the actual conditions of the shelters and the evacuation timing in the area. According to local authorities, 20 official TEBs exist, with enough space to accommodate the entire population (Fig. 9). The authors used the agent-based tsunami evacuation model (MAS et al. 2013a) to investigate a tsunami inundation and the resident evacuation behavior.

A detailed description of the assumptions and constraints for each simulated case can be found in MAS et al. (2013a). In addition to the casualty estimate, more detailed and interesting information is found for the TEBs regarding their capacities and the numbers of evacuees that arrived. At 13 of the 20 evacuation buildings, demand exceeded the available capacity; at the other seven buildings, capacity exceeded the demand, and available space remained. As a consequence, for a total of 4370 residents and buildings with a total capacity for 7930 people, the spatial characteristics of each shelter location produced an imbalance in the preference and number of evacuees during the simulated event. This situation may raise new issues during evacuations, such as

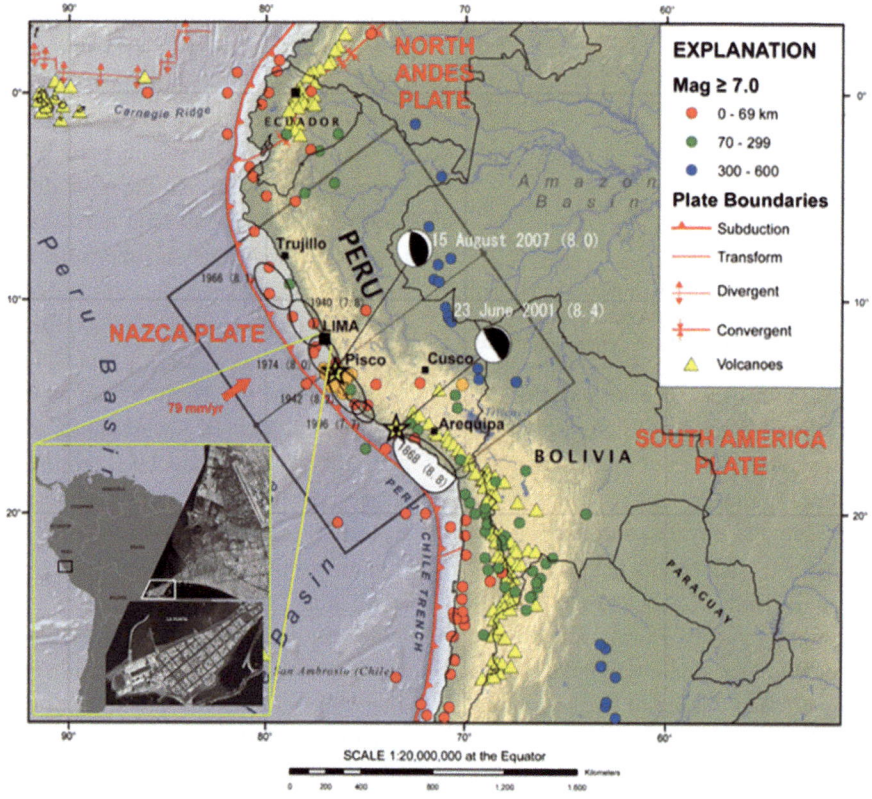

Figure 8
The location of La Punta in Peru and the tectonic settings of the earthquakes in the surrounding regions (modified from YAMAZAKI and ZAVALA 2013; MAS et al. 2013a)

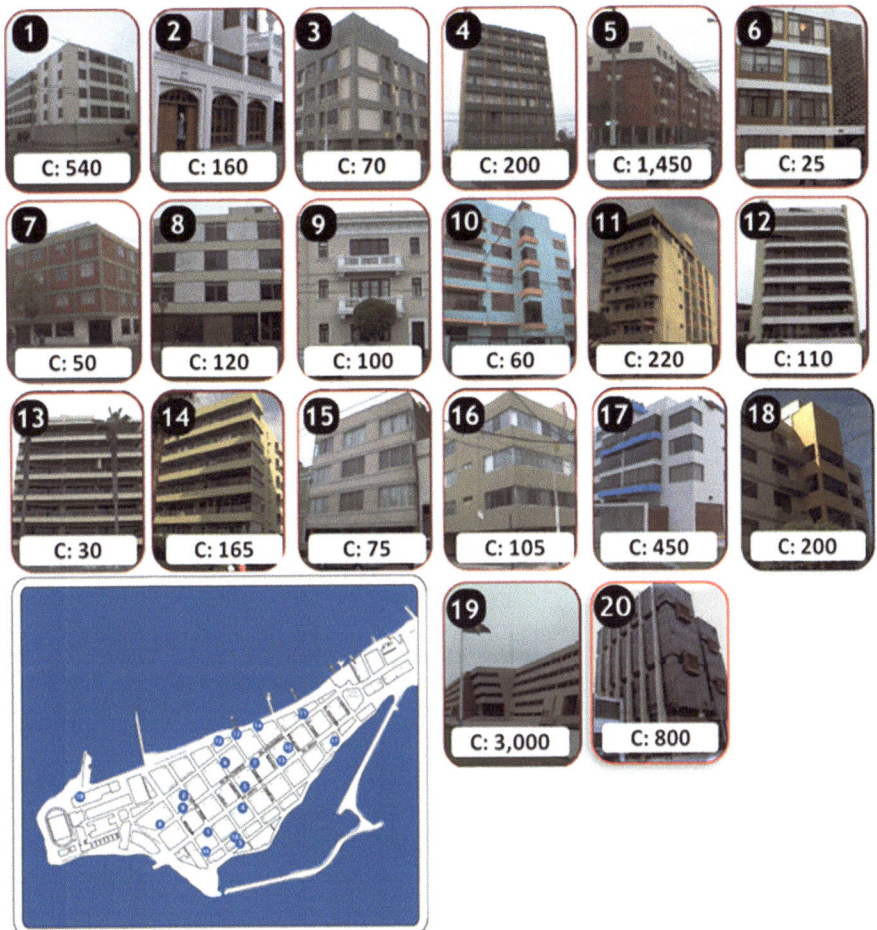

Figure 9
Tsunami evacuation buildings in La Punta. The 20 buildings can accommodate a total of 7930 evacuees in the district of 4370 residents. The *bottom-left inset* shows the locations of these buildings

conducting a secondary evacuation—as in the case of Natori in Japan discussed previously—that delays evacuation and may cause a higher number of fatalities. To inform the local authorities and stakeholders of these results, the capacity–demand rate of the shelters was mapped. The capacity demand index was constructed to represent and easily communicate the spatial issues of the availability of shelters in La Punta (Fig. 10).

The tsunami evacuation ABM was used to reveal the necessity of vertically directed evacuations, particularly in low-lying areas such as La Punta. The outcomes of this study contributed to the identification of risks that could not be identified using static approaches (e.g., GIS least-cost distance analysis, shelter location–allocation solutions, or a direct comparison of the available space versus the number of residents without considering the spatiotemporal issues). The dynamics of a tsunami evacuation simulation are valuable characteristics that should be explored and applied in tsunami mitigation and evacuation planning.

4. Challenges and Future Perspectives

Each time a tsunami occurs, lessons are gathered and shared; unfortunately, not all lessons are fully learned. Similarly, tsunami evacuation research has substantially improved since the 2004 IOT. The future of tsunami evacuation research, as seen by the authors, concerns the comprehensive geophysics of

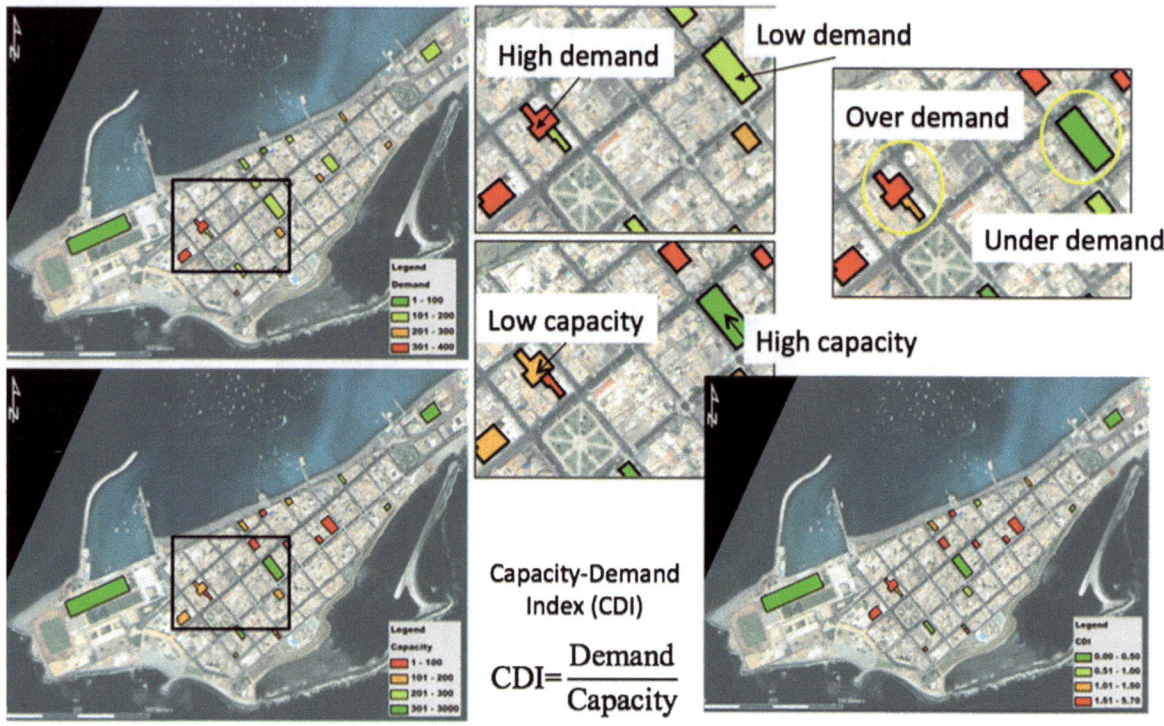

Figure 10
The *top-left inset* shows the evacuee demand for shelter, while the *bottom-left inset* shows the capacity of shelters. The *black square* marks the zoomed-in area with the high/low demand and low/high capacity example of over/under demand evaluations. The *bottom-right* picture is the CDI mapping result for outreach purposes

tsunamis and the physical and psychological traits of the evacuees, all of which would be built into an integrated modeling technique. Efforts to integrate tsunami simulations and social simulations using agent-based modeling (MAS *et al.* 2012, 2013a) and to use supercomputers (WIJERATHNE *et al.* 2013) to bridge gaps between geoscience and social science connect risk assessments with risk management. With these tools, evacuation issues can be understood in a more comprehensive manner. Challenges and emerging issues worth our attention are as follows:

1. Human behavior: Modeling human behavior is not an easy task, but it is not impossible because humans do not behave randomly (KENNEDY 2012). Psychological parameters (VORST 2010), such as evacuees' cognitive and emotional behaviors that affect decision-making or levels of stress and panic should be included. However, capturing individual characteristics in mathematical equations is difficult (PAN *et al.* 2007). Human behavior in tsunami evacuations has been studied using questionnaires. Survey responses provide insight into people's risk perceptions (CHARKNOL and TANABORIBOON 2006; BIRD 2009; GIERLACH *et al.* 2010) and their experiences in real evacuations in post-tsunami events (LACHMAN *et al.* 1961; KATADA *et al.* 2005; SAITO 1990; MAS *et al.* 2011). The statistical outcomes of preference and revealed surveys are incorporated into the agent-based model presented here to stochastically assess uncertainties in human behavior. In the future, we should explore artificial intelligence and cognitive science models in conjunction with agent-based approaches for tsunami evacuation modeling. Agent-based modeling practitioners use these fields quite often to model human behavior (WRAY and LAIRD 2003; SHENDARKAR and VASUDEVAN 2006; DUNIN-KEPLICZ and VERBRUGGE 2010).

2. Verification and validation (V&V) of models: Because of the nature of ABMs, which are based

on simulations rather than equations that can be tested in the laboratory or analytically solved, the process of V&V is difficult (ORMEROD and ROSEWELL 2009). However, replication may aid the model verification process. Specifically, if two distinct implementations of a conceptual model are able to produce the same results, then that outcome supports the hypothesis that the original model correctly implemented the conceptual model (RAND and WILENSKY 2006). Another way to confirm the "inner validity" of the model is by "alignment", which refers to using a different programming language and ABM toolkit to re-implement the model (CASTLE and CROOKS 2006). For validation, real-world data must be collected for comparison with model output; this can be done through controlled evacuation drills. Finally, aggregated data on tsunami evacuations or evacuation drills, such as the number of fatalities in an area or the number of evacuees in shelters, are available, but the validity of the dynamics of the evacuees from the starting point to their shelter and the accuracies of the behavioral models used for agents are uncertain.

3. Use of earthquake disaster "big data": To address the challenges in the two items above, human behavioral models and real-world data can be collected from mobile technologies, which record user locations, speeds, directions, etc. This information was tracked during the 2011 GEJT from mobile phones, car navigation systems and social media. This massive amount of data is known as "Shinsai big data" or earthquake disaster big data. These data can elucidate what occurred on the ground on the day of the tsunami, characteristics of evacuee behaviors, and issues in the process of mobilization; in other words, the collective mind of the society at risk can be explored to calibrate behavioral models and to validate evacuation models. Additionally, virtual big data can also be created to gather additional information on human behavior. Here, we introduce the concept of "virtual big data", which is the information gathered via cloud gaming (LIU et al. 2014) to create a library of human behavior in evacuation scenarios. In the future, real-time big data may be used to support the evacuation of residents during an event.

Other expectations in the field of agent-based modeling that apply to tsunami evacuation models of this nature are discussed by HELBING and BALIETTI (2011), who envision a new way of performing research using supercomputers in data-rich situations. These authors foresee a massive simulation platform with various types of data (i.e., demographic, socio-economic, and geographic) and simulation approaches (i.e., agent-based and equation-based) in large-scale environments within the next 10–15 years (HELBING and BALIETTI 2010).

5. Conclusions

The 2004 IOT and the 2011 GEJT were the two most destructive tsunamis in recent years. Both events emphasize the necessity for effective evacuation plans and rapid evacuation behavior. In addition, the events provided large amounts of data that have been or will be used to produce, verify, validate and improve models to represent the evacuation of populations. Tsunami evacuation models have been developed using several techniques; here, we discussed the agent-based modeling approach because we considered it to be suitable for exploring human behavior and rapid low-level environmental changes using available high-resolution data. An agent-based model was applied to assess tsunami risk and evacuation scenarios in Indonesia, Thailand, Peru and Japan. As described in the case studies, agent-based models benefit tsunami mitigation and evacuation planning by describing the individuality of the evacuees and allowing for the observation of emergent behavior within the dynamics of agent interactions. Agent-based models are flexible and provide a natural description of a particular system. From the perspective of tsunami hazard mitigation, the model presented here provides estimates of casualties, an analysis of evacuee behavior in a two-step evacuation process, identifies bottlenecks, uncovers limitations in shelter capacity and evaluates the use of vehicles in evacuations. All of these outcomes are associated with evacuation planning and cannot be observed solely through evacuation drills or questionnaire surveys. Finally, several challenges to agent-based modeling exist. Past evacuation simulations for tsunamis were unable to model large-scale scenarios and various human traits; present research is considering much finer levels of detail in simulations with a huge amount of data using high-performance computational techniques. Future research should focus on comprehensive and integrated simulations by incorporating complex agent behavior. Engineering, social, psychological and educational sciences should work together to effectively understand, build, apply and share evacuation simulations and outcomes.

Acknowledgments

This research was financially supported by Grant-in-Aid for Scientific Research (Project numbers: 25242035), SATREPS Peru project, CREST project of JST, and IRIDeS grant. In addition, we offer a special thanks to the two anonymous reviewers and the editor in charge of this paper from whom we have received valuable comments that improved the quality of this publication.

Open Access This article is distributed under the terms of the Creative Commons Attribution 4.0 International License (http://creativecommons.org/licenses/by/4.0/), which permits unrestricted use, distribution, and reproduction in any medium, provided you give appropriate credit to the original author(s) and the source, provide a link to the Creative Commons license, and indicate if changes were made.

REFERENCES

ABUSTAN, M. S., HARADA, E., and GOTOH, H. (2012). *Numerical simulation for evacuation process against tsunami disaster at Teluk Batik in Malaysia by multi-agent DEM model.* Proceedings of Coastal Engineering, JSCE, *3*, 56–60.

BERNARD, E., MOFJELD, H. O., TITOV, V. V., SYNOLAKIS, C. E., and GONZÁLEZ, F. I. (2006). *Tsunami: scientific frontiers, mitigation, forecasting and policy implications.* Philosophical Transactions of The Royal Society. Series A, Mathematical, Physical, and Engineering Sciences, *364*(1845), 1989–2007. doi:10.1098/rsta.2006.1809.

BIRD, D.K. (2009). *The use of questionnaires for acquiring information on public perception of natural hazards and risk mitigation a review of current knowledge and practice*, Natural Hazards and Earth System Sciences, Vol. 9, pp. 1307–1325.

CASTLE, C. J. E., and CROOKS, A. T. (2006). Principles and Concepts of Agent-Based Modelling for Developing Geospatial Simulations (No. 110) (Vol. 44, pp. 1–62).

CHARKNOL, T., and TANABORIBOON, Y. (2006). *Tsunami Evacuation Behavior Analysis.* IATSS Research, *30*(2), 83–96.

CLERVEAUX, V. I., KATADA, T., and HOSOI, K. (2008). *Tsunami Scenario Simulator: A Tool for ensuring effective disaster management and coastal evacuation in a multilanguage society.* Science of Tsunami Hazards, *27*(3), 48–71.

DEWI, R. S. (2012). *A Gis Based Approach of an Evacuation Model for Tsunami Risk Reduction.* Journal of Integrated Disaster Risk Management, *2*(2), 1–32. doi:10.5595/idrim.2012.0023.

DUNIN-KEPLICZ, B., and VERBRUGGE, R. (2010). Teamwork in Multi-Agent Systems. A Formal Approach (p. 246). Wiley.

FRASER, S., LEONARD, G. S., MATSUO, I., and MURAKAMI, H. (2012). Tsunami evacuation: Lessons from the Great East Japan earthquake and tsunami of March 11th 2011, GNS Science Report (pp. 1–12).

FREIRE, S., AUBRECHT, C., and WEGSCHEIDER, S. (2012). When the Tsunami Comes to Town – Improving Evacuation Modeling by Integrating High-resolution Population Exposure. In Proceedings of the 9th International ISCRAM Conference (pp. 1–5). Vancouver, Canada.

FUJIOKA, M., ISHIBASHI, K., KAJI, H., and TSUKAGOSHI, I. (2002). *Multi agent Simulation Model for Evaluating Evacuation Management System Against Tsunami Disaster.* Journal of Architecture, Planning and Environmental Engineering, *562*, 231–236. Retrieved from http://ci.nii.ac.jp/naid/110004660530/.

GIERLACH, E., BELSHER, B. E., and BEUTLER, L. E. (2010). *Cross-cultural differences in risk perceptions of disasters.* Risk analysis: an official publication of the Society for Risk Analysis, Vol. 30, pp. 1539–1549.

GONZALEZ-RIANCHO, P., AGUIRRE-AYERBE, I., ANIEL-QUIROGA, I., ABAD, S., GONZALEZ, M., LARREYNAGA, J., GAVIDIA, F., GUTIERREZ, O.Q., ALVAREZ-GOMEZ, J.A., and MEDINA, R. (2013). *Tsunami evacuation modelling as a tool for risk reduction: application to the coastal area of El Salvador.* Natural Hazards and Earth System Sciences, *13*, 3249–3270. doi:10.5194/nhess-13-3249-2013.

GOTO, Y., AFFAN, M., AGUSSABTI, NURDIN, Y., YULIANA, D. K., and ARDIANSYAH. (2012). *Tsunami Evacuation Simulation for Disaster Education and City Planning.* Journal of Disaster Research, *7*(1), 92–101. Retrieved from http://www.fujipress.jp/finder/xslt.php?mode=present&inputfile=DSSTR000700010010.xml.

HELBING, D., BUZNA, L., JOHANSSON, A., and WERNER, T. (2005). *Self-Organized Pedestrian Crowd Dynamics: Experiments, Simulations, and Design Solutions.* Transportation Science, *39*(1), 1–24. doi:10.1287/trsc.1040.0108.

HELBING, D. and BALIETTI, S. (2010) From social simulation to integrative system design. Visioneer white paper. http://www.visioneer.ethz.ch.

HELBING, D., and BALIETTI, S. (2011). How to Do Agent-Based Simulations in the Future: From Modeling Social Mechanisms to Emergent Phenomena and Interactive Systems Design (No. 024) (pp. 1–55).

HOPPE, M., and MARHADIKO, H. S. (2009). 30 Minutes in the City of Padang: Lessons for Tsunami Preparedness and Early Warning from the Earthquake on September 30, 2009. Working Document No. 25. GTZ-GITEWS.

IMAMURA, F., SUZUKI, T., and TANIGUCHI, M. (2001). *Development of a Simulation Method for the Evacuation from the Tsunami and*

Its Application to Aonae, Okushiri Is., Hokkaido. Journal of Japan Society for Natural Disaster Science, 20(2), 183–195. Retrieved from http://sciencelinks.jp/j-east/article/200201/000020020101A0917476.php.

IMAMURA, F., MUHARI, A., MAS, E., PRADONO, M. H., SUGIMOTO, M., and POST, J. (2012). *Tsunami Disaster Mitigation by Integrating Comprehensive Countermeasures in Padang City, Indonesia*. Journal of Disaster Research, 7(1). Retrieved from http://www.fujipress.jp/finder/xslt.php?mode=present&inputfile=DSSTR000700010006.xml.

Japan Meteorological Agency (JMA). (2013). Lessons learned from the tsunami disaster caused by the 2011 Great East Japan Earthquake and improvements in JMA's tsunami warning system October 2013 Japan Meteorological Agency (pp. 1–13). Retrieved from: http://www.data.jma.go.jp/svd/eqev/data/en/tsunami/LessonsLearned_Improvements_brochure.pdf.

JOHNSTON, K. M. (2013). Agent Analyst. Agent-Based Modeling in ArcGIS. (Esri Press, Ed.) (pp. 1–559). California, USA.

KAHOKU SHIMPO (2011). August 3, 2011 published testimony. Retrieved from: http://ameblo.jp/yume-zuki/entry-10987309620.html.

KATADA, T., ASADA, J., KUWASAWA, N., and OIKAWA, Y. (2000). *Development of practical scenario simulator for dissemination of disaster information*. Journal of Civil Engineering Information Processing System, 9, 129–136.

KATADA, T., KUWASAWA, N., KANAI, M., and HOSOI, K. (2004). *Disaster Education for Owase citizen by using Tsunami Scenario Simulator and evaluation of that method*. Sociotechnica, 2, 199–208. Retrieved from http://www.jstage.jst.go.jp/article/sociotechnica/2/0/199/_pdf.

KATADA, T., KODAMA, M., KUWASAWA, N., and KOSHIMURA, S. (2005). *Issues of resident's consciousness and evacuation from the tsunami - From questionnaire survey in Kesennuma city, Miyagi Pref. After the Earthquake of Miyagiken-oki, 2003 -* Proceedings of the Japan Society of Civil Engineers, 789(2), 93–104. Retrieved from http://www.jstage.jst.go.jp/article/jscej/2005/789/789_93/_pdf.

KATADA, T., and KUWASAWA, N. (2006). *Development of tsunami comprehensive scenario simulator for risk management and disaster education*. Transactions of the Japan Society of Civil Engineers (D), 62(3), 250–261.

KATO, S., SHIMOZONO, T., and OKAYASU, A. (2009). *Hybrid Simulation for Tsunami Evacuation in Consideration of Individual Behaviors*. Journal of Japan Society of Civil Engineers, Ser. B2 (Coastal Engineering), 65(1), 1316–1320. doi:10.2208/kaigan.65.1316.

KENNEDY, W.G. (2012) Modelling Human Behavior in Agent-Based Models. In HEPPENSTALL, A. J., CROOKS, A. T., SEE, L. M., and BATTY, M. (2012). Agent-Based Models of Geographical Systems (pp. 1–746). Springer.

KIETPAWPAN, M. (2008). Simulation Approach to Evaluating the Effectiveness of a Tsunami Evacuation Plan for Patong Municipality, Phuket, Thailand. Prince of Songkla University.

LACHMAN, R., TATSUOKA, M., and BONK, W. J. (1961). *Human behavior during the tsunami of May 1960*. Science, 133, 1405–1409. Retrieved from http://www.ncbi.nlm.nih.gov/pubmed/13758063.

LÄMMEL, G., RIESER, M., NAGEL, K., TAUBENBÖCK, H., STRUNZ, G., GOSEBERG, N., SCHLURMANN, T., KLÜPFEL, H., SETIADI, and N. BIRKMANN, J. (2010). Emergency Preparedness in the case of a Tsunami - Evacuation Analysis and Traffic Optimization for the Indonesian city of Padang. In W. W. F. KLINGSCH, C. ROGSCH, A. SCHADSCHNEIDER, and M. SCHRECKENBERG (Eds.), Pedestrian and Evacuation Dynamics 2008 (pp. 171–182). Berlin Heidelberg: Springer. Retrieved from http://www.springerlink.com/index/R0887H6872071Q07.pdf.

LIU, R., DU, J., and ISSA, R.A.R. (2014). *Cloud-based deep immersive game for human egress data collection: a framework*, Journal of Information Technology in Construction (ITcon), Special Issue BIM Cloud-Based Technology in the AEC Sector: Present Status and Future Trends, Vol. 19, pg. 336-349, http://www.itcon.org/2014/20.

MAS, E., IMAMURA, F. and KOSHIMURA, S. (2011). Tsunami Risk Perception Framework for the Start Time Evacuation Modeling, in XXV IUGG General Assembly International Association of Seismology and Physics of Earths Interior, (Melbourne, Australia).

MAS, E., SUPPASRI, A., IMAMURA, F., and KOSHIMURA, S. (2012). *Agent-based Simulation of the 2011 Great East Japan Earthquake / Tsunami Evacuation: An Integrated Model of Tsunami Inundation and Evacuation*. Journal of Natural Disaster Science, 34(1), 41–57. Retrieved from http://www.jsnds.org/contents/jnds/34_1_3.pdf.

MAS, E., ADRIANO, B., and KOSHIMURA, S. (2013a). *An Integrated Simulation of Tsunami Hazard and Human Evacuation in La Punta, Peru*. Journal of Disaster Research, 8(2), 285–295. Retrieved from http://www.fujipress.jp/finder/xslt.php?mode=present&inputfile=DSSTR000800020008.xml.

MAS, E., SUPPASRI, A., SRIVIHOK, P., and KOSHIMURA, S. (2013b). Feasibility of Evacuation at the Pakarang Cape in Thailand based on Tsunami Inundation Model and Human Evacuation Simulation. In 10th International Conference on Urban Earthquake Engineering (pp. 1–6). Tokyo, Japan.

MEGURO, K., and ODA, K. (2005). *Development of Evacuation Simulator for Tsunami Disaster Mitigation*. Production Research, 57(4), 343–347.

MUNADI, K., NURDIN, Y., DIRHAMSYAH, M., and MUCHALIL, S. (2012). Multiagent based Tsunami Evacuation Simulation: A Conceptual Model. In The Proceedings of 2nd Annual International Conference Syiah University 2012 & 8th IMT Uninet Biosciences Conference (pp. 254–259). Banda Aceh, Indonesia.

MUHARI, A., IMAMURA, F., KOSHIMURA, S., and POST, J. (2011). *Examination of three practical run-up models for assessing tsunami impact on highly populated areas*. Natural Hazards and Earth System Science, 11(12), 3107–3123. doi:10.5194/nhess-11-3107-2011.

MUHARI, A., IMAMURA, F., SUPPASRI, A., MAS, E., and KOSHIMURA, S. (2012). *Tsunami arrival time characteristics of 2011 East Japan tsunami revealed from eyewitness, evidences and numerical simulation*. Journal of Natural Disaster Science, 34(1), 91–104.

MURAKAMI, H., TAKIMOTO, K., and POMONIS, A. (2012). Tsunami Evacuation Process and Human Loss Distribution in the 2011 Great East Japan Earthquake - A Case Study of Natori City, Miyagi Prefecture -. In 15th World Conference on Earthquake Engineering (pp. 1–10). Retrieved from http://www.iitk.ac.in/nicee/wcee/article/WCEE2012_1587.pdf.

NGUYEN, T. N. A., ZUCKER, J. D., NGUYEN, H. DU, ALEXIS, D., and VO DUC, A. (2012a). A Hybrid Macro-Micro Pedestrians Evacuation Model to Speed Up Simulation in Road Networks. In AAMAS 2011 Workshop (pp. 371–383).

NGUYEN, T. N. A., ZUCKER, J. D., NGUYEN, M. H., ALEXIS, D., and NGUYEN, H. P. (2012b). *Simulation of emergency evacuation of pedestrians along the road networks in Nhatrang city*. IEEE, 1–6.

Nozawa, S., Watanabe, K., and Kondo, A. (2006). *Development of Evacuation Simulation Model for Tsunami Disaster*. Geographical Information Systems Association, *15*, 483–486.

Ormerod, P., and Rosewell, B. (2009). Validation and Verification of Agent-Based Models in the Social Sciences. In F. Squazzoni (Ed.), Epistemological Aspects of Computer Simulation in the Social Sciences (pp. 130–140). Berlin: Springer.

Pan, X., Han, C. S., Dauber, K., and Law, K. H. (2007). *A multi-agent based framework for the simulation of human and social behaviors during emergency evacuations*. Ai & Society, *22*(2), 113–132. doi:10.1007/s00146-007-0126-1.

Park, S., van de Lindt, J. W., Gupta, R., and Cox, D. (2012). *Method to determine the locations of tsunami vertical evacuation shelters*. Natural Hazards, 891–908. doi:10.1007/s11069-012-0196-3.

Pulido, N., Tavera, H., Aguilar, Z., Nakai, S., and Yamazaki, F. (2013). *Strong Motion Simulation of the M8.0 August 15, 200, Pisco Earthquake; Effect of a Multi-Frequency Rupture Process*. Journal of Disaster Research, *8*(2), 235–242.

Rand, W., and Wilensky, U. (2006). Verification and Validation through Replication: A Case Study Using Axelrod and Hammond's Ethnocentrism Model. In North American Association for Computational Social and Organization Sciences (NAACSOS) (pp. 1–6). Retrieved from http://ccl.northwestern.edu/papers/naacsos2006.pdf.

Saito, T. (1990). *Questionnaire Survey of Human Behaviors and Consciousness on the Tsunami of the 1989 Sanriku-Oki Earthquake*, Journal of Japan Society for Natural Disaster Science, Vol. 9, No.2, pp. 49–63.

Saito, T., and Kagami, H. (2004). Simulation of Evacuation Behavior from Tsunami utilizing Multi Agent System. In 13th World Conference on Earthquake Engineering (pp. 1–10). Vancouver, B.B., Canada. Retrieved from http://www.iitk.ac.in/nicee/wcee/article/13_612.pdf.

Shendarkar, A., and Vasudevan, K. (2006). Crowd Simulation for Emergency Response using BDI Agent Based on Virtual Reality. In L. F. Perrone, F. P. Wieland, J. Liu, B. G. Lawson, D. M. Nicol, and R. M. Fujimoto (Eds.), Proceedings of the 2006 Winter Simulation Conference (pp. 545–553).

Shuto, N. (2009). Tsunami Research - Its Past, Present and near Future -. In Proceedings of the Sixth International Workshop on Coastal Disaster Prevention (pp. 1–24). Bangkok, Thailand.

Simonovic, S. P., and Ahmad, S. (2005). *Computer-based Model for Flood Evacuation Emergency Planning*. Natural Hazards, *34*(1), 25–51. doi:10.1007/s11069-004-0785-x.

Sugimoto, T., Murakami, H., Kozuki, Y., and Nishikawa, K. (2003). *A Human Damage Prediction Method for Tsunami Disasters Incorporating Evacuation Activities*. Natural Hazards, *2035*(29), 585–600.

Suppasri, A. (2010). Tsunami Risk Assessment to Coastal Population and building in Thailand. PhD thesis. Tohoku University.

Suppasri, A., Shuto, N., Imamura, F., Koshimura, S., Mas, E., and Yalciner, A. C. (2012a). *Lessons Learned from the 2011 Great East Japan Tsunami: Performance of Tsunami Countermeasures, Coastal Buildings, and Tsunami Evacuation in Japan*. Pure and Applied Geophysics, *170*(6–8), 993–1018. doi:10.1007/s00024-012-0511-7.

Suppasri, A., Koshimura, S., Imai, K., Mas, E., Gokon, H., Muhari, A., and Imamura, F. (2012b). *Damage characteristics and field survey of the 2011 Great East Japan Tsunami in Miyagi Prefecture*. Coastal Engineering Journal, *54*(1), 1250005–1. doi:10.1142/S0578563412500052.

Suppasri, A., Muhari, A., Ranasinghe, P., Mas, E., Imamura, F., and Koshimura, S. (2014). Damage and Reconstruction After the 2004 Indian Ocean Tsunami and the 2011 Tohoku Tsunami in Tsunami Events and Lessons Learned edited by Kontar, Y., Santiago-Fandino, V., Takahashi, T. Ed. Springer Netherlands, 2014.

Suzuki, T., and Imamura, F. (2005). *Simulation model of the evacuation from a tsunami in consideration of the resident consciousness and behavior*. Journal of Japan Society for Natural Disaster Science, *23*(4), 521–538. Retrieved from http://sciencelinks.jp/j-east/article/200509/000020050905A0347269.php.

Synolakis, C. E., and Bernard, E. N. (2006). *Tsunami science before and beyond Boxing Day 2004*. Philosophical Transactions. Series A, Mathematical, Physical, and Engineering Sciences, *364*(1845), 2231–65. doi:10.1098/rsta.2006.1824.

Takagi, H., Mas, E., and Koshimura, S. (2014). Analysis of the evacuation behavior in Natori, Yuriage during the Great East Japan Earthquake Tsunami. In Annual Meeting of the Tohoku Branch Technology Research Conference, Japan Society of Civil Engineers (Vol. 3, pp. 2011–2012).

Usuzawa, H., Imamura, F., and Shuto, N. (1997). Development of the method for evacuation numerical simulation for tsunami events. In Annual Meeting of the Tohoku Branch Technology Research Conference, Japan Society of Civil Engineers (pp. 430–431). Retrieved from http://library.jsce.or.jp/jsce/open/00322/1997/1997-0430.pdf (In Japanese).

Vorst, H. C. M. (2010). *Evacuation models and disaster psychology*. Procedia Engineering, *3*, 15–21. doi:10.1016/j.proeng.2010.07.004.

Watanabe, K., and Kondo, A. (2009). *Development of Tsunami Evacuation Simulation Model to Support Community Planning for Tsunami Disaster Mitigation*. Journal of Architecture Planning, *74*(637), 627–634.

Watts, J. M. (1987). *Computer Models for Evacuation Analysis*. Fire Safety Journal, *12*, 237–245.

Wijerathne, M. L. L., Melgar, L. A., Hori, M., Ichimura, T., and Tanaka, S. (2013). *HPC Enhanced Large Urban Area Evacuation Simulations with Vision based Autonomously Navigating Multi Agents*. Procedia Computer Science, *18*, 1515–1524. doi:10.1016/j.procs.2013.05.319.

Wood, N. J., and Schmidtlein, M. C. (2012). *Community variations in population exposure to near-field tsunami hazards as a function of pedestrian travel time to safety*. Natural Hazards. doi:10.1007/s11069-012-0434-8.

Wray, R. E., and Laird, J. E. (2003). Variability in Human Behavior Modeling for Military Simulations. In Behavior Representation in Modeling & Simulation Conference (pp. 1–10).

Yamazaki, F., and Zavala, C. (2013). *SATREPS Project on Enhancement of Earthquake and Tsunami Disaster Mitigation Technology in Peru*. Journal of Disaster Research, *8*(2), 224–234.

(Received April 30, 2014, revised May 9, 2015, accepted May 12, 2015, Published online May 20, 2015)

Pure and Applied Geophysics

Interrogation of the Megathrust Zone in the Tohoku-Oki Seismic Region by Waveform Complexity: Intraslab Earthquake Rupture and Reactivation of Subducted Normal Faults

SEMECHAH K. Y. LUI,[1] DON HELMBERGER,[1] SHENGJI WEI,[1,4] YIHE HUANG,[2] and ROBERT W. GRAVES[3]

Abstract—Results from the 2011 Mw 9.1 Tohoku-Oki megathrust earthquake display a complex rupture pattern, with most of the high-frequency energy radiated from the downdip edge of the seismogenic zone and very little from the large shallow rupture. Current seismic results of smaller earthquakes in this region are confusing due to disagreements among event catalogs on both the event locations (>30 km horizontally) and mechanisms. Here we present an in-depth study of a series of intraslab earthquakes that occurred in a localized region near the downdip edge of the main shock. We explore the validity of 1D velocity model and refine earthquake source parameters for selected key events by performing broadband waveform modeling combining regional networks. These refined source parameters are then used to calibrate paths and further simulate secondary source properties, such as rupture directivity and fault dimension. Calculation of stress changes caused by the main event indicate that the region where these intraslab events occurred are prone to thrust events. This group of intraslab earthquakes suggest the reactivation of a subducted normal fault, and are potentially useful in enhancing our understanding on the downdip shear zone and large outer-rise events.

Key words: Waveform modeling, intraslab earthquakes, fault reactivation.

Electronic supplementary material The online version of this article (doi:10.1007/s00024-015-1042-9) contains supplementary material, which is available to authorized users.

[1] Seismological Laboratory, Division of Geological and Planetary Sciences Caltech, California Institute of Technology, 1200 E. California Blvd, MS 252-21, Pasadena, CA 91125, United States. E-mail: klui@caltech.edu

[2] Department of Geophysics, Stanford University, Stanford, CA 94305, United States.

[3] U.S. Geological Survey, Pasadena, CA 91106, United States.

[4] *Present Address*: Earth Observatory of Singapore (EOS), Nanyang Technological University, Singapore 639798, Singapore.

1. Introduction

1.1. Background: Complexity of the Tohoku-Oki Seismic Region

The Mw 9.1 Tohoku-Oki earthquake in 2011 devastated the Northeast coastline of Japan. Analysis of various data sets (i.e. regional and teleseismic broadband seismographic net- works, geodetic networks, ocean-bottom measurements, etc.) demonstrates a unique and complex rupture pattern which indicates varying mechanical and frictional properties along the thrust zone (FUJIWARA et al. 2011; IDE et al. 2011; ITO et al. 2011; SIMONS et al. 2011; YUE and LAY 2011; WEI et al. 2012). Especially intriguing is the difference of frequency content in radiated energy emanating from various parts of the rupture zone. HUANG et al. (2012) shows that the high-frequency radiations in the deeper part are at least partially caused by asperities that have hosted earthquakes before, but the exact mechanism that has caused the variation of such energy concentration is currently unclear. Since the mainshock, the region has also experienced a sharp increase in aftershock activity. These seismic events span a wide range in size and depth. Current national and international earthquake catalogs rely mainly on travel-time data to determine origin time and spatial location. Unfortunately, their results often do not agree, where the hypocentral locations of earthquakes reported by different networks can vary by over 20 km laterally, and by over 10 km vertically depending on the catalog (ZHAN et al. 2012). Furthermore, due to the lack of regional stations on the Pacific side, the resolution decreases with distance from the Japan coast. Such discrepancy creates difficulty in constructing a coherent picture of the thrust zone.

Among the numerous aftershocks, there are a number of earthquakes that occurred along the downdip region of the rupture zone (Fig. 1). Long-period source inversions by the National Research Institute for Earth Science and Disaster Prevention (NIED) placed them at a depth greater than 40 km. According to the slab models used in the JMA catalog and proposed by ZHAN et al. (2012), such focal depths place these seismic events below the slab interface. Despite varying magnitudes, their focal mechanisms, as well as their arriving SV and SH pulses, are akin to one another (Fig. 2). In an earlier study by KATO and IGARASHI (2012), a group of downdip compressional earthquakes located off the coast of Miyagi, Iwate, and Fukushima is also delineated. These events are generally believed to be the outcome of the static stress increase generated by abrupt slip termination during the main shock (LIN and STEIN 2004). The source mechanism and rupture characteristics of these intraslab events are of interest because they occurred beneath the megathrust interface where interesting energy radiation pattern was observed during the mainshock. Within this group of intraslab earthquakes, the largest is a Mw 7.1 aftershock occurred on April 7, 2011, a month after the Mw 9.1 earthquake. Based on detailed seismic tomographic estimation of the 3D velocity structure, NAKAJIMA et al. (2011) identified a low-velocity zone in the focal zone of this aftershock. Hence this event was hypothesized as a possible reactivation of a buried hydrated normal fault that was historically formed in the outer-rise region prior to subduction (Fig. 1 inset). OHTA et al. (2011) also proposed a coseismic displacement model for the same aftershock using regional high-rate GPS data. With a non-linear inversion approach, they estimated a rectangular fault with uniform slip. Even though the results indicated intraslab earthquake characteristics, it was difficult to identify the true rupture plane because the displacement at each GPS site shows similar pattern for both nodal planes. Based on the southward dipping alignment of the aftershock pattern, both NAKAJIMA et al. (2011) and OHTA et al. (2011) considered this event to have occurred on the east-dipping fault plane.

These intriguing observations, as well as the lack of a comprehensive understanding of these intraslab

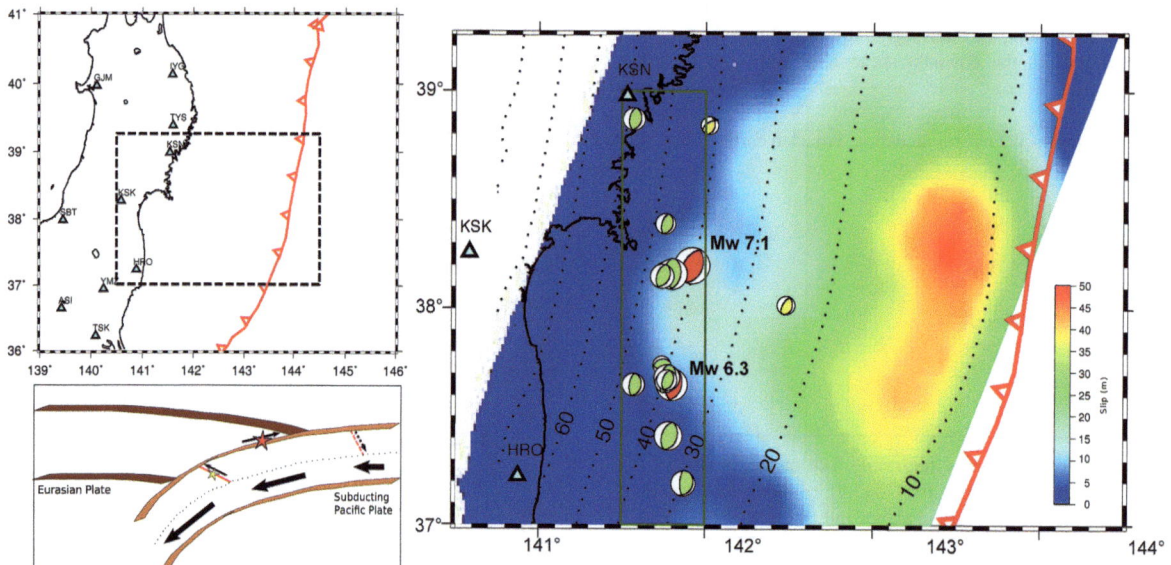

Figure 1

Top left map of the Tohoku-Oki region. *Dashed rectangular box* is the area of the map shown on the *right*. *Right* seismic events in the region 141.5 to 142E and 37 to 39N. Focal Mechanisms marked in *red* represent the Mw 7.1 aftershock and E1 (2011/08/19, Mw 6.3) analyzed in this study. FM marked in *green* are other seismic events with similar strike, dip and hypocenter depth to the two larger events. FM in *yellow* are similar events found outside of the *rectangular region*. Triangles in *cyan* are the three closest F-net stations to E1. The *color scale* represents the slip distribution of the Mw 9.1 main shock in 2011 (WEI et al. 2012). *Dotted lines* are the slab contours. *Bottom left* schematic diagram illustrating the formation of normal fault in the outer rise region which later undergoes subduction and reactivation after the Mw 9.1 Tohoku-Oki event

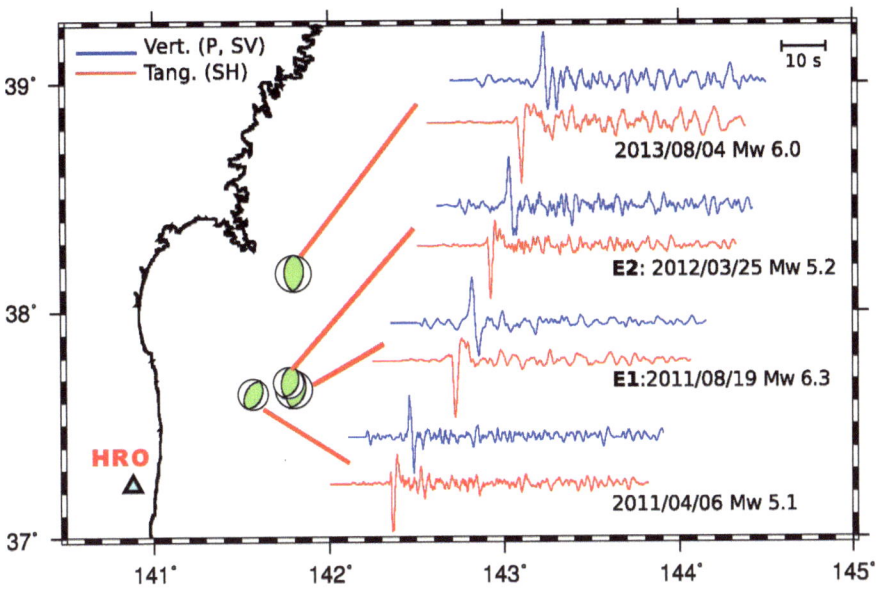

Figure 2
A number of seismic events occurred in the downdip region of the Tohoku-Oki rupture region have displayed very similar focal mechanism and waveform complexity. Waveform (in velocity) shown here are examples recorded by the F-net station HRO: (*blue*) vertical component showing the behavior of P and SV phases, (*red*) tangential component showing the behavior of SH phase. Note the similarity in recording from events labeled E2 and E1

events, have prompted us to investigate this group of earthquakes. In particular, we explore the scope of details that we can extract from existing data, and, by modeling these intraslab events, we test the hypothesis of weakened zones present inside the slab. We also study their correlation with shear stress change in the region caused by the mainshock.

1.2. Overview of Our Study

In order to study the structure of the shear zone and to resolve the source properties of these earthquake in local scale, it requires the analysis of broadband data recorded by regional seismic stations. Therefore we first focus on the validation of seismic velocity models. In particular, we evaluate how well a simplified 1D velocity model can represent the heterogeneous 3D Earth. Based on another aftershock in the downdip region with well-determined source parameters, such detailed comparison allows us to establish which station paths are 1D-like.

With validated velocity models for these selected stations, we then explore the possibility of modeling the rupture characteristics of intraslab events, particularly to identify the true rupture plane. This section focuses on two intraslab earthquakes within the group. One of which occurred on August 19, 2011 with Mw 6.3 (named E1, Fig. 2), while the other occurred 7 months later on March 25, 2012 with Mw 5.2 (named E2). According to the JMA catalog, these two earthquakes have almost identical hypocenters, with a lateral separation of less than 4 km and a 1.8-km difference in depth. Their seismic waveform data also have very similar shapes and frequency content as displayed in Fig. 2. Thus, there is strong evidence indicating that the two earthquakes have occurred at almost the same hypocenter location with similar focal mechanism, but with a difference in moment magnitude. Looking closely at both sets of data, we discover suggestive features of E1 rupturing unilaterally. To verify our hypothesis, E2 is used as the empirical Green's functions to simulate E1 (HARTZELL 1978). Using Taylor-series expansion in time domain, E1 is modeled as the summation of a line of point sources (SONG and HELMBERGER 1996). This search procedure allows the determination of the focal plane on which the earthquake occurred, and gives a robust estimation of fault finiteness. These results, together with the other intraslab events in the region with similar mechanisms, suggests a line of weakness at least

150 km long, which is consistent with the estimation in an earlier study by KATO and IGARASHI (2012).

As an extension from our modeling result, in the last part of the paper, we address the effect of shear stress change caused by the mainshock. While previous studies focus mostly on the shear stress change in the regional scale (KATO et al. 2011; TODA et al. 2011), we study locally the area beneath the downdip edge of the main shock rupture region in order to verify the causal relationship between these intraslab events and the megathrust. Without assuming specific fault geometries or friction coefficients, we evaluate faulting mechanisms that are possible in this location.

2. Methods and Results

2.1. Validity of 1D Seismic Velocity Models

In general, the Earth at teleseismic distances is assumed to be a 1D layered structure, which has velocity varying with depth and does not contain any subducted slab structures, though in reality it is far more complex. In particular, the Japan region involves a complicated tectonic setting, with several tectonic plates, as well as multiple sedimentary basins on land and offshore. With the accessibility to the enormous data set from several major regional seismic networks in Japan, we possess ample resources to characterize waveforms that provide information on earthquake source parameters and rupture properties. Since our emphasis is the detail of the intraslab events in a localized megathrust zone, we begin with testing the validity of 1D velocity model (See Table S1, available in the electronic supplement to this article) by comparing data with 1D and 3D waveform synthetics respectively. In this particular study region, we focus on regional data collected from seismic networks in Japan (F-net, K-net and Kik-net), with most stations within 500 km from the earthquake sources. Here the Green's functions were computed for 1D velocity model listed in Table S1, using a frequency-wave number integration method (ZHU and RIVERA 2002). For 3D synthetics calculation, the 3D Japan Integrated Velocity Structure Model (JIVSM; KOKETSU et al. 2008) is used, which includes a slab structure. We apply the staggered-grid finite difference technique to model 3D wave propagation (GRAVES 1996), with a grid spacing of 0.4 km and

synthetic waveform frequency up to of 0.25 Hz. Focal mechanism inversions in this part of the study is done with the Cut-and-Paste (CAP) method using 1D Green's functions, which has the advantage of performing inversion on selected portions of Pnl and surface waves with timing shifts allowed among segments (ZHAO and HELMBERGER 1994; ZHU and HELMBERGER 1996). We performed a detailed comparison between the synthetic waveforms generated with a 1D velocity model and those from the 3D calculation, to establish which station paths are 1D-like.

The comparison between 1D and 3D synthetics for these paths indicates that within this proximal distance, 1D velocity structure surprisingly has as good resemblance of the real Earth just as the 3D model (Fig. 3). The 3D synthetics are systematically shifted to arrive earlier by 2 s, which implies a generally faster 1D medium. Nonetheless, for synthetic seismograms filtered up to 0.25 Hz, major features such as the SV arrival are captured by both 1D and 3D models. For regional stations within 200 km that have high cross correlations (cc) between 3D synthetics and data, comparison between 1D synthetics and data also show similar results

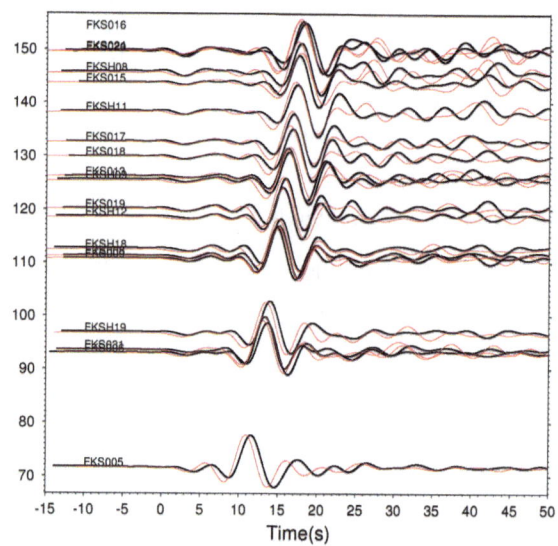

Figure 3
Comparison between 1D (*marked in red*) and 3D (*marked in black*) synthetic waveforms of event E1. Vertical velocity waveform is shown here. Stations selected are between the azimuthal range of 240° and 270°, at the nearest distances. A bandpass filter with cutoff frequencies 0.03 and 0.25 Hz is applied to the synthetics. Also, waveforms in the plot are normalized and the 3D synthetics are systematically shifted to arrive earlier by 2 s

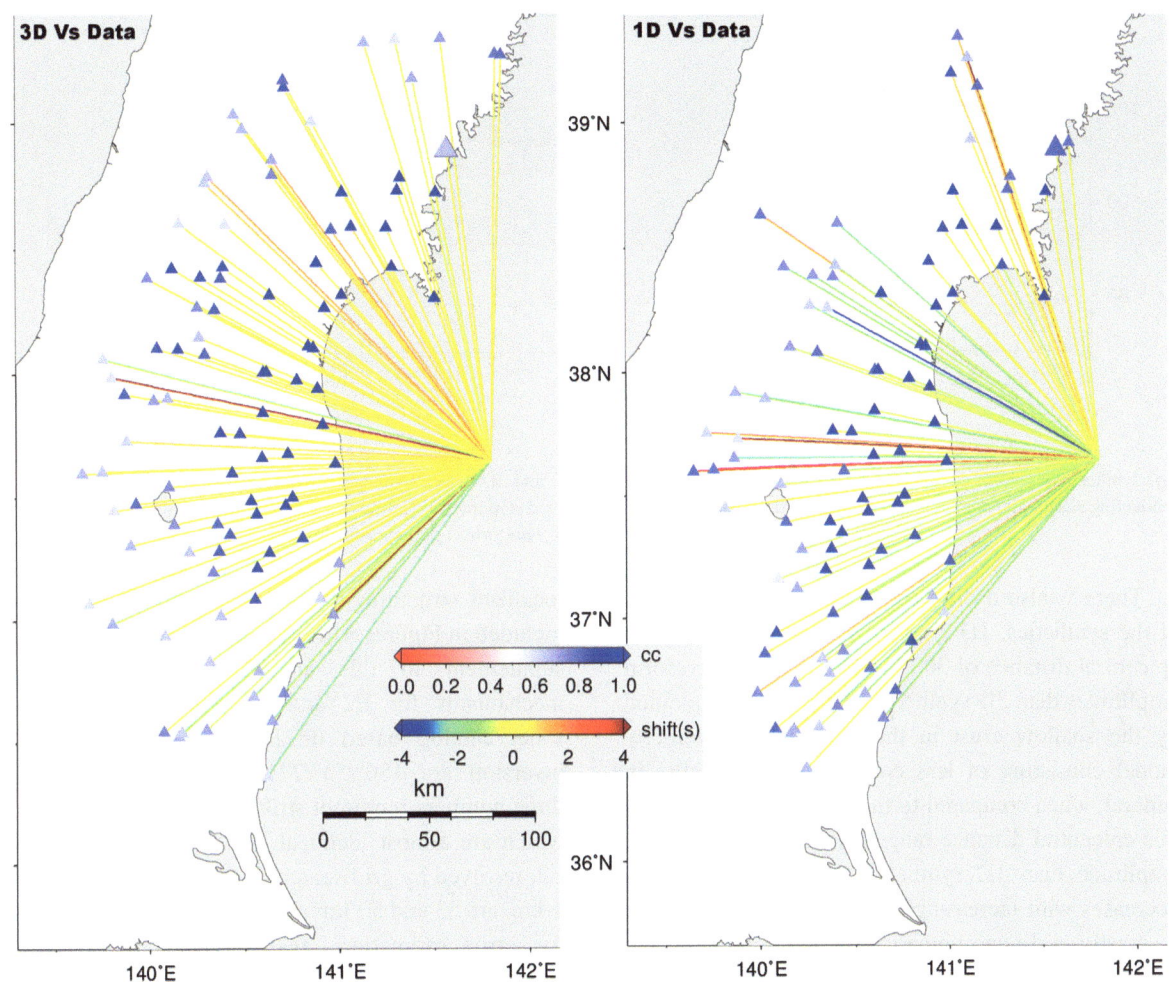

Figure 4
Left stations (*triangles*) are colored by the cross-correlation coefficients (cc) between 3D synthetics and waveform data (vertical component) of event E1. Only stations with >70 % cc values are plotted. *Colored lines* indicate time-shift value to align the data and synthetics, ranging from −4 to 4 s, with positive value representing faster synthetics. *Right* similar to figure on the *left*, except that this is showing the cross-correlation between 1D synthetics and waveform data

(Fig. 4). Disregarding the systematic 2-second shift of 3D synthetics, both models have comparable time shifts values. Note that the 3D synthetics still contain an additional bias relative to the 1D synthetics, i.e. more yellow on average. However, much of the difference between 1D and 3D appears to be caused by the shallow structure, in that data from stations to the northwest are quite late (green) for 1D synthetics while slightly early (gold) for the 3D model. This feature can be expected based on shallow structures embedded in the 3D models, i.e. see Figure S13 of WEI et al. (2012). This comparison is complicated for several reasons: (1) origin times and locations are fixed by the seismic model, (2) the fits displayed are sometimes different for stations where 3D models have strong effects on SV arrivals, and (3) directivity effects are affected by the structure as discussed later. Besides 1D and 3D comparison, we further analyze the similarities among synthetics of different 3D velocity models. Two models are used here—JIVSM and another 3D velocity model from NIED with no subducting slab structure. Results indicate that within 100 km from the epicenter, synthetics generated with the velocity model without a subducting slab are similar to those generated with velocity model having a subducting slab (Figs. 4, 5).

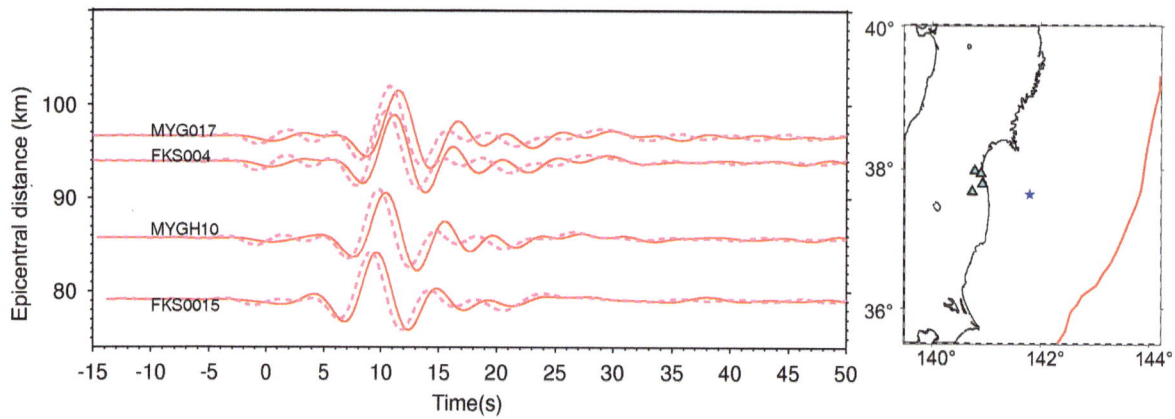

Figure 5
Left synthetic waveform comparison of event E1 generated from two different 3D velocity models. *Solid lines* represent the model with a subducting slab, while *dotted lines* represent model with no slab. *Right* selected stations are within 100 km from the epicenter (*cyan triangles*), with azimuth between 270° and 300°. The *blue star* is the epicenter of E1

There is also a difference in the wave amplitude of the synthetics. 1D synthetics at stations within an epicentral distance of 300 km have higher waveform amplitudes than 3D synthetics. This can be explained by the shallow crust in the 3D (JIVSM) velocity model consisting of less consolidated materials. In general, when compared to the data, 3D synthetics at this epicentral distance range also have more similar amplitude than 1D synthetics, but the discrepancy decreases with increasing epicentral distance.

Lastly, a direct comparison is made between the 1D and 3D synthetics by running 1D CAP inversions with 3D synthetics as data. A substantial number of stations have cc > 70 % for significant phases (P and S). Given such results, we are confident in using the 1D CAP results in our following analysis.

2.2. Source Mechanism and Rupture Characteristics of Selected Intraslab Events

We perform point-source inversions on two selected events, E1 and E2, and compare their focal mechanisms (Fig. 6). Here we use CAP to perform inversion on waveforms from all F-net stations within an epicentral distance of 500 km. Results indicate that the CAP analysis can resolve not only similar focal mechanism as reported by JMA, but also very consistent focal mechanisms for waveform filtered at different frequency bands, even up to 1 Hz. This serves as a useful tool in studying complicated localized structures. (Detailed inversion results are included in Figures S1 to 4, available in the electronic supplement to this article) Furthermore, focal mechanisms for E2 and E1 resolved by JMA's F-net catalog based on long-period point source inversion are 186°/53°/77° and 183°/53°/81° (the three numbers represent strike/dip/rake) respectively, which are almost identical. Focal depths of E1 and E2, resolved by grid-search analysis with grid size of 1 km, are 53 and 50 km respectively (See Figure S5A for error estimation, available in the electronic supplement to this article). They are slightly shallower than F-net catalog's result, but the 3-km difference reinforces the idea that the two events are indeed in close proximity. Moreover, the time-shift values associated with the segments of the arriving P wave are also very similar between E1 and E2 (less than 2 s, Fig. 6), which indicates a very short relative separation. This result reinforces the lateral separation of less than 4 km recorded by the JMA catalog, in which both epicenters are approximately 70 km to the east of the coastline and 180 km from the Japan Trench.

Given the almost identical hypocenter location and well-resolved source parameters, we are prompted to study the actual data in detail. In terms of frequency content, waves traveling to the south are almost identical for both earthquakes. On the other hand, for northern stations, E1 in general has lower frequency than does E2 (Fig. 7a). It is clear that E1

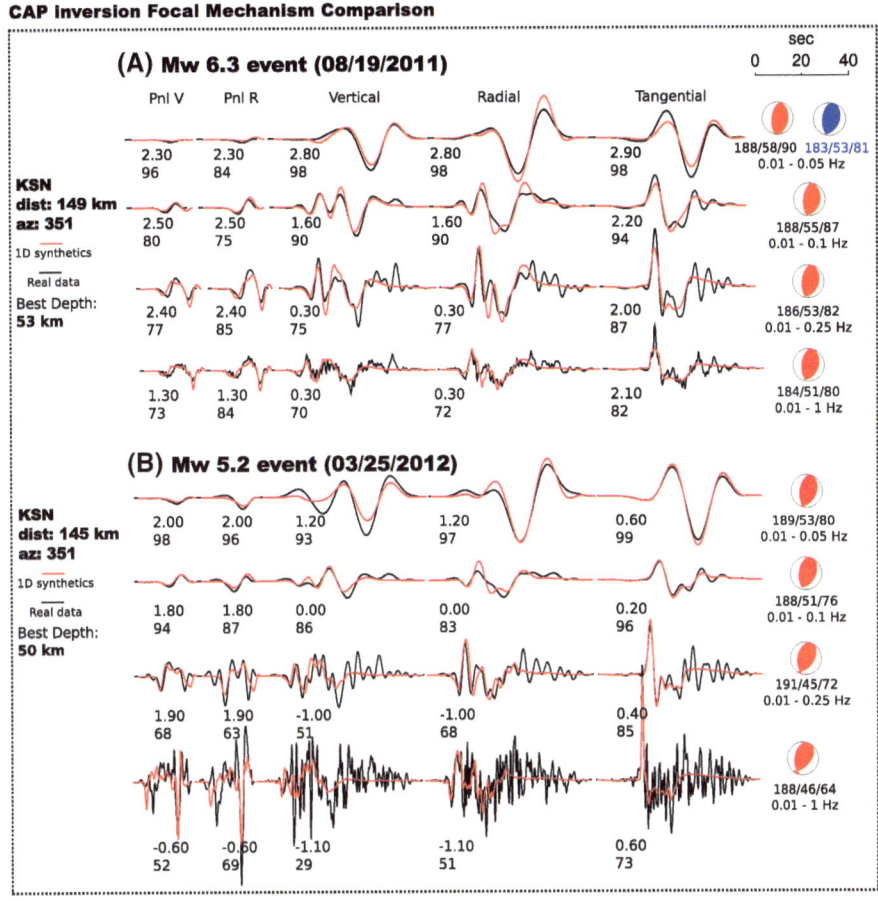

Figure 6
a CAP inversion of E1 waveform data (bandpass filter applied from *top to bottom* up to 20, 10, 4, and 1 s) at F-net station KSN. *Black lines* are real data and *red* are 1D synthetics. Focal spheres in *red* are CAP results, and the one in *blue* is inversion result from JMA F-net catalog. The three-number sets indicate strike/dip/rake values. The two numbers below the waveforms are the relative time shift (*top*) and the cross correlation values (*bottom*). **b** CAP inversion of E2 waveform data at the same station, also up to 1 s

has a wider arriving SV pulse than E2. Furthermore, there are distinctive differences in SH and SV amplitude ratio between stations in opposite directions away from the epicenter. A comparison of SV amplitude ratio among all F-net stations displays an increasing trend from north to south (Fig. 7b). Assuming both earthquakes being highly similar in nature, such observations strongly indicate directivity of E1 rupturing to the south. With both E1 and E2 being intraslab events, they could rupture on either of the focal planes. We therefore search for the actual rupture plane by using E2 to generate the Empirical Green's Functions (EGFs). The moment ratio between E1 (Mw 6.3) and E2 (Mw 5.2) is approximately 45. We therefore discretize the fault into a line of 45 elements, each represented as a point source of E2. Each point source position is then varied by a small time variance depending on their shift in horizontal and vertical direction from the original point source (see Appendix). Assuming the two earthquakes began their rupture at the same spot, E1 can be treated as the summation over all elements. We simulate four simple scenarios, with rupture on each of the two auxiliary focal planes, directing to north or south respectively. The fault geometries are based on CAP inversion with data up to 0.25 Hz. Strike/Dip values are 186°/53° for plane 1 and 19°/37° for conjugate plane 2.

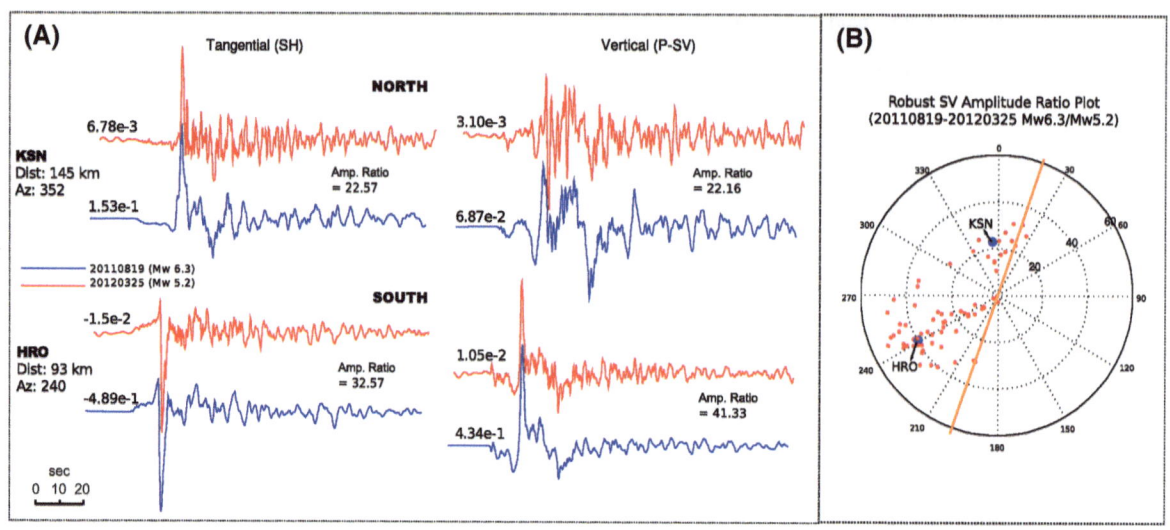

Figure 7

a Broadband waveform comparison (in displacement) for two similar earthquakes (E1 and E2) with an order of difference in Mw at the same two stations. KSN is a station to the north while HRO is a station to the south. E1 has signal in *blue* and E2 in *red*. Numbers stated in the beginning of the waveform are the displacement amplitudes (in cm). Amplitude ratios between the two events (SH on left column; SV on right column) are also shown. **b** Amplitude ratio of SV waves as a function of azimuth with each *dot* representing one station. Here the scale on the concentric circles is the amplitude ratio. Two stations in (**a**) are highlighted in *blue*. The *orange line* indicates the strike angle of E1

With fixed rupture direction and fault geometry, we search for a wide range of rupture velocity and rupture length to obtain the simulation with the lowest misfit error between E1 broadband data and the corresponding EGFs (see Table S2 and S3, available in the electronic supplement to this article). The misfit error is the summation of l^2 norm, weighted by their corresponding cc. Here we explore both displacement and velocity data and find consistent results, with the same best rupture velocity resolved and a rupture length difference of 1 km. Results indicate that E1 can be represented by 45 point sources, each 0.18 km apart, rupturing diagonally to the south on the plane dipping 37° to the east, with a rupture velocity at 4.5 km/s over a distance of 8 km (Fig. 8). We consider 4.5 km/s a reasonable rupture speed, given the event occurred inside the mantle where shear wave speed is 4.5 km/s or higher. Assuming a square rupture dimension with diagonal length of 8 km ($\sim 5.6 \times 5.6$ km^2), E1 has generated an offset of approximately 3 m. For data and EGF comparisons of different station components, see Figure S6 and S7 (available in the electronic supplement to this article).

2.3. Intraslab Thrust Events and Stress Change from the Mw 9.1 Main Shock

Since March 11, 2011, numerous studies have been focusing on the state of stress changes in the seismogenic zone (HASEGAWA et al. 2011; KATO et al. 2011; TODA et al. 2011; ZHANG et al. 2008). The series of intraslab events observed occurring along this long weakened zone inside the slab also seems to suggest a causal effect. With well resolved source mechanism and rupture characteristics of E1, we extend our study to explore whether the faulting of E1, or other intraslab thrusting events in neighboring region with similar mechanism, is directly related to the Mw 9.1 main shock. Here we calculate the static stress change induced by the slip of the mainshock. We use the slip distribution obtained by HUANG et al. (2013), who in their simulation reproduce the final slip distribution model and stress drop of the main shock. Their result is consistent with the general final slip inferred for this earthquake (see Figure S8, available in the electronic supplement to this article). With calculated stress drops and based on the Coulomb model, we model

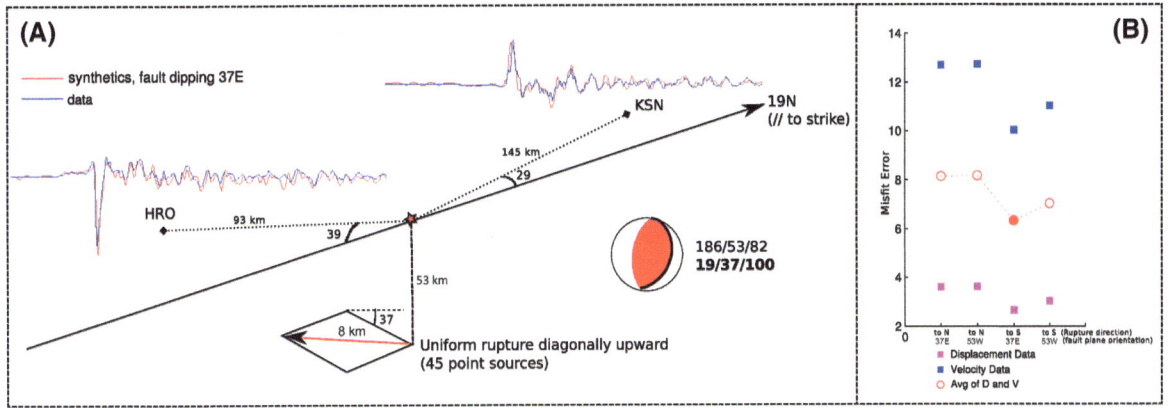

Figure 8

a Geometrical setting of the study on secondary source parameters: Here we simulate E1 waveform (*blue*) using a summation of 45 point sources equivalent to E2, each with a small amount of time shift calculated using power series expansion based on general ray theory. Four rupture patterns are tested and the one with highest cross-correlation (cc) value between data and synthetic is shown, with estimated rupture direction and total rupture length. Focal mechanism of E1 is also shown here, with *black line* on the focal sphere indicating the preferred rupture fault plane. **b** Comparison of misfit error among four rupture simulations. Rupture toward the South on the fault plan dipping 37°E has the lowest average misfit error (*filled red circle*) for both displacement and velocity data. The misfit error is the l^2 norm weighted by their corresponding cc value. Detailed quantitative comparison is shown in Table S2 and S3 (available in the electronic supplement to this article)

the differential stress $(\sigma_1 - \sigma_2)$ caused by the megathrust rupture (Fig. 9, top) to the surrounding area, where σ_1 and σ_2 are the maximum and minimum stresses in principal directions. The maximum shear stress $(\sigma_1 - \sigma_2)/2$ is concentrated at the portion of the megathrust approximately 90 km from the trench (0 km), which also coincides with the downdip end of the rupture plane in the model proposed by WEI et al. (2012).

The effect of maximum shear stress varies along area adjacent to the megathrust (Fig. 9, bottom). Depending on specific locations, it favors different faulting mechanisms, so either normal and thrust events could possibly be triggered, both below and above the megathrust interface. Our analysis shows that the region between 0 and 5 km below the downdip tip of the interface is favorable for thrusting events (resolved as negative dip angles in the simulation) on fault dipping between 20° to 50°. The dip angles of the intraslab events discussed here lie well within this range. Our result is consistent with the hypothesis that this sequence of intraslab aftershocks, which possibly ruptured a continuous weakening zone inside the slab, are the results of the stress change induced by the main shock.

3. Discussion

The epicenters of E1 and E2 are only 50 km south of the Mw 7.1 intraslab aftershock that took place on April 7, 2011. Within the region, there is also a series of events that possess focal depth within the intraslab range and fault-plane orientation similar to that of E1, with strike and dip values ± 10°. Altogether, they form a narrow line of events between latitude 37° to 39° from north to south (Fig. 1). Interestingly, intraslab events with such features are almost completely absent from neighboring area. Such events also did not exist in this region before the Mw 9.1 event. These evidence all point to the possibility that this group of earthquakes have occurred on a north–south striking fault dipping about 35° eastward, which is hypothesized to be a weak hydrated zone reactivated as a thrust fault after March 11, 2011 (NAKAJIMA et al. 2011), spanning a distance of 150 km. Since the occurrence of the Mw 9.1 earthquake, numerous studies are prompted to study the physical mechanism of the megathrust that drives such unique rupture pattern. So far it is still proven very difficult to solve the problem. Given the intriguing coincidence that these intraslab events are located in close proximity to the source of high-

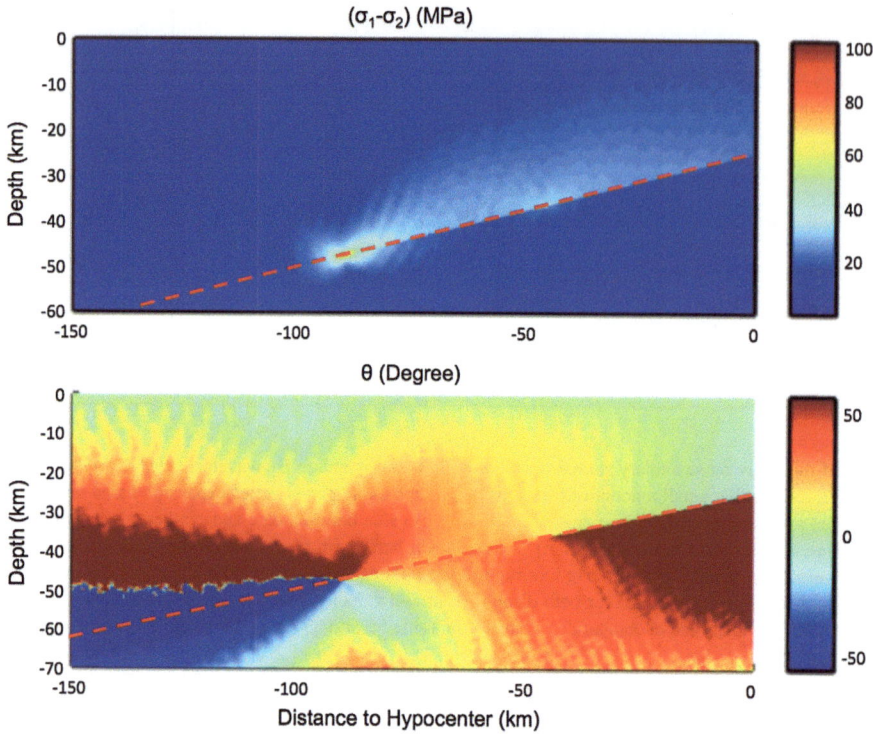

Figure 9
Top Differential stress ($\sigma_1 - \sigma_2$) caused by the Mw 9.1 earthquake to neighboring region. HUANG et al. (2013) assumes a megathrust dipping 14° (*dashed red line*). *Bottom* mechanism of earthquakes as a possible outcome produced by this differential stress. Positive (*warm color*) degrees imply normal faulting, while negative (*cold color*) degrees suggest thrust faulting

frequency radiation during the main event, these intraslab earthquakes in the downdip area can be valuable assets in enhancing our understanding of this local region.

Similar to the forward modeling technique implemented by TAN and HELMBERGER (2010), our study also uses the EGF approach and simulate earthquake ruptures. We show that using two co-located seismic events, the details of fault rupture can be extracted. Assuming this group of intraslab earthquakes occur on the same continuous weakened zone, the dip angle resulted from our inversion (37°E) is in fact very similar to the estimates in studies that are based on the spatial distribution of aftershocks. OHTA et al. (2011) simulated the Mw 7.1 aftershock with the assumption of a fault plane dipping 35.3°E. NAKAJIMA et al. (2011) discovered that the angle between this fault plane and the dip of the slab surface is approximately 60°. Among typical subduction models of this region, the slab at this particular distance from the trench is dipping about 35° westward. This implies that the fault plane on which E1 occurred is dipping about 35°E. These intraslab events are all located within 10 km from the plate interface, thus it is important to resolve for the dimension and directivity of these ruptures in order to estimate the extent of disruption these smaller events can have on the megathrust shear zone.

In fact, the high resolution of the directivity study can be a useful tool in refining the source locations, especially for earthquakes lacking azimuthal coverage of seismic stations. According to JMA, the epicentral location of the E1 and E2 are laterally separated by less than 4 km. Thus, using CAP inversion method, one should expect similar time shift between synthetics and data for both events. However, for waveform filtered up to 0.1 Hz, the resulting time shift of the S arrival differs by as much as 2.5 s between E1 and E2 (Figs. 6, 10a). Such a discrepancy does not necessarily imply incorrect epicenter locations. In particular, we find that in a long-period inversion, one should consider using the centroid

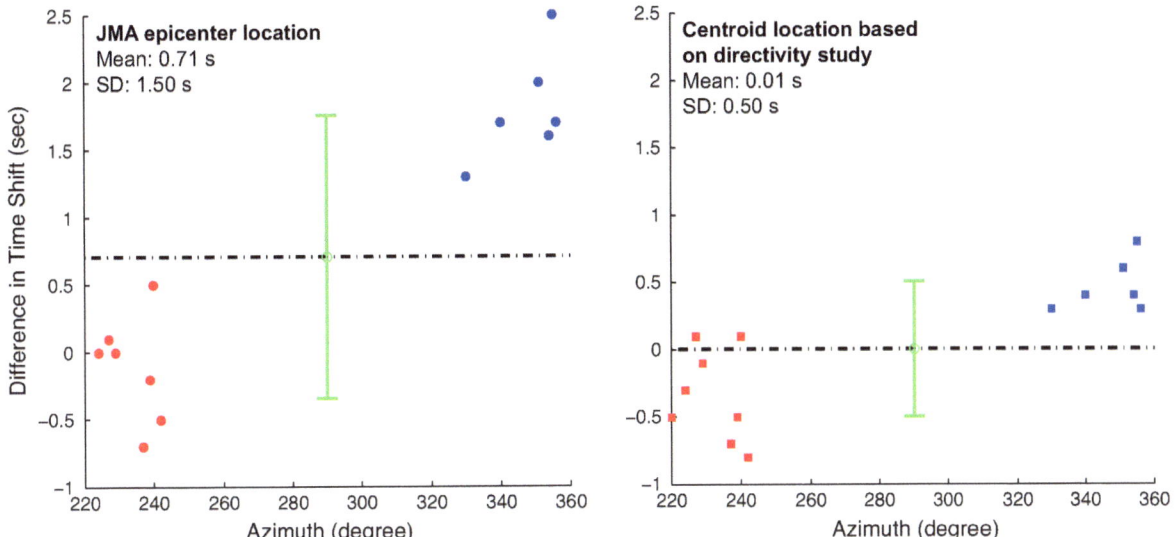

Figure 10
Comparison of the difference in time shift values of E1 and E2 with different epicentral location: *Left* JMA location; *Right* new centroid location based on our directivity study. This is shown that if we use the centroid location for long-period inversion, the difference in time shift is significantly lower

location rather than the epicentral location. Hence we refine the centroid location based on the resolved rupture dimension, and the inversion result indicates a much smaller time shift difference between the two events (Fig. 10b).

It is also interesting to address the origin of these weak lineate zones inside the slab. One possible explanation is that the earthquake is a reactivation of a buried fault formed at the outer rise of the trench as normal faulting prior to subduction. The fault is dipping 37°, which is approximately equivalent to a 60°-dipping normal fault given the slab at this position is subducting at 30° westward (see Fig. 1 inset). Hence, the intraslab events on the fault may serve as useful tools in understanding historical and future outer rise events. The 1933 Sanriku earthquake is one of the several outer-rise earthquakes in the region that has been studied. The depth extend of this earthquake is not well resolved, but there is hypothesis that the rupture is a normal faulting on a 45°-dipping plane that ruptures the oceanic plate, with a dimension of 185 by 100 km^2 (KANAMORI 1971). This outer-rise event occurred over three decades after the 1896 Meiji-Sanriku underthrust earthquake in the region exactly adjacent to the 1896 rupture zone, and both earthquakes generated huge tsunamis. There are also other studies on similar doublets but with a much shorter time separation, in the scale of days (AMMON and LAY 2008; HINO et al. 2009). Nonetheless, our current understanding for large outer rise events is still limited due to their sparse occurrences. Therefore studying the extent of subducted faults formed in the outer rise could provide a better understanding to this phenomenon and the potential risk of tsunami hazards in the area (LAY et al. 2011).

4. Conclusion

The unique rupture pattern of the Tohoku-Oki megathrust earthquake has prompted this study of the shear zone. We lay out a three-step study that focus on exploring the source and rupture mechanism of smaller aftershocks inside the subducting slab, which are beneath the downdip edge of the main shock rupture. Our results indicate that within 150 km proximity to epicentral location, 1D and 3D velocity model show comparable validity in modeling seismic events up to 1 Hz. Using forward modeling approach with EGFs, we are able to simulate the rupture of a

Mw 6.3 intraslab event and resolve for the fault plane without relying on the spatial distribution of other aftershocks. The fault geometry and focal mechanism of this earthquake is also found to be the predicted outcome due to the shear stress change caused by the main shock. Based on the resolved dip angle, together with other similar neighboring intraslab events, a weakened zone of up to 150 km is believed to have reactivated, which can be a long normal fault previously formed in the outer rise region. Further study on the downdip region through these intraslab events may shed light on understanding historical and future outer-rise events, as well as the potential cause of high-frequency energy radiation during the Mw 9.1 main shock.

Acknowledgments

This study was supported by NSF Grant EAR-1142020. Broadband and Strong-motion waveforms were obtained from F-net, K-net, and Kik-Net of NIED. The Generic Mapping Tools (GMT) were used for creating some of the in-text figures. We would like to thank Hiroo Kanamori and Zhongwen Zhan for their insightful comments.

Appendix: Generating Empirical Green's Functions

To generate empirical Green's Functions using E2, based on the generalized ray theory (HELMBERGER 1983), the characteristic travel time of a generalized ray in a layered half-space is given by:

$$t_0 = p_0 r + \sum_i \eta_i d_i \quad (1)$$

where r is the source-receiver distance, η_i the vertical slowness of the ray in each layer, and d_i the vertical distance of the ray segment in each layer. For two very close sources, the paths to the receiver will be highly similar in shape and differ only by a small time shift dt_0. This time variance (dt_0) can be approximated by using Taylor series expansion for t_0 around the position of the point source (r, h).

$$\partial t_0 = \frac{\partial t_0}{\partial r} dr + \frac{\partial t_0}{\partial h} dh \quad (2)$$

$\partial t_0 / \partial r$ is essentially p_0, which is treated as a constant here. $\partial t_0 / \partial h = -\varepsilon \eta_s$, where $\varepsilon = 1$ for down-going rays and $\varepsilon = -1$ for up-going rays. η_s, which equals $[(1/v_s^2) - p_0^2]^{1/2}$, is the vertical slowness of the ray p_0 in the source region. The velocity in the source region is represented by v_s. p_0 and η_s in this study are numerical estimation from synthetics generated at different depths based on the 1D velocity model used for CAP inversion.

Here we assume E1 to be a finite-fault earthquake which is 45 times larger than E2 in moment magnitude. Thus, we discretize the rupture region into a line of 45 elements, each represented as an E2 point source. The total response ($R(t)$) of E1 at the receiver can then be represented by a summation of the 45 rays, each properly lagged in time according to the relative position from the reference point source.

$$R(t) = \sum_{i=1}^{45} E2_i(t - dt_{0i}) \quad (3)$$

Since we assume four rupture scenarios, diagonally northward and southward on the two auxiliary focal planes, there is a set of four empirical Green's Functions $R(t)$ generated, which is then compared with the data obtained to determine rupture directivity.

REFERENCES

AMMON, C. J., KANAMORI. H., and LAY, T (2008), *A great earthquake doublet and seismic stress transfer cycle in the central Kuril Islands*, Nature, 451(7178): 561–565.

FUJIWARA, T., KODAIRA, S., NO, T., KAIHO, Y., TAKAHASHI, N., and KANEDA, Y. (2011), *The 2011 Tohoku-oki earthquake: Displacement reaching the trench axis*, Science, 334(6060): 1240–1240.

GRAVES, R. W. (1996), *Simulating seismic wave propagation in 3d elastic media using staggered- grid finite differences*, Bulletin of the Seismological Society of America, 86(4): 1091–1106.

HARTZELL, S. H. (1978), *Earthquake aftershocks as green's functions*, Geophysical Research Letters, 5(1):1–4.

HASEGAWA, A., YOSHIDA, K., and OKADA, T. (2011), *Nearly complete stress drop in the 2011 Mw 9.0 off the Pacific coast of Tohoku earthquake*, Earth, planets and space, 63 (7):703–707.

HELMBERGER, D. V. (1983), Theory and application of synthetic seismiograms, Earthquakes: Observation, Theory and Interpretation, H. Kanamori (Editor), pages 173–222.

HINO, R., AZUMA, R., ITO, Y., YAMAMOTO, Y., SUZUKI, K., TSUSHIMA, H., SUZUKI, S., MIYASHITA, M., TOMORI, T., ARIZONO, M. (2009), *Insight into complex rupturing of the immature bending normal fault in the outer slope of the Japan trench from aftershocks of the 2005 Sanriku earthquake (mw = 7.0) located by ocean bottom seismometry*, Geochemistry, Geophysics, Geosystems, *10*(7).

HUANG, Y., MENG, L., and AMPUERO, J.-P. (2012), *A dynamic model of the frequency- dependent rupture process of the 2011 Tohoku-Oki earthquake*, Earth Planets Space, *64* (12):1061–1066.

HUANG, Y., AMPUERO, J.-P., and KANAMORI, H. (2013), *Slip-weakening models of the 2011 Tohoku-Oki earthquake and constraints on stress drop and fracture energy*, Pure and Applied Geophysics, pages 1–14.

IDE, S., BALTAY, A., and BEROZA, G. C. (2011), *Shallow dynamic overshoot and energetic deep rupture in the 2011 Mw 9.0 Tohoku-Oki earthquake*, Science, *332* (6036): 1426–1429.

ITO, Y., TSUJI, T., OSADA, Y., KIDO, M., INAZU, D., HAYASHI, Y., TSUSHIMA, H., HINO, R., and FUJIMOTO, H. (2011), *Frontal wedge deformation near the source region of the 2011 Tohoku-Oki earthquake*, Geophysical Research Letters, *38* (7).

KANAMORI, H. (1971), *Seismological evidence for a lithospheric normal faulting the Sanriku earthquake of 1933*, Physics of the Earth and Planetary Interiors, *4* (4): 289–300.

KATO, A. and IGARASHI, T. (2012), *Regional extent of the large coseismic slip zone of the 2011 Mw 9.0 Tohoku-Oki earthquake delineated by on-fault aftershocks*, Geophysical Research Letters, *39* (15).

KATO, A., SAKAI, S., and OBARA, K. (2011), *A normal-faulting seismic sequence triggered by the 2011 off the pacific coast of Tohoku earthquake: Wholesale stress regime changes in the upper plate*, Earth, planets and space, *63* (7): 745–748.

KOKETSU, K., MIYAKE, H., FUJIWARA, H., and HASHIMOTO, T. (2008), *Progress towards a japan integrated velocity structure model and long-period ground motion hazard map*, In Proceedings of the 14th World Conference on Earthquake Engineering, pages S10–038.

LAY, T., AMMON, C. J., KANAMORI, H., KIM, M. J., and XUE, L. (2011), *Outer trench-slope faulting and the 2011 Mw 9.0 off the pacific coast of Tohoku earthquake*, Earth, planets and space, *63* (7): 713–718.

LIN, J. and STEIN, R. S. (2004), *Stress triggering in thrust and subduction earthquakes and stress interaction between the southern San Andreas and nearby thrust and strike-slip faults*, Journal of Geophysical Research: Solid Earth (1978–2012), *109* (B2).

NAKAJIMA, J., HASEGAWA, A., and KITA, S. (2011), *Seismic evidence for reactivation of a buried hydrated fault in the pacific slab by the 2011 M9.0 Tohoku earthquake*, Geophysical Research Letters, *38* (7).

OHTA, Y., MIURA, S., OHZONO, M., KITA, S., IINUMA, T., DEMACHI, T., TACHIBANA, K., NAKAYAMA, T., HIRAHARA, S., SUZUKI, S., ET. AL. (2011), *Large intraslab earthquake (2011 April 7, M 7.1) after the 2011 off the Pacific coast of Tohoku earthquake (M 9.0): Coseismic fault model based on the dense GPS network data*, Earth, planets and space, *63* (12): 1207–1211.

SIMONS, M., MINSON, S. E., SLADEN, A., ORTEGA, F., JIANG, J., OWEN, S. E., MENG, L., AMPUERO, J.-P., WEI, S., CHU, R., et al. (2011), *The 2011 magnitude 9.0 Tohoku-Oki earthquake: Mosaicking the megathrust from seconds to centuries*, Science, *332* (6036): 1421–1425.

SONG, X. J. and HELMBERGER, D. V. (1996), *Source estimation of finite faults from broadband regional networks*, Bulletin of the Seismological Society of America, *86* (3): 797–804.

TAN, Y. and HELMBERGER, D. V. (2010), *Rupture directivity characteristics of the 2003 Big Bear sequence*, Bulletin of the Seismological Society of America, *100* (3): 1089–1106.

TODA, S., STEIN, R. S., and LIN, J. (2011), *Widespread seismicity excitation throughout central Japan following the 2011 M = 9.0 Tohoku earthquake and its interpretation by coulomb stress transfer*, Geophysical Research Letters, *38* (7).

WEI, S., GRAVES, R., HELMBERGER, D. V., AVOUAC, J.-P., and JIANG, J. (2012), *Sources of shaking and flooding during the Tohoku-Oki earthquake: A mixture of rupture styles*, Earth and Planetary Science Letters, *333*: 91–100.

YUE, H. and LAY, T. (2011), *Inversion of high-rate (1 SPS) GPS data for rupture process of the 11 march 2011 Tohoku earthquake (Mw 9.1)*, Geophysical Research Letters, *38* (7).

ZHAN, Z., HELMBERGER, D., SIMONS, M., KANAMORI, H., WU, W., CUBAS, N., DUPUTEL, Z., CHU, R., TSAI, V. C., AVOUAC, J.-P., et al. (2012), *Anomalously steep dips of earthquakes in the 2011 Tohoku-Oki source region and possible explanations*, Earth and Planetary Science Letters, *353*: 121–133.

ZHANG, Z., CHEN, J. Y., and LIN, J. (2008), *Stress interactions between normal faults and adjacent strike-slip faults of 1997 Jiashi earthquake swarm*, Science in China Series D: Earth Sciences, *51* (3): 431–440.

ZHAO, L.-S. and HELMBERGER, D. V. (1994), *Source estimation from broadband regional seismograms*, Bulletin of the Seismological Society of America, *84* (1): 91–104.

ZHU, L. and HELMBERGER, D. V. (1996), *Advancement in source estimation techniques using broadband regional seismograms*, Bulletin of the Seismological Society of America, *86* (5): 1634–1641.

ZHU, L. and RIVERA, L. A. (2002), *A note on the dynamic and static displacements from a point source in multilayered media*, Geophysical Journal International, *148* (3): 619–627.

(Received September 3, 2014, revised December 29, 2014, accepted January 16, 2015, Published online March 20, 2015)

Rapidness and Robustness of Finite-Source Inversion of the 2011 M_w 9.0 Tohoku Earthquake by an Elliptical-Patches Method Using Continuous GPS and Acceleration Data

THOMAS ULRICH[1] and HIDEO AOCHI[1]

Abstract—The kinematic rupture process of the 2011 M_w 9.0 Tohoku earthquake is inverted with an elliptical-patches method, using a genetic algorithm, for the purpose of rapid and robust estimation of the source parameters of a mega-earthquake. We use the ground-displacement field provided by a continuous GPS network and the ground-velocity field recorded by acceleration networks. In addition to the typical inversion procedure in which a data duration long enough to cover the whole rupture process is used, inversions based on shorter signals, giving an incomplete view over the ground shaking sequence, are also carried out. How fast can a robust estimation of the source parameters be obtained? Using the elliptical approximation, we find that robust solutions of $M_w \sim$ 9.0-earthquakes are rather quickly obtained regardless of the frequency band and the elliptical patch description. It is also confirmed that, because of the absence of off-shore recording stations on the east side of the fault, some uncertainties in the rupture process cannot be completely removed. In fact, at the very low frequencies considered, the western part of the fault (deeper portion) contributes more significantly to the recorded signals than does the other end close to the trench. This problem also prevents refinement of the description of the rupture process, in particular when using more than one ellipse.

Key words: Kinematic inversion, Tohoku, Elliptical patch method.

1. Introduction

The great Tohoku earthquake occurred on the 11th March 2011 with an exceptional moment magnitude (M_w) of 9.0. It is one of the largest earthquakes instrumentally recorded. A huge tsunami was triggered by this megathrust earthquake, with disastrous consequences: more than 15,000 fatalities, about 130,000 collapsed buildings and the meltdown of three nuclear reactors. The observation network in Japan being extremely dense, the ground motions generated by the earthquake were recorded by hundreds of seismometers and GPS stations. This is a rare opportunity for scientists to study such an extreme event and to better understand the complex mechanisms of the subduction zone.

The rupture process of an earthquake is a complex phenomenon, which depends on the initial stress distribution applied along the fault and the heterogeneity of the fault plane (KANAMORI and STEWART 1978; AKI 1979). Many researchers have analysed different observations of the rupture process of the 2011 Tohoku earthquake and numerous articles have already been published (IDE *et al.* 2011; OZAWA *et al.* 2011; KOKETSU *et al.* 2011; SUZUKI 2011; SIMONS 2011 and many others). Such analyses are essential in understanding precisely the complex dynamic process of the earthquake and to assess the seismic hazard from the future possible events as well. Despite several decades of seismology, predicting the next big earthquake is a great challenge. Earthquakes remain unpredictable (GELLER 2011). Nevertheless, the analysis of the cumulative slip distribution over a fault permits drawing some conclusions on its future activity. The areas presenting a slip deficit could be roughly identified as the areas having accumulated much stress (NISHIMURA *et al.* 2004; HASHIMOTO *et al.* 2009). On the other side of the Pacific trench, the 2010 Maule, Chile, mega-earthquake occurred in such a seismic gap (e.g. RUIZ *et al.* 2012).

Traditionally in seismology, the multitime-window linearwaveform inversion procedure (HARTZELL and HEATON 1983) is used to obtain the spatial heterogeneity of the earthquake source. Namely, the fault is spatially discretized into small rectangular subfaults, each characterized by a set of parameters

---

[1] Seismic and Volcanic Risks Unit (RSV), Risks and Prevention Division (DRP), Bureau de Recherches Géologiques et Minières (BRGM), 3 avenue Claude Guillemin, BP 36009, 45060 Orléans Cedex 2, France. E-mail: t.ulrich@brgm.fr

(amount of slip, rupture time and rise time, for example, as a minimum set), obtained by linear inversion. MAI and BEROZA (2003) presented a procedure based on spatial Fourier-decomposition for analysing the finite source characterization. MAI (2013) also pointed out that source inversions based on many parameters often have no unique solution, and that the obtained solution may contain significant uncertainty. On the other hand, it is known that strong ground motion areas (SMGAs) can be identified as a small portion of the ruptured fault (e.g., KAMAE et al. 1998; IRIKURA and MIYAKE 2010; KURAHASHI and IRIKURA 2011). Recently DI CARLI et al. (2010) were able to reconstruct the rupture process of the 2000 Tottori, Japan, earthquake with only a few parameters. The geometrical heterogeneity on the fault is characterized as one or two ellipses in both kinematic and dynamic source models, following the method proposed by VALLÉE and BOUCHON (2004). In their dynamic inversion, initial stress and frictional parameters are obtained. The solutions are physically consistent, but the method is very computationally intensive. In the kinematic inversion, elliptical patches of slip are inverted. In order to obtain physically-consistent solutions, additional constraints on the parameters are often necessary. Since then, this so-called elliptical patch method (EPM) has been investigated and developed, in particular by: RUIZ and MADARIAGA (2011) for the 2007 Michilla, Chile, earthquake; RUIZ et al. (2012) for the 2010 Maule, Chile, earthquake; and RUIZ et al. (2013) for the deep 2008 Iwate, Japan, earthquake.

In parallel, from the dynamic point of view, IDE and AOCHI (2005) proposed to express the fault heterogeneity as a superposition of different scales of heterogeneity, called "multi-scale heterogeneity". Based on this concept, AOCHI and IDE (2011) and IDE and AOCHI (2013) were able to dynamically simulate the growth of the rupture process of the 2011 Tohoku earthquake, beginning with a small rupture and finishing with an M_w 9 event. In their modelling, the presence of a large ellipse is assumed, in addition to the smaller circular heterogeneities reflecting the past seismicity in this region. We think that the conception of the multi-scale heterogeneity is consistent with the kinematic description of SMGAs or the EPM presented above. The dynamic models explored by AOCHI and IDE (2011) and IDE and AOCHI (2013) were not constrained nor calibrated with the observed data. Therefore, in this study, we are interested to know if multi-scale properties of the fault heterogeneity can be identified from inversions based on the ellipsoidal expression of the source (DI CARLI et al. 2010). We also aim to investigate the sensitivity of the inversion procedure when using such expressions, as some constraints on the parameters are often necessary and this problem has not been explored in previous works.

2. Model Description and Inversion Procedure

We characterize the source process with only a few parameters as previously proposed for other earthquakes (e.g., DI CARLI et al. 2010; RUIZ and MADARIAGA 2011). On the assumed fault plane, one or more ellipse(s) are assumed. In this study, we began with one ellipse defined by seven parameters (Table 1). The ellipse geometry is constrained by the lengths of its two principal semi-axes a and b, and by the coordinates of its centre (x_c, y_c). We assumed that the principal axis of the ellipse is aligned with the strike direction of the fault. As the hypocentre and the ellipse centre are not necessarily coincident, source directivity effects are implicitly considered. Three slip profiles are tested: a Gaussian profile, as used in DI CARLI et al. (2010); an elliptical profile as proposed by Sergio Ruiz (personal communication, 2013); and a uniform profile (Fig. 1). All the profiles are characterized by the maximum slip D_m. Because of the difference in the slip distribution, different profiles characterized by the same maximum slip D_m do not lead to an identical magnitude. Finally, the kinematics of the rupture are constrained by a constant rupture speed V_r and by the rise time t_r.

Table 1

The inverted parameters and their range of variation, explored by the GA

	a (km)	b (km)	x_c (km)	y_c (km)	V_r (m/s)	t_r (s)	D_m (m)
Range min	0	0	−200	−100	1,000	5	0
Range max	300	100	200	100	3,300	40	50

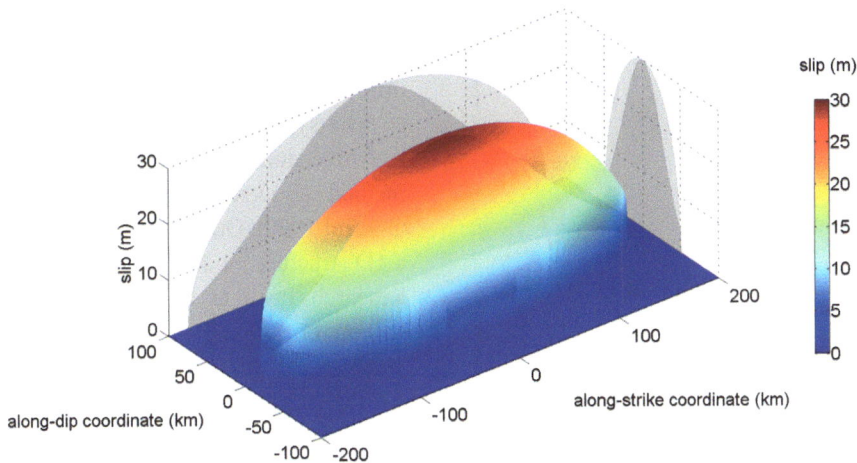

Figure 1
Schematic illustrations of the Gaussian and elliptic profiles of slip distribution, depicted for the same maximum displacement. The elliptic profile presents a more extended and flatter peak and decreases more sharply around its edge

As the model formulation is non-linear, standard linear-optimization techniques cannot be used. A global search algorithm has to be used instead. All the studies previously cited have used the neighbourhood algorithm (NA, e.g., SAMBRIDGE 1999). Nevertheless, in the present study we choose to use a genetic algorithm (GA, e.g., GOLDBERG 1989). This is because from our experience, GAs are able to provide reliable solutions after fewer forward calculations than the NA. Moreover, because of their simple formulation, GAs can be used on highly non-linear problems that require hundreds of thousands of iterations without a reduction in speed, contrary to the NA, which slows down because of its complex geometrical formulation.

The GA mimics the process of natural evolution to find an optimum solution to any usually complex non-linear problem. Following the genetics terminology, a chromosome is a vector of variables to be optimized. An initial set of chromosomes is randomly generated, and the chromosomes evolve and multiply with the classical basic operations of genetics: selection, crossing-over and mutation.

The parameterization of the optimisation algorithm, which permits tuning the trade-off between convergence speed and parameters space exploration, is primordial for maximizing the chance of convergence to the optimum solution. For instance, a low rate of mutation or crossover, or a smaller population size, permits rapid converge to a solution, but the quality is not ensured because the parameter space may not be sufficiently explored. To invert our seven parameters, we create an initial population of one thousand individuals and let it evolve. It is important to generate an initial population large enough so that the parameter space can be sufficiently explored. After several tests, we found that 125 generations were enough to obtain a converged solution of sufficient quality. Finally, the mutation and crossover rate are set respectively to 2 % and 10 %, and pairwise tournaments are considered for the selection process. The parameters and their corresponding search range are displayed in Table 1.

3. Data Processing

Two types of data are investigated to constrain the inversion. Firstly, ground displacements (three components) continuously recorded by twelve selected GPS stations of GEONET (Geospatial Information Authority of Japan: GSI) are used (see Fig. 2). Ten stations along the coast line are selected to ensure a good overall coverage along the fault strike. Two further inland stations (020930 and 020935) are also used to improve the resolution toward the dip direction of the fault. The data are filtered by a low-pass with cut-off frequency of 0.05 Hz (20 s): this

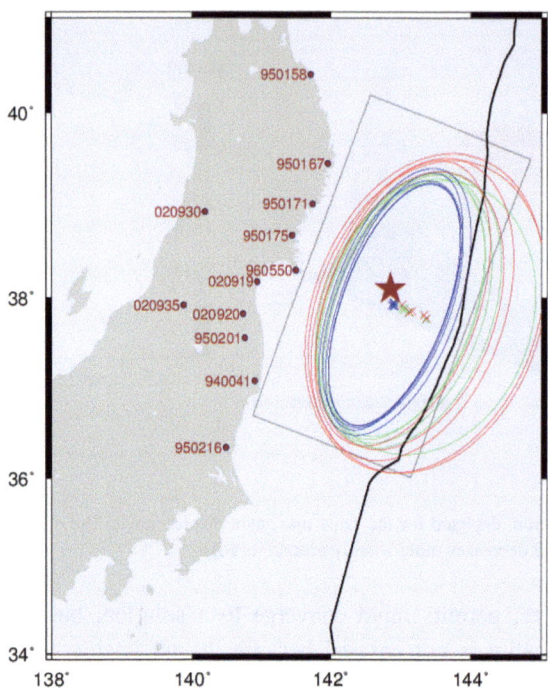

Figure 2
Fault geometry of the five best models independently obtained for the three shapes of slip profiles: Gaussian (*red*), elliptic (*green*) and uniform (*blue*). The *black rectangle* shows the prescribed fault area: only the portion of the ellipse inside this delimited area is considered for the calculation of the synthetic ground motions

relatively low value is used because the simple model cannot recreate waves of higher frequencies.

Secondly, the use of the accelerations recorded by the strong ground motion networks K-NET and KiK-net of the National Research Institute for Earth Science and Disaster Prevention, Japan, is tested. Since local site effects were extensively observed, we select twelve stations (Fig. 4) from the Kik-net and preferably use the downhole records because most of the records at depth did not suffered significantly from these site effects. A rather similar geographical coverage of the stations is considered for both data sets, to allow an unbiased comparison of the inversions results. The acceleration data are first integrated and then processed by a band-pass filter between 50 s and 100 s.

Both GPS and accelerations are (re)sampled at 1.0 s. The GPS signals are compared over 250 s beginning at 14:46:23 (JST), which we call hereafter the rupture triggering time. In fact, our preliminary runs showed that this duration ensures a good trade-off in the optimisation effort between the final displacement field and the transient phase. The same time window is also adopted for the accelerations.

The synthetic seismograms are calculated before the inversions by using the discrete-wavenumber method (BOUCHON 1981; code Axitra written by Coutant 1990). Our analysis being at low frequencies, a 1D structure model (routinely used by the National Research Institute for Earth Science and Disaster Prevention, Japan) is sufficient. A potential fault plane of dimensions 400 km × 200 km, centred on the JMA hypocentre (142.861°E, 38.103°N, 23.7 km depth) and with fault strike of 201°N and dip of 9°, according to the focal mechanism of Global CMT solution, is considered. The fault plane is discretized in 5775 double-couple sources evenly distributed every 4 km along the strike and dip directions.

To evaluate the fit of the synthetic seismograms $y_{i,n}^{syn}(t)$ to the observations $y_{i,n}^{obs}(t)$ at station n, the misfit function ε is computed according to the following equation:

$$\varepsilon = \sum_{n=1}^{n_{max}} \varepsilon_n = \sum_{n=1}^{n_{max}} \frac{\sum_t \sum_{i=1}^{3} \left(y_{i,n}^{obs}(t) - y_{i,n}^{syn}(t)\right)^2}{\sum_t \sum_{i=1}^{3} \left\{y_{i,n}^{obs}(t)\right\}^2}, \quad (1)$$

where i denotes the three components ($i = 1, 2$ and 3) and ε_n is the misfit at station n. As the residual at each station is normalized by the amplitudes of each component, all the stations are equally taken into account. In all the inversions, absolute time is used.

4. Results

We first show the inversion results using the displacement field recorded by continuous GPS stations. Table 2 presents the results of ten independent inversions conducted for different shapes of the slip distribution (Gaussian, elliptical and uniform). Since the optimisation procedure is stochastic, it is important to check if independent inversions, based on different initial populations, converge to a similar solution. The parameters of the best model of each inversion, the corresponding misfit value and the obtained moment magnitude (M_w) are summarized in this table.

Table 2

The best models obtained in ten independent inversions for each shape profile (Gaussian, elliptic, uniform) with a low-pass filter (cut-off frequency at 0.05 Hz) on the ground displacements from continuous GPS

Shape	Run id	a (km)	b (km)	x_c (km)	y_c (km)	V_r (m/s)	t_r (s)	D_m (m)	Misfit	M_w
Elliptic	1	165	88	16	26	1,737	43	45	0.57	9.1
	2	181	116	20	32	1,870	44	30	0.58	9.0
	3	183	145	17	66	1,783	42	35	0.58	9.1
	4	167	124	19	54	1,755	44	43	0.59	9.1
	5	176	104	17	32	1,836	44	36	0.57	9.1
	6	172	99	18	25	1,830	45	36	0.56	9.0
	7	167	122	14	49	1,729	43	40	0.60	9.1
	8	**169**	**89**	**17**	**23**	**1,836**	**43**	**41**	**0.56**	**9.1**
	9	189	103	18	24	1,825	42	30	0.59	9.0
	10	172	133	18	54	1,839	45	36	0.58	9.1
Mean		174	112	17	38	1,804	43	37	0.58	9.06
σ		8	19	2	16	49	1	5	0.01	0.02
Uniform	1	180	108	18	32	1,916	44	23	0.59	9.0
	2	172	79	15	11	1,829	42	27	0.58	9.0
	3	**156**	**67**	**18**	**12**	**1,793**	**45**	**38**	**0.54**	**9.0**
	4	157	66	18	11	1,701	42	38	0.55	9.0
	5	159	73	18	9	1,734	44	32	0.56	9.0
	6	184	114	16	40	1,918	44	23	0.60	9.0
	7	149	99	17	39	1,664	43	37	0.62	9.1
	8	178	132	17	53	1,938	43	23	0.60	9.0
	9	180	110	18	32	1,923	43	22	0.60	9.0
	10	166	67	15	10	1,787	45	34	0.56	9.0
Mean		168	91	17	25	1,820	44	30	0.58	9.03
σ		12	24	1	16	101	1	7	0.03	0.02
Gaussian	1	204	134	17	44	1,965	44	28	0.57	9.0
	2	177	108	18	32	1,774	43	39	0.57	9.1
	3	191	128	19	41	1,871	43	31	0.57	9.0
	4	180	94	15	22	1,729	41	41	0.58	9.1
	5	198	142	18	55	1,906	43	32	0.57	9.0
	6	183	96	13	21	1,852	44	38	0.57	9.0
	7	179	80	15	10	1,688	43	42	0.58	9.0
	8	202	142	15	51	1,962	44	29	0.57	9.0
	9	212	142	18	43	2,024	43	24	0.60	9.0
	10	**193**	**116**	**16**	**35**	**1,946**	**46**	**34**	**0.56**	**9.0**
Mean		192	118	16	35	1,872	43	34	0.58	9.04
σ		12	23	2	14	111	1	6	0.01	0.02

The best model obtained for each profile is highlighted in bold. For each shape profile, the average and standard deviation for each parameter are also indicated

First of all, we find that different assumptions on the shape of final slip fit the data similarly well. The best model is obtained for a uniform profile, but its misfit differs by only 4 % with respect to the best models obtained for an elliptic or a Gaussian profile. This difference is insignificant, particularly when compared with the variability over the misfits obtained from independent inversions with similar profiles. It is important to note that it is the simplest model (uniform profile) which best explains the data. We also observe in Table 2 that the solutions obtained in the independent runs do not present significant differences in terms of misfit. For the elliptical slip profile, the mean relative difference over the misfit is only 1.3 %, with a maximum of 7 %. This confirms that the optimisation algorithm is correctly tuned.

In Table 2 some of the obtained parameters (e.g., geometry and slip amount) differ from one solution to another, although the difference in terms of misfit is not significant and the obtained moment magnitude does not change by much, i.e., the problem has multiple solutions. Thus, it is important to scrutinize all the possible solutions for which the misfit is not significantly different from the very best model. This allows us to quantify the uncertainties in the solution. Figure 2 presents the fault geometry of the five best models obtained for each profile from Table 2. Note that the portion outside the prescribed fault area (black rectangular) is not used in the calculation of the ground motions: the source area considered in the synthetic motions is the intersection between the ellipse and the black rectangle only. This situation occurs for many solutions at the top edge of the fault, thus, solutions apparently presenting different geometries can be practically very similar. This can partly explain why the two geometric parameters along the dip direction (b and y_c) show considerable dispersion. The fact that many solutions tend to extend towards the shallower part is in accordance with the occurrence of a large tsunami. Besides, a good agreement is confirmed for the ruptured dimension (a) and position (x_c) along the longitudinal (strike) direction. Finally, we can notice that the ruptured dimensions obtained with the uniform profile are systematically smaller than the ones obtained with the other profiles, while the obtained moment magnitudes are the same. Thus, in this inversion, the seismic moment, i.e., the total slip multiplied by rupture area, is a key factor.

On the other hand, the other obtained model parameters show quite good convergence. For an elliptical profile, an average rupture speed of 1,804 m/s is obtained, with a standard deviation of 49 m/s. This value is closely related to the total duration of the rupture process, i.e., with the longer axis of the fault geometry. This relatively slow rupture velocity corresponds to the long duration of this earthquake of about 2 min. An average rise time of 43 s is obtained for the three slip profiles with very little dispersion. Finally, the maximum displacement D_m is also quite well constrained with an average value of 37 m (elliptical profile). This average value is a little smaller than the maximum values obtained in other studies of about 45 m [for instance, (SUZUKI et al. 2011) present a model with a peak slip of 48 m near the trench, and KOKETSU et al. (2011) show a peak slip of about 40 m]. This tendency to underestimate is probably due to the supposed shape of the slip distribution, which is too smooth, even with a Gaussian shape, to localize any huge displacement.

Figure 3 shows the comparison of observed and synthetic ground motions of the best model among all 30 inversions. As the final displacement at the later phase is significant, the inversion procedure looks for the best parameter in order to minimize the residual on that stage. This is why all the inversions obtain a robust estimation of the magnitude. The transient part is also quite well modelled, the fits being satisfactory as a first approximation. Nevertheless, our analyses are based on a very low frequency band. An ellipse model is sufficient to briefly describe the macroscopic features of the earthquake but it is too simple to reveal details of the complex rupture process.

In order to quantify the sensitivity of the inversion process to the final displacement level, we now apply a Butterworth band-pass filter between 0.005 and 0.05 Hz (20–200 s). Table 3 shows the best obtained parameters in ten independent inversions, and Fig. 4 compares the ruptured area of the best models obtained (blue) against those obtained with the low-pass filter (green). The obtained ellipses are similar to those obtained previously but are shifted toward the island of Honshu. The inverted peak slip is on average reduced by a factor two compared with what was obtained with the low-pass filter. However, the difference in the obtained magnitude is reasonable, an average value of 8.9 being obtained with the band-pass filter. A comparison of the ground motions with the synthetic signals generated with the model 4 of Table 3 is shown in Fig. 5. The synthetics fit quite well the observations for this frequency range, but they cannot explain the oscillations at higher frequencies. Thus, the use of the final displacement is not absolutely necessary to obtain reasonable solutions since the parameters obtained in both cases are coherent.

Next we carry out the same inversion process but based on the velocity waveforms (processed by a 0.01–0.02 Hz band-pass) recorded by the acceleration networks. In fact, the continuous GPS signals, once differentiated and filtered, could have been used as the input velocity signals, but we chose to use the data from the acceleration networks to show how different types of data can constrain the inversions. The results of ten independent inversions assuming an elliptic profile of final displacements are summarized in

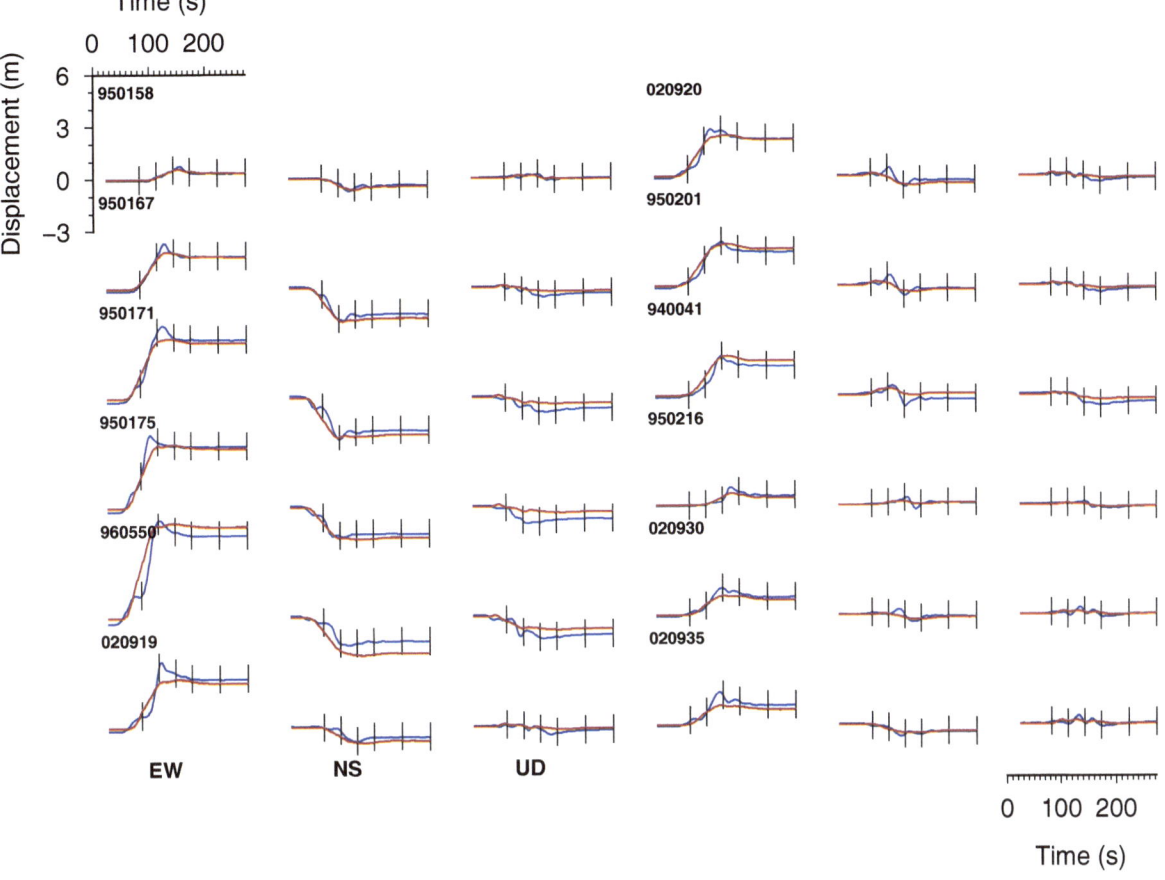

Figure 3

Comparison of the synthetic ground displacement (*red lines*) with the observation (*blue lines*) for model 3 of the uniform sub dataset (best of all 30 models in Table 2). The twelve stations are plotted in Fig. 2. A low-pass filter with a cut-off frequency of 0.05 Hz is used. *Vertical lines* indicate 60, 90, 120, 150, 200 and 250 s

Table 3

The best models obtained in ten independent inversions assuming an elliptic shape profile of displacement for a band-pass filter (0.005–0.05 Hz) on the ground displacements

Shape	Run id	a (km)	b (km)	x_c (km)	y_c (km)	V_r (m/s)	t_r (s)	D_m (m)	Misfit	M_w
Elliptic	1	180	101	25	−11	1,850	40	18	4.28	8.9
	2	167	101	23	−9	1,771	39	20	4.27	8.9
	3	**157**	**94**	**24**	**−5**	**1,686**	**41**	**26**	**4.21**	**8.9**
	4	186	97	27	−14	1,883	41	18	4.32	8.9
	5	169	105	28	−7	1,826	41	19	4.28	8.9
	6	200	103	23	−8	1,950	42	17	4.36	8.9
	7	177	110	24	−11	1,846	38	16	4.33	8.9
	8	186	100	26	−10	1,885	41	18	4.29	8.9
	9	172	106	26	−6	1,816	40	19	4.28	8.9
	10	173	122	25	3	1,818	38	18	4.38	8.9
Mean		177	104	25	−8	1,833	40	19	4.30	8.91
σ		12	8	2	5	71	1	3	0.05	0.01

The best model obtained is highlighted in bold. The average and standard deviation over each parameter are also indicated the obtained parameters fit globally well the data, particularly for the horizontal components. Nevertheless, some ground motions are underestimated for the most distant stations (e.g., CHBH14, IBRH14 and IBRH19). This is mainly because the obtained rupture area is short along the strike. Besides, we find that the model does not fit sufficiently well many of the vertical components. It could be a consequence of the way of computing the misfit function [Eq. (1)] since the vertical amplitudes are often much smaller than the horizontal ones. However, it can be also the consequence of the use of shorter periods than in the previous inversions.

5. Discussion

5.1. Sensitivity Tests

In the previous inversions, some model parameters are robustly obtained, but others are not. In order to better understand the influence of each parameter, different sensitivity analyses are carried out. First, let us study the position of the ellipse along the dip direction (y_c). For that purpose, let us investigate model 8 (elliptic slip profile) of Table 2, this model having a relatively narrow profile. Figure 7 shows the ground displacements for various dip positions keeping the other model parameters unchanged. A deeper location of the rupture area leads to larger final displacements at the selected stations than a shallower slip. In fact, the shallow areas of the fault are farther from the stations than the deep areas. This should be put in perspective with the fact that the effect of a shallower slip is more limited in space and has, therefore, less influence on distant locations. This explains why the rupture area is better constrained at depth (close to the land), and the uncertainty remains for the shallower part.

Next we test the sensitivity of the parameters t_r (rise time) and V_r (rupture velocity). We again use model 8 of Table 2, varying t_r between 10 and 50 s, and V_r between 1,250 and 2,500 m/s, respectively (Fig. 8). Shorter t_r and faster V_r lead to a steeper slope in the rise of displacements (namely shorter transition time), and may reproduce an overshoot earlier to the finite final displacement level.

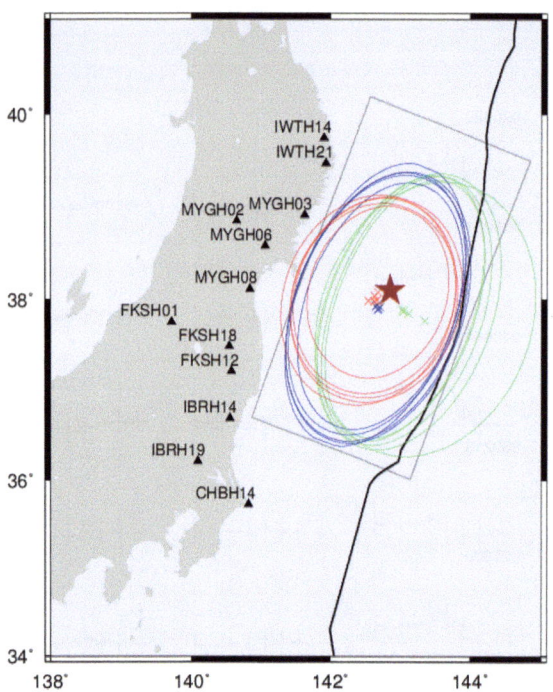

Figure 4
Fault geometry of the five best models independently obtained assuming an elliptic profile and (1) using the GPS displacement data filtered by a 0.05 Hz low-pass filter (*green*) and a 0.05-0.005 Hz band pass filter (*blue*), (2) using the ground-motion velocity filtered by a 0.02-0.01 Hz band-pass filter (*red*). The twelve ground motion stations are located on the map by the *black triangles*

Table 4. The geometry of the five best solutions is shown in Fig. 4 (red ellipses) with the previous inversion results (green and blue), and a comparison of the ground motions for the best solution are illustrated in Fig. 6. The geometry is well resolved in the ten inversions. The obtained ruptured area is mostly circular, being much less extended along the strike direction than in the solution obtained from the ground displacements. The maximum slip location (centre of ellipse) coincides with the hypocentre. The rupture is slower than in the previous inversions and the fault dimension is shorter. The total rupture duration is then kept unchanged. In addition, the obtained rise time is smaller than in the previous results. In the following section, we study the sensitivity of these parameters. Finally, regardless of the discrepancy in some obtained parameters using displacement or velocity ground motions, it is interesting to note that both data sets give the same magnitude of M_w 9. Figure 6 confirms that

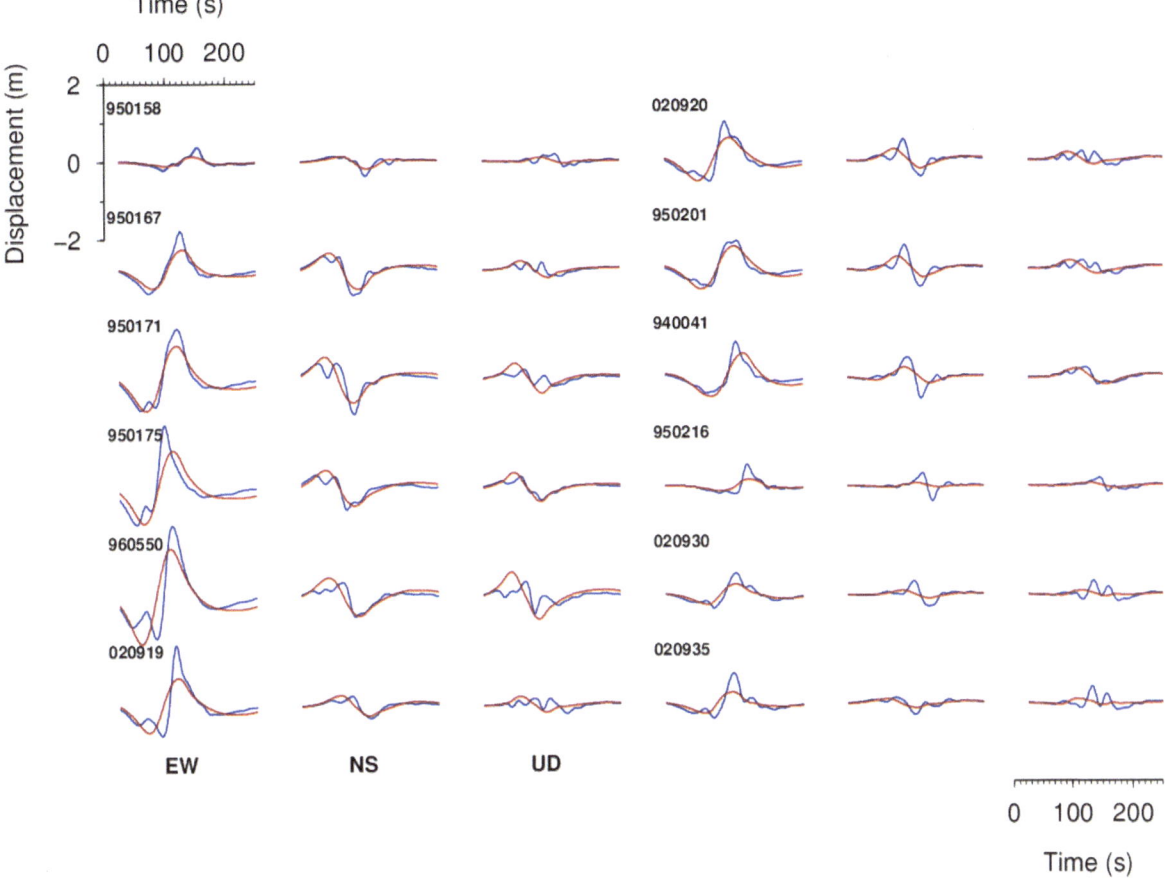

Figure 5
Comparison of the ground displacement between synthetic (*red*) and observation (*blue*) for model 4 of Table 3. A band-pass filter between 0.005 and 0.05 Hz is applied

Table 4

The best models obtained in ten independent inversions assuming an elliptic shape profile, with a band-pass filter (0.01–0.02 Hz) on the ground velocities from the acceleration data

Shape	Run id	a (km)	b (km)	x_c (km)	y_c (km)	V_r (m/s)	t_r (s)	D_m (m)	Misfit	M_w
Elliptic	1	129	111	19	−22	1,368	28	29	6.53	9.0
	2	145	115	24	−23	1,532	40	29	6.74	9.0
	3	**121**	**100**	**14**	**−16**	**1,300**	**29**	**36**	**6.44**	**9.0**
	4	131	127	20	−33	1,502	38	28	6.65	9.0
	5	129	116	22	−22	1,502	42	32	6.57	9.0
	6	128	106	18	−12	1,409	39	38	6.48	9.0
	7	107	87	8	−14	1,094	15	48	6.46	9.0
	8	121	103	20	−15	1,374	37	39	6.50	9.0
	9	131	107	22	−22	1,432	36	31	6.51	9.0
	10	133	111	23	−29	1,468	37	29	6.60	9.0
Mean		127	108	19	−21	1,398	34	34	6.55	9.00
σ		10	11	5	7	129	8	6	0.09	0.02

The best model obtained is highlighted in bold. The average and standard deviation over each parameter is also indicated

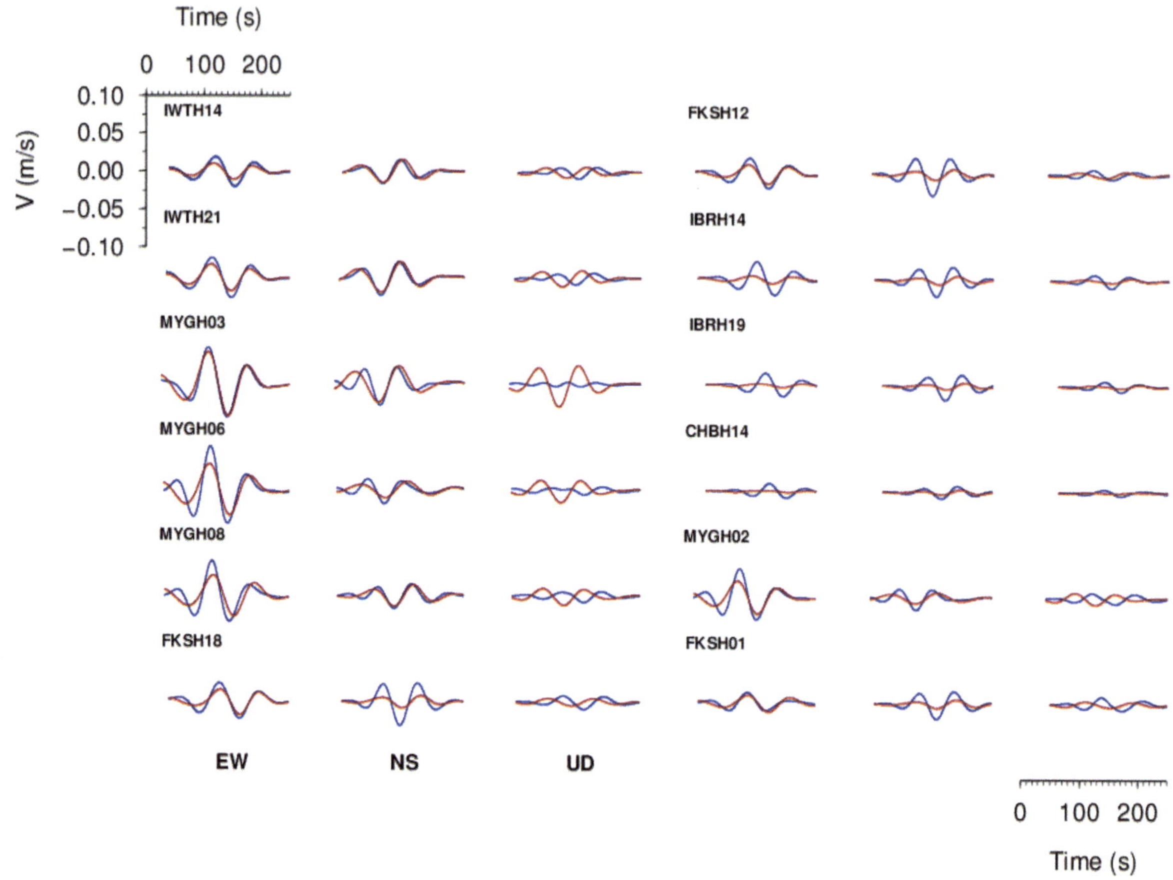

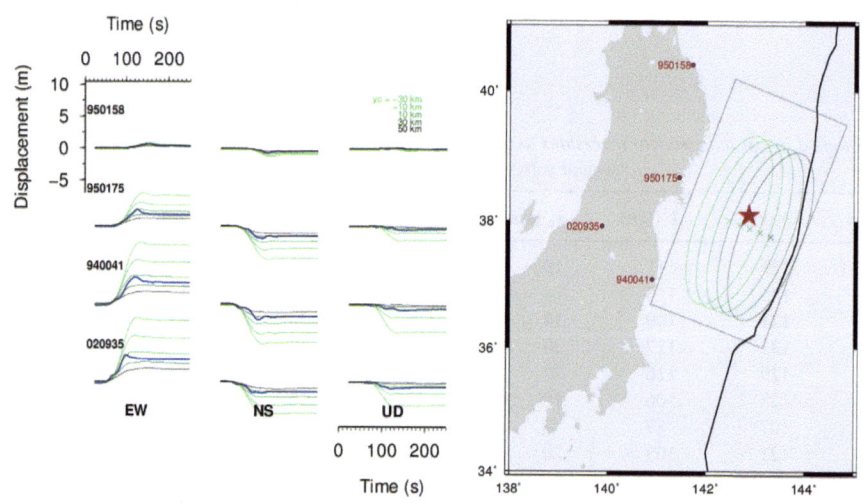

Figure 6
Comparison of the synthetic ground displacement (*red lines*) with the observation (*blue lines*) for model 3 of Table 4. The twelve stations are plotted in Fig. 4. A 0.02–0.01 Hz band-pass filter is used

Figure 7
Sensitivity of the position of the ellipse along the dip direction. The ground displacements are compared at four selected stations. Five models, shown in *different colours*, are tested, with y_c between −30 and 50 km. Observations are shown in *blue*

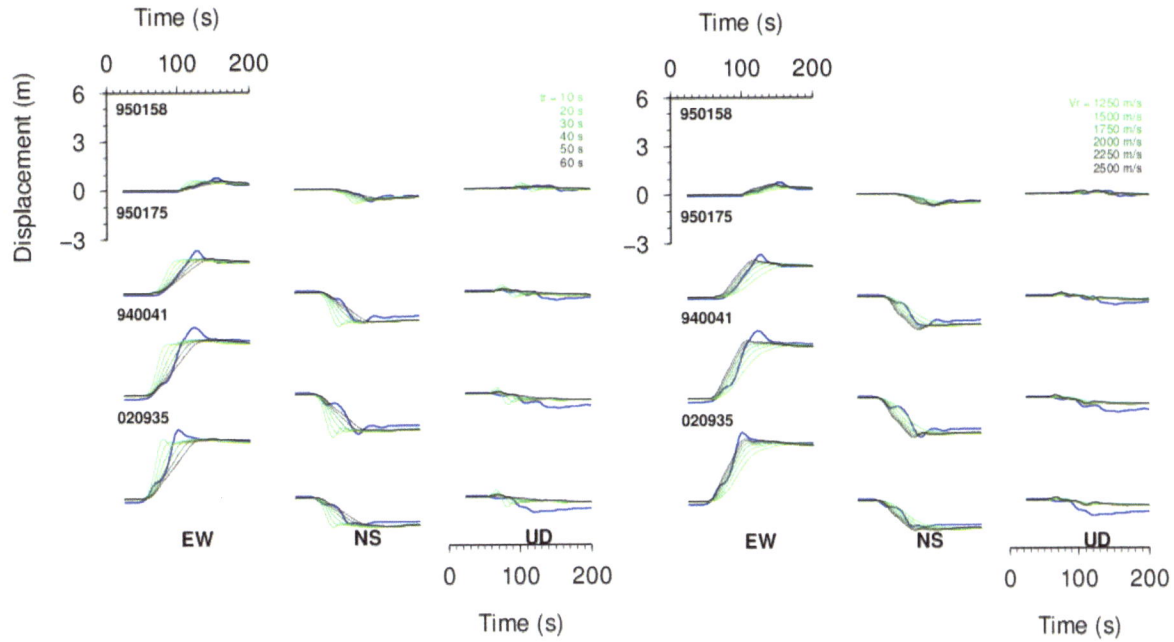

Figure 8
Sensitivity test for the parameters, t_r (rise time) on the *left* and V_r (rupture velocity) on the *right*. The comparisons of ground displacement at the four selected stations are displayed. Observations are shown in *blue*

5.2. Possibility of Two Ellipses Solutions

We saw that the inversion method based on elliptical patches permits a general characterization of the earthquake by a single large patch. On the other hand, Ruiz et al. (2012) were able to characterize the rupture process of the 2010 Maule, Chile, earthquake with two elliptical patches. As mentioned in the introduction, Aochi and Ide (2011) and Ide and Aochi (2013) propose for the 2011 Tohoku earthquake several heterogeneities of different scales. Thus, we try to invert the Tohoku rupture process by using two elliptical patches. Each ellipse is parameterized by the same seven parameters previously used for the single ellipse, in addition to two more parameters necessary to localize the starting point of the rupture initiating on the second ellipse. This point has to be located inside both ellipses. In total, the model is described by 16 parameters. The rupture starts from the hypocentre, and propagates inside the first ellipse. When the rupture reaches the initiating point of the second ellipse, the rupture of this ellipse is triggered.

The three components of the displacement field from the GPS data are again used to constrain the inversion. They are processed by a 20 s low-pass filter. Because of the two-ellipse parameterization, longer rupture process can be generated. We compare the ground motions over a slightly increased duration (300 s instead of 250 s previously) in order to be sure that the contribution of both ellipses is fully accounted for in the comparison. The five best models of the inversions carried out with two ellipses are shown in Fig. 9. The misfits are significantly improved: up to 30 % lower than the best model using a single ellipse. The obtained models globally present the similar geometrical features as the geometry of the two ellipses. The first ellipse is roughly circular and partly located in the deepest area of the fault. The second ellipse is always shallower and more extended in the strike direction, resulting in a flatter geometry. The rupture initiation location of the second ellipse is always close to the hypocentre. The parameterization by two elliptical patches clearly becomes a much more complex problem, whose solutions must be carefully considered. Despite the good agreements between the synthetic ground motions and the observations, the models are physically inconsistent, and not in the

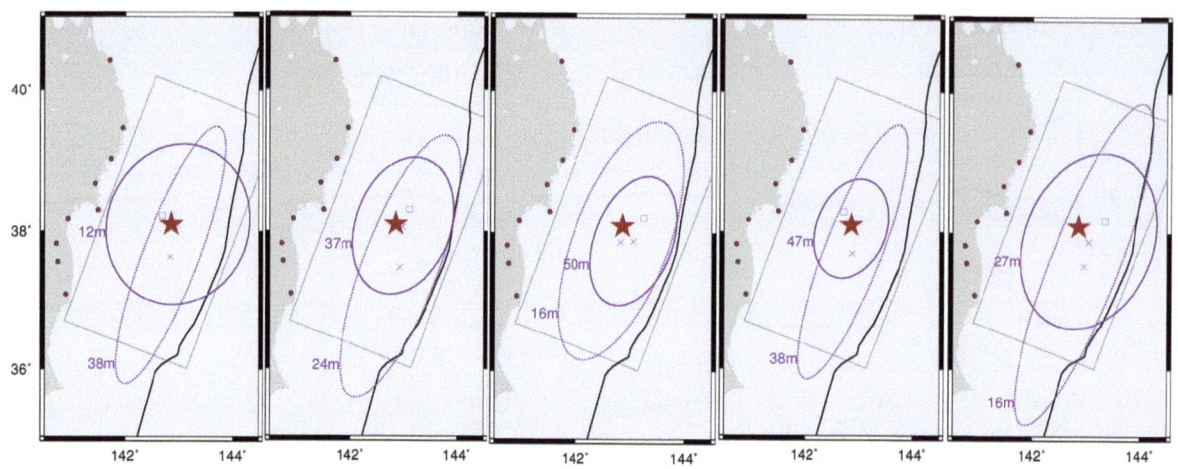

Figure 9
Fault geometry of the five best models independently obtained using two elliptical patches. The *first ellipse* is displayed with a *solid line*, whereas the *second ellipse* is drawn with a *dotted line*. The *maximum slip* is indicated near each ellipse. The *square* represents the rupture-initiation location from which the rupture starts on the second ellipse

way we expected. In fact, their moment magnitudes are extremely large, between 9.6 and 9.95. Again, the limited influence of the shallowest part of the fault on the distant stations for the low frequencies considered here could be the reason of these discrepancies. In fact, with a single-ellipse parameterization, the upper part of the fault, in spite of its low influence on the signals, is well constrained, because of the imposed slip geometry over the elliptic patch. On the other hand, when a second ellipse is introduced, and the first ellipse, more deeply located, recreates most of the signal, the second ellipse is mostly located in the upper part of the fault, which does not contribute much to the signals but significantly increases the magnitude. Thus, the uneven distribution of the stations, preventing uniform constraint of the slip over the whole fault, and the low frequencies considered do not allow us to obtain easily a refined model of the rupture process. Finally, inversions with the same parameterization and a further constraint on the magnitude ($M_w < 9.05$) are tested to see if a physically consistent two ellipses description can be obtained. All obtained solutions present the same features: the first patch is similar to the one obtained in the case of a single ellipse and the second ellipse presents a quasi-null peak slip. This tends to suggest that the Tohoku earthquake cannot be explained by the two-ellipse parameterization.

5.3. "Pseudo Real-Time" Inversion

One major problem highlighted during the 2011 Tohoku earthquake is that a reliable magnitude was not estimated rapidly enough by the Earthquake Early Warning System (EEWS). An estimation of 7.2 was firstly announced 8.6 s after the triggering of the system (14:46:48 JST) and was then improved with time until a magnitude of about 8, based on the method of JMA-displacement magnitude (HOSHIBA et al. 2011), was computed. The EEWS worked well for the first estimation of the earthquake location and ground-motion prediction: our inversion actually uses the location of the hypocentre obtained by the EEWS. However, it is important to obtain rapidly a reliable magnitude estimate and a first finite-fault model. After the Tohoku earthquake improved procedures for the fast estimation of the magnitude have been proposed. For example, COLOMBELLI et al. (2012) suggest some modifications to the classical methods of EEWSs in order to minimize the saturation effect for large magnitudes. Using an approach based on the gradual increase of the P-wave time windows, a magnitude estimate of 8.4 is obtained 35 s after the detection of the first P waves (about 55 s after the triggering of the event). This estimation is still not large enough, but is, nevertheless, better than the one given by

JMA at that time. In WRIGHT *et al.* (2012) the use of continuous GPS data was investigated: assuming a very simple "static" model a magnitude 8.8 could be obtained about 100 s after the earthquake onset. In this study, we investigate the possibility of using the single-ellipse model to give a good first approximation of the rupture process using continuous GPS data. We are then interested in estimating the robustness of our inversion procedure with respect to the duration of the data length we use. Let us vary the time windows width from 60 to 250 s, namely until 14:47:23 to 14:50:33 (JST). The initial time of the windows corresponds to the time when the rupture was triggered (14:46:23 JST). Windows durations of 60, 90, 120, 150, 200 and 250 s are investigated. These time stamps are shown in Fig. 3 with grey vertical lines. For each duration, ten independent inversions are conducted, resulting in a large number of models (about 300,000), which give a good overview of the parameter space.

The inversion results are summarized in Fig. 10. For each time-window duration, we plot the relative difference in misfits with respect to the best model obtained as a function of the resultant magnitude. First, we can observe that when the duration increases, the magnitude is better constrained, namely, the solution converges to a narrower range. For instance, when using a data length of 60 s, the inversion estimates the magnitude between 8.05 and 9.35 with a margin of 20 % in the misfits between runs. This estimation is surprisingly good because at that time only a portion of the entire fault has been ruptured and some of the stations are only showing the arrival of the first waves (only a few centimetres of displacement can be observed, as seen in Fig. 3). Mechanically, the final magnitude could not be playing a role yet. The slow growth of this rupture (e.g., IDE and AOCHI 2013) may appear as slow progress of a large rupture. As a result, the inversion with a few parameters using continuous GPS data (if available in real time) can quickly provide a reliable magnitude of this earthquake, even before the end of the rupture process. Nevertheless, it would be necessary to apply the method on other "moderate" events as well, in order to verify its applicability within EEWSs.

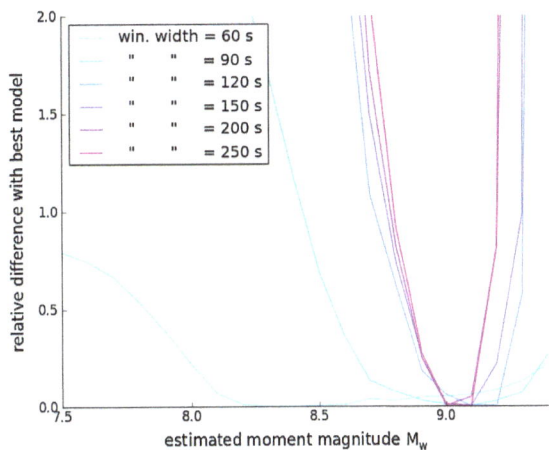

Figure 10
Relative difference of the best model for an obtained moment magnitude in the inversion process. We test six different data lengths, from 60 to 250 s, for the inversion

6. Summary

The elliptical patch method combined with a genetic algorithm has been applied to determine the rupture process of the 2011 Tohoku earthquake. The displacement field recorded by a continuous GPS network and the velocity field integrated from acceleration data are used, using absolute time and various durations for the time window. An ellipse described by seven parameters is robustly obtained in terms of the seismic moment and duration of the rupture. Despite the profusion of data from this earthquake, our very low-frequency inversions are impacted by the absence of off-shore instruments: in particular, the location of the ruptured area is not fully constrained.

Various inversions with different data sets and various frequency ranges led to estimates of magnitude 9, in accordance with the other studies. An inversion with two elliptical patches is tested, but no satisfactory solution is obtained. This implies that no second ellipse (heterogeneity) exists at the same scale as the first one for the 2011 Tohoku earthquake. It is difficult to identify in the current framework a multi-scale description of the rupture, composed of several patches, because of the limited influence of the distant sources on the filtered signals. Finally, the possibility of applying the method to obtain rapidly a reliable estimate of the finite-source parameters is investigated, by varying the duration of the data window. The method is quite efficient, as an estimate of moment magnitude 9 is inferred after only 90 s (75 s after the first arrival of signals at any stations) and the solution is rapidly obtained. Consequently, the procedure has promise within early warning systems for mega-earthquakes.

Acknowledgments

We thank Prof. Takuya Nishimura for providing us GPS data in the framework of the French-Japanese ANR-JST joint program DYNTOHOKU (2011–2013). We used the data from the National Institute for Earth Science and Disaster Prevention, Japan and Geospatial Information Authority of Japan. Discussions with Prof Raul Madariaga and Dr. Sergio Ruiz were very fruitful. This is a contribution to the French national project S4 (Subduction: Slow & Standard Seismology, 2012-2014, ANR-2011-BS56-017) supported by the Agence National de la Recherche. We also benefit from funding from the European Seventh Framework Programme project MARsite for the methodology development. Some calculations were carried out at the French national supercomputing centre GENCI-CINES (Grant c2013-046700). Finally, we thank two anonymous reviewers for their careful comments on an earlier version of this article and John Douglas for proofreading.

REFERENCES

AKI, K. (1979), *Characterization of barriers of an earthquake fault*, J. Geophys. Res., 84, 6140–6148.

AOCHI, H. and S. IDE (2011), *Conceptual multi-scale dynamic rupture model for the 2011 Off-the-Pacific-Coast-of-Tohoku earthquake*, Earth Planets Space, 63, 761–765.

BOUCHON, M. (1981), *A simple method to calculate Green's functions for elastic layered media*, Bull. Seismol. Soc. Am. 71, 959–971.

COLOMBELLI, S., A. ZOLLO, G. FESTA and H. KANAMORI (2012), *Early magnitude and potential damage zone estimates for the great Mw 9 Tohoku-Oki earthquake*, Geophys. Res. Lett., 39, L22306, doi:10.1029/2012GL053923.

DI CARLI, S., C. FRANCOIS-HOLDEN, S. PEYRAT, and R. MADARIAGA (2010), *Dynamic inversion of the 2000 Tottori earthquake based on elliptical subfault approximations*, J. Geophys. Res. 115, B12328, doi: 10.1029/2009JB006358.

GELLER, R. J. (2011), *Shake-up time for Japanese seismology*, Nature, 472, 407–409, doi:10.1038/nature10105.

GOLDBERG, D. (1989), Genetic Algorithms in Search, Optimization, and Machine Learning, Addison-Wesley Professional, ISBN 978-0201157673.

HARTZELL, S. H. and T. H. HEATON (1983), *Inversion of strong ground motion and teleseismic waveform data for the fault rupture history of the 1979 Imperial Valley, California, earthquake*, Bull. Seismol. Soc. Am., 73, 1553–1583.

HASHIMOTO, C., A. NODA, T. SAGIYA and M. MATSU'URA (2009) *Interplate seismogenic zones along the Kuril–Japan trench from GPS data inversion*. Nature Geosci. 2, 141–144.

HOSHIBA, M., K. IWAKIRI, N. HAYASHIMOTO, T. SHIMOYAMA, K. HIRANO, Y. YAMADA, Y. ISHIGAKI and H. KIKUTA (2011), *Outline of the 2011 off the Pacific coast of Tohoku Earthquake (Mw9.0)—Earthquake Early Warning and observed seismicity intensity*, Earth Planets Space, 63, 547–551.

IDE, S. and H. AOCHI (2013), *Historical seismicity and dynamic rupture process of the 2011 Tohoku-Oki earthquake*, Tectonophys., 600, 1–13, doi:10.1016/j.bbr.2011.03.031.

IDE, S. and H. AOCHI (2005), *Earthquakes as multiscale dynamic ruptures with heterogeneous fracture surface energy*, J. Geophys. Res., 110, B11303, doi:10.1029/2004JB003591.

IDE, S., A. BALTAY and G.C. BEROZA (2011), *Shallow dynamic overshoot and energetic deep rupture in the 2011 Mw 9.0 Tohoku-Oki earthquake*. Science 332, 1426–1429, doi:10.1126/science.1207020

IRIKURA, K., and H. MIYAKE (2010), *Recipe for predicting strong ground motion from crustal earthquake scenarios*, Pure Appl. Geophys., 168, 85–104, doi:10.1007/s00024-010-0150-9.

KAMAE, K., K. IRIKURA and A. PITARKA (1998), *A technique for simulating strong ground motion using hybrid Green's function*, Bull. Seism. Soc. Am., 88, 357–367.

KANAMORI, H. and G.S. STEWART (1978), *Seismological aspects of the Guatemala earthquake of February 4, 1976*, J. Geophys. Res., 83, 3427–3434.

KURAHASHI, S. and K. IRIKURA (2011), *Source model for generating strong ground motions during the 2011 off the Pacific coast of Tohoku earthquake*, Earth Planets Space, 63, 571–576.

KOKETSU, K., Y. YOKOTA, N. NISHIMURA, Y. YAGI, S. MIYAZAKI, K. SATAKE, Y. FUJII, H. MIYAKE, S. SAKAI, Y. YAMANAKA, AND T. OKADA. (2011), *A unified source model for the 2011 Tohoku earthquake*. Earth and Planetary Science Letters 310 (3–4), 480–487, doi:10.1016/j.epsl.2011.09.009.

MAI, P. M. (2013), *Uncertainty quantification in earthquake source inversions: The Source Inversion Validation (SIV) project*, Geophys. Res. Abst., 15, EGU2013-3596.

MAI, P.M., and G.C. BEROZA (2003), *A hybrid method for calculating near-source, broadband seismograms from an extended source*, Phys. Earth Planet. Int., 137, 183–199.

NISHIMURA, T., T. HIRASAWA, S. MIYAZAKI, T. SAGIYA, T. TADA, S. MIURA and K. TANAKA (2004), *Temporal change of interplate coupling in northeastern Japan during 1995–2002 estimated from continuous GPS observations*. Geophys. J. Int. 157, 901–916

OZAWA, S., T. NISHIMURA, H. SUITO, T. KOBAYASHI, M. TOBITA, and T. IMAKIIRE (2011), *Coseismic and postseismic slip of the 2011 magnitude-9 Tohoku-Oki earthquake*, Nature, 475, 373–376, doi:10.1038/nature10227.

RUIZ, S., MADARIAGA, R., ASTROZA, M., SARAGONI, G.R., LANCIERI, M., VIGNY, C., and CAMPOS, J. (2012), *Short period rupture process of the 2010 Mw 8.8 Maule earthquake in Chile*, Earthquake Spectra 28 Issue S1

RUIZ, S. AND R. MADARIAGA (2011), *Determination of the friction law parameters of the Mw 6.7 Michilla earthquake in northern Chile by dynamic inversion*, Geophys. Res. Lett., *38*, L09317, doi:10.1029/2011GL047147.

RUIZ, S. and R. MADARIAGA (2013), *Kinematic and dynamic inversion of the 2008 Northern Iwate earthquake*. Bull. Seismol. Soc. Am., *103*, 694–708.

SAMBRIDGE, M. (1999), *Geophysical inversion with a Neighbourhood Algorithm, Searching a parameter space*, Geophys. J. Int. *138*, 479–494.

SIMONS, M., S. E. MINSON, A. SLADEN, F. ORTEGA, J. JIANG, S. E. OWEN, L. MENG, J.-P. AMPUERO, S. WEI, R. CHU, D. V. HELMBERGER, H. KANAMORI, E. HETLAND, A. W. MOORE and F. H. WEBB (2011), *The 2011 Magnitude 9.0 Tohoku-oki earthquake: Mosaicking the megathrust from seconds to centuries*, Science, *332*, 1421, doi:10.1126/science.1206731.

SUZUKI, W., S. AOI, H. SEKIGUCHI, and T. KUNUGI (2011), *Rupture process of the 2011 Tohoku-Oki mega-thrust earthquake (M9.0) inverted from strong-motion data*, Geophysical Research Letters *38*, L00G16, doi:10.1029/2011GL049136.

VALLÉE, M., and M. BOUCHON (2004), *Imaging coseismic rupture in far field by slip patches*, Geophys. J. Int., *156*, 615–630.

WRIGHT, T. J., N. HOULIÉ, M. HILDYARD, and T. IWABUCHI (2012), *Real-time, reliable magnitudes for large earthquakes from 1 Hz GPS precise point positioning: The 2011 Tohoku-Oki (Japan) earthquake*, Geophys. Res. Lett., *39*, L12302, doi:10.1029/2012GL051894.

(Received November 7, 2013, revised April 10, 2014, accepted April 29, 2014, Published online May 20, 2014)

Parallel Implementation of Dispersive Tsunami Wave Modeling with a Nesting Algorithm for the 2011 Tohoku Tsunami

Toshitaka Baba,[1,3] Narumi Takahashi,[1] Yoshiyuki Kaneda,[1] Kazuto Ando,[1] Daisuke Matsuoka,[1] and Toshihiro Kato[2]

Abstract—Because of improvements in offshore tsunami observation technology, dispersion phenomena during tsunami propagation have often been observed in recent tsunamis, for example the 2004 Indian Ocean and 2011 Tohoku tsunamis. The dispersive propagation of tsunamis can be simulated by use of the Boussinesq model, but the model demands many computational resources. However, rapid progress has been made in parallel computing technology. In this study, we investigated a parallelized approach for dispersive tsunami wave modeling. Our new parallel software solves the nonlinear Boussinesq dispersive equations in spherical coordinates. A variable nested algorithm was used to increase spatial resolution in the target region. The software can also be used to predict tsunami inundation on land. We used the dispersive tsunami model to simulate the 2011 Tohoku earthquake on the Supercomputer K. Good agreement was apparent between the dispersive wave model results and the tsunami waveforms observed offshore. The finest bathymetric grid interval was 2/9 arcsec (approx. 5 m) along longitude and latitude lines. Use of this grid simulated tsunami soliton fission near the Sendai coast. Incorporating the three-dimensional shape of buildings and structures led to improved modeling of tsunami inundation.

Key words: 2011 Tohoku tsunami, Boussinesq model, Dispersion, Soliton fission, Parallel computation.

1. Introduction

On 11 March 2011, a large, interplate earthquake between the Pacific and North American lithospheric plates occurred in the Japan Trench subduction zone. The Japan Meteorological Agency estimated the moment magnitude of the earthquake to be 9.0. The coastal region of eastern Japan was strongly shaken for 4–5 min, and that movement was followed by the devastating Tohoku tsunami. The tsunami completely destroyed many coastal cities along the eastern coast of Japan. Clear signals of the tsunami were recorded by tsunami gauges around the world during the event (Hayashi *et al.* 2011; Maeda *et al.* 2011). Many videos were also taken by evacuees and public agencies during the inundation of the coastal region. Tsunami height surveys were conducted at more than 5,900 points along the coast after the event (Mori *et al.* 2012).

Frequency dispersion is apparent in the far-field records of the 2011 Tohoku tsunami (Løvholt *et al.* 2012; Kirby *et al.* 2013). Dispersion of a tsunami occurs because water waves with different wavelengths travel at different speeds. Improvements in offshore tsunami observation technology have facilitated many observations of tsunami wave frequency dispersion in the open ocean in recent years (Horillo *et al.* 2006; Saito *et al.* 2010). Another remarkable phenomenon associated with tsunami dispersion occurred in the near-field region on the shallow, gentle slope along the Sendai coast. That phenomenon, called "tsunami soliton fission" (Shuto 1985), is characterized by split, short-period waves (or "undular bores") around the tsunami crest caused by a combination of wave nonlinearity and dispersion. The wave front of a tsunami propagating on a shallow, gentle slope becomes steep because of wave nonlinearity effects. When the wave front becomes sufficiently steep, the effect of wave dispersion begins to cause fission of the wave (Madsen *et al.* 2008). Analysis of a video of the Tohoku tsunami (Murashima *et al.* 2012) has

---

[1] Japan Agency for Marine-Earth Science and Technology (JAMSTEC), Yokohama Institute for Earth Sciences, 3173-25 Showa-machi, Kanazawa-ku, Yokohama, Kanagawa 236-0001, Japan. E-mail: baba.toshi@tokushima-u.ac.jp

[2] NEC Corporation, 5-7-1 Shiba, Minato-ward, Tokyo 108-8001, Japan.

[3] *Present Address*: Institute of Technology and Science, The University of Tokushima, 2-1 Minami-jyousanjima-cho, Tokushima 770-8506, Japan.

indicated that the wavelengths of the split waves ranged from 100 m to several hundred meters, and the amplitudes were several meters. If soliton fission occurs, the leading wave is amplified dramatically, the result being a larger tsunami force on coastal structures.

It is, therefore, essential to include the characteristics of tsunami soliton fission in tsunami modeling, but commonly used tsunami simulation models based on nonlinear and/or linear long-wave equations cannot reproduce tsunami soliton fission, because a tsunami soliton results from the combined effects of nonlinearity and dispersion. A Boussinesq-type approach that includes a dispersion term in the long-wave equations is an appropriate method for simulating tsunami soliton fission (MATSUYAMA et al. 2007, MURASHIMA et al. 2010; SON et al. 2011; ZHOU et al. 2011). Such simulation requires many computational resources. Rapid progress is being made, however, in parallel computing technology, an efficient way of solving the Boussinesq equations (SITANGGANG and LYNETT 2005).

In this study, we initially developed a new parallelized software by using message-passing interface (MPI) and open multi-processing (OpenMP) libraries; the software solves nonlinear Boussinesq-type dispersive equations with nested bathymetric grids. SITANGGANG and LYNETT (2005) constructed a parallelized scheme for Boussinesq modeling on a uniform finite-difference grid by domain decomposition. ZHOU (2011) investigated a nested approach for Boussinesq modeling, but their approach did not include parallel technology. The software described in this paper enables parallel computation to be used simultaneously with a nesting algorithm. Implementation of this software enabled us to perform a large-scale dispersive tsunami modeling study with high spatial resolution and within a reasonable time. We then applied the new software to a modeling study of the 2011 Tohoku tsunami on the Supercomputer K, which at the time this paper was written was the fastest computer in Japan, and the fastest computer in the world from June 2011 to June 2012. The Supercomputer K is equipped with 82,944 computation processors for a total of 663,552 cores; it has achieved a Linpack performance of 10.51 petaflops with a high computing efficiency ratio of 93.2 % (FUJITSU 2012).

2. Parallelized Dispersive Tsunami Software

To perform large-scale dispersive tsunami wave modeling, we started with the URS Corporation and Geoscience Australia (URSGA) software provided by JAKEMAN et al. (2010). The numerical scheme used by URSGA is an explicit leapfrog difference method that solves either the linear or nonlinear long-wave equations in spherical coordinates. This scheme is based on the uniform finite-difference scheme of SATAKE (2002). Nonlinear terms in the model are approximated with upwind finite differences, and linear terms are approximated by two-point centered finite differences. The URSGA model uses a variable nested algorithm that enables the spatial resolution of the study region to be easily increased. The ratio of the grid spacing of the parent and child nested grids is 3:1.

SAITO et al. (2010) reported a practical method for solving the linear Boussinesq equations derived from PEREGRINE (1972) with a uniform finite-difference scheme. By incorporating the method of SAITO et al. (2010), we attempted to include the dispersion term in existing URSGA software. The URSGA software treats depth-averaged velocities (u and v) as dependent variables. Unfortunately, this software caused numerical instability during dispersive tsunami calculations in areas where the topography changed suddenly, for example at a continental shelf or trench axis. Conserved variables, however, have the property of being able to resolve discontinuities. ROEBER et al. (2010) have described the transformation of physical to conserved variables in the Boussinesq equations. We therefore used conserved, depth-integrated variables to ensure stability during computation. This modification worked successfully and enabled us to avoid numerical instability during dispersive modeling. The governing equations used in the new software are expressed as:

$$\frac{\partial M}{\partial t} + \frac{1}{R\sin\theta}\frac{\partial}{\partial \varphi}\left(\frac{M^2}{d+h}\right) + \frac{1}{R}\frac{\partial}{\partial \theta}\left(\frac{MN}{d+h}\right)$$
$$= -\frac{g(d+h)}{R\sin\theta}\frac{\partial h}{\partial \varphi} - fN$$
$$- \frac{gn^2}{(d+h)^{7/3}}M\sqrt{M^2+N^2}$$
$$+ \frac{d^2}{3R\sin\theta}\frac{\partial}{\partial \varphi}\left[\frac{1}{R\sin\theta}\left(\frac{\partial^2 M}{\partial \varphi \partial t}+\frac{\partial^2(N\sin\theta)}{\partial \theta \partial t}\right)\right]$$
(1)

$$\frac{\partial N}{\partial t} + \frac{1}{R \sin\theta} \frac{\partial}{\partial \varphi}\left(\frac{MN}{d+h}\right) + \frac{1}{R}\frac{\partial}{\partial \theta}\left(\frac{N^2}{d+h}\right)$$
$$= -\frac{g(d+h)}{R}\frac{\partial h}{\partial \theta} + fM$$
$$- \frac{gn^2}{(d+h)^{7/3}} N\sqrt{M^2 + N^2}$$
$$+ \frac{d^2}{3R}\frac{\partial}{\partial \theta}\left[\frac{1}{R\sin\theta}\left(\frac{\partial^2 M}{\partial \varphi \partial t} + \frac{\partial^2 (N\sin\theta)}{\partial \theta \partial t}\right)\right] \quad (2)$$

$$\frac{\partial h}{\partial t} = -\frac{1}{R\sin\theta}\left[\left(\frac{\partial M}{\partial \varphi} + \frac{\partial (N\sin\theta)}{\partial \theta}\right)\right] \quad (3)$$

where h is the water height from the sea surface at rest, t is time, φ and θ are the longitude and co-latitude, respectively, R is the earth's radius, d is the water depth, and the variables M and N are depth-integrated quantities equal to $(d+h)u$ and $(d+h)v$, respectively, along longitude and latitude lines, respectively. g is the gravitational constant, f is the Coriolis parameter, and n is Manning's roughness coefficient. In calculation of the dispersion terms [the final terms on the right-hand sides of Eqs. (1) and (2)], the fact that we used the static water depth (d) means that we ignored the dispersion effect on land to avoid risk of numerical instability caused by complex propagation during inundation. To solve these equations, we used the leapfrog, staggered-grid, finite-difference calculation scheme shown in the Appendix.

We used a domain decomposition method for parallel implementation of the Boussinesq model. We divided the finite-difference grid points of the nested grid into multiple rectangular sub-domains, each of the same size, the number of sub-domains being equal to the number of computation nodes (Fig. 1). An important aspect of decomposing the domain is load balancing. All computation nodes must have equal or almost equal amounts of data to be processed. This condition was achieved in the software by dividing all of the nested grids into the same number of sub-domains. Each sub-domain of the nested grids was associated with one computation node. For example, computation node No. 1 computed the No. 1 sub-domain of nested grid A, which was followed by computation of the No. 1 sub-domain of nested grid B, which was then followed by computation of the No. 1 sub-domain of nested grid C, and so forth. Multi-thread processing with OpenMP was also incorporated into the software for acceleration of loop calculations in the sub-domains. To calculate the variables h, M, and N in Eqs. (1)–(3) at a grid point, we must refer to data at the surrounding grid points. The data needed to calculate the variables at the edges of the sub-domains were acquired from the adjoining sub-domains. This data exchanges between the sub-domains were parallelized by the MPI point-to-point communication routines. To facilitate communication between the nested grids, data at the edges of a child nested grid were over-written by interpolated data from the parent nested grid. All of the data of the child nested grid were re-sampled to match the resolution of the parent nested grid, to enable copying of them to the parent nested grid (Fig. 2). This two-way inter-grid communication enabled seamless propagation of the tsunami between the nested grids. The inter-grid communications were implemented by using the MPI collective communication routines. Figure 3 shows a flowchart of the parallel dispersive computation with the nesting algorithm. We call the new parallelized software JAGURS-D, which stands for the Japan Agency for Marine-Earth Science and Technology (JAMSTEC) improvements of the URSGA software for dispersive tsunamis.

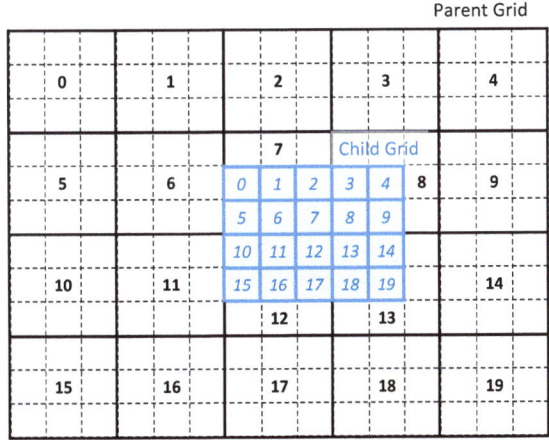

Figure 1
Nested gridding scheme and domain decomposition for parallel computation. *Numbers* indicate MPI ranks assigned for calculation of *rectangular sub-domains*

3. Tsunami Model Validation and Parallel Performance

For the purpose of validation of the dispersive tsunami model developed in this study, we used

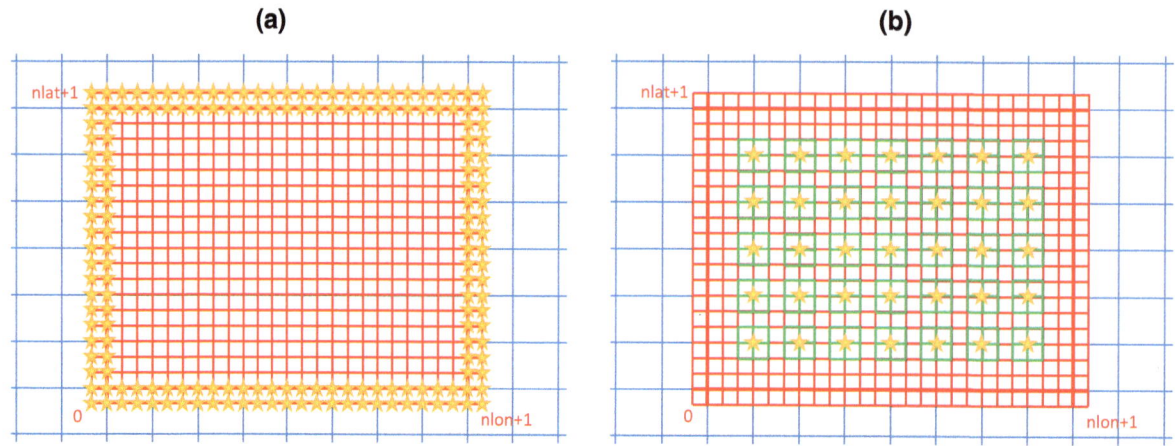

Figure 2
Inter-grid communications **a** from parent to child grids and **b** from child to parent grids. *Blue and red lines* indicate parent and child grid cells, respectively. The *stars* are points for transferred data. For data transfer from the parent to child, the data along the edges of the child grid were linearly interpolated by using the parent data, and copied to the child grid. For data transfer from the child to parent grids, we took an average of nine points of the child data, indicated by *green* in (**b**), to match the resolution of the parent grid

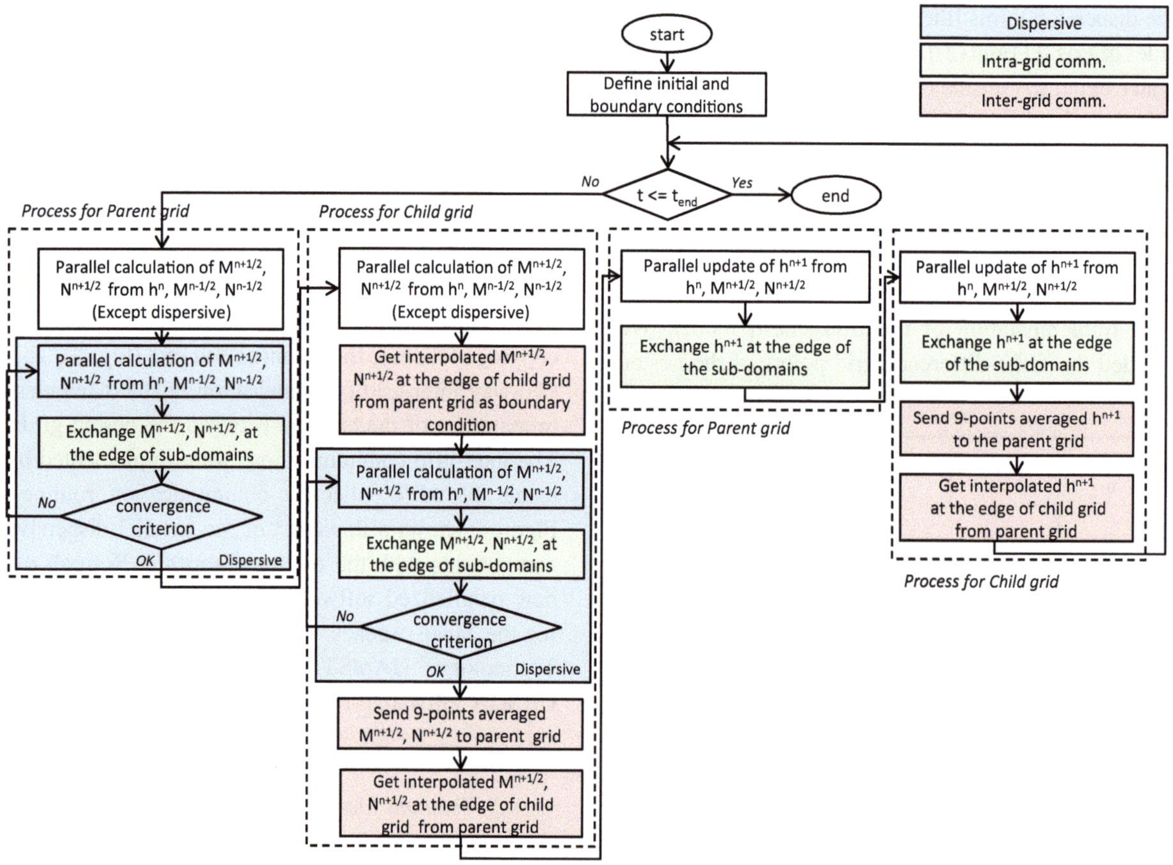

Figure 3
Flowchart of parallel dispersive tsunami calculation with the nesting algorithm. This example uses two nesting grids, parent and child grids, similar to the scheme shown in Fig. 1

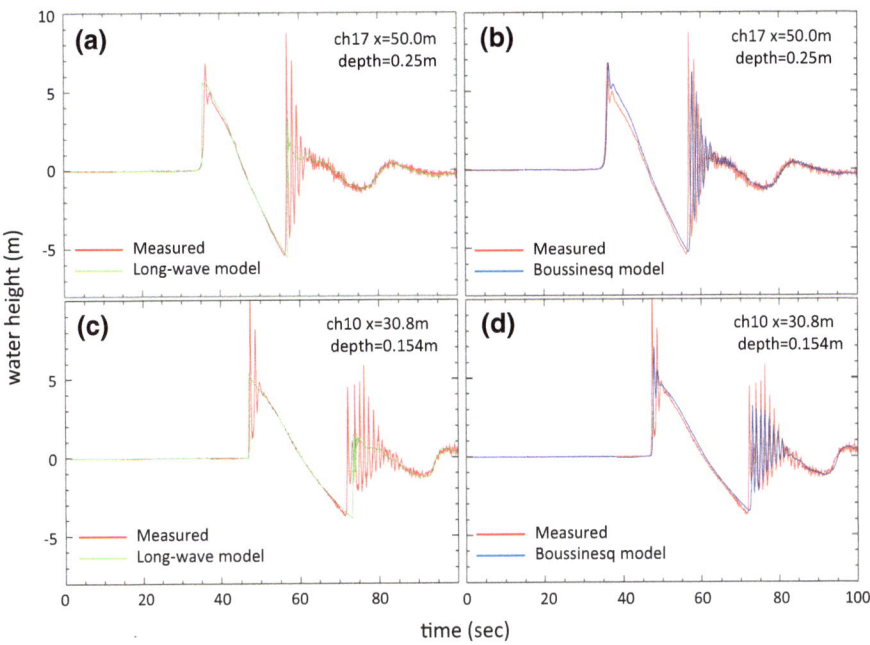

Figure 4
Time histories of water surface elevation calculated with the Boussinesq (*blue*) and long-wave (*green*) models compared with actual fluctuations (*red*) recorded in the wave flume experiment of MATSUYAMA et al. (2007)

results from MATSUYAMA et al. (2007). They conducted a large wave flume experiment to investigate the nature of tsunami soliton fission. The wave flume was 205 m long, 3.4 m wide, and had a maximum depth of 4.0 m (Figs. 1, 2 in MATSUYAMA et al. 2007). We used water surface waveforms recorded during the experiment for which the input wave period was 20 s and the wave amplitude was 0.03 m on a slope gradient of 1/200 (Fig. 5 in MATSUYAMA et al. 2007). Soliton fission occurred at a point 50 m from the modeled shoreline (wet and dry boundary) in that experiment. We constructed a topographic model that simulated the wave flume topography from distances of 80 m to −10 m from the modeled shoreline, with a grid interval of 0.1 m. We have given the recorded waveform time series at a point 80 m from the modeled shoreline as the boundary condition in the calculation. It should be noted that we modified the software from spherical coordinates to Cartesian coordinates to avoid cancellation of significant digits. A single topographic grid with no nesting was applied with a time step of 0.005 s and a Manning's coefficient of 0.025. The integral time was set to 300 s.

Figure 4 shows a comparison of measured (MATSUYAMA et al. 2007) and calculated waveforms at two points 50.0 and 30.8 m from the modeled shoreline in the wave flume experiment. The conventional non-linear long-wave model (Fig. 4a, c) predicted the timing of tsunami arrival and the characteristics of the waveform, except for the soliton fission waves. It could not, however, predict any component of the soliton fission wave. After the wave front became steep, because of nonlinear effects, the wave front propagated without fission in the long-wave model. In contrast, the Boussinesq model developed in this study (Fig. 4b, d) simulated well the time histories of water surface fluctuations recorded in the wave flume experiment, including the soliton fission wave. We can therefore assert that our dispersive tsunami model can be used to investigate issues related to tsunami soliton fission phenomena.

Next, we verified correct operation of the nesting algorithm. We simulated the tsunami caused by the 2011 Tohoku earthquake, using the dispersive model with five nested grids shown in Fig. 5. The finest nesting grid was located on the Sendai coast, where soliton fission was observed during the 2011 Tohoku tsunami. This grid was divided into grid points separated by approximately 2/9 arcsec (about 5 m). The total number of grid points was 21 million. The

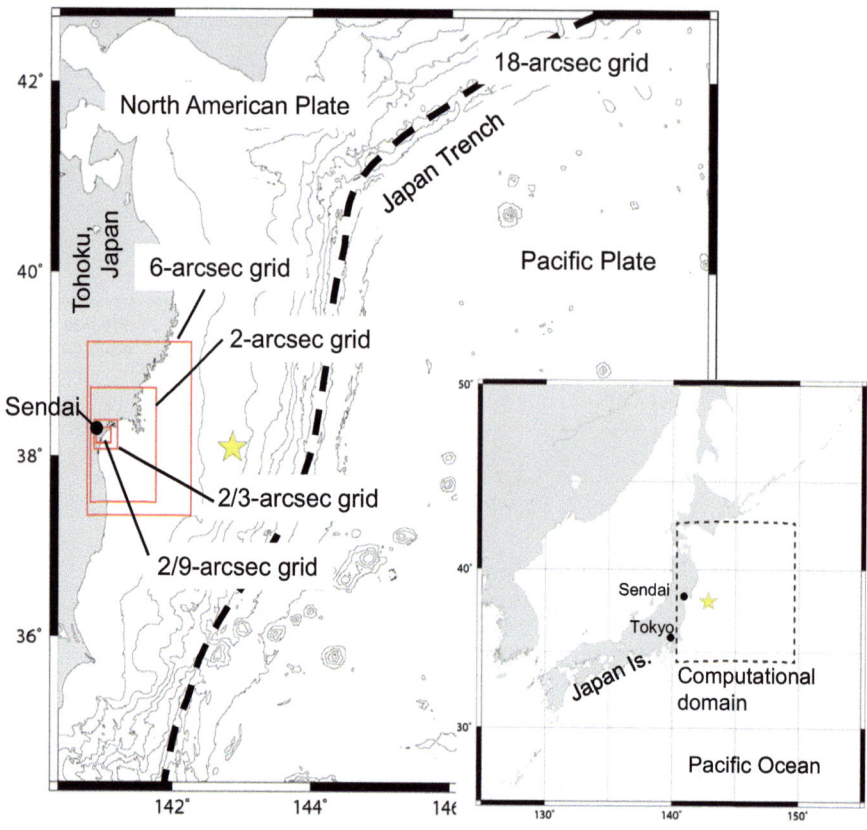

Figure 5
Bathymetry of the computational domain (18-arcsec grid spacing) and *outlines* of nested grids (*rectangles outlined in red*). The depth contour interval is 1,000 m. The *dashed line* is the axis of the Japan Trench. The *star* indicates the epicenter of the 2011 Tohoku earthquake determined by the Japan Meteorological Agency

time step was set to 0.1 s to satisfy the stability condition.

Figure 6 is a sequential time series of maps that indicate sea surface height around the boundary of the nesting grids. Analysis of these maps revealed that the calculated tsunami propagated seamlessly beyond the boundaries of the nesting grids. This result means that the algorithm implemented with our software can successfully achieve transfer of variables between the nesting grids needed to solve the dispersive equations under the parallelized computational scheme.

We also performed a parallel performance test of JAGURS-D on the K computer using the set up described above. We used the MPI+OpenMP mixed parallel model with automatic parallelization by the compiler. We repeated the calculations by changing number of nodes (12, 24, 48, 96, and 192) and measured elapsed time. One node of the K computer consists of eight processing cores. Figure 7 shows the acceleration ratio achieved in the test. Addition of up to 192 nodes efficiently upgraded the speed of computation.

4. Simulation Model for the 2011 Tohoku Tsunami

Many source models have been suggested for the 2011 Tohoku earthquakes, on the basis of seismic data (AMMON *et al.* 2011; YAGI AND FUKAHATA 2011), tsunami data (IMAMURA *et al.* 2011; SAITO *et al.* 2011; Satake *et al.* 2013), and crustal deformation data (SUITO *et al.* 2011; ITO *et al.* 2011). GOTO *et al.* (2012) and BABA *et al.* (2014) used the tsunami source model proposed by IMAMURA *et al.* (2011) for simulation of the near-field tsunami. GRILLI *et al.* (2013) simulated two forms of the Tohoku tsunami derived from a source inverted from teleseismic waves (SHAO *et al.*

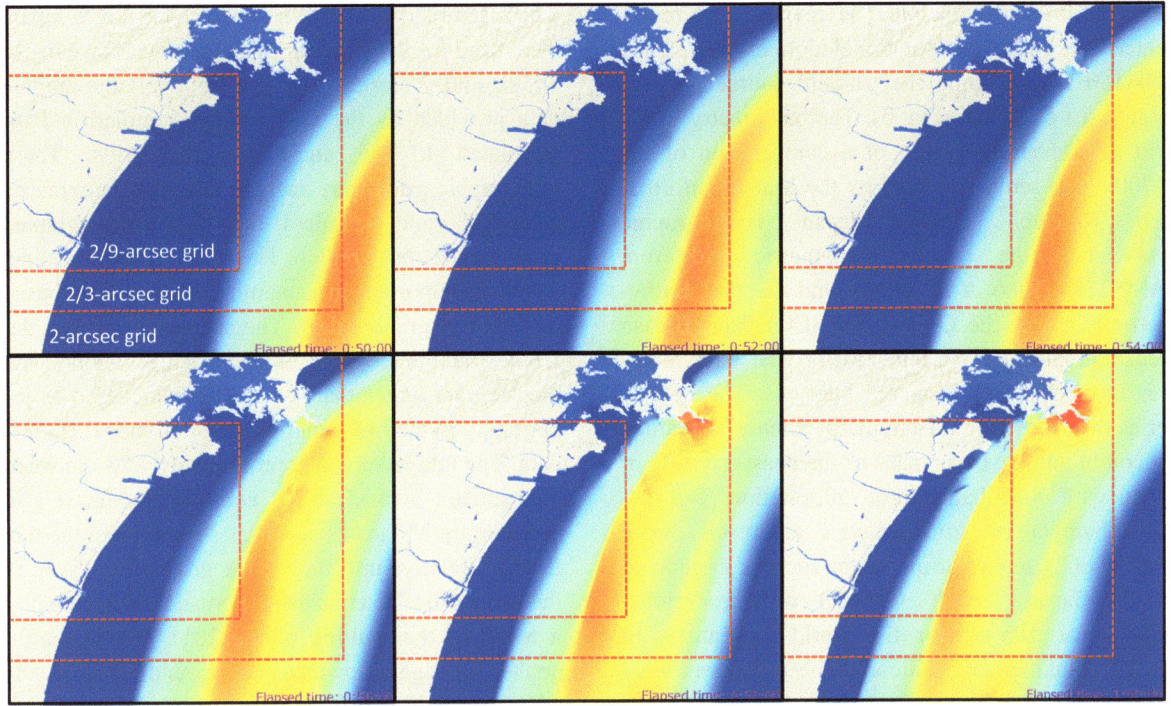

Figure 6
Successive views of water surface elevation during the 2011 Tohoku tsunami, near the Sendai coast, calculated by use of the dispersive tsunami software. *Red dotted lines* indicate the boundaries of nesting grids. The calculated tsunami propagated seamlessly beyond the nesting boundaries

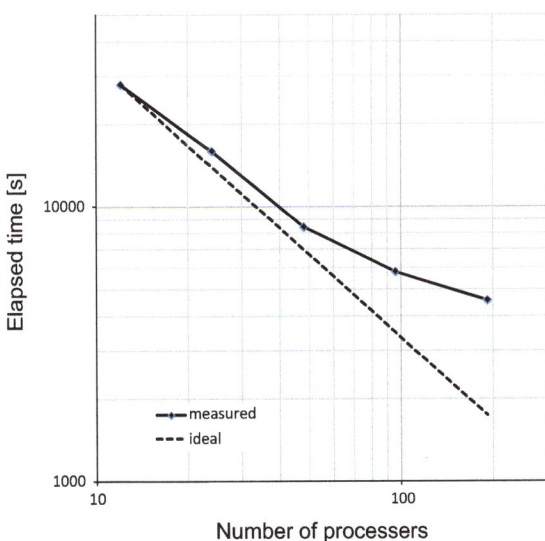

Figure 7
Parallel performance test of JAGURS on the K computer

2011) and their own source based on geodetic data, including sea floor displacements. We used the 2011 Tohoku tsunami source obtained by SAITO *et al.* (2011) in this study, because it is the only model derived from linear dispersive tsunami equations. They inverted the observed tsunami waveforms for the 2011 tsunami source with dispersive Green's functions calculated in a uniform finite-difference scheme. In general, we did not use the dispersive equations but, instead, used linear long-wave equations in tsunami inversion analysis on the assumption that dispersion effects are negligible in the near-field. This assumption is valid when the wavelength of the tsunami source is long enough to be comparable with the water depth. However, the offshore tsunami gauges recorded very short-wavelength tsunami waves during the 2011 tsunami. In this case, the linear long-wave equations could not correctly simulate the short-wavelength tsunami waves, the result being an incorrect image of the tsunami source after inversion analysis. Accordingly, we selected the Saito's model derived from dispersive Green's functions to implement the dispersive tsunami modeling.

For modeling of the 2011 Tohoku tsunami, five bathymetric nested grids, shown in Fig. 5, were defined for our calculation. The coarsest grid represented the entire computational domain (34–43°N, 140–150°E), including the tsunami source and the target area of Sendai (Fig. 5). The bathymetry in the grid was obtained by use of a combination of the M7000 map series provided by the Marine Information Research Center, Japan Hydrographic Association, the Tohoku bathymetric grid from JAMSTEC (KIDO et al. 2011), and General Bathymetric Chart of the Oceans (GEBCO) data (British Oceanographic Data Centre 2010). The bathymetry was interpolated to 18 arcsec intervals. The M7000 series is a set of digital bathymetric contours obtained by combining the basic maps of the coastal waters of Japan with other bathymetric information. The Tohoku bathymetric grid includes all results from JAMSTEC's multi-narrow beam surveys conducted in the Japan Trench. GEBCO provides global bathymetry datasets for the world's oceans with spatial resolution of 30 arcsec. These datasets were subsampled and then interpolated to make nested grids with spacings of 6, 2, 2/3, and 2/9 arcsec for the nesting scheme.

For the land area, we re-sampled Geospatial Information Authority of Japan (GSI) data to produce topographic grids. The 50-m-interval topographic data assembled by the GSI, which covers all of Japan, were used for the topography in grids with spacing of 18, 6, and 2 arcsec. The 5-m interval topographic data provided by the GSI were re-sampled and interpolated to 2/3 and 2/9 arcsec grids. These topographic grids were merged with the bathymetric grids to yield seamless bathymetric–topographic grids for the entire region. The shape of the coastline, which is important in tsunami modeling, was based on GSI topographic data. The shapes of tsunami defense facilities, for example sea walls and breakwaters larger than 7.5 m, were included as topography in the finest digital elevation model (DEM) data. The tide level was approximately −25 cm when the tsunami arrived at the coast in the finest grid (gridded by 2/9 arcsec spacing). We imitated the tide level by relative uplifting of the ground by 25 cm in the simulation. To consider the crustal deformation as a result of the faulting of the 2011 Tohoku earthquake, we again lowered the ground level in the 2/9 arcsec grid by 35 cm, on the basis of NISHIMURA et al. (2011).

We performed a nonlinear dispersive simulation of the tsunami generated by the 2011 Tohoku earthquake by use of the dataset described above. A sponge buffer zone (CERJAN et al. 1985) was applied

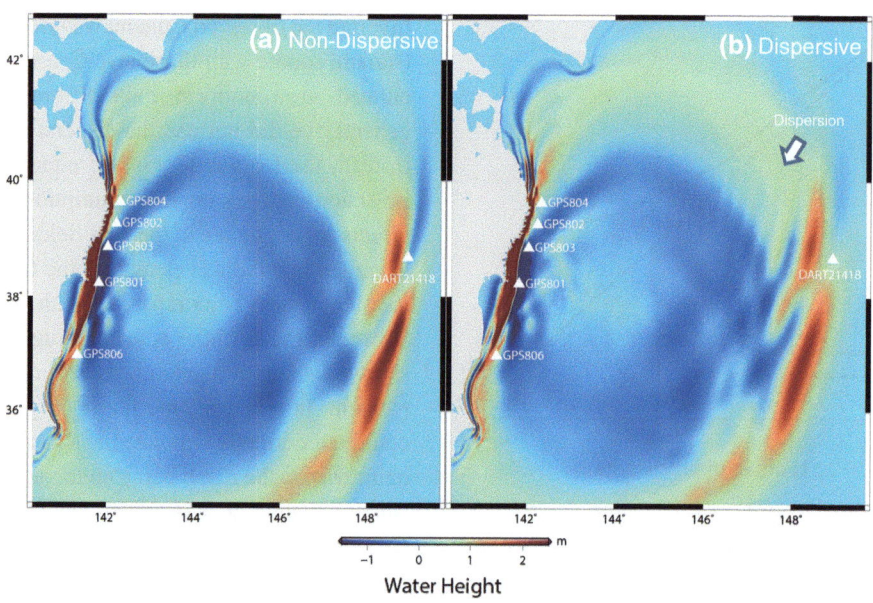

Figure 8
Sea-surface fluctuations 30 min after the earthquake occurred, simulated with nonlinear long-wave equations (a) and nonlinear dispersive wave equations (b). *Triangles* indicate locations of tsunami gauges. Tsunami waveforms are compared in Fig. 9

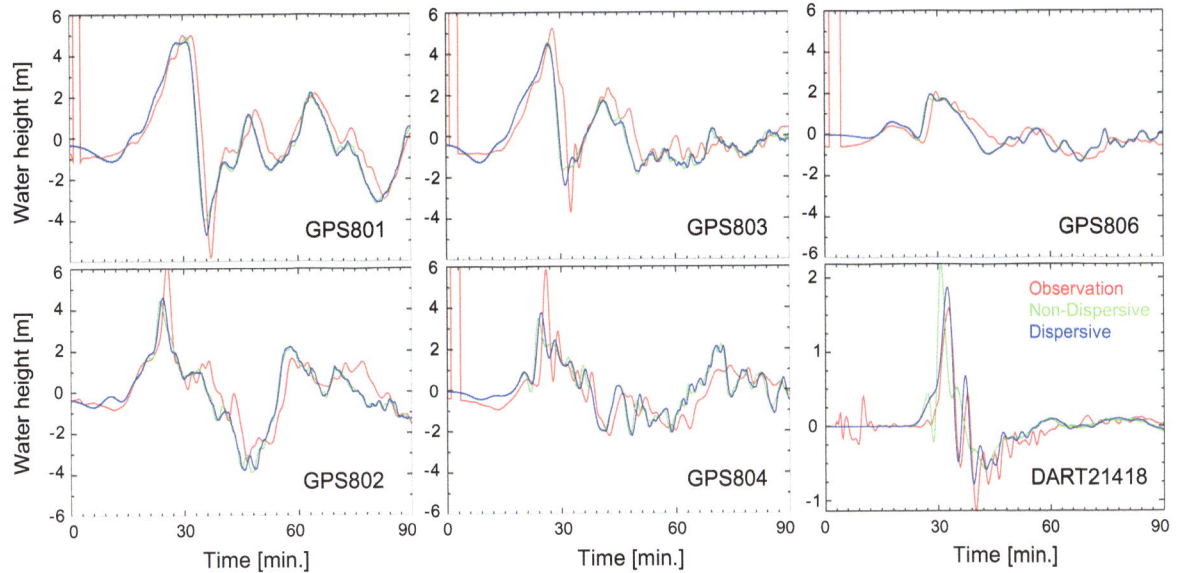

Figure 9
Comparisons of tsunami waveforms determined from observations (*red*), the non-dispersive model (*green*), and the dispersive model (*blue*) at the offshore stations. The locations of the stations are shown in Fig. 8. *Horizontal axis* is elapsed time after the 2011 Tohoku earthquake. The records from stations *GPS801*, *GPS803*, *GPS806*, and *GPS804* lack data at the beginning

at the 60 grid points surrounding the simulation model to avoid reflection of short-period tsunami waves at the outer boundary of the calculation region. For the sea–land boundary, we used a moving boundary so that the tsunami could inundate the land. A uniform Manning's coefficient of 0.025 s m$^{-1/3}$ was used for the whole computation region. The time step was set to be 0.1 s, to satisfy the stability condition for the finite difference algorithm. The integral time was 3 h, including the arrival time of the major tsunami waves at the target region. We implemented this calculation on 192 nodes (1,536 cores) of the K computer. The computation results were produced after an elapsed time of approximately 9.5 h. Non-dispersive modeling, based on the nonlinear long-wave equations, was also performed for comparison.

5. Results and Discussion

Ocean-bottom pressure gauges and global positioning system (GPS) buoys around Japan successfully documented the 2011 Tohoku tsunami offshore (MAEDA et al. 2011; HAYASHI et al. 2011). These offshore gauges were able to provide tsunami waveforms free from such complicating effects as nonlinearity, reflection, and refraction near the coast. These data were therefore useful for validating our dispersive tsunami simulation. Figure 8 compares sea-surface fluctuations between the long-wave and dispersive models. In the region near buoy DART21418, frequency dispersion was apparent in the dispersive simulation (arrow in Fig. 8). Simulated and observed tsunami waveforms are compared in Fig. 9. In the case of the tsunami waveform observed at DART21418, the first tsunami wave was followed by several short-period waves during the time interval from 30 to 60 min after the time of origin of the earthquake. The dispersive simulation produced similar results. This similarity reflects the fact that the source model of SAITO et al. (2011) used dispersive tsunami equations and was adjusted to reproduce the observed data at buoy DART21418. The point is that the short-period wave train following the first tsunami appeared only in the dispersive simulation, but not in the non-dispersive simulation. In contrast, the discrepancy between the non-dispersive and dispersive models was not apparent in the tsunami waveforms recorded by GPS buoys near the Japanese coast. This information is important for determining whether dispersive or non-dispersive

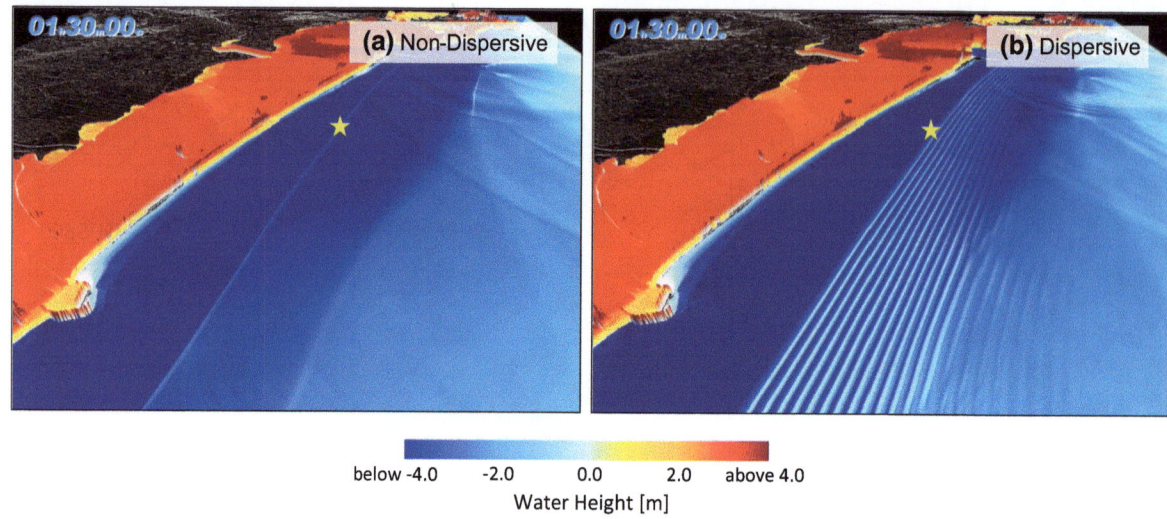

Figure 10
Sea-surface fluctuations of the 2/9-arcsec grid near the Sendai coast 90 min after the earthquake occurred, simulated by use of nonlinear long-wave equations (**a**) and the nonlinear dispersive wave equations (**b**). The *star* indicates the location of the waveform of sea-surface fluctuations plotted in Fig. 11

equations should be used to create Green's functions for source inversion analysis with the tsunami waveforms.

These two models provided quite different images of the area very close to the Sendai coast (Fig. 10). The dispersive model successfully predicted the occurrence of tsunami soliton fission approximately 90 min after the earthquake, when the second large tsunami wave was approaching the coast. The computed wavelength of each split wave was approximately 200 m. Figure 11 shows a comparison of the non-dispersive and dispersive simulations of the tsunami waveforms near the coast. We were able to count 13 split waves in the tsunami waveforms obtained with the dispersive model. The maximum peak-to-trough amplitude of the soliton fission was approximately 3 m. The period of that wave was approximately 13 s. These features of the soliton fission waves were consistent with reports from helicopter observations (MURASHIMA *et al.* 2010). This dispersive model, which used a fine topographic grid interval of 0.22 arcsec (approx. 5 m), simulated the soliton fission waves of the 2011 Tohoku tsunami quantitatively near the coast. We stress that dispersive modeling is essential for simulating the characteristic tsunami phenomena near a coast and the frequency dispersion often observed in the open ocean.

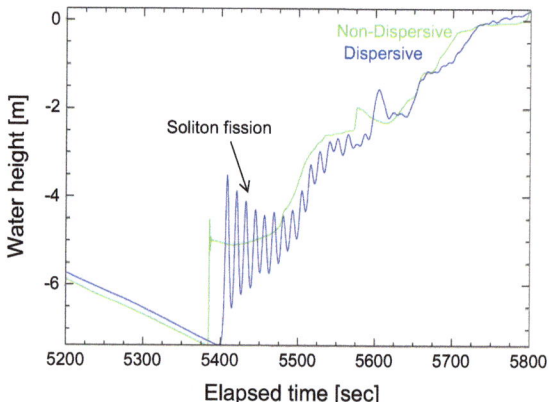

Figure 11
Tsunami waveforms derived from non-dispersive (*green*) and dispersive (*blue*) models at the point indicated by the *star* in Fig. 10

We mapped the differences between the simulated maximum tsunami height in the two models (Fig. 12). Although we were able to describe the tsunami soliton fission for the second tsunami wave as described above, the maximum tsunami height was produced during the first tsunami wave, when the soliton fission wave was less apparent in the calculation. However, weak soliton fission occurred at a point very close to the coast for the first tsunami wave. The occurrence of this soliton fission was accompanied by a large-amplitude tsunami wave near

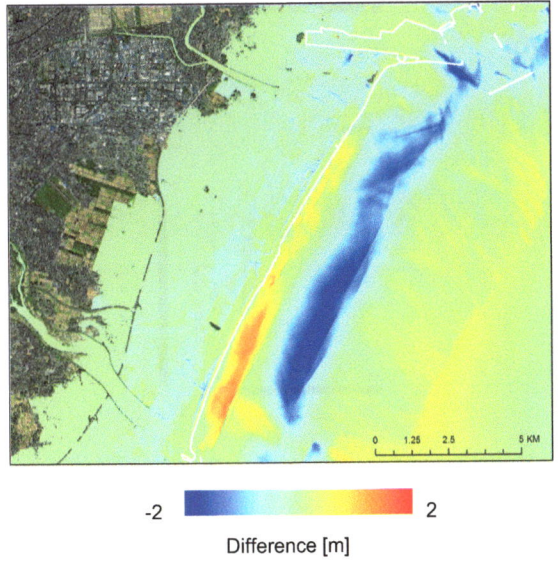

Figure 12
Difference between maximum tsunami height in the non-dispersive and dispersive models for the 2/9-arcsec grid. *Positive values* indicate greater height simulated by the dispersive model. The *white line* is the coastline

the coast. This fact is apparent in Fig. 12, which depicts the difference between the simulated maximum height of the tsunami in the dispersive and non-dispersive models. A positive (greater height in the dispersive model) band of approximately 1 m is apparent along the coastline. In contrast, a negative (smaller height in the dispersive model) band is apparent offshore of the positive band. The amplitude of the nonlinear long-wave tsunami becomes large because of the shoaling effect as the wave approaches the coast. The dispersion term also works to depress the amplification because of wave nonlinearity. Consequently, the dispersive model may produce a tsunami of smaller amplitude than the nonlinear long-wave model when soliton fission does not occur.

How does the dispersive model make a difference in terms of tsunami inundation on land? Comparison of the calculated results revealed that the dispersive model resulted in maximum inundation slightly smaller than the long-wave model (Fig. 12). But the countermeasures taken against a tsunami disaster would make the difference negligible. This small difference may be related to the fact that the maximum inundation was recorded during the first tsunami wave, when soliton fission was much less apparent than it was during the second wave. We therefore do not believe the tsunami derived from the dispersive simulation to produce the maximum inundation on land differs from that of the long-wave model. A previous study that investigated the effect of using a dispersive model to simulate inundation during the 2004 Sumatra tsunami (SHIGIHARA et al. 2006) revealed that the dispersive model and long-wave models produced similar inundation results. However, for the 1983 Nihonkai-Chubu earthquake, a dispersive model resulted in a much larger inundation (IWASE 2005). It would be inappropriate to conclude how much use of a dispersive model affects the simulated inundation. The effect may differ on case-by-case basis. With the dataset we used, inclusion of dispersion in the model had little effect on the maximum simulated inundation height on land.

In this study, we used a source model inverted from only offshore tsunami waveforms observed by ocean-bottom pressure gauges and GPS buoys (IMAMURA et al. 2011). These offshore tsunami gauges are able to detect a tsunami before it arrives on the coast and are therefore useful for early prediction of tsunami. In fact, several algorithms for estimating the source of a tsunami on a real-time basis have been investigated by inverting offshore data (TSUSHIMA et al. 2012; TAKAGAWA and TOMITA 2012). Our concern is how accurately the characterization of a tsunami source based on inversion of offshore tsunami data describes the tsunami and inundation near the coast with the methodology used in this study. MORI (2011) measured numerous tsunami inundation heights in the coastal region after the 2011 Tohoku tsunami. Our simulation results are compared with their survey results in Fig. 13. The predicted and observed tsunami heights are positively correlated (Fig. 13c). We also used Aida's method (AIDA 1978) to quantitatively validate the numerical simulation. AIDA (1978) defined two indices, the geometric mean K and geometric standard deviation κ, that can be used to evaluate the reproducibility of numerical simulations of tsunami events. For the data in Fig. 13, the calculated K and κ values were 0.94 and 1.28, respectively.

However, this model seems to systematically underestimate tsunami heights along the coastline (Fig. 13b). We took into account highly accurate DEM data in the tsunami simulation, but large, strong

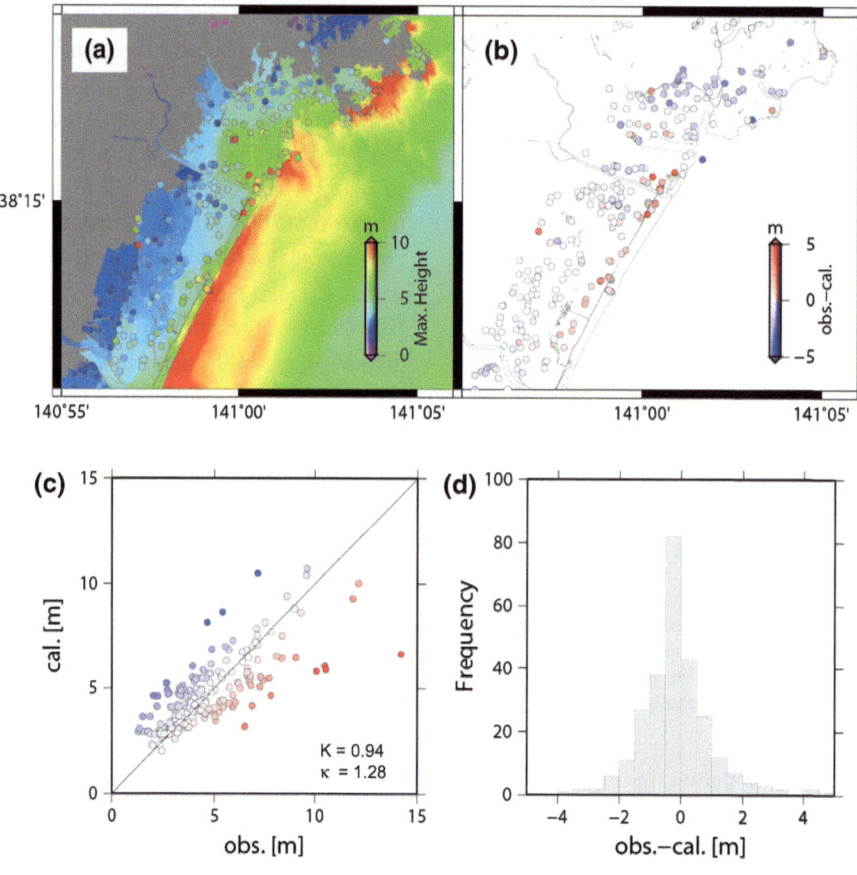

Figure 13
Comparison of measured inundation heights and simulations with the dispersive model using conventional DEM. **a** *Colored map* showing the maximum inundation height in the 2/9-arcsec grid obtained with the simulation. *Colored circles* are derived from a field survey (MORI 2011). **b** The map shows differences between observed and calculated heights at specific points. **c** Relationship between observed and calculated values. The *straight line* indicates equality of calculated and observed values. K and κ are the geometric mean and geometric standard deviation, respectively, of the reproduction indices, proposed by AIDA (1978). **d** Histogram of deviations of the calculated values from the observations

buildings should, similar to sea walls, afford direct protection against an incoming tsunami. We inferred that incorporating three-dimensional (3D) shapes of buildings and structures may lead to improved modeling of tsunami inundation in the coastal region. Lidar measurements are being carried out along the Japanese coast by the GSI. Lidar collects reflections with high spatial resolution from the ground surface, and reflections from such elevated surfaces as roads, bridges, the roofs of buildings, and the tops of trees. We therefore embedded the 3D building data derived from lidar measurements as topographic highs in the dispersive tsunami model to reproduce tsunami barriers in the coastal area. We repeated the tsunami calculation after replacing topographic data from the DEM only with data that included 3D building information embedded in the DEM (Fig. 14). The predicted tsunami became large in front of the buildings along the coastline. The predicted maximum inundations reproduced observations better than the maximum inundations obtained with the DEM model. The values of K and κ were improved to 0.97 and 1.27, respectively. These values satisfy the adequacy criteria for tsunami numerical modeling established by the Japan Society of Civil Engineers (2002) ($0.95 < K < 1.05$, $\kappa < 1.45$). The 3D building data helped to improve the accuracy of the simulated inundations. We conclude that the measured heights were simulated well by this method of calculation, which used 3D building data and tsunami source information acquired with

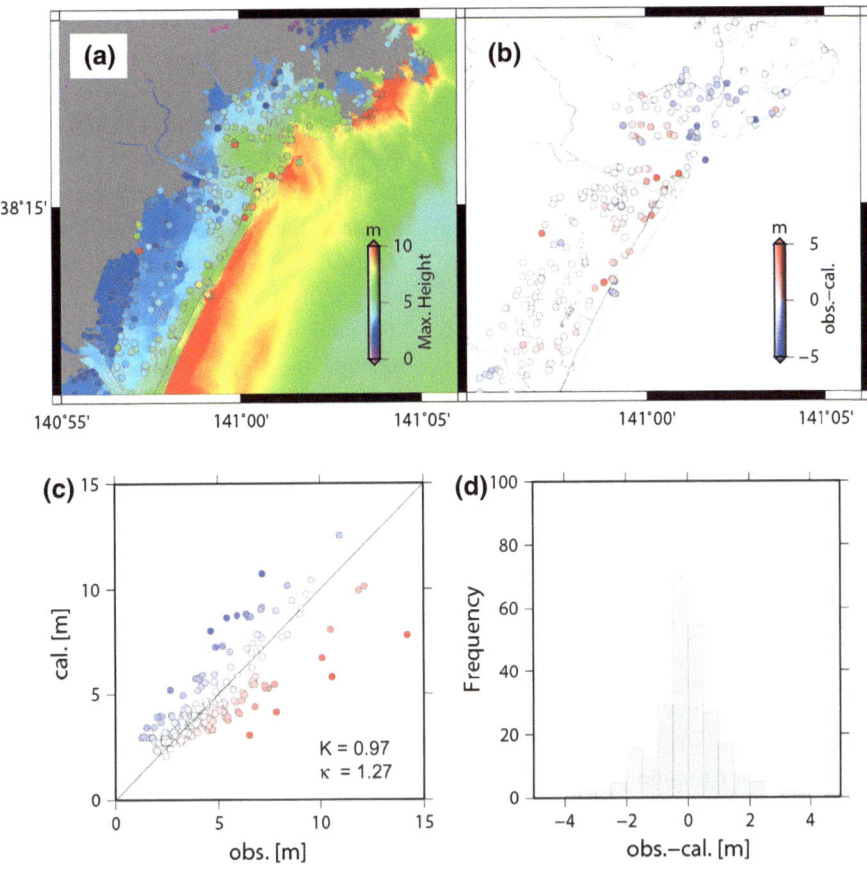

Figure 14
Comparison of measured inundation heights and simulations with the dispersive model using three-dimensional shapes of buildings embedded in the DEM. The *panels* are explained in the caption to Fig. 13

offshore data only without the coastal data. In this study, we successfully developed high-speed software for tsunami propagation and inundation. It would now be very desirable to develop a practical and real-time method that can accurately characterize tsunami sources by analysis of offshore tsunami data for mitigation of future tsunami damage.

During the tsunami simulations, we developed two ideas for further improving the accuracy of tsunami models. The first idea relates to topographic data. We applied the highly accurate DEM and 3D building data on land, and used highly reliable data for the shape of the sea bottom. There were, however, no measured data for such inland waters as rivers, ponds, and small channels, so topographic shape was generated by interpolating surrounding data. The reproducibility of the observed tsunami height was relatively poor in these areas in this simulation. The second idea concerns the finite-difference scheme in the current software. The current model uses a low-order upwind differencing method to calculate the nonlinear term, but that method is associated with significant numerical dissipation when the flow curvature is large (Matsuyama et al. 2010; Son et al. 2011). The result may be an obstructive factor that prevents amplification of the height of the tsunami because of nonlinear and dispersion effects. We are attempting to solve these problems so the model can predict tsunami characteristics more accurately with high speed and high resolution.

6. Conclusions

In this study we developed new software for dispersive tsunami wave modeling. The software solves the nonlinear Boussinesq dispersive equations in a finite-difference scheme with variable nested grids. The software was fully parallelized with the MPI and OpenMP libraries so that large-scale dispersive modeling of the 2011 Tohoku tsunami was possible on the K Supercomputer. A clear discrepancy was apparent from comparison of tsunami waveforms derived from dispersive and non-dispersive simulations at the DART21418 buoy located in the deep ocean. Tsunami soliton fission near the coast recorded by helicopter observations was accurately reproduced by the dispersive model with the high-resolution grids. This calculation scheme, with incorporation of 3D data on building shapes and tsunami source characteristics independently retrieved from offshore data alone, satisfied the adequacy criteria for the prediction of tsunami traces on land.

Acknowledgments

Dr Hong Kie Thio, Dr Phil Cummins, and Dr David Burbidge kindly provided us with the URSGA tsunami software. The wave flume experimental data was provided by the Central Research Institute of the Electric Power Industry. The 2011 Tohoku tsunami data for the GPS buoy were provided by the Ministry of Land, Infrastructure, Transport and Tourism, and DART data came from the Pacific Marine Environmental Laboratory. The DEM and 3D building data were provided by the Geospatial Information Authority of Japan. The digital data of the initial sea-surface displacement of the 2011 Tohoku tsunami were provided by Dr Tatsuhiko Saito. Comments from Dr Takayuki Miyoshi and Dr Takane Hori were very useful for improving the manuscript. We thank anonymous reviewers for constructive comments. This research was implemented in the Strategic Programs for Innovative Research, Field 3, and was also partially supported by the Science and Technology Research Partnership for Sustainable Development. Some figures were prepared by use of Generic Mapping Tools (WESSEL and SMITH 1998), Seismic Analysis Code (GOLDSTEIN et al. 2007), and ArcGIS.

Appendix: Finite-difference Scheme for Nonlinear Dispersive Equations

The finite difference calculation was performed in the staggered-grid system shown in Fig. 15. Because the integration over time was solved with a leapfrog method, the water height (h) was defined at time $t = n\Delta t$, and the depth-integrated quantities (M, N) were defined at $t = (n - 1/2)\Delta t$, where Δt is the time step and $n = 1, 2, 3 \ldots$. In the Appendix, φ and θ indicate the longitude and co-latitude, respectively, R is the earth's radius, d is the water depth, and D is total depth, that is $d + h$. g is the gravitational constant, f is the Coriolis parameter, and n is the Manning's roughness coefficient. We considered the finite-difference form of the dispersion term, the final term on the right-hand side of Eq. (1).

$$\frac{d^2}{3R \sin\theta} \frac{\partial}{\partial\varphi} \left[\frac{1}{R \sin\theta} \left(\frac{\partial^2 M}{\partial\varphi \partial t} + \frac{\partial^2 (N \sin\theta)}{\partial\theta \partial t} \right) \right]$$
$$= \frac{d^2}{3R^2 \sin^2\theta} \frac{\partial}{\partial t} \left(\frac{\partial^2 M}{\partial\varphi^2} + \frac{\partial^2 (N \sin\theta)}{\partial\varphi \partial\theta} \right), \quad (4)$$

where

$$\frac{\partial^2 M}{\partial\varphi^2} = \frac{M_{i+1,j} - 2M_{i,j} + M_{i-1,j}}{\Delta\varphi^2}, \quad (5)$$

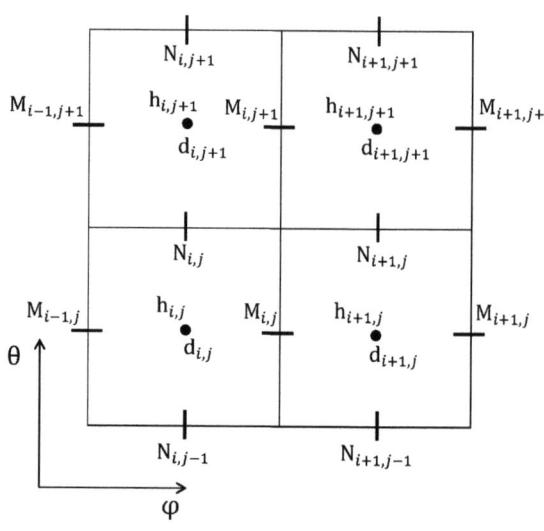

Figure 15
Staggered-grid system for the finite-difference simulation of nonlinear dispersive equations

$\Delta\varphi$ is the grid size along a longitude line. We introduced

$$U = M_{i+1,j} - 2M_{i,j} + M_{i-1,j}, \qquad (6)$$

to write

$$\frac{\partial}{\partial t}\left(\frac{\partial^2 M}{\partial \varphi^2}\right) = \frac{\partial}{\partial t}\left(\frac{U}{\Delta\varphi^2}\right) = \frac{U^{n+\frac{1}{2}} - U^{n-\frac{1}{2}}}{\Delta\varphi^2 \Delta t}. \qquad (7)$$

The other term of Eq. (4) can be expressed as:

$$\frac{\partial^2 (N\sin\theta)}{\partial\varphi\partial\theta} = \frac{\partial}{\partial\varphi}\left(\frac{\bar{N}_{i,j+1}\sin(\theta+\Delta\theta) - \bar{N}_{i,j}\sin(\theta)}{\Delta\theta}\right)$$
$$= \frac{\bar{N}_{i,j+1}\sin(\theta+\Delta\theta) - \bar{N}_{i-1,j+1}\sin(\theta+\Delta\theta) - \bar{N}_{i,j}\sin(\theta) + \bar{N}_{i-1,j}\sin(\theta)}{\Delta\theta\Delta\varphi}, \qquad (8)$$

where $\Delta\theta$ is the grid size along a latitude line. Because of the staggered grid system:

$$\bar{N}_{i,j} = \frac{N_{i,j} + N_{i+1,j} + N_{i,j-1} + N_{i+1,j-1}}{4}. \qquad (9)$$

By defining:

$$\bar{V} = \bar{N}_{i,j+1}\sin(\theta+\Delta\theta) - \bar{N}_{i-1,j+1}\sin(\theta+\Delta\theta) - \bar{N}_{i,j}\sin(\theta) + -\bar{N}_{i-1,j}\sin(\theta), \qquad (10)$$

we can write:

$$\frac{\partial}{\partial t}\left(\frac{\partial^2(N\sin\theta)}{\partial\varphi\partial\theta}\right) = \frac{\partial}{\partial t}\left(\frac{\bar{V}}{\Delta\theta\Delta\varphi}\right) = \frac{\bar{V}^{n+\frac{1}{2}} - \bar{V}^{n-\frac{1}{2}}}{\Delta\theta\Delta\varphi\Delta t}. \qquad (11)$$

Accordingly, the dispersion term of Eq. (1) can be expressed by using Eqs. (7) and (11) in finite-difference form as:

$$\frac{d^2}{3R\sin\theta}\frac{\partial}{\partial\varphi}\left[\frac{1}{R\sin\theta}\left(\frac{\partial^2 M}{\partial\varphi\partial t} + \frac{\partial^2(N\sin\theta)}{\partial\theta\partial t}\right)\right]$$
$$= \frac{\bar{d}_{i,j}^2}{3R^2\sin^2\theta\Delta t}\left(\frac{U^{n+\frac{1}{2}} - U^{n-\frac{1}{2}}}{\Delta\varphi^2} + \frac{\bar{V}^{n+\frac{1}{2}} - \bar{V}^{n-\frac{1}{2}}}{\Delta\theta\Delta\varphi}\right), \qquad (12)$$

where also, because of the staggered grid system:

$$\bar{d}_{i,j} = \frac{d_{i,j} + d_{i+1,j}}{2} \qquad (13)$$

Finally, Eq. (1) can be written in finite-difference form as:

$$M^{n+\frac{1}{2}} = M^{n-\frac{1}{2}} - \frac{1}{R\sin\theta}\frac{\Delta t}{\Delta\varphi}\left(\frac{M_{i+1,j}^{n-\frac{1}{2}^2}}{\bar{D}_{i+1,j}} - \frac{M_{i,j}^{n-\frac{1}{2}^2}}{\bar{D}_{i,j}}\right)$$
$$- \frac{1}{R}\frac{\Delta t}{\Delta\theta}\left(\frac{M_{i,j+1}^{n-\frac{1}{2}}\bar{N}_{i,j+1}^{n-\frac{1}{2}}}{\bar{D}_{i,j+1}} - \frac{M_{i,j}^{n-\frac{1}{2}}\bar{N}_{i,j}^{n-\frac{1}{2}}}{\bar{D}_{i,j}}\right)$$
$$+ \frac{g\bar{D}_{i,j}}{R\sin\theta}\frac{\Delta t}{\Delta\varphi}\left(h_{i+1,j}^n - h_{i,j}^n\right)$$
$$- f\bar{N}_{i,j}^{n-\frac{1}{2}}\Delta t - \frac{gn^2}{\bar{D}_{i,j}^{7/3}}M_{i,j}^{n-\frac{1}{2}}\sqrt{M_{i,j}^{n-\frac{1}{2}^2} + \bar{N}_{i,j}^{n-\frac{1}{2}^2}}\Delta t$$
$$+ \frac{\bar{d}_{i,j}^2}{3R^2\sin^2\theta}\left(\frac{U^{n+\frac{1}{2}} - U^{n-\frac{1}{2}}}{\Delta\varphi^2} + \frac{\bar{V}^{n+\frac{1}{2}} - \bar{V}^{n-\frac{1}{2}}}{\Delta\theta\Delta\varphi}\right), \qquad (14)$$

where, again, because of the staggered grid system:

$$\bar{D}_{i,j} = \frac{d_{i,j} + d_{i+1,j} + h_{i,j} + h_{i+1,j}}{2} \qquad (15)$$

It should be noted that the 2nd and 3rd terms on the right-hand side of Eq. (14) were approximated with upwind finite differences.

Similarly, we considered the finite difference form for the dispersion term (the final term on the left-hand side) of Eq. (2):

$$\frac{d^2}{3R}\frac{\partial}{\partial\theta}\left[\frac{1}{R\sin\theta}\left(\frac{\partial^2 M}{\partial\varphi\partial t} + \frac{\partial^2(N\sin\theta)}{\partial\theta\partial t}\right)\right]$$
$$= \frac{d^2}{3R^2}\frac{\partial}{\partial t}\left[\frac{\partial}{\partial\theta}\left(\frac{1}{\sin\theta}\frac{\partial M}{\partial\varphi}\right) + \frac{\partial}{\partial\theta}\left(\frac{1}{\sin\theta}\frac{\partial N\sin\theta}{\partial\theta}\right)\right] \qquad (16)$$

where:

$$\frac{\partial}{\partial\theta}\left(\frac{1}{\sin\theta}\frac{\partial M}{\partial\varphi}\right) = \frac{\partial}{\partial\theta}\left(\frac{1}{\sin\theta}\frac{\bar{M}_{i+1,j} - \bar{M}_{i,j}}{\Delta\varphi}\right)$$
$$= \frac{1}{\Delta\theta\Delta\varphi}\left(\frac{\bar{M}_{i+1,j} - \bar{M}_{i,j}}{\sin\theta} - \frac{\bar{M}_{i+1,j-1} - \bar{M}_{i,j-1}}{\sin(\theta - \Delta\theta)}\right), \qquad (17)$$

and because of the staggered grid system:

$$\bar{M}_{i,j} = \frac{M_{i,j} + M_{i,j+1} + M_{i-1,j} + M_{i-1,j+1}}{4}. \qquad (18)$$

By further defining:

$$\bar{U} = \frac{\bar{M}_{i+1,j} - \bar{M}_{i,j}}{\sin\theta} - \frac{\bar{M}_{i+1,j-1} - \bar{M}_{i,j-1}}{\sin(\theta - \Delta\theta)}, \qquad (19)$$

we can write:

$$\frac{\partial}{\partial t}\left[\frac{\partial}{\partial \theta}\left(\frac{1}{\sin\theta}\frac{\partial M}{\partial \varphi}\right)\right] = \frac{\partial}{\partial t}\left(\frac{\bar{U}}{\Delta\theta\Delta\varphi}\right) = \frac{\bar{U}^{n+\frac{1}{2}} - \bar{U}^{n-\frac{1}{2}}}{\Delta\theta\Delta\varphi\Delta t}.$$

(20)

The other term of Eq. (16) can be expressed as:

$$\frac{\partial}{\partial \theta}\left(\frac{1}{\sin\theta}\frac{\partial N \sin\theta}{\partial \theta}\right)$$

$$= \frac{\partial}{\partial \theta}\left(\frac{1}{\sin\theta}\frac{N_{i,j+1/2}\sin(\theta + \Delta\theta/2) - N_{i,j-1/2}\sin(\theta - \Delta\theta/2)}{\Delta\theta}\right)$$

$$= \frac{1}{\Delta\theta}\frac{\partial}{\partial \theta}\left(\frac{N_{i,j+1/2}\sin(\theta + \Delta\theta/2) - N_{i,j-1/2}\sin(\theta - \Delta\theta/2)}{\sin\theta}\right)$$

$$= \frac{1}{\Delta\theta^2}\left(\frac{N_{i,j+1}\sin(\theta + \Delta\theta) - N_{i,j}\sin(\theta)}{\sin\left(\theta + \Delta\theta/2\right)}\right.$$

$$\left. - \frac{N_{i,j}\sin(\theta) - N_{i,j-1}\sin(\theta - \Delta\theta)}{\sin\left(\theta - \Delta\theta/2\right)}\right).$$

(21)

By defining:

$$V = \frac{N_{i,j+1}\sin(\theta + \Delta\theta) - N_{i,j}\sin(\theta)}{\sin\left(\theta + \Delta\theta/2\right)}$$
$$- \frac{N_{i,j}\sin(\theta) - N_{i,j-1}\sin(\theta - \Delta\theta)}{\sin\left(\theta - \Delta\theta/2\right)},$$

(22)

and we can write:

$$\frac{\partial}{\partial t}\left(\frac{\partial}{\partial \theta}\left(\frac{1}{\sin\theta}\frac{\partial N \sin\theta}{\partial \theta}\right)\right) = \frac{1}{\Delta\theta^2}\frac{\partial V}{\partial t} = \frac{V^{n+\frac{1}{2}} - V^{n-\frac{1}{2}}}{\Delta t \Delta\theta^2}.$$

(23)

By using Eqs. (20) and (23), the dispersion term of Eq. (2) can be expressed in finite-difference form as:

$$\frac{d^2}{3R}\frac{\partial}{\partial \theta}\left[\frac{1}{R\sin\theta}\left(\frac{\partial^2 M}{\partial\varphi\partial t} + \frac{\partial^2(N\sin\theta)}{\partial\theta\partial t}\right)\right]$$
$$= \frac{\bar{d}_{i,j}^2}{3R}\left(\frac{\bar{U}^{n+\frac{1}{2}} - \bar{U}^{n-\frac{1}{2}}}{\Delta\theta\Delta\varphi\Delta t} + \frac{V^{n+\frac{1}{2}} - V^{n-\frac{1}{2}}}{\Delta t \Delta\theta^2}\right),$$

(24)

where because of the staggered grid system:

$$\bar{d}_{i,j} = \frac{d_{i,j} + d_{i,j+1}}{2}.$$

(25)

Finally, Eq. (2) can be written in finite-difference form as:

$$N^{n+\frac{1}{2}} = N^{n-\frac{1}{2}} - \frac{1}{R\sin\theta}\frac{\Delta t}{\Delta\varphi}\left(\frac{\bar{M}_{i,j+1}^{n-\frac{1}{2}}N_{i,j+1}^{n-\frac{1}{2}}}{\bar{D}_{i,j+1}} - \frac{\bar{M}_{i,j}^{n-\frac{1}{2}}N_{i,j}^{n-\frac{1}{2}}}{\bar{D}_{i,j}}\right)$$
$$+ \frac{g\bar{D}_{i,j}}{R}\frac{\Delta t}{\Delta\theta}\left(h_{i,j+1}^n - h_{i,j}^n\right) + f\bar{M}_{i,j}^{n-\frac{1}{2}}\Delta t$$
$$- \frac{gn^2}{\bar{D}_{i,j}^{7/3}}N_{i,j}^{n-\frac{1}{2}}\sqrt{\bar{M}_{i,j}^{n-\frac{1}{2}2} + N_{i,j}^{n-\frac{1}{2}2}}\Delta t$$
$$+ \frac{\bar{d}_{i,j}^2}{3R^2}\left(\frac{\bar{U}^{n+\frac{1}{2}} - \bar{U}^{n-\frac{1}{2}}}{\Delta\theta\Delta\varphi} + \frac{V^{n+\frac{1}{2}} - V^{n-\frac{1}{2}}}{\Delta\theta^2}\right),$$

(26)

where, because of the staggered grid system:

$$\bar{D}_{i,j} = \frac{d_{i,j} + d_{i,j+1} + h_{i,j} + h_{i,j+1}}{2}.$$

(27)

It should also be noted that the 2nd and 3rd terms on the right-hand side of Eq. (26) are approximated with upwind finite differences.

Equation (3) can be written in the finite-difference form as:

$$h^{n+1} = h^n - \frac{\Delta t}{R\sin\theta}\left[\left(\frac{M_{i,j}^{n+\frac{1}{2}} - M_{i-1,j}^{n+\frac{1}{2}}}{\Delta\varphi}\right)\right.$$
$$\left. + \frac{\left(N_{i,j}^{n+\frac{1}{2}}\sin(\theta) - N_{i,j-1}^{n+\frac{1}{2}}\sin(\theta - \Delta\theta)\right)}{\Delta\theta}\right].$$

(28)

At time $t = n\Delta t$, we calculate $M^{n+1/2}$ and $N^{n+1/2}$ by substituting h^n, $M^{n-1/2}$, and $N^{n-1/2}$ into Eqs. (14) and (26). These equations are solved by use of an iterative method (Gauss–Seidel method) (PRESS et al. 1986). The calculated $M^{n+1/2}$ and $N^{n+1/2}$ are substituted into Eq. (28) to obtain h^{n+1}. Then h^{n+1}, $M^{n+1/2}$, and $N^{n+1/2}$ are used to solve Eqs. (14), (26), and (28) at the next time step.

REFERENCES

AIDA, I., *Reliability of a tsunami source model derived from fault parameters*, J. Phys. Earth, 26, 57–73, 1978.

AMMON C.J., T. LAY, H. KANAMORI, and M. CLEVELAND. 2011. *A rupture model of the 2011 off the Pacific coast of Tohoku Earthquake*, Earth Planets Space, 63, 693–696, 2011.

BABA, T., N. TAKAHASHI, Y. KANEDA, Y. INAZAWA and M. KIKKOJIN, Tsunami inundation modeling of the 2011 Tohoku earthquake using three-dimensional building data for Sendai, Miyagi Prefecture, Japan, in V. S.-FANDIÑO et al. (ed.): Tsunami Events and

Lessons Learned, Advances in Natural and Technological Hazards Research, SPRINGER, 35, 89–98, doi:10.1007/978-94-007-7269-4_3, 2014.

British Oceanographic Data Centre, GEBCO (General Bathymetric Chart of the Oceans). 2010. http://www.gebco.net/data_and_products/gridded_bathymetry_data, 2010.

CERJAN, C., D. KOSLOFF, R. KOSLOFF, and M. RESHEF, A nonreflecting boundary condition for discrete acoustic and elastic wave equations, Geophysics, 50, 705–708, doi:10.1190/1.1441945, 1985.

Fujitsu Corporation, press release, http://www.fujitsu.com/global/news/pr/archives/month/2011/20111102-02.html, 2012.

GOLDSTEIN, P., D. DODGE, M. FIRPO, L. MINNER, J. E. TULL, D. HARRIS, and W. C. TAPLEY, SAC—Seismic Analysis Code, http://www.iris.edu/manuals/sac/manual.html, 2007.

GOTO, K., K. FUJIMA, D. SUGAWARA, S. FUJINO, K. IMAI, R. TSUDAKA, T. ABE and T. HARAGUCHI, Field measurements and numerical modeling for the run-up heights and inundation distances of the 2011 Tohoku-oki tsunami at Sendai Plain, Japan, Earth Planets Space, 64, 1247–1257, 2012.

GRILLI, S.T., J.C. HARRIS, T.S.T. BAKHSH, T.L. MASTERLARK, C. KYRIAKOPOULOS, J.T. KIRBY and F. SHI, Numerical simulation of the 2011 Tohoku tsunami based on a new transient FEM co-seismic source: Comparison to far- and near-field observations, Pure Appl. Geophys., 170, 1333–1359, 2013.

HAYASHI, Y., H. TSUSHIMA, K. HIRATA, K. KIMURA, and K. MAEDA, Tsunami source area of the 2011 off the Pacific coast of Tohoku Earthquake determined from tsunami arrival times at offshore observation stations, Earth Planets Space, 63, 809–813, 2011.

HORILLO, J., KOWALIK, Z., SHIGIHARA, Y., Wave dispersion study in the Indian Ocean-tsunami of December 26, 2004, Marine Geodesy, 29, 149–166, 2006.

IMAMURA, F., S. KOSHIMURA, T. OIE, Y. MABCHI, and Y. MURASHIMA, Tsunami simulation for the 2011 off the Pacific coast of Tohoku Earthquake (Tohoku University model ver. 1.0), 12 pp., 2011.

ITO, T., K. OZAWA, T. WATANABE, and T. SAGIYA. Slip distribution of the 2011 off the Pacific coast of Tohoku Earthquake inferred from geodetic data, Earth Planets Space, 63, 627–630, 2011.

IWASE, H., Development of numerical model including dispersion effect from the tsunami source to the coastal area (in Japanese), PhD. thesis, 166p., 2005.

JAKEMAN, J.D., O.M., NIELSEN, K., VANPUTTEN, R., MLECZEKO,, D. BURBIDGE, and N. HORSPOOL, Towards spatially distributed quantitative assessment of tsunami inundation models, Ocean Dynamics, doi:10.1007/s10236-010-0312-4, 2010.

Japan Society of Civil Engineers, Tsunami Assessment Method for Nuclear Power Plants in Japan, JSCE, 321p., 2002.

KIDO, Y., T. FUJIWARA, T. SATAKI, M. KINOSHITA, S. KODAIRA, M. SANO, Y. ICHIYAMA, Y. HANAFUSA, S. TSUBOI. 2011. Bathymetric feature around Japan Trench obtained by JAMSTEC multi narrow beam survey, MIS036-P58 (in Japanese), Japan Geoscience Union Meeting, 2011.

KIRBY, J. T., F. SHI, B. TEHRANIRAD, J. C. HARRIS, and S. T. GRILLI, Dispersive tsunami waves in the ocean: Model equations and sensitivity to dispersion and Coriolis effects, Ocean Modelling, 62, 39–55, 2013.

LØVHOLT, F., G. KAISER, S. GLIMSDAL, L. SCHEELE, C. B. HARBITZ, and G. PEDERSEN, Modeling propagation and inundation of the 11 March 2011 Tohoku tsunami, Nat. Hazards Earth Syst. Sci., 12, 1017–1028, doi:10.5194/nhess-12-1017-2012, 2012.

MADSEN, P.A., D.R. FUHRMAN, and H.A. SCHÄFFER, On the solitary wave paradigm for tsunamis, J. Geophys. Res., 113, C12012, doi:10.1029/2008JC004932, 2008.

MAEDA, T., T. FURUMURA, S. SAKAI, and M. SHINOHARA, Significant tsunami observed at ocean-bottom pressure gauges during the 2011 off the Pacific coast of Tohoku Earthquake, Earth Planets Space, 63, 803–808, 2011.

MATSUYAMA, M., M. IKENO, T. SAKAKIYAMA, and T. TAKEDA, A study of tsunami wave fission in an undistorted experiment, Pure Appl. Geophys., 164, 617–631, doi:10.1007/s00024-006-0177-0, 2007.

MORI, N., T. Takahashi and The 2011 Tohoku Earthquake Tsunami Joint Survey Group, Nationwide survey of the 2011 Tohoku earthquake tsunami, Coastal Engineering Journal, 54, 1–27, 2012.

MURASHIMA, Y., S. KOSHIMURA, H. OKA, Y. MURATA, K. FUJIMA, H. SUGINO, Y. IWABUCHI, Numerical simulation of soliton fission in 2011 Tohoku tsunami using nonlinear dispersive wave model, (in Japanese with English abstract), J. Japan Society of Civil Engineers (B2), 68, I_206–I_212, 2012.

MURASHIMA, Y., S. KOSHIMURA, H. OKA, Y. MURATA, and F. IMAMURA, Development of the practical river run-up model of tsunami based on non-linear dispersive wave theory (in Japanese), J. Japan Society of Civil Engineers (B2), 66, 201–205, 2010.

NISHIMURA, T., H. MUNEKANE, and H. YARAI, The 2011 off the Pacific coast of Tohoku Earthquake and its aftershocks observed by GEONET, Earth Planets Space, 63, 631–636, 2011.

PEREGRINE, H., Equations for water waves and the approximations behind them, edited by R. E. MEYER, pp. 95–121, Waves on Beaches and Resulting Sediment Transport, Academic Press, New York, 1972.

PRESS, W. H., B. P. FLANNERY, S. A. TEUKOLSKY, and W. T. VETTERLING, Numerical Recipes, Cambridge Univ. Press, Cambridge, 1986.

ROEBER, V., CHEUNG, K. F., and KOBAYASHI, M. H., Shock-capturing Boussinesq-type model for nearshore wave processes, Coastal Engineering, 57, 407–423, 2010.

SAITO T., K. SATAKE, and T. FURUMURA, Tsunami waveform inversion including dispersive waves: the 2004 earthquake off Kii Peninsula, Japan, J. Geophys. Res., 115, B06303, doi:10.1029/2009JB006884, 2010.

SAITO, T., Y. ITO, D. INAZU, and R. HINO, Tsunami source of the 2011 Tohoku-Oki earthquake, Japan: Inversion analysis based on dispersive tsunami simulations, Geophys. Res. Lett., 38, L00G19, doi:10.1029/2011GL049089, 2011.

SHAO, G, X. LI, C. JI, and T. MAEDA, Focal mechanism and slip history of 2011 Mw 9.1 off the Pacific coast of Tohoku earthquake, constrained with teleseismic body and surface waves, Earth Planets Space, 63, 559–564, 2011.

SATAKE, K. Tsunamis, in International Handbook of Earthquake and Engineering Seismology, (eds. LEE,W.H.K., KANAMORI, H., JENNINGS, P.C., and KISSLINGER, C.), Academic Press, 81A, 437–451, 2002.

SATAKE, K., Y. FUJII, T. HARADA, and Y. NAMEGAYA, Time and space distribution of coseismic slip of the 2011 Tohoku earthquake as inferred from tsunami waveform Data, Bull. Seismol. Soc. Am., 103, 1473–1492, doi:10.1785/0120120122, 2013.

SHIGIHARA, Y., and FUJIMA K., Wave dispersion effect in the Indian ocean tsunami, J. Disaster Research, 1, 142–147, 2006.

SHUTO, N., The Nihonkai-Chubu earthquake tsunami on the north Akita coast (in Japanese), Costal Engin. Japan JSCE 28, Tokyo, Japan, 255–264, 1985.

SITANGGANG K. I. and LYNETT, P., *Parallel computation of a highly nonlinear Boussinesq equation model through domain decomposition*, Int. J. Numer. Fluids, *49*, 57–74, 2005.

SON, S., LYNETT, P., KIM, D.-H., *Nested and multi-physics modeling of tsunami evolution from generation to inundation*, Ocean Modell. *38*, 96–113, doi:10.1016/j.ocemod.2011.02.007, 2011.

SUITO, H., T. NISHIMURA, M. TOBITA, T. IMAKIIRE, and S. OZAWA, *Interplate fault slip along the Japan Trench before the occurrence of the 2011 off the Pacific coast of Tohoku Earthquake as inferred from GPS data*, Earth Planets Space, *63*, 615–619, 2011.

TAKAGAWA, T., and T. TOMITA, *Tsunami source inversion with time evolution and real-time estimation of permanent deformation at observation points*, J. Japan Society of Civil Engineers (B2), *68*, I_311–I_315, 2012.

TSUSHIMA, H., R. HINO, Y. TANIOKA, F. IMAMURA, H. FUJIMOTO, *Tsunami waveform inversion incorporating permanent seafloor deformation and its application to tsunami forecasting*, J. Geophys. Res., *117*, B03311, doi:10.1029/2011JB008877, 2012.

WESSEL, P., SMITH, W.H.F., *New, improved version of generic mapping tools released*, EOS Trans., AGU79, 579, 1998.

YAGI, Y. and Y. FUKAHATA, *Rupture process of the 2011 Tohoku-oki earthquake and absolute elastic strain release*, Geophys. Res. Lett., *38*, L19307, doi:10.1029/011GL048701, 2011.

ZHOU, H., C. W. MOORE, Y. WEI, and V. V. TITOV, *A nested-grid Boussinesq-type approach to modeling dispersive propagation and runup of landslide-generated tsunamis*, Nat. Hazards Earth Syst. Sci., *11*, 2677–2697, doi:10.5194/nhess-11-2677-2011, 2011.

(Received February 19, 2014, revised February 3, 2015, accepted February 4, 2015, Published online February 21, 2015)

Pure Appl. Geophys. 172 (2015), 3473–3491
© 2015 Springer Basel
DOI 10.1007/s00024-015-1051-8

Pure and Applied Geophysics

Investigation of Hydrodynamic Parameters and the Effects of Breakwaters During the 2011 Great East Japan Tsunami in Kamaishi Bay

CEREN OZER SOZDINLER,[1] AHMET CEVDET YALCINER,[2] ANDREY ZAYTSEV,[3,4] ANAWAT SUPPASRI,[5] and FUMIHIKO IMAMURA[5]

Abstract—The March 2011 Great East Japan Tsunami was one of the most disastrous tsunami events on record, affecting the east coast of Japan to an extreme degree. Extensive currents combined with flow depths in inundation zones account for this devastating impact. Video footage taken by the eyewitnesses reveals the destructive effect and dragging capability of strong tsunami currents along the coast. This study provides a numerical modeling study in Kamaishi Bay, calculating the damage inflicted by tsunami waves on structures and coastlines in terms of the square of the Froude number Fr^2; and also other calculated hydrodynamic parameters, such as the distribution of instantaneous flow depths, maximum currents and water surface elevations that occurred during this catastrophic tsunami. Analyses were performed by using the tsunami numerical modeling code NAMI DANCE with nested domains at a higher resolution. The effect of the Kamaishi breakwater on the tsunami inundation distance and coastal damage was tested by using the conditions of "with breakwater," "without breakwater," and "damaged breakwater." Results show that the difference between the hydrostatic pressure on the seaward side of the breakwater and the leeward side of the breakwater is quite high, clarifying conditions contributing to failure of the breakwater. Lower water surface elevations were calculated in the case of a breakwater existing at the entrance, a partly valid condition for the damaged breakwater case. The results are different for current velocities and Fr^2_{max} in the "with breakwater" condition due to the concentration of energy through the breakwater gaps.

Key words: Tsunami, inundation, current velocity, water surface elevation, flow depth, Froude number, tsunami damage, dynamic tsunami input, march 11 2011 Great East Japan Tsunami.

1. Introduction

Tsunamis have the power to impart extensive and destructive effects on coastal and marine structures. When tsunamis reach land, the effects of hydrodynamic parameters become substantial. The main reason for this devastating damage is the occurrence of extensive currents combined with flow depths in inundation zones. The 2011 Great East Japan (GEJE) Tsunami, triggered by an enormous magnitude 9.0 earthquake and ground accelerations up to 3 g, was the fifth largest earthquake recorded in the last 2,000 years (JMA 2011). The event was dramatic evidence of the tsunami phenomenon, affecting the entire east coast of Japan and destroying many coastal settlements with extremely long inundation distances, excessive water levels and strong tsunami currents with high dragging capabilities. There is much video footage of the destructive tsunami effects taken by eyewitnesses who managed to survive by escaping to higher elevations in concrete buildings.

In addition to the extreme damage along the eastern coast of Japan, the GEJE Tsunami also had far-field effects along the coasts of Hawaii and Northwest America. The maximum tsunami runup height was measured at 40.5 m (Coastal Engineering Committee 2011) and the inundated area was estimated to be as large as 507 km² (SUPPASRI *et al.* 2011, 2012; Iwate-Miyagi-Fukushima Province 2011; Geospatial Information Authority 2011; KOIZUMI 2011), primarily overwhelming the area from the towns of Taro to Kesennuma along the east coast of Japan.

---

[1] Department of Geophysics, Bogazici University Kandilli Observatory and Earthquake Research Institute, Cengelkoy-Uskudar, 34684 Istanbul, Turkey. E-mail: ceren.ozer@boun.edu.tr

[2] Department of Civil Engineering, Ocean Engineering Research Center, Middle East Technical University, 06800 Ankara, Turkey. E-mail: yalciner@metu.edu.tr

[3] Special Research Bureau for Automation of Marine Researches, Far Eastern Branch of Russian Academy of Sciences, Uzhno-Sakhalinsk Gorkiy str. 25, 693013 Yuzhno-Sakhalinsk, Russia. E-mail: aizaytsev@mail.ru

[4] Nizhny Novgorod State Technical University, Nizhny Novgorod 24 Minin Street, 603950 Nizhnii Novgorod, Russia.

[5] International Research Institute of Disaster Science, Tohoku University, 468-1 Aoba, Aramaki, Aoba, Sendai 980-0845, Japan. E-mail: suppasri@irides.tohoku.ac.jp; imamura@irides.tohoku.ac.jp

Hydrodynamic demand (HD), considered to be the tsunami damage level, is defined as a dimensionless parameter that represents the proportionality of the drag force with respect to the hydrostatic force on the structure (OZER SOZDINLER et al. 2014). Assuming similar shaped dragged objects, HD is actually the square of the Froude number Fr^2, which is an instantaneous parameter occurring at any location during tsunami inundation, depending on the instantaneous values of the current velocity and the flow depth at the location (OZER 2012). The purpose of this study is to investigate the behavior of tsunami hydrodynamic parameters such as Fr^2, maximum values of water surface elevations, flow depths, and currents, specifically for the case of the coastal protection structures in Kamaishi Bay, and, by doing so, determine levels of damage and areas prone to tsunamis in residential regions.

In recent years, advances in laboratory equipment, hardware and related software have facilitated measurement and evaluation of tsunami damages on structures. Among numerous studies, Japanese studies are the most prevalent since Japan has experienced many notable tsunami events throughout history. SHUTO (2009) stated that tsunami-induced currents may be the main reason for damage to coastal structures and defined four principle types of tsunami damage due to strong currents as: (1) erosion of soil embankments near underpasses or bridges by concentrated water currents; (2) scour and destruction of a structure toe by strong currents parallel to long structures; (3) erosion of soil embankments by overflowing tsunamis water; (4) damage to quay wall toes due to the waterfall that occurs when water returns and hits the sea bottom during a tsunami drawdown. SHUTO (2009) also exemplifies the effects of concentrated current velocity through a railroad, and in bays and harbors, through the case of a gravity-type quay wall. Understanding the effects of overflow during tsunami inundation is also essential, especially on coastal embankments.

Conducting hydraulic experiments is an alternative effective method of investigating tsunami forces exerted onto structures during tsunami inundation. One study to mention such experiments is FUJIMA (2009), which tries to understand the characteristics of the time history of wave pressure and total force exerted onto structures. He performed several experiments by varying the scale of the buildings, distance from the shoreline and the type of incident wave strike. In all cases, the incident wave broke in the shallow zone and impacted a vertical seawall. Current velocity was measured by propeller current meter, wave force on the model structures was measured by a load cell, and wave pressure was measured by pressure gauges. FUJIMA (2009) mentioned an empirical formula, proposed by ASAKURA et al. (2000), to calculate the maximum tsunami force on structures by integrating the envelope of the maximum standing-wave pressure. A hydrostatic form of the formula for estimating tsunami force was proposed. After evaluating the results, FUJIMA (2009) suggested that the tsunami damage estimation based on only the inundation depth may be inaccurate for structures far from a shoreline since the results are either overestimated or underestimated. FUJIMA (2009) also provided the tsunami force on structures by calculating the hydrodynamic force (drag force) on the exposed buildings. Drag force formulated as a function of inundation depth, velocity and the drag coefficient is $F_D = \frac{C_D}{2} \rho B (h_i u_i^2)_m$ where u_i is the velocity at a point, h_i is the inundation depth at that point, ρ is the density of water, B is the width of the model, and C_D is the drag coefficient depending on the shape of the area of the structure exposed to the waves defined as functions of $\frac{h_{im}}{D}$. It was indicated that, since the maximum inundation depth h_{im} and the maximum velocity u_{im} may not occur at the same time, it is preferred to estimate the maximum momentum flux.

Surge front tsunami force is an important concept that can be studied in order to determine tsunami damage. ARIKAWA (2009) studied surge front tsunami force using physical laboratory experiments. Wooden and concrete walls were examined to predict their behavior in case of failure processes in full-scale experiments under the load of a breaking tsunami. The experiments were conducted in a large wave flume, called the Hydro-Geo Flume, at the Port and Airport Research Institute (PARI) in Japan. The flume is 184 m in length, 3.5 m in width and 12 m in depth, having a piston-type wave generator capable of generating a 2.5 m high tsunami. The experimental results show that the wooden wall was fully destroyed by a tsunami of 2.5 m high measured at 30 m offshore. The

Figure 1
The Location of Kamaishi city and a satellite image of Kamaishi Bay taken from Google Earth

tsunami also damaged a concrete wall, though partly, leaving a damaged hole near the bottom of the wall. The failure occurred in the form of bending or punching shear when the strength of the wall was small; the failure mode shifted from local damage to wide destruction depending on the strength of the wall. From these results, it is obvious that examining the failure processes of walls is highly essential for designing seawalls and breakwaters that can resist tsunami forces.

On the other hand, a new measure called tsunami fragility has been used to estimate structural tsunami damage and fatalities by gathering remote satellite sensing data, field survey information, numerical modeling and historical data analysis using geographical information systems (GIS) (KOSHIMURA et al. 2009). Tsunami fragility can be defined as the probability of structural damage or the fatality ratio related to the hydrodynamic parameters of tsunami inundation. Several empirical approaches exist for correlating tsunami hazard and vulnerability (KOSHIMURA et al. 2009). However, their findings are generally based on the inception of local aspects of tsunami damage. Therefore, it is difficult to identify the vulnerability quantitatively. To estimate vulnerability, it is necessary to consider several uncertain sources, such as the hydrodynamic parameters of tsunami inundation, structural characteristics, population, land use and any other site conditions. Recently, fragility analysis has been used and developed via application to the 2011 GEJE by many researchers (e.g., SUPPASRI et al. 2012, 2013; CHARVET et al. 2014; LEELAWAT et al. 2014).

The 2011 GEJE Tsunami caused extreme damage along the east coast of Japan. Kamaishi is one of the coastal cities subjected to extreme tsunami inundation, with strong currents and high flow depths. It is located on the northeast coast of Japan, on Honshu, and is protected from the sea by a natural harbor (Fig. 1). Kamaishi is the birthplace of modern iron manufacturing in Japan, and because of the iron and steel industries, it has a population of nearly 40,000. According to historical records, the 1896 Meiji Sanriku tsunami was so devastating for Kamaishi that the city lost more than 75 % of its population.

For the mitigation of tsunami disasters, a tsunami breakwater was constructed at the Kamaishi Bay entrance in 1978–2008. The city was protected by two breakwaters, 670 and 990 m long, at the entrance of the bay, with a 6-m crest elevation and a 300-m gap distance (Fig. 1). These breakwaters were built at a depth of 63 m, becoming the world's deepest breakwaters; thusly, they made it into the Guinness World Records in 2010.

Figure 2
Tsunami damage in Kamaishi city after the GEJE tsunami (courtesy of Prof. Dr. Ahmet Cevdet Yalciner). These impressive frames reveal the extent of this devastating tsunami, as almost all of the light timber houses were destroyed; the concrete buildings were heavily damaged; huge tankers were dragged towards the land and grounded

Tsunami amplitudes, as measured by a GPS buoy installed and operated by PARI, Japan, reached 6.7 m at 24 km offshore of Kamaishi at a water depth of 400 m. It is reported that at least four out of 69 designated evacuation sites in the city were inundated by the tsunami (KAMAISHI PORT OFFICE 2011). As seen from the photos in Fig. 2, most of the timber-framed buildings were dragged; a huge tanker was carried onto the quay, and the foundations of buildings on the waterfront were highly scoured.

2. Technical Background

2.1. Tsunami Numerical Code NAMI DANCE

In this study, the tsunami numerical modeling code NAMI DANCE (NAMI DANCE 2011; OZER 2012) was used. It was developed by following the similar computational procedure of TUNAMI N2 (TUNAMI-N2 2001; GOTO et al. 1997; SHUTO et al. 1990; IMAMURA 1989, 1995) in C++ programming

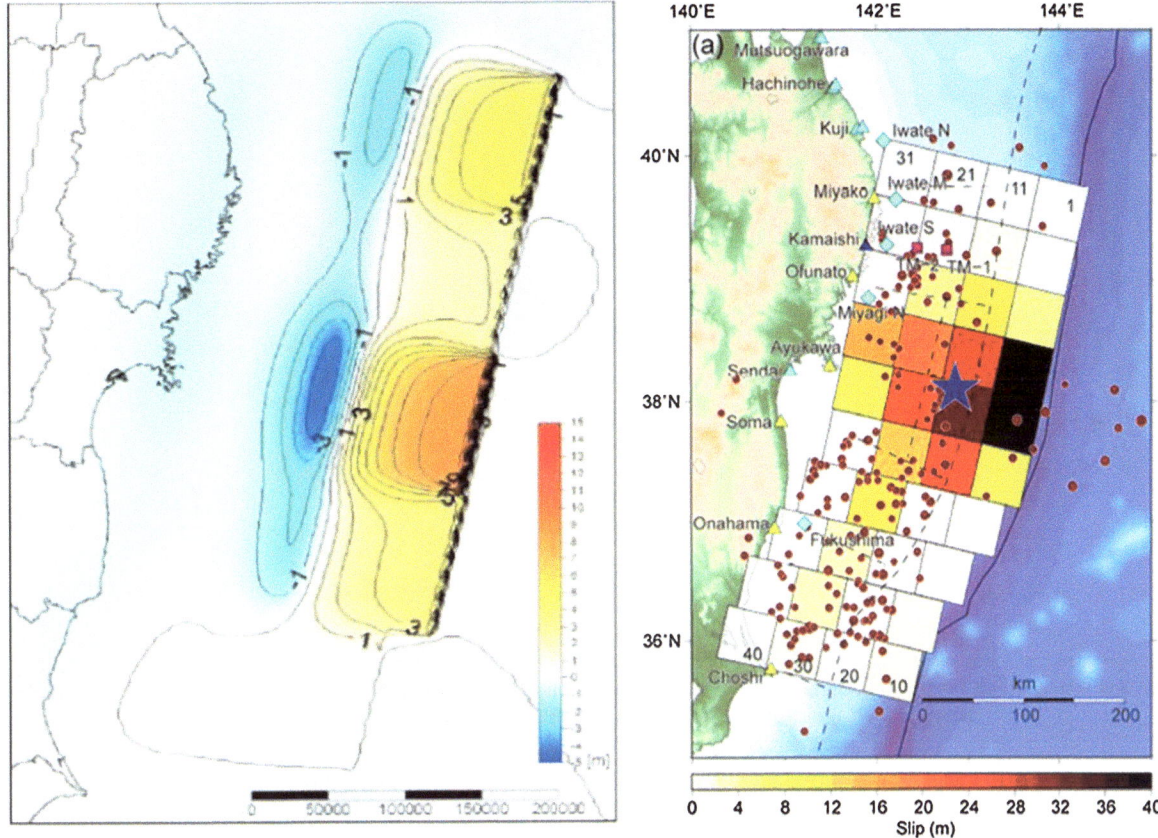

Figure 3
Source models for the 2011 Great East Japan Earthquake. Left is the Tohoku University-Imamura source (IMAMURA et al. 2011); Right is the Fujii-Satake source (FUJII et al. 2011)

language as a user-friendly code for tsunami simulations and visualizations, and it has been applied to several tsunami events (YALCINER et al. 2001, 2002, 2003, 2004; KURKIN et al. 2003; ZAHIBO et al. 2003; ZAITSEV et al. 2002; YALCINER and PELINOVSKY 2007; ZAITSEV et al. 2008; OZER et al. 2008, 2011; YALCINER et al. 2010). The code calculates the principal tsunami hydrodynamic parameters, namely water surface elevation, current velocities and their directions, flow depth, and the Froude number in selected output time intervals throughout the study domain with either cartesian or spherical coordinate system. Moreover, the model can make the calculations using either static source data created as an initial wave or dynamic source data (time history of water surface fluctuation) inputted as a line from an arbitrary location. The improved code version was applied here, calculating land inundation more precisely without instability by using very fine bathymetric/topographical data.

2.2. Comparison of the NAMI DANCE Calculation Results with Wave Records from the Kamaishi Offshore GPS Buoy

Before starting the model application on the test basins, NAMI DANCE results were verified with real-world measurement data in order to ensure geophysical reality. For this purpose, wave records obtained from the PARI GPS buoy, located 24 km offshore of Kamaishi city at a depth of 400 m, during the 2011 GEJE tsunami were used to compare with the NAMI DANCE results. Ground deformation models developed by IMAMURA et al. (2011) and FUJII et al. (2011) were used as the tsunami sources for the GEJE simulations (Fig. 3). The slip distribution given

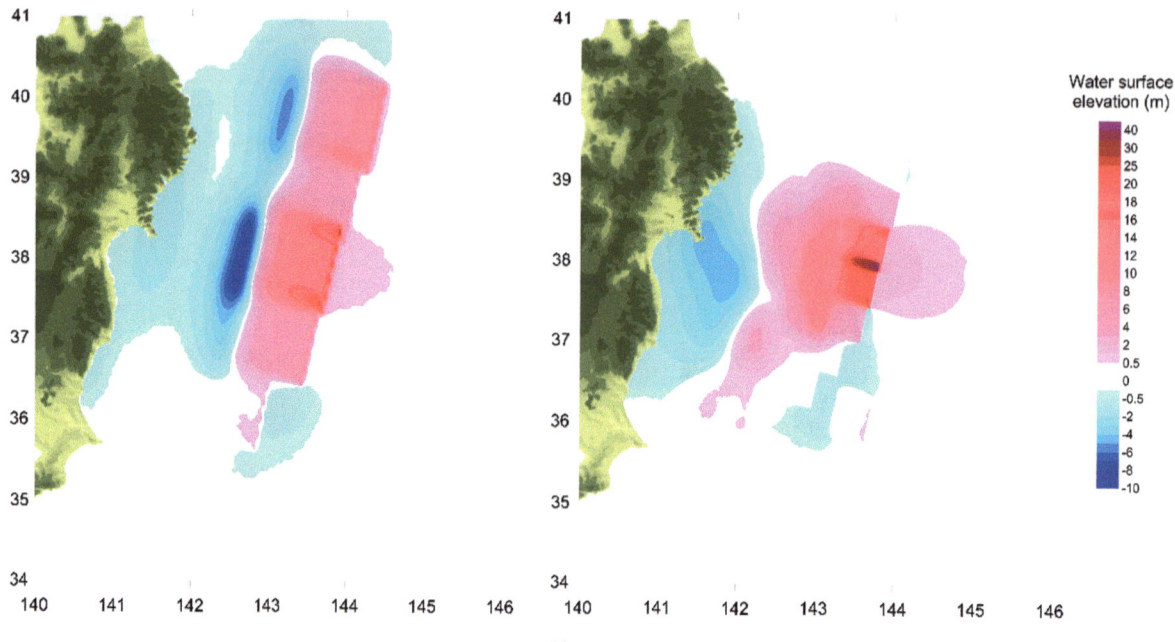

Figure 4
The computed initial water surfaces for the Great East Japan Tsunami using the tsunami source models provided by IMAMURA et al. (2011) (*left*) and FUJII et al. (2011) (*right*) given in Fig. 3. The computed tsunami sources indicate that the Imamura source model gives higher descending initial sea surface while the Fujii-Satake model provides higher ascending parts

in FUJII et al. (2011) was then improved and published in KOKETSU et al. (2011) and SATAKE et al. (2013), including tsunami waveforms at more locations along the eastern Japanese coasts. Both of the models used a tsunami waveform inversion method, while the Imamura model used tsunami height, inundation area and land uplift/subsidence. The Fujii-Satake model described in KOKETSU et al. (2011) was also taken into consideration; however, FUJII et al. (2011) slip distribution was used in our analyses.

The ground deformation models are digitized and converted to tsunami sources in several segments. The segments are then combined to form one single tsunami source. After this procedure, the initial water surfaces of the 2011 GEJE tsunami wave are computed by NAMI DANCE for both slip models (Fig. 4). Even if there are discrepancies between the tsunami sources, both source models are acceptable since they were determined using seismic event ground motion data and by comparing the observed and computed peak coastal tsunami amplitudes.

The tsunami wave record from the Kamaishi GPS buoy officially provided by PARI is given in Fig. 5. As seen from the record, the first peak occurred at around 15:10 local time, about 25 min after the earthquake time 14:46 (Japan Standard Time). The first peak, which was very steep and high, was approximately 6.7 m. The wave periods were not regular for the first and third waves. Then, the wave period was 55 min, in a regular pattern (TAKAHASHI et al. 2011). The average sea level rise in the record is about 55 cm.

Bathymetry of the study area closest to the east coast of Japan was obtained from the GEBCO 30 arcsec database. The best possible, finer grid size of bathymetry was selected in order to achieve a better match between numerical results and recorded data. Two different simulations with two different tsunami source models (i.e., Imamura and Fujii-Satake) given in Fig. 4 were performed to compute water surface elevations at the Kamaishi GPS buoy. Comparison of the numerical results of both simulations is given in Fig. 5. The results reveal that the computed water surface elevations of both source models fit quite well with the wave record for both wave amplitudes and periods. For the first descending wave, numerical results calculated using the Imamura source model gave higher values of recession. In this respect, the

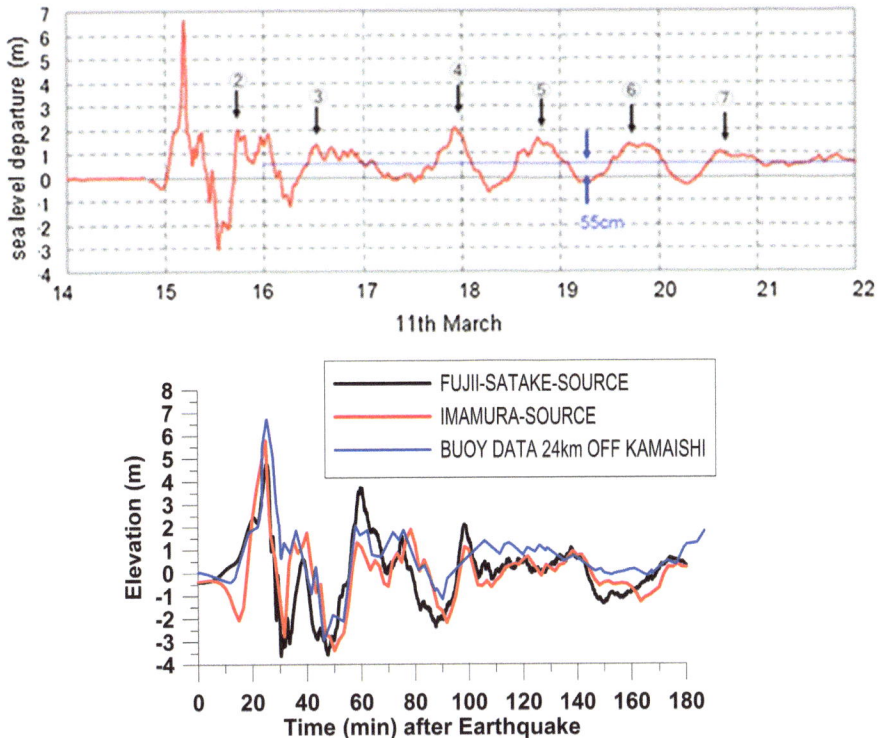

Figure 5
PARI GPS buoy wave record after the earthquake, 24 km offshore of Kamaishi, and comparison of the measured and computed data using both the Fujii-Satake and Imamura sources

first wave calculated using the Fujii-satake source model fits better with the GPS buoy record. After the first wave, the calculated waves seem to be not so compatible with the wave record in some parts of the wave train. This result enables us to conclude that either of these models can be used in this study to calculate tsunami hydrodynamic parameters in Kamaishi Bay.

3. Model Setting

In order to compute and compare the distribution of tsunami hydrodynamics as flow depth, current velocities and Froude numbers in Kamaishi, tsunami simulation was performed for Kamaishi Bay and for the city with fine grid bathymetry and topography. Well-known video footage taken from the roof of a three-storey building at the north of the bay reveals the devastating effect of tsunami inundation. As seen from the photos in Fig. 1, this concrete building was highly scoured during tsunami inundation and other light timber houses were totally swept away. It is thus essential to include the locations of the buildings into the topography with their heights. Therefore, the locations of the buildings and roads together with the coastline of Kamaishi city were specified from satellite imagery, meticulously digitized and added to the topographic data and digital data of the study domain was generated with 3 m spatial grid resolution in order to achieve more realistic results (Fig. 6). The tsunami breakwaters of Kamaishi are also inserted in the bathymetric data by assuming its crest elevation as 6 m.

The tsunami source proposed by IMAMURA et al. (2011) was used as the input of the near-field modeling from the source to the Kamaishi area. The time histories of water level fluctuations at the entrance of Kamaishi Bay, as obtained from near-field modeling, is inputted as the forcing function into the fine grid model of Kamaishi bay for the computation of inundation and nearshore tsunami parameters.

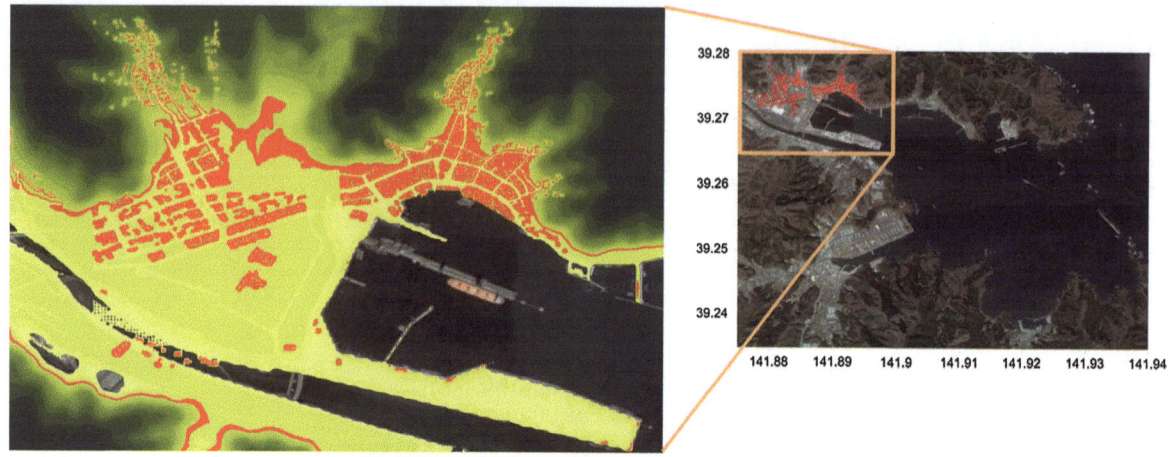

Figure 6
Digitized bathymetry and topography of Kamaishi Bay and City. Roads and buildings delineating in the study domain were digitized from the satellite images and combined with the topographic data

During the GEJE tsunami, the Kamaishi breakwater was heavily damaged by the first 9-m wave, which left Kamaishi defenseless (ONISHI 2011). The waves deflected from the breakwater are also thought to have contributed to the high amplitudes in the northern part of the city (see the digitized topography in Fig. 6, on the left). Eyewitnesses in Ryo-ishi, a small village at the north end of Kamaishi claimed that the waves reflected from the breakwaters significantly increased the damage to houses. ARIKAWA et al. (2012) examined the stability of breakwaters under tsunami overflow by doing hydraulic model experiments and conducting numerical simulations with Kamaishi Bay breakwaters as the model in order to understand the failure mechanisms of the trunk of breakwaters. Due to the experimental results, they assumed that the major cause of breakwater collapse is the difference of water level at the two sides of the breakwaters, strong overflow and scouring. In addition to that study, ARIKAWA and SHIMOSAKO (2013) included the countermeasure of tsunami scouring and the resiliency of protective structures through these experiments.

In order to analyze the effects of the Kamaishi breakwater on tsunami inundation and damage, the analyses of tsunami inundation were focused on three conditions, depending on the presence and state of the breakwater: "with breakwater," "without breakwater" and with "damaged breakwater," conditions that may reflect the real scenario and provide comparison of the breakwater's performance. A damaged breakwater is represented to reflect a submerged breakwater without a crown wall.

Analyses were performed in two phases as (1) investigation of the effect of the breakwater on near shore tsunami hydrodynamic parameters and its performance on protecting the bay, and (2) investigation of the behavior of hydrodynamic parameters around the buildings at the North end of the bay. Figure 7 shows the distribution of numerical gauge points in these two focused areas. The sections were determined along the direction of incoming waves in order to investigate the change of hydrodynamic parameters on the seaward side and the protected harbor side. On the other hand, the numerical gauge points at the residential area were selected among those at the front of the buildings and along the roads.

3.1. Investigation of Hydrodynamic Parameters Around the Kamaishi Breakwaters

As failure of the Kamaishi breakwater had serious consequences, the reasons for the breakwater collapse are discussed in terms of the differences in hydrodynamic parameters on the seaward and leeward sides of the breakwater. Sections 1 and 4, shown in Fig. 7, were selected for these comparisons.

Water surface elevations on the seaward and leeward sides of the breakwater were computed and given in Fig. 8. The plots show that water surface

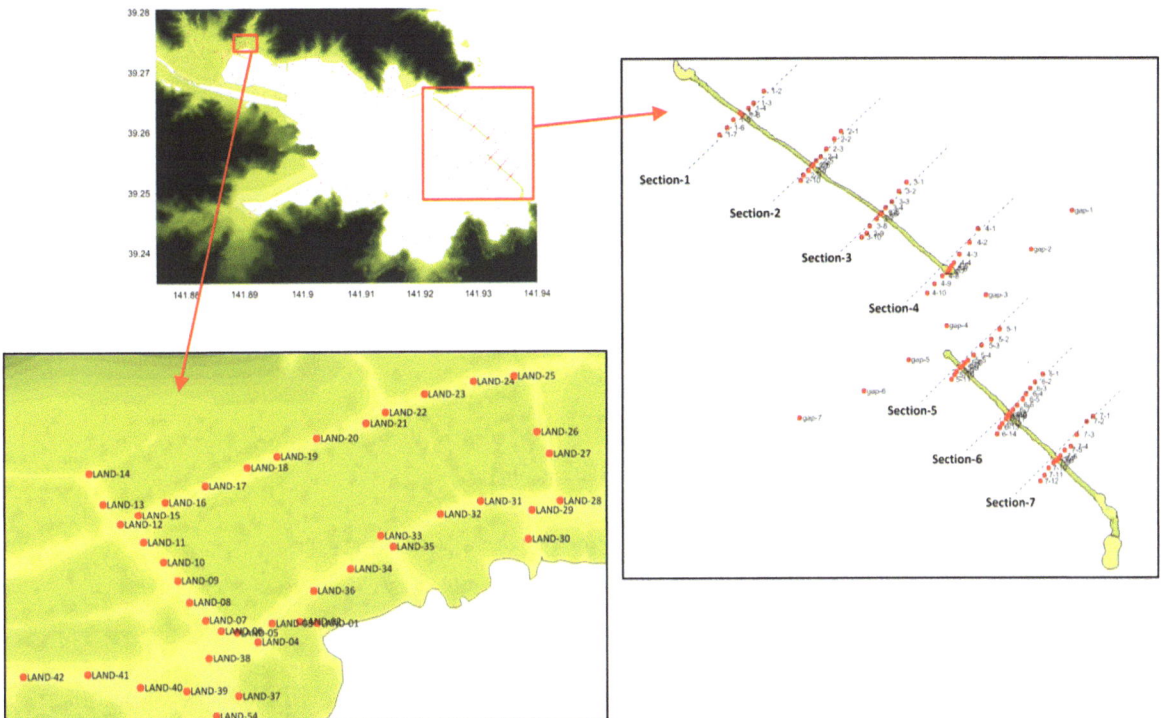

Figure 7
The numerical gauge points around the buildings along the roads at the north end of Kamaishi Bay and the sections crossing the two breakwaters

elevation on the seaward side is considerably higher than its values on the leeward side. Especially for Sect. 1, when the first wave hits the breakwater, water elevation on the seaward side is almost four times of that on the leeward side in about 10 min. This situation tends to cause significant differences in the hydrostatic pressures between the two sides of the breakwater. Therefore, the resisting forces on the leeward side become inadequate against the sliding and overturning forces, and hence the collapse due to the sliding and overturning becomes inevitable. Figure 8 also indicates that the difference of water elevation at the two sides of the breakwater decreases at the toe. This result corresponds well with the expectations since the waves pass from the gap and therefore cannot raise the water elevation on the seaward side of Sect. 4. However, the seaward side of the breakwater at Sect. 1 is directly exposed to the tsunami attack without the passage of waves through any gap.

The wave overtopping during the tsunami attack was investigated by plotting the water levels at the gauge points located at the breakwater crest. Figure 9 shows the water surface fluctuation on the crest of the breakwater at Sects. 1, 2 and 4 during the simulation. The results show the first wave hits the breakwater at around the 30th minute of the simulation through Sects. 1 and 2, and 1 min later at Sect. 4. The water elevation reaches approximately 8.5 m at about the 5th minute after overtopping, and the water passage continues with a 0.5 m flow depth at the top of breakwater. The second wave hits at around 62nd minute, again with an approximately 8-m water level.

The role of breakwater gap on the change of tsunami hydrodynamic parameters was investigated in this study by examining the ratio of maximum current velocity between with and without breakwater cases. As seen from Fig. 10, the current velocity is concentrated through the gap of breakwater that may instantaneously cause high values of currents in inundation area. However, the occurrence times of maximum current velocity in these two cases may not occur concurrently. Similarly, Fig. 11 shows the ratio of maximum Fr^2 between with and without

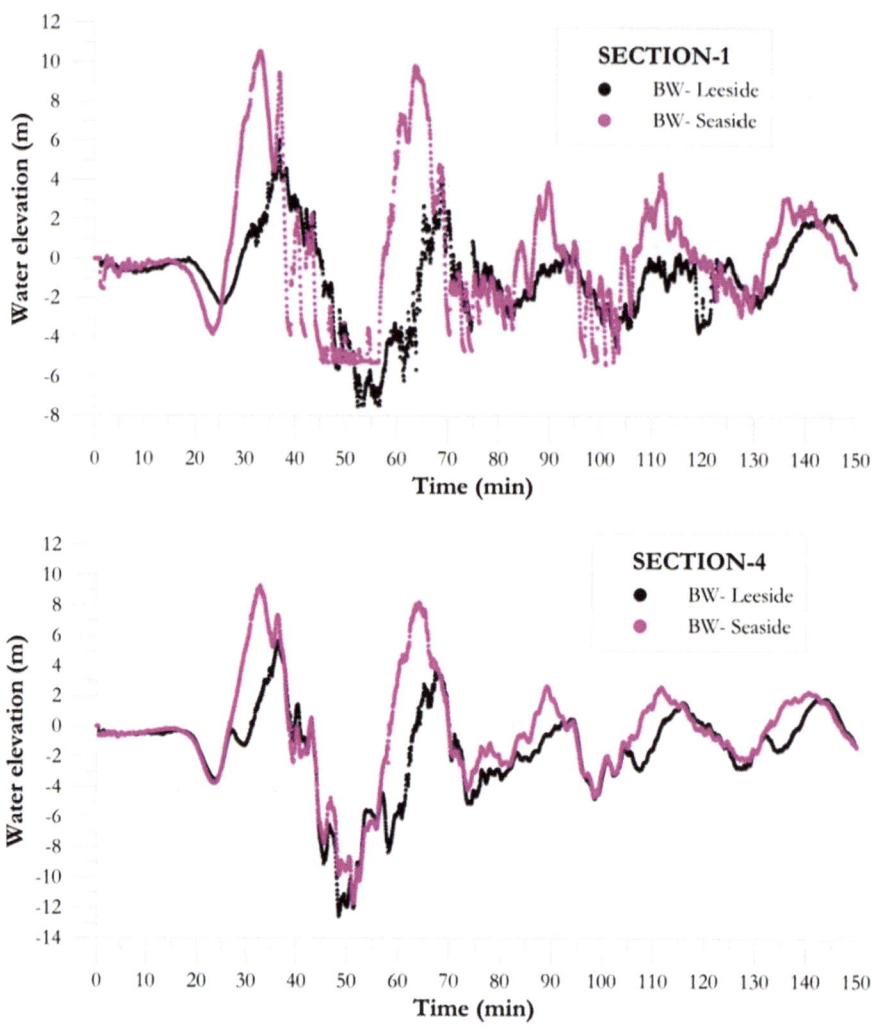

Figure 8
Comparison of water surface elevations on the seaward and leeward sides of the breakwater during the simulation time

breakwater cases. The figure indicates that the existence of Kamaishi breakwater leads to higher values of Fr^2 at some locations on land. This means that in case of protection with breakwater, higher damage may be expected at some locations in inundation area due to high current velocities that may occur instantaneously and be directed by the existence of breakwater.

The change in current velocity at numerical gauge points along the gap for the with- and without breakwater conditions are highlighted in Fig. 12. The results presented in this figure also support the presence of a concentrated distribution of current velocity through the gap. The plots reveal that the current velocities at the numerical gauge points Gap_3 and Gap_4 are fairly higher when the breakwater exists. This is due to the intrusion of water through the gap in a concentrated manner even if the waves overtop. The difference of current velocities between the with- and without breakwater conditions decreases at the gauge point Gap_5 since this point is farther from the gap entrance.

3.2. *The Effect of the Kamaishi Breakwater on Hydrodynamic Parameters in Residential Areas*

In light of the previous results, changes in hydrodynamic parameters also require evaluation within the inundation area in regards to the three

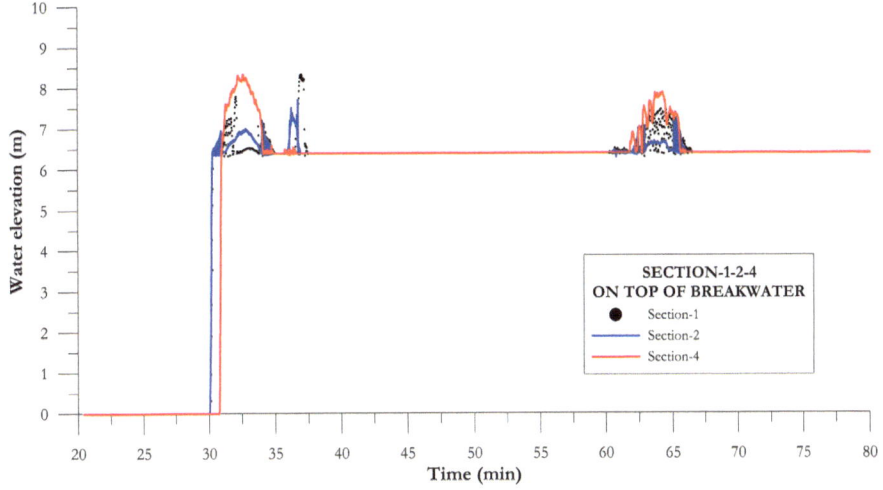

Figure 9
Water surface fluctuations at the top of breakwater along Sects. 1, 2 and 4 during simulation

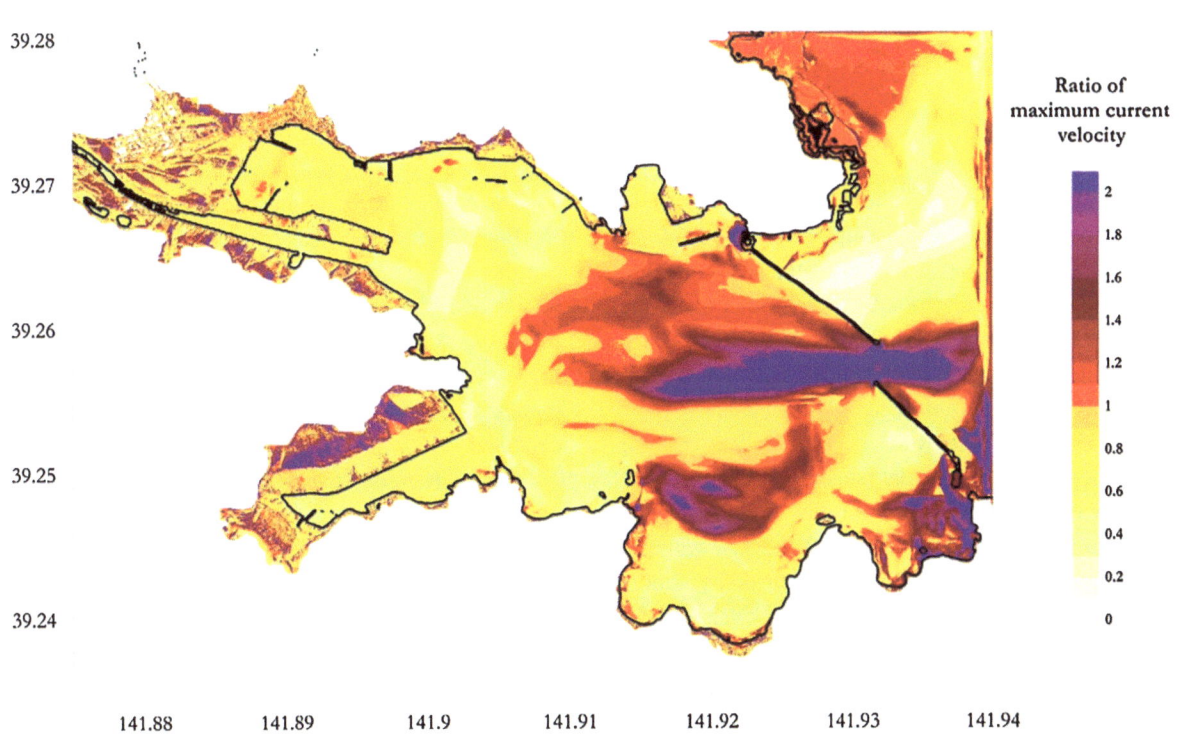

Figure 10
The ratio of maximum current velocity between with- and without breakwater cases

conditions: with breakwater, without breakwater and with damaged breakwater.

Water level fluctuations in the residential area are shown in Fig. 13 at three selected gauge points, Land_04, Land_08 and Land_20. Land_04 is located at the entrance of a road parallel to the incoming wave direction near the shoreline, Land_08 is along that road and Land_20 is selected on another road

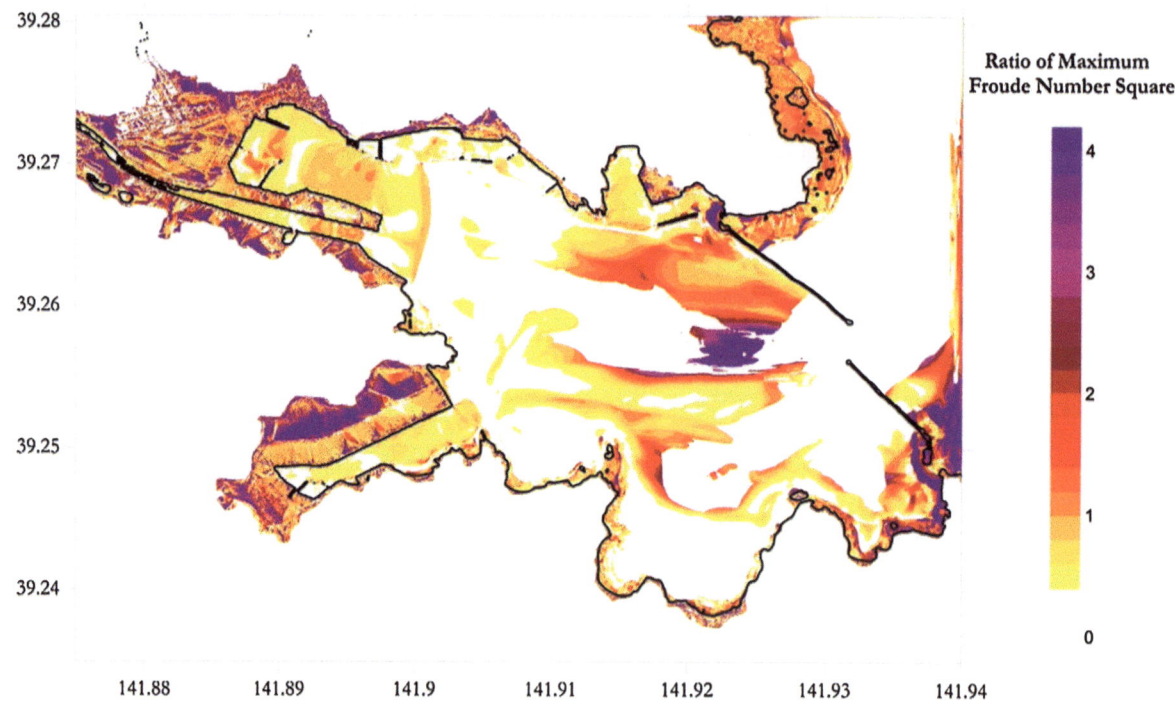

Figure 11
The ratio of maximum Fr^2 between with- and without breakwater cases

parallel to the shoreline. The well-known Kamaishi footage was taken near the gauge point Land_04.

Figure 13 shows that higher water elevations are observed at these gauge points. On the other hand, the lowest values occur when the bay is protected by a breakwater. The results also indicate that, although it is not as effective as in a non-damaged breakwater case, a damaged breakwater can partly decrease the water elevations and prevent tsunami damage, even if the top part of the breakwater has collapsed. The results also show that the arrival of the first wave to the locations selected on land was reduced by almost 5 min in the case of not having a breakwater at the entrance of the bay.

The current velocities at the same gauge points are given in Fig. 14 for the three breakwater cases. It is obvious from the results that when the waves first inundate, the existence of a breakwater causes instantaneous high current velocities at the locations in the direction of the incoming wave. For the locations along the road parallel to the shoreline, no significant change of current velocities is observed.

The change of Fr^2 at these gauge points is plotted in Fig. 15 for the same breakwater conditions. Similar to the change of current velocities, higher values of Fr^2 are observed at the locations along the incoming wave direction in residential areas if the breakwater exists. These results are directly consistent with the outcomes for current velocities, as expected.

A similar study assessing the Kamaishi breakwater performance during this tsunami was performed by PARI (TAKAHASHI et al. 2011) with numerical modeling. It was indicated in TAKAHASHI et al. (2011) that the breakwaters maintained their function until the peak and could delay tsunami arrival by about 4 min and reduce tsunami runup about 50 %, which also supports the outcomes of our study. Different from TAKAHASHI et al. (2011), our manuscript used indepth topographic data, including sea structures, buildings and roads in fine resolution for investigating the change in the maximum wave amplitude, current velocity, square of the Froude Number, and flow depth in inundation zones.

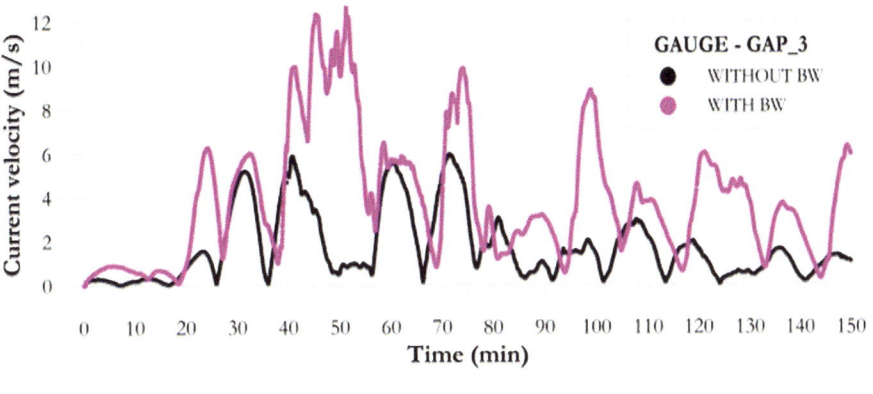
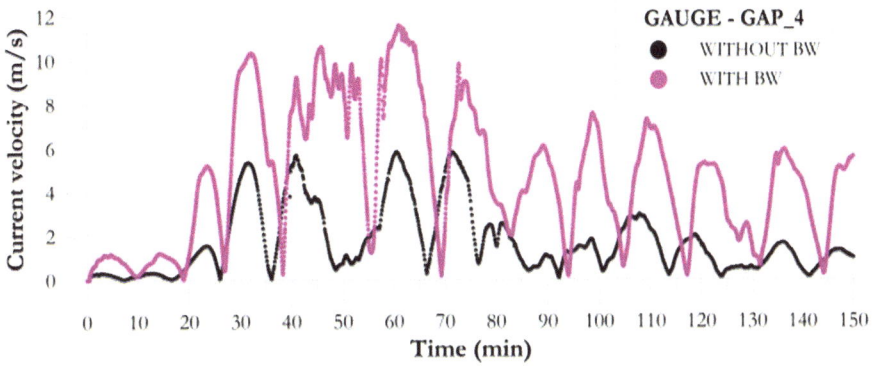
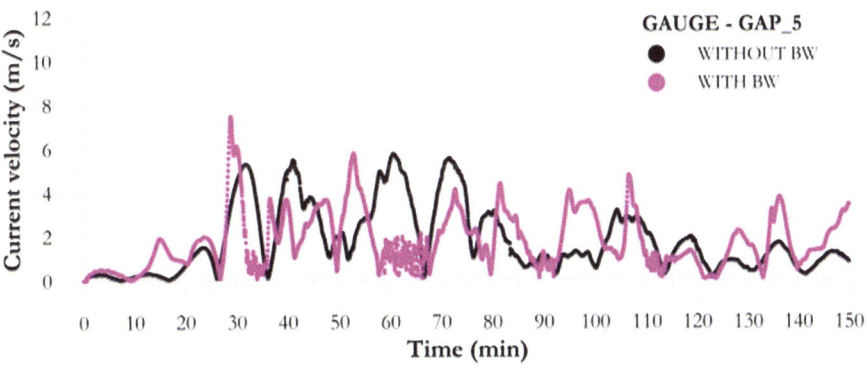

Figure 12
Change in current velocity during the simulations at the gauge points located through the breakwater gap for with- and without breakwater cases

3.3. The Investigation of Hydrodynamic Parameters in Case of Full Protection of Breakwater Without Overtopping

The previous section investigated the change of hydrodynamic parameters in relation to with whether the bay was protected by a breakwater or not. However, wave overtopping is allowed, and the differences in current velocities and Fr^2 between with- and without a breakwater cases can not reflect the actual effect of breakwaters resulting in concentrated energy penetrating into the bay through the breakwater gap. For this reason, the hydrodynamic parameters are investigated in case of high crested breakwaters that do not allow overtopping. The crest height of a breakwater is assigned 20 m, although it is not realistic.

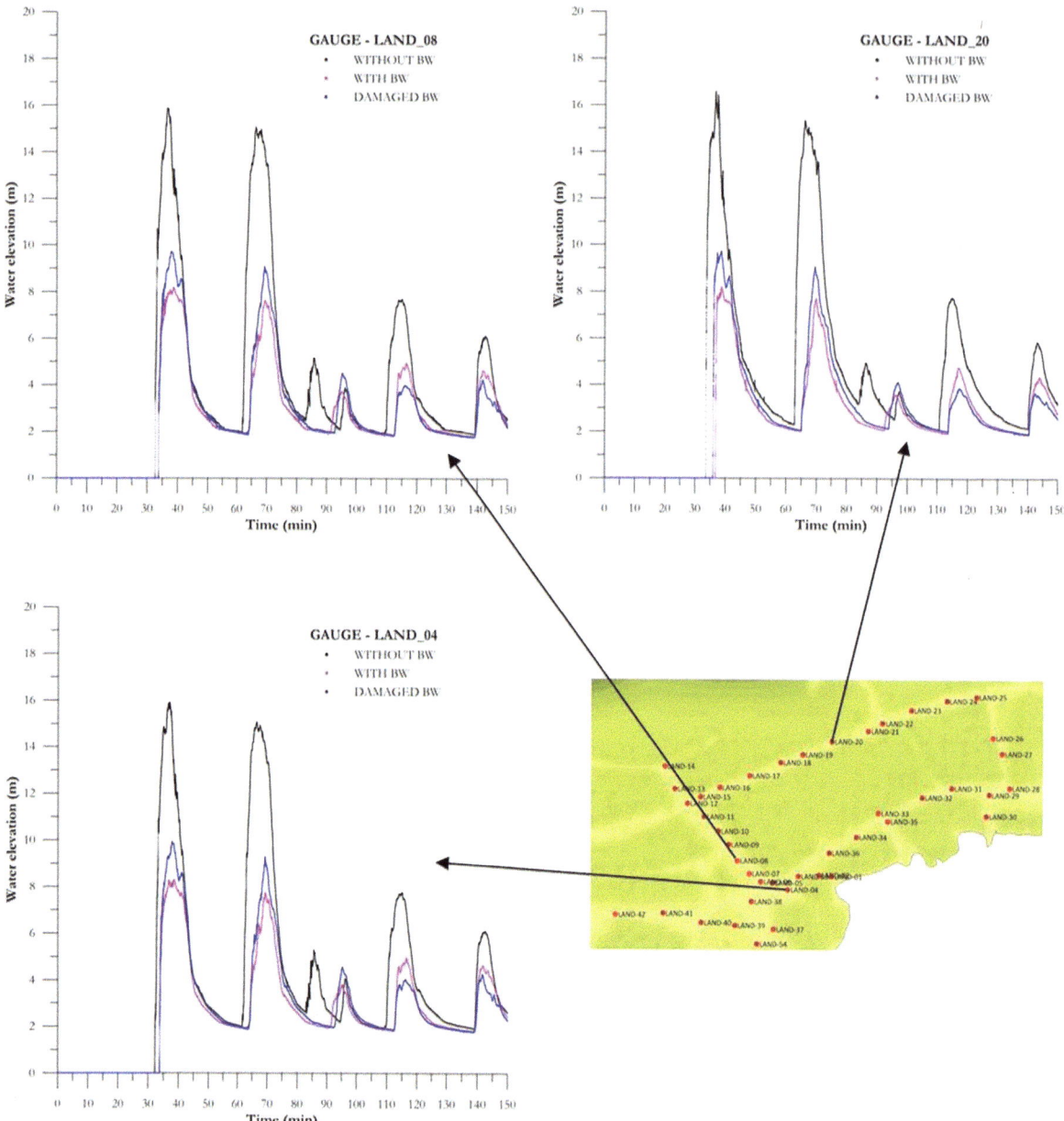

Figure 13
Water level fluctuations during the simulations at the three selected gauges for with-, without- and damaged breakwater cases

Figure 16 shows the current velocity ratio between the two breakwater cases with a 20-m crest height that does not allow overtopping and the existence of no breakwater. The distribution of current velocity in the figure indicates that the fully concentrated energy due to the breakwater results in higher currents in some locations on the land. This means that some parts of the residential area become vulnerable to tsunami attacks due to the effect of breakwaters and the gap between them at the entrance of the bay.

Similarly, Fig. 17 shows the ratio of maximum Fr^2 between the cases of full breakwater protection with no overtopping and no breakwater protection. The plots reveal that, when the breakwater does not allow overtopping, the concentration of wave energy

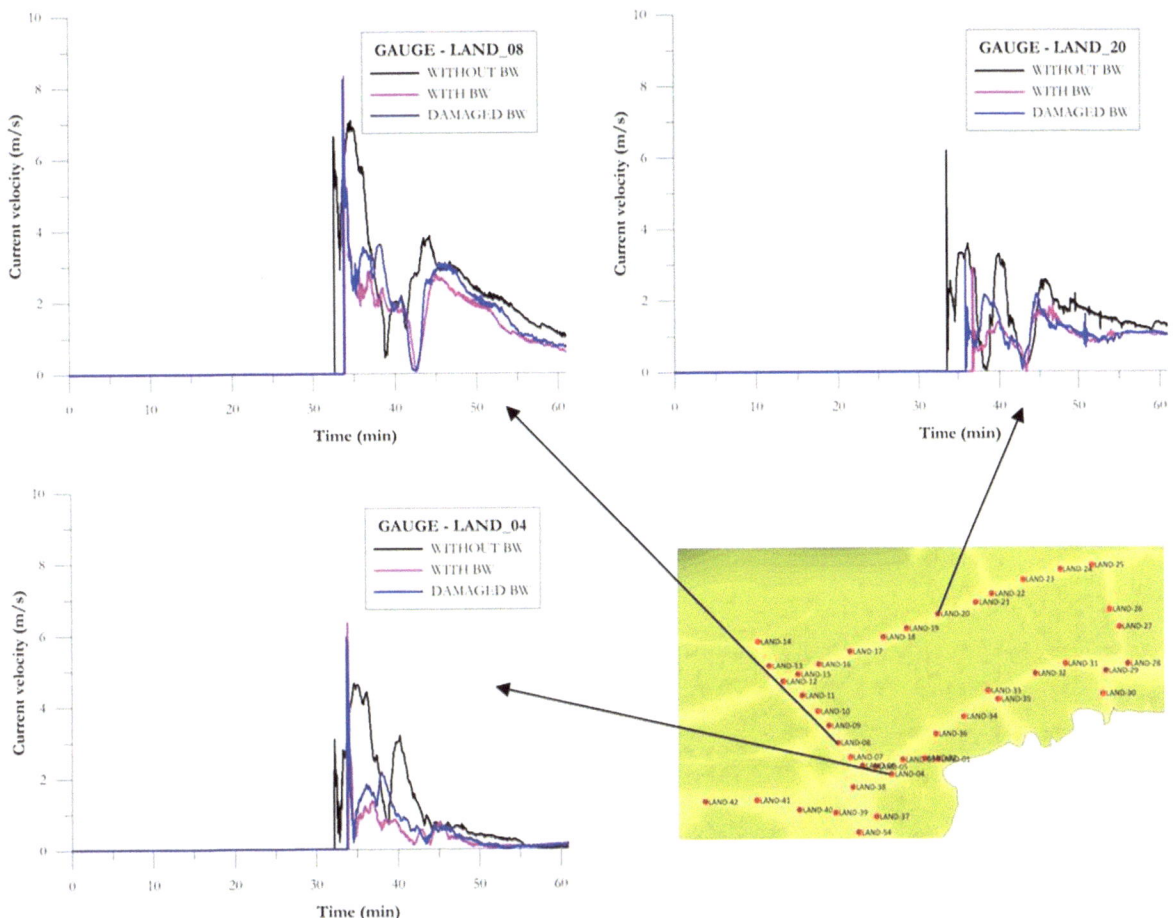

Figure 14
Current velocities changes during the simulation at three selected gauge points for with-, without- and damaged breakwater cases

through the breakwater gap is more substantial and that the results indicate much higher values of Fr^2 at some locations in residential area occurring instantaneously.

4. Evaluation of the Results and Conclusion

This study investigated the hydrodynamic parameters in inundated zones of Kamaishi city and determined the performance of coastal protection structures during the 2011 GEJE tsunami. The results enabled us to make significant inferences for the changes in Froude Number, currents, flow depths and water levels during tsunami inundation. Besides this, it is possible to comment on the damage to the Kamaishi breakwater by focusing on the hydrodynamic parameters around these structures and also assess the reasons for the excessive currents at some locations within the residential areas.

The analyses of tsunami inundation were performed in three conditions (with breakwater, without breakwater and with damaged breakwater) for understanding the effects of the Kamaishi breakwater on tsunami inundation and damage. Results for the with breakwater scenario indicate that for about 10 min, the hydrostatic pressure on the leeward side of the breakwater was dramatically less than the values on the seaward side due to the enormous raise of water surface elevation there. The disequilibrium of hydrostatic pressure causes inevitable sliding and overturning, which is the cause of the breakwater collapse. This huge instability in hydrostatic pressure was not determined at the toe of the breakwater since the waves could pass from the

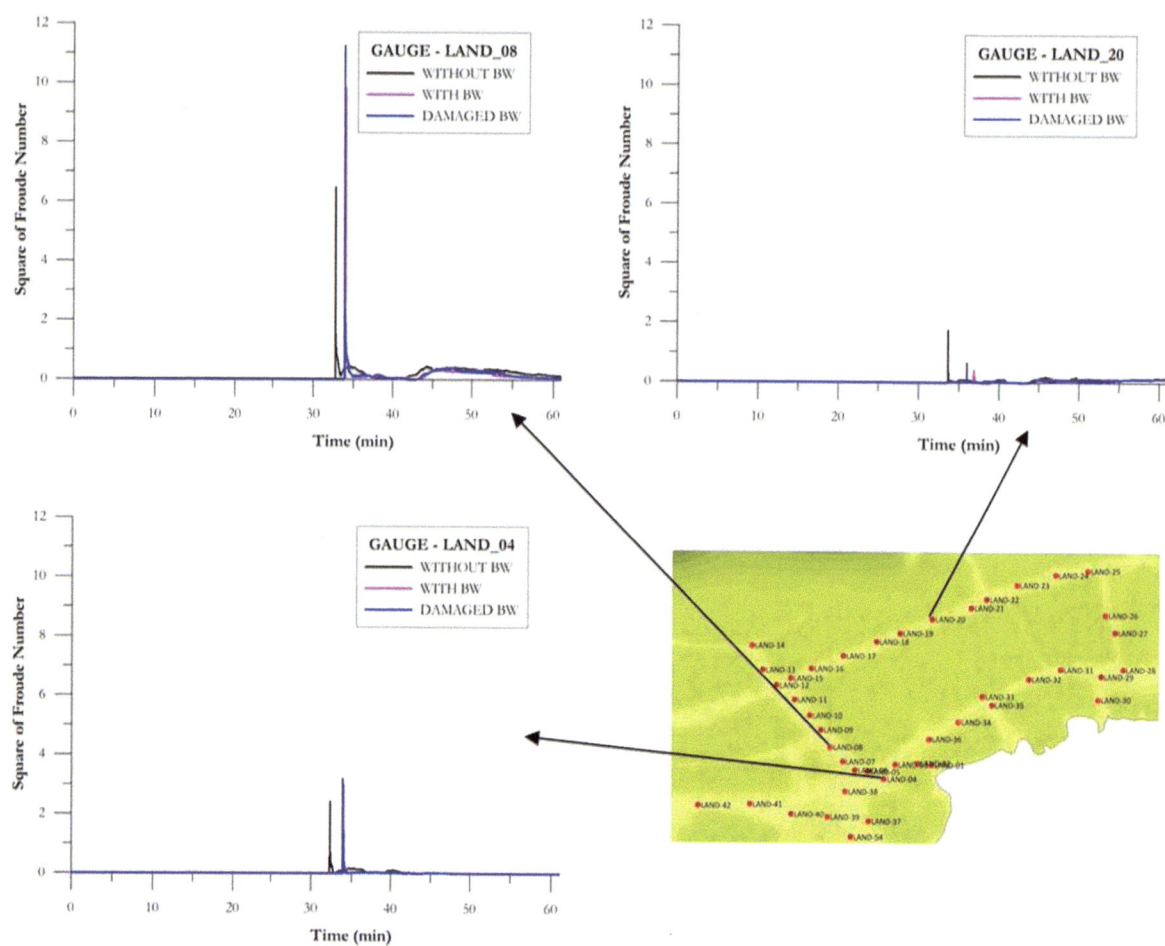

Figure 15
The change of Fr^2 during the simulation at three selected gauge points for with-, without- and damaged breakwater cases

breakwater gap and, therefore, the water elevation could not rise. This result is also supported by the hydraulic model experiments performed in the PARI wave flume (ARIKAWA et al. 2012) and also by several eyewitness videos recorded in Kamaishi (The New York Times 2011).

Wave overtopping during the tsunami attack was also investigated. The water surface elevations at selected cross-sections along the top of the breakwater show that the first wave hit the structure at around the 30th minute of the simulation, reaching its maximum value of around 8.5 m at about the 5th minute after overtopping. The second wave was calculated at the 62nd minute, again with an approximately 8-m water level. During overtopping, the flow depth was calculated as 0.5 m at the top of the breakwater.

The comparison of with-, without- and damaged-breakwater cases shows that lower water levels occur if a breakwater exists. Also, the damaged breakwater can partly protect the inundation area with lower calculated water levels than can the case with no breakwater. Regarding the current velocities, the situation is a bit different from the water levels in that instantaneous high current velocities occurred at the locations in the direction of the incoming wave in the case of an existing breakwater. There was no significant change in current velocities for the locations along the road parallel to the shoreline at the north end of Kamaishi city.

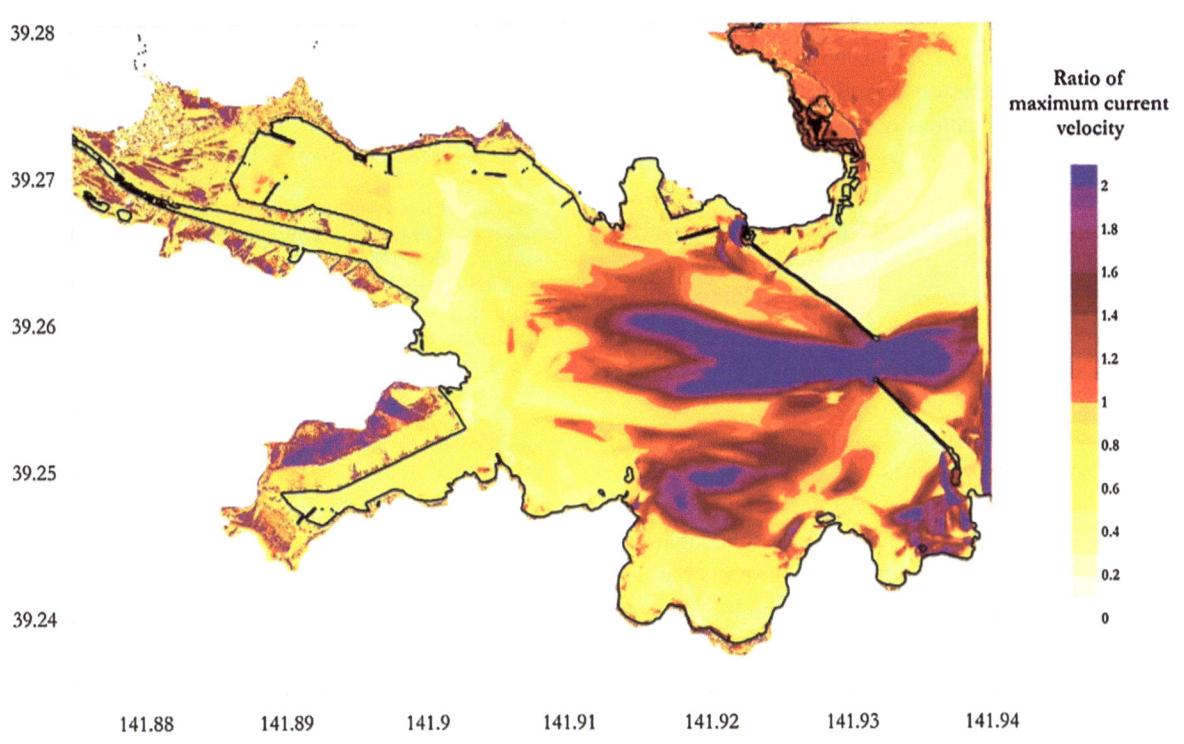

Figure 16
The ratio of maximum current velocity of a 20-m crest height (no overtopping) between the with breakwater and without breakwater cases

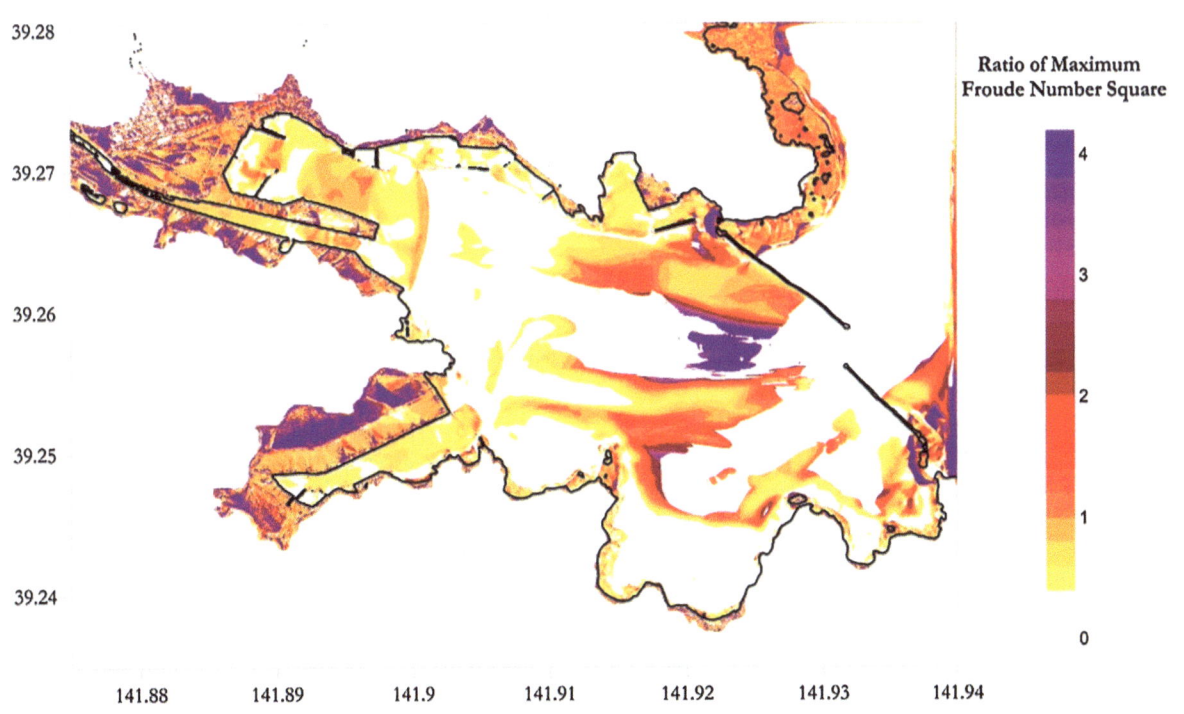

Figure 17
The ratio of maximum Fr^2 of a 20-m crest height (no overtopping) between the with breakwater and without breakwater cases

In case of not allowing the wave overtopping, the results show that the fully concentrated energy due to the existence of breakwater results in higher currents and Fr^2 in some locations on the land, meaning that some residential areas may be exposed to tsunami damage due to the effect of a breakwater with no overtopping and the gap between them at the entrance of the bay.

According to an overall evaluation of the results, the similarity between the pattern of the square of the Froude Number and current velocity reveals that these parameters must be included in tsunami hazard assessment studies together with water elevations. Depending on the outcomes, it can be easily stated that when a breakwater is constructed for the protection of a bay or a semi-enclosed area against tsunamis, changes in tsunami flow patterns should be assessed by mathematical modeling at the design stage.

Acknowledgments

The authors would like to acknowledge the partial support of the Scientific and Technological Research Council of Turkey (TÜBİTAK) research grant no: 108Y227, "Model Development and Risk Analysis for Tsunamis in the Black Sea and Mediterranean (MORAT)" and by European Commission-funded research project "Tsunami Risk ANd Strategies For the European Region (TRANSFER)." The authors also acknowledge the precious colleagues from PARI who provided the Kamaishi GPS buoy tsunami wave record.

REFERENCES

ARIKAWA T. and SHIMOSAKO K. (2013). Failure Mechanism of Breakwaters Due To Tsunami; a Consideration to the Resiliency, Proceedings of 6th Civil Engineering Conference in the Asian Region, August 20–22, 2013, Jakarta, Indonesia.

ARIKAWA, T., SATO, M., SHIMOSAKO, K., HASEGAWA, I., YEOM, G-S. and TOMITA T. (2012). Failure Mechanism of Kamaishi Breakwaters Due to the Great East Japan Earthquake Tsunami, Proceedings of the 33rd International Conference on Coastal Engineering (ICCE2012), Santander, Spain, July 1–6, 2012.

ARIKAWA, T. (2009). *Structural Behavior under Impulsive Tsunami Loading*, Journal of Disaster Research, Vol.*4* No.6, Dec. 2009, pp. 377–381.

ASAKURA, R., IWASE, K., IKEYA, T., TAKAO, M., KANETO, T., FUJII, N. and OMORI, M. (2000). *An Experimental Study on Wave Force Acting on On-shore Structures Due to Overflowing Tsunamis*, Proc. of Coastal Engineering, JSCE, Vol. *47*, pp. 911–915 (in Japanese).

CHARVET, I., SUPPASRI, A., IMAMURA, F. (2014). *Empirical Fragility Analysis of Building Damage Caused by the 2011 Great East Japan Tsunami in Ishinomaki City using Ordinal Regression, and Influence of Key Geographical Features*. Stoch Env Res Risk A., doi:10.1193/053013EQS138M.

Coastal Engineering Committee, Japan (2011). Tsunami height distribution, http://www.coastal.jp/ttjt/index.php?%E7%8F%BE%E5%9C%B0%E8%AA%BF%E6%9F%BB%E7%B5%90%E6%9E%9C.

FUJII, Y., SATAKE, K., SAKAI, S., SHINOHARA, M, and KANAZAWA, T. (2011). *Tsunami source of the 2011 off the Pacific coast of Tohoku Earthquake*, Earth Planets Space, *63*, 815–820, doi:10.5047/eps.2011.06.010.

FUJIMA, K. (2009). Necessity of advanced Tsunami Damage Index, Proceedings of the 6th International Workshop on Coastal Disaster Prevention, Bangkok, Thailand, December 1–2, 2009, pp. 61–70.

Geospatial Information Authority, Japan (2011). Tsunami inundation map, Retrieved from http://www.gsi.go.jp/kikaku/kikaku60003.html.

GOTO, C., OGAWA, Y., SHUTO, N., IMAMURA, F. (1997). IUGG/IOC TIME Project: Numerical Method of Tsunami Simulation with the Leap-Frog Scheme, Intergovernmental Oceanographic Commission of UNESCO, Manuals and Guides # 35, Paris, 4 Parts.

IMAMURA, F., KOSHIMURA, S., MURASHIMA, Y., AKITA, Y., SHINTANI, Y. (2011). The 2011 East Japan off the Pacific coast Earthquake and Tsunami, Tohoku University Source Model version 1.1, Sendai, Miyagi, Japan: Disaster Control Research Center (in Japanese).

IMAMURA, F. (1989). Tsunami Numerical Simulation with the Staggered Leap-Frog Scheme (Numerical Code of TUNAMI-N1), School of Civil Engineering, Asian Inst. Tech. and Disaster Control Research Center, Tohoku University.

IMAMURA, F. (1995). Numerical Method of Tsunami Simulation with the Leap-Frog Scheme, Part 2: Propagation in the Ocean in the Spherical Coordinates, TUNAMI-F1 and its Program List, Sendai-Japan.

Iwate-Miyagi-Fukushima Province, Japan (2011), Observed Tsunami around Iwate-Miyagi-Fukushima Province, Retrieved from http://www.jma.go.jp/jma/en/2011_Earthquake/2011_Earthquake_Tsunami.pdf.

JMA (2011), Japan Meteorological Agency, Retrieved from http://www.jma.go.jp/jma/en/2011_Earthquake.html.

Kamaishi Port Office (2011), Retrieved from http://www.pa.thr.mlit.go.jp/kamaishi/bousai/index.html.

KOIZUMI T. (2011). "The Tohoku Tsunami of 11th March: the Response of Tsunami Early Warning Systems and the Tsunami Analysis"—Takeshi Koizumi (JMA); UNESCO/IOC Tsunami and Civil Protection Workshop: "Tsunami hazard in the Northeastern Atlantic, the Mediterranean and connected seas (NEAM region)–A challenge for Science and Civil protection"; 15–16 June, JRC-Ispra, Italy.

KOKETSU, K., YOKOTA, Y., NISHIMURA, N., YAGI, Y., MIYAZAKI, S., SATAKE, K., FUJII, Y., MIYAKE, H., SAKAI, S., YAMANAKA, Y. and OKADA, T. (2011). "*A Unified Source Model for the 2011 Tohoku*

Earthquake", Earth and Planetary Science Letters, doi:10.1016/j.epsl.2011.09.009, Vol. *310* (2011), pp. 480–487.

KOSHIMURA, S., NAMEGAYA, Y. and YANAGISAWA, H. (2009). *"Tsunami fragility—A new measure to identify tsunami damage"*, Journal of Disaster Research, Vol.*4* No.6, 2009, pp. 404–409.

KURKIN A.A., KOZELKOV A.C., ZAITSEV A.I., ZAHIBO N., and YALCINER A.C. (2003). *Tsunami risk for the Caribbean Sea Coast, Izvestiya*, Russian Academy of Engineering Sciences, vol. *4*, pp. 126–149 (in Russian).

LEELAWAT, N, SUPPASRI, A., CHARVET, I., IMAMURA, F. (2014). *Building damage from the 2011 Great East Japan tsunami: quantitative assessment of influential factors*, Natural Hazards, vol. *73* Issue 2, pp. 449–471.

NAMI DANCE (2011). Manual of Numerical Code NAMI DANCE, Published in http://namidance.ce.metu.edu.tr.

New York Times (2011). http://www.nytimes.com/video/world/asia/100000001150840/japans-failed-breakwaters.html, November 3, 2011.

ONISHI, N. (2011). Japan Revives a Sea Barrier That Failed to Hold, an article in The New York Times, http://www.nytimes.com/2011/11/03/world/asia/japan-revives-a-sea-barrier-that-failed-to-hold.html?_r=1&emc=eta1, Nov. 2, 2011.

OZER SOZDINLER, C., YALCINER, A.C., ZAYTSEV, A. (2014). *Investigation of Hydrodynamic Parameters in Inundation Zones with Different Structural Layouts*, J. of Pure Appl. Geophys., 2014, doi:10.1007/s00024-014-0947-z.

OZER, C. (2012). Tsunami Hydrodynamics in Coastal Zones, PhD. Thesis, Middle East Technical University, Ankara, Turkey, 2007, 124 pages.

OZER, C. and YALCINER, A.C. (2011). Sensitivity Study of Hydrodynamic Parameters during Numerical Simulations of Tsunami Inundation, Pure Appl. Geophys., Springer Basel AG, doi:10.1007/s00024-011-0290-6.

OZER, C., KARAKUS, H., YALCINER, A.C. (2008). Investigation of Hydrodynamic Demands of Tsunamis in Inundation Zone, Proceedings of 7th International Conference on Coastal and Port Engineering in Developing Countries, Dubai, UAE, February 24–28, 2008.

SATAKE, K., FUJII, Y., HARADA, T., NAMEGAYA, Y. (2013). *Time and Space Distribution of Coseismic Slip of the 2011 Tohoku Earthquake as Inferred from Tsunami Waveform Data*, Bulletin of the Seismological Society of America, Vol. *103*, No. 2B, pp. 1473–1492, May, 2013, doi:10.1785/0120120122.

SHUTO, N. (2009). *Damage to Coastal Structures by Tsunami-Induced Currents in the Past*, Journal of Disaster Research, Vol.*4*, No.6, 2009, pp. 462–468.

SHUTO, N., GOTO C., IMAMURA, F. (1990). *Numerical Simulation as a Means of Warning for Near-field Tsunamis*, Coastal Engineering in Japan, Vol. *33*(2), pp. 173–193.

SUPPASRI, A., MUHARI, A. and IMAMURA, F. (2011). JST-JICA and RISTEK Indonesia: Guide Book of the 2011 Tsunami Field Trip in Sendai, 12 pages.

SUPPASRI, A., KOSHIMURA, S., IMAI, K., MAS, E., GOKON, H., MUHARI, A. and IMAMURA, F. (2012). *Damage Characteristic and Field Survey of the 2011 Great East Japan Tsunami in Miyagi Prefecture*, Coastal Engineering Journal, Vol. *54*(1) Special Anniverary Issue on the 2011 Tohoku Earthquake Tsunami, 1250008.

SUPPASRI, A., SHUTO, N., IMAMURA, F., KOSHIMURA, S., MAS, E. and YALCINER, A.C. (2013). *Lessons Learned from the 2011 Great East Japan Tsunami: Performance of Tsunami Countermeasures, Coastal Buildings and Tsunami Evacuation in Japan*, Pure and Applied Geophysics, Vol. *170*(6–8), 993–1018.

TAKAHASHI, S., KURIYAMA, Y., TOMITA, T., KAWAI, Y., ARIKAWA, T., TATSUMI, D., NEGI, T. (2011). Urgent Survey for 2011 Great East Japan Earthquake and Tsunami Disaster in Ports and Coasts—Part I (Tsunami), An English Abstract of Technical Note of Port and Airport Research Institute, No: 1231, April 28, 2011.

TUNAMI-N2 (2001). "Tsunami Modelling Manual (Tunami Model)" by Imamura, F., Yalciner, A. C. and Ozyurt, G.

YALCINER, A.C., SYNOLAKIS, C.E., ALPAR, B., BORRERO, J., ALTINOK, Y., IMAMURA, F., TINTI, S., ERSOY, Ş., KURAN, U., PAMUKCU, S., KANOGLU, U. (2001). Field Surveys and Modeling 1999 Izmit Tsunami, International Tsunami Symposium ITS 2001, Paper 4–6, Seattle, Washington, pp. 557–563.

YALCINER, A.C., ALPAR,B., ALTINOK,Y., OZBAY,I., IMAMURA,F (2002). *Tsunamis in the Sea of Marmara: Historical Documents for the Past, Models for Future*, Marine Geology, 2002, *190*, pp. 445–463.

YALCINER, A.C., PELINOVSKY, E., SYNOLAKIS C., OKAL, E (2003). NATO SCIENCE SERIES, Submarine Landslides and Tsunamis, Publisher: Kluwer Academic Publishers, Netherlands (Editors; Yalciner, A.C., Pelinovsky, E., Synolakis, C., Okal, E.), 329 Pages, ISBN: 1-4020-1348-5 (HB), ISBN: 1-4020-1349-3 (PB).

YALCINER, A., PELINOVSKY, E., TALIPOVA, T., KURKIN, A., KOZELKOV, A. and ZAITSEV, A (2004). Tsunamis in the Black Sea: Comparison of the Historical, Instrumental, and Numerical Data, J. Geophys. Res., AGU, V 109, C12023, doi:10.1029/2003JC002113.

YALCINER, A.C. and PELINOVSKY, E. (2007). *A Short Cut Numerical Method for Determination of Resonance Periods of Free Oscillations in Irregular Shaped Basins*, Ocean Engineering, Vol. *34*, Issues 5–6, April 2007, pp. 747–757.

YALCINER, A.C., OZER, C., KARAKUS, H., ZAYTSEV, A., GULER, I. (2010). Evaluation of Coastal Risk at Selected Sites against Eastern Mediterranean Tsunamis, Proceedings of 32nd International Conference on Coastal Engineering (ICCE 2010), Shanghai, China, June 30–July 5, 2010.

ZAHIBO, N., PELINOVSKY, E.,YALCINER, A.C., KURKIN, A., KOZELKOV, A., ZAITSEV, A. (2003). *Modeling the 1867 Virgin Island Tsunami*, Journal of Natural Hazards and Earth System Sciences, European Geosciences Union, *3*, June-2003, pp. 367–376.

ZAITSEV, A.I., KOZELKOV, A.C., KURKIN, A.A.,PELINOVSKY, E.N., TALIPOVA, T.G., YALCINER, A.C. (in Russian order of alphabet) (2002). *Tsunami Modeling in Black Sea, published in Izvestiya of Russian Academy of Engineering Sciences*, Series: Applied Mathematics and Mechanics, 2002, Vol. *3*, pp. 27–45.

ZAITSEV, A., YALCINER, A.C., PELINOVSKY, E., KURKIN, A., OZER, C., INSEL, I., KARAKUS, H., OZYURT, G. (2008). Tsunamis in Eastern Mediterranean, Histories, Possibilities and Realities, Proceedings of 7th International Conference on Coastal and Port Engineering in Developing Countries, Dubai, UAE, February 24–28, 2008.

(Received September 1, 2014, revised January 31, 2015, accepted February 5, 2015, Published online March 14, 2015)

Pure Appl. Geophys. 172 (2015), 3493–3508
© 2015 Springer Basel (outside the USA)
DOI 10.1007/s00024-015-1127-5

Pure and Applied Geophysics

On the Leading Negative Phase of Major 2010–2014 Tsunamis

MARIE C. EBLÉ,[1] GEORGE T. MUNGOV,[2,3] and ALEXANDER B. RABINOVICH[4,5]

Abstract—Time series observations from instruments sited in the deep ocean and along coastal margins of the Pacific during major tsunami events in the years since 2010 were systematically processed. Examination of these records during four events, 2010 Chile (Maule), 2011 East Japan (Tohoku), 2012 Haida Gwaii and 2014 Chile (Iquique), show the prevalence of a small negative phase leading the first major positive tsunami wave, a phase that is not typically reproduced by current modelling approaches. We present leading negative phase signatures in examples from the more than 40 deep-ocean bottom pressure and approximately 200 tide gauge records investigated for this study. High sampling rate time series (15-s) were given greater weight in our investigation than the more readily available 1-min series. Careful investigation of tsunami arrival at each deep-ocean site highlights the role filtering techniques may play in misleading researchers or masking specific tsunami features such as the leading negative phase that is the basis of this study. The main focus of this investigation is to characterise the scale and repeatability of the phenomenon in support of recent similar findings rather than to provide a definitive explanation as to the cause. In general, our findings are in good agreement with and support the theoretical results of WATADA *et al.* (J Geophys Res Solid Earth 119:4287–4310, 2014).

Key words: Tsunami leading negative phase, 2010 Chile (Maule), 2011 Tohoku, 2012 Haida Gwaii and 2014 Chile (Iquique) tsunamis, DART, coastal tide gauges.

---

[1] Pacific Marine Environmental Laboratory, National Oceanic and Atmospheric Administration, 7600 Sand Point Way NE, Seattle, WA 98115, USA. E-mail: marie.c.eble@noaa.gov

[2] National Centers for Environmental Information, National Oceanic and Atmospheric Administration, Boulder, CO 80305-3328, USA.

[3] Cooperative Institute for Research in Environmental Sciences, University of Colorado at Boulder, 216 UCB, Boulder, CO 80309-0216, USA.

[4] Department of Fisheries and Oceans, Institute of Ocean Sciences, 9860 West Saanich Road, Sidney, BC V8L 4B2, Canada.

[5] P.P. Shirshov Institute of Oceanology, Russian Academy of Sciences, 36 Nakhimovsky Prosp., Moscow 117997, Russia.

1. Introduction

The existence of tsunami waves has been known and chronicled throughout history yet knowledge of fine scale details has historically been limited by sparseness of high-quality deep-ocean data and by low-quality coastal tide gauge records (DUNBAR *et al.* 2008). The 2004 Sumatra tsunami served as impetus for a focused effort on the acquisition of measurements to understand the tsunami as a destructive natural hazard. As a direct result, major tsunamis occurring in the years since 2009 in the Pacific Ocean have been recorded by an unprecedented number of instruments deployed in the deep ocean and along coastal margins (EBLÉ *et al.* 2011). By focusing on analysis of these data sets, RABINOVICH *et al.* (2013b) established a correlation between mean energy decay rates and the frequency content of three Pacific Basin tsunamis. During the course of their investigation, the authors noted a specific feature present in bottom pressure and coastal tide gauge records of the 2010 Chile (Maule) tsunami (RABINOVICH *et al.* 2013a). The feature, a negative phase preceding the main tsunami wave crest, became evident in residual (after subtraction of predicted tides) cabled Canadian North-East Pacific Underwater Networked Experiments (NEPTUNE-Canada) bottom pressure recorders installed seaward of the southwest coast of Vancouver Island (THOMSON *et al.* 2011; RABINOVICH and EBLÉ 2015) and autonomous Deep-ocean and Assessment of Tsunami (DART) bottom pressure records (Fig. 1a) (RABINOVICH *et al.* 2013a). The same feature was subsequently identified by the authors in the records of most tide gauges along the coasts of British Columbia and Washington (Fig. 1b) (RABINOVICH *et al.* 2013a). Specifically, a 1- to 8.5-cm trough, negligibly small in comparison with the tides (typical

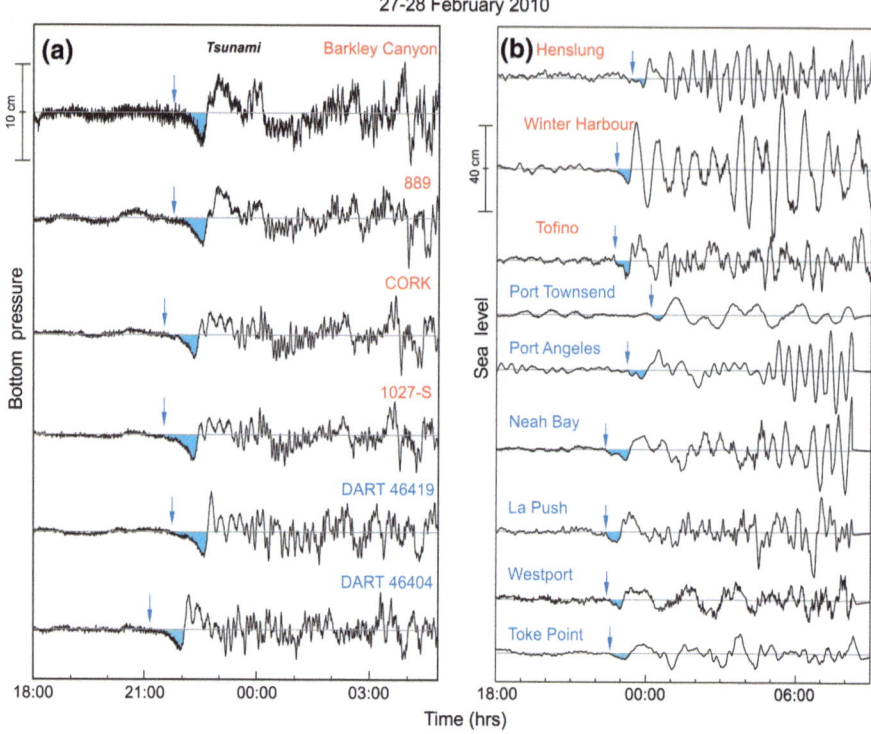

Figure 1
The 27 February 2010 Chile (Maule) tsunami extracted **a** from open-ocean NEPTUNE-Canada and DART bottom pressure records off the coast of southwest Vancouver Island, and **b** from coastal sea level records on the coast of British Columbia (BC) and Washington State (WA). The names of NEPTUNE and BC stations are denoted in *red*, while DART and WA stations are in *blue*. The time series were de-tided by removing predicted tides and high-pass filtered with 4-h Kaiser-Bessel window. *Blue shaded* area indicates the negative initial phase of incoming tsunami waves; *blue arrows* denote the arrival of the tsunami signal (modified from RABINOVICH et al. 2013a)

range of ~3 m), arrived ahead of the frontal major tsunami crest wave.

The presence of a trough preceding the main tsunami wave crest is conceptually not new. The Roman historian Ammianus Marcellinus provided a detailed account of the AD365 earthquake and tsunami that devastated Alexandria and the eastern Mediterranean. His vivid description includes tsunami generation followed by an initial drawdown of water that enticed thousands to their death (KELLY 2004). This and other historical accounts (e.g. VIANA-BAPTISTA and SOARES 2006; ATWATER et al. 2005), of course, refer to the well-understood leading trough introduced by the mechanism of tsunami generation. The initial dipole source structure oriented along the mainland coast with the subsidence zone on the continental side and the uplift zone on the oceanic side is typical for thrust fault earthquakes in major subduction zones of the ocean. Consequently, a positive tsunami wave propagates to the open ocean, while a negative wave moves toward the coast. Typical examples are the 2004 Sumatra tsunami (RABINOVICH and THOMSON 2007) or the 2011 East Japan (Tohoku) tsunami (SAITO et al. 2011; SONG et al. 2012).

What is relatively new, however, is introduction of a minor leading negative feature unrelated to the generation mechanism. The notion of such a leading small negative phase not associated with the tsunami source, eluded discernment as a unique phenomenon deserving of distinction. EBLÉ et al. (2012) identified the feature in DART and tide gauge records during the 2011 East Japan (Tohoku) event. WATADA et al. (2012, 2014) too identified the feature and went further by proposing a physical explanation. The same feature was again observed by SHEVCHENKO et al. (2013) at DART stations 21416 and 21419 near the Kuril Islands in the northwest Pacific during the 2010

Chile (Maule) tsunami. Still, the feature is most commonly attributed to shortcomings during data processing. For this reason, presentation of a leading small negative phase preceding a major wave crest throughout the Pacific Basin by RABINOVICH et al. (2013a) was met first with the reviewers' skepticism. To address the suggestion that this finding might be an artefact of analysis (e.g. associated with filtering of the records, THOMSON and EMERY 2014), the authors (RABINOVICH et al. 2013a) selected six DART stations and de-tided the data without applying any type of filter. A leading small trough remained a resultant feature in all records examined.

RABINOVICH et al. (2013a) proposed that the leading trough of the 2010 tsunami originated with a negative ocean bottom depression along the outer boundary of the source region near northern Chile. However, the National Earthquake Information Center (NEIC) and other models of the 2010 Chile (Maule) earthquake source area did not include such a depression. Moreover, further estimation based on the duration of the trough (≥ 40 min) indicated that if there were a correlation between the trough and such a depression, the extent would supposedly need to be on order several 100 km, i.e., size of the entire source area, a scale considered unrealistic. At the same time, preliminary analysis of DART observations during the 2011 East Japan (Tohoku) and the 2010 Chile (Maule) tsunamis by EBLÉ et al. (2012), MUNGOV et al. (2013), and RABINOVICH et al. (2013b) show the presence of the same leading phase feature unassociated with source. Specifically, it became clear that the negative phase observed during the 2010 Chile (Maule) tsunami was not related to specific properties of the source area but was determined to display a more general character likely associated with particular properties of tsunami physics. WATADA et al. (2012, 2014) and WATADA (2013) indicated that both for the 2010 Chile (Maule) and 2011 East Japan (Tohoku), the initial negative phase was not observed near the tsunami sources. Recent numerical simulations by TSAI et al. (2013), INAZU and SAITO (2013), WATADA (2013), and ALLGEYER and CUMMINS (2014) show that wave distortion and travel time delays between observed and modelled results are, at least in part, attributable to tsunami mass loading and water column stratification. Specifically, 1-dimensional computations by WATADA et al. (2014) for a spherical Earth model (PREM) with a 4-km ocean layer, which takes into account these effects, indicates that the leading negative phase is related primarily to the elasticity of the solid earth, as well as to the compressibility of seawater and to the gravity potential changes associated with the mass motion during tsunami propagation. WATADA et al. (2014) and ALLGEYER and CUMMINS (2014) show improved agreement between observed and synthetic distant tsunami waveforms, thereby highlighting the need for additional physics in shallow water modelling approaches to reproduce observed waveforms and more accurately forecast arrival times.

The purpose of the current study is to present observations of a leading negative phase to show the repeatability of the phenomenon during large and small events alike, both in the open ocean and on the coast, and to illustrate the hallmark signature despite scale. The events considered here: 2010 Chile (Maule), 2011 East Japan (Tohoku), 2012 Haida Gwaii British Columbia, and 2014 Chile (Iquique), have epicentres located in three distinctly different seismically active regions of the Pacific Ocean Basin (Fig. 2). All were recorded by a large number of

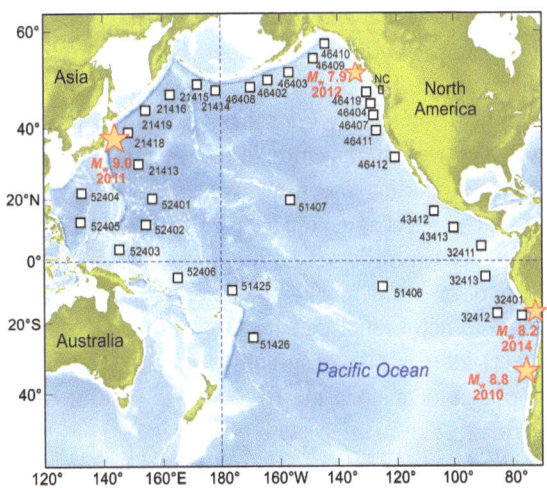

Figure 2
Map of the Pacific Ocean showing the locations of the $M_w = 8.8$ 2010 Chile (Maule), $M_w = 9.0$ 2011 East Japan (Tohoku), $M_w = 7.7$ 2012 Haida Gwaii, and $M_w = 8.2$ 2014 Chile (Iquique) earthquake epicentres (*stars*) and positions of the deep-ocean DART stations (*circles*). The *rectangle* labelled 'NC' near the Haida Gwaii epicentre denotes the location of the NEPTUNE-Canada geophysical cabled bottom stations

deep-ocean DART systems as well as by hundreds of coastal stations, distributed throughout the Pacific Ocean. The M_w 8.8 2010 Chile (Maule) earthquake occurred offshore south-central Chile and was the largest in the Southern Hemisphere since 1960 (DELOUIS et al. 2010). The resultant tsunami was recorded by more than 200 coastal tide gauges and by many deep-ocean instruments throughout the Pacific Ocean. The M_w 9.0 2011 East Japan (Tohoku) earthquake was centred offshore northeastern Honshu, Japan and was the strongest ever instrumentally recorded (SIMMONS et al. 2011; SAITO et al. 2011). Approximately, 250 coastal tide gauges and numerous bottom pressure gauges at autonomous and cabled observatory sites recorded the tsunami (e.g. SONG et al. 2012; SAITO et al. 2011; BORRERO and GREER 2013; MATSUMOTO and KANEDA 2013; FINE et al. 2013; RABINOVICH et al. 2013b). The M_w 7.7 2012 Haida Gwaii earthquake was the second strongest instrumentally recorded in Canadian history and was unique for this region in that it was a thrust fault event (CASSIDY et al. 2013). Many near-field and distant tide gauges, DARTs and NEPTUNE stations in operation at the time recorded the tsunami (FINE et al. 2015). The M_w 8.2 2014 Chile (Iquique) earthquake generated a tsunami offshore the northern coast of Chile that was observed at coastal tide gauge and DART locations throughout the Pacific Ocean. The complete 2014 Chile observation data set will be examined for the feature when all bottom pressure units that were in operation at the time are recovered by NDBC during station maintenance.

Our greatest attention is paid to the analysis of deep-ocean records, since, as previously mentioned, these have much higher precision, exhibit a higher signal to noise ratio, and are free from the distortion effects of coastal topography. Sited at depths of thousands of metres, the BPR detects pressure fluctuations of tsunamis, with wavelengths on order hundreds of kilometres, but effectively filter out lower period waves. The DART locations at which high-resolution (15-s) observations were obtained and used for analysis are listed in Table 1 and shown in Fig. 2. The position provided is that of the bottom pressure unit associated with each DART station. Changes in location may occur and are due to hardware redeployment during station maintenance operations. For investigative comparison and to complete a synoptic picture, we take advantage of coastal tide gauge observations that, for each event, are available from a great number of island and mainland sites. Results presented here show the consistent presence of a negative leading phase not associated with the source of tsunami generation.

2. Instrumentation and Data Processing

An extensive network of coastal tide gauges in the Pacific Basin include many that have been in continuous operation for more than 100 years along the Pacific coast of North America. Until the late twentieth century, observations were typically pen-and-paper analogue marigrams in which time and wave height resolutions were low. Since network inception, infrastructure has been upgraded, most recently to digital technology, to improve station to shore communication, robustness, and to provide higher precision and shorter time resolution than predecessor units. At the same time, basic sea level measurements provided by each tide gauge, as well as station locations, have remained constant. The higher sampling rate has proven particularly important in exposing localised tsunami effects across a suite of events. Measurements from tide gauges are reported every 1 min instead of previous 6-min intervals and are made available via a web service (ALLEN et al. 2008). Of particular significance, a backup or secondary unit is housed alongside the primary 1-min report unit for the purpose of providing higher (15-s) observations from a bottom pressure sensor. These highest resolution data are not automatically reported but, instead, are only available on demand upon direct, manual communication with tide gauge hardware. Fifteen-second data have proven especially important to the investigation of the spatial extent of the basin wide negative leading wave for all events considered and presented in the following sections. While 1-min data are suitable for confirming tsunami impact for warnings, the relative scale of the leading negative phase to that of the tides coupled with siting in complex environmental regimes makes recognition of the feature highlighted in this study challenging. Even in 15-s residual records, the feature is easy to

Table 1

Locations of bottom pressure recorder units that provided observations at the DART stations listed during the 2010 Chile (Maule), 2011 East Japan (Tohoku), 2012 Haida Gwaii, and 2014 Chile (Iquique) tsunamis

DART	2010 Chile (Maule)		2011 East Japan (Tohoku)		2012 Haida Gwaii		2014 Chile (Iquique)	
	Lat	Lon	Lat	Lon	Lat	Lon	Lat	Lon
21401	–	–	42.62	152.58	–	–	–	–
21413	30.52	152.12	30.52	152.12	30.52	152.11	30.52	152.11
21414	48.94	178.25	48.94	178.25	48.94	178.25	48.95	178.26
21415	50.18	171.85	50.18	171.85	50.18	171.00	–	–
21416	48.04	163.49	48.05	163.51	48.05	163.51	–	–
21418	38.71	148.69	38.71	148.69	38.71	148.69	–	–
21419	44.30	155.74	44.46	155.74	44.46	155.74	44.46	155.74
32401	–	–	−19.30	−74.75	–	–	−20.47	−73.43
32411	4.92	−90.69	4.94	−90.65	–	–	5.01	−90.84
32412	−17.99	−86.39	−17.99	−86.39	–	–	−17.98	−86.34
32413	–	–	7.40	−93.50	–	–	−7.40	−93.50
43412	16.07	−107.00	16.03	−107.00	–	–	16.07	−107.00
43413	10.84	−100.08	10.84	−100.08	–	–	10.84	−100.14
46402	51.07	−164.01	51.07	−164.01	51.07	−164.01	51.07	−164.02
46403	52.64	−156.93	52.65	−156.92	52.65	−156.92	–	–
46404	45.86	−128.77	45.86	−128.77	45.86	−128.77	–	–
46407	42.59	−128.90	42.59	−128.90	42.59	−128.90	42.66	−128.81
46408	49.62	−169.31	49.62	−169.31	49.62	−169.31	49.67	−169.89
46409	55.30	−148.49	55.30	−148.49	55.30	−148.49	–	–
46410	57.63	−143.80	57.63	−143.80	57.63	−143.80	–	–
46411	39.33	−127.01	39.32	−126.99	39.32	−126.99	39.35	−127.02
46412	32.46	−120.56	32.46	−120.56	32.46	−120.56	32.46	−120.57
46419	48.48	−129.36	48.76	−129.61	48.76	−129.61	48.77	−129.63
51406	−8.49	−125.02	−8.48	−125.03	–	–	–	–
51407	19.64	−156.52	19.62	−156.51	19.62	−156.51	19.59	−156.59
51425	−9.50	−176.25	−9.50	−176.23	−9.50	−176.23	−9.51	−176.24
51426	−23.01	−168.11	−22.99	−168.10	–	–	–	–
52401	19.26	155.76	–	–	19.25	155.78	19.26	155.77
52402	11.74	154.22	11.88	154.11	11.88	154.11	11.87	154.04
52403	5.00	145.60	4.03	145.60	–	–	4.05	145.59
52404	–	–	–	–	20.94	132.30	20.79	132.34
52405	12.88	132.33	12.88	132.33	12.88	132.33	12.99	132.18
52406	–	–	−5.30	165.01	–	–	−5.29	165.00
54401	−33.00	−172.99	−33.01	−172.99	–	–	–	–
55012	–	–	−15.80	158.40	–	–	–	–
55013	–	–	−46.67	161.00	–	–	–	–
55015	−46.92	160.56	–	–	–	–	–	–
55023	–	–	−14.80	153.58	–	–	–	–

discount as a filtering effect or difficult to discern if not looking specifically for it.

In the deep ocean, a network of several tens of DART systems were in operation throughout the Pacific Ocean during the four events presented here (MOFJELD 2009; MUNGOV et al. 2013). Developed and maintained by the U.S. National Oceanographic and Atmospheric Administration (NOAA), DART technology was designed to capture direct observations of tsunami waves where signals are free from distorting coastal and topographic effects for research and tsunami warning applications (e.g. USLU et al. 2011; WEI et al. 2013; RABINOVICH and EBLÉ 2015). The system measures and records temperature and absolute bottom pressure in terms of oscillating crystal frequency counts at a sampling interval $\Delta t = 15$ s. In normal or standard monitoring mode, one message encoded with four 15-s observations is communicated from the bottom pressure recorder each hour to the surface buoy where it is stored until six messages are collected. Four times each day, a 6-h packet is then telemetered to shore as a measure of system health. When tsunami detection criteria are met, specifically a rapid change of pressure from that predicted across two samples, the system immediately transmits "raw" 15-s data for several minutes and then switches transmissions to 1-min averages until what is known as event mode terminates, approximately 4-h later (MOFJELD, 2009). All 15-s pressure and temperature data, including those transmitted in real-time during an event, are stored on flash memory internal to the bottom pressure recorder for download following instrument recovery one to 2 years after deployment (MUNGOV et al. 2013). More detailed descriptions of the DART technology are provided by EBLÉ and GONZÁLEZ (1991) and MOFJELD (2009).

Time series of tsunami observations from coastal tide gauges and deep-ocean DART systems during all events investigated as part of this study were processed in the same manner. High-resolution, 15-s, time series observations retrieved from the internal flash storage cartridge of 24 DART bottom pressure recorders and manual downloaded by remote communication with many coastal tide gauges are weighted most heavily in our investigation. An approach based exclusively on tidal analyses as described by PARKER (2007) was taken so as to avoid introducing filtering or other processing artefact into the residual time series. Such artefacts (Fig. 3) are likely magnified precisely at the location where the leading negative phase might be expected. As described by MUNGOV et al. (2013), a customised version of the FOREMAN et al. (2009) tidal harmonic analyses package was developed to allow greater flexibility in accommodating time series longer than 1 year. Based on FOREMAN et al. (2009), non-linear time variations in the astronomical parameters are considered such that up to 69 tidal harmonic constituents may be specified for tidal analysis. The standard limit of 39 tidal harmonic constituents as specified in the widely used traditional FOREMAN (1977) method does not adequately account for tidal contributions.

Input time series were first quality controlled for gaps or outliers and de-trended before analyses for tidal harmonic constituents at respective locations. Selection of tidal constituents was, in general, based on the Rayleigh criterion (FOREMAN 1977; FOREMAN et al. 2009) but since the purpose of this investigation is not tidal in nature, a greater number of constituents were solved for than indicated when series length might otherwise be considered insufficient (MUNGOV et al. 2013). At each location, predicted tides computed at every time step were subtracted from de-trended observations to generate residual time series. It should be noted that trends in the records are typically of a long time scale, due primarily to instrument equilibration when first deployed so de-trending does not alter or otherwise affect the tsunami frequency band. Following the work of MUNGOV et al. (2013), spectral energy reduction of the main diurnal and semi-diurnal tidal frequency bands provides the

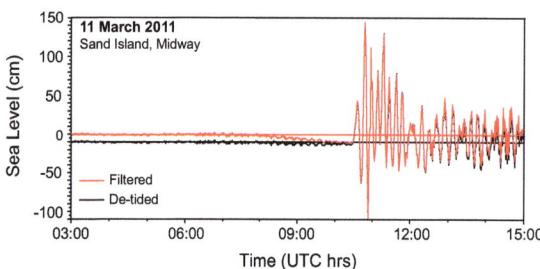

Figure 3
Example of signal distortion after filtering Sand Island, Midway tide gauge observations. The *red line* is the filtered time series and the *black line* shows the same original observed record after de-tiding

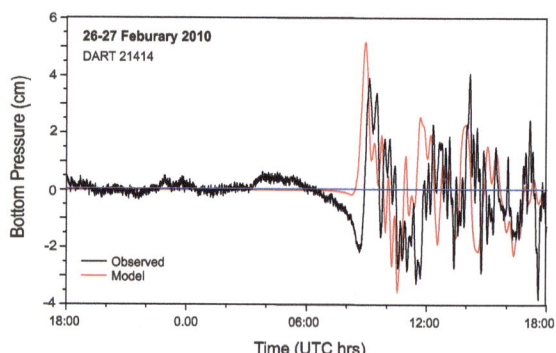

Figure 4
Comparison of the time series observed at DART 21414 during the 2010 Chile (Maule) tsunami with the modelled series generated by the shallow water wave MOST model for the same location

Table 2

The leading negative phase (LNP) determined from observations at each of the bottom pressure recorder units at the identified DART stations during each of four tsunami events. DARTs are sorted by distance from source (Δ Source). Earthquake epicentre is provided in the header

2010 Chile (Maule) 136.122°N 72.898°W			2011 East Japan 38.297°N 142.373°E			2012 Haida Gwaii 52.788°N 132.101°W			2014 Chile (Iquique) 19.612°S 70.769°W		
DART	Δ Source (km)	LNP	DART	Δ Source (km)	LNP	DART	Δ Source (km)	LNP	DART	Δ Source (km)	LNP
32412	2413	−0.0009	21418	553	0	46419	481	0	32401	294	0.0000
32411	4908	−0.0035	21401	989	0	46404	808	0	32412	1651	0.0000
43413	5929	−0.0074	21413	1243	0	46410	950	0	32413	2802	0.0000
51406	6086	−0.0131	21419	1308	0	46409	1107	−0.001	32411	3500	0.0000
43412	6801	−0.0144	21416	2021	0	46407	1159	0	43413	4660	NaN
54401	8728	−0.0026	21415	2674	0	46411	1547	0	43412	5594	−0.0001
51426	8981	−0.0033	52405	2986	NaN	46403	1669	−0.0003	46412	7830	−0.0012
46412	9071	−0.0209	21414	3092	−0.0026	46402	2185	−0.0023	46411	8761	−0.0015
55015	9379	0	52402	3150	−0.0025	46412	2442	−0.0013	46407	9101	−0.0014
46411	10029	−0.014	52403	3809	−0.0036	46408	2594	−0.0042	46419	9551	0.0000
46407	10406	−0.0166	46408	4000	−0.0059	21414	3451	−0.0039	51407	10313	−0.0022
51425	10572	−0.0043	46402	4373	−0.0036	21415	3859	−0.003	51425	11254	0.0000
46404	10668	−0.0171	46403	4847	−0.0046	51407	4241	−0.0025	46402	11904	−0.0023
51407	10726	−0.0111	46409	5355	−0.003	21416	4452	−0.0025	46408	12295	−0.0025
46419	10911	−0.0144	52406	5363	−0.0025	21419	5171	NaN	21414	13153	−0.0001
46410	12305	−0.0147	46410	5593	−0.0025	21418	6031	NaN	52406	13334	−0.0010
46409	12402	−0.0197	55023	5993	−0.0023	21413	6444	−0.0007	21419	14930	−0.0013
46403	12733	−0.0193	51407	6189	−0.0101	52401	7127	−0.0007	52401	15162	−0.0021
46402	13096	−0.0059	55012	6218	NaN	52402	7874	−0.0007	52402	15170	NaN
46408	13384	−0.0229	46419	6807	−0.0084	51425	8072	−0.0007	21413	15566	−0.0020
21414	14224	−0.024	51425	6816	−0.0056	52404	8545	−0.0005	52403	15726	−0.0019
52402	14591	NaN	46404	7012	−0.0082	52405	9248	−0.0003	52405	17475	−0.0013
21415	14705	−0.0262	46407	7178	−0.0086	21401	–	–	52404	17617	−0.0014
52403	14791	−0.0065	46411	7501	−0.0076	32401	–	–	21401	–	–
52401	14936	−0.0068	46412	8409	−0.0105	32411	–	–	21415	–	–
21416	15283	−0.0268	51426	8529	−0.0095	32412	–	–	21416	–	–
21413	15826	−0.0208	54401	9146	NaN	32413	–	–	21418	–	–
21419	15846	−0.0255	55013	9592	0	43412	–	–	46403	–	–
21418	16363	NaN	43412	10626	−0.0101	43413	–	–	46404	–	–
52405	16411	−0.0039	51406	10817	−0.0088	51406	–	–	46409	–	–
21401	–	–	43413	11565	−0.0106	51426	–	–	46410	–	–
32401	–	–	32413	12346	−0.0111	52403	–	–	51406	–	–
32413	–	–	32411	12760	−0.0096	52406	–	–	51426	–	–
52404	–	–	32412	14805	−0.0108	54401	–	–	54401	–	–
52406	–	–	32401	15872	−0.0127	55012	–	–	55012	–	–
55012	–	–	52401	–	–	55013	–	–	55013	–	–
55013	–	–	52404	–	–	55023	–	–	55023	–	–
55023	–	–									

primary measure of residual series quality (e.g. RABINOVICH et al. 2011). Data processing is conducted with great care to sharply cut the tidal frequencies that dominate each observational time series record. A thorough discussion of data processing is included in MUNGOV et al. (2013).

3. Leading Negative Phase

Data recorded by deep-ocean DART and coastal tide gauge stations during each of four events (2010–2014) include a small but distinguishable leading negative phase just prior to arrival of the first

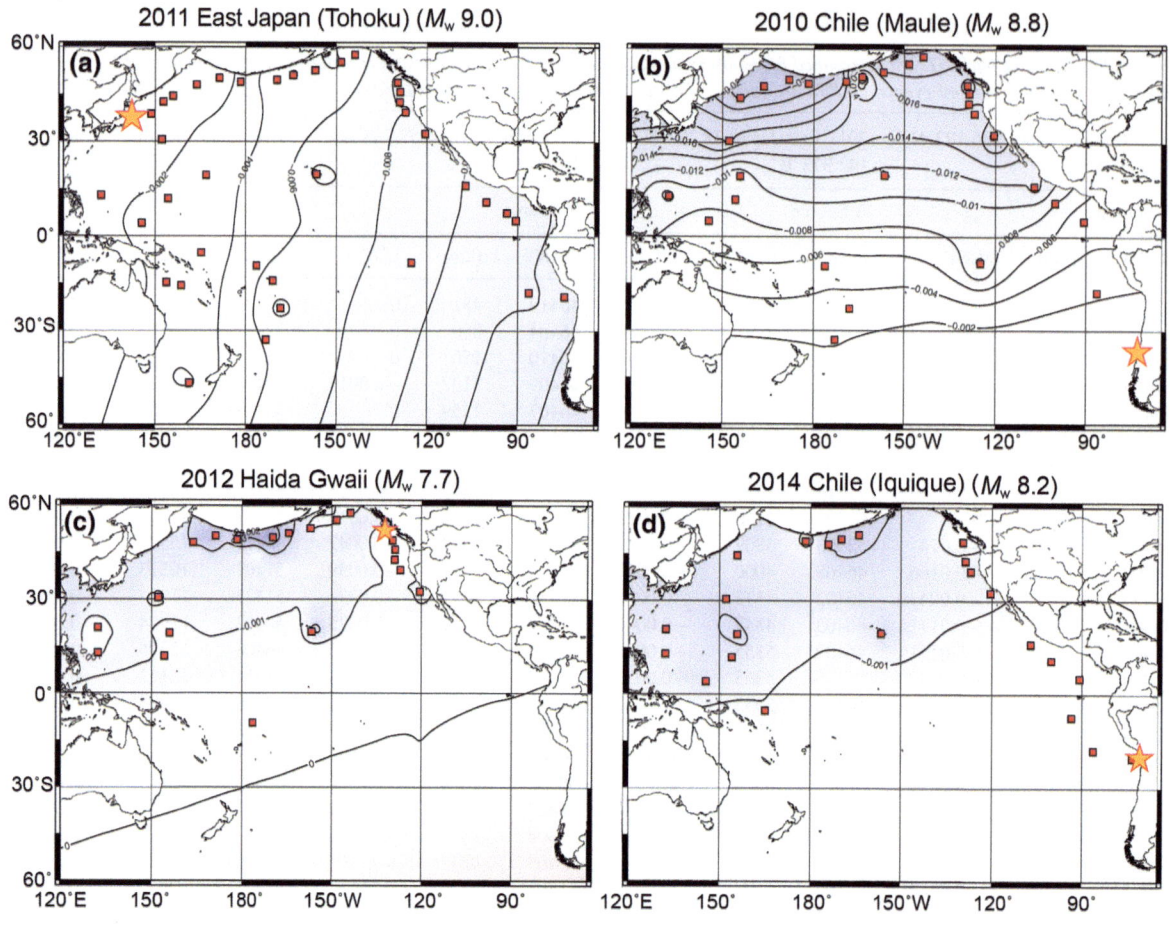

Figure 5
Contours of the minimum negative wave phase observed at the deep-ocean sites marked by red squares during the **a** 2011 East Japan (Tohoku) tsunami **b** 2010 Chile (Maule) tsunami **c** 2012 Haida Gwaii tsunami, and **d** 2014 Chile (Iquique) tsunami. *Colour* gradation from the *pale red shade* to the *purple hue* in all frames visually shows the intensification of the leading negative phase with increasing propagation distance from each respective source. Contour units are metres

tsunami wave crest in most time series examined. The leading negative phase is typically not reproduced by tsunami models commonly used for forecasting but is well defined in the 2010–2014 observations (Fig. 4). The signature feature was recorded throughout the Pacific at locations both nearby and far away from each respective source of generation. The shape tends to be consistent from record to record across all events: a gradual downward descending concave profile followed by an abrupt increase in sea level. Of significance, the negative phase intensifies with tsunami wave propagation, deepening and becoming more pronounced as the distance between the tsunami and generating source increases, regardless of the specific mechanism that initiated ground motion.

Contours of the negative phase minima observed in 15-s deep-ocean DART records, listed in Table 2, illustrate this effect (Fig. 5). The pale red shading marks a featureless region nearest the source that gradually deepens to purple along the far reaches of the basin relative to each event, where the feature is most pronounced. From event to event, the contour patterns (Fig. 5) are consistent with one another in that the feature is not present, or at least not distinguishable, nearest the tsunami generating source (in agreement with the findings of FINE et al. 2015 and WATADA et al. 2014). The deepest shade of purple appears along coasts furthest from the three largest events; 2010 Chile (Maule) (Fig. 5b), 2011 East Japan (Tohoku) (Fig. 5a), and 2014 Chile (Iquique)

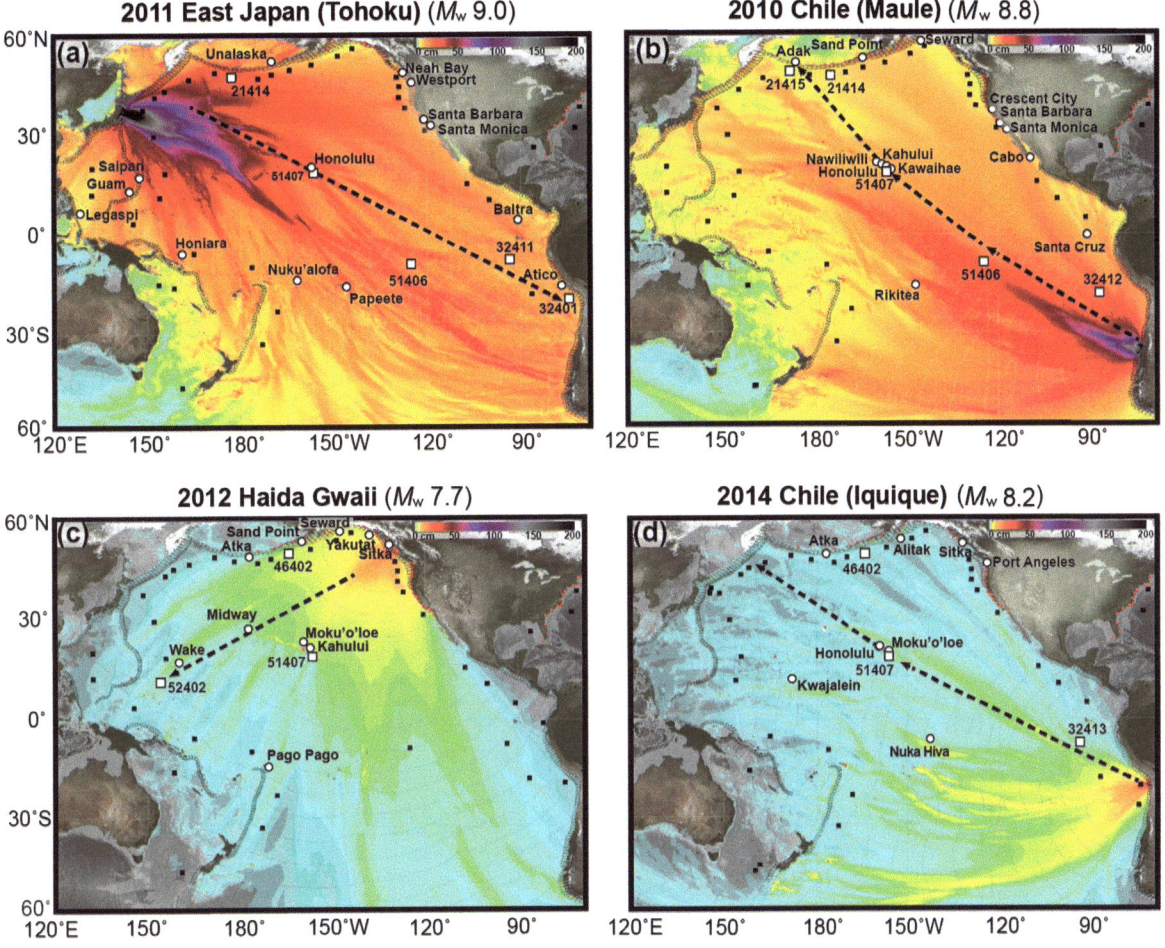

Figure 6
Modeled maximum tsunami wave amplitude for **a** 2011 East Japan (Tohoku) tsunami **b** 2010 Chile (Maule) tsunami **c** 2012 Haida Gwaii tsunami, and **d** 2014 Chile (Iquique) tsunami. Deep-ocean DART stations are identified by *black squares*. *Smaller red squares* show the locations at which tsunami forecast model grids provide operational forecasts. *Larger white squares* identify the DART sites at which observed leading negative phases in deep-ocean observations along the *dashed transect line* are shown in Fig. 7. Observations of the leading negative phase at the tide gauge stations marked by the *white circles* appear in Figs. 8 thru 11

(Fig. 5d). It is noted here that while no 15-s deep-ocean observations were retrieved in the vicinity of the 2010 Chile (Maule) source, the conclusion that an initial negative phase was not a near-source feature is supported by the observations of WATADA et al. (2012, 2014), WATADA (2013) and ALLGEYER and CUMMINS (2014) and by the absence of such a feature in coastal sea level observations in the near-field area. Contours of the feature in observations recorded during 2012 Haida Gwaii (Fig. 5c) show the deepest colouring along the Aleutians to the north of the generation source. Although the feature is identifiable in most of the available records (Table 2), high-resolution (15-s) data from 9 of 10 possible DART stations furthest from the 2012 Haida Gwaii source in the South Pacific were not available so are not represented in results.

Comparison of maximum amplitude plots (Fig. 6) with leading negative phase contours (Fig. 5) show that, for each event, as maximum amplitudes decrease with distance from the source, the leading negative phase deepens. Instruments in the deep ocean and along the coasts nearest the source measured the largest tsunami waves yet the negative leading phase is most pronounced furthest from each respective tsunami source (Table 2; Figs. 5, 6). The

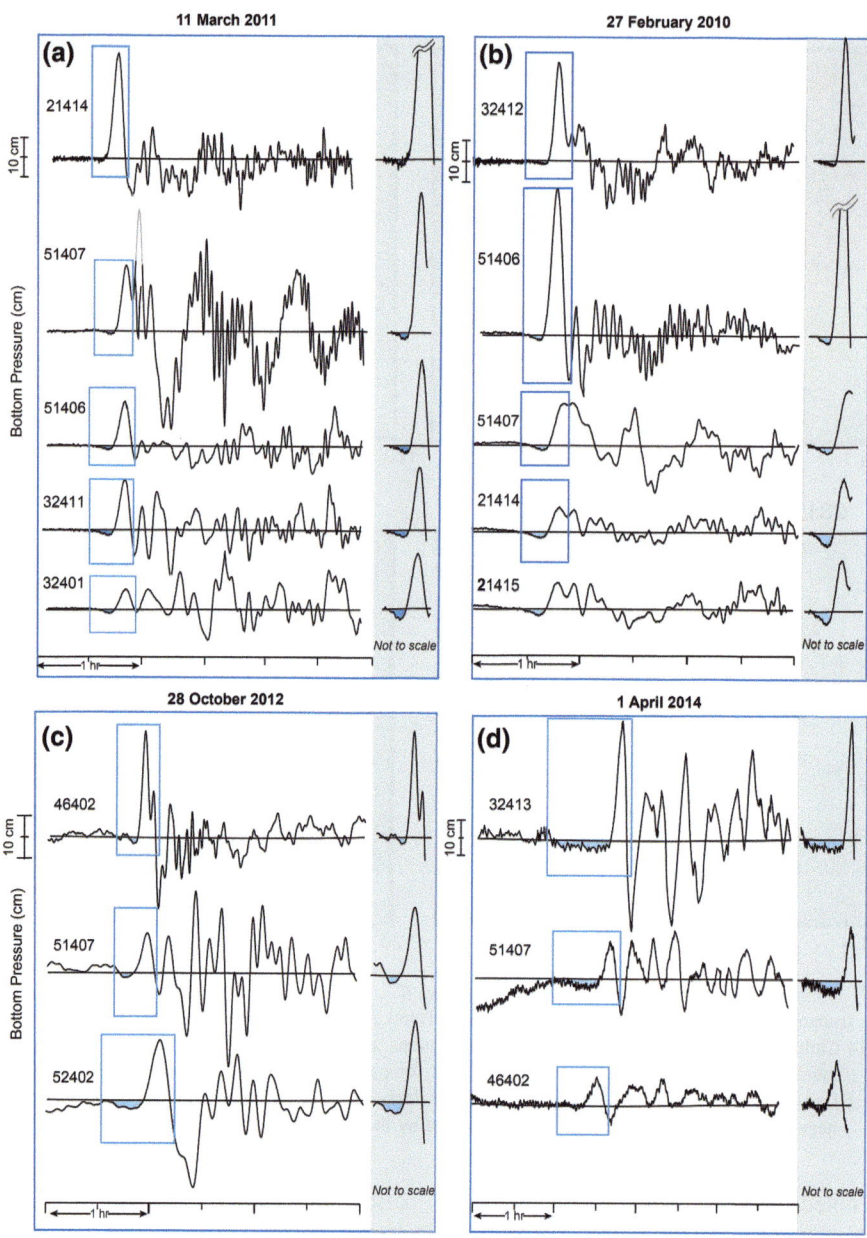

Figure 7
The tsunami extracted in DART records during **a** the 2011 East Japan (Tohoku), **b** the 2010 Chile (Maule), **c** the 2012 Haida Gwaii, and 2014 Chile (Iquique) tsunami events. The time series were de-tided by removing predicted tides. The *Blue shaded area* highlights the leading negative phase marking arrival of tsunami waves

leading negative phase is apparent in 27 of 30 available 2010 Chile (Maule) observations, in 24 of the 35 available observations recorded during the 2011 East Japan (Tohoku) tsunami, and in 15 of 23 high-resolution observations recorded during the 2014 Chile (Iquique) tsunami and recovered to date

(Table 2). Observations recorded during the smallest, 2012 Haida Gwaii event, show a less pronounced leading negative phase than those observed during the 2010 Chile (Maule), 2011 Japan East Japan (Tohoku), and 2014 Chile (Iquique). Still, following greater data processing care and scrutiny, a leading

27 - 28 February 2010

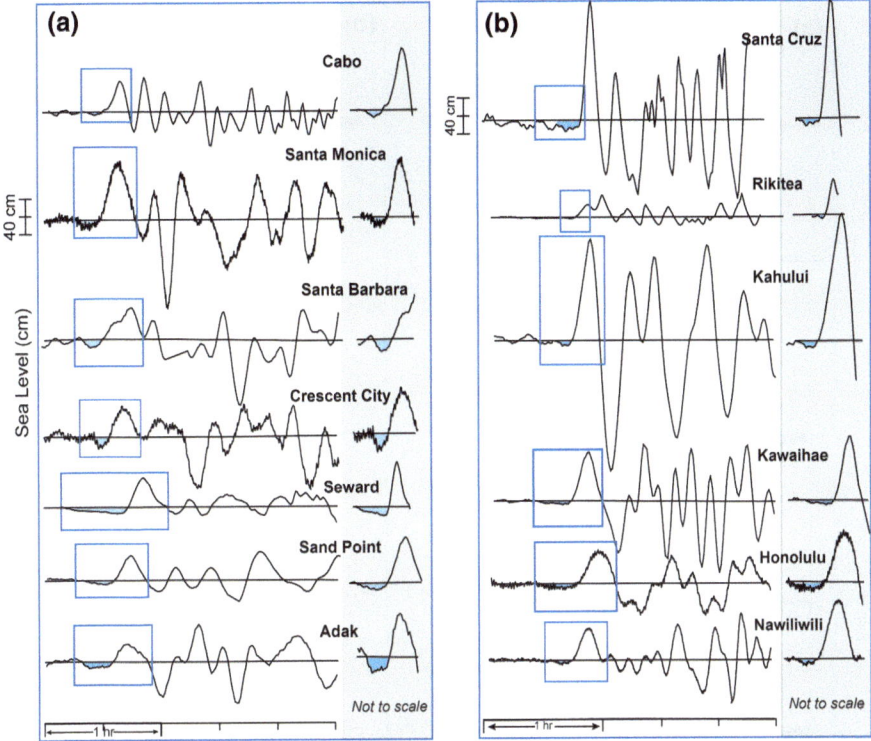

Figure 8
The leading negative phase in de-tided observations recorded during the 2010 Chile (Maule) tsunami by the tide gauges at locations marked by *white circles* in Fig. 6b. The section of signal that includes the leading negative phase is shown at a larger scale within the *grey shaded plot* area at **a** continental coastal locations and **b** Pacific island locations. Plot scales of the two frames are independent of one another but are consistent within each frame

negative phase was clearly discernible in 15 of 22 available DART high-resolution observations recorded during this event.

Observations at DART stations along the dashed transect lines that are superimposed on the maximum amplitude plots of Fig. 6 provide graphical illustration of the feature's signature in deep-ocean records and the deepening effect with distance (Fig. 7). Transects were chosen with data availability in mind. A path that included the greatest number of DART observations and provided observations of the propagating wave was selected. The five time series shown for each of the two largest events, 2011 East Japan (Tohoku) (Fig. 7a) and 2010 Chile (Maule) (Fig. 7b), show deepening of the phase as the tsunami propagates along the respective transects. The three time series presented for 2012 Haida Gwaii (Fig. 7c) and 2014 Chile (Iquique) (Fig. 7d) each show the leading negative phase clearly. The deepening effect is apparent but is not as evident as in Fig. 7a and b. The lesser amount of available high-resolution data effectively limited the number of DART candidate sites.

Great numbers of coastal tide gauges throughout the Pacific Basin recorded the tsunamis generated by all four events. The leading negative phase is discernible in many de-tided records but variations in maximum amplitude and site specific background noise associated with local seiching and infragravity waves make graphical comparison of the scale challenging. Coastal tide gauge observations recorded during 2010 Chile (Maule), 2011 East Japan (Tohoku), 2012 Haida Gwaii, and 2014 Chile (Iquique) are shown in Figs. 8, 9, 10, and 11. Stations at which observations demonstrated the leading negative phase feature at a scale conducive to grouping for comparative purposes were preferentially selected for

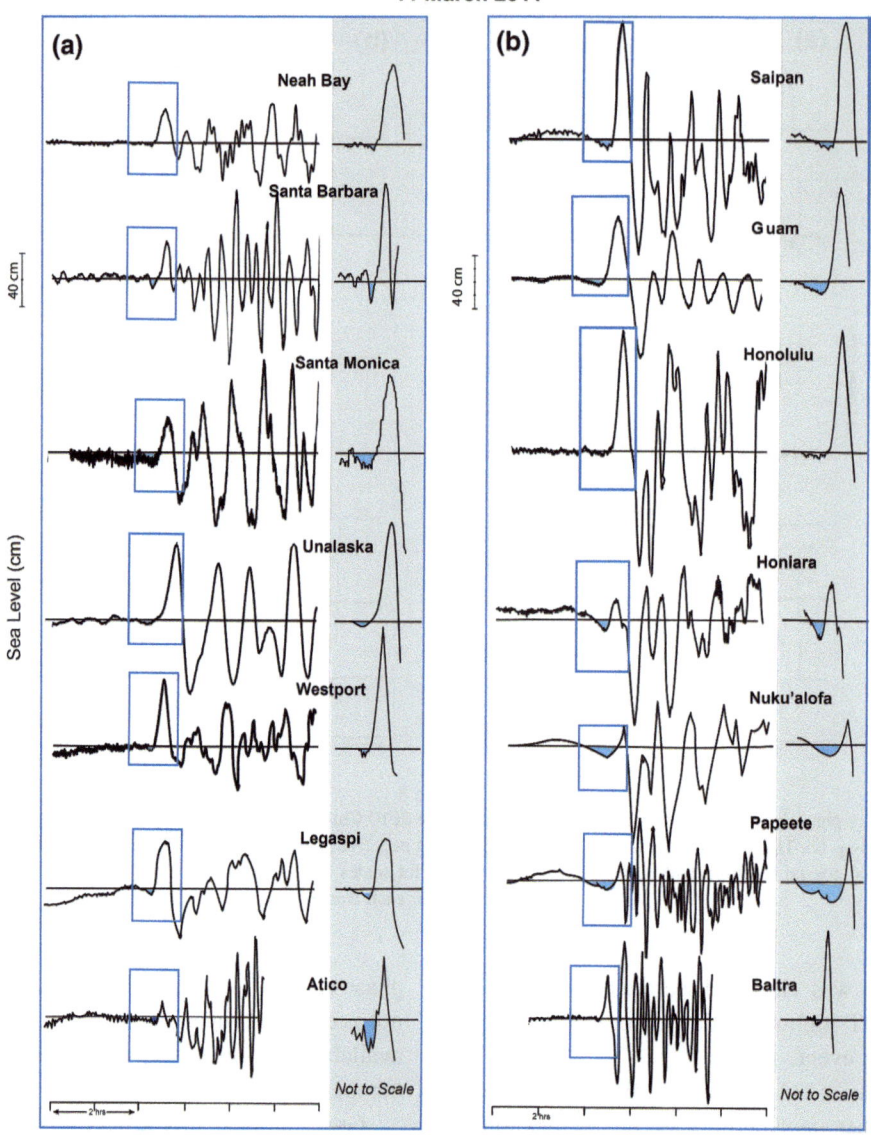

Figure 9
The leading negative phase in de-tided observations recorded during the 2011 East Japan (Tohoku) tsunami by the tide gauges at locations marked by *white circles* in Fig. 6a. The section of signal that includes the leading negative phase is shown at a larger scale within the *grey shaded* plot area at **a** continental coastal locations and **b** Pacific island locations. Plot scales of the two frames are independent of one another but are consistent within each frame

display. Grouped observations, those within each frame, are plotted at the same vertical and horizontal scales in order of increasing distance from the respective source. The shaded region to the right of each frame provides a close-up look at the region bounded by the blue inset box.

During the 2010 Chile (Maule) tsunami, coastal observations along the Mexico, U.S. West Coast, and Alaska coastlines (Fig. 8a) and at island stations, most in Hawaii (Fig. 8b), indicate a leading negative phase that exhibits the same signature as seen in deep-ocean records (Fig. 7). The observations in Fig. 8 show the presence of the leading negative phase but do not clearly display the expected deepening with distance from the source. De-tided coastal observations recorded during the 2011 East Japan

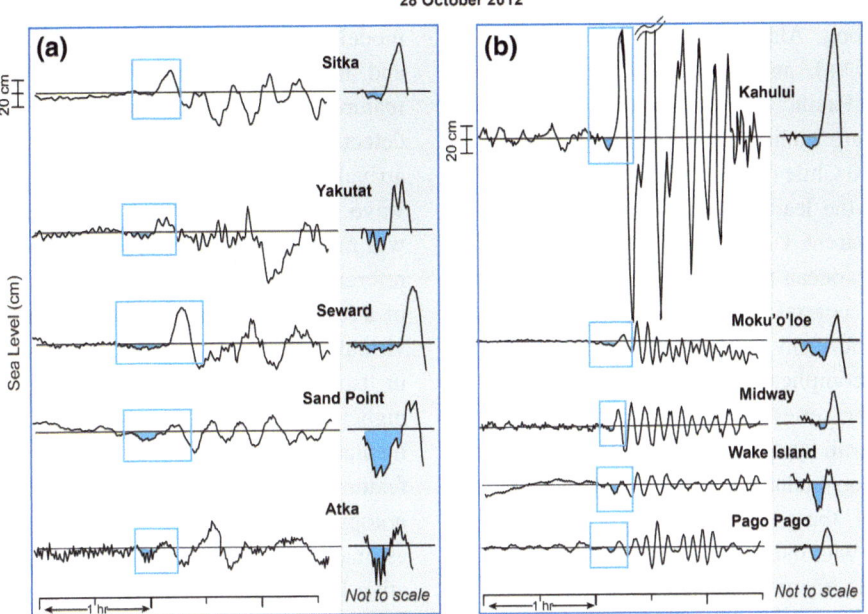

Figure 10
The leading negative phase in de-tided observations recorded during the 2012 Haida Gwaii tsunami by the tide gauges at locations marked by *white circles* in Fig. 6c. The section of signal that includes the leading negative phase is shown at a larger scale within the *grey shaded* plot area at **a** continental coastal locations and **b** Pacific island locations. Plot scales of the two frames are independent of one another but are consistent within each frame

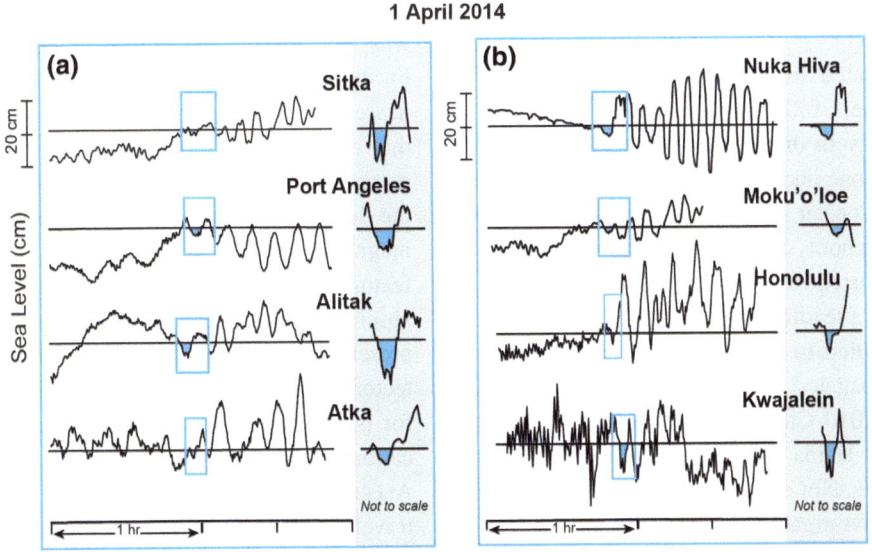

Figure 11
The leading negative phase in de-tided observations recorded during the 2010 Chile (Iquique) tsunami by the tide gauges at locations marked by *white circles* in Fig. 6d. The section of signal that includes the leading negative phase is shown at a larger scale within the *grey shaded* plot area at **a** continental coastal locations and **b** Pacific island locations. Plot scales of the two frames are independent of one another but are consistent within each frame

(Tohoku) tsunami at seven locations along the U.S. west coast, along Alaska, Peru, and Philippines coastlines (Fig. 9a), and at seven island locations throughout the Pacific (Fig. 9b) also clearly reveal the same signature leading negative phase but, as was the case for 2010 Chile (Maule) observations (Fig. 8), a deepening of the leading negative phase with distance is not apparent. Unlike DART systems that are sited in the deep ocean to provide the purest possible tsunami signal, coastal tide gauges are installed in complex near-shore environments where site-specific processes may complicate the signal.

Residual time series recorded during the smaller 2012 Haida Gwaii and 2014 Chile (Maule) events (Figs. 10, 11) too include a leading negative phase. During both of these smaller events, the leading negative phase was most easily discernible in north Pacific coastal records, notably along Alaska (Fig. 10a) and at the island stations shown in Figs. 10b, 11b. For these events, a low signal to noise ratio is common and makes discernment of the leading negative phase more difficult than in records with a strong tsunami signal [eq. Kahului (Fig. 10)].

4. Discussion and Conclusions

The accelerated instrumentation of coastal and deep-ocean waters following the 2004 Sumatra tsunami led directly to discovery of a feature that has, until recently, gone undetected. Our analyses of high-resolution deep-ocean and coastal tide gauge datasets collected throughout the Pacific Basin during four events that occurred in 2010–2014 consistently show the presence of a leading negative phase. The feature is clearly identified in 15-s DART records of tsunamis that propagated across the Pacific Ocean following the 2010 Chile (Maule), 2011 East Japan (Tohoku), 2012 Haida Gwaii, and 2014 Chile (Iquique) earthquakes. Not only is this feature present, but observations investigated during each of the four events show an increase in effect with increasing distance from the source of tsunami generation. The feature has likely been previously missed in plots due to axis scaling, has been masked by low resolution sampling, or has been discounted as an artefact of filtering. The fact that this leading negative phase appears more common than not has implications for modelling of tsunami waves for real-time forecasting and in research applications. Normally, the main feature of incoming tsunami waves, simplifying their detection, is a rather sharp ("frontal") first wave arrival followed by an abrupt change in the in situ wave field properties. The observation of this feature was made a posteriori here and by other investigators referenced above. The very smooth initial signature of a leading negative phase has, until recently, gone virtually undetected and is not typically reproduced in tsunami model results. Careful processing of higher-resolution data from more and more instrumentation shows that routine identification of this feature is possible. Improved ocean sensing technologies coming online are expected to enable robust estimates of the observed effect, in this case, a negative leading phase not associated with the source of tsunami generation and increase the probability of discovering additional information about the properties of propagating tsunami waves.

Following careful analyses and examination, comparison of the records from various coastal and deep-sea stations revealed that the observed small but consistent negative phase was specifically related to the leading edge of the incoming tsunami waves and was not associated with background oscillations. Further, the absence of the leading negative phase nearest each of the four sources lead us to conclude that the generating mechanisms are not in any way responsible for the observed feature; the presence of the feature appears independent of tsunami source magnitude and location. The magnification of the feature away from the source coupled with the absence of the feature near the source in records presented here supports an explanation other than tectonic. Inclusion of physics in models to account for tsunami mass loading and water column stratification has been shown by Watada et al. (2014) and Allgeyer and Cummins (2014) to improve modelled travel time accuracy and to more closely reproduce observed tsunami waveforms. The complement of empirical results presented here with the numerical results of Watada et al. (2014) show clearly that a leading negative phase is a real and identifiable feature that should not be summarily dismissed nor attributed solely to source characteristics.

Acknowledgments

The authors thank the National Data Buoy Center and the National Ocean Service for their dedication in securing quality observations, the National Tsunami Warning Center for providing many coastal data sets, and Lindsey Wright and NOS Pacific, especially Caleb Gostnell, for the many hours invested in calling tide gauge backup units and manually downloading the 15-s data that are the basis of our investigations. We also thank Karen Birchfield for her assistance with graphics and the reviewers for taking the time to provide candid and greatly appreciated comments for manuscript improvement. The contribution of Alexander Rabinovich was partially funded under NOAA/PMEL Contract Number WE-133R-13-SE-1659 and by the Russian Science Foundation Grant 14-50-00095. This is PMEL contribution 4099.

REFERENCES

ALLEN, A. L., DONOHO, N. A., DUNCAN, S. A., GILL, S. K., MCGRATH, C. R., MEYER, R. S., and SAMANT, M.R.N. (2008), *NOAA's National Ocean Service Supports Tsunami Detection and Warning through Operation of Coastal Tide Stations.* Solutions to Coastal Disasters *2008*: pp. 1–12; doi:10.1061/40978(313)1

ALLGEYER, S., and CUMMINS, P. (2014), *Numerical tsunami simulation including elastic loading and seawater density stratification*, Geohys. Res. Lett., *41*, 2368-2375; doi:10.1002/2014GL059348.

ATWATER, B. F., MUSUMI-ROKKAKU, S., SATAKE, K., YOSHINOBU, T., KAZUE, U., and YAMAGUCHI, D. K. (2005). *The Orphan Tsunami of 1700 – Japanese Clues to a Parent Earthquake in North America*. U.S. Geological Survey Professional Paper 1707. United States Geological Survey-University of Washington Press. 98 pp. ISBN 978-0295985350.

BORRERO, J. C., and GREER, S. D. (2013), *Comparison of the 2010 Chile and 2011 Japan tsunamis in the far field*, Pure Appl. Geophys., *170*, 1249-1274; doi:10.1007/s00024-012-0559-4.

CASSIDY, J. F., ROGERS, G. C., and HYNDMAN, R. D. (2013), *An overview of the October 28, 2012 Mw 7.7 earthquake in Haida Gwaii, Canada: a tsunamigenic thrust event along a predominantly strike-slip margin*, Pure Appl. Geophys., *171*, 3457–3465; doi:10.1007/s00024-014-0775-1.

DELOUIS, B., NOCQUET, J. M., and VALLE, E. M. (2010), *Slip distribution of the February 27, 2010 $M_w = 8.8$ Maule earthquake, central Chile, from static and high-rate GPS, InSAR, and broadband teleseismic data*, Geophys. Res. Lett., *37*, L17305, doi:10.1029/2009GL043899.

DUNBAR, P. K., STROKER, K. J., BROCKO, V. R., VARNER, J. D., MCLEAN, S. J., TAYLOR, L. A., EAKINS, B. W., CARIGNAN, K. S., and WARNKEN, R. R. (2008), *Long-term tsunami data archive supports tsunami forecast, warning, research and mitigation*, Pure Appl. Geophys. *165*, 2275-2291.

EBLÉ, M. C., and GONZÁLEZ, F. I. (1991), *Deep-ocean bottom pressure measurements in the Northeast Pacific*, J. Atmos. Ocean. Tech. *8*(2), 221–233.

EBLÉ, M. C., TITOV, V., DENBO, D., MOORE, C., MUNGOV, G., and BOUCHARD, R. (2011), *Signal-to-noise ratio and the isolation of the 11 March 2011 Tohoku tsunami in deep-ocean tsunameter records*, OCEANS '11 MTS/IEEE KONA, http://www.oceans11mtsieeekona.org/

EBLÉ, M., MUNGOV, G., RABINOVICH, A., HARRIS, E., TITOV, V. (2012), *Spatial and Temporal Characterization of the 11 March 2011 Tsunami*. AGU Fall Meeting, San Francisco, 3-7 December 2012, Abstract NH11C-1563.

FINE, I. V., KULIKOV, E. A., and CHERNIAWSKY, J. Y. (2013), *Japan's 2011 tsunami: Characteristics of wave propagation from observations and numerical modelling*. Pure Appl. Geophys., *170*, 1295-1307; doi:10.1007/s00024-012-0555-8.

FINE, I. V., CHERNIAWSKY, J. Y., THOMSON, R. E., RABINOVICH, A. B., and KRASSOVSKI, M. V. (2015), *Observations and numerical modeling of the 2012 Haida Gwaii Tsunami off the coast of British Columbia*, Pure Appl. Geophys. *172* (3-4), 699-718; doi:10.1007/s00024-014-1012-7.

FOREMAN, M. G. G. (1977, revised 2004). *Manual for Tidal Heights Analysis and Prediction*. Pacific Marine Science Report. 77-10. Institute of Ocean Sciences, Patricia Bay, 58 pp. http://www.pac.dfo-mpo.gc.ca/science/oceans/tidal-marees/index-eng.htm.

FOREMAN, M. G. G., CHERNIAWSKY, J. Y., and BALLANTYNE, V. A. (2009), *Versatile Harmonic Tidal Analysis: Improvements and Applications*. J. Atmos. Oceanic Technol. *26*, 806–817. doi:http://dx.doi.org/10.1175/2008JTECHO615.1

INAZU, D., and SAITO, T. (2013), *Simulation of distant tsunami propagation with a radial loading deformation effect*, Earth Planets Space *65*, 835–842.

KELLY, G. (2004). *Ammianus and the Great Tsunami*. Journal of Roman Studies *94*, 141–167; doi:10.2307/4135013. JSTOR 4135013.

MATSUMOTO, H., and KANEDA, Y. (2013), *Some features of bottom pressure records at the 2011 Tohoku earthquake - Interpretation of the far-field DONET data*. Proceed. 11th SEGJ Intern. Symp.

MOFJELD, H. O. (2009), *Tsunami measurements*. In The Sea, Volume *15*: *Tsunamis* (eds. A. Robinson and E. Bernard), (Harvard University Press, Cambridge, MA, 2009) pp. 201–235.

MUNGOV, G., EBLÉ, M., and BOUCHARD, R. (2013). *DART tsunameter retrospective and real-time data: A reflection on 10 years of processing in support of tsunami research and operations*. Pure Appl. Geophys., *170*, 1369-1384; doi10.1007/s00024-012-0477-5.

PARKER, B. (2007), *Tidal Analysis and Prediction*, NOAA Special Publication NOS CO-OPS *3*; 378 pp.

RABINOVICH, A. B., and EBLÉ, M. C. (2015), *Deep ocean measurements of tsunami waves*, Pure Appl. Geophys., *2015*, 172 (this issue); doi: 10.1007/s00024-015-1058-1.

RABINOVICH, A. B. and THOMSON R. E. (2007), *The 26 December 2004 Sumatra tsunami: Analysis of tide gauge data from the World Ocean Part 1. Indian Ocean and South Africa*. Pure Appl. Geophys *164* (2/3), 261-308.

RABINOVICH, A. B., STROKER, K., THOMSON, R., and DAVIS E. (2011), *DARTs and CORK in Cascadia Basin: High-resolution observations of the 2004 Sumatra tsunami in the northeast Pacific,*

Geophys. Res. Lett. *38*, L08607, 5 pp., doi:10.1029/2011GL047026.

Rabinovich, A. B., Thomson, R. E. and Fine I. V. (2013a). *The 2010 Chilean tsunami off the west coast of Canada and the northwest coast of the United States.* Pure Appl. Geophys., *170*, 1529-1565.

Rabinovich, A. B., Candella, R. N., and Thomson, R. E., (2013b), *The open ocean energy decay of three recent trans-Pacific tsunamis.* Geophys. Res. Lett., *40*, doi:10.1002/grl.50625.

Saito, T., Ito, Y. Inazu, D., and Hino, R. (2011), *Tsunami source of the 2011 Tohoku-Oki earthquake, Japan: Inversion analysis based on dispersive tsunami simulations*, Geophys. Res. Lett, 38, L00G19, doi:10.1029/2011GL049089.

Shevchenko G., Ivelskaya T., Loskutov A. and Shishkin A. (2013) *The 2009 Samoan and 2010 Chilean tsunamis recorded on the Pacific coast of Russia*, Pure Appl. Geophys. *170*, 1511-1527.

Simons, M., et al. (2011), *The 2011 magnitude 9.0 Tohoku-Oki earthquake: Mosaicking the megathrust from seconds to centuries*, Science, *332* (6036), 1421–1425; doi:10.1126/science.1206731.

Song, Y. T., Fukumori, I. Shum, C. K., and Yi, Y. (2012), *Merging tsunamis of the 2011 Tohoku-Oki earthquake detected over the open ocean,* Geophys. Res. Lett., *39*, L05606, doi:10.1029/2011GL050767.

Thomson, R. E., Fine, I. V., Rabinovich, A. B., Mihaly, S. F., Davis, E. E., Heesemann, M., and Krassovski, M. V. (2011), *Observations of the 2009 Samoa tsunami by the NEPTUNE-Canada cabled observatory: Test data for an operational regional tsunami forecast model,* Geophys. Res. Lett., *38*, L11701, doi:10.1029/2011GL046728

Thomson, R. E. and Emery, W. J. (2014), *Data Analysis Methods in Physical Oceanography: 3rd Edition.* Elsevier Science, Amsterdam, London, New York (August 2014), 716 pp.

Tsai, V. C., Ampuero, J.-P., Kanamori, H., and Stevenson D. J. (2013), *Estimating the effect of Earth elasticity and variable water density on tsunami speeds,* Geophys. Res. Lett., *40*, 492–496, doi:10.1002/grl.50147.

Uslu, B., Power, W., Greenslade, D., Titov, V.V., and Eblé, M. (2011), *The July 15, 2009 Fiordland, New Zealand tsunami: Real-time assessment.* Pure Appl. Geophys., *168* (11), doi:10.1007/s00024-011-0281-7, 1963–1972

Viana-Baptista M. A., and Soares P. M. (2006), *Tsunami propagation along Tagus estuary (Lisbon, Portugal) preliminary results.* Science of Tsunami Hazards; 24(5):329 Online PDF. Accessed 2015-05-9. Archived 2009-05-27.

Watada, S. (2013), *Tsunami speed variations in density-stratified compressible global oceans,* Geophys. Res. Lett., *40*, 4001–4006, doi:10.1002/grl.50785.

Watada, S., Kusumoto, S., Fujii, Y., and Satake, K. (2012), *Cause of delayed first peak and reversed initial phase of distant tsunami.* AGU Fall Meeting, San Francisco, 3-7 December 2012, Abstract NH43B-1649.

Watada, S., Ksumoto, S. and Satake, K. (2014), *Traveltime delay and initial phase reversal of distant tsunamis coupled with the self-gravitating elastic Eart.* J. Geophys. Res. Solid Earth, *119*, 4287–4310, doi:10.1002/2013JB010841.

Wei, Y., Chamberlin, C. Titov, V., Tang, L., and Bernard, E. N. (2013), *Modeling of the 2011 Japan tsunami – Lessons for near-field forecast.* Pure Appl. Geophys., *170* (6–8), 1309–1331; doi:10.1007/s00024-012-0519-z.

(Received October 1, 2014, accepted June 16, 2015, Published online July 11, 2015)

The Great 2006 and 2007 Kuril Earthquakes, Forearc Segmentation and Seismic Activity of the Central Kuril Islands Region

B. V. Baranov,[1] A. I. Ivashchenko,[1] and K. A. Dozorova[1]

Abstract—We present a structural study of the Central Kuril Islands forearc region, where the great megathrust tsunamigenic earthquake (M_w 8.3) occurred on November 15, 2006. Based on new bathymetry and seismic profiles obtained during two research cruises of R/V *Akademik Lavrentiev* in 2005 and 2006, ten crustal segments with along-arc length ranging from 30 to 100 km, separated by NS- and NW–trending transcurrent faults were identified within the forearc region. The transcurrent faults may serve as barriers impeding stress transfer between the neighboring segments, so that stress accumulated within separate forearc segments is usually released by earthquakes of moderate-to-strong magnitudes. However, the great November 15, 2006 earthquake ruptured seven of the crustal segments probably following a 226-year gap since the last great earthquake in 1780. The geographic extent of earthquake rupture zones, aftershock areas and earthquake clusters correlate well with forearc crustal segments identified using the geophysical data. Based on segmented structure of the Central Kuril Islands forearc region, we consider and discuss three scenarios of a great earthquake occurrence within this area. Although the margin is segmented, we suggest that a rupture could occupy the entire seismic gap with a total length of about 500 km. In such a case, the earthquake magnitude M_w might exceed 8.5, and such an event might generate tsunami waves significantly exceeding in height to those produced by the great 2006–2007 Kuril earthquakes.

Key words: Island arc segmentation, Central Kuril Islands, transcurrent faults, subduction, earthquakes.

1. Introduction

On November 15, 2006, at 11:14 UTC, a great megathrust tsunamigenic earthquake of magnitude $M_w = 8.3$ (http://www.globalcmt.org) occurred in the Central Kuril Islands region in a well-established

---

Topical issue of PAGEOPH "Tsunami Science: Ten Years after the 2004 Sumatra Tsunami".

[1] P. P. Shirshov Institute of Oceanology, Russian Academy of Sciences, 36, Nakhimovskiy Prospect, 117997 Moscow, Russia. E-mail: aii_imgg@mail.ru

seismic gap within an active subduction zone. The epicenter of the mainshock was located on continental slope of the Kuril-Kamchatka deep-water trench, approximately 90 km to the southeast from Simushir Island (Fig. 1). Two months later, on January 13, 2007, at 04:23 UTC, a second great earthquake of magnitude $M_w = 8.1$ (http://www.globalcmt.org) occurred within the outer rise of the Central Kuril Islands region. The epicenter of its mainshock was located on the oceanic side of the Kuril-Kamchatka trench, approximately 100 km to the east from the epicenter of the first event. Both earthquakes generated destructive tsunami waves in coastal regions of the Central Kuril Islands (Levin et al. 2008; MacInnes et al. 2009), but only moderate-amplitude tsunamis were produced around the Pacific Ocean (Rabinovich et al. 2008; ITDB/WLD 2014; NGDC/WDS 2014).

This seismic doublet occupies a specific place in the seismic history of the region because no one great earthquake had been observed in the central part of the Kuril Islands Arc during the period of instrumental seismic observations. Generation and propagation of tsunami waves by both earthquakes were studied by many investigators (Fujii and Satake 2008; Koshimura et al. 2008; Laverov et al. 2009; Levin et al. 2008; Lobkovsky et al. 2008, 2009, 2010; Rabinovich et al. 2008; Tanioka et al. 2008; and others). The 2006–2007 Kuril earthquake sequence provided an opportunity to investigate the process of tectonic stress transfer from one large event to another and to reveal specific character of faulting for interplate and intraplate events and their implications for seismic hazard (Ammon et al. 2008; Steblov et al. 2008; Lay et al. 2009; Raeesi and Atakan 2009; Ogata and Toda 2010).

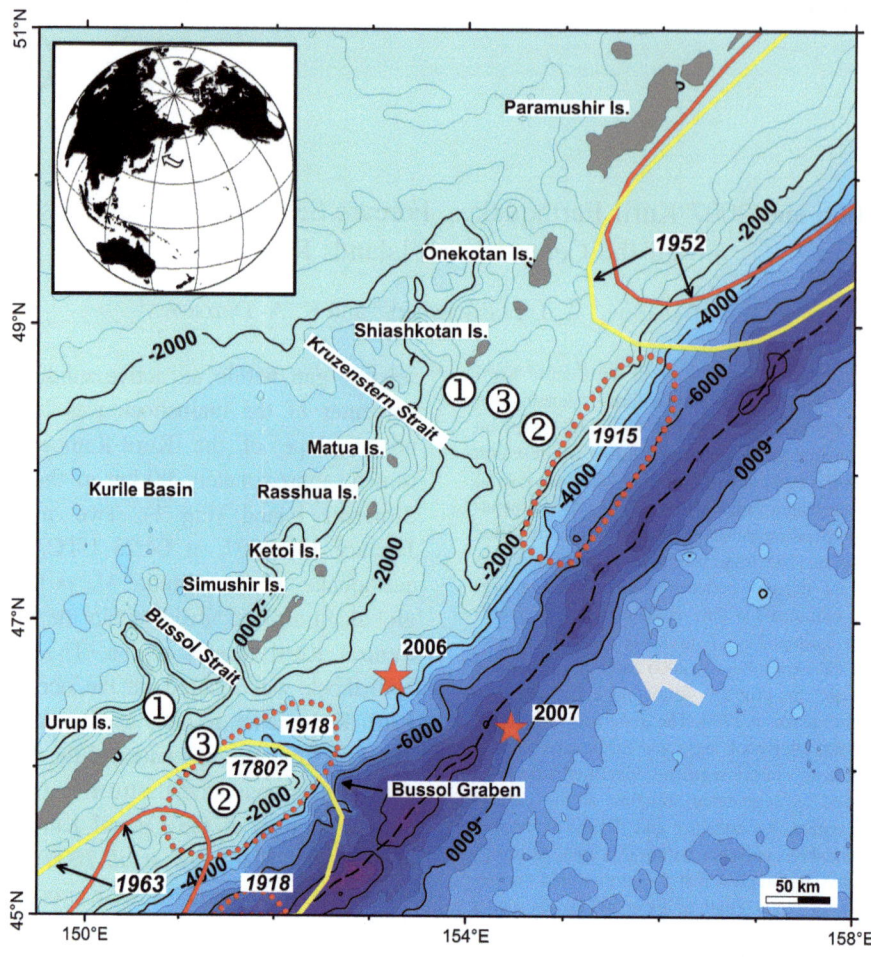

Figure 1
Main features of the Central Kuril Islands region. *Open arrow* within the *inset* shows general location of the study area; *1* inner (volcanic) arc; *2* outer arc (Vityaz Ridge); *3* inter-arc Middle-Kuril trough. *Thick closed contours* delineate source zones of the great historical earthquakes: *red*—according to FEDOTOV and CHERNYSHEV (2002), *dotted* where uncertain; *yellow* estimated in present work based the on updated aftershock locations. A source zone of the great 1780 earthquake is not shown. The Central Kuril Islands seismic gap was suggested to be between rupture zones of the great 1952 and 1963 earthquakes (FEDOTOV 1965, 1968). *Two red stars* denote the epicenters of the great 2006 and 2007 Kuril earthquakes. *Thick gray arrow* indicates direction of motion of the Pacific plate relative to the North American plate at a rate of 80 mm/year according to the NUVEL-1 model (DEMETS et al. 1990). Contour interval of bathymetry is 500 m, after (SMITH and SANDWELL 1997). *Dashed line* marks axis of the deep Kuril Trench

In this paper, we examine the segmented structure of the Central Kuril Islands forearc region and its relation to seismic activity and current seismic potential. Subdivision of the forearc region into separate segments (crustal blocks) is based on the seismic reflection and bathymetric surveys data obtained during two research cruises of R/V *Akademik Lavrentiev* in 2005 and 2006. We identify ten crustal segments (blocks) with along-arc length ranging from 30 to 100 km, separated by NS– and/or NW–striking transcurrent faults. Seismic activity of the Central Kuril Islands region is investigated using the following earthquake catalogues (see references): New Catalog (through 1977); ISC-GEM (1913–2009); EHB (1960–2008), ISC (2009–2014); ANSS ComCat (1973–2014); and CMT (1976–2014), with a corresponding time span shown within brackets. The EHB is a groomed version of the ISC Bulletin (ENGDAHL et al. 1998) supported before a new ISC location algorithm was introduced since

2009 (BONDÁR and STORCHAK 2011). We compare the seismicity to the geographic extent of earthquake rupture zones, aftershock areas and distribution of earthquake clusters and show a correlation with forearc crustal segments (blocks) identified using the geophysical data.

2. Tectonic Setting

The central segment of the Kuril Islands Arc (the Central Kurils) is located between the Bussol and Kruzenstern Straits and includes the relatively large Simushir Island and several smaller islands: Ketoi, Rasshua, Matua, and others (Fig. 1). The large Paramushir Island is located northeast of the Central Kurils, while Urup Island borders this region in the southwest. From the southeast, the region is flanked by the Kuril-Kamchatka Trench, and from the northwest, it borders the deep Kuril Basin of the Okhotsk Sea.

The Central Kuril Islands region consists of the inner (volcanic) and outer (Vityaz Ridge) arcs separated by the inter-arc trough (Fig. 1). The volcanic arc consists of an Upper Oligocene (?)—Middle Miocene green-tuff overlain by Middle Pliocene and Quaternary volcanic and volcanogenic-sedimentary rocks of the island-arc series (SERGEEV 1976). Ten recent active volcanoes are known within the Central Kuril Islands region, and the volcanic front is located at a distance of about 170 km from the Kuril Trench (AVDEIKO et al. 1992).

The outer (tectonic) arc is represented by the submerged Vityaz Ridge, which consists of two separate branches—northeastern and southwestern, and is missing between Simushir Island and Rasshua Island (Fig. 1). Depth of the ridge top varies from 150–200 to 900–1000 m. Dredging on the Vityaz Ridge recovered volcanic rocks belonging to several complexes (LELIKOV et al. 2008). The Late Cretaceous volcanites and intrusive rocks form a single volcano–plutonic complex that constitutes the basement of the structure. Cenozoic volcanics belong to the Eocene, Late Oligocene, Miocene, and Pliocene–Pleistocene (?) complexes.

The Vityaz Ridge is separated from the volcanic arc by the inter-arc Middle-Kuril trough, which is covered by sediments up to 2 km in thickness to the southwest and to the northeast (SERGEEV 1976). Within the Central Kuril Islands region, the Bussol submarine canyon is a remarkable graben oriented transverse to the island arc (VASILIEV and SUVOROV 1979). KIMURA (1986) suggested, that it was formed by southwest displacement of southern Kuril sliver including southwestern branch of the Vityaz Ridge. This displacement could occur due to the shear component of motion caused by oblique subduction of the Pacific plate beneath the southwestern Kuril Islands Arc. At least one more transverse fault is identified within the Central Kuril Islands forearc region; it is marked by the Kruzenstern Strait (SERGEEV 1976) (Fig. 1).

Seismic refraction data indicate the crust beneath the northern extremity of Urup Island is 30-km thick; under the Bussol Strait it is 28-km thick, and beneath Simushir and Matua Islands—25-km thick (ZLOBIN and ZLOBINA 1991). Although the crust structure of the central part of the Kuril Islands Arc is in general similar to that of its flanks, the central segment differs from the flanks by some geological and geophysical features. This is primarily manifested in strong volcanic activity of the rear part of the islands: the linear density of volcanoes per 100 km on the southeastern slope of the Kuril Basin between the Bussol and Kruzenstern Straits is 1.7–4.8 times higher as compared with that in the frontal (oceanic) part of the volcanic arc; the heat flow in this area amounts to 100 mW/m^2 (AVDEIKO et al. 1992).

Based on geophysical investigations during two aforementioned research cruises, KULINICH et al. (2007) have suggested that forearc area within the Central Kuril Islands region is a zone of intense extension and destruction of the basement and crust within the study area. They consider an area from Simushir Island to the trench as an imposed rifting zone, which is transverse to the arc. The rupture of the Vityaz Ridge was incomplete, as the ridge is traceable in the basement as a chain of hidden swells along the entire structure. In contrast, the inter-arc Middle-Kuril trough does not exist as a single negative structure. Therefore, the "seismic gap" has a more complicated structure than was previously believed.

3. Historical Seismicity

Seismicity of the Kuril Islands Arc is dominated by subduction of the Pacific plate beneath the Okhotsk plate at a rate of about 8 cm/year in N60°W direction (DeMets et al. 1990). Interaction of overriding and subducting plates is responsible for a high rate of observed seismic activity (Fig. 1). However, the central part of the Kuril Islands Arc was characterized by relatively lower seismic activity as compared with its southern and northern flanks. Due to the historic absence of great earthquakes here, the Central Kuril Islands region between rupture zones of the great 1952 M_w 9.0 and 1963 M_w 8.5 earthquakes was the first time identified by Fedotov (1965) as a seismic gap, i.e. as a likely place for occurrence of a next great earthquake ($M \geq 7\frac{3}{4}$) in future. A number of authors following Fedotov (1965, 1968) also considered this part of the Kuril subduction zone as a seismic gap (Kelleher and McCann 1976; McCann et al. 1979; Lay et al. 1982; Nishenko 1991), yet noting its seismic history as quite unclear (McCann et al. 1979; Nishenko 1991).

Russian and Japanese historical sources mentioned a great earthquake that occurred near Central Kuril Islands on June 29, 1780 and caused tsunami wave run-ups on coasts of Urup (up to 10–12 m), Simushir and Ketoi islands (Soloviev and Ferchev 1961; Iida et al. 1967; New Catalog 1982). The southwestern edge of the rupture area for this event probably overlapped with a rupture zone of the great 1963 M_w 8.5 earthquake (Fukao and Furumoto 1979; Beck and Ruff 1987; Hatori 2007); position of its northeastern limit was not determined (Fig. 1).

Two other great historical seismic events are known to have occurred in the southwestern margin of study area, namely the seismic doublet of September 7, 1918 (M_w 8.2) and November 8, 1918 (M_w 7.6) (New Catalog 1982; Pacheco and Sykes 1992); source zones of both shocks are also assumed to be partially overlapped with a rupture zone of the great 1963 earthquake (Beck and Ruff 1987; Hatori 2007). The first earthquake caused tsunami waves with run-ups of 12 m onto the coast of Urup Island (Iida et al. 1967). From the estimated location of the tsunami source, the rupture area of the 1918 M_w 8.2 earthquake continued to the northeast from Bussol Graben (Fedotov 1965; Fukao and Furumoto 1979) (Fig. 1). One recent 1991 M_w 7.6 seismic event also partially ruptured the 1963 source area (Perez 2000) suggesting a relatively short recurrence time interval of large earthquakes in the southwestern part of the Central Kuril Islands seismic gap.

The rupture zone of the May 1, 1915 M_w 7.8 earthquake is located to the southwest from the source area of the great 1952 Kamchatka earthquake (New Catalog 1982; ISC-GEM 2013) and it was considered by Fedotov (1965, 1968) as a part of seismic gap similar to a source zone of the 1918 M_w 8.2 earthquake.

The extent of the Central Kuril Islands seismic gap was estimated from 450 to 550 km, which approximately corresponds to rupture length of a great megathrust earthquake with moment magnitude $M_w \sim 8.8$–9.0 (Wells and Coppersmith 1994; Blaser et al. 2010; Strasser et al. 2010). According to the forecasting method proposed by Fedotov (1965, 1968), an earthquake of maximum magnitude $M_{Smax} = 8.2$ was expected to occur within the Central Kuril Islands seismic gap, and probability of its occurrence was estimated, among other parameters, for consecutive 5-year-long periods. For the last period (2001–2005) preceding occurrence of the 2006 M_w8.3 earthquake, probability of occurrence of an earthquake with $M_w \geq 7.75$ within the Central Kuril Islands seismic gap was estimated as 12 % (Fedotov and Chernyshev 2002; Fedotov 2005). This long-term forecast is considered to be successful after occurrence of the great 2006 Kuril earthquake (Fedotov et al. 2007, 2012).

4. Data

To study the tectonic structure of the Central Kuril Islands seismic gap, the Russian Academy of Sciences undertook the *Kurile-2005* and *Kurile-2006* marine expeditions (cruises 37 and 41 of the R/V *Akademik Lavrentiev*) in the Central Kuril Islands region in August–September, 2005 and 2006. The main attention was focused on identifying zones of transverse faults, which are thought to bound seismotectonic blocks and to affect rupture zones of great earthquakes and play an important role in preparation and initiation of such events (LOBKOVSKY et al. 1991; COLLOT et al. 2004). The purpose of our geophysical studies was to reveal the reason for this long-lived seismic gap, to estimate its seismic potential and expected tsunami hazard from possible occurrence of great megathrust earthquake within the studied seismic gap.

The cruises set up three regional, 560–650 km long, geophysical profiles (single channel seismic profiling, bathymetric, magnetic and gravimetric surveys, seismological observations) across the entire seismic gap; several study areas were investigated in more detail (Fig. 2, inset). Data obtained in cruises 37 and 41 were partly published (LAVEROV et al. 2006; KULINICH et al. 2007; IVANENKO et al. 2008; KOVACHEV et al. 2009).

The new data reported here consist of 6320 line-km of air gun seismic reflection profiles from two cruises of the R/V Akademik Lavrentiev: cruise 37 (4390 line-km of single channel digital data), and cruise 41 (1930 line-km of single channel digital data). The seismic system used consisted of one or

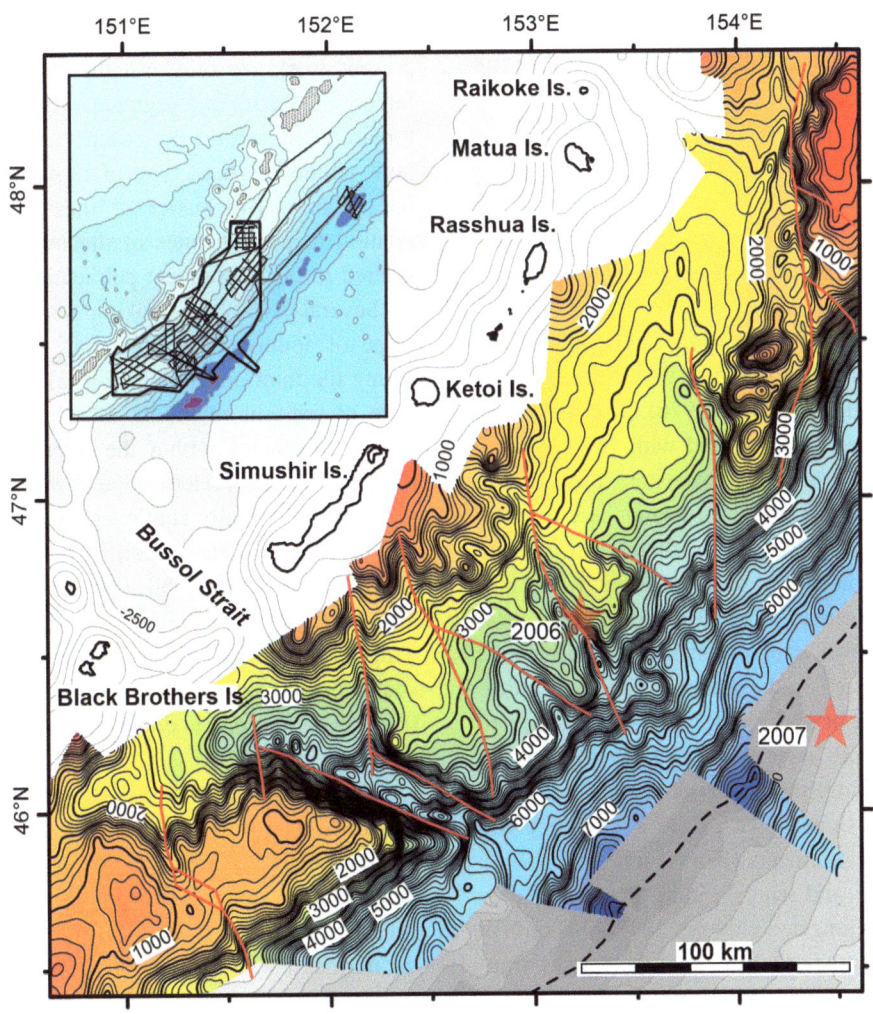

Figure 2
Bathymetric map of the Central Kuril Islands forearc region composed of the data obtained during 37 and 41 cruises of the R/V *Akademik Lavrentiev* in 2005 and 2006, respectively. Contour interval is 100 m. *Inset* shows general location of the map (*closed contour*) and tracks of the geophysical surveys performed during the cruises. *Thin red lines* designate transcurrent faults interpreted using data of geophysical investigations. *Two red stars* denote the epicenters of the great 2006 and 2007 Kuril earthquakes (see also Fig. 1). *Dashed line* marks the trench axis

two air guns and the streamer with total active length of 42 m. New bathymetric data were also collected with a single beam echo sounder operating at a frequency of 12.4 kHz.

Seismicity of the Central Kuril Islands region for the time period through 2014 was investigated using the different available earthquake catalogues (see references): *New Catalog* (through 1977); ISC-GEM (1913–2009); EHB (1960–2008), ISC (2009–2014); ANSS ComCat (1973–2014); and CMT (1976–2014) for time spans shown within brackets. The main subject of our study was a search of correlation between seismic activity and tectonic pattern. We mostly examine the distribution of seismic events with magnitude $M_w \geq 5$ and source depth $H < 105$ km as they provide the most valuable information for understanding the interaction of tectonic plates within subduction zones.

5. Transcurrent Faults and Forearc Segmentation

Two main zones of transcurrent faults, Bussol and Kruzenstern, were distinguished within the Central Kuril Islands region (SERGEEV 1976). The first zone, the Bussol Graben, is associated with Bussol Strait, and the second one—with Kruzenstern Strait; these straits separate the central Kuril Islands Arc from its southwestern and northeastern flanks. In present work, seismic reflection profiles together with compiled bathymetric map allowed us to identify new, previously unknown transcurrent faults. These faults are well manifested in the bottom relief and correspond to canyons or scarps (Fig. 2). They are located within a sea depth interval from 1500 to 5000 m, have a length of 60 through 80 km, and are traced from the upper slope down to the accretion prism made up by soft sediments (KLAESCHEN et al. 1994). Because canyons and scarps within the forearc area are presumably elongated either in NNW–NS or NW direction, we distinguish here two systems of transcurrent faults by their orientation (Fig. 3).

On seismic profiles the transcurrent faults are associated with the scarps and canyon walls; the last cut sedimentary cover or the exposed acoustic basement on the sea floor. In case of their overlapping by sediments, the faults were associated with steep walls of blocks within the acoustic basement. Considered here structure of the study area is highly generalized and includes only major faults greater than 50 km in length. By using single-channel seismic profiling method for investigation of heterogeneous sedimentary structure within the forearc area, one can recognize only sub-surface fault structure. Gravimetric and magnetic data obtained in cruises 37 and 41 of R/V *Akademik Lavrentiev* suggest that transcurrent faults are crustal ones (KULINICH et al. 2007; IVANENKO et al. 2008). Refraction profiles within the frontal and rear parts of the Central Kurils evidence that transcurrent faults cut the crust and may penetrate down to the mantle (ZLOBIN et al. 2008). In present work we only use these main transcurrent faults for separating segments (blocks) within the forearc crust and do not touch their kinematics, i.e. fault parameters and type of movements along the fault planes. Kinematics of these faults and their relation to geodynamics of the region will be considered in the consequent publication.

Several NNW-NS-striking canyons cross the slope of southwestern branch of the Vityaz Ridge facing the inter-arc trough (Fig. 2) and show a presence of transcurrent faults. The biggest one (V1 in Fig. 3) originates within the inter-arc trough, then crosses the northwestern slope of the Vityaz Ridge as a canyon, and finally continues on southeastern slope of the ridge as a steep scarp with a height of about 1000 m (Fig. 4, profs. 50–1 and 04–1, respectively). The fault has en-echelon structure on the Vityaz Ridge top and divides top surface of the ridge into southwestern and northeastern steps (Fig. 4, prof. 03–3). The first one is downshifted 300 m relative to the other.

Vityaz Ridge is truncated on the northeast by the Bussol Graben, which represents the most spectacular example of NW-striking faults. This structure separates southwestern part of the Vityaz Ridge with water depth of about 1000 m from the forearc area to the northeast with depth of about 3000 m. Maximum depth of the Bussol Graben is 6000 m and, consequently, its southwestern and northeastern flanks have the heights of 5000 and 3000 m, correspondingly; slope steepness reaches 45°. Graben flanks correspond to almost straight-line faults of NW strike (B1 and B2 in Figs. 3, 5).

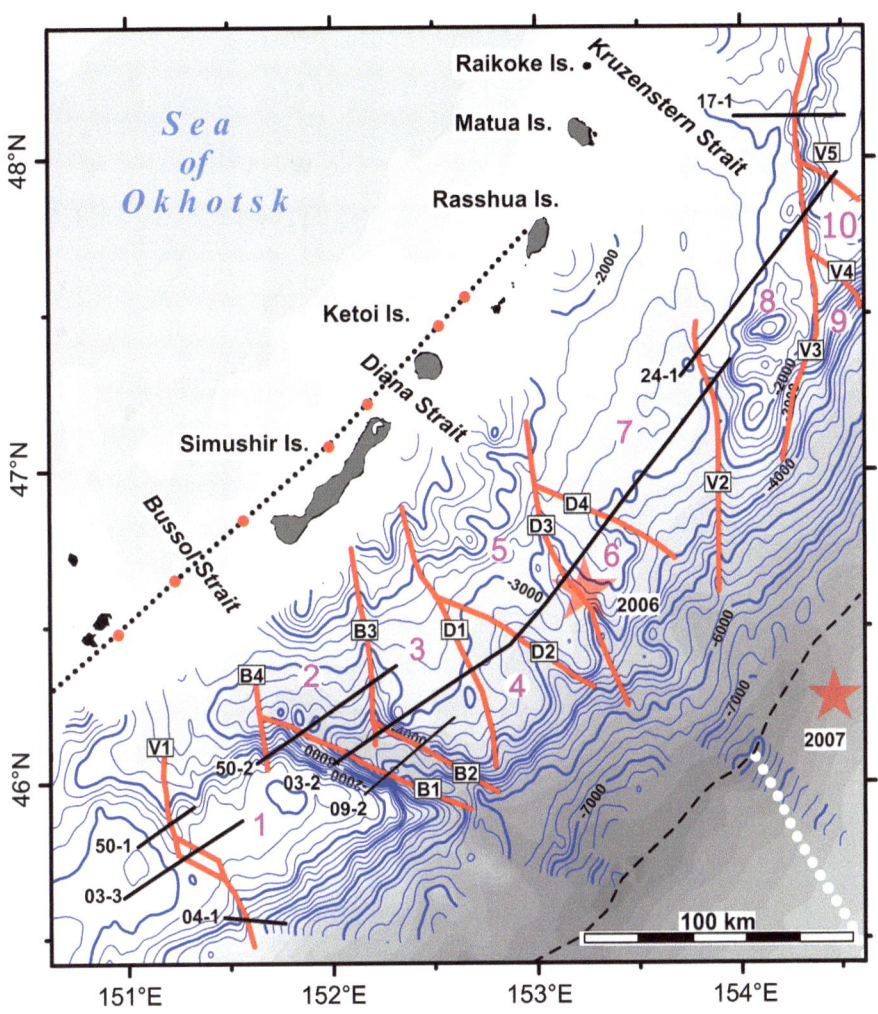

Figure 3
Location of the main identified transcurrent faults (*red solid lines*) in the Central Kuril Islands forearc region based on the data of geophysical surveys. *Letters* and *numbers within boxes* designate faults associated with the Vityaz Ridge (V1–V5), Bussol Graben (B1–B4) and Diana Strait (D1–D4). *Numbers within circles* (1–10) mark separate forearc crustal blocks. *Numbered black straight lines* denote one-channel seismic reflection profiles shown in details in Figs. 4, 5, 6, and 7. *Dotted line* marks the deep seismic refraction profile and locations of the supposed crustal faults (*red filled circles*) according to ZLOBIN et al. (2008). Contour interval is 250 m. See Fig. 2 for other legend notations

The Bussol Graben slightly widens in northwestern direction where it intersects the Middle-Kuril inter-arc trough. The northeastern flank of the graben is limited by NS-striking fault. The fault is confined to a canyon that continues to the Simushir Island and was mapped up to contour of 1500 m (B3 in Figs. 3, 5). According to the deep refraction data (ZLOBIN et al. 2008), one of the crustal faults shifting the Moho discontinuity in the rear of Simushir Island is located on the continuation of transcurrent fault B3. Another NS-striking fault (B4 in Figs. 3, 5) limits south-western flank of the Bussol Graben and apparently is traced into the Bussol Strait; probably another crustal fault exists on its continuation as well (ZLOBIN et al. 2008). We have no data on the intersection of Bussol Graben with the Kuril Trench.

Between southwestern and northeastern branches of the Vityaz Ridge, the forearc has a block structure manifested in both the bottom relief and acoustic basement (Figs. 3, 6). The blocks are limited by NS–trending faults or by a combination of NS– and NW–trending faults. Two NS-trending faults are

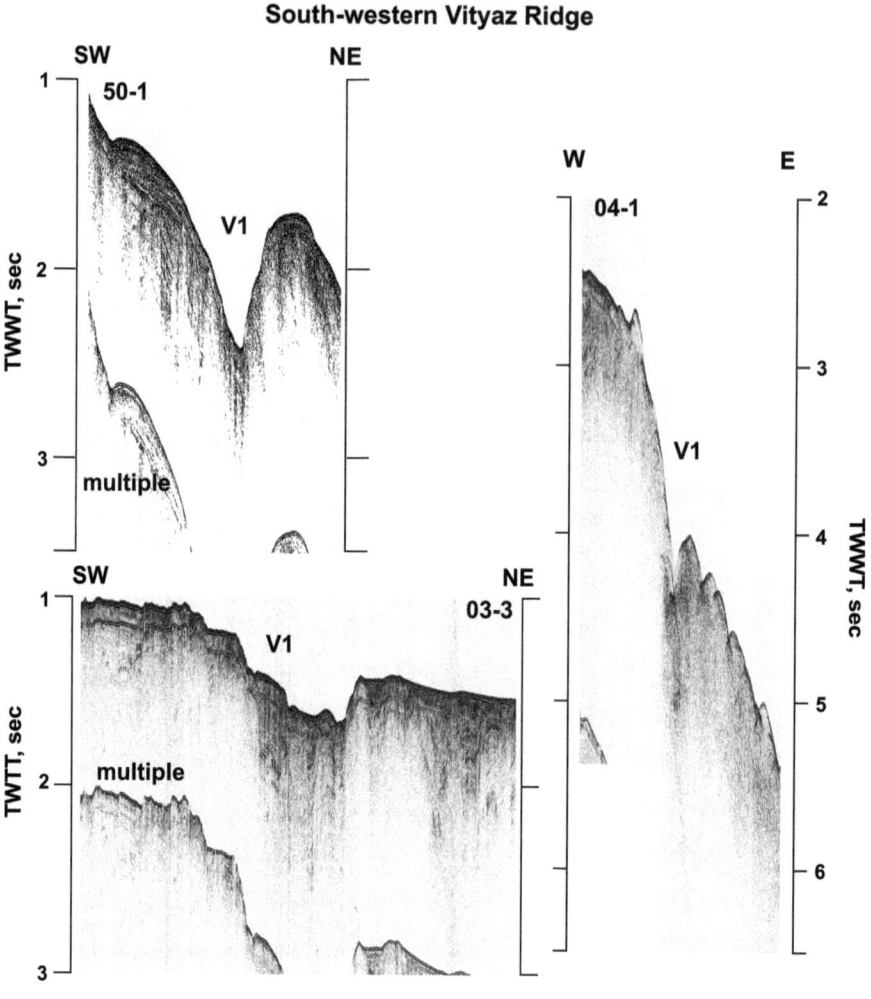

Figure 4
Seismic reflection data obtained along the seismic profiles crossing the transcurrent fault V1 that breaks southwestern branch of the Vityaz Ridge. See Fig. 3 for location of the Vityaz Ridge and position in plan of the seismic profiles and the fault. Acronym TWTT, a label of the vertical axis, stands for "two-way travel time" of the acoustical pulse, in s

manifested as a canyon in the bottom relief and as a scarp limiting the tilted block of the acoustic basement (D1 and D3 correspondingly in Figs. 3, 6). The canyon begins from a depth of about 4500 m and extends to 2000 m contour line.

The first of the two NW-striking faults identified in this area corresponds to a canyon traced in a depth interval of 2500–4500 m (D2 in Figs. 3, 6). The second one is represented by a scarp with a height increasing towards the Kuril Trench up to 400 m; the fault limits the tilted block of the acoustic basement (D4 in Figs. 3, 6). Sedimentary horizons above the tilted block are deposited parallel to the acoustic basement which may evidence active tectonics. The eastern part of the forearc area is characterized by a simple relief of bottom and acoustic basement within a zone between the scarp and northwestern branch of the Vityaz Ridge (Fig. 6). According to available data, there are no transcurrent faults here.

The northeastern branch of the Vityaz Ridge consists of two ridges (Fig. 3). The ridge, only 200 m deep in the north, steps down to the southwest to a depth 2500 m. Steepness of NS slopes, their straightness and character of contact with sedimentary cover near the ridge foot give evidence to suppose that the slopes are fault controlled. Several

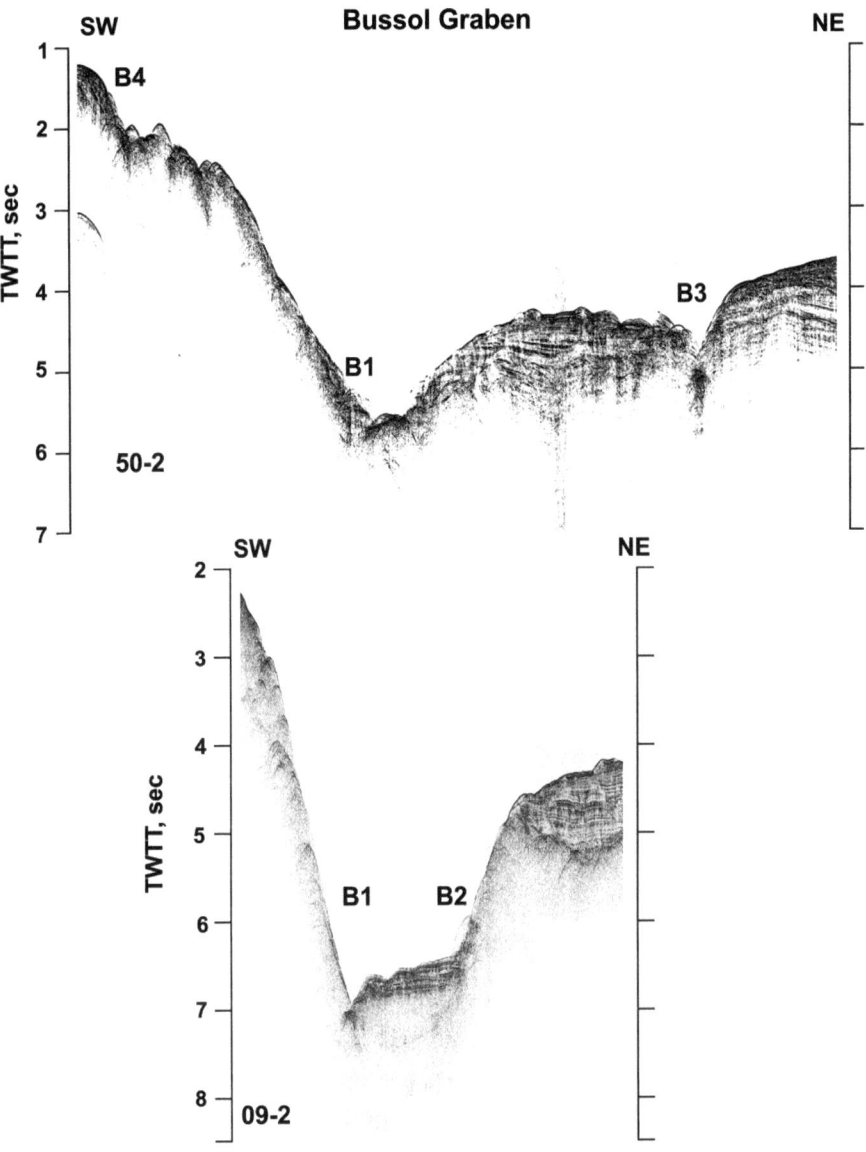

Figure 5
Seismic reflection data showing structure of the Bussol Graben in its northwestern (profile 50–2) and southeastern (profile 09–2) parts and locations of transcurrent faults B1–B4. See Fig. 1 for location of the Bussol Graben, Fig. 3 for position in plan of the seismic profiles and faults, and Fig. 4 for a note

faults can be identified. They form NS-striking zone that separates both steps from each other and the ridge itself from the forearc area (V2, V3 in Figs. 3, 7). Two NW-striking faults, which are conjugated to the upper scarp can be identified as well (V4, V5 in Figs. 3, 7). On regional seismic profiles, beyond the area of detailed investigations, one can see that the Vityaz Ridge slope facing the deep-water trench consists of blocks separated by faults, but their strike remains unclear.

Based on locations of the transcurrent faults we have identified ten forearc crustal segments (Fig. 3). As far as the segments are bordered by NS-striking faults or by a combination of NS- and NW-trending faults, they have rectangular and triangular shape, correspondingly. Below we consider a correlation

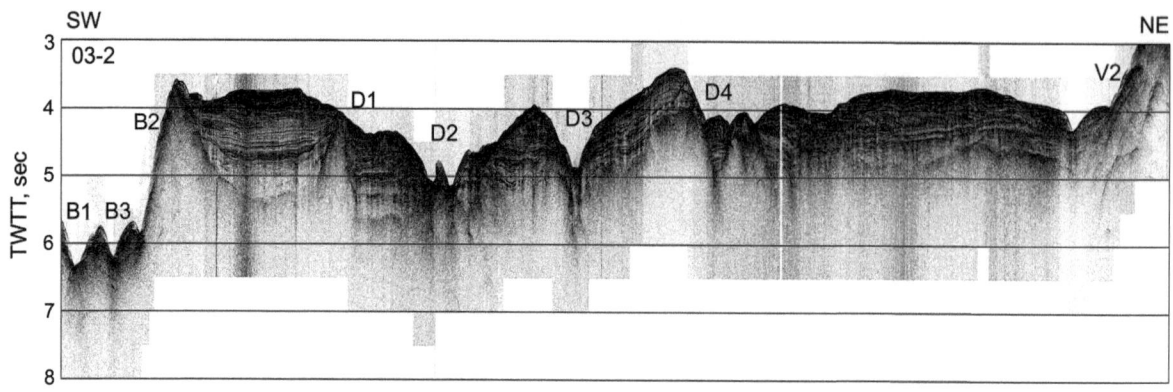

Figure 6
Seismic reflection data showing along-strike surface structure of the Central Kuril Islands forearc region and locations of transcurrent faults (B1, B2, B3, etc.) on the seismic profile 03–2 stretched between the southwestern and northeastern branches of the Vityaz Ridge. See Fig. 3 for position in plan of the seismic profile and faults, and Fig. 4 for a note

between the suggested forearc structural pattern of the Central Kuril Islands forearc region and a distribution of background seismicity. Rupture zones of the great 2006 and 2007 Kuril earthquakes will be also considered in this context.

6. Recent Seismic Activity

To evaluate background seismicity of the Central Kuril Islands region and to determine aftershock zones of the great 2006–2007 Kuril earthquakes, we investigated the distribution of earthquakes ($M_w \geq 4.5$, $H < 105$ km) that occurred within the region during the time period from 1977 to the end of 2014 using presumably the EHB (2009), ANSS ComCat (2014), and CMT (2014) catalogs. We have assumed that the best hypocenter locations are obtained for the EHB (1977–2008), ISC (2009–2012.09), and ANSS ComCat (2012.10–2014) catalogs, while the best estimates of magnitude M_w and 'type-of-motion' in the source are provided by the CMT catalog (1977–2014). No one CMT solution is available for the data year 1976 within the study area. Historical seismicity for the time period 1913–1976 was shortly examined based on the newest global ISC-GEM catalog (2013) containing uniform M_w estimates for all earthquakes. In case of ambiguous solution for any old great event, we preferred data from the New Catalog (1982) as they were corrected using available tsunami data.

We have divided the history into two time periods: (1) from 1977 to the occurrence of the great November 15, 2006 Kuril earthquake (for background seismicity); and (2) from the occurrence of the 2006 mainshock up to the end of 2014 (for aftershock activity following both great events). Hereafter, magnitudes M_w of major earthquakes ($M_w \geq 7$) are indicated according to the CMT catalog (since 1976) or to the ISC-GEM catalog (before 1976).

6.1. Background Seismicity: 1977–2006

During the time period preceding occurrence of the great 2006 Kuril earthquake, distribution of earthquakes within the studied seismic gap and its vicinity was not uniform (Fig. 8a–d). At first, we shortly examine historical earthquakes within the study area for the period since 1913 through 1976 using the newest uniform ISC-GEM (2013) catalogue that is considered to be complete for $M_w \geq 7.0$ since 1918 and for $M_w \geq 6.5$ since 1955 (Fig. 8a). During this 64-year period, the great 1915 $M_w 7.8$ and 1918 $M_w 8.2$ earthquakes occurred within the study area, and two even greater events, the 1952 $M_w 9.0$ Kamchatka earthquake and 1963 $M_w 8.5$ Kuril earthquake, occurred northwest and southeast of it, respectively. Note that epicenter of the great 1918 event is shown according to the New Catalog (1982), not the ISC-GEM (2013) catalogue, where it is

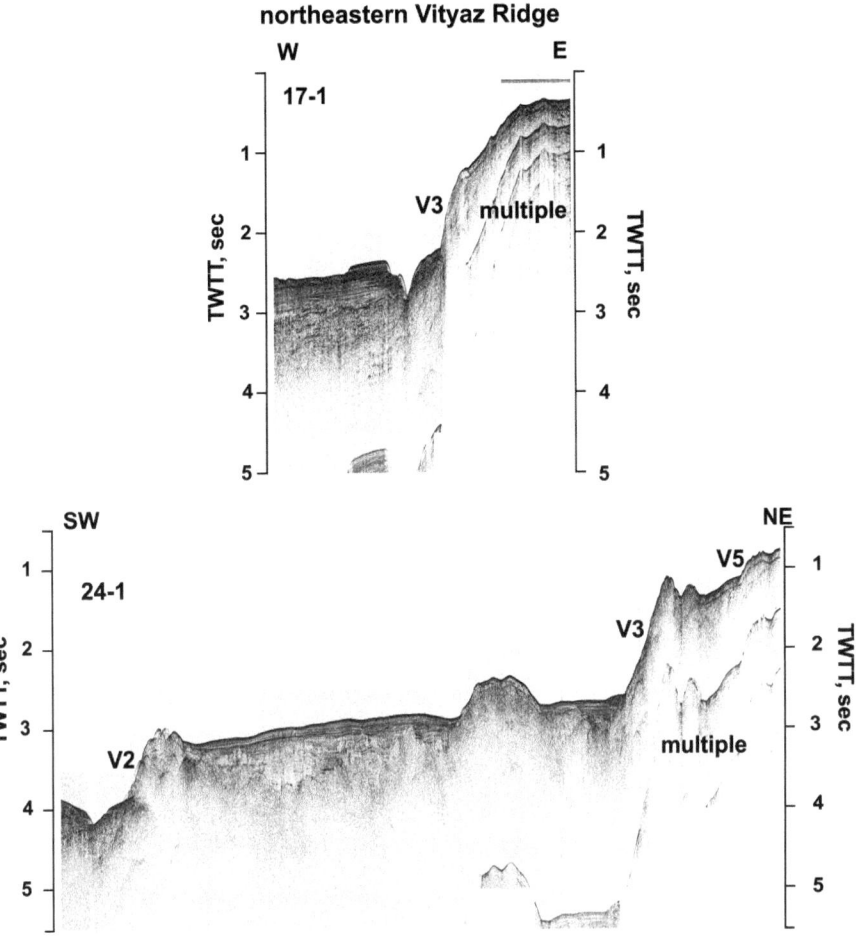

Figure 7
Seismic reflection data obtained along the seismic profiles crossing transcurrent faults V2, V3, and V5 that break northeastern branch of the Vityaz Ridge. See Fig. 3 for location of the Vityaz Ridge and position in plan of the seismic profiles and faults, and Fig. 4 for a note

located in the Sea of Okhotsk, which contradicts the data of tsunami observations. The most important features of the observed seismicity are as follows: (1) in space, epicenters of strong (M_w 6+) shallow earthquakes are mostly concentrated along the southwestern and northeastern branches of the Vityaz Ridge and its likely continuation within the basement in central part of the study area; several moderate (M_w 5+) to strong earthquakes occurred within the oceanic slope of the trench on traverse between the Bussol and Kruzenstern Straits; (2) in time, the bursts of seismic activity and occurrences of major earthquakes ($M_w \geq 7$) within the epicentral area of the future great 2006 and 2007 events were observed during the periods of 1922–1927, 1951–1955, 1963–1964, and 1971–1974, suggesting their relation, at least, to the great 1952 and 1963 seismic events.

To examine the background seismicity, we choose the 30-year time period from 1977 to the occurrence of the great November 15, 2006 Kuril earthquake and use the EHB catalog for obtaining earthquake locations and the CMT catalog for obtaining estimates of M_w and type-of-motion in the source. For comparison with the preceding time period (1913–1976, Fig. 8a), the background seismicity is shown in Fig. 8b for earthquakes with $M_w \geq 5.5$ and $H < 105$ km. The distribution of moderate-to-strong earthquakes within central part of the study area for two observational periods is similar except the low activity of oceanic slope of the trench during the years 1977–2006.

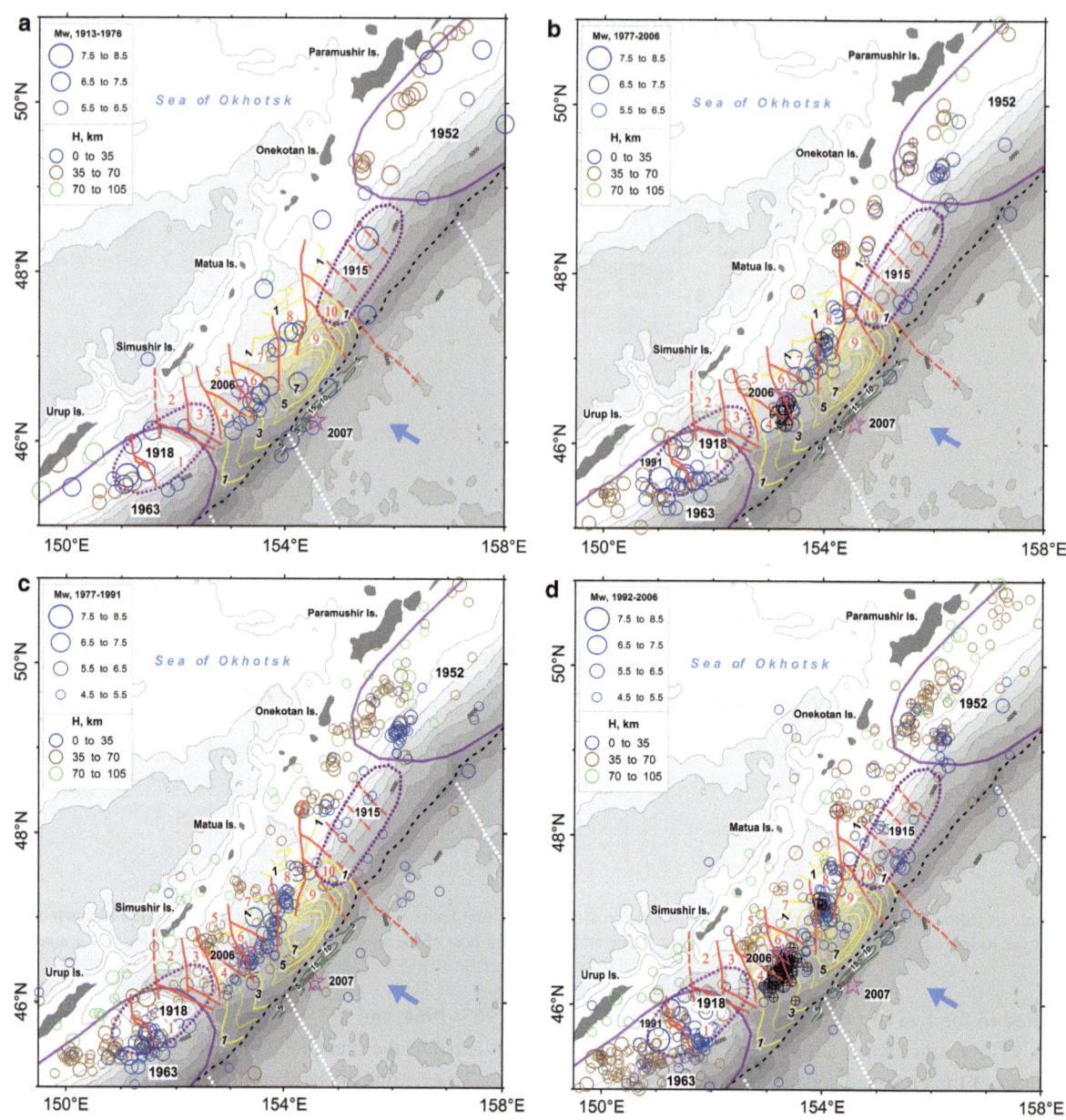

Figure 8
Epicenters of earthquakes ($H < 105$ km) occurred in the Central Kuril Islands region during the time periods (indicated within legend): **a** 1913–1976, $M_w \geq 5.5$ (catalog GEM); **b** 1977–2006, $M_w \geq 5.5$ (catalog EHB); **c** 1977–1991, and **d** 1992–2006, both for $M_w \geq 4.5$ (catalog EHB). Hereafter, open magenta stars indicate the epicenters of the great 2006 and 2007 Kuril earthquakes, and *yellow* and *dark green solid lines* show distribution of slip (in m) in the sources of both events, respectively, according to JI (2006, 2007). The legend gives a classification of shocks by moment magnitude M_w and source depth H. Foreshocks recorded 26.09.2006 through 15.11.2006 are *highlighted in black* with a cross (for $H < 35$ km), and in *dark brown* with a cross (for $35 \leq H < 70$ km). *Solid magenta contours* mark source zones of the great 1952 and 1963 earthquakes, *dotted* where uncertain (1915 and 1918 earthquakes). *Solid red lines* show identified transcurrent faults, *dashed* where assumed, and *dotted white lines* fracture zones within the oceanic crust according to HILDE *et al.* (1977). *Red numbers* indicate separate crustal segments (blocks). *Thick blue arrow* indicates direction of motion of the Pacific plate relative to the North American plate according to the NUVEL-1 model (DEMETS *et al.* 1990). Contour interval of bathymetry is 1000 m, after (SMITH and SANDWELL 1997). *Dashed black line* marks axis of the deep Kuril Trench

More detailed picture of the background seismicity is given by distribution of earthquakes with $M_w \geq 4.5$ within two 15-year time intervals: 1977–1991 (Fig. 8c), and 1992–2006 (Fig. 8d). The highest rate of seismicity was observed at the end of 1991, when ten shocks of magnitude $M_w \sim 6$ and greater were recorded during 2 weeks; these events were mostly aftershocks of the largest background earthquake within the studied area (22.12.1991, $M_w 7.6$). The epicenter of the mainshock was located near the southwestern boundary of the seismic gap (Fig. 8d), and a rupture zone inferred from the aftershock distribution was slightly elongated in southeast direction towards the Kuril trench. In general, maximum rates of background seismicity were observed within a broad zone of the forearc slope between the axis of the Vityaz Ridge and 5000 m contour line (Fig. 8c, d). The earthquakes are rare beneath the trench axis and in oceanic slope of the trench, where epicenters are distributed within a broad stripe of about 90 km in width. The seismic activity here is conditioned by bending of the oceanic plate before its subduction in the trench.

In Fig. 8c, d clusters of earthquake epicenters are clearly seen. The first one is located within the southwestern branch of the Vityaz Ridge. The cluster has a distinct northeastern boundary, which strikes approximately in NW direction across the terminus of the southwestern Vityaz Ridge, probably due to transcurrent NW-trending faults. Seismic activity was weak within the Bussol Graben during both time periods (1977–2006). During the second time period (1992–2006), two distinct earthquake clusters formed within central part of the study area (Fig. 8d). The first cluster crosses lower regions of the forearc crustal segments (blocks) 4–6 bordered by a system of NW-striking transcurrent faults, while the second one corresponds to the crustal segment 8 bordered by two NS-striking faults. Note that foreshock activity observed 1.5 months prior to occurrence of the great 2006 earthquake, and highlighted in blue in Fig. 8b, d, was only located within these clusters. Both clusters were even observed during the first considered time span (1913–1976); however, during the interval of 1977–1991 they were not separated from each other (see Fig. 8a, c, respectively). Three earthquake clusters are identified in northeastern part of the study area (Fig. 8c, d). The first, southern one is associated with supposed NW-striking faults crossing the northeastern branch of the Vityaz Ridge. Two other clusters are observed beneath the forearc region in front of Onekotan Island and seem to be associated with unknown transverse fault crossing the northernmost part of the Vityaz Ridge and bordering the northeastern boundary of suggested seismic gap. A noticeable shift of epicenters, which follow the general strike of the Vityaz Ridge, towards the islands is clearly seen in the northeastern part of the study area. It can be explained by decrease in the dip angle of Benioff seismic zone northeast of the Kruzenstern Strait supposed by TARAKANOV et al. (1983).

The CMT catalog (available at http://globalcmt.org/) contains centroid moment tensor solutions of 667 earthquake sources within the study area for the time period from 1977 to 2014. Using the CMT catalog, we determined the 'type-of-motion' in the earthquake source and classified seismic events into three types according to this parameter: reverse fault (R), normal fault (N), and strike-slip fault (S) events. There are CMT solutions of 324 earthquake sources with $H < 105$ km for the time period of 1977–2006, and they are classified as follows: R type—284 events (88 %); N—27 (8 %); and S—13 (4 %). Therefore, the reverse faulting is a highly dominating type of source rupture for shallow background seismicity within the Central Kuril Islands region. For earthquakes with a source depth greater than 100 km, however, the normal faulting becomes prevailing.

In July 2006, a gradual increase in seismic activity of the studied area began, which continued to the end of September 2006. Most of the recorded shocks that time concentrated in a relatively small area closer to the trench in forearc segments 4–6 and 8 (Fig. 8d). The September 28, 2006 $M_w 5.9$ earthquake was the largest event in this group and a precursor of the September–October 2006 earthquake swarm observed in these area. An unusual burst (swarm) of seismic activity occurred here from September 30 to October 1, 2006, i.e. 1.5 months before the mainshock of the great 2006 earthquake (Fig. 9). The epicenters of these shocks are highlighted in black with a cross in Fig. 8b, d. A series of about 60 earthquakes was recorded that time, with the largest

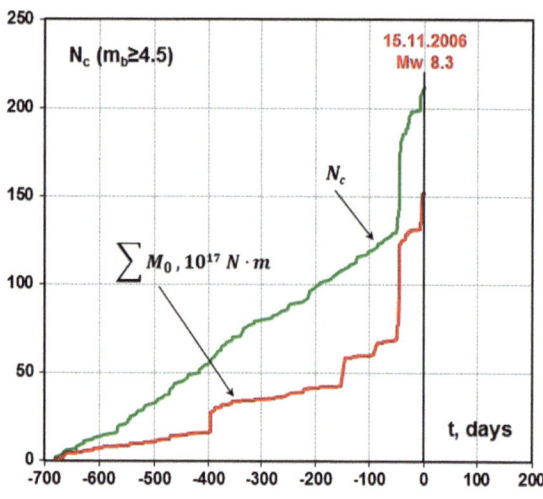

Figure 9
Seismic activity rate within the Central Kuril Islands region since 01.01.2005 preceding occurrence of the great November 15, 2006 M_w 8.3 Kuril earthquake. *Green line* denotes cumulative number N_c of shocks of magnitude $m_b \geq 4.5$ (by ANSS ComCat catalog), and *red line* shows cumulative released seismic moment M_0 within the study area. Note a sharp increase of both parameters about 1.5 month (~45 days) prior to occurrence of the November 15, 2006 main shock. See details in the text

one of $M_w = 6.6$ located directly on the NW-striking transcurrent fault. All the recorded events were shallow ($H = 10$–40 km) and mostly formed a compact group of epicenters within lower regions of forearc segments 4–6 bordered by transcurrent faults. Since October 2 until the November 15, 2006, a gradual decrease in seismic activity was observed, although the general distribution of recorded earthquakes in space was similar to that before the swarm. In this time period, two small clusters of shocks located within the forearc segments 4–6 and near the border of northeastern branch of the Vityaz Ridge were identified.

6.2. 2006 and 2007 Mainshocks and Aftershocks

To investigate the distribution of seismicity following main shocks of the great 2006 and 2007 Kuril earthquakes we examined the seismic events with $M_w \geq 4.5, H < 105$ km observed within the study area during four time stages: (1) from 15 November 2006 to 13 January 2007, i.e. between two main shocks occurrences (about 2 months); (2) from 13 January 2007 to 15 March 2007, i.e. 2 months after the 2007 main shock occurrence; (3) from 15 March 2007 to the end of 2010 (about 4 years); and (4) 2011–2014 (4 years). For the stages 1 and 2 we used the EHB catalog, and for the stages 3 and 4—the ANSS ComCat catalog, because the EHB catalog only exists up to the end of 2008.

6.2.1 1st Stage: 15/11/2006–13/01/2007

The mainshock of the great November 15, 2006 M_w8.3 earthquake occurred on the megathrust; its epicenter was located near the boundary fault separating crustal blocks 5 and 6 (Fig. 10a). The first-day aftershocks form two clusters: the first one occupies parts of the forearc area, while the second cluster is confined to the Kuril Trench and its oceanic slope. There is a distinct gap between both clusters about 25–50 km in width elongated parallel to the oceanward slope of the Vityaz Ridge. Just a few aftershocks were located within the mentioned gap. The relative position and size of both clusters are illustrated in Fig. 11 (red line) using a projection across the Kuril Trench.

After the mainshock occurrence, contours of the aftershock area changed (Fig. 10a). Within 2 months, the first cluster (beneath the island arc slope) expanded to the southwest and northeast, where some transcurrent faults were activated, as well as to the northwest up to Simushir and Ketoi Islands due to slow slip migration into the deeper regions of subduction zone (35–70 km). The aftershock area delineated by the first cluster occupied six crustal segments during the first day, but seven segments during the next 2 months (segments 4–9 and 4–10 in Fig. 10a, respectively). The second cluster (beneath the oceanic slope of the trench) extended mostly in southwestern and northeastern directions, although not far beyond the closest fracture zones within the oceanic plate.

According to the CMT solution, the November 15, 2006 Kuril earthquake was a great low-angle megathrust event, which is typical for subduction zones. Slip distribution in the source was estimated from inversion of teleseismic body and surface waves (JI 2006; AMMON et al. 2008; LAY et al. 2009; BABA et al. 2009; RAEESI and ATAKAN 2009; and others), GPS data (STEBLOV et al. 2008), and tsunami

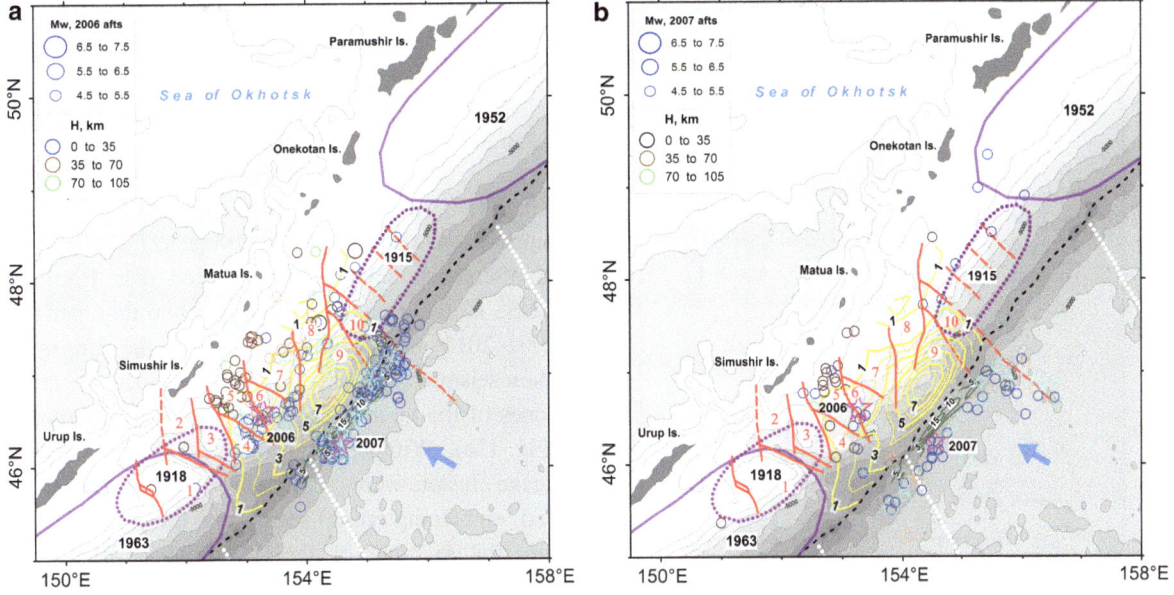

Figure 10
Epicenters of the great 2006 M_w8.3 (**a**) and 2007 M_w8.1 (**b**) Kuril earthquakes (*open magenta stars*), and their 2-month aftershocks (*circles*) according to the EHB catalog. One-day aftershocks are highlighted in *cyan* (for $H < 35$ km), and in *peach* (for $35 \leq H < 70$ km). The legend gives a classification of earthquakes by moment magnitude M_w and source depth H. See other legend notation in Fig. 8

observations (FUJII and SATAKE 2008; TANIOKA et al. 2008). Using the slip distribution according to JI (2006), we can see that few aftershocks occurred in the region of highest slip (greater than 4–5 m) that usually is referred to as 'asperity' (Fig. 10a). Similar picture was observed for many earthquakes around the world and is interpreted as implying that asperities break completely and thus cannot produce aftershocks (DAS and HENRY 2003). The type of motion in sources of the aftershocks occurred beneath the island arc slope and beneath the oceanic slope of the Kuril trench was quite different. In the first case, reverse faults (thrusts) were prevailing, but normal faults dominated in the second case; in both cases rupture planes were oriented mostly parallel to the trench (AMMON et al. 2008; LAY et al. 2009; LOBKOVSKY et al. 2009). Similar distribution of type of motion in the aftershock sources after great megathrust earthquakes is also known for other subduction zones and is usually explained by stress transfer (AMMON et al. 2008; LAY et al. 2009; LOBKOVSKY et al. 2009; OGATA and TODA 2010).

6.2.2 2nd Stage: 13/01/2007–15/03/2007

The mainshock of the great January 13, 2007 M_w8.1 Kuril earthquake was located under the oceanic slope of the Kuril Trench at a distance of about 20 km from the trench axis (Fig. 10b). The first-day aftershocks were also confined to the oceanic plate, maximum activity was observed at a distance of 25–30 km oceanward from the trench axis (Fig. 11, blue line), and few aftershock clusters were located in the southwestern and northeastern margins of the source area, in vicinity of two fracture zones within the oceanic plate. A remarkable chain of aftershock epicenters bordering the rupture zone of the 2007 event from the northeast coincides with NW-striking fault suggested from the bathymetric data (Fig. 10b). During the next 2 months the aftershock area slightly expanded in the southwestern, along-trench direction only. As for diffused seismic activity located within the forearc area at this stage, we associate this observation with the megathrust 2006 Kuril event rather than with the intraplate 2007 event.

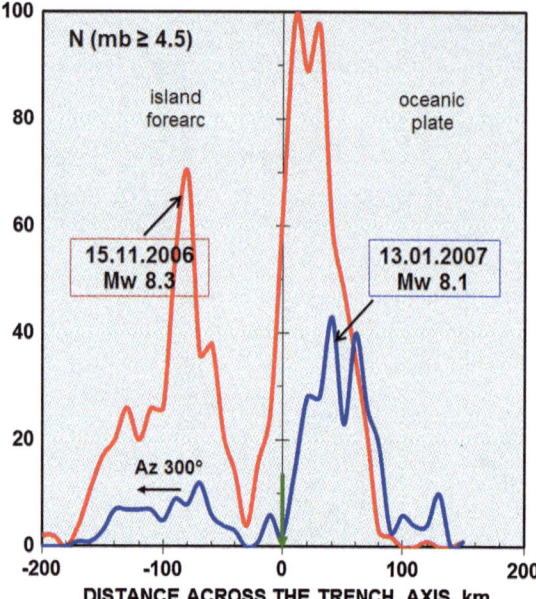

Figure 11
Distribution of 2-month aftershocks ($m_b \geq 4.5$) of the great 2006 and 2007 Kuril earthquakes across the Kuril Islands Arc slope, deep Kuril Trench, and incoming oceanic plate. Number of shocks is shown as a function of the distance D across the trench axis with a step $\Delta D = 10$ km. Note that maximum aftershock activity following both great events was observed within the oceanic plate, although at different distances from the trench. See details in the text

According to the CMT solution, type of motion within the source of the January 13, 2007, $M_w 8.1$ earthquake was a normal fault, which represents a typical response of the oceanic plate within the outer rise to the great megathrust earthquake (JI 2007; AMMON et al. 2008; LAY et al. 2009; OGATA and TODA 2010). Again, similar to the case of 2006 event, we can see few aftershock locations within the region of highest slip in the source of 2007 event, according to JI (2007) (Fig. 10b). Very similar focal mechanism, i.e. normal fault, was obtained for most of aftershocks of the 2007 event with both focal planes being oriented parallel to the trench. Data about detailed structure of the oceanic plate in the study area are absent. However, we can suggest the existence of normal faults parallel to the trench axis as such kind of structural pattern was established, for example, for the oceanic slope offshore the Kamchatka Peninsula (SELIVERSTOV 2009). Few events had focal planes oriented perpendicular to the trench, with normal or strike-slip type of motion in the sources located near the northeastern and southwestern borders of the January 2007 source zone (see Fig. 1 in LAY et al. 2009).

6.2.3 3rd Stage: 15/03/2007–31/12/2010

After 15/03/2007, aftershocks continued filling the rupture zones of both great earthquakes. For analysis of aftershock activity, we selected data from the ANSS ComCat catalog (2014) for two time periods: 2007–2010 and 2011–2014. During the 3rd stage, main seismic activity was mostly confined to rupture zones of the great 2006 and 2007 Kuril earthquakes (Fig. 12a). Within rupture area of the 2006 earthquake clusters were mainly observed in southwestern part, embracing crustal segments 4 through 5. Within the rupture zone of the 2007 event, aftershocks concentrated predominantly in the northeastern margin near the fracture zone within the oceanic plate. Outside the 2006 rupture zone, increase in seismic activity was observed beneath the southwestern Vityaz Ridge (segment 1 in Fig. 12a), and in the vicinity of the supposed NW-trending faults northeast of the source region. The largest for this time period aftershock with $M_w 7.4$ occurred on 15/01/2009 within the oceanic plate near the northeastern terminus of the 2007 rupture zone. Its hypocenter was located at a depth of 25–30 km and the focal mechanism evidences a compressional event with a rupture plane oriented along the Kuril trench and steeply dipping to the northwest (LAY et al. 2009).

6.2.4 4th Stage: 01/01/2011–31/12/2014

Since 2012, however, the pattern of seismic activity within the study area abruptly changed. Number of shocks inside rupture zones of the great 2006 and 2007 Kuril earthquakes significantly decreased. At the same time, seismic activity revived noticeable near the assumed boundaries of the Central Kuril Islands seismic gap, relatively far from rupture zones of both great events (Fig. 12b). From February 25 through 29, 2012, 12 shallow earthquakes occurred near the northeastern boundary of the seismic gap, and half of them were of magnitude $M_w > 5$. In March–April 2012, more events occurred here, and the largest one was of $M_w = 5.6$. From June 7

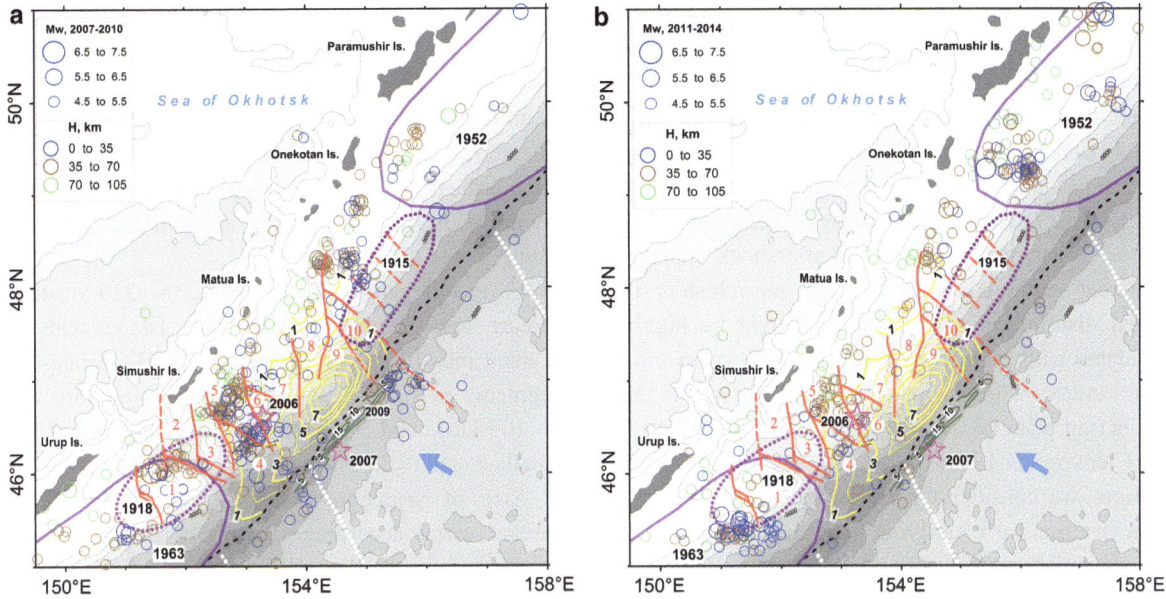

Figure 12
Epicenters of earthquakes ($M_w \geq 4.5, H < 105$ km) occurred in the Central Kuril Islands region during the time periods (indicated within legend): **a** 2007–2010; and **b** 2011–2014, based on the ANSS ComCat catalog. The legend gives a classification of shocks by moment magnitude M_w and source depth H. See other legend notation in Fig. 8

through 14, 2012, a burst of seismic activity was observed near the southwestern boundary of the seismic gap, when 10 shallow earthquakes with magnitude of up to $M_w = 5.6$ occurred here (Fig. 12b). A new episode of seismic activity was observed in 2013, when several shocks with $M_w \geq 5$ and one shock with $M_w > 6$ occurred here.

We conclude that all earthquakes, which occurred within the studied area since 30/09/2006 belong to a unified seismic process. The process started with the earthquake swarm of 30/09/2006–01/10/2006 that represented foreshocks of the great 2006 Kuril earthquake and was concentrated within crustal segments 4–6 and 8 (Fig. 8d). A rupture zone of the 2006 earthquake also occupied these segments and finally expanded up to the eastern boundary of segment 10.

The $M_w 8.1$ earthquake of January 13, 2007, occured within the oceanic slope of the Kuril trench. The November 15, 2006, $M_w 8.3$ earthquake caused abrupt decrease in compression stress within the bending oceanic plate and promoted the January 13, 2007, $M_w 8.1$ normal-fault earthquake (LAY et al. 2009; OGATA and TODA 2010). Character of displacements in sources of the 13/01/2007 and 15/01/2009 earthquakes occurred within the oceanic plate, correlates well with a type of stress existing within the plate when it bends before subduction. In the upper part of the oceanic plate extension conditions result in normal faulting and in the lower part compressional conditions cause reverse faulting.

7. Discussion

7.1. Segmentation of the Central Kuril Islands Forearc

Most forearcs and active continental margins are structurally segmented. Boundaries between segments are usually established on the base of a number of geological and seismological parameters summarized in (CARR et al. 1974; CHEN et al. 2011). Geological parameters include: (1) changes in the strike and displacement of the axes of deep-sea trenches; (2) changes in the strike and displacement of geological structures and lines of active volcanoes; (3) transcurrent faults; (4) differences in the directions or rates of long-term permanent vertical motion.

Seismological parameters are: (1) amounts of strain accumulation and release; (2) patterns and rates of seismic activity; (3) focal mechanism orientations; (4) recurrence intervals for megathrust ruptures; (5) lateral boundaries of the sources of great earthquakes; (6) changes in the strike and dip angles of subduction zones, and (7) anomalies of seismic velocities. It is supposed that rupture zones and aftershock areas of the great earthquakes as well as gaps/clusters in spatial distribution of earthquakes might be highly correlated with segmented structure of the overriding plate and/or with main tectonic features of the subducting plate (BARRIENTOS and WARD 1990; BECK and CHRISTENSEN 1991).

Here we consider the transcurrent faults as the main criterion for identifying the crustal segments. It was shown above that NS and NW-striking transcurrent faults are widely spread within the Central Kuril Islands forearc area. The single channel seismic reflection method does not allow estimating a depth of faulting. Such estimates were obtained, for instance, in the work (COLLOT et al. 2004). The authors performed investigation of transcurrent faults of the northern Ecuador—southwest Colombia margin using multi-channel seismic survey, and showed that transcurrent faults are deep-seated and can be traced down to the roof of subducting oceanic plate. The transcurrent faults within the Central Kuril Islands forearc area are deep-seated based on analysis of P wave velocity distribution with depth from a 250 km-long deep seismic sounding profile along the Kuril Islands Arc from Urup Island in the south up to Rasshua Island in the north (ZLOBIN et al. 2008). The obtained V_P distribution, which is highly heterogeneous, allows identifying of, at least, five crustal blocks (segments) separated by faults that are traced down to the Moho.

Only the Bussol faults can be confidently traced to the Kuril Trench. On the other hand, the segmentation of the Pacific plate, which is subducting under the Kuril-Kamchatka island arc, is determined by the occurrence of a number of fracture zones that are almost normal to the strike of the given subduction zone. These fracture zones have been recognized from relative displacements of the Mesozoic magnetic lineaments (HILDE et al. 1977). Investigations of the oceanward slope of the Kuril trench (GNIBIDENKO et al. 1981) showed that fracture zones are manifested in both the bottom topography and the acoustic basement as linear scarps or basement highs 1–1.5 km in amplitude. Two faults within the island arc slope (the above-mentioned Bussol Graben and one fault in the northeastern Vityaz Ridge) are located on the continuation of these fracture zones under the overriding plate.

From our geophysical surveys, the 330 km-long forearc area in the Central Kuril Islands can be divided into 10 crustal segments (Fig. 3). Triangular segments 3, 4 and 5 have the largest along-arc size ($\sim$100 km), and a rectangular segment 9 has the smallest size (<30 km).

Size of segments identified within the island forearc varies according to the criteria used. Based on changes in the dip of the subducting plate (from 10°N to 45°S), only five segments were identified in a zone of the Nazca Plate subduction under South America along the whole length of over 6000 km (BARAZANGI and ISACKS 1976). Size of segments appears to be much smaller if they are identified based on rupture zones and/or aftershock areas of great earthquakes, data of neotectonics (including coseismic horizontal and vertical displacements), and morphological features. For instance, size of segments within the Nankai arc varies from 100 to 150 km with average value equal to 135 km (ANDO 1975). Within the Hokkaido area, these values are from 115 to 260 km with an average size of 160 km (HIRATA et al. 2009).

The along-arc size of segments identified based on earthquake distribution was as follows: 150–500 km along the Sumatra–Andaman subduction zone (BILHAM 2005); 100–750 km within the Aleutians (LU and WYSS 1996; NISHENKO and JACOB 1990; SHENNAN et al. 2009), and 100–300 km along the Kuril–Kamchatka Arc (LOBKOVSKY et al. 1991). Based on the established transcurrent faults, two segments with along-arc size 260 and 83 km were identified within the northern Ecuador–southwest Colombia subduction margin (COLLOT et al. 2004).

If the geological data are used (paleoshore line positions of modern or fossil coral reefs), estimated segment size appears to be even smaller. For instance, along-arc size of forearc segments identified within the Mariana, New Hebrides and Tonga island arcs is 30–100 km (DICKINSON 2001; DICKINSON and BURLEY 2007; TAYLOR et al. 1990, 1995; TAYLOR and BLOOM 1977). The same method allows identifying of 16 forearc segments for Solomon Islands Arc with a size from 30 to 130 km and average value of 75 km (CHEN et al. 2011).

Our geophysical investigations of the Central Kuril Islands forearc region allow obtain very similar results. We conclude that the 330 km-long forearc area in the Central Kuril Islands region to be divided into 10 crustal segments with a maximum along-arc segment size not exceeding 100 km, and with an average segment size of 33 km.

7.2. Relationships Between Forearc Segmentation and Seismic Activity

Based on the instrumental and historical data it was suggested that within the Kuril-Kamchatka subduction zone average duration of seismic cycle, i.e. the time interval T between two successive earthquakes of $M \geq 7.7$ in any place of zone follows normal distribution with parameters: $T_1 = 140 \pm 60$ or $T_2 = 120 \pm 50$ years (FEDOTOV 1968, 2005; FEDOTOV et al. 1980; FEDOTOV and CHERNYSHEV 1990). Arc segments where no earthquake of $M \geq 7.7$ occurred during 80 years or longer were considered as seismic gaps, i.e. as expected places of the next large earthquake occurrence in future.

Large earthquakes did not occur near the edges of the Central Kuril Islands seismic gap during at least 90–100 years (see Fig. 1). There are no instrumental data about such events in the central part of the seismic gap. Russian and Japanese historical sources mention the large tsunamigenic earthquake may have occurred in front of Urup or Simushir Islands on 29 June 1780 (New Catalog 1982; HATORI 2007), but so far the location of its rupture area is unknown. Based on the observed attenuation of tsunami run-ups along the Japanese and Kuril Islands coasts, HATORI (2007) suggested that a source area of the 1780 event was very similar to source zones of the great 1918 and 1963 Kuril earthquakes. If we assume occurrence of the 1780 earthquake within the Central Kuril Islands seismic gap, then at the moment of the great 2006 Kuril earthquake occurrence, a quiescence period within seismic gap is estimated as 226 years, which significantly exceeded suggested average duration of seismic cycle (T_1 and/or T_2) for the Kuril-Kamchatka subduction zone.

Within the Central Kuril Islands seismic gap, transcurrent faults separate the forearc area into segments with along-arc size from 30 to 100 km, which could serve as barriers that prevent stress transferring between the neighboring segments. Perhaps the long recurrence interval reflects the time required for the system to reach a state when simultaneous and synchronous activation of several segments (blocks) becomes likely. In fact, the great 2006 $M_w 8.3$ Kuril earthquake ruptured a subduction zone interface of 250 km in length, so that accumulated stress was released simultaneously within seven segments (4–10) composing a rupture zone of this event (Fig. 10a). According to the finite fault model developed for this event by JI (2006), the rupture initiated on the megathrust beneath the boundary between segments 5 and 6, i.e. within the region of low slip, and propagated mostly within the near-trench region parallel to the trench axis where the region of highest slip was located. As well, the coseismic slip occurred simultaneously within seven crustal segments (blocks) 4 through 10, however the large transverse faults in front of Bussol and Kruzenstern Straits apparently prevented further spread of the rupture beyond these limits (Fig. 10a). If we use the source model for the 2006 event from LAY et al. (2009), we arrive at the same conclusion as the two models are similar.

Slip distribution models for the great 2006 Kuril earthquake were also obtained using inversion of the data of instrumental records: GPS (STEBLOV et al. 2008); and tsunami waves (FUJII and SATAKE 2008; TANIOKA et al. 2008) (Fig. 13). In addition, in Fig. 13 the slip distribution model obtained by IOKI and TANIOKA (2011) for the great 1963 $M_w 8.5$ Kuril earthquake using inversion of available tsunami records is presented. For the 2006 event, slip distributions in the source inferred from teleseismic and tsunami records are similar in size of the rupture area and value of maximum slip (7–8 m), although differ in location of the region of highest slip; this discrepancy is probably explained by uncertainties in tsunami travel times at long distances due to imperfect bathymetry data. However, the slip model

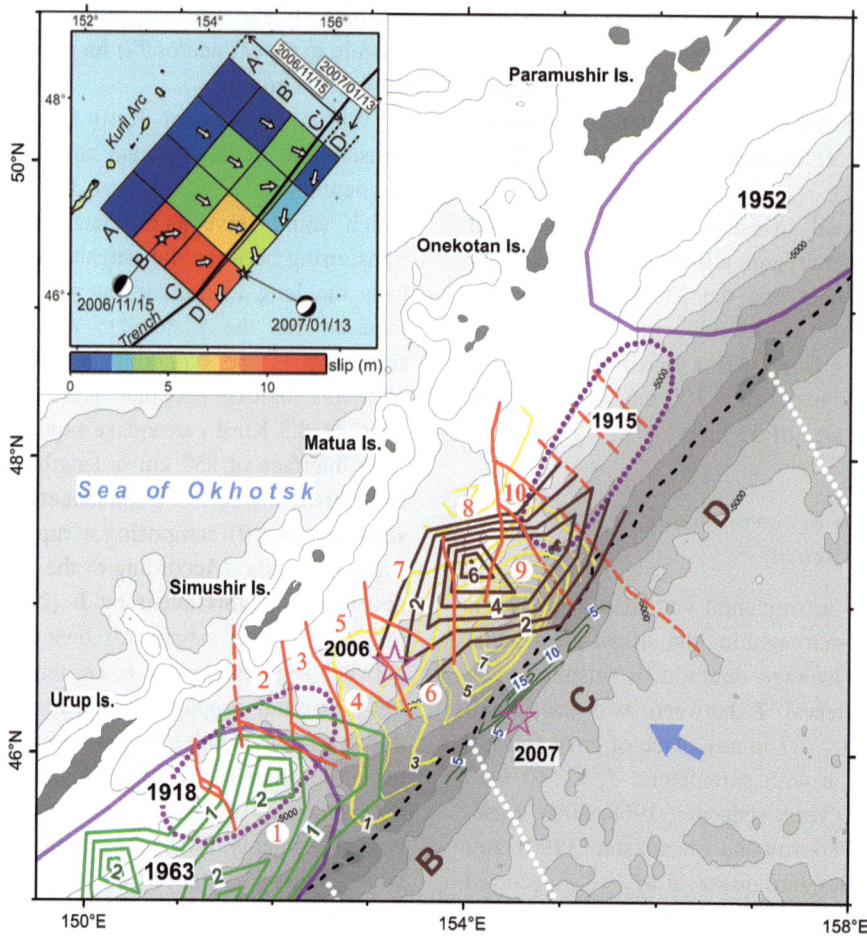

Figure 13
Coseismic slip (in m) in the source of the great 1963 M_w8.5 Kuril earthquake (*solid green lines*) obtained by IOKI and TANIOKA (2011) and in the source of the great 2006 M_w8.3 Kuril earthquake (*solid dark brown lines*) obtained by FUJII and SATAKE (2008). Both slip models were inferred using inversion of the recorded tsunami data; here we interpolate slip distributions based on the table data presented in these works. *Inset* shows the distribution of slip in the source of the great 2006 Kuril event obtained using the data of regional GPS observations (STEBLOV et al. 2008). Note that for the 2006 Kuril event, regions of highest slip in three slip models are shifted relative to each other. *Letters B, C,* and *D* denote segments within the subducting oceanic plate. See other legend notation in Fig. 8

obtained from GPS data (STEBLOV et al. 2008) (see Fig. 13, inset), shows the region of highest slip much closer to the epicenters of 2006 and 2007 events, within segments 4–6, and almost between the rupture zones of 1963 and 2006 events inferred from tsunami data. The discrepancy between different slip models needs to be clarified. Nevertheless, all obtained slip models evidence coherent failure of several segments of the overriding plate during the great megathrust earthquakes.

Similar scenarios of occurrence of the great earthquakes are known as well for other subduction zones. The Aleutian Island Arc may be considered as an example. The arc consists of adjacent blocks with linear along-arc size from tens to hundred kilometers (GEIST et al. 1988). Boundaries between neighboring block are represented by fault-governed deep canyons cutting the frontal (southern) part of the island arc transversally to its general strike. Two great earthquakes that occurred here in 1957 and 1965 had anomalously long rupture zones (~800 and 600 km, relatively), which embraced several segments (blocks) of the arc. One recent example of similar earthquake scenario is the tragic 2004 Sumatra–

Andaman event with its rupture zone embraced from 9 to 12 blocks and a length of about 1300 km (AMMON et al. 2005).

7.3. On the Long-Term Earthquake Forecast for the Central Kuril Islands Region

A rupture zone of the great 2006 $M_w8.3$ Kuril earthquake had a length of about 250 km and did not cover the whole extent of the long-lived Central Kuril Islands seismic gap (450–550 km). The similar or even larger in size northeastern part of the seismic gap was considered as a likely place of the next great earthquake occurrence (FEDOTOV et al. 2007; LAY et al. 2009). As well, southwestern part of the seismic gap was also included in the long-term forecasting scheme (FEDOTOV et al. 2007, 2012). To date, the long-term forecast for the Kuril-Kamchatka subduction zone is published for the 5-year-long period since September 2013 through August 2018 (FEDOTOV and SOLOMATIN 2015). Two likely places of a large earthquake ($M \geq 7.7$) occurrence were distinguished within the Kuril-Kamchatka subduction zone: (1) an area offshore the southeastern coast of Kamchatka Peninsula, and (2) Central Kuril Islands seismic gap, from northern tip of Urup Island to southern tip of Paramushir Island, excluding the rupture zone of the great 2006 $M_w8.3$ Kuril earthquake (Fig. 14, inset). Integrated probability of a great earthquake occurrence during the mentioned 5-year time period is equal for both places: ~ 33 % for the first place, and ~ 35 % for the second place.

According to this forecasting scheme, the entire Kuril-Kamchatka seismogenic zone is subdivided into 20 along-arc segments of 100–150 km in length (FEDOTOV 2005). It is assumed that a large earthquake of $M \geq 7.7$ may have occurred within the Kuril-Kamchatka zone with a 100 % probability during any 5-year-long time interval. In case of random occurrences, each of 20 segments has a 5 % probability to be ruptured during a 5-year-long time interval. However, different segments show different rates of seismic activity depending on current stage of seismic cycle for a given segment and may have different probabilities of failure in a large earthquake. There are no definite criteria to determine which segment among others to be ruptured first. From their experience, FEDOTOV et al. (2007) noted that one of three segments with largest estimated probabilities of failure was usually ruptured in a large earthquake first.

In our opinion, alternative long-term earthquake forecasts for the studied seismic gap are also possible (BARANOV et al. 2013). We believe that one important aspect of long-term forecasting is as follows. According to the "keyboard model" of great earthquakes (LOBKOVSKY et al. 1991), during the mainshock and aftershocks occurrences, the release of accumulated tectonic stress within ruptured blocks never exceeds 20–30 % of the critical failure stress, especially in the rear part of the forearc wedge (LOBKOVSKY et al. 2004). Thus, after the great earthquake occurrence, the block is still being under significant horizontal compression, i.e. permanent initial stress of about 200–500 bars exists any time within it, while slow additional loading and quick partial unloading of the block are superimposed on the permanent stress state. Due to existence of significant initial stress and corresponding initial strain, we need to revise our estimates of accumulated elastic energy as compared with relatively undeformed state, which leads to important conclusion: recurrence interval of great earthquakes can vary significantly (decreasingly). We believe this consideration must be taken into account in long-term forecasting schemes.

The Aleutian Island arc represents an illustrative example for making such a conclusion. Estimated recurrence interval for great earthquakes here ($M_w \geq 7.8$) varied from 50 to 103 years being of 80 years on the average (DAVIES et al. 1981). Using this background, the following long-term earthquake forecast for the period from 1983 to 2003 was constructed (JACOB 1984): next great earthquakes could occur within three arc segments determined as seismic gaps by SYKES et al. (1981) with a probability from 99 to 30 %. In addition, a probability of great earthquake occurrence within arc segments containing rupture zones of the great 1957, 1964 and 1965 earthquakes was estimated to be in a range from 17 to 9 %. However, this long-term forecast failed because two great earthquakes, which happened in 1986 ($M_w \geq 8.0$) and in 1996 ($M_w \geq 7.9$), were confined to rupture zones of the 1957 event (JOHNSON et al. 1994; TANIOKA and GONZÁLEZ 1998), and the 2003

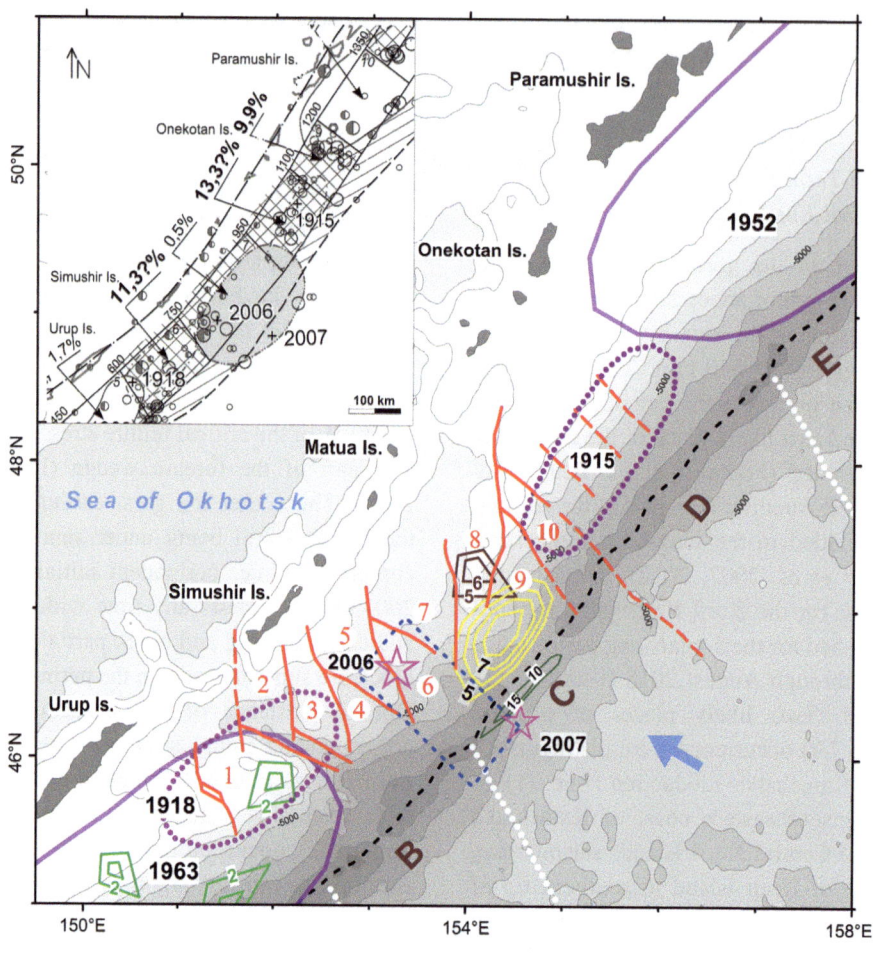

Figure 14
Supposed lithospheric segments within the Central Kuril Islands forearc region (A–E) based on locations of source zones of the great historical earthquakes, distribution of asperities (regions of highest slip) along the forearc region, and fracture zones within the oceanic plate. *Dotted blue line* delineates region of highest slip in the source of the 2006 Kuril earthquake based on inversion of GPS data (STEBLOV et al. 2008). *Inset* shows probabilities of occurrence of $M_w \geq 7.7$ earthquake during 09/1913–08/1918 within defined along-arc segments 5 through 10, according to FEDOTOV and SOLOMATIN (2015). See details in the text. See other legend notation in Fig. 8

event ($M_w \geq 7.8$)—to a rupture zone of the great 1965 earthquake (RUPPERT et al. 2007). The estimated duration of seismic cycle, at least for the Central Aleutians, appeared to be of 30–40 years, i.e. twice shorter than it was supposed.

Another example is the great 1906 $M_w 8.4$ earthquake within Ecuador–Colombia subduction zone when a synchronous rupture of three asperities occurred, and later, within decades from each other, these asperities ruptured again as separate great earthquakes (COLLOT et al. 2004).

As it was shown above, seismicity pattern within a rupture zone of the 2006 Kuril earthquake dramatically changed after 2011. Before the end of this year, aftershocks concentrated beneath southwestern and northeastern margins of rupture zone (Fig. 12a), however, beginning since 2012 the number of aftershocks abruptly decreased and nearly approached to background level (Fig. 12b). The change in seismic activity probably evidences that rupture zone of the 2006 event is already unloaded to initial stress, the latter is suggested to be 20–30 % lower than the critical failure stress accumulated before the 2006 event. Occurrences of earthquake swarms in 2012 near southwestern and northeastern edges of the seismic gap just in the same manner as it was observed before 2006 earthquake (Fig. 8d) can indicate that reorganization of tectonic stress after the great 2006 event has already completed, and stress accumulates again within the entire seismic gap.

The occurrence of great earthquakes within the Central Kuril Islands forearc region is closely related to its segmented structure. We can see here segments of two kinds: (a) larger lithospheric segments within the subducting oceanic plate separated by old fracture zones from each other and indicated by brown letters B, C, D, and E in Fig. 14; and (b) smaller crustal segments within the overriding plate separated by the transcurrent faults and denoted by red numbers 1, 2, ..., 10 in Fig. 14. Both fracture zones and transcurrent faults serve as barriers preventing stress transfer between neighboring segments and thus controlling stress accumulation and release. The along-arc size of the crustal forearc segments is 30–100 km which corresponds to the earthquake magnitude $M_w = 6.8$–7.7, whereas segments of oceanic lithosphere are 150–180 km in extent along the arc, and, therefore, are capable to produce megathrust earthquakes of magnitude $M_w = 8.0$–8.1 (BLASER et al. 2010). If two lithospheric segments rupture simultaneously, for example, C and D, then rupture area can reach 320–350 km in length which corresponds to earthquake magnitude $M_w = 8.5$–8.6. And in case of rupture embracing three such segments (B, C, and D, i.e. the entire Central Kuril Islands seismic gap), the source extent will be about 500 km, which corresponds to earthquake magnitude $M_w \approx 8.9$.

The most likely place for a great earthquake occurrence within the Central Kuril Islands region is oceanward slope of the northeastern Vityaz Ridge between Kruzenstern and Third Kuril Straits, where the last great earthquake occurred in 1915 (segment D and probably some parts of adjacent segments C or/and E). The second likely place is, probably, the source zone of the great 1918 earthquake, as indicated by FEDOTOV and SOLOMATIN (2015). This segment of subduction zone seems to have a very short seismic cycle $T = 60 \pm 15$ years (based on occurrences of 1780, 1843, 1918, and 1963 great earthquakes).

At the present time, we cannot give any forecast in time. Too many factors should be taken into account, however, they mostly are poorly understood.

8. Conclusions

The occurrence of the great November 15, 2006 $M_w 8.3$ Kuril earthquake reflects the segmented crustal structure of the forearc area in this region. From the structural data obtained in two marine expeditions, the Central Kuril Islands forearc region is divided into 10 crustal segments (blocks) separated by transcurrent faults. The along-arc size of identified forearc segments varies from 30 to 100 km, i.e. somewhat smaller than those reported for most other convergent margins. The transcurrent faults serve as barriers that prevent stress transferring between the neighboring segments. Therefore, stress accumulated within separate segments is mainly released by earthquakes of moderate-to-strong ($M_w \sim 5/7$) magnitudes.

The distribution of observed historical earthquakes within the Central Kuril Islands forearc region for the last ~ 100 years is in a good agreement with segmented crustal structure of the overriding plate. The geographic extent of earthquake rupture zones, aftershock areas, and seismicity clusters correlate well with our tectonically defined forearc segments (see Figs. 8a–d, 10a–b, 12a–b). Based on aftershock and slip distributions obtained by inversion of teleseismic and tsunami data, a rupture zone of the great 2006 Kuril earthquake having a length of about 250 km was located in a central part of the entire seismic gap with a total extent of 450–550 km. The remaining parts of the seismic gap situated to the southwest and, especially, to the northeast of the 2006 rupture area are now considered as possible places of the next great earthquake occurrence (FEDOTOV et al. 2007, 2012; LAY et al. 2009; LOBKOVSKY et al. 2009).

Aftershock activity after the great 2006 Kuril earthquake was observed during a rather short time period. Reorganization of the stress to a new state seems to be completed before the 2012, when an increase in seismic activity was observed near the edges of the Central Kuril Islands seismic gap (see Fig. 12b). We can suggest, following LOBKOVSKY et al. (2004), that release of accumulated tectonic stress during the 2006 main shock and aftershock stages was incomplete, as it usually did not exceed 20–30 % of the critical failure stress. Based on this approach, we suggest three possible scenarios of the great earthquake occurrence within the Central Kuril Islands region.

The most probable scenario implies the occurrence of a great, $M_w \sim 8.0$-8.1 earthquake to the northeast of the 2006 rupture area, between Kruzenstern and Third Kuril Straits, where a source zone of the great 1915 earthquake was located (Fig. 14).

The second possible scenario implies the occurrence of a great $M_w \sim 8.0$ earthquake to the southwest of the 2006 rupture zone, but the present state of the forearc area in this part of the Central Kuril Islands seismic gap is unclear. The short duration of seismic cycle $\sim 60 \pm 15$ years for earthquakes of $M_w \geq 7.7$ in this area supports the second scenario as the last great event occurred here in 1963, more than 50 years ago. However, Ioki and Tanioka (2011) suggested that rupture areas of the great 1963 and 2006 earthquakes are adjacent and do not overlap each other, which makes the second scenario doubtful.

The third possible scenario implies that two lithospheric segments B, C or C, D rupture simultaneously in a great earthquake of magnitude $M_w \sim 8.5$-8.6. This scenario has a low probability, because accumulated stress and strain within segment C were partly released during the great 2006 Kuril earthquake. It should be noted that the megathrust event of $M_w 8.5$-8.6 might generate a tsunami waves significantly exceeding in height the waves produced by the great 2006–2007 Kuril earthquakes. In addition, the destructive tsunami waves can penetrate through deepwater Bussol and Kruzenstern straits into the Okhotsk Sea and reach the Eastern Sakhalin coasts, where intensive oil-and-gas production activity is developed. Therefore, simulation of hazardous tsunami generation in result of great megathrust earthquake occurrence within the Central Kuril Islands seismic gap is undoubtedly a task of vital importance.

We believe that new bathymetry and seismic reflection data presented here can help to better understand the segmented structure of the Central Kuril Islands forearc region and eventually resolve between possible scenarios of the great earthquake occurrences here.

Acknowledgments

We acknowledge the Editor, Alexander Rabinovich, for his positive and constructive comments and suggestions. We are deeply grateful to Ray Wells for patiently correcting the manuscript and for his valuable suggestions, which significantly enhanced the manuscript. We thank the anonymous reviewer for his very helpful criticism and comments. This work was supported by the Russian Science Foundation, RSF Grant # 14-50-00095.

REFERENCES

Ammon, C.J., Ji, C., Thio, H.-K., Robinson, D., Ni, S., Hjorleifsdottir, V., Kanamori, H., Lay, T., Das, S., Helmberger, D., Ichinose, G., Polet, J., and Wald, D. (2005), *Rupture process of the 2004 Sumatra–Andaman earthquake*, Science *308*, 1133–1139.

Ammon, C.J., Kanamori, H., and Lay, T. (2008), *A great earthquake doublet and seismic stress transfer cycle in the Central Kuril Islands*, Nature *451*, 561–565.

Ando, M. (1975), *Source mechanisms and tectonic significance of historical earthquakes along the Nankai Trough, Japan*, Tectonophysics *27*, 119–140.

ANSS ComCat (2014), Composite Earthquake Catalog, Northern California Earthquake Data Center. http://www.quake.geo.berkeley.edu/cnss/.

Avdeiko, G.P., Antonov, A.Y., Volynets, O.N., et al. (1992), Submarine Volcanism and Zonality of the Kuril Islands Arc (Nauka, Moscow 1992), 528 p. **(in Russian)**.

Baba, T., P.R. Cummins, H.K. Thio, and H. Tsushima (2009), *Validation and joint inversion of teleseismic waveforms for earthquake source models using deep ocean bottom pressure records: a case study of the 2006 Kuril megathrust earthquake*, Pure Appl. Geophys. *166*, 55–76.

Baranov, B.V., Lobkovsky, L.I., and Dozorova, K.A. (2013), *Probability of occurrence of a very strong earthquake in the Central Kurils Region*, Dokl. Earth Sci. *448*, 206–208.

Barazangi, M., and Isacks, B.L. (1976), *Spatial distribution of earthquakes and subduction of the Nazca plate beneath South America*, Geology *4*, 686–692.

Barrientos, S.E., and Ward, S.N. (1990), *The 1960 Chile earthquake: inversion for slip distribution from surface deformation*, Geophys. J. Int. *103*, 589–598.

Beck, S.L., and Ruff, L.J. (1987), *Rupture process of the great 1963 Kurile Islands earthquake sequence: asperity interaction and multiple event rupture*, J. Geophys. Res. *92*, 14,123–14,138.

Beck, S.L., and Christensen, D.H. (1991), *Rupture process of the February 4, 1965, Rat Islands earthquake*, J. Geophys. Res. *96*, B2, 2205-2221.

Bilham, R.A. (2005), *Flying start, then a slow slip*, Science *308*, 1126–1127.

Blaser, L., F. Krüger, M. Ohrnberger, and F. Scherbaum (2010), *Scaling relations of earthquake source parameter estimates with special focus on subduction environment*. Bull. Seism. Soc. Am. *100*, 2914–2926.

Bondár, I. and D. Storchak (2011), *Improved location procedures at the International Seismological Centre*, Geophys. J. Int. *186*, 1220–1244.

Carr, M.J., Stoiber, R.E. and Drake, C.L. (1974), The segmented nature of some continental margins, In: The Geology of Continental Margins (ed. Burke, C.A., and Drake, C.), (Springer, New York 1974), pp. 105–114.

CHEN, M.-C., FROHLICH, C., TAYLOR, F.W., BURR, G., and VAN UFFORD, A.Q. (2011), *Arc segmentation and seismicity in the Solomon Islands Arc, SW Pacific*, Tectonophysics 507, 47–69.

CMT (2014), Global CMT Web Page. http://www.globalcmt.org/.

COLLOT, J.-Y., MARCAILLOU, B., SAGE, F., MICHAUD, F., AGUDELO, W., CHARVIS, PH., GRAINDORGE, D., GUTSCHER, M.-A., and SPENCE, G. (2004), *Are rupture zone limits of great subduction earthquakes controlled by upper plate structures? Evidence from multichannel seismic reflection data acquired across the northern Ecuador-southwest Colombia margin*, J. Geophys. Res. *109*, B11103.

DAS, S., and C. HENRY (2003), *Spatial relation between main earthquake slip and its aftershock distribution*, Rev. Geophys. *41* (3), 1013.

DAVIES, J., SYKES, L., HOUSE, L., and JACOB, K. (1981), *Shumagin seismic gap, Alaska Peninsula: history of great earthquakes, tectonic setting and evidence for high seismic potential*, J. Geophys. Res. *86*, 3821–3855.

DEMETS, C., GORDON, R.G., ARGUS, D.F., and STEIN, S. (1990), *Current plate motions*, Geophys. J. Int. *101*, 425–478.

DICKINSON, W.R. (2001), *Paleoshoreline record of relative Holocene sea levels on Pacific islands*, Earth-Sci. Revs. *55*, 191–234.

DICKINSON, W.R., and BURLEY, D.V. (2007), *Geoarchaeology of Tonga: geotectonic and geomorphic controls*, Geoarchaeology *22*, 229–259.

EHB (2009), International Seismological Centre, EHB Bulletin, Thatcham, UK. http://www.isc.ac.uk.

ENGDAHL, E.R., R.D. VAN DER HILST, and R. BULAND (1998), *Global teleseismic earthquake relocation with improved travel times and procedures*, Bull. Seism. Soc. Am., *88*, 722–743.

FEDOTOV, S. A. (1965), *Regularities of the distribution of strong earthquakes in Kamchatka, the Kurile Islands, and northeastern Japan*, Trudy Inst. Fiz. Zemli, Akad. Nauk SSSR, *36*, 66–93 **(in Russian)**.

FEDOTOV, S.A. (1968), The seismic cycle, possibility of the quantitative seismic zoning, and long-term seismic forecasting, In: Seismic Zoning in the USSR (ed. S.V. Medvedev), (Nauka, Moscow 1968), pp. 133–166 **(in Russian)**.

FEDOTOV, S.A. (2005), Long-term earthquake forecasting for the Kuril–Kamchatka Arc (Nauka, Moscow 2005), 303 p. **(in Russian)**.

FEDOTOV, S.A., and CHERNYSHEV, S.D. (2002), *A long-term earthquake forecast for the Kuril–Kamchatka Arc: accuracy of forecast for the period 1986–2000, development of the method, and a new forecast for the period 2001–2005*, Volcanol. Seismol. *6*, 3–24, **(in Russian)**.

FEDOTOV, S.A., SOLOMATIN, A.V., and CHERNYSHEV, S.D. (2007), *A long-term earthquake forecast for the Kuril–Kamchatka Arc for the period 2006–2011 and approved forecast of the $M_s = 8.2$ Middle Kuril earthquake of November 15, 2006*, J. Volcanol. Seismol. *1*, 143–163.

FEDOTOV, S.A., SOLOMATIN, A.V., and CHERNYSHEV, S.D. (2012), *A long-term earthquake forecast for the Kuril–Kamchatka Arc for the period from September 2011 to August 2016. The likely location, time, and evolution of a next great earthquake with $M \geq 7.7$ in Kamchatka*, J. Volcanol. Seismol. *6*, 65–88.

FEDOTOV, S.A., and SOLOMATIN, A.V. (2015), *The long-term earthquake forecast for the Kuril-Kamchatka island arc for the September 2013 to August 2018 period; the seismicity of the arc during preceding deep-focus earthquakes in the Sea of Okhotsk (in 2008, 2012, and 2013 at $M = 7.7$, 7.7, and 8.3)*, J. Volcanol. Seismol. *9*, 65–80.

FUJII, Y., and SATAKE, K. (2008), *Tsunami sources of the November 2006 and January 2007 great Kuril earthquakes*, Bull. Seism. Soc. Am. *98*, 1559–1571.

FUKAO, Y., and FURUMOTO, M. (1979), *Stress drops, wave spectra and recurrence intervals of great earthquakes—Implications of the Etorofu earthquake of 1958 November 6*, Geophys. J. Roy. Astron. Soc. *57*, 23–40.

GEIST, E.L., CHILDS, J.R., and SCHOLL, D.W. (1988), *The origin of summit basins of the Aleutian Ridge: implications for block rotation of an arc massif*, Tectonics 7, 327–341.

GNIBIDENKO, H.S., SVARICHEVSKY, A.S., SEDELNIKOVA, S.P. and ZHIGULEV, V.V. (1981), *The structure of Tuscarora fracture zone, northwestern Pacific*, Geo-Mar. Lett. *1*, 221–224.

HATORI, T. (2007), *Magnitudes and the source areas of historical tsunamis from South Kurile to East Hokkaido*, Historical Earthquake 22, 151–155 **(in Japanese)**.

HILDE, T.W.C., UYEDA, S., and KROENKE, L. (1977), *Evolution of the Western Pacific and its margin*, Tectonophysics *38*, 145–165.

HIRATA, K., K. SATAKE, Y. TANIOKA, and Y. HASEGAWA (2009), *Variable tsunami sources and seismic gaps in the southernmost Kuril Trench: A review*, Pure Appl. Geophys. *166*, 77–96.

IIDA, K., COX, D.C., and PARARAS-CARAYANNIS, G., (1967), Preliminary catalog of tsunamis occurring in the Pacific Ocean, Data Rep. 5, HIG-67-10 (Hawaii Inst. Geophys., Univ. Hawaii, Honolulu).

IOKI, K., and Y. TANIOKA (2011), *Slip distribution of the 1963 Great Kurile Earthquake estimated from tsunami waveforms*, Pure Appl. Geophys. *168*, 1045–1052.

ISC (2014), International Seismological Centre, ISC Bulletin. http://www.isc.ac.uk/iscbulletin/.

ISC-GEM (2013), The ISC-GEM Global Instrumental Earthquake Catalogue (1900–2009). http://www.isc.ac.uk/iscgem/.

ITDB/WLD (2014), Integrated Tsunami Database for the World Ocean. 2000 BC to present, CD-ROM, Tsunami Laboratory, ICMMG SD RAS, Novosibirsk. 2014. Web-version: http://tsun.sscc.ru/nh/tsunami.php.

IVANENKO, A.N., FILIN, A.M., GORSHKOV, A.G., and SHISHKINA N.A. (2008), *New data about the structure of the anomalous magnetic field in the central part of the Kuril-Kamchatka Island Arc*, Oceanology *48*, 554–568 **(in Russian)**.

JACOB, K.H. (1984), *Estimates of long-term probabilities for future great earthquakes in the Aleutians*, Geophys. Res. Lett. 11, 295–298.

Ji, C. (2006), Rupture process of the 2006 Nov 15 magnitude 8.3 Kuril Islands earthquake (revised). http://earthquake.usgs.gov/eqcenter/eqinthenews/2006/usvcam/finite_fault.php.

Ji, C. (2007), Rupture process of the 2007 Jan 13 magnitude 8.1 Kuril Islands earthquake (revised). http://earthquake.usgs.gov/eqcenter/eqinthenews/2007/us2007xmae/finite_fault.php).

JOHNSON, J.M., TANIOKA, Y., RUFF, L.J., SATAKE, K., KANAMORI, H., and SYKES, L.R. (1994), *The 1957 great Aleutian earthquake*, Pure Appl. Geophys. *142*, 3–28.

KELLEHER, J., and MCCANN, W. (1976), *Buoyant zones, great earthquakes and unstable boundaries of subduction*, J. Geophys. Res. *81*, 4885–4896.

KIMURA, G. (1986), *Oblique subduction and collision: forearc tectonics of the Kuril arc*, Geology 14, 404–407.

KLAESCHEN, D., BELYKH, I., GNIBIDENKO, H., PATRIKEYEV, S., and VON HUENE, R. (1994), *Structure of the Kuril Trench from seismic reflection records*, J. Geophys. Res. *99*, 24,173–188.

KOSHIMURA, S., HAYASHI, Y., MUNEMOTO, K., and IMAMURA, F. (2008), *Effect of the Emperor seamounts on trans-oceanic* propagation of the 2006 Kuril Island earthquake tsunami, Geophys. Res. Lett. *35*, L02611.

KOVACHEV, S.A., KUZIN, I.P., and LOBKOVSKY, L.I. (2009), *Marine seismological observations in the Central Kurile segment before catastrophic earthquakes on November, 2006 (M = 8.3) and January, 2007 (M = 8.1)*, Izvestiya, Phys. Solid Earth *45*, 777–793.

KULINICH, R.G., KARP, B.YA., BARANOV, B.V., LELIKOV, E.P., KARNAUKH, V.N, VALITOV, M.G., NIKOLAEV, S.M., KOLPASHCHNIKOVA, T.N., and TSOI, I.B. (2007), *Structural and geological characteristics of a "seismic gap" in the central part of the Kuril Island Arc*, Russian J. Pac. Geol. *1*, 3–14.

LAVEROV, N.P., LAPPO, S.S., LOBKOVSKY, L.I., BARANOV, B.V., KULINICH, R.G., and KARP, B.YA. (2006), *The central Kurils "gap": structure and seismic potential*, Dokl. Earth Sci. *409*, 787–790.

LAVEROV, N.P., LOBKOVSKY, L.I., LEVIN, B.W., RABINOVICH, A.B., KULIKOV, E.A., FINE, I.V., and THOMSON, R.E. (2009), *The Kuril tsunamis of November 15, 2006, and January 13, 2007: two trans-pacific events*, Dokl. Earth Sci. *426*, 658–664.

LAY, T., KANAMORI, H., and RUFF, L. (1982), *The asperity model and the nature of large subduction zone earthquakes*, Earthq. Predict. Res. *1*, 3–71.

LAY, T., KANAMORI, H., AMMON, C.J., HUTKO, A.R., FURLONG, K., and RIVERA, L. (2009), *The 2006–2007 Kuril Islands great earthquake sequence*, J. Geophys. Res. *114*, B11308.

LELIKOV, E.P., EMEL'YANOVA, T.A., and BARANOV, B.V. (2008), *Magmatism of the submarine Vityaz Ridge (Pacific slope of the Kuril Island Arc)*, Oceanology *48*, 239–249.

LEVIN, B.V., KAISTRENKO, V.M., RYBIN, A.V., NOSOV, M.A., PINEGINA, T.K., RAZZHIGAEVA, N.G., SASOROVA, E.V., GANZEI, K.S., IVEL'SKAYA, T.N., KRAVCHUNOVSKAYA, E.A., KOLESOV, S.V., EVDOKIMOV, YU.V., BOURGEOIS, J., MACINNES, B., and FITZHUGH, B. (2008), *Manifestations of the tsunami on November 15, 2006, on the Central Kuril Islands and results of the runup heights modeling*, Dokl. Earth Sci. *419*, 335–338.

LOBKOVSKY, L.I., KERCHMAN, V.I., BARANOV, B.V., and PRISTAVAKINA, E.I. (1991), *Analysis of seismotectonic processes in subduction zones from the standpoint of a keyboard model of great earthquakes*, Tectonophysics *199*, 211–236.

LOBKOVSKY, L.I., NIKISHIN, A.M., and KHAIN, V.E. (2004), Current problems of geotectonics and geodynamics (Scientific World, Moscow 2004), 612 p. **(in Russian)**.

LOBKOVSKY, L.I., KULIKOV, E.A, RABINOVICH, A.B., IVASHCHENKO, A.I., FAIN, I.V., and IVEL'SKAYA, T.N. (2008), *Earthquakes and tsunamis (November 15, 2006, and January 13, 2007) in the Central Kuril Islands region: a justified prediction*, Dokl. Earth Sci. *419*, 320–324.

LOBKOVSKY, L.I., RABINOVICH, A.B., KULIKOV, E.A, IVASHCHENKO, A.I., FINE, I.V., THOMSON, R.E., IVELSKAYA, T.N., and BOGDANOV, G.S. (2009), *The Kuril earthquakes and tsunamis of 15 November, 2006 and 13 January, 2007: observations, analysis and numerical modeling*, Oceanology *49*, 166–181.

LOBKOVSKY, L.I., MAZOVA, R.KH., KISEL'MAN, B.A., and MOROZOVA, A.O. (2010), *Numerical simulation and spectral analysis of the November 15, 2006 tsunami in the Kurile-Kamchatka region*, Oceanology *50*, 449–458.

LU, Z., and WYSS, M. (1996), *Segmentation of the Aleutian plate boundary derived from stress direction estimates based on fault plane solutions*, J. Geophys. Res. *101*, 803–816.

MACINNES, B.T., PINEGINA, T.K., BOURGEOIS, J., RAZZHIGAEVA, N.G., KAISTRENKO, V.M., and KRAVCHUNOVSKAYA, E.A. (2009), *Field survey and geological effects of the 15 November 2006 Kuril tsunami in the Middle Kuril Islands*, Pure Appl. Geophys. *166*, 9–36.

MCCANN, W.R., NISHENKO, S.P., SYKES, L.R., and KRAUSE, J. (1979), *Seismic gaps and plate tectonics: seismic potential for major boundaries*, Pure Appl. Geophys. *117*, 1082–1147.

NGDC/WDS (2014), National Geophysical Data Center/World Data Service. Global Historical Tsunami Database (NGDC, NOAA). doi:10.7289/V5PN93H7.

New Catalog of Strong Earthquakes in the USSR from Ancient Times through 1977 (1982) (eds. Kondorskaya, N.V., Shebalin N.V.), (WDC-A for Solid Earth Geophysics, Boulder 1982).

NISHENKO, S.P. (1991), *Circum-Pacific seismic potential, 1989–1999*, Pure Appl. Geophys.*135*, 169–259.

NISHENKO, S.P., and JACOB, K.H. (1990), *Seismic potential of the Queen Charlotte–Alaska–Aleutian Seismic Zone*, J. Geophys. Res. *95*, 2511–2532.

OGATA, Y., and S. TODA (2010), *Bridging great earthquake doublets through silent slip: on- and off-fault aftershocks of the 2006 Kuril Island subduction earthquake toggled by a slow slip on the outer rise normal fault of the 2007 great earthquake*, J. Geophys. Res. *115*, B06318.

PACHECO, J.F., and L.R. SYKES (1992), *Seismic moment catalog of large shallow earthquakes, 1900–1989*, Bull. Seism. Soc. Am. *82* (3), 1306–1349.

PEREZ, O.J. (2000), *Kuril Islands Arc: two seismic cycles of great earthquakes during which the complete history of seismicity ($M_S \geq 6$) is observed*, Bull. Seism. Soc. Am. *90*, 1096–1100.

RABINOVICH, A.B., LOBKOVSKY, L.I., FINE, I.V., THOMSON, R.E., IVELSKAYA, T.N., and KULIKOV, E.A. (2008), *Near-source observations and modeling of the Kuril Islands tsunamis of 15 November 2006 and 13 January 2007*, Adv. Geosci. *14*, 105–116.

RAEESI, M., and K. ATAKAN (2009), *On the deformation cycle of a strongly coupled plate interface: the triple earthquakes of 16 March 1963, 15 November 2006, and 13 January 2007 along the Kurile subduction zone*, J. Geophys. Res. *114*, B10301.

RUPPERT, N.A., LEES, J.M., and KOZYREVA, N.P. (2007), *Seismicity, earthquakes and structure along the Alaska–Aleutian and Kamchatka–Kurile subduction zones: a review*, In: Volcanism and subduction: The Kamchatka Region (Geophys. Monogr. Ser. *172*, AGU), pp. 129–144.

SELIVERSTOV, N.I. (2009), Geodynamics of the junction zone between the Kuril-Kamchatka and Aleutian Island Arcs (Kamchatka State Univ., Petropavlovsk-Kamchatsky 2009), 191 p. **(in Russian)**.

SERGEEV, K.F. (1976), Tectonics of the Kuril Islands System (Nauka, Moscow 1976), 240 p. **(in Russian)**.

SHENNAN, I., BRUHN, R., and PLAFKER, G. (2009), *Multi-segment earthquakes and tsunami potential of the Aleutian megathrust*, Quaternary Sci. Revs. *28*, 7–13.

SMITH, W.H.F., and SANDWELL, D.T. (1997), *Global sea floor topography from satellite altimetry and ship depth sounding*, Science *277*, 1956–1962.

SOLOVIEV, S.L., and FERCHEV, M.D. (1961), *Summary of data on tsunamis in the USSR*, Bull. Counc. Seismol. *9*, 1–37, **(in Russian)**. Eng. Transl. by W.G. Van Campen (Hawaii Inst. Geophys., Transl. Ser. *9*, 1961).

STEBLOV, G.M., KOGAN M.G., LEVIN B.V., VASILENKO N.F., PRYTKOV A.S., and FROLOV D.I. (2008), *Spatially linked asperities of the 2006-2007 great Kuril earthquakes revealed by GPS*, Geophys. Res. Lett. *35*, L22306.

STRASSER, F.O., M.C. ARANGO, and J.J. BOMMER (2010), *Scaling of the source dimensions of interface and intraslab subduction-zone earthquakes with moment magnitude*, Seismol. Res. Lett. *81*, 941–950.

SYKES, L.R., KISSLINGER, J.B., HOUSE, L., DAVIS, J.N., and JACOB, K.H. (1981), Rupture zones and repeat times of great earthquakes along the Alaska–Aleutian Arc, 1784–1980, In: Earthquake Prediction: Int. Review Maurice Ewing Series, *4*, (eds. Simpson, D.W., Richards, P.G.), (AGU, Washington, D.C. 1981), pp. 73–80.

TANIOKA, Y., and GONZÁLEZ, F.I. (1998), *The Aleutian earthquake of June 10, 1996 (M_w 7.9) ruptured parts of both the Andreanof and Delarof segments*, Geophys. Res. Lett. *25*, 2245–2248.

TANIOKA, Y., HASEGAWA, Y., and KUWAYAMA, T. (2008), *Tsunami waveform analyses of the 2006 underthrust and 2007 outer-rise Kuril earthquakes*, Adv. Geosci. *14*, 129–134.

TAYLOR, F.W., and BLOOM, A.L. (1977), Coral reefs on tectonic blocks, Tonga Island Arc, Proc. 3rd Int. Coral Reef Symp. 2, (Univ. of Miami, Miami), pp. 275–281.

TAYLOR, F.W., EDWARDS, R.L., WASSERBURG, G.J., and FROHLICH, C. (1990), *Seismic recurrence intervals and timing of aseismic subduction inferred from emerged corals and reefs of the Central Vanuatu (New Hebrides) Frontal Arc*, J. Geophys. Res. *95*, 393–408.

TAYLOR, F.W., BEVIS, M.G., SCHUTZ, B.E., KUANG, D., RECY, J., CALMANT, S., CHARLEY D., REGNIER, M., PERIN, B., JACKSON, M., and REICHENFELD, C. (1995), *Geodetic measurements of convergence at the New Hebrides island arc indicate arc fragmentation caused by an impinging aseismic ridge*, Geology *23*, 1011–1014.

VASILIEV, B.I., and SUVOROV, A.A. (1979), Geological structure of the Bussol underwater valley (Kuril Islands Arc), In: New data on the geology of the Far Eastern seas (ed. Vasiliev, B.I.), (FESC USSR Academy of Sciences, Vladivostok 1979), pp. 58–68 **(in Russian)**.

WELLS, D.L., and K.J. COPPERSMITH (1994), *New empirical relationships among magnitude, rupture length, rupture width, rupture area, and surface displacement*, Bull. Seism. Soc. Am. *84*, 974–1002.

ZLOBIN, T.K., and ZLOBINA, L.M. (1991), *Crustal structure of the Kuril Islands system*, Pacific Geology *6*, 24–35 **(in Russian)**.

ZLOBIN, T.K., LEVIN, B.W., and POLETS A.YU. (2008), *First results of the comparison of catastrophic Simushir earthquakes on November 15, 2006 (M = 8.3) and January 13, 2007 (M = 8.1) with the deep structure of the Earth's crust in the Central Kuril Islands*, Dokl. Earth Sci. *420*, 615–619.

(Received November 29, 2014, revised June 4, 2015, accepted June 5, 2015, Published online June 26, 2015)

Pure Appl. Geophys. 172 (2015), 3537–3555
© 2015 Springer Basel
DOI 10.1007/s00024-015-1077-y

Pure and Applied Geophysics

Estimation of Extreme Sea Levels for the Russian Coasts of the Kuril Islands and the Sea of Okhotsk

GEORGY SHEVCHENKO[1] and TATIANA IVELSKAYA[2]

Abstract—Extreme sea levels arising from the combination of tides, storm surges, seasonal oscillations and tsunamis were estimated by the joint probability method for the coast of the Sea of Okhotsk and the Pacific coast of the Kuril Islands. The sea-level observations at 10 coastal tide gauges were examined. The tidal heights at most stations are about 1.5–2 m, and only at Magadan are they much larger (about 5 m). Storm surges have the largest heights for the central Kuril Islands (Matua and Iturup islands), while at the North and South Kuril Islands the surge heights are the smallest. The recurrence of tsunami heights of various probabilities was estimated for each station. The influence of tides and storm surges on the tsunami risk assessment for the Pacific coast of the Kurile Islands was found to be relatively small. For the coast of the Sea of Okhotsk, the contribution of tides and surges is the primary influence, especially for return periods less than 100 years. For longer return periods, tsunamis play the major role in forming the extreme levels (similar to the Russian coast of the Sea of Japan, e.g., RABINOVICH *et al.* 1992).

1. Introduction

The intensive development of coastal areas and the construction of complex and expensive structures (such as nuclear power stations) increase the risk of flooding (or draining) and require the precise estimation of extreme high (or low) sea levels. The tragic effects of the 2011 Tohoku tsunami on the Fukushima-Daiichi nuclear power plant is the most spectacular example of the importance of this problem. Damage to facilities for the extraction and transportation of oil and natural gas on the northeastern shelf of Sakhalin Island can lead to large-scale environmental disaster in the western part of the Sea of Okhotsk.

The sea-level variations near the coast are related to various factors: tides, atmospheric pressure and wind stress changes that cause non-tidal oscillations including storm surges, seasonal oscillations, and (in some rare cases) seismically generated tsunamis. The coincidence of these factors produces abrupt fluctuations of sea level and can lead to severe damage. The most important factor is the increase of tsunami and storm surge heights by the coincidence with high tide. For example, on November 5, 1952, the catastrophic Kamchatka tsunami occurred at Severo-Kurilsk, Paramushir Island, during the high water of spring tide and this additional 1 m was responsible for many victims (RABINOVICH and SKRIPNIK 1984). The probability of tsunami and tide superposition is very important both for the Tsunami Warning System (TWS) and long-term tsunami risk estimations (MOFJELD *et al.* 2007; KARIM and ISMAIL 2010; CONSOLI *et al.* 2014). Time of arrival of the 2010 Chilean tsunami at the Pacific coast of Russia took place during high tide; this coincidence was taken into account when the Russian TWS declared a tsunami alarm. Happily, the first arriving waves were not the strongest, and the maximum wave arrived later coinciding with the close-to-zero tide. This tsunami reached the coast of southern California during low tide. However, strong tsunami-caused oscillations lasted more than 8 h, so maximum impact was observed about 7 h later at high tide (WILSON *et al.* 2010).

The coincidence of a tsunami and a significant storm surge is rare, and the probability of their superposition is very small. Usually it does not get taken into account when estimating tsunami risk. However, such examples are also described in the scientific literature. One of the most illustrative

[1] Institute of Marine Geology and Geophysics, FEB RAS, Yuzhno-Sakhalinsk, Russia. E-mail: g.shevchenko@imgg.ru
[2] Sakhalin Tsunami Warning Center, Yuzhno-Sakhalinsk, Russia.

examples is the coincidence of the Sumatra tsunami of 26 December 2004, which reached the Atlantic coast of North America at the time of a severe storm. According to the tide gauge records, waves from the two events coalesced along the shores of Maine and Nova Scotia on 27 December where they produced damaging waves with heights in excess of 1 m (THOMSON et al. 2007). The tsunami waves were identified in almost all outer tide gauges from Florida to Nova Scotia with maximum tsunami heights estimated to be 32–39 cm (northern regions) and 15–33 cm (southern region).

Another example of a tsunami with a strong storm coincidence was a weak Simushir event on 15 January 2009 (USGS earthquake magnitude M_w 7.4) at Kurilsk (Iturup Island) (SHEVCHENKO et al. 2011). The superposition of these two factors produced an anomalous enhancement of resonant oscillations in Kitoviy Bay with a period of about 20 min; their heights exceeded 1 m. Also, the 2010 Chilean tsunami was strengthened by surge in some harbors of southern California (WILSON et al. 2010).

For various adverse factors to coincide, a fairly long period of time is needed. Correctly estimating the probability of such a coincidence from observations of sea level is very difficult. PUGH and VASSIE (1978) suggested the joint probability method (JPM) to estimate the probability of extreme sea-level heights formed by superposition of tides and storm surges. RABINOVICH and SKRIPNIK (1984) and RABINOVICH and SHEVCHENKO (1990) generalized this method to include multiple factors and to estimate the tsunami risk combined with tides and meteorological sea-level oscillations. It is evident that the relative role of these factors (sea-level constituents) depends sufficiently on particular physical and geographical conditions of the specific region.

Extreme sea levels of rare recurrence arising from combination of tides, storm surges, seasonal oscillations and tsunamis were estimated by this method (JPM) for the Russian coast of the Sea of Japan (RABINOVICH et al. 1992). Following this work, this study has carried out similar estimates for the coasts of the Sea of Okhotsk and the Pacific coast of the Kuril Islands. We assumed that for the Pacific coast, in contrast to the Sea of Japan, the corrections associated with tides and surges are likely to be relatively small, since the expected height of tsunami runups is quite high. Nevertheless, the tragic 1952 event at Severo-Kurilsk demonstrates that the secondary factors should be taken into account. For the Sea of Okhotsk, the contribution of tides and surges can be essential.

2. Data and Methods

Hourly tide gauge sea-level values collected by the Sakhalin Hydrometeorological Service at eight stations (Table 1; Fig. 1) were used for the present analysis. Additional tsunami records were obtained by coastal tsunameters of the Sakhalin Tsunami Warning System (TWS) during recent years. The length of sea-level series is of primary importance for the study of meteorological oscillations, especially for the statistic of storm surges. The shortest lengths were at Magadan and Nabil Bay (Katangli—shorter than 20 years). All available data were carefully edited and corrected, and then used to estimate the probability distribution of tidal and storm surge (detided residual) components. Non-systematic gaps in the data do not affect this kind of analysis (PUGH and VASSIE 1978).

The joint probability method (JPM) was proposed by PUGH and VASSIE (1978) to investigate combining probability functions of surge and tide, and to estimate extreme sea levels based on relatively short observation series (the method was applied for drilling platforms in the North Sea). Based on RABINOVICH and SHEVCHENKO (1990) and RABINOVICH et al. (1992), we briefly describe the expanded

Table 1

The list of sea-level data

Station	Observation period (years)	Duration (years)
Burevestnik	1964–1984	21
Kurilsk	1960–1998	39
Matua	1960–1982	23
Severo-Kurilsk	1967–1989	23
Yuzhno-Kurilsk	1958–1998	41
Malokurilsk	1970–2013	43
Magadan	1977–1988	12
Nabil	1960–1964; 1987–1997	16
Poronaisk	1965–1998	34
Korsakov	1948–1992	45

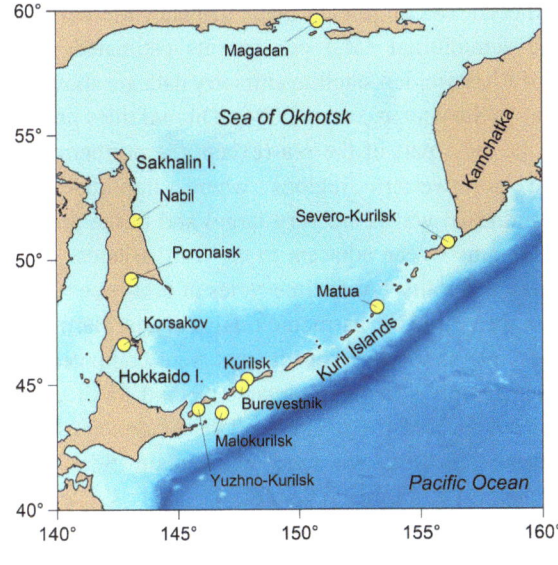

Figure 1
The location of tide gauges used in this study

version of this method, which can be used to estimate tsunami risk in combination with other factors.

Sea-level elevation (ζ) at any time (t) may be considered as the sum of a few individual components

$$\zeta(t) = \zeta_0 + \zeta_t + \zeta_s + \zeta_m + \zeta_{ts} \qquad (1)$$

where ζ_0 is the mean sea level, ζ_t is the tides, ζ_s the seasonal oscillations, ζ_m the meteorologically induced sea-level component (including storm surges), and ζ_{ts} the tsunami (very rare event). The meteorological component remains after subtraction of mean levels, tides and seasonal oscillations from the original records (for this component sometimes we use the term "residual fluctuations").

The representation of the original sea level in the form of a linear combination of the individual components (1) is based on the assumption that the influence of the nonlinear interaction is small. This assumption is usually true, except for some shallow seas or bays (GERMAN and LEVIKOV 1988). The examined stations are not located in shallow-water regions; however, the presence or absence of nonlinear effects has not been specifically tested. At the same time, we can mention that previous studies of storm surges in these regions (RABINOVICH and SKRIPNIK 1984; SHEVCHENKO 1997a, b) did not reveal indications of nonlinear interaction of tides and meteorologically induced sea-level oscillations.

In such a case, the probability density P_ζ of sea-level oscillations caused by several various uncorrelated factors may be represented as:

$$P_\zeta(y) = \int\limits_{-\infty}^{+\infty} P_1(x_1) \int\limits_{-\infty}^{+\infty} P_2(x_2) \cdots \int\limits_{-\infty}^{+\infty} P_{N-1}(x_{N-1})$$
$$\times P_N(y - x_1 - x_2 - \cdots - x_{N-1})\mathrm{d}x_1 \mathrm{d}x_2 \ldots \mathrm{d}x_N, \qquad (2)$$

where P_i are the probability density of each type of variation, x_i and y are their sea-level values, and N is the number of components (RABINOVICH and SHEVCHENKO 1990).

Based on Eq. (2) and accounting for Eq. (1), the probability of total sea-level height h can be expressed as the sum of probabilities of all possible combinations of individual components. The probability of the sea level exceeding height h is:

$$F(h) = \int\limits_h^\infty P_\zeta(y)\mathrm{d}y \qquad (3)$$

The corresponding return period may be calculated as (PUGH and VASSIE 1978):

$$T(h) = \frac{1}{nF(h)} \qquad (4)$$

where n equals to a number of samples in a year (for hourly sea-level series $n = 8766$). Therefore, the problem of estimating the probability function and return periods of extreme sea levels reduces to the study of the probability densities of each individual component separately (their discretization using a class interval of 5 cm to group the levels) and then estimation of their joint probability based on expressions (2–3) wherein integration was replaced by summation.

For evaluating sea-level components, it is important to consider their physical properties. Tides are a deterministic process: based on a 1-year observational series, it is possible with high precision to estimate the amplitudes and phases of tidal constituents, and to predict hourly tidal sea levels to determine extreme high tide values for any reasonable period (ZETLER and FLICK 1985). Seasonal oscillations are quasi-deterministic. Their amplitudes and phases can change from 1 year to another but in general they are relatively consistent. Tidal and seasonal components may be combined (we included the annual Sa and semiannual Ssa components, as well as the diurnal and semidiurnal constituents); their possible extreme values can be estimated accurately. Tsunami and storm surges are stochastic processes. To estimate their extreme heights it is possible to use the methods of extreme statistics (GUMBEL 1958; GERMAN and LEVIKOV 1988):

$$1 - P(h) = \exp(-\exp(-y)) \qquad (5)$$

where $P(h)$ is the probability of the sea level exceeding height h and y is a reduced variable with linear dependency of h and y:

$$h = a\, y + b \qquad (6)$$

The third nonlinear Gumbel distribution corresponding to the physical limitations of natural disasters is probably more physically reasonable. But for our stations, a linear relationship was in good agreement with observations, which is typical for relatively short periods of observation (GERMAN and LEVIKOV 1988). The coefficients a and b are determined by the least squares method from the ranked series of maxima with empirical probabilities

$$1 - P_i = i/N + 1, \qquad (7)$$

where N is a total number of maxima and i is a number of the maximum in the ranked set. Annual maxima of residual (de-tided) sea-level series were used for estimation of extreme storm surges. For tsunami, which are the rare events, we used an additional coefficient which is an average frequency of tsunamis $f = N/T$ where N is a number of events and T is a period of observations.

An analysis of individual components and their extreme values is a subject of independent interest. That is why we consider below the concise description of the analysis of individual components, as well as the combination of various components.

3. Tidal Analysis and Estimations of Extreme Tidal Levels

Large tidal amplitudes and their strong spatial variability are particular features of the Sea of Okhotsk. The amplitude maps of major diurnal (K_1) and semidiurnal (M_2) constituents estimated from TOPEX/Poseidon satellite altimetry data are shown in Fig. 2 (SHEVCHENKO et al. 2004). Diurnal tides prevail in the most part of the sea (except the northernmost and northwestern regions where amplitudes of semidiurnal waves are very large) and in the area of the Pacific Ocean adjacent to the Kuril Islands.

High quality 1-year time series of hourly sea-level data were used for harmonic tidal analysis. Harmonic constants of 67 tidal constituents were computed for each station by the least squares method. In practice, it is not necessary to evaluate all these harmonics. For future estimates, we used eight main tidal constituents (4 diurnal Q_1, O_1, P_1 and K_1 and 4 semidiurnal N_2, M_2, S_2, and K_2), and amplitudes of other tidal waves do not exceed 1.5 cm for all stations except Magadan. Very strong tides are observed in this part of the Sea of Okhotsk; thus we used 26 harmonics for that station. To describe the spatial tidal variability in the study area, we used the root mean square (RMS) tidal amplitude ("mean amplitude" in the following text) (SHEVCHENKO et al. 2004):

$$A_T = \sqrt{\sum_{i=1}^{8} H_i^2}, \qquad (8)$$

where H_i are the amplitudes of the aforementioned eight major tidal constituents. Mean amplitude A_T related to zero mean sea level is smaller than 40 cm at the coast of Sakhalin Island (Nabil, Poronaisk, and Korsakov); it is within 40–50 cm at the coasts of the South and Middle Kuril Islands, and exceeds 64 cm in the region adjacent to the North Kuril Islands (Table 2). Magadan stands out among all stations; the mean tidal amplitude at this station reaches 144 cm. As can be seen in Fig. 2, tides are very high in the northern part of the Sea of Okhotsk.

Using harmonic constants (seasonal constituents Sa and Ssa were included), hourly time series of predicted tides for a period of 101 years (2000–2100) were simulated for all stations. Figure 3 presents the corresponding time series for four stations: Severo-Kurilsk, Kurilsk, Nabil and Magadan. Well-expressed tidal variations with 19-year period (more accurately 18.61 years, the tidal nodal cycle) were found in annual maxima and minima of tidal levels at various

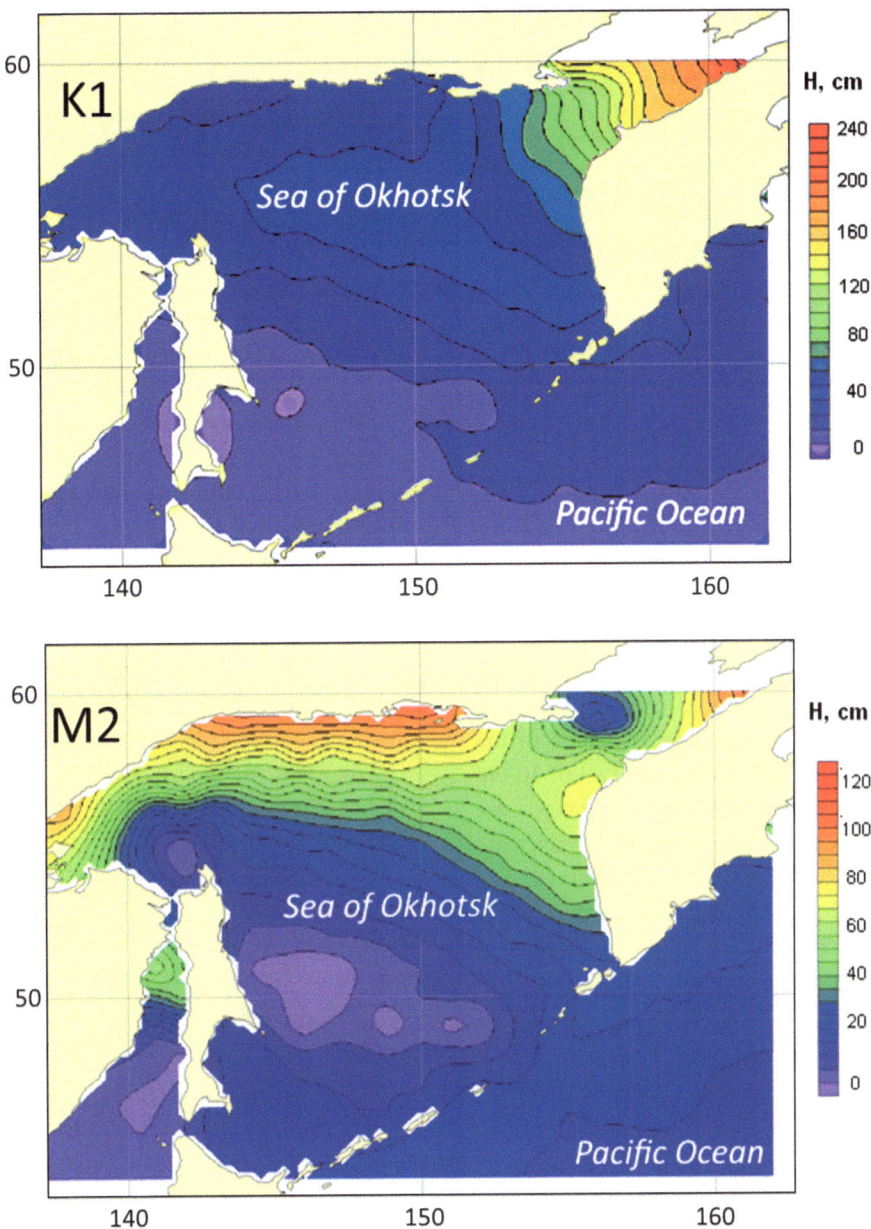

Figure 2
Spatial distribution of major diurnal (K_1-*top*) and semidiurnal (M_2-*bottom*) tidal constituents in the Sea of Okhotsk and adjacent areas calculated from TOPEX/Poseidon satellite altimetry data (from SHEVCHENKO *et al.* 2004)

stations. However, this cycle is not clearly expressed at Magadan, where a cycle with a period of 4.4 years is more significant. This is typical for regions where the semidiurnal K_2 wave plays an important role. Plots for all other stations are very similar to the one at Kurilsk. Maximum (Table 2) and minimum values from the 100-year predicted time series may be considered as the upper and lower limit of tidal level which cannot be exceeded (the highest and lowest astronomic tidal levels).

The nodal cycle is very important: for example, at Severo-Kurilsk the difference of annual maxima in the year of maximum nodal tide (2025) and minimum nodal tide (2015) is 16 cm. Therefore, the 19-year

Table 2

Statistical characteristics of sea-level tidal oscillations relative to mean sea level: maximum and minimum tidal levels, form factor (R) and mean tidal amplitude (A_T)

Station	Max (cm)	Min (cm)	R	A_T (cm)
Burevestnik	63.4	−107.6	1.3	48.7
Kurilsk	68.7	−72.9	1.8	35.6
Matua	70.2	−102.3	1.9	45.8
Severo-Kuriksk	88.0	−144.4	2.1	64.1
Yuzhno-Kurilsk	54.3	−100.1	1.0	45.6
Malokurilsk	55.8	−90.6	1.0	41.1
Magadan	213.6	−280.2	0.5	144.6
Nabil	64.4	−78.2	5.3	37.6
Poronaisk	73.7	−74.3	1.0	37.9
Korsakov	80.0	−75.7	1.5	39.1

interval is necessary for correct generation of the probability density function (PDF). Predicted time series for the interval (we used 2000–2018, although we could take any other interval with such duration) were used to calculate PDF (histograms) with a class interval of 5 cm. We considered the histogram for each station to be the discrete approximation of the tidal PDF. Histograms with a class interval of 10 cm for Nabil, Severo-Kurilsk and Magadan stations are shown in Fig. 4. Some differences in PDF shapes are associated with the relative role of diurnal and semidiurnal constituents. Such a tidal regime is related to a "form factor" (it defines the interannual and seasonal changes of tidal range and is very important for estimation of tidal PDF)

$$R = \frac{H_{O_1} + H_{K_1}}{H_{M_2} + H_{S_2}}. \quad (9)$$

Computed values of form factor for different stations are presented in Table 2. Diurnal tides strongly dominate at Nabil, semidiurnal tides prevail at Magadan, and a mixed regime is at Severo-Kurilsk and the other stations.

4. Seasonal Variations

Seasonal changes of mean sea level in the Sea of Okhotsk are related to several factors: water density and circulation, discharge from the Amour River and other rivers, long period tides, wind forcing, atmospheric pressure, etc. The winter maximum in sea level is a specific feature of seasonal variations in this region (POEZZHALOVA and SHEVCHENKO 1997; SEDAEVA and SHEVCHENKO 2001) which is associated with the winter amplification of cyclonic circulation over the Sea of Okhotsk and with the prevailing winds.

The whole series of monthly mean sea levels were used to compute the amplitudes and phases of annual (Sa) and semiannual (Ssa) seasonal constituents (Table 3). Individual yearly series were also processed to estimate variations in the respective parameters; the maximum amplitudes are also presented in Table 3. Computed values of seasonal constants are not as consistent as they are in the Sea

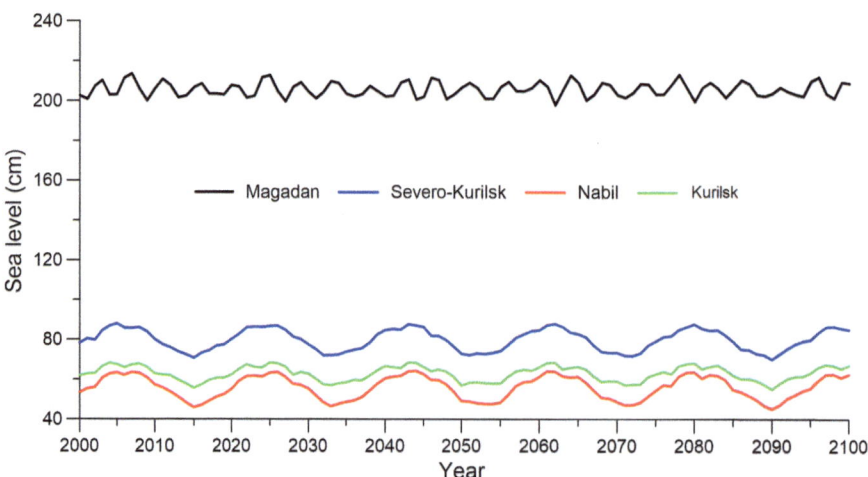

Figure 3
Annual maximum tidal sea levels predicted for the period of 2000–2100 for stations Magadan, Severo-Kurilsk, Nabil and Kurilsk

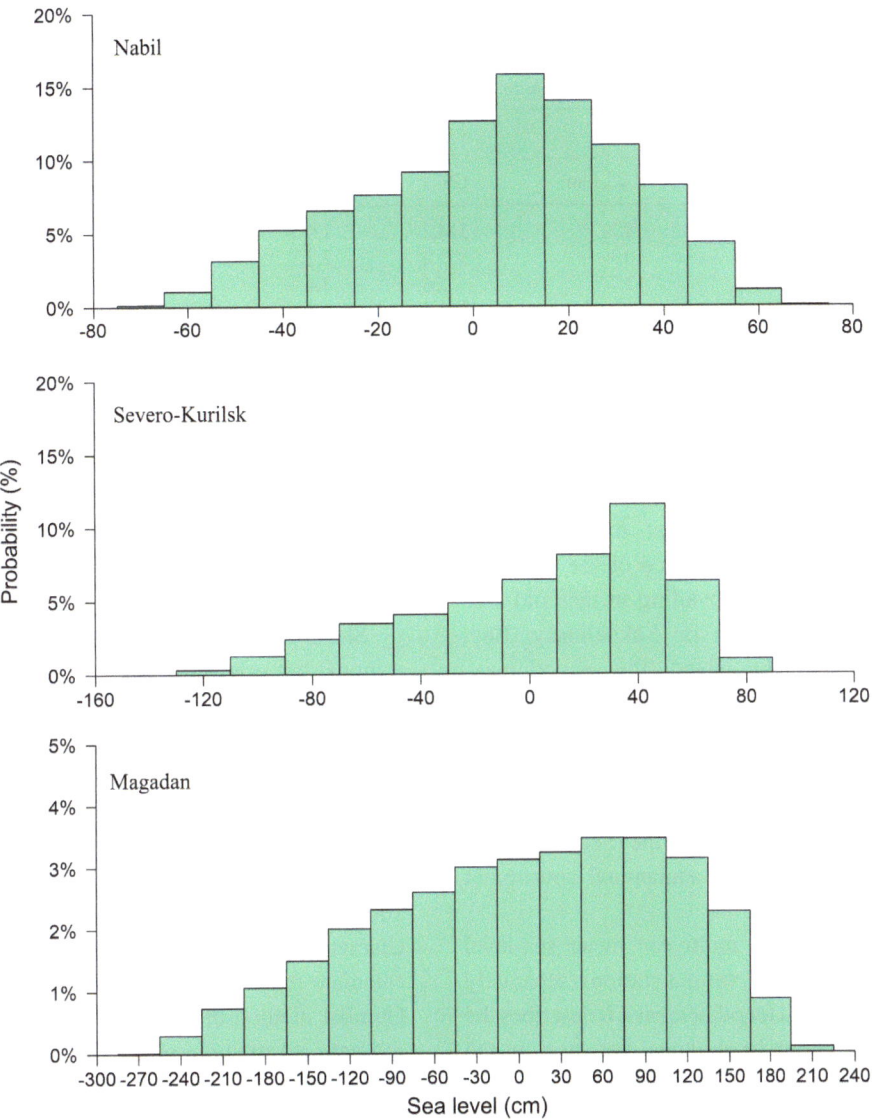

Figure 4
Probability density functions of tidal sea-level oscillation calculated from predicted tidal series for period 2000–2018 for stations Nabil, Severo-Kurilsk and Magadan

of Japan (RABINOVICH et al. 1992). This is primarily due to significant variations in the position and intensity of the Aleutian Low, which determines the wind flows over the study region and the winter circulation (winter maximum of sea level) in the Sea of Okhotsk and adjacent regions (SEDAEVA and SHEVCHENKO 2001). Also, the amplitude of the annual harmonic varies considerably in space: it ranges from 3.4 cm (Magadan) to 8.9 cm (Severo-Kurilsk) and phase shift between these stations reaches 100°. The seasonal form factor, similar to that for ordinary tides,

$$F = \frac{H_{Sa}}{H_{Ssa}}, \qquad (10)$$

varies from 1.1 (Magadan) to 3.0 (Nabil), suggesting the annual cycle prevails in seasonal variations in most of the study region with the maximum sea level in December–January (POEZZHALOVA and SHEVCHENKO 1997). The winter sea level

Table 3

Statistical characteristics of seasonal sea-level oscillations: mean (A) and maximum (A_{max}) amplitudes, mean phase (G) of annual (Sa) and semiannual (Ssa) harmonics, and seasonal form factor F (10)

Station	Sa			Ssa			F
	A (cm)	A_{max} (cm)	G (°)	A (cm)	A_{max} (cm)	G (°)	
Burevestnik	3.5	8.2	314.7	1.9	4.4	19.7	1.8
Kurilsk	5.0	10.3	325.2	2.2	6.1	351.4	2.3
Matua	6.1	11.7	5.6	3.4	7.9	24	1.8
Severo-Kurilsk	8.9	13.9	359.4	3.7	7.4	7.5	2.4
Yuzhno-Kurilsk	4.7	9.9	303.7	3.3	6.8	21.9	1.4
Malokurilsk	6.1	9.4	318.6	2.4	5.6	8.4	2.5
Magadan	3.4	9.9	259.8	3.1	6.8	322.5	1.1
Nabil	7.3	10.9	319.9	2.4	6.1	281.7	3.0
Poronaisk	4.8	8.5	281.3	2.4	3.5	284.6	2.0
Korsakov	6.1	10.7	325.7	3.8	6.4	332.1	1.6

maximum on the northern coast of the Sea of Okhotsk wanes under the influence of the northerly and north-westerly winds prevailing in the cold season. There are anomalies in Sakhalinsky Bay, northward from Sakhalin Island (the second maximum in summer due to the discharge of the Amur River) and in the area adjacent to the South Kuril Islands and the northern coast of Hokkaido Island (a summer maximum induced by the Soya Warm Current bringing warm water from the Sea of Japan) (ROMANOV et al. 2004). The semiannual constituent prevails in these areas.

Figure 5 presents the multiyear mean sea-level month-to-month changes at various stations (separately for the Kuril Islands and for other areas). It describes the general character of sea-level annual variations in the region under study. The results of analysis show that seasonal variations play an important role in general sea-level variability in the Sea of Okhotsk and make significant contribution to their extreme heights.

The mean values of seasonal harmonics were used to calculate the PDF in combination with tides. Thus, Fig. 4 shows the joint PDF of tidal and seasonal oscillations. Normally, maximum semidiurnal tides occur near the spring (March) and autumn (September) equinoxes when the S_2 and K_2 constituents are in phase; diurnal (or mixed) tides are enhanced in December–January and June–July when K_1 and P_1 are in phase (ZETLER and FLICK 1985). For the main part of the Sea of Okhotsk, the winter maximum of tides coincided with the seasonal maximum, creating the total sea level maximum at this time.

5. Storm Surges and Meteorologically Induced Oscillations

Storm surges are one of the most widespread and ruinous hazards for coastal areas of the Sea of Okhotsk (LYUBITSKY et al. 2013). Statistical properties of storm surges, including the estimation of mean number of events exceeding particular thresholds, were studied by SHEVCHENKO (1997a, b). The thresholds were the following: 30 cm (*surge phenomenon*), 50 cm (*strong surge*), 75 cm (*dangerous surge*) and 100 cm (*catastrophic surge*). These are important characteristics of marine disaster, which were used to calculate the recurrence of storm surge heights. The Gumbel method of extreme statistics was applied to estimate storm surge probability. The computed values of surge heights with probability 0.04, 0.02 and 0.01 (i.e., with return periods of 50, 100 and 200 years) are given in Table 4. The accuracy of these values depends on the observation series length. For Nabil (total length of 16 years) and Magadan (12 years) the series lengths were too short to estimate correctly the storm surge heights for the return period of 100 and 200 years. For other stations, the observation periods were long enough to evaluate the surge heights of rare recurrence.

The most dangerous surges in the Sea of Okhotsk are observed in shallow Sakhalinsky Bay in the northern part of Sakhalin Island (LYUBITSKY et al. 2013); however, there are no tide gauge observations at this region and it is not considered in the present paper. From those stations that were examined, the

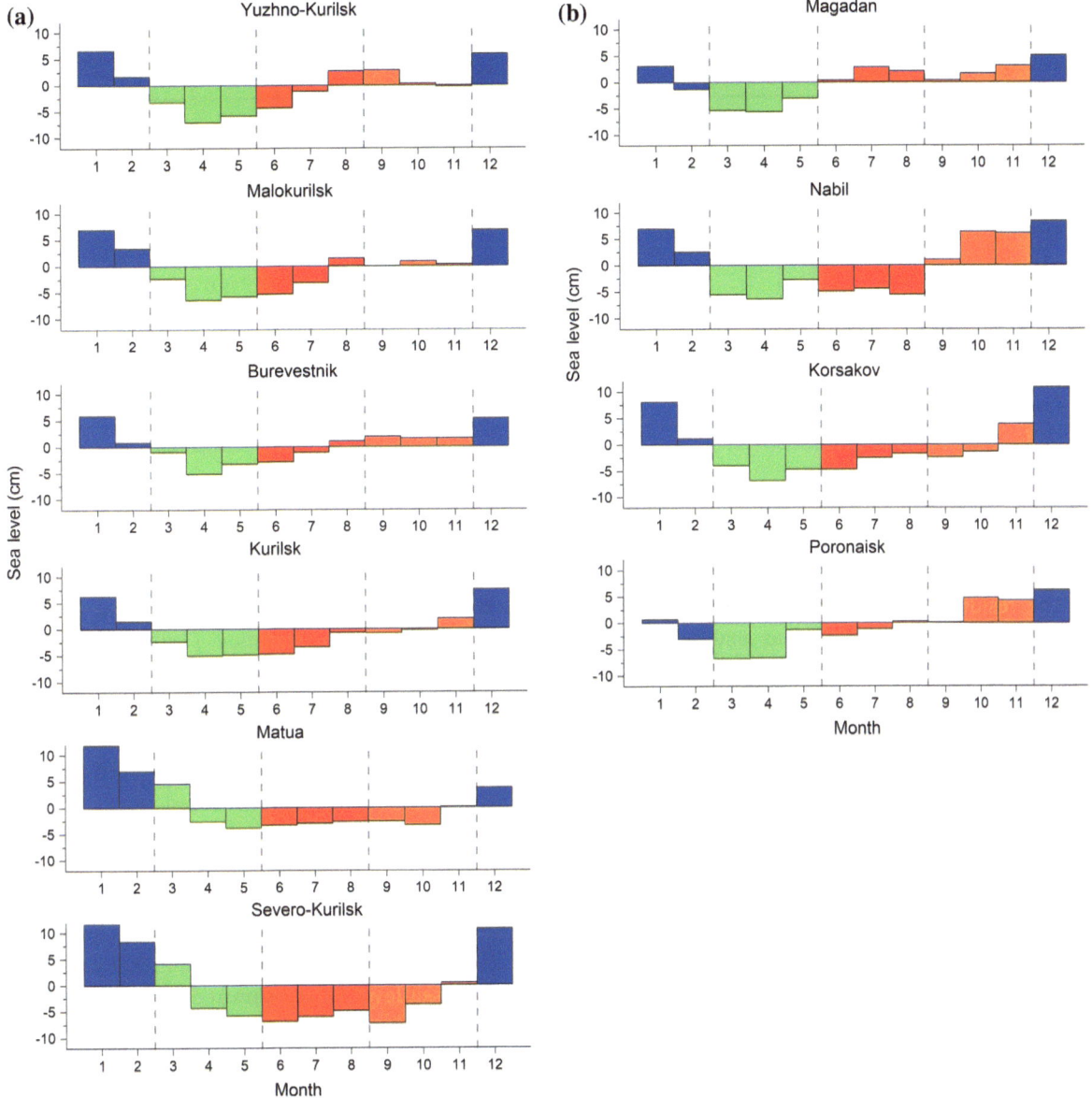

Figure 5
Multiyear mean sea-level changes for the stations located on the coast of the Kuril Islands (a) and the Sea of Okhotsk (b)

largest storm surge heights were recorded on the coast of Matua Island, the Okhotsk Sea (Kurilsk) and the Pacific (Burevestnik) coast of Iturup Island. The large storm surge height at Matua is a puzzle because the physical and geographical conditions of the island (small size, deep and short shelf) do not look conducive for surge formation. This is an interesting scientific problem but from the applied point of view surges at this island pose no threat because there is no resident population at the present time. The strongest storm surge at the Kuril Islands (204 cm height relative to the mean sea level) occurred on December 3–4, 1971 in the area of Kitoviy Bay (the Okhotsk Sea coast of Iturup Island). This surge caused a great flood in the town of Kurilsk and led to extensive damage.

The weakest storm surges were observed on the South and North Kuril Islands: in the southwestern

Table 4

Statistical characteristics of storm surges: observed maximum (relative to mean sea level) and calculated extreme surge heights for return periods of 50, 100 and 200 years

Station	Observed maximum (cm)	Return period (year)		
		50	100	200
Burevestnik	130.6	132.3	147.4	152.6
Kurilsk	204.5	164.4	181.6	188.8
Matua	166.5	182.2	200.0	207.8
Severo-Kurilsk	68.1	73.4	78.0	82.6
Yuzhno-Kurilsk	75.5	79.7	86.5	93.5
Malokurilsk	61.9	63.4	68.1	72.8
Magadan	81.2	88.3	93.6	98.9
Nabil	84.0	102.4	107.8	112.7
Poronaisk	112.2	115.5	126.2	136.9
Korsakov	88.7	78.6	84.7	90.9

flank on Kunashir (Yuzhno-Kurilsk) and Shikotan (Malokurilsk) Islands, and in the northeastern flank on Paramushir Island (Severo-Kurilsk).

The residual series (after subtracting tides and seasonal variations) were used to analyze meteorologically induced sea levels and storm surges, and to identify their extrema. However, for correct estimation of the probability (or return periods) of the highest sea levels produced by the superposition of surges and tides, it is necessary to take into account not only extreme values but also ordinary meteorologically induced background oscillations. For this purpose, we used hourly residual series for the entire observation period at each station to construct the probability density functions in the same way it was done for tides (Sect. 3). Figure 6 presents some examples of the respective distributions. About 100,000 individual sea-level values were used to obtain PDF for Magadan and for many other stations. The probability density functions look similar to the Gaussian function, except their very tails. Actually, the correct estimation of the "tail" (extreme values) of PDF of the residual oscillations is the key question of the joint probability method.

TAWN (1992) performed some revision of the joint probability method. He recommended the parametric smoothing and extrapolation of surge distribution in the range of small probabilities. In his opinion, this JPM modification is important in the case of relatively short periods of sea-level observations (less than 10 years) and if the role of tides is smaller in comparison with residual oscillations. We analyzed sufficiently long series of sea-level observations, and tides in the Sea of Okhotsk are very high.

In addition to these specific values, we used the results of extreme statistics (Table 4); the extrapolated sea-level elevations exceeded the observed maximum (to a value corresponding to a return period of 500 years) and these values supplemented the actual PDFs. Then, we used these modified distributions of meteorologically induced oscillations in expression (2) to estimate the joint probability of these oscillations and tides. However, we found no significant effect in the range of return periods of 50–200 years. The most significant corrections were at the Nabil for periods of more than 1000 years. The greatest sea-level value estimated without extrapolation had an extremely small probability 3.2×10^{-9}, which corresponds to a return period of about 35,000 years. For stations with longer series of observations, this effect was even less.

An example demonstrating the combined effect of the high tide and surge is presented in Fig. 7. Fortnightly, sea-level records (November 1990) at Kurilsk (the Okhotsk Sea side of Iturup Island) show the initial series and two individual components: tidal and meteorological (non-tidal). Two surges were recorded at this station during this time interval: a moderate surge on 5 November (37 cm) and a strong surge on 10 November (60 cm). The first surge coincided with the high tide, and the total sea level reached an extreme value of 92 cm. The second event

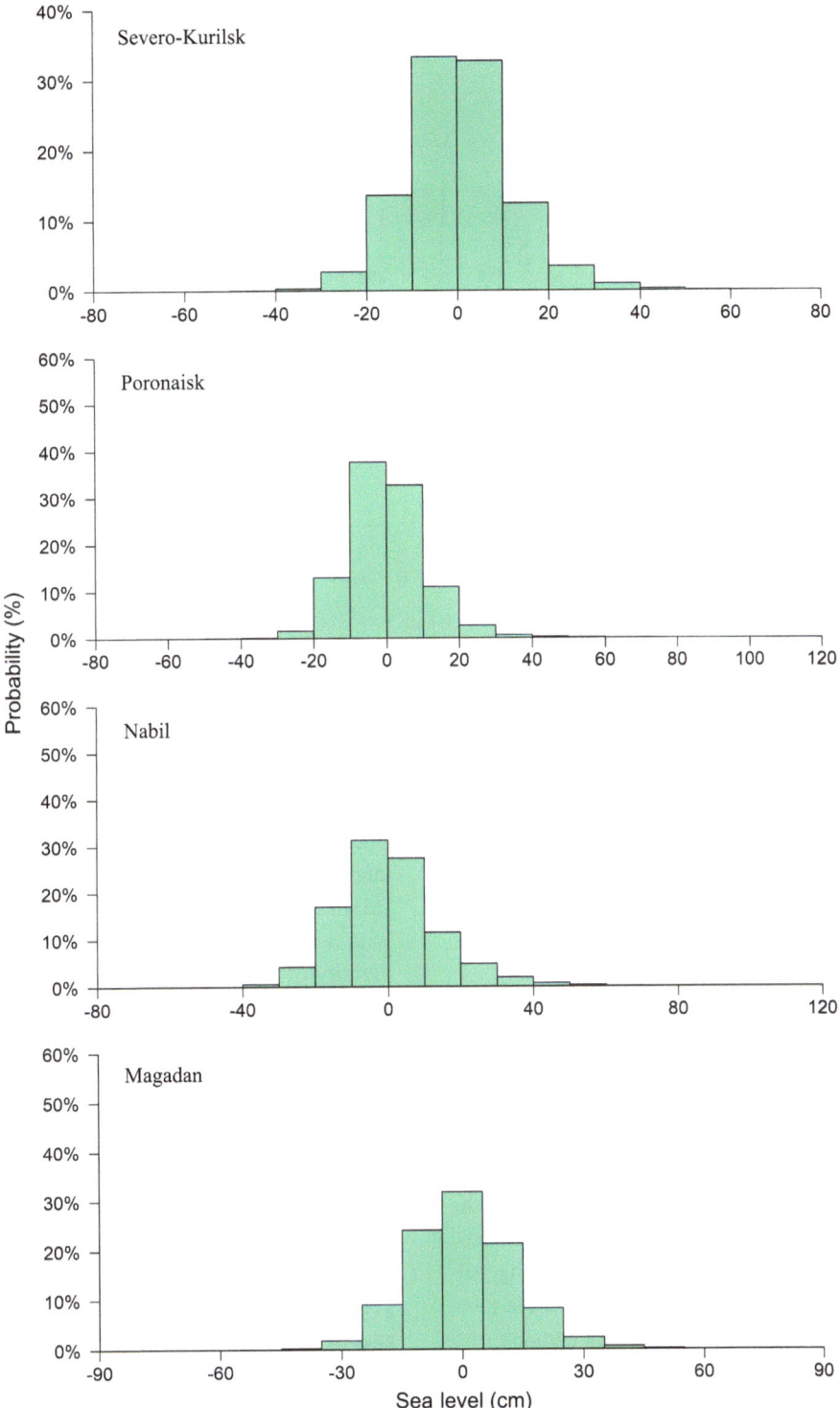

Figure 6
Probability density functions of residual (de-tided) sea-level oscillation calculated from the whole series for stations Nabil, Severo-Kurilsk, Poronaisk and Magadan

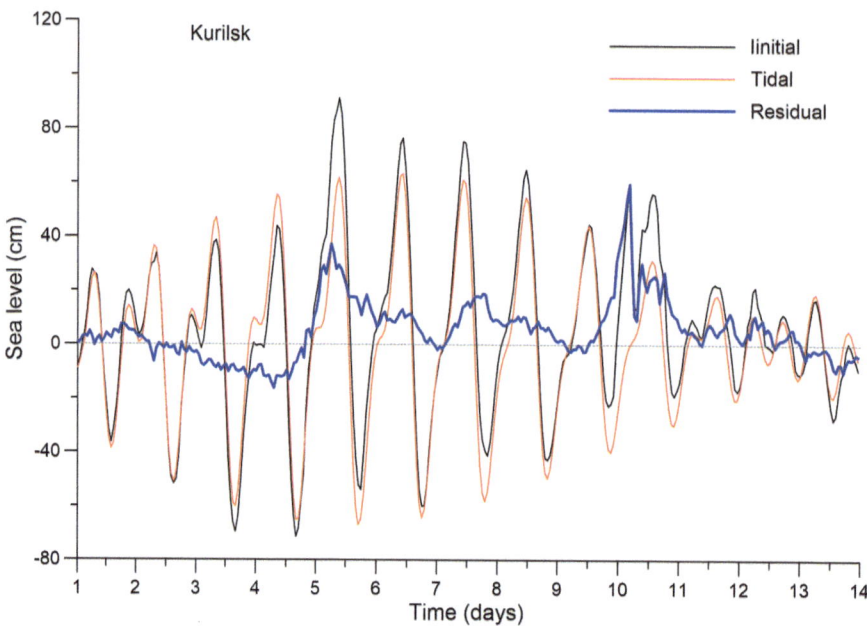

Figure 7
Initial, tidal and residual (de-tided) sea levels at Kurilsk (November 1–14 1990)

occurred during zero tide, and the total sea level was significantly lower than 5 days earlier (despite the fact that the surge was substantially higher). This example shows the importance of tides on the formation of extreme sea levels during surge events.

6. Tsunami

Strong earthquakes occur relatively seldom in the Sea of Okhotsk, and most of them have a large focal depth. Therefore, tsunamis are rarely generated within the sea. The historical tsunami catalog (SOLOVIEV and GO 1974) shows only one weak local tsunami generated off the northern coast of Hokkaido Island. However, the continental slope adjacent to the Kuril Islands and Kamchatka is one of the most seismically active regions in the world. Strong tsunamis are frequently induced in this region; the most destructive of them was the Kamchatka tsunami of 5 November 1952 generated by an M_w 9.0 earthquake offshore of southern Kamchatka. This tsunami destroyed the town of Severo-Kurilsk and a number of smaller settlements on the Pacific coast of the Kamchatka Peninsula and the North Kuril Islands (KAISTRENKO and SEDAEVA 2001).

The Simushir (Kuril Islands) earthquake (M_w 8.3) of 15 November 2006 occurred in the central part of the Kuril Islands (LOBKOVSKY et al. 2009) with tsunami run ups up to 15–20 m. Fortunately, this tsunami seriously impacted only the uninhabited islands of Simushir, Matua, Ketoy, and others (MACINNES et al. 2009). Several dangerous tsunamis with wave heights of >5 m were generated in the vicinity of the South Kuril Islands; in particular, the Shikotan tsunami of 5 October 1994 (KAISTRENKO 2014).

Strong remote earthquakes can also generate tsunamis that can badly affect settlements along the coast of the Kuril Islands. For example, the Chilean tsunami of 22 May 1960, generated by the strongest instrumentally recorded earthquake of M_w 9.5, produced tsunami waves with a trough-to-crest height of 5–6 m on this coast and was one of the most dangerous for the Pacific coast of Russia. This tsunami caused low-frequency waves that penetrated into the Sea of Okhotsk and were actually considerably stronger than tsunamis caused by the Kuril Islands earthquakes. The only tide gauge that was in operation in the Kuril Islands during the 1960 event was Yuzhno-Kurilsk which recorded a maximum wave height of 2.6 m (Fig. 8). However, this is an underestimate of the wave height because the sea level had dipped below the minimum limit of the tide gauge. The estimated wave height is about 3.1 m. At Magadan and on the eastern coast of Sakhalin Island, quality tide gauge records were obtained in May 1960 (the records obtained at different tide gauges are presented in Fig. 8). The maximum height of 2.5 m is observed at the Magadan tide gauge (that is higher than 2.2 m specified by SOLOVIEV and GO 1974). The lowest wave height is observed at Nabil (0.8 m). The first waves and those with maximum amplitude of the 1960 Chilean tsunami reached the Kuril Islands when the tide was falling. Waves with maximum amplitude occurred at the time of low tide, so tides reduced the tsunami heights at Yuzhno-Kurilsk and on the coasts of the Sea of Okhotsk.

Approximately 50 years later, the Chilean tsunami of 27 February 2010, generated by an M_w 8.8 earthquake off northern Chile, was also potentially dangerous. Wave heights of about 2 m were observed on the Pacific coast of the Kuril Islands (SHEVCHENKO et al. 2013). The first waves of the 2010 Chilean

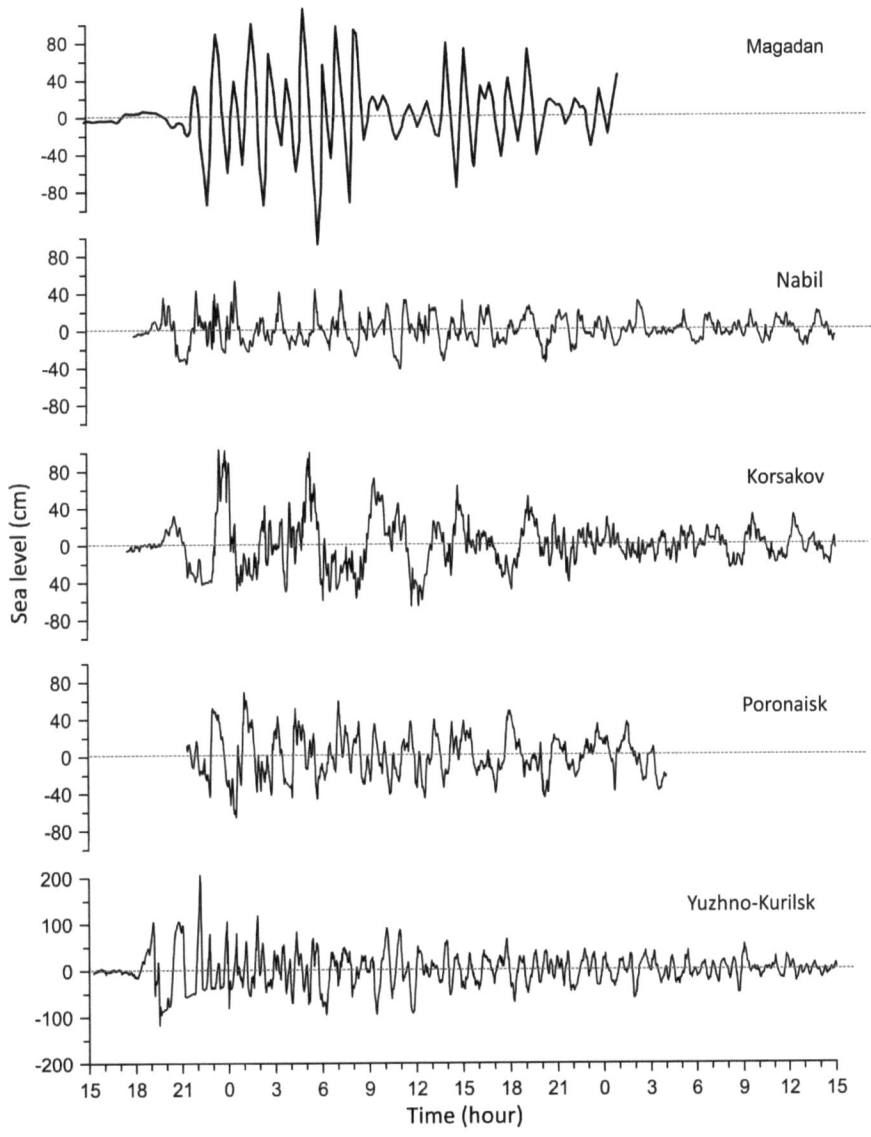

Figure 8
Two-day records (May 23–25, 1960, UTC time) of the Great Chilean tsunami recorded at stations in the Russian Far East

tsunami reached the coasts of Kamchatka and the Kuril Islands at the time of high tides. However, the first waves were not the greatest along these coasts except at the oceanic coast of Shikotan Island. Waves with the maximum amplitude occurred mainly during a falling tide and in some areas (for example, at Malokurilsk and Shikotan) tides led to reduced cumulative heights. However, in general, tides did not produce significant effect on heights of the 2010 Chilean tsunami.

Tsunami records obtained at stations Severo-Kurilsk and Malokurilsk on the Pacific coast of the Kuril Islands and Korsakov, Poronaisk and Magadan in the Sea of Okhotsk (de-tided series) are presented in Fig. 9. In general, the 2010 tsunami on the coast of the Russian Far East was much weaker than the 1960 Chilean tsunami. The ratio of maximum wave heights was about 1:3 which is in a good agreement with results obtained for the western coasts of USA and Canada (RABINOVICH et al. 2013). The signal frequencies were also significantly higher. Tsunami height at Magadan was considerably larger than on the coast of Sakhalin Island, similar to the 1960 event.

The correct evaluation of tsunami risk is a difficult problem due to the small number of observed events. The total number of recorded tsunamis on the Pacific coast of the Kuril Islands is ~10–15 at various stations, and on the coast of the Sea of Okhotsk this number ranges from 5 to 8. Therefore, in assessing tsunami heights of rare recurrence the common method is to use numerical modeling (e.g., MOFJELD et al. 1999; KULIKOV et al. 2005). Using this method, the probability of tsunamis with extreme

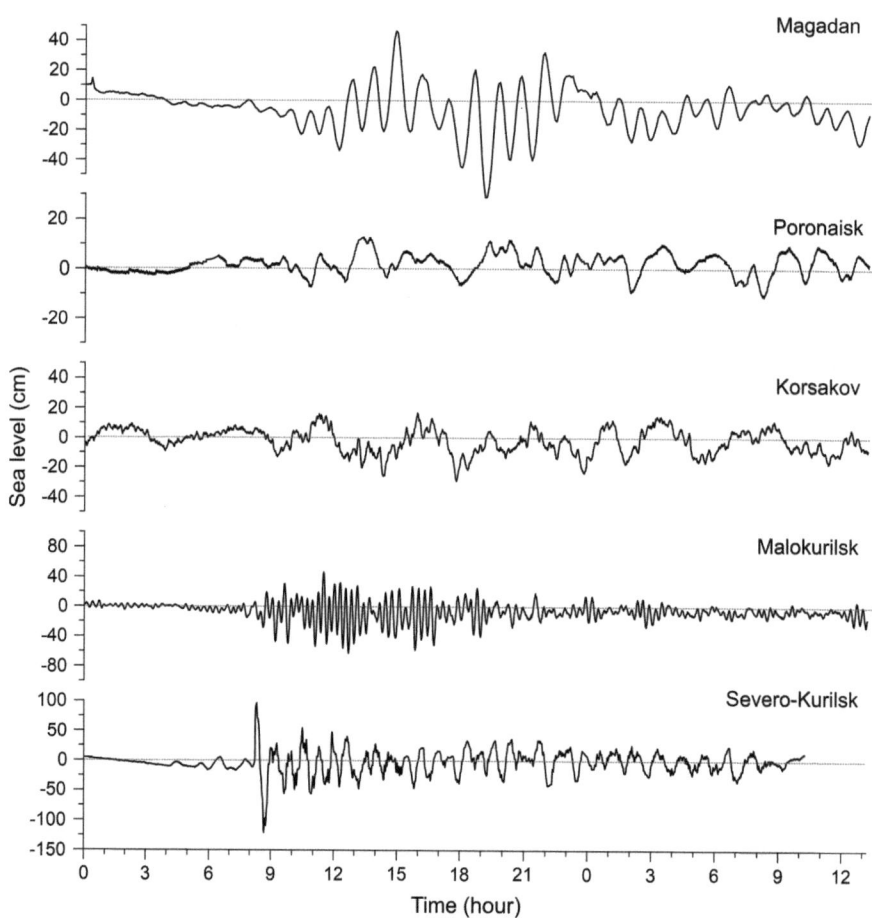

Figure 9
One and a half day records (February 28–March 1, 2010) of the Chilean tsunami recorded at the stations in the Russian Far East

heights is estimated based on the probability of the respective strong earthquakes in the study area. The methods of extreme statistics (GUMBEL, 1958) are also used, but calculation errors in that case are usually large (KULIKOV et al. 2005; KAISTRENKO, 2014).

An estimation of the tsunami risk for the coasts of Sakhalin and the Kuril Islands was made by RABINOVICH and SHEVCHENKO (1990), KHRAMUSHIN and SHEVCHENKO (1994) and KAISTRENKO (2014). In accordance to this method, the function describing the tsunami recurrence with wave height exceeding 0.1 m may be represented as (GO et al. 1985; KAISTRENKO 2014):

$$\frac{N}{T} = Ae^{-\frac{h}{h^*}}, \qquad (11)$$

for any coastal site. Here, N is the number of tsunamis with wave height $h \geq 0.1$ m recorded at the given site during T years, A is the coefficient determined by the frequency of large tsunamis in the study area, and the parameter h^* characterizes the relative local topographic amplification of tsunamis at a particular site. The Eq. (11) is based on the assumption that a tsunami is a Poisson process. The validity of this assumption for the Pacific Ocean was demonstrated by GEIST et al. (2009). It is important that the correlation radius of tsunamis is large and, therefore, parameter A varies insignificantly within the specific region (KAISTRENKO 2014).

Data on tsunami heights on a particular coast of the Russian Far East can be divided into two groups. The first group is the instrumental measurements of tsunamis by coastal tide gauges or other tsunami recorders (tsunameters) installed in recent years for the purposes of the Russian TWS. The second group includes data on tsunami runup and inundation collected during field surveys in tsunami-exposed zones.

In the present work, we primarily used data from the first group. The examination of tsunamis in combination with tides, storm surges and other types of sea-level oscillations is more difficult for the data from the second group. It is known that tsunami heights estimated based on the instrumental data are usually lower than the tsunami runup estimates in the adjacent parts of the coast. Thus, we obtain the lower estimates which relate to the location of a particular tide gauge. However, actual tsunamis are characterized by a large variability of wave heights along the coast related to detailed properties of local topography. For *local tsunami-zoning*, i.e., for creating maps of tsunami risk (maximum tsunami runups and inundation zones), the local topographic effects are crucial. Numerical modeling can be used for this purpose (KHRAMUSHIN and SHEVCHENKO 1994).

The maximum observed tsunami heights (runup values for Pacific coast of Kuril Islands) and calculated heights for return periods of 50, 100 and 200 years are presented in Table 5. For various parts of the coast of the Sea of Okhotsk, the highest recorded tsunamis are the same: 1952 Kamchatka, 1960 Chile, 1963 Urup, 2006 Simushir, 2010 Chile and 2011 Tohoku. However, not all of these tsunamis were recorded at all stations. Thus, at Nabil, due to the short observation period, only two of the listed prominent tsunamis were recorded along with a few weaker events including the 1994 Shikotan tsunami. The Korsakov permanent tide gauge was closed in 1992. However, the records of the 2010 and 2011 tsunamis were obtained by newly installed TWS tsunameters (SHEVCHENKO et al. 2013). Therefore, the frequency of recorded tsunamis is not the same at various stations, although the actual frequency of large tsunamis is the same for the entire Sea of Okhotsk.

For the Kuril Islands, our resulting values are relatively high and are in a good agreement with the estimates of KAISTRENKO (2014). The exception is the

Table 5

Statistical characteristics of tsunami waves: observed maximum trough-to-crest wave height and calculated extreme tsunami heights for return periods of 50, 100 and 200 years

Station	Observed maximum (m)	Return period (year)		
		50	100	200
Burevestnik	3.0	3.2	4.0	4.9
Kurilsk	1.0	0.8	1.1	1.4
Matua	3.5	4.0	5.3	6.6
Severo-Kurilsk	11.0	5.9	9.6	13.3
Yuzhno-Kurilsk	4.6	3.5	4.6	5.7
Malokurilsk	4.6	4.2	5.4	6.7
Magadan	2.5	1.8	2.5	3.2
Nabil	0.8	0.8	1.1	1.4
Poronaisk	1.3	0.9	1.5	2.1
Korsakov	1.5	1.2	1.6	1.9

Kurilsk station on the Okhotsk coast of Iturup Island. The maximum recorded tsunami height at this station (1 m, the 1960 Chilean tsunami) was the lowest among all stations (together with Nabil on the northeastern coast of Sakhalin Island). Accordingly, the calculated tsunami height at Kurilsk is lower than the storm surge height, while at Nabil these values are similar. For other stations on the Sea of Okhotsk coasts, the calculated tsunami heights are substantially higher than the heights of storm surges, and the difference increases with the return period.

The sampled tsunami PDF was calculated as the difference in values of the probability distribution function with a class interval of 5 cm. 10 cm was the minimum value and the maximum was the height corresponding to a return period of 500 years. Probability of the value 0 cm was defined from the condition that the sum of all the probabilities equals 1.

7. Estimations of Extreme Sea Levels using JPM

The joint probability method provides realistic estimates of extreme sea level in those cases when it is critical to take into account the possible combination of several unfavorable factors. This applies to coastal structures, such as ports, water intakes and other objects that are built in the coastal zone. Of particular importance are objects whose failure or damage could create serious environmental consequences. For the Sea of Okhotsk, these are the objects on the northeastern shelf of Sakhalin Island that provide production and transportation of oil and gas. This approach can also be used to evaluate the potential impact on marine water intakes of nuclear power plants. The requirements of the International Atomic Energy Agency (IAEA) to assess events with a probability of 10^{-6} (http://www.slideshare.net/atomlibrary/iaea-safety-standard-no-ssg2) are in good agreement with the probability of coincidence of a tsunami with the highest high tide or with a strong storm surge. JPM is also very efficient when the observational series are relatively short; i.e., not sufficient to have observed floods associated with the highest combination of several factors.

The main outcome of this study is the estimate of maximum sea level at individual tide gauges located

Table 6

JPM calculated extreme sea levels (cm) estimated as combination of tide+surge

Station	Return period (year)		
	50	100	200
Burevestnik	178	183	187
Kurilsk	228	238	246
Matua	213	218	223
Severo-Kurilsk	145	148	151
Yuzhno-Kurilsk	114	117	120
Malokurilsk	104	107	110
Magadan	262	267	271
Nabil	138	141	144
Poronaisk	157	164	170
Korsakov	135	141	147

Table 7

JPM calculated extreme sea levels (m) estimated as combination of tide+surge+tsunami

Station	Return period (year)		
	50	100	200
Burevestnik	3.4	4.1	4.9
Kurilsk	2.3	2.4	2.5
Matua	4.1	5.3	6.6
Severo-Kurilsk	5.9	9.6	13.3
Yuzhno-Kurilsk	3.5	4.6	5.7
Malokurilsk	4.2	5.4	6.7
Magadan	2.7	2.9	3.4
Nabil	1.4	1.4	1.5
Poronaisk	1.6	1.7	2.2
Korsakov	1.4	1.6	2.0

on the coast of the Sea of Okhotsk and on the Pacific coast of the Kuril Islands with return periods of 50, 100 and 200 years. These values are produced by superposition of tide+surge (Table 6) and tide+surge+tsunami (Table 7).

The maximum heights of tidal oscillations at most stations are of the same order (Table 2), ranging from 54 cm at the South Kuril Islands to 88 cm at the North Kuril Islands. These values determine the tides contribution in the JPM estimations of the extreme sea-level heights (tide+surge). The Magadan station is the only one where the positive deviation of tides is more than 2 m. The extreme sea level at this station is significantly higher than at other stations when considering tide plus storm surge. Possible sea-level heights of more than 2 m for a return period of

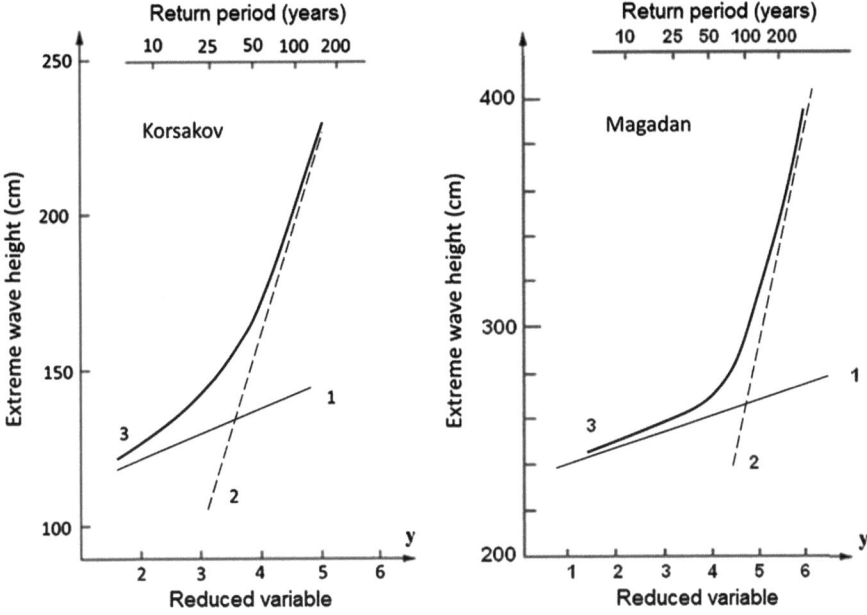

Figure 10
Calculated extreme sea levels at Korsakov and Magadan: *1* JPM estimates of tide+surge; *2* tsunami wave heights estimated based on the Gumbel's extreme statistics method; *3* JPM estimates of tide+surge+tsunami

100 years were found to be expected at stations Matua and Kurilsk; anomalously strong storm surges caused high extreme total sea levels at these stations.

Our calculations (Table 7) showed that the assessments of possible tsunami heights for stations located on the Pacific coast are only insignificantly changed by the influence of tides and storm surges. An increase of only a few centimeters was found at these stations for the recurrence period of 50 years.

In contrast, for most of the stations on the coast of the Sea of Okhotsk, the contribution of tidal and meteorological components is more significant, especially for the return period of 50 years or less (the corresponding examples for Korsakov and Magadan are shown in Fig. 10). With increased return periods, the contribution of these components to the overall assessment becomes smaller, except Kurilsk, where the role of tsunamis is minor. This is due to the fact that at this station, the maximum height of the surge in 1971 exceeded 2 m, while the maximum observed tsunami height (for the 1960 Chilean event) was only about 1 m. The most significant contribution of the tidal component was found at Magadan, where tides are very high. However, tsunamis in this area are also significant. Thus, the total extreme height level of sites examined in the Sea of Okhotsk is found at Magadan.

Specifically, it was found that the highest sea levels constructed by superposition of tides, storm surges and tsunamis are on the Pacific coast of the Kuril Islands, and at Magadan. The lowest sea levels are on the coast of Sakhalin Island. Another interesting and new result is an estimation of extreme heights and relative importance of individual sea-level components. Different components were found to play an essential role in various parts of the study area. Tides are especially strong in the northern part of the Sea of Okhotsk (Magadan). Tsunamis at this station are also the largest in comparison with the stations located on the coast of Sakhalin Island. Tsunamis prevail on the Pacific coast of the Kuril Islands. The relative role of storm surges is the most significant at stations Kurilsk and Matua.

8. Conclusions

The joint probability method provided realistic extreme sea levels based on observational time series (some were relatively short) from 10 tide gauges located on the Pacific coast of the Kuril Islands and on the coast of the Sea of Okhotsk. Several different factors (tides, seasonal oscillations, storm surges and tsunamis) were taken into account.

The absolute estimates of the total maximum sea levels for various stations at the Russian Far East coast with return periods of 50, 100 and 200 years (Tables 6, 7) are the main result of our study. Specifically, it was found that the highest sea-level elevations (without taking tsunamis into account) are found at Magadan on the northwestern coast of the Sea of Okhotsk, where very high tides occur, and at Matua and Kurilsk in the Kuril Islands where high storm surges take place. The lowest non-tsunami sea-level oscillations are at Malokurilsk and Yuzhno-Kurilsk in the South Kuril Islands.

However, taking the tsunami into account significantly changes the entire picture. The highest sea-level elevations in that case were found to occur on the Pacific coast of the Kuril Islands. The relative role of other sea-level components is negligibly small. For the coast of the Sea of Okhotsk, we found the maximum overall sea-level height to be at Magadan; both tsunamis and tides are the highest in this region of the Sea of Okhotsk. The lowest sea-level elevations were found on the coast of Sakhalin Island (Nabil, Korsakov and Poronaisk). Tsunami influence is minor at Kurilsk, where the maximum observed tsunami height was about one half of the maximum storm surge.

In summary, we can say that for the coast of the Sea of Okhotsk, for short return periods, the extreme heights are determined by storm surges, tides and seasonal oscillations for all examined stations. For longer return periods, tsunamis begin to play an essential role in the formation of extreme sea levels (similar to the Russian coast of the Sea of Japan, RABINOVICH et al. 1992). In our estimates, we have not considered an important factor that is the duration of the tsunamis and storm surges. This is the subject of a separate study.

Tsunamis are among the world's most destructive natural hazards. To mitigate the loss of life and property, the possible impact of tsunamis must be taken into account prior to major development or construction; particularly in seismically active regions of the ocean coast. Tsunamis and earthquakes are constraints to economic development of the Kuril Islands and other coastal areas of Russia's Far East. At the same time, we should remember that tsunamis are not an isolated process but they occur on top of other sea-level oscillations that can significantly increase their negative effect. The results presented in this study may be useful in evaluating the relative role of various sea-level components, and in planning safe construction of new or existing facilities on the coast of the Sea of Okhotsk and Pacific coasts of the Kuril Islands.

Acknowledgments

We are grateful to Artem Loskutov (IMGG FEB RAS, Yuzhno-Sakhalinsk, Russia) for his help with the figures preparation, and to Fred Stephenson (IOS, Sidney, BC, Canada) for productive comments and editing the text. We are also quite thankful to our reviewers Eric Geist (USGS, Menlo Park, California, USA) and Paul Whitmore (WC/ATWC, Palmer, AK, USA) for their thorough evaluation and helpful comments and suggestions that significantly improved this manuscript. This study for was partly supported by the RFBR Grant 13-05-00936-a, and by a Grant of the Russian Academy of Sciences (Far East Branch) No 15-I-1-045 e.

REFERENCES

CONSOLI, S., RECUPERO, D., and ZAVARELLA, V. (2014). *A survey of tidal analysis and forecasting methods for tsunami detection*, Science of Tsunami Hazards, *33*(1), 1–57.

GEIST, E. L., PARSONS, T., TEN BRINK, U. S., and LEE H. J. (2009). *Tsunami probabilities*, In: The Sea, Vol.15, Tsunamis (Eds. A.Robinson and E. Bernard), Harvard University Press, Cambridge, USA, pp. 93–135.

GERMAN, V.H. and LEVIKOV, S.P. (1988). *Probability Analysis and Numerical Modelling of Sea Level Oscillations*. Hydrometeoizdat, Leningrad, 231 p. (in Russian).

GO, CH. N., KAISTRENKO, V. M. and SIMONOV, K. V. (1985). *A two-parameter scheme for tsunami hazard zoning*, Marine Geodesy 9(4), 469–476.

GUMBEL, E. J. (1958). *Statistics of extremes*, New York: Columbia University Press, 375 pp.

KAISTRENKO, V. (2014). *Tsunami recurrence function: structure, methods of creation, and application for tsunami hazard estimates*, Pure Appl. Geophys., *171*, 3527–3538.

KAISTRENKO, V. and SEDAEVA, V. (2001). *The 1952 North Kuril Tsunami: new data from archives*. In: Tsunami Research at the

End of a Critical Decade (Ed. G. Hebenstreit), Springer, Dordrecht, 91–102.

KARIM, M.F. and ISMAIL, A.I.M. (2010). *Estimation of expected maximum water level due to tide and tsunami interaction along the coastal belts of Pelang Island in peninsular Malaysia* Science of Tsunami Hazards, 29(3), 127–138.

KHRAMUSHIN, V.N. and SHEVCHENKO, G.V. (1994). *A method of detailed tsunami zoning for coastal Aniva Bay*. Oceanology, 34(2), 192–197.

KULIKOV, E.A., RABINOVICH, A.B., and THOMSON, R. E. (2005). *Estimation of tsunami risk for the coasts of Peru and Northern Chile*. Natural Hazards, 35, 185–209.

LOBKOVSKY, L.I., RABINOVICH, A.B., KULIKOV, E.A., IVASHCHENKO, A.I., FINE, I.V., THOMSON, R.E., IVELSKAYA, T.N., and BOGDANOV, G.S. (2009). *The Kuril earthquakes and tsunamis of November 15, 2006, and January 13, 2007: Observations, analysis, and numerical modeling*, Oceanology, 49(2), 166–181.

LYUBITSKY, Yu.V., SHEVCHENKO, G.V., and ELISOV V.V. (2013). *Storm surges*. In: *World Ocean. Geology and Tectonics of the Ocean. Catastrophic Events in the Ocean*. Scientific World, Moscow, pp. 559–575 (in Russian).

MACINNES, B.T., PINEGINA, T.K., BOURGEOIS, J., RAZHIGAEVA, N.G., KAISTRENKO, V.M. and KRAVCHUNOVSKAYA, E.A. (2009). *Field survey and geological effects of the 15 November 2006 Kuril tsunami in the middle Kuril Islands*, Pure Appl. Geophys., 166, 9–36.

MOFJELD, H. O., GONZALEZ, F. I., and NEWMAN, J. C. (1999). *Tsunami prediction in U.S. coastal regions*, Coastal and Estuarine Studies, 56, American Geophysical Union, 353–375.

MOFJELD, H.O., GONZÁLEZ, F.I., TITOV, V.V., VENTURATO, A.J. and NEWMAN, J.C. (2007). *Effect of tides on maximum tsunami wave heights: probability distributions*. Atmosph. Oceanic Techn., 24, 117–123.

POEZZHALOVA, O.S. and SHEVCHENKO, G.V. (1997). *Seasonal oscillations of sea level in the Sea of Okhotsk*, Proceedings and abstracts of the 12th International Symposium on the Okhotsk Sea & Sea Ice, 2-5 February 1997, Mombetsu, Hokkaido, Japan, 248–255.

PUGH, D.T. and VASSIE, J.M. (1978). *Extreme sea levels from tide and surge probability*, Proc. 16th Coast. Eng. Conf., Hamburg, 911–930.

RABINOVICH, A.B. and SKRIPNIK, A.V. (1984). *Consideration of tidal and meteorological sea level oscillations in estimates of tsunami-risk in the area of Severo-Kurilsk*, In: Non-Stationary Wave Processes on the Shelf of the Kuril Islands, FED USSR Acad. Sciences, Vladivostok, pp. 81–92 (in Russian).

RABINOVICH, A.B. and SHEVCHENKO, G.V. (1990). *Estimations of extreme sea level heights as superposition of tides, storm surges and tsunamis. Summary*. In: *Tsunamis: Their Science and Hazard Mitigation*. Proc. Intern. Tsunami Symposium. July 31- August 3, 1989. Novosibirsk, pp. 201–205.

RABINOVICH, A.B., SHEVCHENKO, G.V., and SOKOLOVA, S.E. (1992). *An estimation of extreme sea levels in the northern part of the Sea of Japan*, La mer., 30, 179–190.

RABINOVICH, A.B., THOMSON, R.E., and FINE, I.V. (2013). *The 2010 Chilean tsunami off the west coast of Canada and the northwest coast of the United States*. Pure Appl. Geophys., 170, 1529–1565.

ROMANOV, A.A., SEDAEVA, O.S., and SHEVCHENKO, G.V. (2004). *Seasonal and tidal variations of the sea level between Hokkaido and Sakhalin Islands based on satellite altimetry and coastal tide gauge data*, Pacific Oceanography, 2(1–2), 117–125.

SEDAEVA, O. and SHEVCHENKO, G. (2001). *Investigation of seasonal fluctuations of sea level and atmospheric pressure in the area of Kuril ridge*. Proceedings and abstracts of the 16th International Symposium on the Okhotsk Sea & Sea Ice, 4–8 February 2001, Mombetsu, Hokkaido, Japan, 339–342.

SHEVCHENKO, G.V. (1997a). *Statistical characteristics of storm surges in the southern part of Sakhalin Island*, Proceedings of the Russian Geographical Society, 129(3), 94–107 (in Russian).

SHEVCHENKO, G.V. (1997b). *Storm surges on the Kuril Islands*. In: *Tsunami and Accompanied Events*. IMGG FEB RAS, Yuzhno-Sakhalinsk, 106–116 (in Russian).

SHEVCHENKO, G., ROMANOV, A., and BOBKOV, A. (2004). *Definition of spatial variability of tide characteristics in the Sea of Okhotsk from satellite altimetry data*, Proc. Third PICES Workshop on the Okhotsk Sea and Adjacent Areas. Sidney, B.C., Canada, PICES Scientific report, 26, 33–43.

SHEVCHENKO, G.V., CHERNOV, A.G., KOVALEV, P.D., KOVALEV, D.P., LIKHACHEVA, O.N. LOSKUTOV, A.V. and SHISHKIN, A.A. (2011). *The tsunamis of January 3, 2009 in Indonesia and of January 15, 2009 in Simushir as recorded in the South Kuril Islands*, Science Tsunami Hazards 30(1), 43–61.

SHEVCHENKO, G., IVELSKAYA, T., LOSKUTOV, A., and SHISHKIN, A. (2013). *The 2009 Samoan and 2010 Chilean Tsunamis Recorded on the Pacific Coast of Russia*, Pure Appl. Geophys., 170, 1511–1527.

SOLOVIEV, S.L. and GO. CH.N. (1974). *Catalogue of Tsunamis on the Western Shore of the Pacific Ocean*. Nauka Publ. House, Moscow, 310 p. (in Russian; English translation; Canadian Transl. Fish. Acuatic Sci., No. 5078, Ottawa, 439 pp., 1984).

TAWN, J.A. (1992). Estimating probabilities of extreme sea levels. Journal of the Royal statistical society. Series C (Applied statistics), 41(1), 77–93.

THOMSON, R. E., RABINOVICH, A. B. and KRASSOVSKI M. V. (2007). *Double jeopardy: Concurrent arrival of the 2004 Sumatra tsunami and storm-generated waves on the Atlantic coast of the United States and Canada*, Geophys. Res. Lett., 34, L15607, doi:10.1029/2007GL030685.

WILSON, R.I., DENGLER, L.A., LEGG, M.R., LONG, K. AND MILLER, K.M. (2010), *The 2010 Chilean tsunami on the California coastline*, Seism. Res. Lett. 81(3), 545–546.

ZETLER, B.D., and FLICK, R.E. (1985). *Predicted extreme high tides for mixed-tide regimes*, Phys. Oceanogr. 15, 357–369.

(Received January 18, 2015, accepted April 6, 2015, Published online April 19, 2015)

ns of the Relationship Between Coral Damage and Tsunami Dynamics; Case Study: 2009 Samoa Tsunami

Derya I. Dilmen,[1,2] Vasily V. Titov,[1,2] and Gerard H. Roe[2]

Abstract—On September 29, 2009, an Mw = 8.1 earthquake at 17:48 UTC in Tonga Trench generated a tsunami that caused heavy damage across Samoa, American Samoa, and Tonga islands. Tutuila island, which is located 250 km from the earthquake epicenter, experienced tsunami flooding and strong currents on the north and east coasts, causing 34 fatalities (out of 192 total deaths from this tsunami) and widespread structural and ecological damage. The surrounding coral reefs also suffered heavy damage. The damage was formally evaluated based on detailed surveys before and immediately after the tsunami. This setting thus provides a unique opportunity to evaluate the relationship between tsunami dynamics and coral damage. In this study, estimates of the maximum wave amplitudes and coastal inundation of the tsunami are obtained with the MOST model (Titov and Synolakis, J. Waterway Port Coast Ocean Eng: pp 171, 1998; Titov and Gonzalez, NOAA Tech. Memo. ERL PMEL 112:11, 1997), which is now the operational tsunami forecast tool used by the National Oceanic and Atmospheric Administration (NOAA). The earthquake source function was constrained using the real-time deep-ocean tsunami data from three DART® (Deep-ocean Assessment and Reporting for Tsunamis) systems in the far field, and by tide-gauge observations in the near field. We compare the simulated run-up with observations to evaluate the simulation performance. We present an overall synthesis of the tide-gauge data, survey results of the run-up, inundation measurements, and the datasets of coral damage around the island. These data are used to assess the overall accuracy of the model run-up prediction for Tutuila, and to evaluate the model accuracy over the coral reef environment during the tsunami event. Our primary findings are that: (1) MOST-simulated run-up correlates well with observed run-up for this event ($r = 0.8$), it tends to underestimated amplitudes over coral reef environment around Tutuila (for 15 of 31 villages, run-up is underestimated by more than 10 %; in only 5 was run-up overestimated by more than 10 %), and (2) the locations where the model underestimates run-up also tend to have experienced heavy or very heavy coral damage (8 of the 15 villages), whereas well-estimated run-up locations characteristically experience low or very low damage (7 of 11 villages). These findings imply that a numerical model may overestimate the energy loss of the tsunami waves during their interaction with the coral reef. We plan future studies to quantify this energy loss and to explore what improvements can be made in simulations of tsunami run-up when simulating coastal environments with fringing coral reefs.

Key words: Tsunami, Samoa, Tutuila, MOST, coral, reef, 2009, Tonga.

1. Introduction

Large tsunamis can wreak devastation upon the near-shore environment. There is abundant documentation of the impacts on the subaerial portion of that environment, but much less on the impacts on the submarine portion. In many tropical settings, coral reefs form an important component of the submarine environment, being the cornerstone of the local ecosystems, as well as shaping the near-shore bathymetry. There is thus the potential for two-way interactions between reefs and tsunamis. The reef bathymetry influences the tsunami dynamics; and tsunami events may cause significant damage to fragile coral structures. In this study, we report on a unique opportunity to document tsunami-related damage, and to evaluate whether the damage can be straightforwardly related to particular aspects of the tsunami dynamics.

On 29 September 2009 at 17:48 UTC, an Mw 8.1 earthquake occurred along Tonga-Kermadec Trench. A complicated fault rupture produced bottom deformations and resulted in tsunami waves that generated localized run-ups exceeding 17 m on the island of Tutuila. These waves claimed 34 lives (out of total 192 deaths for the event) and caused extensive damage around the island. The tsunami was detected by coastal tide gauges and offshore sea-level sensors

---

[1] Pacific Marine Environmental Laboratory, NOAA Center for Tsunami Research, Seattle, USA. E-mail: dilmen@uw.edu; vasily.titov@noaa.gov
[2] Department of Earth and Space Sciences, University of Washington, Seattle, USA. E-mail: groe@uw.edu

located in Pacific Ocean. The tectonic setting of the Tonga Trench has produced several tsunamis during past hundred years (OKAL et al. 2011).

Following the September 29, 2009 tsunami, field surveys were conducted (FRITZ et al. 2011) to document the relationship between the physical near-shore environment and the tsunami impact. According to survey results, the tsunami produced a maximum run-up of 17 m at Poloa on the western coast of Tutuila, 12 m at Fagasa on the northern coast, and 10 m at Tula on the eastern coast. The survey team recorded large variations in the impacts of the tsunami along the coastal bays: a wide range of tsunami run-ups, wave directions and inundation. The high degree of spatial variability in these various tsunami fields was somewhat of a surprise to scientists studying the event (FRITZ et al. 2011; OKAL et al. 2010; BEAVAN et al. 2010; ROEBER et al. 2010), but was clearly established in the field surveys and confirmed by residents.

The impact of this tsunami on Tutuila has proven unusually hard to simulate in numerical models (BEAVAN et al. 2010; OKAL et al. 2010; ROEBER et al. 2010; FRITZ et al. 2011; ZHOU et al. 2012). The discrepancies between observations and models have been variously attributed to many factors, including the low-resolution bathymetry and topography, wave dispersion effects, the possibility of resonance over the coral reefs, but most importantly, the unusual complexity of the tsunami source mechanism that may have included multiple ruptures of several fault systems at the same time (BEAVAN et al. 2010). We also perform a simulation of this event, and aim to build upon the experience of these earlier studies: we try to eliminate any bathymetric and topographic discrepancies by using a very high-resolution (10 m) dataset (LIM et al. 2009); further, we optimize the tsunami source function by calibrating it with direct tsunami observations. Both far-field pressure sensors (DARTs) and near-field coastal sea-level stations (tide gauges) were used to calibrate the tsunami source for this event. We establish good agreement with the near-field tide gauges (Sect. 3), which means it is unlikely that additional details in the source function would impact the simulation.

One challenge of modeling tsunamis in tropical settings such as this are the pervasive barrier and fringing coral reefs, which create tremendous complexity in bathymetry and topography (Fig. 1). The impact of reefs in tsunami dynamics has been a topic of discussion in the literature. Such analyses point to a complex picture, and conclusions can occasionally appear contradictory. BABA et al. (2008) performed numerical simulations of the 2007 Solomon islands Tsunami to explore the effect of Great Barrier Reef (GBR) on tsunami wave height, using the low-resolution bathymetry and ignoring sea-bottom friction and wave dispersion. The results indicate reefs decrease the tsunami wave height due to the refraction and reflection. KUNKEL et al. (2006) performs 1D and 2D numerical modeling of tsunami run-up for an idealized island with barrier reefs around the island, and shows that coral reefs reduce tsunami run-up by order of 50 %. However the KUNKEL et al. (2006) simulations also suggest the possibility that gaps between adjacent reefs can result in flow amplification and actually increase local wave heights. FERNANDO et al. (2005, 2008) lend support to these numerical results: coral reefs protect coastline behind them but local absences of reefs cause local flow amplification due to gaps. Their results are based on field observations, laboratory measurements (FERNANDO et al. 2008), and interviews done by local people in Sri Lanka after the 2004 Indian Ocean tsunami. However their laboratory simulations treated corals as a submerged porous barrier made of a uniform array of rods, which likely oversimplifies the complex structural distribution of coral reefs.

Other studies find no effect, or even suggest the opposite conclusions (KUNKEL et al. 2006). Based on quantitative field observations of coral assemblages at less than 2 m depth in Aceh after the 2004 Sumatra–Andaman tsunami, BAIRD et al. (2005) conclude that the limit of inundation at any particular location is determined by a combination of wave height and coastal topography, and is independent of the reef quality or development prior to the tsunami. Further, CHATENOUX and PEDUZZI (2007) perform statistical and observational analysis of 56 sites located in Indonesia, Thailand, India, Sri Lanka and Maldives with a coarse resolution bathymetry and qualitative coral damage data. They find that the higher the percentage of the corals, the larger the inundation distances behind coral reef on the coast. Lastly, ROEBER et al. (2010) identify strong correlations

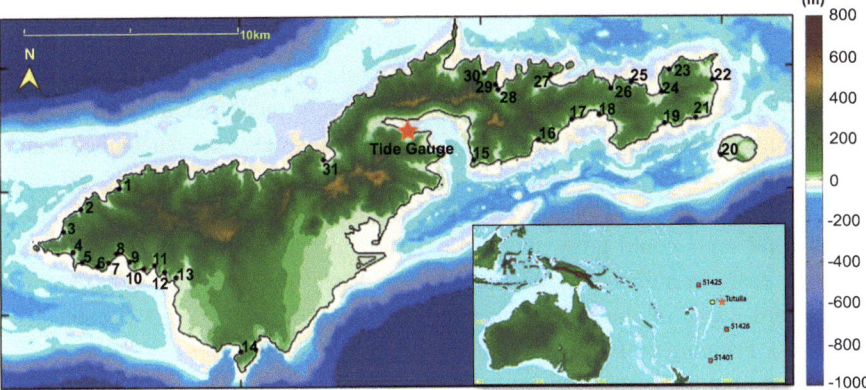

Figure 1
Tutuila island. The *beige* and *light purple colors* show the location of the fringing and barrier reefs. The location of Tutuila island is given as *red star* on the *lower right map*. The *yellow dot* on the same map shows the epicenter of the 2009 Samoa earthquake. In the *inset panel* the location of the DART buoys used in optimizing the earthquake fault source used in the MOST simulations are shown. In the *main panel*, the location of the PagoPago Tide gauge in Tutuila is shown as *red star*

between the high variability of run-up and inundation along bays at Tutuila during the 2009 Samoa tsunami with the geomorphology of the island, and suggest a role for high concentrations of resonance energy within particular bays. All of these locations of high-energy concentration have fringing reefs extending 100–200 m from the shores. Based on their tsunami simulations, they hypothesize that fringing reefs might amplify near-shore tsunami energy and worsen the impact of short-period dispersive waves.

In this paper, wave heights, inundation at the coast, and tsunami wave dynamics are simulated for the island of Tutuila for the 2009 tsunami. The simulations are compared with field observations at the coast and wave pressure gauges (DARTs) located around Tutuila to find a relationship between coral damage and coastal metrics of tsunami dynamics. The results contribute to an ongoing discussion about how tsunami dynamics impact corals and how, in turn, that damage might potentially be used to constrain tsunami simulations.

The remainder of the paper is organized as follows: Section 2 describes the study area, the earthquake and tsunami event, and the observational datasets. Section 3 describes the numerical modeling of the earthquake source and the subsequent tsunami. Section 4 presents an analysis of the relationships among the observations, datasets and simulated tsunami fields. We conclude with a summary and discussion that suggests an outline for future research directions.

2. Study Area and Observations

2.1. Tutuila Island

The study area for this research is Tutuila Island, in American Samoa, the United States' southernmost territory. The Samoa island chain in the central South Pacific Ocean includes five islands, of which Tutuila is the largest and also its center of the government. It is located at roughly 14° south of the equator between longitudes 169° and 173° west (see Fig. 1). The following will summarize some aspects of the geometry of Tutuila island that created unique challenges in modeling of the tsunami.

The island formed in the late Quaternary period, from oceanic crust as the Pacific tectonic plate moved over a hotspot (TERRY et al. 2005). Due to its volcanic formation, it has rocky, steep topography and bathymetry, with narrow valleys that rise from ocean floor (McDOUGALL 1985). The island sits on a shallow submarine platform, which then drops off to a depth of over 3000 m to meet the abyssal plain. Tutuila is approximately 32 km long, with a width that ranges from less than 2 km to a maximum of 9 km. An insular shelf (<100 m depth) with an average width of 4 km extends along the entire north coast and the southwest region of the island (see Fig. 1).

The island is surrounded by fringing and barrier coral reefs, which contain a diversity of coral reef habitats, and coral species. The island has possibly subsided faster than coral reefs could grow upward, leaving former barrier reefs as submerged offshore banks along the seaward edges of the insular shelf (BIRKELAND et al. 2008). Fringing reefs have a width ranging from 0 to 600 m, but 90 % of them are less than 217 m (GELFENBAUM et al. 2011). The barrier reefs are located 2–3 km from the coastline. The total area of coral reefs in the territory of Tutuila is approximately 300 km^2.

2.2. The Samoa Event

The tsunami of September 29, 2009 earthquake was generated at the most active region of deep seismicity of Tonga Trench, and reached the Samoan island chain approximately 20 min later. The tsunami caused devastating property damage and loss of life on Tutuila island, because of its close proximity to the epicenter and the high population density on its coasts.

The cause of the earthquake was the rupture of a normal fault with a moment magnitude of Mw = 8.1 in the outer trench-slope at the north end of the trench, near the sharp bend to the west, followed by two inter-plate ruptures on the nearby subduction zone with moment magnitudes of Mw = 7.8 (LAY et al. 2010). Fault displacements measured by seismic signals (LAY et al. 2010), Global Positioning System (GPS) Stations, and ocean-bottom pressure sensors (BEAVAN et al. 2010) for these three separate faulting events support this picture. These fault displacements led to vertical movement of the seafloor, and created a complex tsunami source mechanism.

2.3. Observations

This particular tsunami afforded a unique opportunity to systematically evaluate the relationship between tsunami dynamics and coral reef damage. Six months prior to the tsunami, NOAA Coral Reef Ecosystem Division (CRED)-certified divers performed comprehensive surveys of the reefs around Tutuila. The survey lines totaled 110 km in length. Observers measured the number of live, dead, and stressed corals, sea cucumbers, and urchins along track lines at the depths of 10–20 m. In the immediate aftermath of the tsunami the divers retraced most of the original survey lines. They documented clear evidence of fresh damage at depths between 10 and 20 m. This depth range was selected because of the location of the fore-reef at these depths, which is where the coral population is a maximum. Damage at depths shallower than 10 m was not recorded during the survey (BRAINARD et al. 2008).

The divers operated a tow board of instruments as it was tugged behind a boat at a depth of about 15 m. Data taken included direct observations from the diver, a downward facing camera and electronic instrumentation, including GPS (Fig. 2). The downward-pointed camera recorded the sea bottom habitat. It also captured images at 15 s intervals (NOAA-PIFSC-CRED unpublished data). Selected images of broken and overturned table corals and broken branching corals are presented in Fig. 2. The survey covered a total of 83 km linear distance within a 5 m horizontal zone either side of the track line. Divers were careful to try to differentiate between damage directly due to the tsunami itself, and land-originating debris entrained into the water.

The damage survey report synthesized the direct observations, aggregating track-line data into groupings based on 31 nearby villages, and reported the total number of damage observations (Table 3). The number of coral damage reports was supplemented with notes. Examples of such notes are: at Onenoa Village, "coral damage was low, with only one damaged tabulate Acropora sighting was recorded between both divers"; and at Amaluia Village it "consisted of isolated sightings of broken branching (species of Pocillopora and Acropora) corals". The survey is not an absolute measure of coral damage and involves a degree of subjectivity: it records the number of observations of coral damage, not the absolute number of damaged corals. It is nonetheless a useful window onto the impact of tsunami dynamics in the immediate aftermath of the event. Since the full coral density of the island is not available, variation

Figure 2
Photographs of the coral survey, methods and typical observations. *Upper left image* shows the NOAA-certified diver surveying a track line with a tow board tugged behind a boat. *Lower left image* shows the instrument suite on tow boards, among which are observer data sheet, gauges and timers, a camera and strobes. *Upper right* shows the table and branching corals that have been overturned; *lower right* shows a table coral that has been broken due to the tsunami. *Images* are taken from NOAA-Marine Debris Division

in coral density might influence the results (NOAA-PIFSC-CRED unpublished data).

The datasets reported by the divers for the number of damaged corals are discontinuous, unevenly and non-normally distributed, and this precludes classifying the coral damage observations with conventional methods such as standard deviation or equal intervals. Instead we used Jenks Natural Breaks classification method, a univariate version of k-means clustering (JENKS 1967) by sorting it from lowest value to highest and looking for large gaps, or natural breaks. This is done by seeking to minimize each class's average deviation from the class mean, while maximizing each class's deviation from the means of the other classes. In other words, the method iteratively seeks to reduce the variance within the same classes and maximize the variance between classes. The final classification in terms of coral damage is (0 no damage, 1–27 low, 38–63 medium, 83–159 high, 310 very high damage). While we felt the Jenks method is most appropriate for these data, our overall conclusions are not sensitive to this choice.

An international tsunami survey team observed and recorded tsunami run-up and inundation on the islands of the Samoan archipelago including Tutuila a week after the tsunami (FRITZ et al. 2011). The surveys followed the tsunami survey protocols reviewed by SYNOLAKIS and OKAL (2005). The team marked the values of run-up at 59 different field locations at Tutuila.

3. Modeling the Event

3.1. The Model Setup

We simulate the 2009 Samoa tsunami using the MOST Model (TITOV and GONZALEZ 1997). MOST is an established tsunami model that has been widely tested and evaluated, and it is used operationally for forecasting (e.g., TITOV 2009) and hazard assessment (e.g., TITOV et al. 2003). There are other numerical tsunami models with alternative dynamical equations and/or numerical schemes. Recognizing the importance for inter-model evaluations (SYNOLAKIS et al.

2008), recent community efforts have focused on using models that satisfy theoretical benchmarks and case study comparisons, such as those proposed by the National Tsunami Hazard Mitigation Program (NTHMP 2012). MOST meets the benchmarks and performs comparably to other tsunami models for the real-world case studies.

The primary metrics for comparison with observations are wave run-up and inundation. MOST solves the shallow water equations with a leapfrog finite difference scheme (TITOV and SYNOLAKIS 1998). We define three, nested bathymetric and topographic grids. The earthquake dislocation is input as the tsunami source; several predetermined tsunami sources were tried in order to optimize the agreement with tide-gauge observations. Regional bathymetry and topography datasets (Table 1) were compiled and provided by National Geophysical Data Center (NGDC) and used to create the three nested grids (resolutions of 360, 60, and 10 m, respectively, see Fig. 3).

3.2. The Choice of Source Function

We simulated the 2009 Samoa Tsunami with a tsunami source function, f, calibrated to direct observations. For this event, several combinations of source functions f have been developed for use in

Table 1

Bathymetry compiled by NGDC to create the three nested grids (LIM et al. 2009)

Source of data	Production date	Data type	Horizontal and vertical datum	Spatial Resolution (m)
NGDC	1962–1998	Single beam echo-sounder	WGS-1984 and MHW	~100
NGDC	2009	Digitized coastline		30
NGDC	1996–2005	Multi-beam swath sonar		30–90
Gaia Geo-Analytical	2008	Estimated depths from satellite imagery		~5
NAVEOCEANO	2006	Bathymetric-topographic data		5
SCSC	2002	Vector data		10
USGS	1996–2006	NED digital elevation model		30

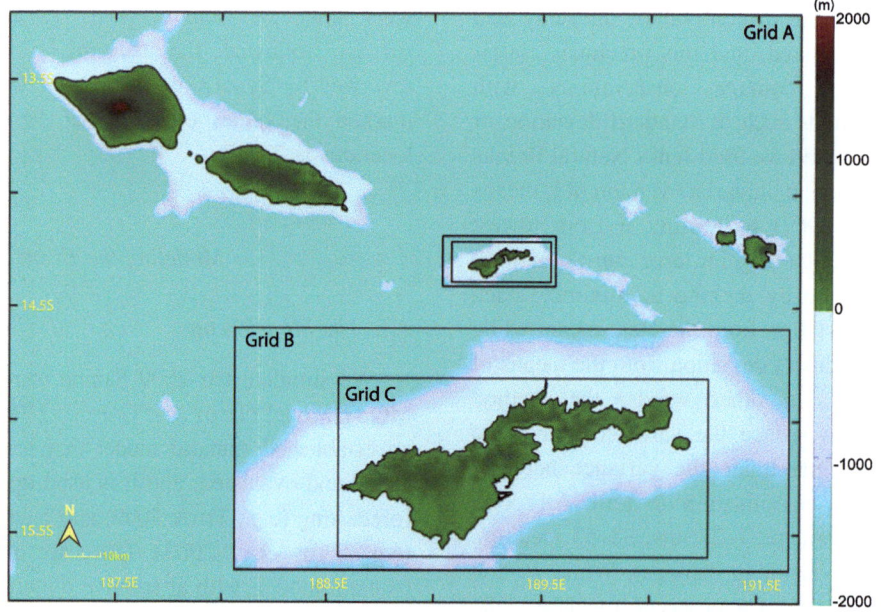

Figure 3
The boundaries of the nested A, B and C grids used in the MOST simulations

previous works (ZHOU et al. 2012; and TANG et al. 2009). Earthquakes are modeled as a combination of "unit sources", S_1, S_2, S_3 and so on. Each unit source is a reverse thrust of a given strike, dip, and depth, and each has a moment magnitude of 7.5 (GICA et al. 2008). The parameters for these unit sources were chosen according to the inversion results of the method described in GICA et al. (2008). The tsunami source function f, is converted into an initial wave height using the elastic model of GUSYAKOV (1972). This assumes the rupture of rectangular fault planes causes vertical displacements of the sea floor, and that the initial water level movement is equal and instantaneous to the corresponding vertical sea-bottom displacement. The inversion finds the linear combination of unit sources that best matches the DART buoy data (PERCIVAL et al. 2009).

We tested four previously optimized source function f. Our choice of source function f was based on optimizing the agreement to the PagoPago tide gauge data. Of the source functions we considered, one significantly underestimated wave heights, the other three sources all performed comparably and performed well: for the first four waves, they all matched the tide-gauge wave amplitudes to within about 10 %, and the timing of crests and troughs to within 20 min. In order to make sure that our choice of source function was not being overfit to a single data point (PagoPago), we checked model output at a 'virtual' tide gauge at a model grid point west of the island. At this virtual tide gauge, the time series of wave height was robust to the choice of f. The outlier f for PagoPago remained an outlier, and there was close agreement among the other three fs. Although our results would be similar for any of these three fs, we picked the source function with the best agreement to Pago-Pago (TANG et al. 2009) for which

$$f = 6.45 \times S_1 + 6.21 \times S_2, \quad (1)$$

where the specific parameters of S_1 and S_2 are given in Table 2.

3.3. Evaluation of the Model Results with DART and Tide Gauges

We first compare the tsunami wave amplitudes simulated with MOST to the tide-gauge observations in PagoPago in the near field and three DART Buoys 51425, 51426 and 54410 in the far field regions. For their locations, see Fig. 1. The simulated amplitudes match fairly well with the recorded values, particularly for PagoPago (Fig. 4), and particularly the first half-dozen fluctuations. The DART buoys record high-frequency crustal Rayleigh waves in the hour or so ahead of the arrival of the lower frequency tsunami waves (Fig. 5). Because the DART buoys lie farther from the source than Tutuila, phase discrepancies are expected to appear, which is particularly evident for DART Buoy 52425 (Fig. 5a). The discrepancies between the computed and recorded values at the DART buoys are likely also due to a secondary rupture occurred during the earthquake, which has been characterized in the model as one instantaneous rupture of the source function f. However, because of the excellent agreement with the PagoPago tide-gauge observations, we are confident these discrepancies are negligible for the purpose of run-up and inundation computations around Tutuila.

Table 2

Parameters of the two tsunami unit source functions, S_1 and S_2 (Eq. 1), used to simulate the 2009 Samoa tsunami

Unit Source	Longitude (°E)	Latitude (°S)	Dip (°)	Rake (°)	Strike (°)	Depth (m)	Mw	L (km)	W (km)	Slip (m)	Scaling parameter ()
1	187.2330	16.2754	9.68	90	182.1	5.00	8.1	100	50	1	6.45
2	187.8776	15.6325	57.06		342.4	6.57					6.21

4. Analysis

A comprehensive summary of data and analyses is presented in Table 3. We first evaluate whether there is any clear relationship between the two main observational datasets for this event—the coral damage and tsunami run-up for the 31 village sites. From Fig. 6 it is visually obvious that no such relationship exists ($r = 0.12$). Even excluding outliers (13 m run-up, 100 damage numbers at Vaitogi, 5 m run-up and 310 damage numbers at Fagatele) the data still do not yield a clean story ($r = 0.18$). Across a range of run-ups between 2 and 8 m, coral damage is as likely to be low or very low, as it is to be high or very high.

We are obviously constrained by the limitations of the data that are available for analysis, but on this basis no clear relationship between observed run-up

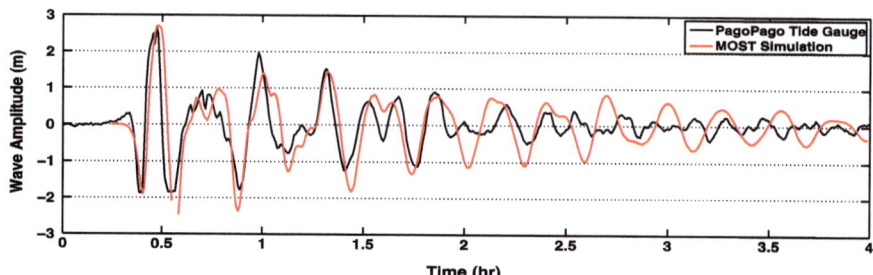

Figure 4
Comparison of observed (*black*) and simulated (*red*) water surface elevations at the PagoPago tide gauge (see Fig. 1) in the 4 h after the rupture at $t = 0$ (17:48:10 UTC, on Sep 29th, 2009). The MOST model estimated a maximum surface elevation of 2.3 m at the tide gauge, which agrees well with the recorded value

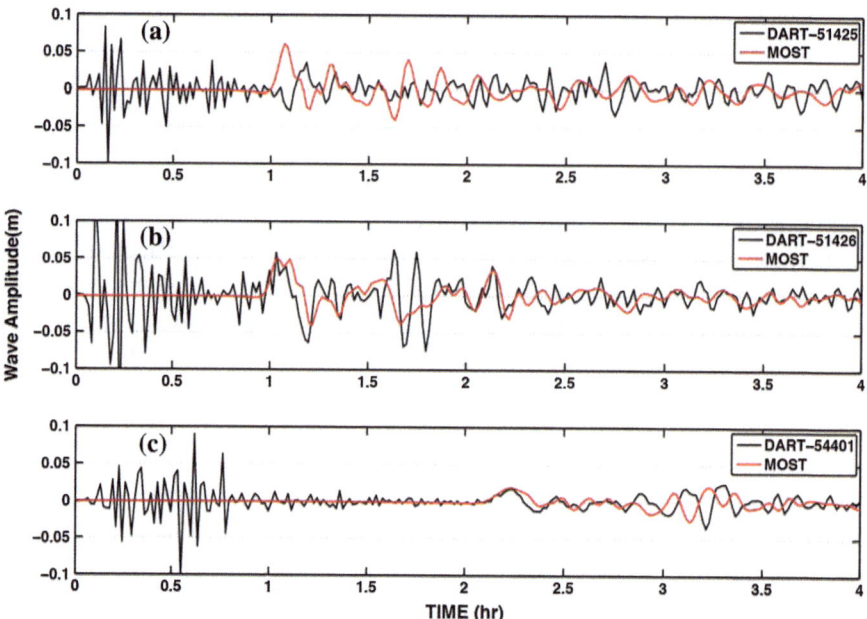

Figure 5
A comparison of the observed (*black*) and simulated (*red*) water surface elevations at three DART buoys, in the 4 h after the rupture at $t = 0$ (17:48:10 UTC on Sep 29th, 2009): **a** Buoy #51425, **b** Buoy #51426, and **c** Buoy #54401. High frequency, crustal Rayleigh waves are seen ahead of the arrival of the lower-frequency tsunami waves. The MOST model estimated maximum surface elevation of 0.05 m at selected DART locations

Table 3

Model, simulated and field run-up, differences, and coral damage at selected 31 villages around Tutuila

Village # (Fig. 1)	Village	Model run-up (m)	Field run-up (m)	Run-up difference (m)	Difference (%)	Average by village (%)	Coral damage	Coral dam. (numbers)
28	Afona1	3.52	4.31	−0.79	−18.33	−3.05	Medium	41
		3.6	4.08	−0.48	−11.76			
		3.9	4.41	−0.51	−11.56			
		3.4	3.75	−0.35	−9.33			
		3.69	3.59	0.1	2.79			
		3.05	2.59	0.46	17.76			
29	Afona2	3	2.25	0.75	33.33	38.43	Medium	41
		3.1	2.16	0.94	43.52			
10	Afao	4.2	5.89	−1.69	−28.69	−28.69	Low-medium	1
6	Agugulu	2.5	6.12	−3.62	−59.15	−59.15	High	86
30	Amalau	2.12	2.93	−0.81	−27.65	−19.45	No data	No data
		2.13	2.4	−0.27	−11.25			
4	Amanave	5.25	7.74	−2.49	−32.17	−32.17	High	151
11	Asilii	6	6.81	−0.81	−11.89	−11.89	Medium	42
12	Amaluia	5.2	5.39	−0.19	−3.53	−3.53	High	116
18	Amaua	3.2	2.91	0.29	9.97	9.97	No damage	0
19	Amouli	3.36	3.38	−0.02	−0.59	1.37	Low	13
		3.1	3	0.1	3.33			
24	Aoa	2.2	2.23	−0.03	−1.35	−1.35	No damage	0
21	Auasi	3.08	3.79	−0.71	−18.73	−18.73	Medium	38
20	Aunu'u	2.14	2.15	−0.01	−0.47	−0.47	High	121
16	Avaio	3.3	3.92	−0.62	−15.82	−3.36	Low	14
		3	2.75	0.25	9.09			
2	Fagailii	5.15	5.82	−0.67	−11.51	−18.82	No data	No data
		4.61	6.24	−1.63	−26.12			
31	Fagasa	4.2	4.13	0.07	1.69	1.69	No data	No data
1	Fagamalo	3	6.39	−3.39	−53.05	−48.71	No damage	0
		3.5	6.79	−3.29	−48.45			
		3.2	5.78	−2.58	−44.64			
14	Fagatele	3.3	4.92	−1.62	−32.93	−32.93	Very high	310
17	Fagaitua	4.75	3.5	1.25	35.71	59.77	No data	No data
		5	2.72	2.28	83.82			
5	Failolo	2.7	6.54	−3.84	−58.72	−58.72	High	159
13	Leone	2.65	2.75	−0.1	−3.64	67.86	High	83–116
		3.2	0.97	2.23	229.9			
		3.75	4.85	−1.1	−22.68			
27	Masefau	4	2.96	1.04	35.14	13.02		
		4.4	4.84	−0.44	−9.09		No data	No data
26	Masausi	3.15	2.79	0.36	12.9	12.9	No data	No data
9	Nua	4.2	4.09	0.11	2.69	2.69	No damage	0
23	Onenoa	2.6	2.74	−0.14	−5.11	−4.75	No damage	0
		2.4	2.51	−0.11	−4.38			
3	Poloa	7.6	17.59	−9.99	−56.79	−44.26	Low	24.00
		4.93	10.04	−5.11	−50.9			
		8	12.99	−4.99	−38.41			
		8.5	12.31	−3.81	−30.95			
25	Sailele	2.25	2.95	−0.7	−23.73	−23.73	Low	1
8	Seetaga	5.57	5.69	−0.12	−2.11	−2.11	Low	5
22	Tula	3.8	9.52	−5.72	−60.08	−40.87	Medium	63.00
		3.8	7.62	−3.82	−50.13			
		3.24	6.93	−3.69	−53.25			
		3.79	3.79	0	0			
7	Utumea	4	4.51	−0.51	−11.31	−11.31	Low	13
15	Vaitogi	1.8	3.36	−1.56	−46.43	−46.43	High	96

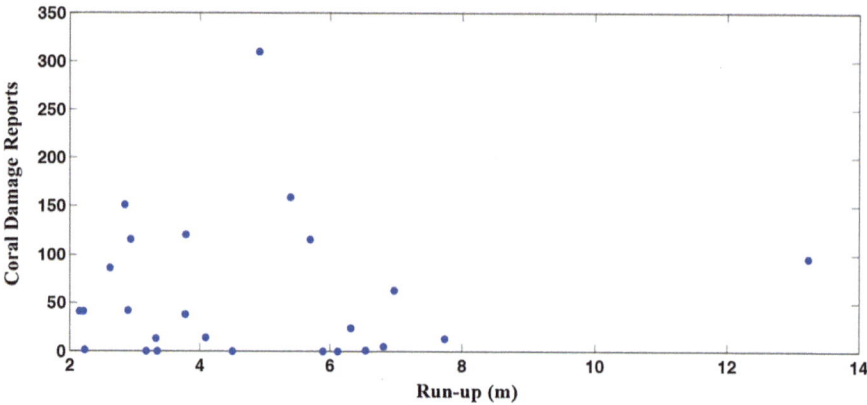

Figure 6
Coral damage vs. observational run-up. The *y-axes* shows coral damage numbers reported by the survey team, and the *y* axes is the average of the run-up observations after aggregating the data into 31 separate village locations (see Table 3; Fig. 1). There is no clear relationship between these two datasets

and coral damage can be inferred. It is not known whether this is because of limitations in the coral dataset (being only a sub-sampling of the reef environment), whether run-up is not the most relevant metric of tsunami dynamics, or whether the occurrence of coral damage is actually driven by many other unknown factors and antecedent conditions.

We next turn to a comparison of the observations with the MOST simulation of the event. We determined the run-up from MOST by taking the highest value of maximum wave amplitudes within a 3×3 grid box (30 m $\times$ 30 m) around each of the 59 run-up data points.

We begin by presenting a comparison of the MOST simulations with observations for two representative locations. For the first (Fig. 7a) in the vicinity of the villages of Amaneve, Failolo, and Agugulu (4, 5, and 6 in Fig. 1), observed run-up averaged 8 m. For these villages, MOST does poorly, underestimating the run-up by an average of 45 %. This was a region where high coral damage was documented. For the second (Fig. 7b), near the villages of Utumea West, Nua and Seetaga (7, 8 and 9 in Fig. 1) average run-up was 4 m. Here MOST does well, simulating run-up to within 2.4 %. The documented coral damage was very low. Even though these two locations are only 3 km apart, these very different run-ups, coral damage, and simulation performance illustrate the complexities of the setting.

Figure 8 shows some of the dynamical fields simulated by MOST for this event: maximum wave amplitude (panel a), maximum current (panel b), maximum flux (panel c) and peak stress (panel d), respectively, together with the coral damage along the survey tracks. From a visual analysis, there are not obvious strong relationships between coral damage and tsunami dynamical fields. Take just two examples, Leone and Fagatele (13 and 14 in Fig. 1): at Fagatele the maximum flux, current and maximum amplitudes are low, but coral damage is very high. In contrast, at Leone, the maximum flux is high with low maximum amplitudes and currents, and high coral damage.

The absence of a statistically significant relationship between the modeled tsunami fields and the observed damage is confirmed by averaging MOST output along each survey track and creating scatter plots of observed damage versus track-averaged model output, for several relevant model fields (see Fig. 11 in "Appendix").

The clearest basis that we identify for comparing the observations and the model is the observed coral damage and run-up, and the simulated run-up, at each of the 31 sites where observations were made.

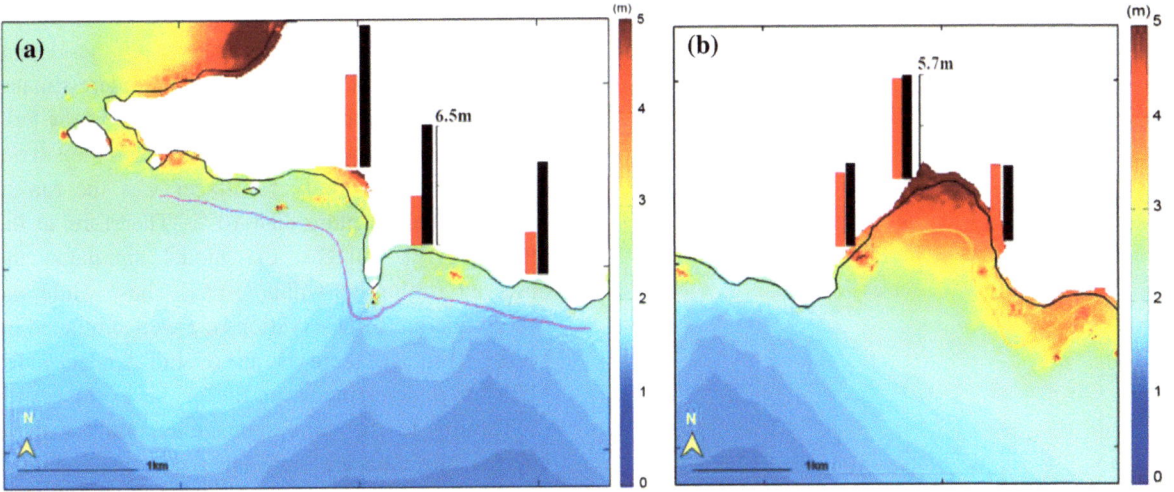

Figure 7
Two examples comparing observed run-up with MOST run-up and maximum wave height. Coastal run-up is shown as *black bars* (observed) and *red bars* (simulated). *Colors* show contours of maximum wave height simulated by MOST. **a** The villages of Amaneve, Failolo and Agugulu (locations 4,5, and 6 in Fig. 1; Table 3); **b** the village of Utumea West, Seetaga and Nua (locations 7, 8, and 9 in Fig. 1; Table 3) The survey tracks nearest these villages are also shown, *color-coded* according to damage scale in Fig. 10

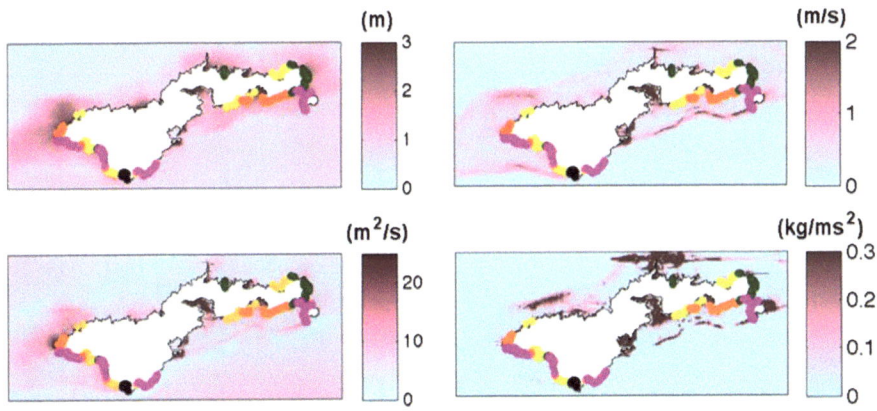

Figure 8
Maximum wave amplitudes, peak currents, peak fluxes and max, stresses calculated from the MOST simulation of the 2009 Tutuila tsunami. Also shown are the post-tsunami survey tracks with coral damage according to the *color scale* in Fig. 10

Table 3 compiles the complete results from all the available observations, We have aggregated the observed and computed run-up values into reports at 31 villages by taking the mean of the total data points at every village.

The complete dataset for the whole island is presented graphically in Figs. 9 and 10. A scatterplot of simulated versus observed run-up correlates at $r = 0.78$, demonstrating significant overall skill for the MOST model (Fig. 9). This overall level of agreement between observations is comparable to other tsunami models and case studies (e.g., NTHMP 2012).

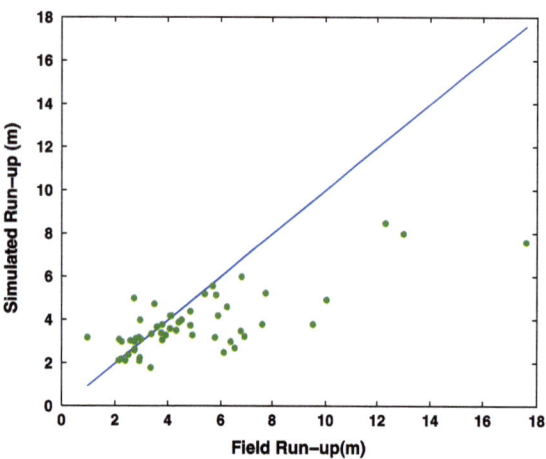

Figure 9
Scatter plot of simulated vs. observed run-up (m). The *blue line* shows the 1-to-1 line. The simulated vs. observed run-up is correlated at $r = 0.78$, but MOST underestimates run-up at many places

Despite this success, it is also clear MOST underestimates run-up in many places. Of the 31 total villages, there are 15 for which MOST underestimates run-up by more than 10 %. Figure 10 shows the bulk of these are on the west side of the island, although not exclusively so. At only 5 villages was the run-up overestimated by more than 10 %. Therefore, at the remaining 11 villages, the model simulated the observed run-up to within 10 %. Thus, while the nearby tide-gauge observations are well simulated by MOST (Fig. 4), there is an overall tendency for MOST to underestimate run-up for this event.

Turning to the coral damage reports for these villages, the data are suggestive of a general relationship with the accuracy of the run-up simulations (Table 3). Of the 15 villages where the model underestimated run-up, the breakdown in terms of coral damage is 8 very high/high, 4 medium, 3 low/very low, 1 no damage, 1 no data. For 11 villages where the model estimates run-up well, the coral damage is 2 very high/high, 1 medium, 3 low/very low, 4 no damage, 1 no data. For the 5 villages where

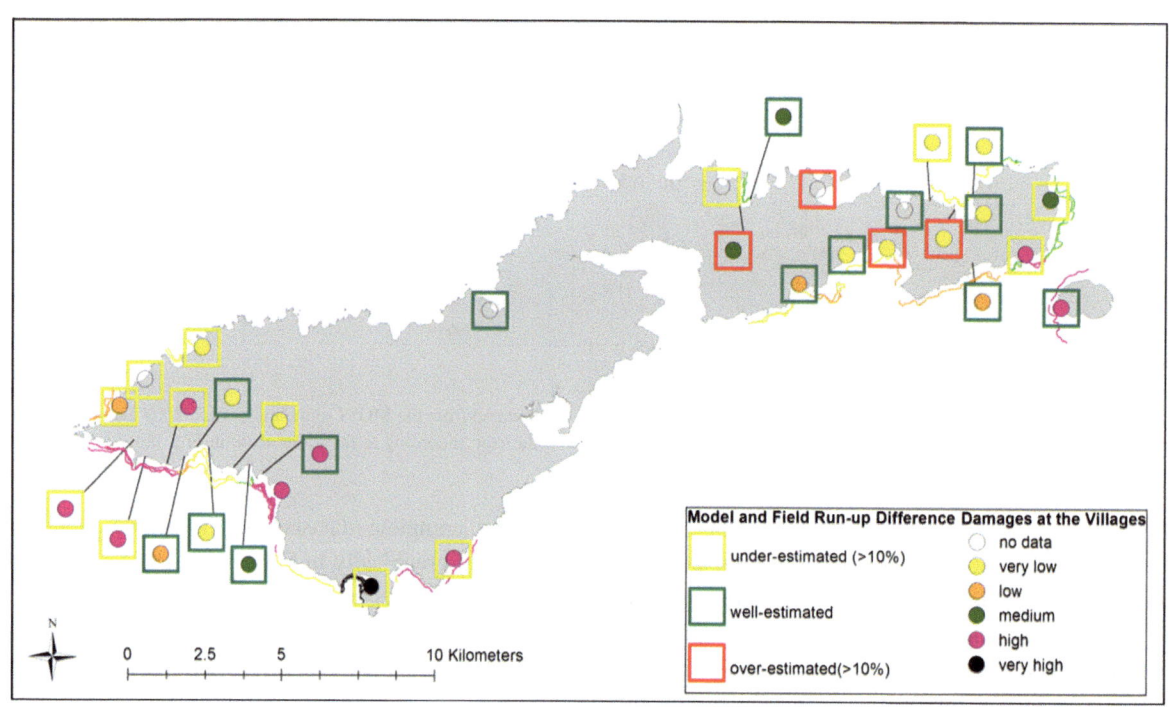

Figure 10
A summary of the comparison between model run-up skill and coral damage. *Rectangles* indicate the difference between modeled and observed run-ups. The *filled dots* indicate the coral damage reports using the qualitative classification described in the text. *Lines* are the color-coded survey track-lines followed to estimate the damage on corals. See also Table 3

run-up is overestimated, limited coral damage data preclude strong interpretation 1 high, 1 medium, 3 no data.

While these associations fall short of a formal, statistically significant correlation, there is an indication that at villages where MOST underestimates run-up, coral damage is much likely to be high or very high. It is interesting to speculate on whether there is a systematic reason for such associations, and whether it can be used to refine run-up dynamics. We turn to this in the discussion.

5. Summary and Discussion

Focusing on the impact of the 2009 Samoa tsunami on Tutuila island, we conducted numerical model simulations of tsunami run-up and inundation at the coastal zones. We performed an integrated analysis to evaluate the relationship between the tsunami hydrodynamics and the coral damage by using numerical modeling and post-tsunami surveys. The results for 31 villages on Tutuila island suggest that, while the numerical model simulates run-up with a high correlation with the observations, there is also a tendency to underestimate run-up in regions of high or very high coral damage, and that run-up tends to be better estimated in locations where coral damage is low.

The dataset synthesized in Table 3 is a preliminary assessment of the damage to the fringing reef coral. Although the 2009 Tutuila tsunami was a one-of-a-kind opportunity to investigate coral damage and tsunami dynamics, the data have some limitations. In particular, the damage assessments inevitably involve a degree of subjectivity and the damage reports have not been normalized to the background coral density. Moreover, the data cover only the east and west sides of Tutuila due to the bad weather conditions that existed on the south and north sides of the island during the surveys.

In the present setting of Tutuila island, the variable simulated run-up differences in our high-resolution tsunami model might be due to sharp changes in bottom roughness values caused by coral reefs. One expects that in reality there are strong spatial variations in roughness values (NUNES and PAWLAK 2008). However, a constant roughness value was defined in the MOST simulations. This is a limitation and a standard practice shared by most current tsunami models. A productive direction for future research would be to implement spatially varying roughness, and to focus on detailed case studies of particular bays and reefs to understand how roughness variations affect run-up. Even though we used a very high-resolution model, for the Tutuila tsunami, there are significant differences from observations, and thus modeling this event remains a big challenge. A model sensitivity analysis simulating the effect of varying the depths where coral reefs exist may better elucidate their role in controlling run-up on the coastlines they shield.

While we have found some intriguing relationships between tsunami dynamics and coral damage, at least in the spatial variations in the skill of the numerical simulations, it is clear these are only tentative gleanings amid a great deal of variability. We did not find any clear relationships between coral damage and other simulated dynamical tsunami fields. Such relationships might be drawn out in more targeted and more detailed simulations, but the real situation is obviously very complicated at small scales, and many factors operate. The failure to establish a stronger connection between the simulated dynamical tsunami fields and coral damage may be because the coral damage dataset was not comprehensive enough; or because the MOST model does not represent the correct spatial scales in roughness or bathymetry, or MOST model does not represent the processes that actually cause damage (for instance, damage may be inflicted on corals by retreating waves carrying debris and sand); or because coral damage is inherently stochastic and unpredictable, and depends unknowably on antecedent conditions. Since so little is known about the damage to coral reefs by tsunamis, more studies are needed to examine the influence of water depth, three-dimensional effects, wave-wave interactions and coral strengths.

The role of reefs in tsunami dynamics remains enigmatic. Given the degree of complexity we found in both observations and simulations, it would, for instance, be hazardous to conclude that reefs provide universal protection for their inshore coastlines.

In some ways it is disappointing to find no correlation between the run-up and coral datasets. On the other hand, it is significant to establish that result, and it also points to new questions. The documentation of the submarine ecological impacts of a tsunami is an important goal in its own right, but the tentative inference from our study (the first of its kind) is that the relationship of coral damage to a variety of tsunami metrics (i.e., observed run-up and inundation, modeled maximum currents, fluxes, and stress) is not a simple one, at least in our setting. Tutuila represented a 'target of opportunity', since the all-important pre-tsunami survey existed. Alternative future surveying strategies range from identifying simpler bathymetric settings, more rigorous reporting protocols, or perhaps more comprehensive surveys in a smaller domain.

Would using a different mathematical model make a difference to our analyses? Some other tsunami models employ full three-dimensional Navier–Stokes equations (ABADIE et al. 2006) or Boussinesq approximations (KIRBY et al. 1998). However, bigger models do not always make for better models, and for long wavelength tsunamis shallow water equations have been shown to perform well in many benchmark problems and other comparisons. It is important to gauge the level of model complexity relative to other aspects of the problem. Key among these are: the uncertainty in the spatial roughness and the mathematical formulation of dissipation at small scales; the detailed bathymetry and coastal geometry at small scales, and how that interacts with the reflection and refraction of multiple wave fronts; and the intrinsic challenge of building numerical models of highly turbulent, chaotic, and one-off events with a single set of governing equations and parameters. As the situation stands, the advantages of each model class can be demonstrated under idealized circumstances, but in inter-comparison studies all perform with comparable skill for real-world events (e.g., NTHMP 2012). As models are developed and refined, continued benchmarking and inter-model comparison will be essential to establish the strengths and ultimate limitations of each approach.

It is likely that the role of reefs depends sensitively on the detailed dynamics of any particular tsunami and the characteristics of the reefs themselves. This study is intended to contribute to the existing body of work on reef-tsunami interactions. It points to the need for more comprehensive research gathering more detailed coral damage data and testing the results with different case studies.

Acknowledgments

This publication makes use of data products provided by the Coral Reef Ecosystem Division (CRED), Pacific Islands Fisheries Science Center (PIFSC), National Marine Fisheries Service (NMFS), National Oceanic and Atmospheric Administration (NOAA), with funding support from the NOAA Coral Reef Conservation Program. NOAA Center for Tsunami Research under TSUNAMI TASK-2 Project, Pacific Marine Environmental Laboratory (contribution number 4368) supports this research. Hermann Fritz generously provided the post-tsunami run-up survey datasets. We are grateful to Randall Leveque, Joanne Bourgeois, Hongqiang Zhou, Yong Wei, Christopher Moore, Marie Eble, Diego Arcas and Lijuan Tang for their endless advice and their help.

Appendix

We construct scatterplots by averaging the MOST output for maximum amplitude, current, flux, and stress along each survey associated with a village (Fig. 1), and plotting against the corresponding observed coral damage numbers. None of the correlations are significant (Fig. 11).

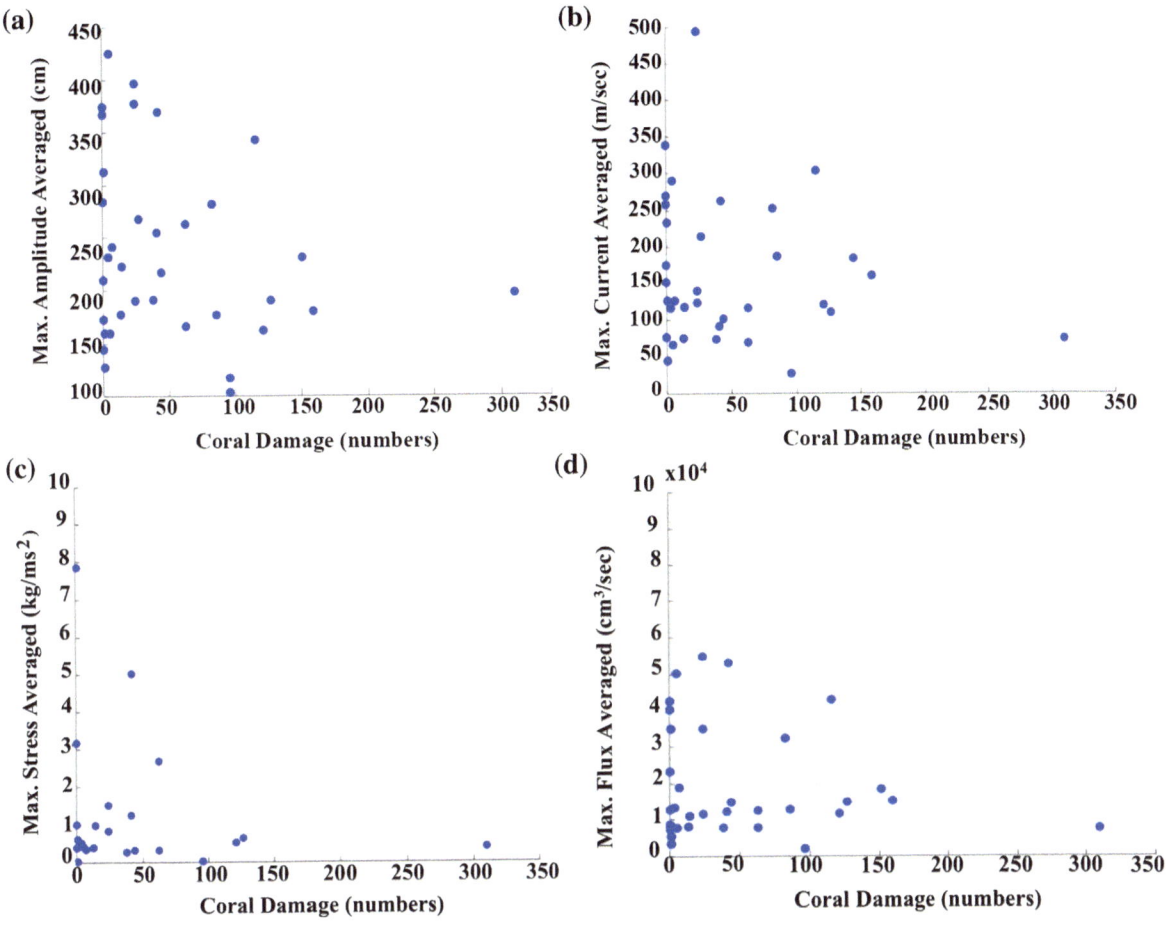

Figure 11
Scatter plots of coral damage (on the *x-axis*) vs. **a** maximum amplitude; **b** maximum current; **c** maximum flux and **d** maximum stress along coral damage track-lines. The dynamical fields are averaged over each track-lines associated with each village (Fig. 1), to evaluate any relationship between coral damage and dynamical fields

References

ABADIE S., GRILLI S., GLOCKNER S., (2006), *A Coupled Numerical Model for Tsunami Generated by Subaerial And Submarine Mass Failures*, In Proc. 30th Intl. Conf. Coastal Eng., San Diego, California, USA. 1420–1431.

BABA T., MLECZKO R., BURBIDGE D., CUMMINS P., (2008), *The Effect of the Great Barrier Reef on the Propagation of the 2007 Solomon Islands Tsunami Recorded in Northeastern Australia*, Pure and Applied Geophysics, Vol. 165, 2003–2018.

BAIRD A. H., CAMPBELL S. J., ANGGORO A. W., (2005), *Acehnese Reefs in the Wake of the Asian Tsunami*, Current Biology, Elsevier, Vol. 15, Issue 21, 1926–1930.

BERGER M. J., and LEVEQUE R. J., (1998), *Adaptive Mesh Refinement Using Wave-Propagation Algorithms for Hyperbolic Systems*. SIAM J. Numer. Anal. 35, 2298–2316.

BRAINARD R., ASHER J., GOVE J., HELYER J., KENYON J., MANCINI F., MILLER J., MYHRE S., NADON M., ROONEY J., SCHROEDER R., SMITH E., VARGAS-ANGEL B., VOGT S., VROOM P., BALWANI S., CRAIG P., DESROCHERS A., FERGUSON S., HOEKE R., LAMMERS M., LUNDBLAD E., MARAGOS J., MOFFITT R., TIMMERS M., VETTER O., (2008), U.S. Dept. of Commerce, National Oceanic and Atmospheric Administration, National Marine Fisheries Service.

BEAVAN J., WANG X., HOLDEN C., WILSON K., POWER W., PRASETYA G., BEVIS M. and KAUTOKE R., (2010), *Near-simultaneous Great Earthquakes at Tongan Megathrust and Outer-rise in September 2009*, Nature., Vol. 466. doi:10.1038/nature09292.

BIRKELAND C., CRAIG P., FENNER F., SMITH L., KIENE W. E. and RIEGL B., (2008), *Geologic Setting and Ecological Functioning of Coral Reefs in AS*. Coral reefs of the USA, Springer Publishers, Chap. 20, Vol. 33, 803.

CHATENOUX B., PEDUZZI P., (2007), *Impacts from the 2004 Indian Ocean Tsunami: Analyzing the Potential Protecting Role of Environmental Features*, Natural Hazards, Vol. 40, 289–304.

DRAPER N. R., SMITH H., (1998), Applied Regression Analysis (3rd ed.). Wiley, New York.

FERNANDO H. J. S., (2005), *Coral Poaching Worsens Tsunami Destruction in Sri Lanka*, Eos, Vol. *86*, No. 33.

FERNANDO H. J. S., SAMARAWICKRAMA S. P., BALASUBRAMANIAN S., HETTIARACHCHIB S. S. L., VOROPAYEVA S., (2008), *Effects of Porous Barriers Such as Coral Reefs on Coastal Wave Propagation*, Journal of Hydro-environment Research, Elsevier, Vol. *1*, Issues 3–4, 187–194.

FRITZ H. M., BORRERO J. C., SYNOLAKIS C. E., OKAL E. A., WEISS R., LYNETT P. J., TITOV V. V., FOTEINIS S., JAFFE B. E., LIU P. E., and CHAN C., (2011), *Insights on the 2009 South Pacific Tsunami in Samoa and Tonga from Field Surveys and Numerical Simulations*, Earth Sci. Rev., Vol. *107*, 66–75. doi:10.1016/j.earscirev.2001.03.004.

GELFENBAUM, G., APOTSOS, A., STEVENS, A. W., JAFFE, B., (2011), *Effects of fringing reefs on tsunami inundation: American Samoa*, Earth-Science Reviews, Vol. *107*, 12–22.

GICA, E., SPILLANE, M. C., TITOV, V. V., CHAMBERLIN, C. D., and NEWMAN, J. C., (2008), *Development of the forecast propagation database for NOAA's Short-term Inundation Forecast for Tsunamis (SIFT)*, NOAA Tech. Memo. OAR PMEL-*139*, 89.

GUSYAKOV, V. K., (1972), *Mathematical problems of geophysics, chapter Generation of tsunami waves and ocean Rayleigh waves by submarine earthquakes*, Vol. *3*, Novosibirsk, VZ SO AN SSSR, 250–272.

JENKS G. F., (1967), *The Data Model Concept in Statistical Mapping*, International Yearbook of Cartography, Vol. *7*, 186–190.

KIRBY J., WEI G., CHEN Q., KENNEDY A., DALRYMPLE R., (1998), FUNWAVE 1.0, Fully Nonlinear Boussinesq Wave Model Documentation and Users Manual. Tech. Rep. Research Report No. CACR-98-06, Center for Applied Coastal Research, University of Delaware.

KUNKEL M., HALLBERG R. W., OPPENHEIMER M., (2006) *Coral reefs reduce tsunami impact in Model Simulations*, Geophysical Research Letters, Vol. *33*, 123612. doi:10.1029/2006GL027892.

LIM E., TAYLOR L. A., EAKINS B. W., CARIGNAN K. S., WARNKEN R. R., GROTHE P. R., (2009), *Digital Elevation Models of Craig, Alaska:Procedures, Data Sources and Analysis*. NOAA Technical Memorandum NESDIS NGDC-27.

LAY T., AMMON C. J., KANAMORI H., RIVERA L., KOPER K. D., HUTKO A. R., (2010), *The 2009 Samoa–Tonga Great Earthquake Triggered Doublet*, Nature, Vol. *466*, 964–968. doi:10.1038/nature09214.

MCDOUGALL I., (1985), *Age and Evolution of the Volcanoes of Tutuila, American Samoa* Pacific Science, Vol. *39*, No:4.

NATIONAL TSUNAMI HAZARD MITIGATION PROGRAM, (NTHMP), (2012). Proceedings and Results of the 2011 NTHMP Model Benchmarking Workshop. Boulder: U.S. Department of Commerce/NOAA/NTHMP; (NOAA Special Report). 436.

NUNES V. and PAWLAK G., (2008), *Observations of bed roughness of a coral reef*, Journal of Coastal Research, *24(2B)*, 39–50, West Palm Beach (Florida), ISSN 0749-0208.

OKADA Y., (1985), *Surface Deformation due to Shear and Tensile Faults in a Half-Space*, Bulletin of the Seismological Society of America, Vol. *75* (4), 1135–1154.

OKAL E., BORRERO J. C., CHAGUE-GOFF C., (2011), *Tsunamigenic Predecessors to the 2009 Samoa Earthquake*, Earth–Science Reviews, Vol. *107*, 128–140.

OKAL E. A., FRITZ H. M., SYNOLAKIS C. E., BORRERO J. C., WEISS R., LYNETT P. J., TITOV V. V., FOTEINIS S., JAFFE B. E., CHAN C., LIU P. E., (2010), *Field Survey of the Samoa Tsunami of 29 September 2009*, Seismol. Res. Lett., Vol. *81* (4), 577–591.

PERCIVAL D. B., ARCAS D., DENBO D. W., EBLE M. C., GICA E., MOFJELD H. O., SPILLANE M. C., TANG L., TITOV V. V., (2009), *Extracting tsunami source parameters via inversion of DART® buoy data*. NOAA Tech. Memo. OAR PMEL-*144*, 22.

ROEBER V., YAMAZAKI Y., CHEUNG K. F., (2010), *Resonance and Impact of the 2009 Samoa Tsunami Around Tutuila, American Samoa*, Geophys. Res. Lett., Vol. *37*, 121604. doi:10.1029/2010GL044419.

SYNOLAKIS C. E., OKAL E. A., (2005), *1992–2002 Perspective on a Decade of Post Tsunami Surveys.*, Adv. Nat. Technol. Hazards, Vol. *23*, 1–30. doi:10.1007/1-4020-3331-1_1.

SYNOLAKIS, C.E., BERNARD E., TITOV V. V., KÂNOĞLU U., and GONZÁLEZ F. I., (2008), *Validation and Verification of Tsunami Numerical Models*. Pure Appl. Geophys., *165*(11–12), 2197–2228.

TANG L., TITOV V. V., CHAMBERLIN C. D., (2009), *Development, testing, and applications of site-specific tsunami inundation models for real-time forecasting*. J. Geophys. Res., *114*, C12025. doi:10.1029/2009JC005476.

TERRY J. P., KOSTASCHUK R. A., GARIMELLA S., (2005), *Sediment deposition rate in the Falefa River basin, Upolu Island, Samoa*, Journal of Environmental Radioactivity, Vol. *86*(1), 45–63.

TITOV, V. V., (2009), *Tsunami Forecasting*. Chapter 12 in The Sea, Vol. *15*: Tsunamis, Harvard University Press, Cambridge, MA and London, England, 371–400.

TITOV, V. V., and SYNOLAKIS C., (1998), *Numerical Modeling of Tidal Wave Runup*, Journal Of Waterway, Port, Coastal, And Ocean Engineering, Vol. *124*(4), 157–171.

TITOV, V. V., and GONZALEZ, F. I., (1997), *Implementation and Testing of the Method of Splitting Tsunami (MOST) Model*, NOAA Tech. Memo. ERL PMEL-*112*, NTIS:PB98-122773.

TITOV, V.V., F.I. GONZÁLEZ, H.O. MOFJELD, and A.J. VENTURATO, (2003), *NOAA TIME Seattle Tsunami Mapping Project: Procedures, Data sources, and Products*. NOAA Tech. Memo. OAR PMEL-*124*, NTIS: PB2004-101635, 21.

ZHOU H., WEI Y., TITOV V. V., (2012), *Dispersive Modeling of the 2009 Samoa Tsunami*, Geophysical Research Letters, Vol. *39*, L16603. doi:10.1029/2012GL053068.

(Received February 19, 2015, revised June 11, 2015, accepted July 24, 2015, Published online August 30, 2015)

Did an underwater landslide trigger the June 22, 1932 tsunami off the Pacific coast of Mexico?

NÉSTOR CORONA[1] and MARÍA-TERESA RAMÍREZ-HERRERA[2,3]

Abstract—On June 22, 1932, a 10- to 12-m-high tsunami wave struck ~60 km off the Mexican Pacific coast. The associated earthquake that apparently produced this tsunami is questionable because of its relatively small magnitude ($M_s = 6.9$) to produce such tsunami heights. Historical documents, survivor testimony, tsunami catalogs, a post-tsunami survey report, together with geomorphological interpretation of the continental shelf and slope, and numerical modeling were combined to characterize the tsunami parameters. Our results suggest that recorded maximum tsunami wave height, horizontal inundation, arrival time, directivity, effects, and damage are compatible with those characteristics related to an underwater landslide tsunami. The associated landslide (slump) is 4.2 km long, 3.9 km wide, 0.448 km thick, and is located in the upper continental shelf of the Armería Canyon. Elucidating the cause and mechanisms of the near-field 1932 tsunami would aid in considering a wider spectrum of tsunami sources in hazard mitigation programs of the Mexican Pacific coast.

Key words: Landslide-tsunami, earthquake, Mexican Pacific coast, historical data, tsunami modeling.

1. Introduction

In recent years, submarine mass movements have been revealed as the source mechanism of a number of large, destructive tsunamis (TAPPIN *et al.* 1999, 2014; RANGUELOV *et al.* 2008; KAWAMURA *et al.* 2012;

Electronic supplementary material The online version of this article (doi:10.1007/s00024-015-1171-1) contains supplementary material, which is available to authorized users.

[1] Centro de Estudios en Geografía Humana, El Colegio de Michoacán A.C., Cerro de Nahuatzen 85, Fracc. Jardines del Cerro Grande, 59370 La Piedad, Michoacán, Mexico. E-mail: corona@colmich.edu.mx

[2] Laboratorio Universitario de Geofísica Ambiental and Instituto de Geografía, Universidad Nacional Autónoma de México, Circuito de la Investigación, Ciudad Universitaria, 04510 Coyoacán, Mexico. E-mail: tramirez@igg.unam.mx; ramirez@seismo.berkeley.edu; ramirezt@berkeley.edu1

[3] Berkeley Seismological Laboratory, University of California Berkeley, 299 McCone Hall, Berkeley, CA 94720, USA.

STRASSER *et al.* 2013; HARBITZ *et al.* 2014; VON HUENE *et al.* 2014). Landslide tsunamis present complex triggering processes, low frequency in human time scale, that make them almost impossible to observe and instrument, thus listing them as one of the oddest natural threats (HARBITZ 2014). Reliable assessment of submarine landslides requires complex studies that involves intense geophysical fieldwork, oceanographic surveys, review of historical archive (when available), detailed bathymetric, and geological data, among others (FAVALLI *et al.* 2009; ZANIBONI *et al.* 2014; HARBITZ 2014). Therefore, numerical modeling is one of the key tools in evaluating submarine landslides as tsunami source mechanism (HARBITZ 2006).

Multiple signatures and important criteria to identify landslide-tsunamis include (1) large run-up values close to the source and focusing effects (ASSIER-RZADKIEAICZ 2000; WARD 2001; SATAKE and TANIOKA 2003; OKAL and SYNOLAKIS 2003, 2004; HARBITZ *et al.* 2006; FRITZ *et al.* 2007); (2) limited far-field effects when compared with earthquake-tsunamis (SATAKE and TANIOKA 2003; OKAL and SYNOLAKIS 2004) due to the dipole nature of landslide tsunamis that decays faster in the far-field (SATAKE and TANIOKA 2003; MOHAMMED and FRITZ 2012); (3) propagation tends to produce radial directivity patterns (BEN-MENAHEM and ROSENMAN 1972; IWASAKI 1997; OKAL 2000; OKAL *et al.* 2003); and (4) landslide tsunamis can propagate and reach coastlines in a very short time because the source is often located in continental margins at shallow waters (ZANIBONI *et al.* 2014), e.g., 1994 Skagway, Alaska, and 2010, Haiti reached the coast in 1 min (KULIKOV *et al.* Kulikov et al. 1996; SYNOLAKIS *et al.* 2000; FRITZ *et al.* 2013). It is also evident that seismicity produced along active margins has a direct influence on landslide generated tsunamis (e.g., 1998 Papua New Guinea, 2011 Tohoku, 2007 Chile, among others).

1.1. Seismic, tsunami, and seafloor setting

The Mexican Pacific coast is a seismically active zone, located parallel to the Middle American Trench (MAT) along the North American (NAP), Cocos (CP), and Rivera plates (RP) convergent margin (Fig. 1) (SINGH *et al.* 1981, 1985; NIXON 1982; EISSLER and MCNALLY 1984; LUHR *et al.* 1985). More than 900 earthquakes magnitude $M_s > 5$ have occurred since 1732–2011 (SERVICIO SISMOLÓGICO NACIONAL 2011), and 70 tsunamis have been recorded (SÁNCHEZ and FARRERAS 1993; NOVOSIBIRSK TSUNAMI LABORATORY 2011). Of those, the March 28, 1787 off the Corralero, Oaxaca coast, the November 16, 1925 off the coast of Zihuatanejo, Guerrero, and the June 22, 1932, off Cuyutlán, Colima, have been the largest and most destructive tsunamis recorded in Mexico. Only the last two events were considered local tsunamis, and the associated earthquakes were not large enough to produce the reported run-ups. This is the reason why the June 22, 1932 tsunami is a unique event which deserves our particular attention.

The large submarine Armería Canyon is an active depositional environment, where terrigenous sediment are transported and deposited by the Armería River (RAMÍREZ-HERRERA and URRUTIA-FUCUGAUCHI 1999). The canyon sediments consist mainly of mud, mudstone, and in smaller quantity sand and sandstone (COULBOURN *et al.* 1982), arranged in

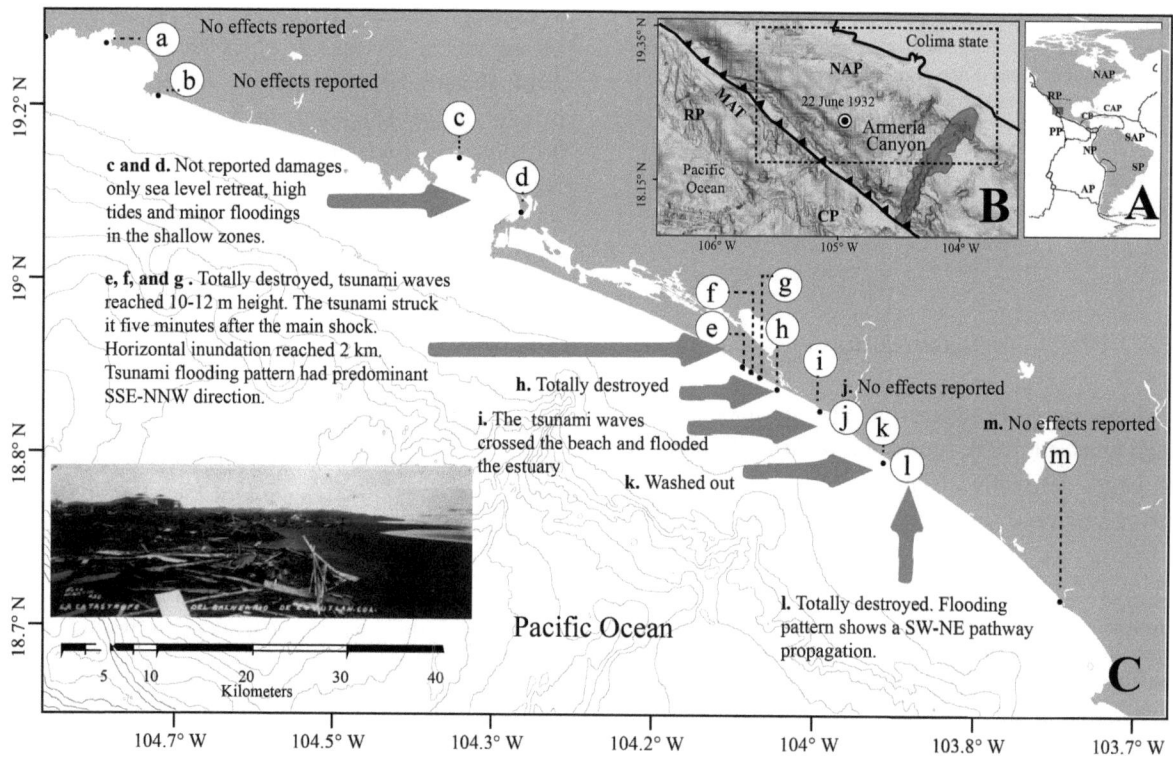

Figure 1

a Regional tectonic setting. **b** Local tectonic setting. *NAP* North American plate, *RP* Rivera plate, *MAT* Mesoamerican trench, *CP* Cocos plate. Bull eye—epicenter of the 22 June, 1932, earthquake (CRUZ and WYSS 1983; CUMMING 1933; SINGH *et al.* 1981; SINGH *et al.* 1985; SERVICIO SISMOLÓGICO NACIONAL 2011). Shaded contour—Armería Canyon. C. June 22, 1932 tsunami spatial behavior. Sites: *a* Cuastecomates, *b* Barra de Navidad, *c* Western Manzanillo Bay, *d* Eastern Manzanillo Bay, *e* Western Cuyutlán, *f* Central Cuyutlán, *g* Eastern Cuyutlán, *h* Palo Verde, *i* El Paraíso, *j* Boca de Pascuales, *k* El Real, *l* El Tecuanillo, and *m* Boca de Apiza

400–600 m thick packages (DIVINS 2003). These factors increase the likelihood of massive rotational landslides (slumps) during seismic events (EINSELE 1992; LEE et al. 2009; YAMADA et al. 2012).

1.2. The June 22, 1932 tsunami

On June 3, 1932, the largest earthquake, $M_s = 8.2$, recorded in México (SINGH et al. 1985) was followed by a 3-m-high tsunami, and two aftershocks, magnitudes $M_s = 7.9$ on June 18, and $M_s = 6.9$ on June 22.

The June 22 earthquake occurred at 7:00 (UTC-6), magnitude $M_s = 6.9$ and epicenter ~ 86 km south off the Bahía de Manzanillo (CUMMING 1933; SINGH et al. 1981, 1985; CRUZ and WYSS 1983; SERVICIO SISMOLÓGICO NACIONAL 2011). Five min. later, approximately, a tsunami struck the coastline from Bahía de Manzanillo (Fig. 1, sites c and d) to Salinas del Guasango [nowadays, El Tecuanillo (Fig. 1, site l)], flooding 2 km inland, and a 10- to 12-m maximum tsunami wave height (MTWH) was recorded at Cuyutlán (EL UNIVERSAL GRÁFICO NEWSPAPER 1932b; EXCÉLSIOR NEWSPAPER 1932b; SOLOVIEV and GO 1975; SALAZAR 1989; FARRERAS and SÁNCHEZ 1991; SÁNCHEZ and FARRERAS 1993; CORONA and RAMÍREZ-HERRERA 2012). The reported fatalities were ~ 50, and 1500 people injured (EL INFORMADOR NEWSPAPER 1932a, 1932b; EL UNIVERSAL GRÁFICO NEWSPAPER 1932a; EXCÉLSIOR NEWSPAPER 1932a; ORDOÑEZ 1933), and major damage and destructive effects in several coastal communities. Among the damages reported standout the following: Palo Verde, El Real, and El Tecuanillo villages were totally washed away (Fig. 1, sites h, k, and l respectively); tsunami waves went beyond the beach and flooded the estuary at Las Boquillas (Fig. 1, site i, (CHÁVEZ 1932; CUMMING 1933)). Figure 1 shows the spatial distribution of the above described historical accounts.

The specific source mechanism of the June 22 tsunami has not been thoroughly defined (SINGH et al. 1985). It has been suggested as a seismic source (SÁNCHEZ and FARRERAS 1993; SINGH et al. 1998; OKAL and BORRERO 2011; NATIONAL GEOPHYSICAL DATA CENTER 2012). Nevertheless, the June 22 earthquake was relatively small ($M_s = 6.9$) to produce the large 10–12 m height tsunami (FARRERAS and SÁNCHEZ 1991; CORONA and RAMÍREZ-HERRERA 2012). For example, the 2012 El Salvador tsunami earthquake showed a much larger magnitude ($M_w = 7.3$) and only produced half the runup, ~ 6 m (BORRERO et al. 2014).

OKAL and BORRERO (2011), based on a seismological re-assessment of the event, proposed the most comprehensive analysis up to now for the source of this tsunami. They proposed a change in earthquake parameters assuming a dip angle of 45°, and reducing the slab rigidity to 1.8×10^{11} dyn cm^{-2}. They recreated a splay fault mechanism and affirmed that the tsunami was a typical Fukao style "tsunami earthquake". However, these authors acknowledged that the seismic data they employed were sparse and suboptimal, and also the applied method M_M (Mantle magnitude) was empirical (OKAL and BORRERO 2011). Moreover, their best fit tsunami model underestimates the tsunami heights, showing 6–7 m tsunami height at Cuyutlán, where coastal sand dunes reach 10 m height; therefore, their modeled tsunami heights do not fit with the reported tsunami flooding and damage at this location. Furthermore, their model cannot explain the apparent tsunami focusing effects. On the other hand, the rule aspect ratio I_2, postulated by OKAL and SYNOLAKIS (2004), which relates the maximum run-up of a tsunami to the effect along the beach and defines that dislocation sources cannot be greater than $I_2 = 1.00 \times 10^{-4}$ (OKAL and SYNOLAKIS 2004); in the case of the June 22 tsunami, 10–12 m run-up maximum values along 60 km of the coastline indicate $I_2 = 1.67 \times 10^{-4}$ and 2.00×10^{-4}, respectively. Thus I_2 rule cannot be satisfied by any of the aforementioned scenarios.

Based on interpretation of historical data, higher resolution bathymetric grids, geomorphological analysis of the Armería Canyon and improved tsunami modeling codes that can resolve the tsunami effects on the shore at higher detail, we propose that the June 22, 1932 tsunami was produced by a submarine landslide. To prove our hypothesis, we analyzed tsunami arrival time to the coast, maximum tsunami wave height spatial distribution along the shoreline, tsunami directivity, and horizontal inundation. All these parameters are directly related to the location of the tsunami source and its mechanism (TAPPIN et al. 2014).

2. Methods

2.1. Geomorphic analysis of topobathymetric data

We used topobathymetric data obtained from the Global Multi-Resolution Topography model compiled by Ryan et al. (2009), and generated a ~200 × 200 m pixel size grid using GeoMapApp 3.0.1 software. This grid is an integrated digital model that summarizes different multibeam data at different spatial resolutions, which provides detailed information about the seafloor in some areas, and reveals important topographical features, such as dunes and swales (See Fig. 1 of the Supplementary Material). Limitations of the model include the presence of a banding effect which makes resolution variable across the study area, primarily on the upper zone of the Armería Canyon (Fig. 2 of the supplementary material). Because multibeam bathymetric coverage is incomplete across the trench slope, the bathymetric grid includes N–S and E–W streaking artifacts in some areas. Therefore, it is likely that the actual number of slumps is underestimated.

Morphological interpretation of the seafloor allowed us to determine slump characteristics such as: location (longitude, latitude), length, width, mean slope, orientation, depth, and distance of slump movement. Bathymetric data were processed using GIS to develop hillshaded relief, slope models, as well as isobath contour lines. We tested different contour intervals, starting at 200 m and up to 20 m equidistance and found that slump features were enhanced at different levels accordingly. To avoid pixel size effect, a smooth-line filter was applied during every single testing step. Figure 3, in the supplementary material, shows the three steps applied during the bathymetric filtering process. This procedure was applied to enhance the bathymetric grid that initially masked detailed information. It should be mentioned that slump features remain coarse and smooth; however the slump head scarp can be distinguished. Based on a general mass failure model, we considered landslides as those areas with rapid changes in slope, with an arch shape facing down-slope (interpreted as head scarp), with extending sub-parallel sidewalls, and in some cases rubble at the base of the landslide body (e.g., Hampton et al. 1996; McAdoo et al. 2000; Lee et al. 2009; Masson et al. 2006).

2.2. Tsunami modeling

The June 22, 1932 tsunami was modeled assuming two consecutive events: (1) Co-seismic deformation resulting from the 6.9 M_s earthquake (Singh et al. 1985; Servicio Sismológico Nacional 2011), and (2) a submarine landslide triggered ~10 s after the earthquake. Triggering time was inferred following data recorded on the arrival for the first P-S waves at Manzanillo station (Singh et al. 1985).

Input data for the earthquake were taken from Singh et al. (1984, 1985), except for the rupture length and width, that were taken from Okal and Borrero (2011). Seafloor was modeled assuming an elastic dislocation model that yields the vertical deformation of the seafloor in the epicentral area as a function of the ground elastic parameters and the fault plane geometry (Okada 1985).

Input data for the slump, i.e., longitude, latitude, length, width, mean slope, orientation, depth, and distance of slide motion, were obtained from morphological interpretation. Bulk density was assumed to be a constant value, 1.906 kg/m^3, consistent with the lithological composition of the Armería Canyon, that is defined by mud, mudstone, and in smaller quantity sand and sandstone (Coulbourn et al. 1982). Finally, the landslide thickness attributes were inferred from the sediment deposit and thickness model proposed by Divins (2003).

Tsunami modeling was carried out using GEOWAVE software. We selected this software because it is a comprehensive tsunami simulation model formed in part by combining the Tsunami Open and Progressive Initial Conditions System (TOPICS) with the fully nonlinear Boussinesq water wave model FUNWAVE [see Watts et al. (2003)]. GEOWAVE simulates tsunami generation, propagation, and inundation using the fourth-order fully nonlinear and fully dispersive Boussinesq wave model with multiple wave dissipation mechanism, wave breaking, and dry land overflow. This software was selected due to its ability to model tsunamis induced from a variety of sources, including submarine landslides. Its capability has been well demonstrated by Watts and Tappin (2012) for several tsunami events, including the 125 ka Alika 2 Hawaii landslide, the 1908 Messina Strait (Italy)

Did an underwater landslide trigger the June 22, 1932 tsunami off the Pacific coast of Mexico?

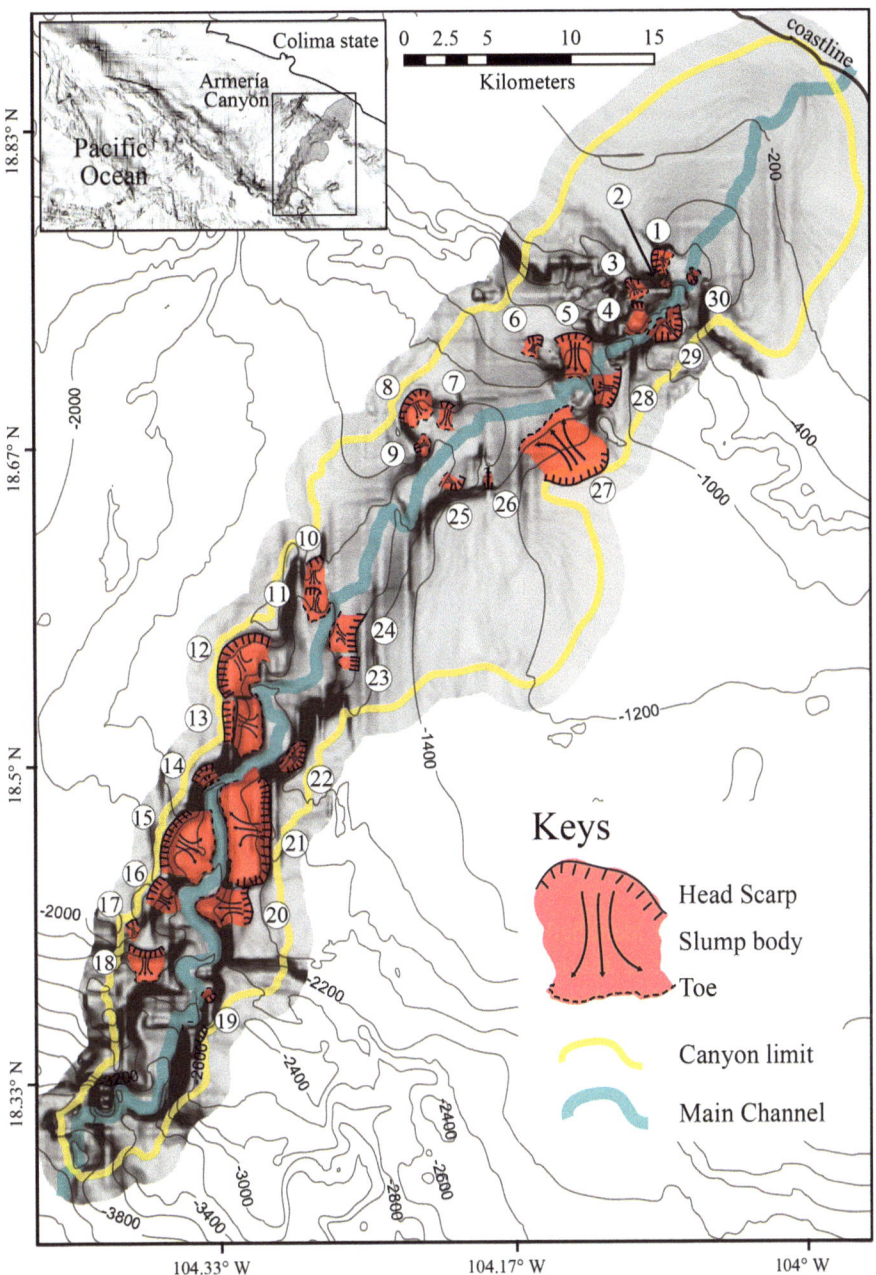

Figure 2
Armería Canyon and interpreted submarine landslides. *Gray shading* indicates seafloor slope, *darker shades* indicate steeper seafloor. Isobaths' contour interval is 200 m. Table 2 shows landslides identification number (1–30) and their dimensions

tsunami, the 1946 Unimal Alaska (US) tsunami, and the 1998 Sissano Papua New Guinea tsunami. In addition, the model has been proven more effective and accurate than the Nonlinear Shallow Water (LYNETT et al. 2003; HARAHAP et al. 2014).

Simulation time was 33.83 min for each model. It required 14 h of computer time. The ratio of computed to simulated time resulted in 24.83. Models were run in a Quad core processor at 2.9 GHz frequency, supported by 8 GB of RAM.

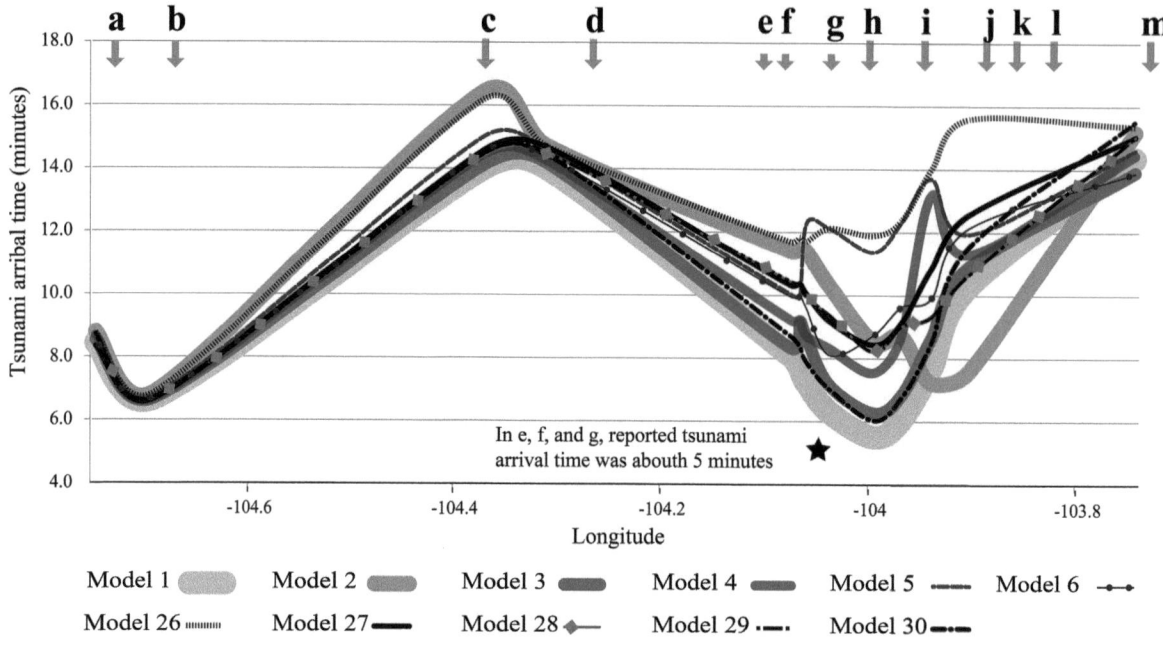

Figure 3
Tsunami arrival times at observed stations. *Dark star* shows documented time-lag after the earthquake

3. Results and discussion

3.1. Landslides characterization

Thirty plausible landslides were identified in the Armería Canyon (see Fig. 2). Most of the landslide features are found by the continental slope, below 600 and to 2200 m depth. Their spatial distribution is similar on both, NW and SE, canyon faces, and thus direction of movement is towards the SE and NW, respectively (Fig. 2). The direction in which the mass of the slump moves and produces a positive wave is of importance. Thus, here the slumps on the SE canyon face are a major hazard to local coastal communities (Table 1).

The slumps' geometry characteristics range from 0.53 to 4.8 km long, 0.38–6 km wide, and 70–480 m thick (see Table 2; Fig. 2). The dimensions of the identified slumps are relatively small at the interface of the shelf and the continental slope, from 600 to 1000 m depth, compared with those at larger depth, from 1400 to 2400 m depth (see Table 2; Fig. 2). Slumps' volume (V) was calculated according TEN BRINKS et al. (2014) approach, where there is a strong correlation between area and volume of the submarine mass failure (SMF); for rotational mass movements (slumps) $V = kA^d$, where A is the area, $k = 0.03$ is a constant in the correlation function, and $d \approx 1.3$ (TEN BRINK et al. 2014). Volume values range from 0.006 to 0.491 Km^3 and the largest volume corresponds to M27 slump. In spite of bathymetric data being relatively coarse in large zones (see Figs. 2, 3 on supplementary material), it was possible to identify typical morphological features revealing the presence of landslides, plausible slump type, which agree with descriptions of mass movement models (HAMPTON et al. 1996; MCADOO et al. 2000; MASSON et al. 2006; LEE et al. 2009).

3.2. Tsunami modeling

A tsunami model was run for every landslide following a dual source: seismic and submarine landslide mechanisms. In total 30 tsunami models were run; however, based on their behavior of the selected variables, only 11 models, 1–6 and 26–30, fit our hypothesis. We excluded those models that either

Table 1

Parameters used in the seismic tsunami model

x_0 (deg.)[a]	y_0 (deg.)[a]	d (km)[b]	Φ (deg.)[b]	λ (deg.)[c]	δ (deg.)[b]	M_s^b	L (km)[c]	W (km)[c]	μ (Pa)
Input TOPICS									
$-104.680°$	$18.739°$	30	$310°$	$90°$	$14°$	6.9	86	41	$4.00E+10$
Output TOPICS									
M_o (J)	λ_0 (km)	η_0 (m)							
$2.13E+20$	-0.1970	0.4545							

Longitude of the earthquake centroid x_0, latitude of the earthquake centroid y_0, centroid depth d, fault strike ϕ, fault rake λ, fault dip δ, earthquake magnitude M_s, fault length along rupture L, fault width across rupture W, and shear modulus μ, seismic moment M_o, characteristic wavelength λ_0, and characteristic tsunami amplitude η_0

[a] Taken from Singh et al. (1984) and Servicio Sismológico Nacional (2011)

[b] According to Singh et al. (1985)

[c] As per Okal and Borrero (2011)

showed: tsunami arrival time (TAT) far away of the expected values, and maximum tsunami wave heights (MTWH) too small (less than 2 m in height) to produce inundation inland (Horizontal Inundation, HI). For these reasons, those models are not included in the discussion. Nonetheless, these models are included in Fig. 4 of the supplementary material which is a summary of all models. We describe in the following sections the results for the different variables included in tsunami modeling.

3.3. Tsunami arrival times (TAT)

Tsunami arrival times showed a wide range of values for each one of the 30 produced models. Models 1–6 and 27–30 showed the smallest TAT values (Fig. 3), because distance from the modeled landslides to the coastline is shorter (see Fig. 2 for landslides' location relative to the coastline). Sites a, b, and c are closer to the seismic source and their modeled TATs range between 8.5 and 10 min. TAT values 4.3 and 9.2 min are observed at sites e, f, g, h, i, and j; while at site m, values are 13.9 min minimum and 17.0 min maximum. The lowest TAT values were observed at: El Paraíso ~5.5 min (site i), Palo Verde ~6.8 min (site h), Boca de Pascuales ~6.9 min (sites j), and Cuyutlán ~7.6 min (sites e, f, and g).

TAT for model 27 is ~7 min; however, the time reported by historical documents is 5 min as shown in Fig. 3 (El Universal Gráfico Newspaper 1932b; Excélsior Newspaper 1932a; Soloviev and Go 1975; Salazar 1989; Farreras and Sánchez 1991; Sánchez and Farreras 1993; Corona and Ramírez-Herrera 2012). Probable explanations for the ~2 min delay in TAT might be attributed to either quality of the bathymetric data, button friction conditions, and/or location and orientation of slump source. Figure 6 on the supplementary materials shows the wave form history of model 27; it describes the times and heights of the tsunami when it reaches the 12 stations.

3.3.1 Maximum tsunami wave heights (MTWH) and horizontal inundation (HI)

Tsunami inundation characteristics depend on a combination of several factors, such as coastal topography, bathymetry, terrain roughness, and the resulted maximum tsunami wave heights, derived from the triggering mechanism generating the tsunami. Based on these factors, only 10 of 30 tested models (models 1, 3, 4, 6, 11, 20, 24, 27, 28, and 29) produced tsunami horizontal inundations in some sectors of the studied coast. Results showed a maximum tsunami wave height (MTWH) concentration pattern on sites e, f, g, h, I, and j, with a gradual lowering towards sites a, b, c, and d, and m (Fig. 4). This behavior can be explained by the maximum horizontal inundation at sites c to l (Fig. 5).

Modeling shows MTWH values <3 m with no inundation at sites a and b, which agrees with historical data on tsunami effects. In addition, the models show 2–5 m MTWHs and 400 m maximum HI in Manzanillo Bay (sites c and d), where no damage was reported. At sites e, f, and g (Cuyutlán),

Table 2

Identified slump parameters

Id	Longitude	Latitude	b^a (km)	T^b (m)	ω^a (km)	d^a (m)	θ^a (deg)	Volume (km^3)
1	−104.079	18.762	1.05	468	1.46	560	19	0.0460
2	−104.078	18.750	0.53	468	0.38	611	9	0.0060
3	−104.095	18.748	1.8	470	1.25	921	6.5	0.0675
4	−104.092	18.736	1.08	448	0.68	832	9	0.0220
5	−104.094	18.728	0.75	448	0.57	932	11	0.0128
6	−104.128	18.714	2.53	448	1.85	1248	9	0.1404
7	−104.154	18.718	0.75	448	0.57	1128	8.5	0.0128
8	−104.202	18.683	0.75	429	0.57	1694	8	0.0128
9	−104.219	18.687	4.8	450	2.4	1730	6	0.3456
10	−104.215	18.666	0.75	445	0.57	1620	8	0.0128
11	−104.277	18.594	3.6	430	1.3	1860	7.5	0.1404
12	−104.318	18.553	2.8	422	4.13	1900	7	0.3469
13	−104.320	18.521	3.1	466	2.6	1928	7	0.2418
14	−104.340	18.495	1.2	450	1.4	2051	8	0.0504
15	−104.349	18.457	3	466	3.5	2400	8	0.3150
16	−104.366	18.432	2.12	540	1.47	2222	17	0.0935
17	−104.382	18.415	1.02	540	0.79	2120	9.9	0.0242
18	−104.374	18.396	1.73	540	2.04	2521	10.2	0.1059
19	−104.339	18.380	0.75	0	0.61	2750	10	0.0137
20	−104.331	18.428	3.5	540	2.02	2370	18	0.2121
21	−104.311	18.464	2.7	466	6	2280	9	0.4860
22	−104.290	18.515	1.2	480	4.2	2220	7	0.1512
23	−104.258	18.553	1.2	432	0.72	1860	9	0.0259
24	−104.267	18.569	1.9	432	1.5	1950	6	0.0855
25	−104.199	18.647	1.4	445	0.83	1749	7.2	0.0349
26	−104.178	18.648	0.96	445	0.46	1481	9	0.0132
27	−104.136	18.667	4.2	448	3.9	1100	8	0.4914
28	−104.111	18.698	1.55	448	2.2	1185	7	0.1023
29	−104.076	18.727	1.68	468	2.17	898	8	0.1094
30	−104.062	18.754	0.83	468	0.73	758	6.5	0.0182

b = Initial landslide length, T = maximum initial landslide thickness, ω = maximum landslide width, d = mean initial landslide depth, θ = Mean initial incline angle

[a] Parameters taken directly from the topobathymetric model (RYAN et al. 2009)

[b] Taken from the sediment deposit and thickness model (DIVINS 2003), V = volume, calculated according to ten BRINK's et al. (2014) model

the highest MTWH values reached 9.8, 8.7, and 8.1 m, respectively (see Fig. 6 on the supplementary materials). HI maximum values were produced in these stations too, ranging from 1000 to 2200 m; and model 27 shows both MTWH and HI maximum values (see Figs. 4, 5). In fact, model 27 results are comparable with those reported by historical data, i.e., a MTWH 8.1–9.8 m (Fig. 4) and a HI ranging from 300 to 2000 m at Cuyutlán (Fig. 5). This area coincides with the maximum reported damage (50 people dead) and tsunami waves of 10–12 m, inundating inland up to 2000 m (EL UNIVERSAL GRÁFICO NEWSPAPER 1932b; EXCÉLSIOR NEWSPAPER 1932a; SOLOVIEV and GO 1975; SALAZAR 1989; FARRERAS and SÁNCHEZ 1991; SÁNCHEZ and FARRERAS 1993; CORONA and RAMÍREZ-HERRERA 2012). Also, neighboring Palo Verde settlement was totally washed away by the tsunami (CHÁVEZ 1932; CUMMING 1933), and here MTWH values reached 4.6 m, and HI reached up to 1100 m. The occurrence of this tsunami at this particular location and at its magnitude has been confirmed by comprehensive evidence of paleotsunami deposits, i.e., sand layers that correspond to the 22 June, 1932 tsunami (RAMÍREZ-HERRERA et al. 2014). Thus, geological data corroborate historical accounts and are in agreement with modeled results.

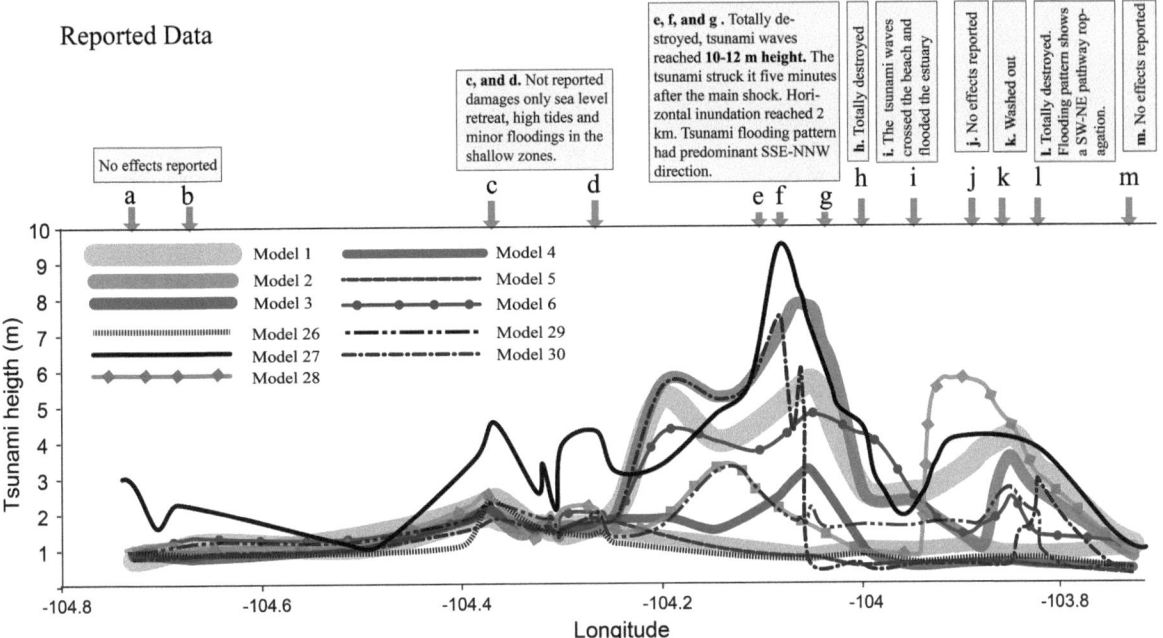

Figure 4
Maximum tsunami wave height models and local effects reported by historical sources. Locations: *a* Cuastecomates, *b* Barra de Navidad, *c* Western Manzanillo Bay, *d* Eastern Manzanillo Bay, *e* Western Cuyutlán, *f* Central Cuyutlán, *g* Eastern Cuyutlán, *h* Palo Verde, *i* El Paraíso, *j* Boca de Pascuales, *k* El Real, *l* El Tecuanillo, and *m* Boca de Apiza

Historical data show that the tsunami went beyond the beach and entered the estuary, located 200 m from the shoreline, at Las Boquillas, site I (CHÁVEZ 1932; Excélsior Newspaper 1932a, 1932b; CUMMING 1933), which is consistent with model 27 showing 300 m HI and 2 m MTWH. This model also shows maximum HI of 300 and 500 m, respectively, with 3–4 m of MTWH at sites El Real and Tecuanillo (j and k). Figure 4 shows the maximum tsunami height pattern along the coastline: Heights are concentrating in a small areas surrounding Cuyutlán, and decay to the East and West. On the other hand, Fig. 5 shows the HI pattern, and how topographic features, i.e., sand dune barriers, affect HI distance at observed sites. These values agree with the described damage reported by CUMMING (1933) in his post-earthquake-tsunami survey and with the historical accounts summarized by CHÁVEZ (1932). Finally, no damage was reported at Boca de Apiza which also corresponds with the modeled 1 m MTWH, and no horizontal inundation at this location.

Tsunami modeling and historical descriptions suggest that the mayor destructive effects were focused along ~40 km of coastline and a maximum tsunami height ~9.8 m, from Cuyutlán to El Tecuanillo (Fig. 1). Using these data, $I_2 = 2.5 \times 10^{-4}$, consequently, the 22 June tsunami clearly exceeds the threshold for seismic dislocation (OKAL and SYNOLAKIS 2004).

3.3.2 Directivity

Tsunami directivity is defined as the maximum flow direction resulting from numerical modeling, serving as a robust parameter to define tsunami source, following IWASAKI (1997), OKAL (2003) and OKAL and SYNOLAKIS (2003). Values of directivity were similar in all modeled scenarios due to the location of the submarine canyon with respect to all studied coastal sites. The main NW–SE directivity at sites a and b is explained by the influence of the seismic source. The predominant directivity at sites e, f, g and h, resulted in

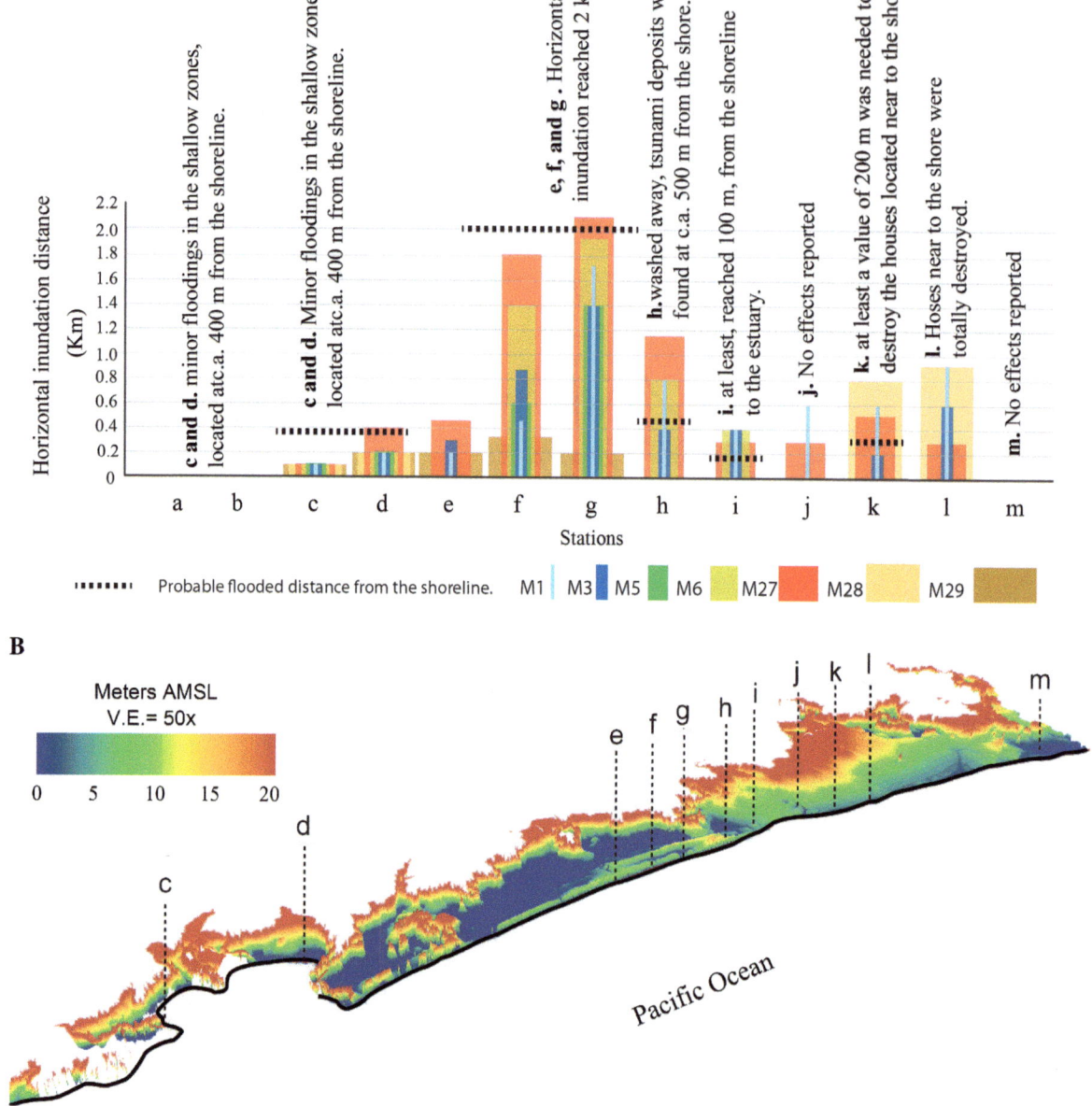

Figure 5
a Plotted values: horizontal inundation with respect to observed effects at each site. **b** Digital elevation model of the coastal area. Irregular topography determines the possibilities of tsunami inundation inland

SE-NW direction. At site i, main directivity was oriented from S to N; and at sites j, k, l and m, was predominantly from SW to NE (see Fig. 2 on supplementary material). A predominant SE-NW direction at Cuyutlán (site h) is consistent with the directivity reported by CUMMING (1933), and the anisotropy of magnetic susceptibility data (116° azimuth) reported for the 1932 tsunami deposits at Palo Verde by RAMÍREZ-HERRERA et al. (2014). Finally, CUMMING (1933) reported that palm trees at Tecuanillo (site 12) were knocked down by the tsunami and laid down in SE-NW direction.

4. Conclusions

We discussed here a line of evidence to support the hypothesis of a submarine landslide that contributed to produce the large tsunami observed on June 22, 1932. First, we found original historical documents of this particular tsunami and its characteristics along the Jalisco and Colima coast. Precise tsunami height was only reported in Cuyutlán (sites e, f, and g), where a plausible 10–12 m waves were observed. Neighboring this location a horizontal inundation reached from 500 to 2000 m (CORONA and RAMÍREZ-HERRERA 2012). By interpreting historical accounts and documents, we inferred the specific tsunami behavior in the rest on the affected sites.

Historical data together with seafloor morphological analysis and numerical modeling allow us to suggest that (1) characteristic morphological elements of submarine landslides are present at Armería Canyon; (2) identified submarine landslides at Armería Canyon are capable of producing tsunamis of variable dimensions; and (3) tsunami modeling results, TAT for each site, MTWH value distribution, horizontal inundation and directivity are consistent with the described effects from historical data for the June 22, 1932 tsunami.

Our results show that model 27 (Fig. 6) of an underwater slump (4.2 km long × 3.9 km wide × 0.480 km thick), of 0.491 km^3 volume, at 1100 m depth, located on the eastern slope of the

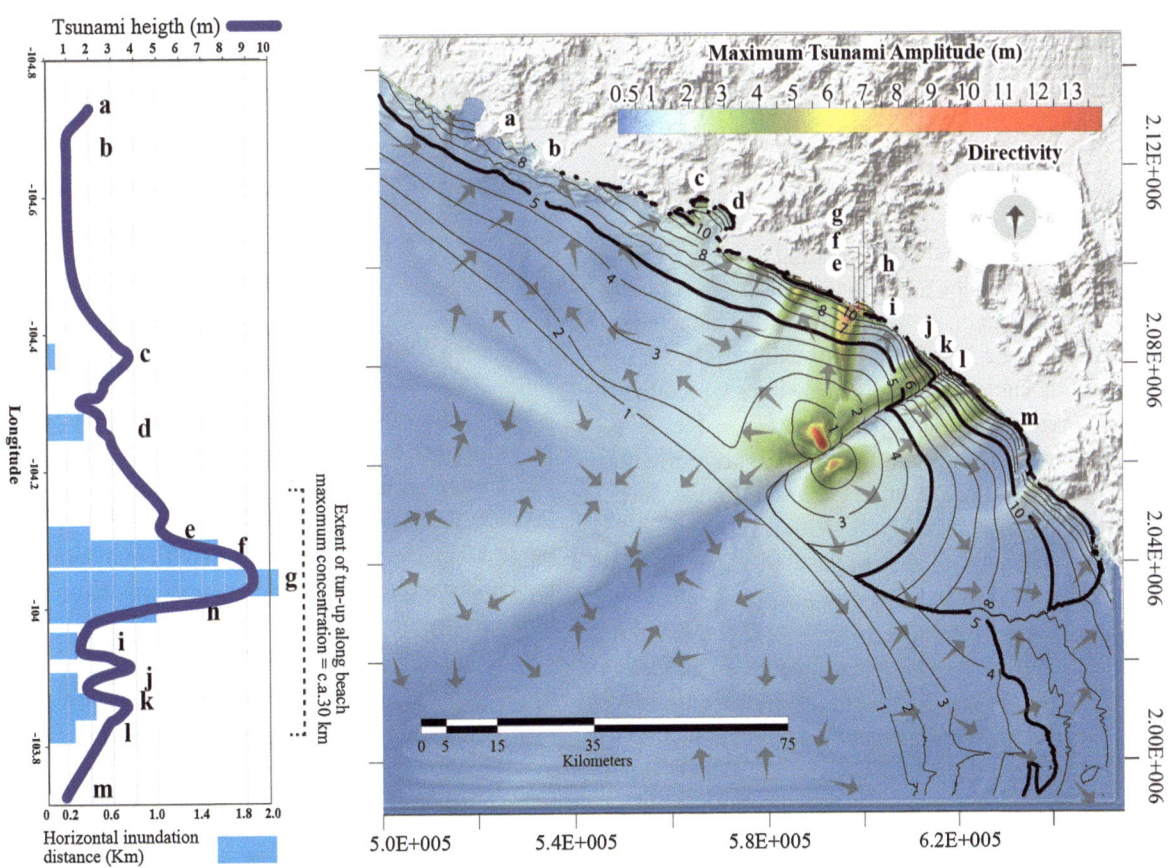

Figure 6
Chart to the *left* displays the modeled maximum wave height along the coast (continuous *bold line*) as well as the *modeled horizontal inundation* at all sites (*bars*) that resulted in model 27. *Graph* at right displays maximum tsunami amplitudes (*gradient scale colors* from *blue* to *red*). *Dark lines* represent isochrones at 1-min intervals, thicker *dark lines* refer to 5-min intervals. *Dark arrows* indicate predominant tsunami directivity

Table 3

Parameters used for tsunami model M27

Parameters	Values
Input TOPICS	
Longitude (W)	104.1363
Latitude (N)	18.6670
Specific density, γ (Kg/m^3)[a]	1.906
Initial landslide length, b (km)[b]	4.1
Maximum initial landslide thickness, T (m)[c]	448
Maximum landslide width, ω (km)[b]	3.9
Mean initial landslide depth, d (m)[b]	1100
Mean initial incline angle, θ (deg.)[b]	8
Output TOPICS	
Slide initial acceleration, a_0 (m/s^2)	0.294
Characteristic distance of slide motion, S_0 (Km)	490
Characteristic time of slide motion, t_0 (s)	40.824
Characteristic wavelength, λ_0 (Km)	8.939
Characteristic tsunami amplitude η_0 (m)	7.1288

[a] Inferred, in agreement with sediment materials COULBOURN et al. (1982)

[b] Parameters taken directly from the topobathymetric model (RYAN et al. 2009)

[c] Taken from the sediment deposit and thickness model (DIVINS 2003)

Armería Canyon (parameters are listed in Tables 3 and Table 1 Supplementary materials), and ~27 km off the coast, can trigger a tsunami of similar characteristics to those of the June 22, 1932 tsunami, described in historical documents and by eyewitness accounts. Model 27 resulted in an aspect ratio $T/b = 1.1$. This value is within the expected values of T, according to WATTS et al. (2005), and other observed slumps, such as the 1998, Papua New Guinea, $T/b = 1.7$ ($b = 4500$ m and $T = 760$), or the 1999 Izmit Bay $T/b = 1$ ($b = 5000$ m and $T = 500$ m).

In the tsunami modeling process we found that the value of thickness were overestimated in models: 1–5, 7, 8, 10, 14, 16–19, 22–26, 28–30 the thickness (see Table 1 on supplementary materials). In these models, data provided by DIVINS et al. (2003) produced overestimated results in the tsunami modeling. In the rest of slumps, their geometry presented an acceptable T/b ratio; values were ranging from 0.09 to 1.6. However, we emphasize that these are inferred landslides and essentially they indicate the scale of the landslides that would be needed for our model to reproduce the coastal effects.

Landslide tsunamis are still underestimated on the Mexican coasts. The Mexican tsunami warning system is limited by considering seismic sources as the only probable source for tsunamis.

Despite the limitations of the employed, relatively coarse, bathymetric grid, we were able to model the landslide tsunami behavior. The core of the discussion was to provide arguments to support the existence of a number of slumps, identified in the Armería Canyon, and to model different scenarios for each one of these landslides to test the hypothesis of a submarine landslide as the triggering mechanism for the June 22, 1932 tsunami. Nevertheless, detailed geological and bathymetric data would contribute to the characterization of the landslide responsible for the June 22, 1932 tsunami.

By providing more scientific information on potential landslide tsunamis on the Mexican Pacific continental slope, hazard levels could be reduced, and detailed data on submarine landslides as tsunamigenic mechanism would aid in our understanding of this phenomenon.

Acknowledgments

N. Corona thanks CONACYT for a Ph.D. scholarship, grant number 216323. M.T. Ramírez-Herrera acknowledges research funding by PAPPIT-UNAM, grant number IN123609 and SEP-CONACYT Ciencia Básica grant number 129456, and DGAPA-PASPA-2015. The authors want to thank PAAG guest editor Hermann M. Fritz and 3 anonymous for their incisive comments and suggestions that helped improve this manuscript.

REFERENCES

ASSIER-RZADKIEAICZ, S., P. HEINRICH, P. C. SABATIER, B. SAVOYE, and BOURILLET, J. F. (2000), *Numerical Modelling of a Landslide-Generated Tsunami: The 1979 Nice Event*, Pure Appl. Geophys. 157, 1707–1727. doi:10.1007/PL00001057.

BEN-MENAHEM, A., and ROSENMAN, M. (1972), *Amplitude patterns of tsunami waves from submarine earthquakes*, J. Geophys. Res. 77, 3097–3128. doi:10.1029/JB077i017p03097

BORRERO, J. C., KALLIGERIS, N., LYNETT, P.J., FRITZ, H.M., NEWMAN, A.V., CONVERS, J.A. (2014), *Observations and Modelling of the*

August 27, 2012 Earthquake and Tsunami affecting El Salvador and Nicaragua, Pure Appl. Geophys., 171, 3421–3435. doi:10.1007/s00024-014-0782-2.

CHÁVEZ, L. (1932), Dramático relato de la reciente catástrofe del balneario de Cuyutlán, El Nuevo Mexicano Newspaper, July 21, p. 2.

CORONA, N., and RAMÍREZ-HERRERA, M. T. (2012), Mapping and historical reconstruction of the great Mexican 22 June tsunami, Nat. Hazards Earth Sys. 12, 1337–1352. doi:10.5194/nhess-12-1337-2012.

COULBOURN, W. T., AUBOUIN, J., VON HUENE, R., AZEMA, J., COWAN, S. S., CURIALE, J. A., DENGO, C. A., FAAS, R. W., HARRISON, W. E., HESSE, R., LADD, J. W., MUZYLOV, N., SHIKI, T., THOMSON, P. R., WESTBERG, M. J., and ORLOFSKY, S. (1982), Initial Reports of the Deep Sea Drilling Project covering Leg 67 of the cruises of the drilling vessel Glomar Challenger, Manzanillo, Mexico to Puntarenas. 67, 5–25.

CRUZ, G., and WYSS, M. (1983), Large earthquakes, mean sea level, and tsunamis along the Pacific Coast of Mexico and Central America, Bull. Seis. Soc. Am. 73, 553–570.

CUMMING, J. L. (1933), Los terremotos de Junio de 1932 en los estados de Colima y Jalisco, Universidad de México, 31-32, 68-104.

DIVINS, D. L., Total Sediment Thickness of the World's Oceans and Marginal Seas (National Geophysical Data Center, Boulder 2003).

EINSELE, G., Sedimentary Basins: Evolution, Facies, and Sediment Budget (Springer-Verlag, Germany, 1992).

EISSLER, H. K., and MCNALLY, K. C. (1984), Seismicity and Tectonics of the Rivera Plate and Implications For the 1932 Jalisco, Mexico, Earthquake, J. Geophys. Res. 89, 4520–4530. doi:10.1029/JB089iB06p04520.

EL INFORMADOR NEWSPAPER (1932a), Cuyutlán a punto de desaparecer bajo las aguas del Océano Pacífico, June 23, pp. 1–2.

EL INFORMADOR NEWSPAPER (1932b), La Catástrofe Ocurrida Anteayer En Cuyutlán No Tiene Precedente, June 24, pp. 1–2.

EL UNIVERSAL GRÁFICO NEWSPAPER (1932a), Colima y Cuyutlán, June 24, pp. 6, 7.

EL UNIVERSAL GRÁFICO NEWSPAPER (1932b), En Cuyutlán no quedó piedra sobre piedra, June 23, p. 15.

EXCÉLSIOR NEWSPAPER (1932a), Cuyutlán arrasado por la invasión del océano: gigantescas olas arrasaron las casas y la gente, June 23, pp. 1, 4.

EXCÉLSIOR NEWSPAPER (1932b), Tres minutos fueron suficientes para arrasar el balneario de Cuyutlán, June 25, pp. 1, 3.

FARRERAS, S. F., and SÁNCHEZ, A. J. (1991), The tsunami threat on the Mexican west coast: A historical analysis and recommendations for hazard mitigation, Nat. Hazards, 4, 301–316. doi:10.1007/bf00162795.

FAVALLI, M., BOSCHI, E., MAZZARINI, F. and PARESCHI, M. T. (2009), Seismic and Landslide Source of the 1908 Straits of Messina Tsunami (Sicily, Italy), Geophys. Res. Lett. 36, L16304. doi:10.1029/2009GL039135.

FRITZ, H. M., HILLAIRE, J. V., MOLIÈRE, E., WEI, Y. and MOHAMMED, F. (2013). Twin tsunamis triggered by the 12 January 2010 Haiti earthquake, Pure Appl. Geophys., 170, 1463–1474. doi:10.1007/s00024-012-0479-3

FRITZ, H. M., KONGKO, W., MOORE, A., MCADOO, B. GOFF, J., HARBITZ, C., USLU, B., KALLIGERIS, N., SUTEJA, D. and KALSUM, K. (2007), Extreme Runup from the 17 July 2006 Java Tsunami, Geophys. Res. Lett. 34, L12602. doi:10.1029/2007GL029404.

HAMPTON, M. A., LEE H. J., and LOCAT, J. (1996), Submarine landslides, Rev. Geophys. 34, 33–59, doi:10.1029/95RG03287.

HARAHAP, I. S. H., and HUAN V. N. P. (2014), Generation, Propagation, Run-Up and Impact of Landslide Triggered Tsunami: A Literature Review, App. Mech. Mater. 567, 724–29. doi:10.4028/www.scientific.net/AMM.567.724.

HARBITZ, C. B., LØVHOLT, F., and BUNGUM, H. (2014). Submarine landslide tsunamis: how extreme and how likely?, Nat. Hazards, 72, 1341–1374. doi:10.1007/s11069-013-0681-3.

HARBITZ, C. B., LOVHOLT, F., PEDERSEN, G., GLIMSDAL, S. and MASSON, D. G. (2006), Mechanisms of Tsunami Generation by Submarine Landslides - a Short Review, Norw. J. Geol. 86, 255–264.

IWASAKI, S. (1997), The wave forms and directivity of a tsunami generated by an earthquake and a landslide, Sci. Tsunami Haz. 15, 23–40.

KAWAMURA, K., SASAKI, T., KANAMATSU, T., SAKAGUCHI, A., and OGAWA, Y. (2012), Large submarine landslides in the Japan Trench: A new scenario for additional tsunami generation, Geophys. Res. Lett. 39, 1–5. doi:10.1029/2011GL050661.

KULIKOV, E. A., RABINOVICH, A. B., THOMSON, R. E., and BORNHOLD, B. D. (1996), The landslide tsunami of November 3, 1994, Skagway harbor, Alaska, J. Geophys. Res. 101, 6609–6615, doi:10.1029/95JC03562.

LEE, H. J., LOCAT, J., DESGAGNÉS, P., PARSONS, J. D., MCADOO, B. G., ORANGE, D. L., PUIG, P., WONG, F. L., DARTNELL, P. and BOULANGER, E., Submarine Mass Movements on Continental Margins, In Continental Margin Sedimentation: From Sediment Transport to Sequence Stratigraphy (eds. NITTROUER, C. A., AUSTIN, J. A., FIELD, M. E., KRAVITZ, J. H., SYVITSKI, J. P. M., and WIBERG, P. L.) (Oxford 2009) pp. 213–274.

LUHR, J. F., NELSON, S. A., ALLAN, J. F., and CARMICHAEL, I. S. E. (1985), Active rifting in southwestern Mexico: Manifestations of an incipient eastward spreading-ridge jump, Geology, 13, 54–57.

LYNETT, P. J., BORRERO,J. C., LIU, P. F., and SYNOLAKIS, E. C., Field Survey and Numerical Simulations: A Review of the 1998 Papua New Guinea Tsunami, In: Landslide Tsunamis: Recent Findings and Research Directions (e. BARDET, J. P., IMAMURA, F., SYNOLAKIS, C. E., OKAL, E. A., and DAVIES, H. L.) (Birkhäuser Basel 2003) pp. 2119–2146.

MASSON, D., HARBITZ, C., WYNN, R., PEDERSEN, G., and LØVHOLT, F. (2006), Submarine landslides: processes, triggers and hazard prediction, P. Roy. Soc. Lon. A Mat. 364, 2009–2039. doi:10.1098/rsta.2006.1810.

MOHAMMED, F. and FRTITZ, H. M. (2012), Physical modeling of tsunamis generated by three-dimensional deformable granular landslides, J. Geophys. Res. 117, C11015, doi:10.1029/2011JC007850.

MCADOO, B., PRATSON, L., and ORANGE, D. (2000), Submarine landslide geomorphology, US continental slope, Mar. Geol. 169, 103–136.

NATIONAL GEOPHYSICAL DATA CENTER, NATIONAL OCEANIC AND ATMOSPHERIC ADMINISTRATION (2012), Tsunami Event Database, http://www.ngdc.noaa.gov/struts/form?t=101650&s=70&d=7 (accessed 20 December 2012).

Nixon, G., T. (1982), *The relationship between Quaternary volcanism in central Mexico and the seismicity and structure of subducted ocean lithosphere*, Geol. Soc. Am. Bull. *93*, 514–523.

Novosibirsk Tsunami Laboratory (2011), Historical Tsunami Database for the World Ocean, Institute of Computational Mathematics and Mathematical Geophysics SB RAS. http://tsun.sscc.ru/nh/tsun_descr.html (accessed 15 September 2011).

Okal, E. A. (2003), *Normal mode energetics for far-field tsunamis generated by dislocations and landslides*, Pure Appl. Geophys. *160*, 2189–2221. doi:10.1007/s00024-003-2426-9

Okada, Y. (1985), *Surface deformation due to shear and tensile faults in a half-space*, Bull. Seismol. Soc. Am. *75*, 1135–1154.

Okal, E. A. (2000), *T waves from the 1998 Sandaun PNG sequence: Definitive timing of the slump, Eos*, Trans. Am. Geophys. Union. *81*, 142.

Okal, E. A. and Synolakis, C.E. (2003), *Field Survey and Numerical Simulations: A Theoretical Comparison of Tsunamis from Dislocations and Landslides*, Pure Appl. Geophys. *160*, 2177–2188.

Okal, E. A., and Borrero, J. C. (2011), *The 'tsunami earthquake' of 1932 June 22 in Manzanillo, Mexico: seismological study and tsunami simulations*, Geophys. J. Int. *187*, 1443–1459. doi:10.1111/j.1365-246X.2011.05199.x.

Okal, E. A. and Synolakis, C.E. (2004), *Source discriminants for near-field tsunamis*, Geophys. J. Int. *158*, 899–912.

Ordoñez, E. (1933), *Seismic activity in Mexico during June, 1932*, B. Seismol. Soc. Am. *23*, 80–82.

Ramírez-Herrera, M.-T., Corona, N., Lagos, M., Černý, J., Goguitchaichvili, A., Goff, J., Chagué-Goff, C., Machain, M. L., Zawadzki, A., Jacobsen, G., Carranza-Edwards, A., Lozano, S., and Blecher, L. (2014), *Unearthing earthquakes and their tsunamis using multiple proxies: the 22 June 1932 event and a probable fourteenth-century predecessor on the Pacific coast of Mexico*, Int. Geol. Rev. 56, 1584–1601. doi:10.1080/00206814.2014.951977.

Ramírez-Herrera, M. T., and Urrutia-Fucugauchi, J. (1999), *Morphotectonic zones along the coast of the Pacific continental margin, southern Mexico*, Geomorphology, *28*, 237–250.

Ranguelov, B., Tinti, S., Pagnoni, G., Tonini, R., Zaniboni, F., and Armigliato, A. (2008), *The nonseismic tsunami observed in the Bulgarian Black Sea on 7 May 2007: Was it due to a submarine landslide?*, Geophys. Res. Lett. *35*, L18613. doi:10.1029/2008gl034905.

Ryan, W. B. F., Carbotte, S. M., Coplan, J. O., O'Hara, S., Melkonian, A., Arko, R., Weissel, R. A., Ferrini, V., Goodwillie, A., Nitsche, F., Bonczkowski, J., and Zemsky, R. (2009), *Global Multi-Resolution Topography synthesis*, Geochem. Geophy. Geosy. *10*, Q03014. doi:10.1029/2008GC002332.

Salazar, J. El Maremoto de Cuyutlán 1932 (Sociedad Colimense de Estudios Históricos, Colima 1989).

Sánchez, A. J., and Farreras, S. F., Catálogo de Tsunamis (Maremotos) en la Costa Occidental de México (Boulder, Colorado, 1993).

Satake, K. and Tanioka, Y. (2003), *The July 1998 Papua New Guinea Earthquake: mechanism and quantification of unusual tsunami generation*, Pure Appl. Geophys. *160*, 2087–2188.

Singh, S. K., Astiz, L., and Havskov, J. (1981), *Seismic gaps and recurrence periods of large earthquakes along the Mexican subduction zone: A reexamination*, B. Seismol. Soc. Am. *71*, 827–843.

Singh, S. K., Pacheco, J. F., and Shapiro, N. (1998), *The earthquake of 16 November, 1925 (Ms = 7.0) and the reported tsunami in Zihuatanejo, Mexico*, Geofis. Int. *37*, Short Note.

Singh, S. K., Ponce, L., and Nishenko, S. P. (1985), *The great Jalisco, Mexico, earthquakes of 1932: Subduction of the Rivera plate*, B. Seismol. Soc. Am. *75*, 1301–1313.

Singh, S. K., Rodríguez, M., and Espindola, J. M. (1984), *A catalog of shallow earthquakes of Mexico from 1900 to 1981*, B. Seismol. Soc. Am. *74*, 267–280.

Soloviev, S. L., and Go, C. N., Catalogue of tsunamis on the eastern shore of the Pacific Ocean, Canadian Translation of Fisheries and Aquatic Sciences No. 5078 (Nauka Publishing House, Moscow 1975).

Servicio Sismológico Nacional (2011), Catalogo de sismos, http://www.ssn.unam.mx/ (accessed 2 November 2011).

Strasser, M., Kölling, M., dos Santos Ferreira, C., Fink, H. G., Fujiwara, T., Henkel, S., Ikehara, K., Kanamatsu, T., Kawamura, K., and Kodaira, S. (2013), *A slump in the trench: Tracking the impact of the 2011 Tohoku-Oki earthquake*, Geology, *41*, 935–938.

Synolakis, C. E. Borrero, J. C., Plafker, G., Yalçiner, A., Greene, G., and Watts, P. (2000) *Modeling the 1994 Skagway, Alaska Tsunami, EOS*, Trans. Am. Geophys. Union *81*, F748 (abstract).

Tappin, D. R., Grilli, S. T., Harris, J. C., Geller, R. J., Masterlark, T., Kirby, J. T., Shi, F., Ma, G., Thingbaijam, K. K. S., and Mai, P. M. (2014), *Did a submarine landslide contribute to the 2011 Tohoku tsunami?*, Mar. Geol. *357*, 344–361. doi:10.1016/j.margeo.2014.09.043.

Tappin, D. R., Matsumoto, T., Watts, P., Satake, K., McMurtry, G. M., Matsuyama, M., Lafoy, Y., Tsuji, Y., Kanamatsu, T., Lus, W., Iwabuchi, Y., Yeh, H., Matsumotu, Y., Nakamura, M., Mahoi, M., Hill, P., Crook, K., Anton, L., and Walsh, J. P. (1999), *Sediment slump likely caused 1998 Papua New Guinea tsunami*, EOS T. Am. Geophys. Un. *80*, 329. doi:10.1029/99EO00241.

ten Brink, U. S., Chaytor, J. D., Geist, E. L., Brothers, D. S., Andrews, B. D. (2014), *Assessment of tsunami hazard to the U.S. Atlantic margin*, Mar. Geol., *353*, 31–54. doi:10.1016/j.margeo.2014.02.011.

von Huene, R., Kirby, S., Miller, J., and Dartnell, P. (2014), *The destructive 1946 Unimak near-field tsunami: New evidence for a submarine slide source from reprocessed marine geophysical data*, Geophys. Res. Lett., *41*, 6811–6818, doi:10.1002/2014GL061759.

Ward, S. N. (2001), *Landslide tsunami*, J. Geophys. Res-Sol. Ea. *106*, 11201–11215.

Watts, P., Grilli, S. T., Kirby, T. J., Fryer, G. J., and Tappin, D. R. (2003), *Landslide tsunami case studies using a Boussinesq model and a fully nonlinear tsunami generation model*, Nat. Hazard Earth Sys. *3*, 391–402.

Watts, P., Grilli, S. T., Tappin, D. R., and Fryer, G. J. (2005). *Tsunami generation by submarine mass failure. II: Predictive equations and case studies*. J. Waterw. Port Coast. Ocean Eng., *131*, 298–310.

Watts, P., and Tappin, D., Geowave Validation with Case Studies: Accurate Geology Reproduces Observations, In Submarine Mass Movements and Their Consequences (e. Yamada, Y., Kawamura, K, Ikehara, K., Ogawa, Y., Urgeles, R., Mosher, D., Chaytor, J., and Strasser, M.) (Springer, Netherlands 2012) pp. 517–524.

YAMADA, Y., KAWAMURA, K., IKEHARA, K., OGAWA, Y., URGELES, R., MOSHER, D., CHAYTOR, J., and STRASSER, M. (Eds.) Submarine Mass Movements and Their Consequences (Springer 2012).

ZANIBONI, F., ARMIGLIATO, A., PAGNONI, G. and TINTI, S. (2014), *Continental Margins as a Source of Tsunami Hazard: The 1977 Gioia Tauro (Italy) Landslide–tsunami Investigated through Numerical Modeling*, Mar. Geol. *357* 210–217. doi:10.1016/j.margeo.2014.08.011.

(Received December 4, 2014, revised August 25, 2015, accepted August 31, 2015, Published online September 21, 2015)

Far-Field Tsunami Impact in the North Atlantic Basin from Large Scale Flank Collapses of the Cumbre Vieja Volcano, La Palma

BABAK TEHRANIRAD,[1] JEFFREY C. HARRIS,[2,4] ANNETTE R. GRILLI,[2] STEPHAN T. GRILLI,[2] STÉPHANE ABADIE,[3] JAMES T. KIRBY,[1] and FENGYAN SHI[1]

Abstract—In their pioneering work, Ward and Day suggested that a large scale flank collapse of the Cumbre Vieja Volcano (CVV) on La Palma (Canary Islands) could trigger a mega-tsunami throughout the North Atlantic Ocean basin, causing major coastal impact in the far-field. While more recent studies indicate that near-field waves from such a collapse would be more moderate than originally predicted by Ward and Day [LØVHOLT et al. (J Geophy Res 113:C09026, 2008); ABADIE et al. (J Geophy Res 117:C05030, 2012)], these would still be formidable and devastate the Canary Island, while causing major impact in the far-field at many locations along the western European, African, and the US east coasts. ABADIE et al. (J Geophy Res 117:C05030, 2012) simulated tsunami generation and near-field tsunami impact from a few CVV sub-aerial slide scenarios, with volumes ranging from 20 to 450 km^3; the latter representing the most extreme scenario proposed by Ward and Day. They modeled tsunami generation, i.e., the tsunami source, using THETIS, a 3D Navier-Stokes (NS) multi-fluid VOF model, in which slide material was considered as a nearly inviscid heavy fluid. Near-field tsunami impact was then simulated for each source using FUNWAVE-TVD, a dispersive and fully nonlinear long wave Boussinesq model [SHI et al. (Ocean Modell 43–44:36–51, 2012); KIRBY et al. (Ocean Modeling, 62:39–55, 2013)]. Here, using FUNWAVE-TVD for a series of nested grids of increasingly fine resolution, we model and analyze far-field tsunami impact from two of ABADIE et al.'s extreme CVV flank collapse scenarios: (i) that deemed the most "credible worst case scenario" based on a slope stability analysis, with a 80 km^3 volume; and (ii) the most extreme scenario, similar to Ward and Day's, with a 450 km^3 volume. Simulations are performed using a one-way coupling scheme in between two given levels of nested grids. Based on the simulation results, the overall tsunami impact is first assessed in terms of maximum surface elevation computed along the western European and African, and US east coasts (USEC). Strong wave elevation decay is predicted over the wide USEC shelf, which is shown to be essentially due to bottom friction effects. We then show more detailed results for the USEC, which is the object of high-resolution tsunami inundation mapping under the auspices of the US National Tsunami Hazard Mitigation Program. In this context, we compare the maximum surface elevation predicted along the coastline for each CVV scenario and show that, besides the initial directionality of the sources, coastal impact is mostly controlled by focusing/defocusing effects resulting from the shelf bathymetric features. A simplified ray-tracing analysis confirms this controlling effect of the wide USEC shelf for incident long waves. Finally, we perform high-resolution (10 m) inundation mapping for the most extreme CVV scenario and show results at one of the most vulnerable and exposed communities in the mid-Atlantic US states, in and around Ocean City, Maryland. Such maps are being generated for all exposed areas of the USEC, to be used in tsunami hazard assessment and mitigation work.

Key words: Tsunami propagation, coastal geohazard, sub-aerial landslide, navier-stokes VOF model, boussinesq wave models, volcano collapse, Cumbre Vieja, La Palma.

1. Introduction

Since 2010, under the auspices of the US National Tsunami Hazard Mitigation Program (NTHMP), the authors have conducted modeling work to gradually develop tsunami inundation maps for the most critical or vulnerable areas of the US east coast (USEC). These first generation maps are constructed as envelopes of maximum inundation caused by the most extreme near- and far-field tsunami sources, both historical and hypothetical, in the Atlantic Ocean basin, without considering their return period or probability. Probabilistic tsunami hazard analyses will be part of future generations of inundation maps. To perform this inundation mapping work, all the relevant extreme tsunami sources in the Atlantic Ocean basin were first identified and parameterized, and then tsunami generation, propagation, and coastal

---

[1] Center for Applied Coastal Research, Department of Civil and Environmental Engineering, University of Delaware, Newark, DE 19716, USA.

[2] Department of Ocean Engineering, University of Rhode Island, Narragansett, RI 02882, USA. E-mail: grilli@oce.uri.edu

[3] Laboratoire SIAME, Université de Pau et des Pays de l'Adour, 64600 Anglet, France.

[4] *Present Address:* Saint-Venant Laboratory for Hydraulics, Université Paris-Est, 78400 Chatou, France.

impact from each of those was simulated to the considered areas of the US coastline. The extreme sources identified and used so far include [see TEN BRINK et al. (2014) for a more comprehensive review]: (i) near-field submarine mass failures on or near the continental shelf break GRILLI et al. (2009, 2015b); (ii) an extreme hypothetical M9 seismic event occurring in the Puerto Rico Trench GRILLI et al. (2010); (iii) a repeat of the historical 1755 M8.9 earthquake occurring in the Azores convergence zone BARKAN et al. (2009); and (iv) a large scale volcanic flank collapse of the Cumbre Vieja Volcano (CVV) in the Canary Archipelago, which is the object of this paper.

Large subaerial landslides are known to occur on the flanks of active volcanos (a.k.a. volcanic flank collapses), because volcanic material continuously accumulates until the slope becomes unstable (HOLCOMB and SEARLE 1991). Large deposits from past landslides have been found on the seafloor surrounding young volcanos in Hawaii (MOORE et al. 1989; ROBINSON and EAKINS 2006) and Réunion Island (COCHONAT et al. 1990; OEHLER et al. 2004). In the Canary Islands, MASSON et al. (2002) identified at least 14 large paleo-landslides, associated with failures that occurred in the last one million years on the flanks of the youngest volcanoes in the islands of El Hierro, La Palma, and Tenerife (with the youngest one, at El Hierro, being only 15,000 years old; Figs. 1, 2). Some of these landslides have volumes of $\mathcal{O}(100 \text{ km}^3)$ or more and could have triggered mega-tsunamis (WARD and DAY 2001). It is believed that such large, potentially catastrophic, events may have occurred in average every 100,000 years in the Canary Archipelago, which is a much longer return period than the typical hundreds of year periodicity of megathrust seismic events in large subduction zones, which have the potential of causing large tsunamis. [For instance, in the Atlantic Ocean basin, GRILLI et al. (2010) estimated 200 and 600 year return periods for M8.7 and M9.1 earthquakes, respectively, in the Puerto Rico Trench.] A long return period, however, does not necessarily mean low risk, particularly regarding tsunami hazard for critical coastal facilities, such as nuclear power plants or large maritime terminals. Therefore, comprehensive tsunami hazard assessment for such critical facilities must consider all the potential extreme tsunami sources in the relevant ocean basin, such as the volcanic flank collapses considered here. More detailed analyses of landslides mechanisms in the Canary Islands and of their recurrence are outside the scope of this paper, but the reader can consult the works of WYNN and MASSON (2003) and HUNT et al. (2011, 2013), for more information. It should be noted that the latter three studies provide evidence for a multi-stage failure, which was not considered here in our extreme scenarios.

In the Canary Islands, CVV (Figs. 1 and 2) is the fastest growing volcano (CARRACEDO et al. 1999), with the potential of causing very large flank collapses and thus of generating a mega-tsunami. Ward and Day were the first to suggest this possibility, by considering an extreme collapse of nearly the entire CVV western flank, with an estimated 500 km^3 volume. Using fairly simple models, they simulated the resulting tsunami generation and propagation and predicted extremely large (kilometer high) near-field waves which, despite significant decay in the far-field, would still be on the order of 10–20 m, when reaching the USEC. Because they used simplified models to estimate coastal impact, which in particular lacked dissipation, they concluded that flow depths on the order of 10–20 m could occur along the USEC, depending on location. Other authors later recognized the potential for the generation of large waves from an extreme CVV collapse, but questioned both Ward and Day's catastrophic landslide scenario and near-shore wave modeling (MADER 2001; PARARAS-CARAYANNIS 2002). In more recent work, Ward and Day's and other similar CVV collapse scenarios were modeled using more accurate models of both landslide and wave generation/propagation (PÉRIGNON 2006; LØVHOLT et al. 2008; ZHOU et al. 2011; ABADIE et al. 2012).

ABADIE et al. (2012) provided a review to date of relevant CVV landslide and tsunami modeling studies. Based on a slope stability analysis of the CVV western flank (RISS et al. 2010), they estimated both the geometry and volume of the most credible extreme flank collapse scenario. Although they found high safety factors, greater than one, in all the cases they analyzed, the lowest one was obtained for an 80-km^3 failure, which they thus deemed the likeliest

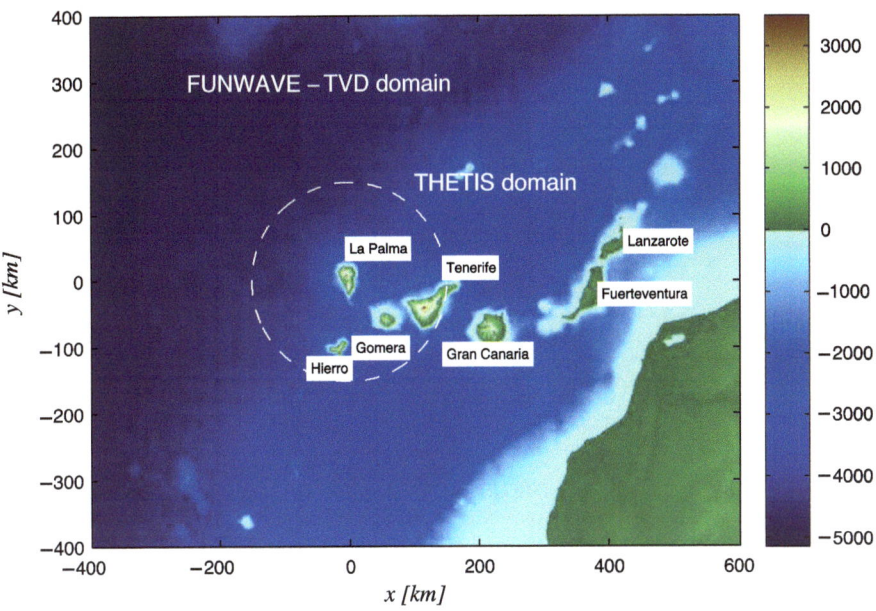

Figure 1
Computational domains covering the Canary Islands, including the higher resolution 3D THETIS domain surrounding La Palma (*dashed circle*) and the 500 m resolution FUNWAVE-TVD domain (*outer box*). Bathymetry (<0) and topography (>0) are represented by the color scale (meter)

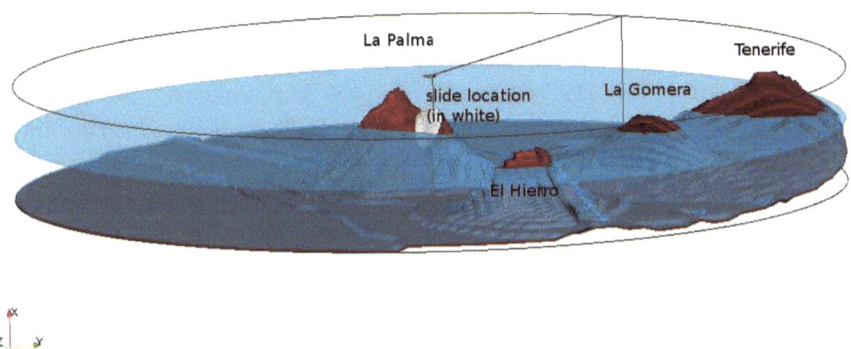

Figure 2
Sketch of THETIS 3D computational domain for the simulation of landslide tsunami generation. The *white area* in La Palma indicates the site of the 80 km^3 subaerial slide. See ABADIE et al. (2012) for details

extreme flank collapse scenario; here we will refer to this case instead as: *extreme credible worst case scenario* or ECWCS, which is more standard terminology in hazard analysis. While the return period for such an event is unknown, as discussed above, it can be estimated to be $\mathcal{O}(100,000)$ years, which is the age of deposits from earlier flank collapses found at the toe of the volcano. despite being stable under present conditions, the CVV western flank could be destabilized by a large earthquake or volcanic eruption, and a slide be triggered. [This mechanism is supported by recent field work; S. Day, personal communication, 2014.]. In addition to their likeliest scenario, to compare their model results to those of earlier studies such as LØVHOLT et al. 's, which considered only the most catastrophic scenario proposed by Ward and Day, ABADIE et al. also modeled a similar extreme CVV western flank collapse scenario, with a volume estimated at 450 km^3, based on higher resolution bathymetry and topography. Finally, they modeled

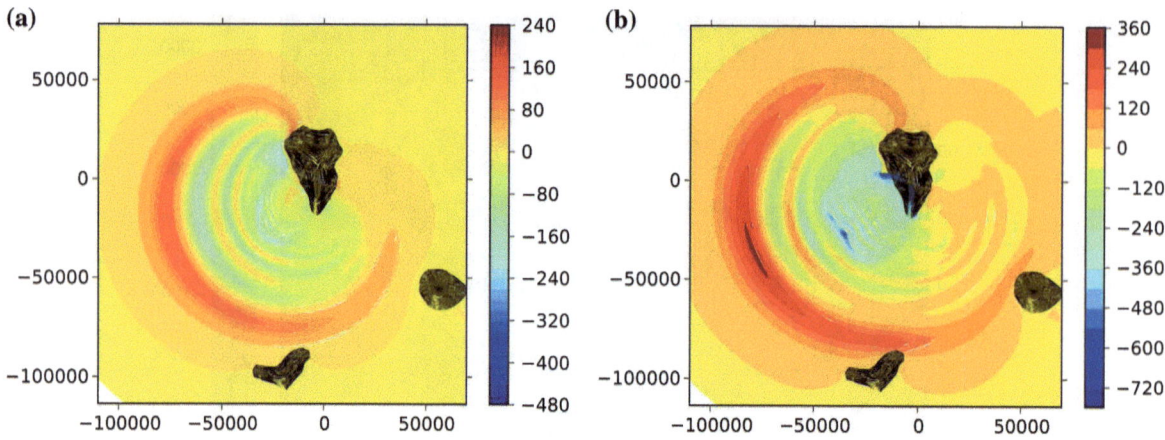

Figure 3
Surface elevations (*color scale* in meter) computed with THETIS for CVV flank collapse scenarios with volume: **a** 80, **b** 450 km^3, at $t = 7.5$ min after the start of the event. Note, x and y axes represent distance in meter

two more scenarios representing partial failure of their likeliest scenario, with 20 and 40 km^3 volumes, respectively.

For simulating these 4 scenarios, ABADIE et al. used the three-dimensional (3D) multi-material Navier-Stokes model THETIS, in which the slide material was modeled as a heavy Newtonian fluid. [THETIS was developed since 1996 by the TREFLE CNRS laboratory, at the University of Bordeaux, France (http://thetis.enscbp.fr).] THETIS has been validated for modeling wave generation by rigid (ABADIE et al. 2010) and deformable (MORICHON and ABADIE 2010) slides; additionally the model has been used to simulate breaking waves (ABADIE et al. 1998; LUBIN et al. 2006). Once the landslide tsunamis were fully generated in the 3D model (which was assessed by quantifying the slide-to-water energy transfer), near-field tsunami propagation within the Canary Islands area was simulated using the two-dimensional (2D) fully nonlinear and dispersive long wave model FUNWAVE-TVD (SHI et al. 2012), in a 500 m resolution Cartesian grid (Fig. 1). The model was initialized 5 min into the event, based on the depth-averaged horizontal velocity and surface elevation computed with THETIS (see details in ABADIE et al. 2012). These simulations predicted initial wave elevations in front of the volcano of up to 1200 m for the 450 km^3 scenario and 800 m for the 80 km^3 scenario (Fig. 3), with the dominant direction of wave propagation towards the far-field being at 24° south of West. For both scenarios, runup was calculated to be over 100 m in the back of the La Palma island and 10–50 m on nearby islands, 10–25 min into the event. Details of near-field impact can be found in ABADIE et al. (2012).

In this paper, our main goal is to model and assess the far-field tsunami impact resulting from the two largest CVV flank collapse scenarios studied by ABADIE et al.: (i) the ECWCS with a 80 km^3 volume; and (ii) the most extreme scenario with a 450 km^3 volume. Consistent with our NTHMP project of developing tsunami inundation maps for the USEC, based on the most extreme tsunami sources identified in the Atlantic Ocean basin, we first focus our work on this area and present in greater details results of tsunami propagation and far-field impact for the most extreme 450 km^3 flank collapse scenario. Besides maximum surface elevation along the entire coastline, we show detailed results in finer nested grids for one of the most exposed areas of the US east coast, near and around Ocean City, MD. Then, for the purpose of comparison, limited results of the 80 km^3 scenario simulations are presented and compared to the former. As an additional goal, we also model the far-field impact of both CVV scenarios on the very exposed, and closer to the source, western European and north African coasts. Here, similar to the near-field tsunami impact study performed by ABADIE et al. (2012), we essentially consider the ECWCS 80 km^3 scenario and provide limited results for the extreme scenario, for comparison.

2. Modeling of Landslide Tsunami Generation

We perform far-field simulations of tsunami propagation and coastal impact for two extreme scenarios of CVV flank collapse: (i) the ECWCS failure with 80 km^3 of slide material; and (ii) the most extreme scenario with 450 km^3. Figures 1 and 2 show the footprint and perspective view, respectively, of the 3D computational domain used by ABADIE et al. in their simulations with THETIS of the volcano collapse and initial tsunami generation. This domain was discretized by a cylindrical mesh (8 km tall and 150 km radius; 300 by 80 stretched grid cells in the radial and vertical directions, and 140 grid cells in the tangential direction). The slide area marked in white on Fig. 2 corresponds to the 80 km^3 volume. Besides the island of La Palma, El Hierro, La Gomera, and Tenerife were included in the domain, since their bathymetry and topography affected the early stages of wave propagation. Similar to GISLER et al. (2006), ABADIE et al. modeled the CVV slide debris flow as an inviscid fluid with a constant 2,500 kg/m^3 density (i.e., corresponding to basalt). Hence, neither basal friction nor resistance to internal deformation were modeled, which they indicated causes more energetic and dynamic slides, likely to generate worst case scenario tsunamis, that are conservative as far as near- and far-field coastal hazard assessment.

Figure 3 shows the surface elevations computed with THETIS at $t = 7.5$ min, for the two slide scenarios. The directivity of tsunami waves is similar in both cases, with a dominant direction of wave propagation at about 24° south of West, but the maximum surface elevation of the generated leading tsunami wave increases and occurs farther away from the volcano for the larger slide volume (as a result of amplitude dispersion effects). The reader will be referred to ABADIE et al. (2010, 2012) for details of the THETIS model features, set-up, and application to the CVV landslide tsunami simulations.

3. Modeling of Near- and Far-field Tsunami Impact

Because THETIS is computationally expensive (the 3D simulations in ABADIE et al. run for a month on a single-core processor, as the cylindrical version was not parallel yet) and the generated landslide tsunami waves quickly become long waves as they propagate away from CVV, ABADIE et al. simulated tsunami impact in the near-field, beyond the 150 km radius of the THETIS domain, with the 2D-horizontal Boussinesq Model (BM) FUNWAVE-TVD, in a 500 m resolution Cartesian grid (Fig. 1). [To correct for earth's sphericity, a transverse secant Mercator projection was used, with its origin located at 28.5 N and 18.5 W.] While still computationally intensive, most simulations with FUNWAVE-TVD reported here typically run for a few hours to half a day on 24 processors (see details below). FUNWAVE-TVD's near-field simulations were initialized using surface elevations (such as shown in Fig. 3) and depth-averaged horizontal velocities computed with THETIS, at 5 min into the event, which ensured that the slide had fully transferred its energy to the water motion and the generated tsunami had not yet reached the neighboring island. Since THETIS and FUNWAVE-TVD have different dimensionality and physics, to provide for a smooth transition of simulations from one model to the other, a filter was applied to THETIS' results to eliminate residual oscillations and vortices in the tail of the generated wave train, near the volcano, that would otherwise perturb FUNWAVE simulations and could possibly trigger instabilities (due to differences in model equations, dimensionality and treatment of horizontal vorticity). By running THETIS for a longer time, ABADIE et al. verified that the filtering method was accurate and did not affect the subsequent near- and far-field wave propagation. Details can be found in the reference.

As they include frequency dispersion effects, BMs simulate more complete physics than models based on the Nonlinear Shallow Water Equations (NSWE), which until recently were traditionally used to model co-seismic tsunami propagation. Dispersion is key to accurately simulate landslide tsunamis, which typically are made of shorter and hence more dispersive waves than for co-seismic tsunamis (WATTS et al. 2003; GRILLI and WATTS 2005; MOHAMMED and FRITZ 2012). However, dispersion is also key to model the coastal impact of any tsunami since dispersive shock waves (a.k.a. undular bores) can be generated near the crest of long waves in increasingly shallow water

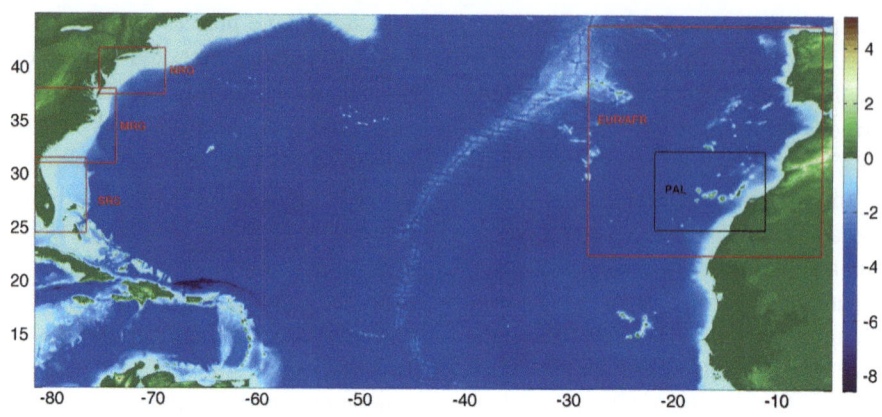

Figure 4
FUNWAVE-TVD simulation grids (see Table 1): (PAL) 500 m local Cartesian grid around La Palma (*black box*); 1 arc-min spherical Atlantic grid (*outer box*); 20 arc-sec spherical regional grids for simulating far-field coastal effects on the USEC (*red boxes*): (NRG) Northern USEC regional, (MRG) Middle USEC regional, (SRG) Southern USEC regional, grids; (EUR/AFR) 15 arc-sec spherical regional grid for simulating far-field coastal effects in western Europe and western Africa. *Color scale* indicates bathymetry (<0) and topography (>0) in km

(MADSEN et al. 2008; GEIST et al. 2009; GRILLI et al. 2012, 2015b). A review of dispersive effects in tsunamis can be found in GLIMSDAL et al. (2013).

FUNWAVE-TVD is based on the equations of SHI et al. (2012) and is a recent improvement of FUNWAVE (WEI et al. 1995), which was originally developed and used to model coastal and nearshore waves, but was later also applied to a variety of tsunami case studies, both landslide and co-seismic (WATTS et al. 2003; DAYS et al. 2005; GRILLI et al. 2007, 2010, 2013; IOUALALEN et al. 2007; Tappin et al. 2008; Karlsson et al. 2009; Tappin et al. 2014). The importance of dispersive effects in far-field tsunami propagation was illustrated in case studies, by running the model in both BM and NSWE modes, by TAPPIN et al. for the 1998 Papua New Guinea landslide tsunami, and by IOUALALEN et al. for the 2004 Indian Ocean and GRILLI et al. (2013) for the 2011 Tohoku, coseismic tsunamis.

FUNWAVE-TVD was developed as a fully nonlinear version in Cartesian coordinates (SHI et al. 2012), but currently is only implemented as a weakly nonlinear approximation in spherical coordinates, including Coriolis effects (KIRBY et al. 2013). Both versions of the model use a combined finite-volume and finite-difference MUSCL-TVD scheme. As in the earlier FUNWAVE version, improved linear dispersion properties are achieved, up to nearly the deep water limit, by expressing the BM equations in terms of the horizontal velocity vector computed at 0.531 times the local depth. Additionally, wave breaking dissipation is adequately modeled by switching from the Boussinesq to the NSWE equations when the local height to depth ratio exceeds 0.8 (which has been shown to closely approximate the physical dissipation in breaking waves). Bottom friction is parameterized as a quadratic term based on a friction coefficient C_d; in all the present simulations, we use the standard value for coarse sand, $C_d = 0.0025$, which only causes moderate dissipation, except for long distances of propagation over wide shallow shelves; hence this value is conservative as far as predicting maximum flow depth at the coastline; see GEIST et al. (2009) and GRILLI et al. (2015b) for a study of the influence of bottom friction on landslide tsunami nearshore propagation and coastal impact. Additional discussions in this respect can be found in KAISER et al. (2011). FUNWAVE-TVD's latest implementation is fully parallelized using MPI, for efficient use on large computer clusters (a nearly 90% scalability is achieved; all simulations reported in this paper for the oceanic propagation were performed using 24 CPUs and those in the many finer coastal nested grids used hundreds of CPUs). FUNWAVE-TVD was fully validated against all of NOAA's National Tsunami Mitigation Program (NTHMP) mandatory benchmarks (TEHRANIRAD et al. 2011).

Table 1

Latitudinal and longitudinal extension, and resolution, for FUNWAVE-TVD simulation grids of Fig. 4

Grid	Latitude	Longitude	Resolution
Local Palma	24.8706°–32.2337° N	−22.0184° to −11.4356°E	500 m
Atlantic Basin	10.0°–45.0° N	−82.0° to −5.0°E	1′
N. US reg.	37.45°–41.7667° N	−75.7° to −69.25°E	20″
Mid US reg.	31°–38° N	−82° to −74°E	20″
S. US reg.	24.5°–31.5° N	−82.0° to −77°E	20″
W. Europe and W. Africa	22.5°–44.0° N	−28.5° to −6.0°E	15″

We simulate the transoceanic tsunami propagation of each CVV scenario using the spherical version of FUNWAVE-TVD, initially in a 1 arc-min Atlantic ocean basin grid; we then pursue simulations in nested regional coastal grids: three 20 arc-sec (about 610 m) resolution grids along the USEC and one 15 arc-sec (about 450 m) resolution grid along the north African and western European coasts (Fig. 4). Following the methodology established in our NTHMP inundation mapping work, simulations of tsunami coastal impact along the USEC are performed in a series of nested Cartesian grids (to use the fully nonlinear implementation of the model), based on results of the 3 regional grids, with a reduction by a factor of 3 or 4 in mesh size, from one coarser grid level down to a finer grid level. The finest grid resolution used in the inundation maps is 10–30 m, depending on the complexity of the coastal features, which typically requires 3 or 4 additional levels of nested grids beyond the 20 arc-sec US regional grids. This will be illustrated in results of inundation mapping around Ocean City, MD, presented later.

Bathymetry in the 1 arc-min Atlantic basin grid is interpolated from the ETOPO-1 database (i.e., 1 arc-sec accurate data). In the US regional grids, bathymetry is interpolated over the continental shelf from NOAA's 3 arc-sec (about 90 m) Coastal Relief Model (CRM) data, and in the African and European regional grid from the 500 m resolution EMODNET data set (http://portal.emodnet-hydrography.eu). For the finer coastal nested grids, bathymetry and topography from the 10 m resolution NOAA tsunami Digital Elevation Maps (DEMs) are used, wherever available, and if not, the similar resolution FEMA DEMs are used. Table 1 gives the extent of the Atlantic ocean basin and regional simulation grids.

Simulations in the 1 arc-min Atlantic ocean basin grid are initialized using the surface elevation and horizontal velocity computed at 20 min into the event in the 500 m grid (Fig. 1), which are shown in Fig. 5 for the 80 and 450 km^3 flank collapse scenarios. At this time, for both sources, the tsunami is made of a large number of quasi-circular concentric waves, with the leading elevation wave being 15–40 m and 40–100 m high for each source, respectively, and the largest values occurring in the predominant direction of propagation, at 24° south of West. The leading crest is followed by a deep trough and a long oscillatory dispersive tail with at least 5 significant waves. This pattern will also be observed in far-field results. A detailed analysis of the features and decay patterns during propagation to the far-field of the tsunami wave train, computed for each source, can be found in ABADIE et al. (2012). For the simulations in the 15 arc-sec regional grid, encompassing western Europe and western Africa, because this grid includes the Canary Islands, computations with FUNWAVE-TVD are simply initialized based on the source computed 20 min into the event, as for the 1 arc-min grid. All the nested grid simulations in 20 arc-sec regional grids and finer Cartesian grids along the USEC are performed using a one-way coupling scheme, which works by computing time series of free surface elevations and currents in a coarser grid level, for a large number of numerical gages (stations) defined along the boundary of the finer grid level. Computations in the finer nested grid level are then performed using these time series as boundary conditions. In this scheme, reflected waves propagating from the area covered by each finer grid are included in the time series computed in the coarser grids, along the finer grid boundaries, thus satisfying an open boundary

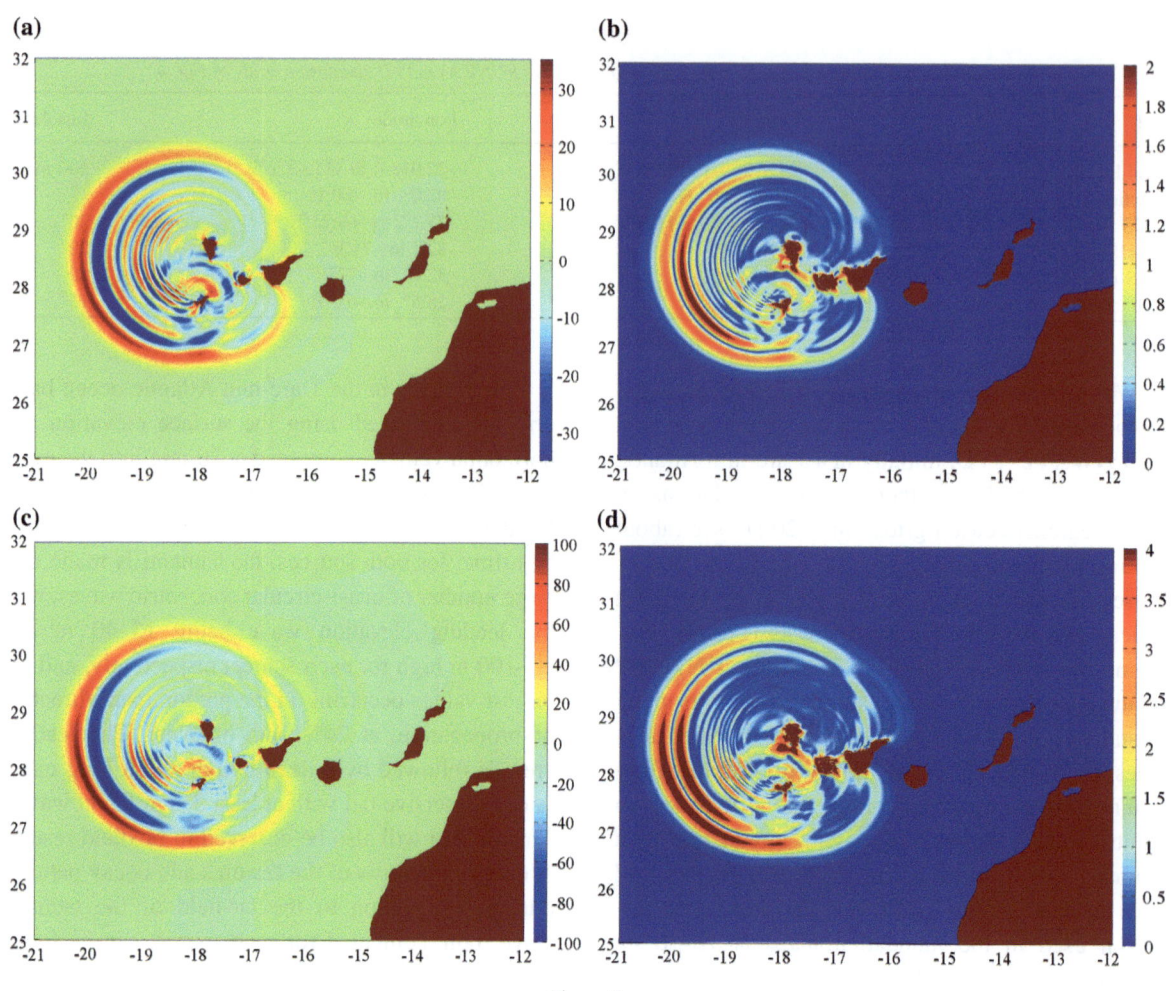

Figure 5

a, **c** Surface elevation (*color scale* in meter) and **b**, **d** horizontal water velocity magnitude (*color scale* in m/s) computed with FUNWAVE-TVD in the 500 m Cartesian grid defined in Fig. 1, at 20 min after the start of the CVV event, for ABADIE et al.'s (2012): **a**, **b** 80 km^3 ; and **c**, **d** 450 km^3 flank collapse scenarios

condition. To reduce reflection in the first coarsest grid level (here either the 1 arc-min Atlantic ocean basin grid for the westward propagation, or the 15 arc-sec grid for the propagation towards western Africa and Europe), 200 km thick sponge (absorbing) layers are specified along all the open boundaries.

4. Transatlantic Tsunami Propagation to the Far-field

Because the overall features and patterns of tsunami waves generated by both CVV flank collapse scenarios are similar, and we focus on the most extreme tsunami sources, in the following, we only detail the propagation to the far-field of the extreme 450 km^3 scenario and its impact along the US east coast; in north Africa and western Europe, consistent with the earlier work of ABADIE et al. (2012), we detail instead the impact of the 80 km^3 scenario. Whenever relevant to the discussion, however, results of the other scenario are also shown for comparison.

Figure 6 shows three snapshots of surface elevation computed in the 1 arc-min grid for the 450 km^3 scenario. In Fig. 6a, 1h20′ into the event, the tsunami has already reached and impacted the western coast of Africa, in the Sahara and Morocco. In Fig. 6b, 2h20′ into the event, the tsunami has reached the

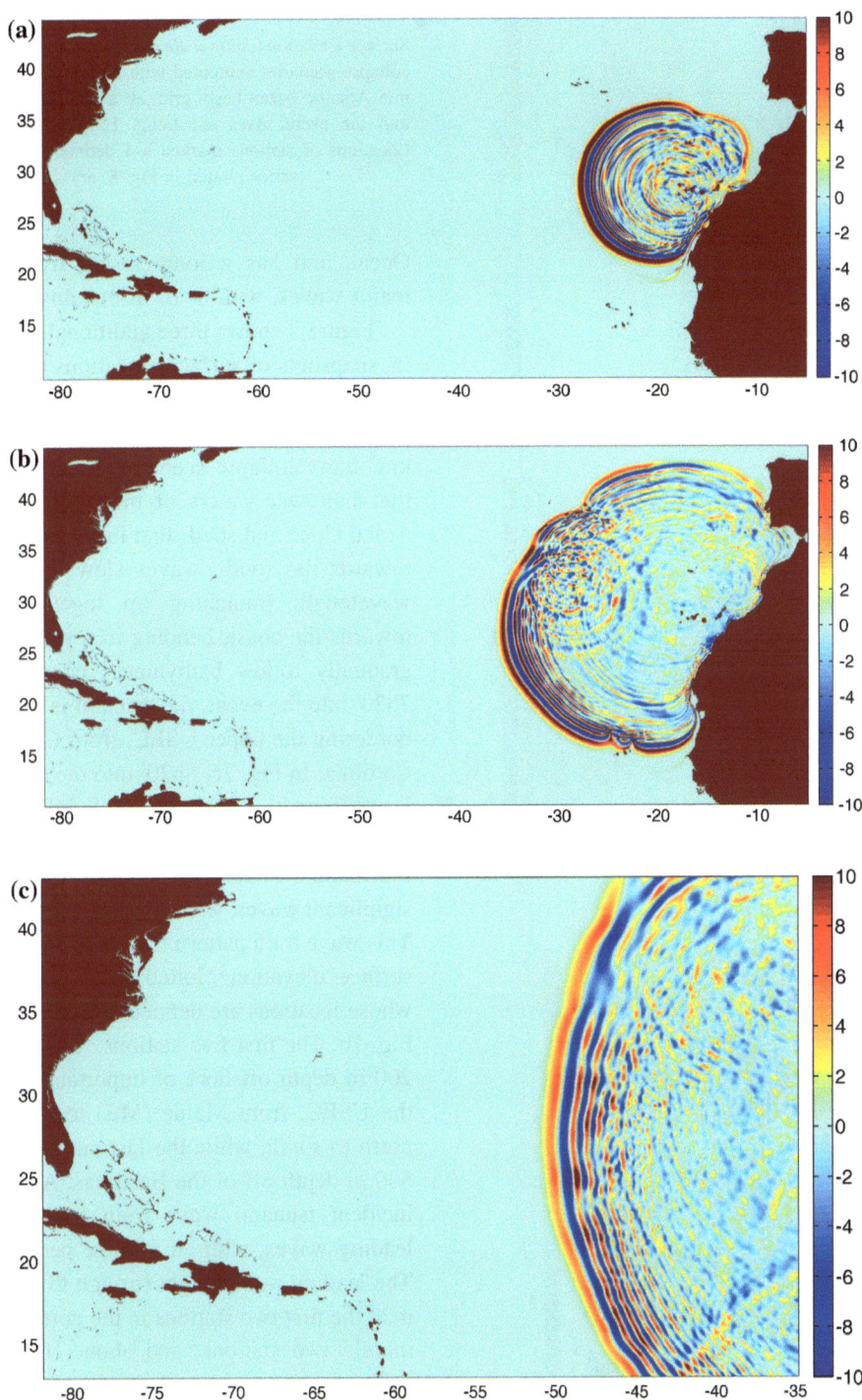

Figure 6
Surface elevation (*color scale* in meter) for the CVV 450 km³ flank collapse scenario, computed with FUNWAVE-TVD in the 1 arc-min Atlantic ocean basin grid, at: **a** 1h20′; **b** 2h20′; and **c** 4h20′, into the event. Axes are Long. E. (deg.) and Lat. N. (deg.)

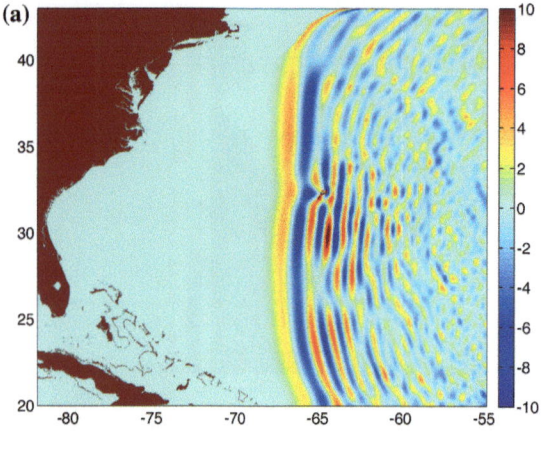

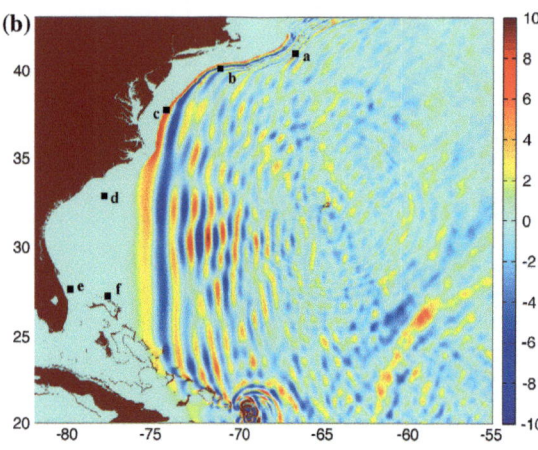

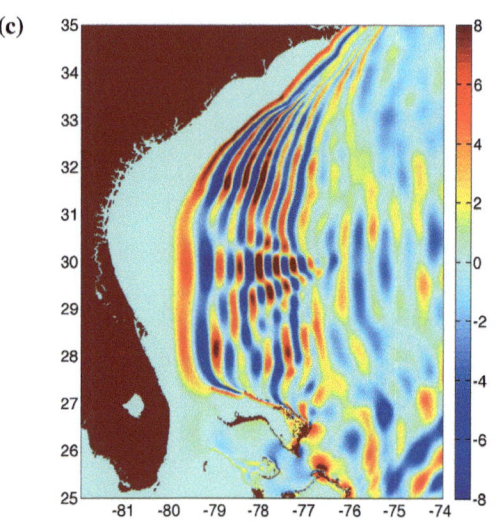

southern part of the western European coast and is about to enter the Mediterranean sea by the Strait of Gibraltar. In Fig. 6c, 4h20′ into the event, the tsunami has propagated about half-way into the Atlantic

Figure 7
Surface elevation (*color scale* in meter) for the CVV 450 km^3 flank collapse scenario, computed with FUNWAVE-TVD in the 1 arc-min Atlantic ocean basin grid, at: **a** 6h20′; **b** 7h20′; and **c** 8h20′, into the event. *Axes* are Long. E. (deg.) and Lat. N. (deg.). Locations of stations marked **a–f** defined in Table 2, with time series plotted in Fig. 8, are shown in **b**

Ocean and has a long-crested front made of 4–5 major waves, with 8–10 m maximum amplitude.

Figure 7 shows three additional, slightly zoomed-in, snapshots of surface elevations computed in the 1 arc-min grid, that detail the far-field tsunami propagation towards the US east coast, for the CVV 450 km^3 flank collapse scenario. As the tsunami reaches the shallower waters of the north American continental slope and shelf, first in the north and gradually towards the south, waves slow down, reduce their wavelength (bunching up together), and refract towards the coast, bending in a way that their crests gradually follow bathymetric contours. In Fig. 7b, 7h20′ into the event, the tsunami is reaching the shelf bordering the upper USEC, from Cape Cod to North Carolina. In Fig. 7c, 8h20′ into the event, the tsunami is reaching the shelf off of South Carolina and is approaching Florida and Georgia. We see that off of Florida, the tsunami dispersive tail has increased to 7 significant waves, with 6 to 8 m maximum amplitude. This wave train pattern is clearer on the time series of surface elevation plotted in Fig. 8, for 6 stations whose locations are defined in Table 2 and marked in Fig. 7b. The first five stations "a–e" are located in a 200 m depth offshore of important cities or areas of the USEC, from Maine (ME) to Florida (FL) from north to south, while the last one "f" is located in a 800 m depth off of the Bahamas. At stations a–e, the incident tsunami wave train has at least 3 large leading waves, with an average period of 9–12 min. The leading wave height (trough to crest) is about 10 m at the first two stations in the north, 18–21 m at the middle two stations, and about 10 m again at the southern station. At the last station f, in deeper water, the wave train is less organized, but the first three waves have an average period of about 13 min and the leading wave height is nearly 14 m.

Figure 9 shows the envelopes of maximum surface elevation computed in the 1 arc-min Atlantic Ocean basin grid, for the 80 km^3 and 450 km^3 CVV

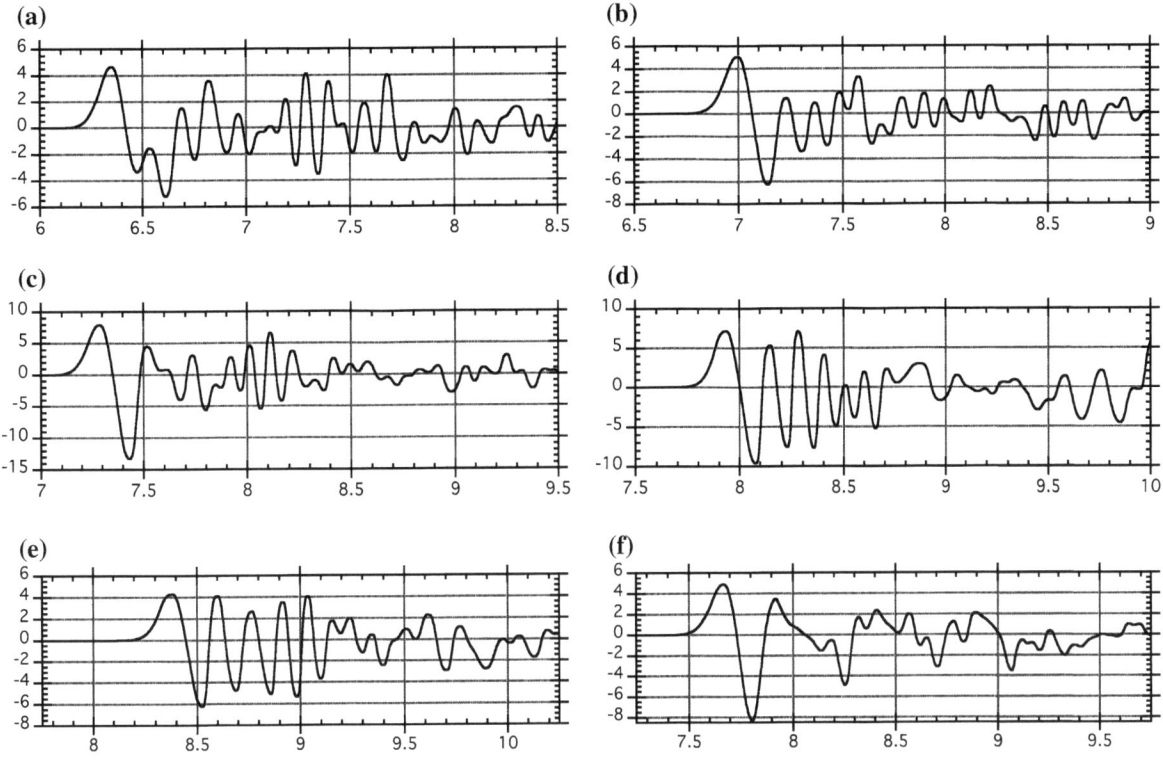

Figure 8
Time series of surface elevation (x-axis in hour and y-axis in meter) for the CVV 450 km³ flank collapse scenario, computed at stations a–f (see Table 2 and Fig. 7b for location) with FUNWAVE-TVD, in the 1 arc-min Atlantic Ocean basin grid

Table 2
Locations of stations used to compute time series of incident tsunami for CVV flank collapse scenarios shown in Fig. 8 (see Fig. 7b for location)

Location	Map index	Latitude (Deg. E)	Longitude (Deg. N)	Depth (m)
Offshore of NH, ME	a	−66.6318	40.9542	200
Offshore of NY, RI	b	−71.1429	40.0837	200
Offshore of NJ, MD	c	−74.3086	37.7094	200
Offshore of SC	d	−77.9096	32.8421	200
Offshore of FL	e	−79.8882	27.5791	200
Bahamas	f	−77.7118	27.1834	800

flank collapse scenarios. In both cases, we see a significant variation (alongshore modulation) in the maximum wave height in the far-field, resulting both from the directionality of the tsunami source (Fig. 5) and wave guiding and focusing/defocusing effects caused by refraction over the ocean bathymetry. The latter effects are particularly important near and over the continental shelf and will be further detailed in the next section. Note that similar strong wave guiding and focusing effects were observed during the deep water propagation of recent extreme tsunamis, for the 2004 Indian Ocean tsunami impact in Somalia, which distance-wise was as far from the earthquake source as the far-field areas considered here are from CVV (FRITZ and BORRERO 2006), and for the Tohoku 2011 tsunami in Crescent City, which was similarly far away from the source in the Japan Trench (GRILLI et al. 2013; KIRBY et al. 2013).

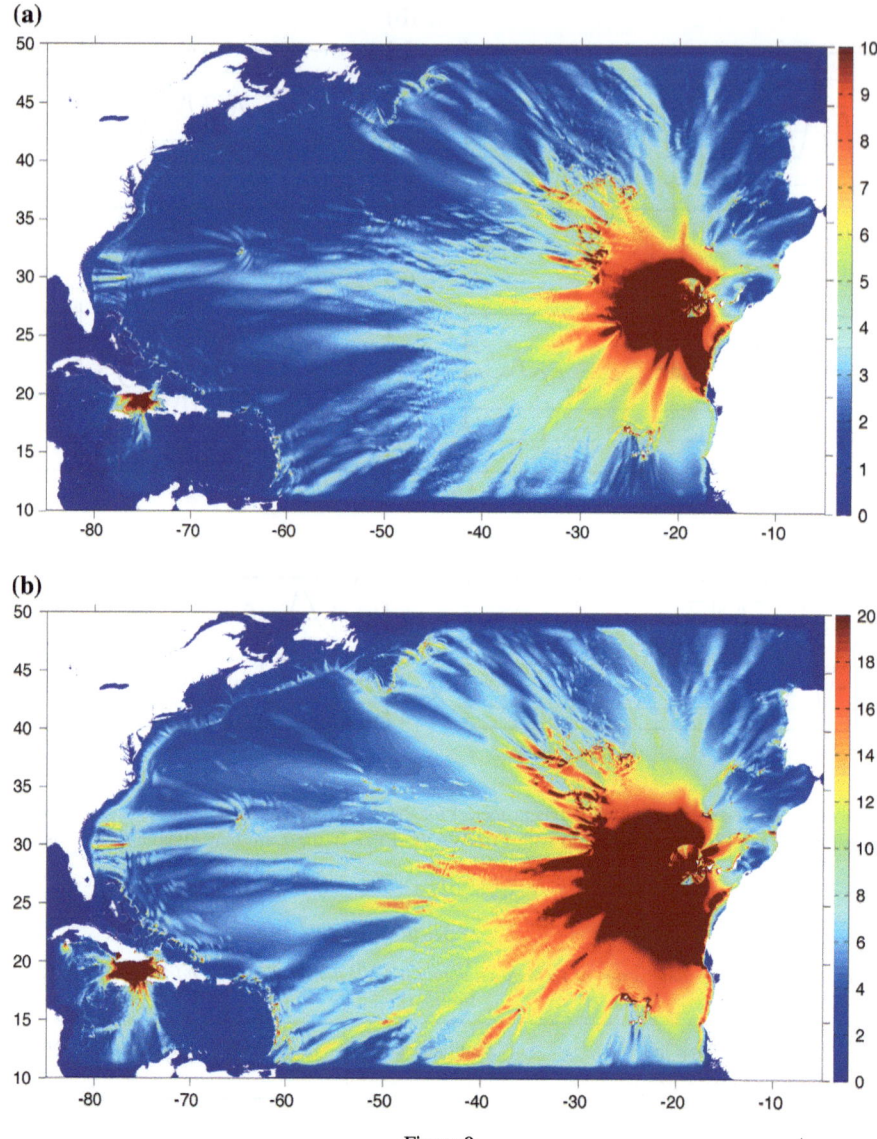

Figure 9
Maximum envelope of surface elevation (*color scale* in meter) computed with FUNWAVE-TVD in 1 arc-min Atlantic Ocean basin grid for the tsunami generated by the: **a** 80 km³; and **b** 450 km³, CVV flank collapse scenarios. *Axes* are Long. E. (deg.) and Lat. N. (deg.). [Note that both figures are plotted with a different *color scale*. Also note the large elevations, in both cases, west of Haiti and south of Cuba are an artifact of numerical simulations; more accurate simulations and details of the extreme CVV scenario impact in this area can be found in GRILLI *et al.* (2015a).]

Besides these modulations, along the USEC, the figures show first increasing maximum surface elevations towards the coast, due to wave shoaling over the continental slope and shelf break, and then decreasing surface elevations over the wide continental shelf, due to dissipation by bottom friction and, closer to shore, by breaking of the steepest waves; this wave decay will be clearer in results of simulations in the regional grids detailed in the next section.

More specifically, regarding variability of tsunami impact in the far-field, for both flank collapse scenarios, we see: (i) in the east, very large tsunami impact in western Africa and significant impact as well in Portugal; (ii) in the west-southwest, which is the sources' dominant direction, large waves propagating towards south America and the eastern Caribbean islands; (iii) in the northwest, large surface elevations in the Grand Banks, off of Newfoundland, but not closer to shore, likely as a result of significant dissipation from wave breaking and subsequent bottom friction over the shallow shelf; and (iv) in the west, at the scale of the plots, the largest waves occurring off of Florida, and these waves remaining large all the way to North Carolina.

5. Far-field Tsunami Features and Coastal Impact

5.1. Instantaneous and Maximum Surface Elevations

Based on the results in the 1 arc-min Atlantic Ocean basin grid, computations were performed by one-way coupling in three 20 arc-sec nested grids covering the entire USEC and one 15 arc-sec nested grid covering the west African and European coasts (Fig. 4; Table 1). The bathymetry and topography in these grids is shown in Fig. 10.

Results in regional grids along the USEC Figure 11a shows the instantaneous surface elevation computed for the extreme CVV 450 km^3 flank collapse scenario in the northern regional grid (N. US reg. grid; Table 1), at 8h10′ into the event, when the long-crested leading elevation wave, more than 5 m high at most locations, is about to impact (from west to east) Montauk, at the eastern extremity of Long Island, NY, Block Island, RI and the islands of Nantucket and Martha's Vineyard, off of Cape Cod, MA (see Fig. 16 for a definition of state borders along the USEC and locations of major coastal cities). One can clearly see that the leading and following few wave crest elevations are modulated along-crest, as a result of bathymetric focusing/defocusing (see Fig. 10a). This is clearer in Fig. 11b, which shows the envelope of maximum surface elevation computed in this grid up to 9h10′. Here, we see a significant alongshore modulation of the maximum surface elevation, as a series of cross-shore stripes, with at the same time a gradual decrease of the wave elevation towards the shore, due to wave breaking and bottom friction dissipations. More specifically, in this figure, waves are being refracted away from the Hudson River Canyon, off of New York City, which makes them focus on both northern New Jersey and Western Long Island. Further east, as observed before, surface elevations are larger off of Montauk and Nantucket, due to underwater ridges. To the south, focusing also occurs towards Atlantic City and Cape May, NJ.

Figures 12 and 13 show similar results for the middle and southern regional grids (Mid. and S. US reg. grids; Fig. 4; Table 1), respectively. In both grids, the incoming tsunami wave train also has a large leading long-crested elevation wave, followed by a train of several large waves. As before, the maximum surface elevation of the leading wave is significantly modulated alongshore but slightly less so off of the Chesapeake Bay. Overall, there is significant wave focusing towards the outer banks (Cape Hatteras; around 35.5 deg. N Lat.) and the Cape Fear area (near Wilmington around 34 deg. N Lat.) in North Carolina, and off of South Carolina (south of Charlestown; around 32 deg. N Lat.), with likely intense breaking nearshore in these locations, causing significant decay in wave elevation. Off of Florida, Fig. 13 shows a very strong alongshore modulation of the maximum surface elevations, with these being largest off of Orlando to Jacksonville (around 30 N Lat.). In all cases, we observe a strong cross-shore decay of surface elevations, which is initiated quite far offshore at around the shelf break (see Fig. 10c), This results from large energy dissipation over the wide shelf, but due to the fairly large depth where this strong decay is initiated, this may be due more to bottom friction than wave breaking dissipation. Energy dissipation in the simulations is analyzed in greater detail below.

Analysis of wave elevation decay in cross-shore transect Here, we verify that the model prediction of large cross-shore wave elevation decay over the shelf, seen in many instances of the above results, is caused by physical rather than numerical dissipation, particularly when water depth is too large for wave breaking to occur. Thus, we compared model results in a cross-shore transect off of Florida (Fig. 13d), where the incident offshore waves exceed 10 m, to an analytical formula based on linear wave theory, predicting the decay of wave amplitude $a(x)$ due to bottom friction, over the actual bathymetry of a

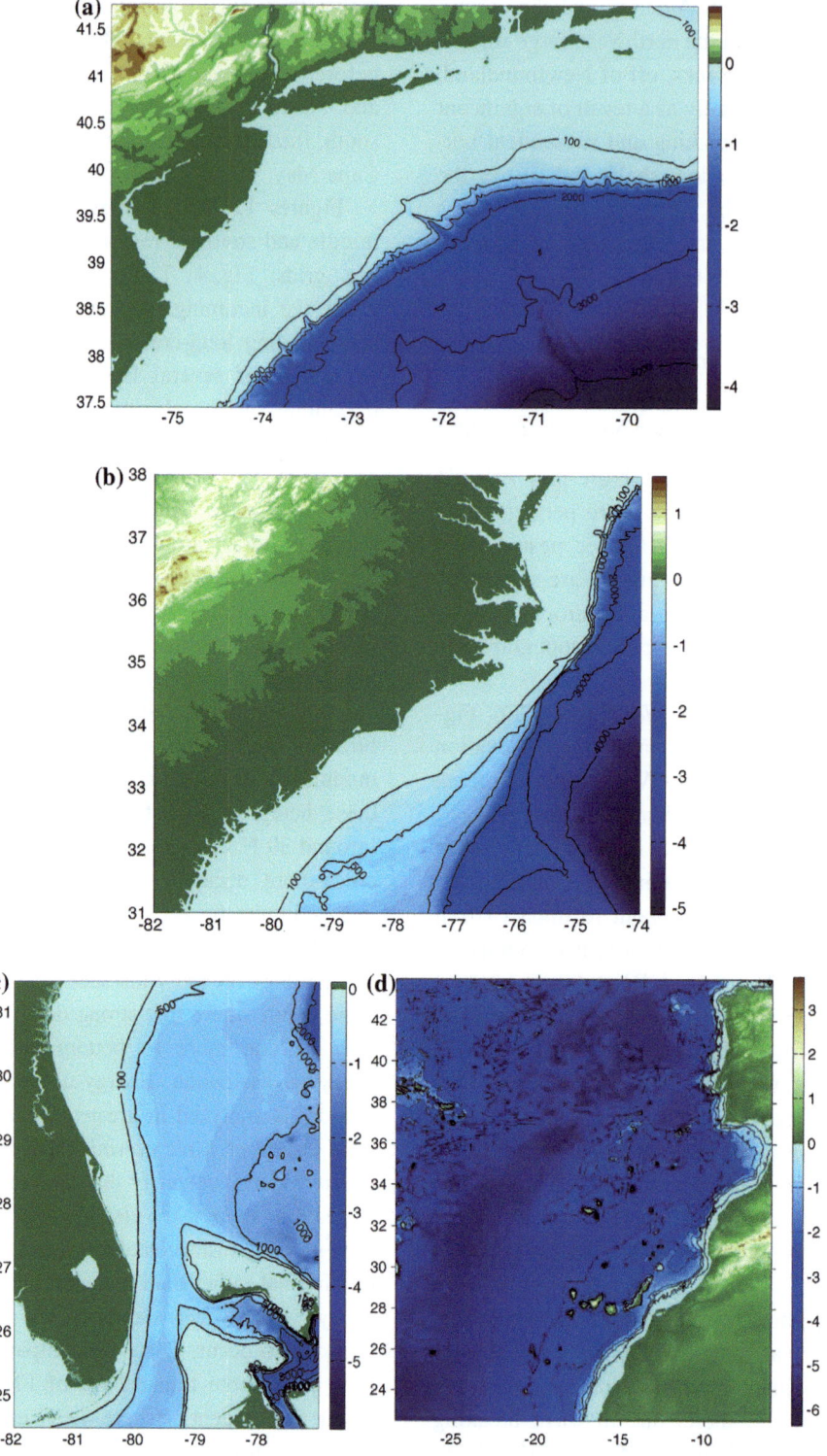

Figure 10
Bathymetry (<0), topography (>0) (*color scale* in km; *contours* in meter) in regional grids (see Fig. 4; Table 1): **a** N. US reg.; **b** Mid. US reg.; **c** S. US reg.; and **d** W. Africa and W. Europe (bathymetric contours are for the same depth sequence as other grids and are not labeled for clarity). *Axes* Long. E. and Lat. N. (deg.)

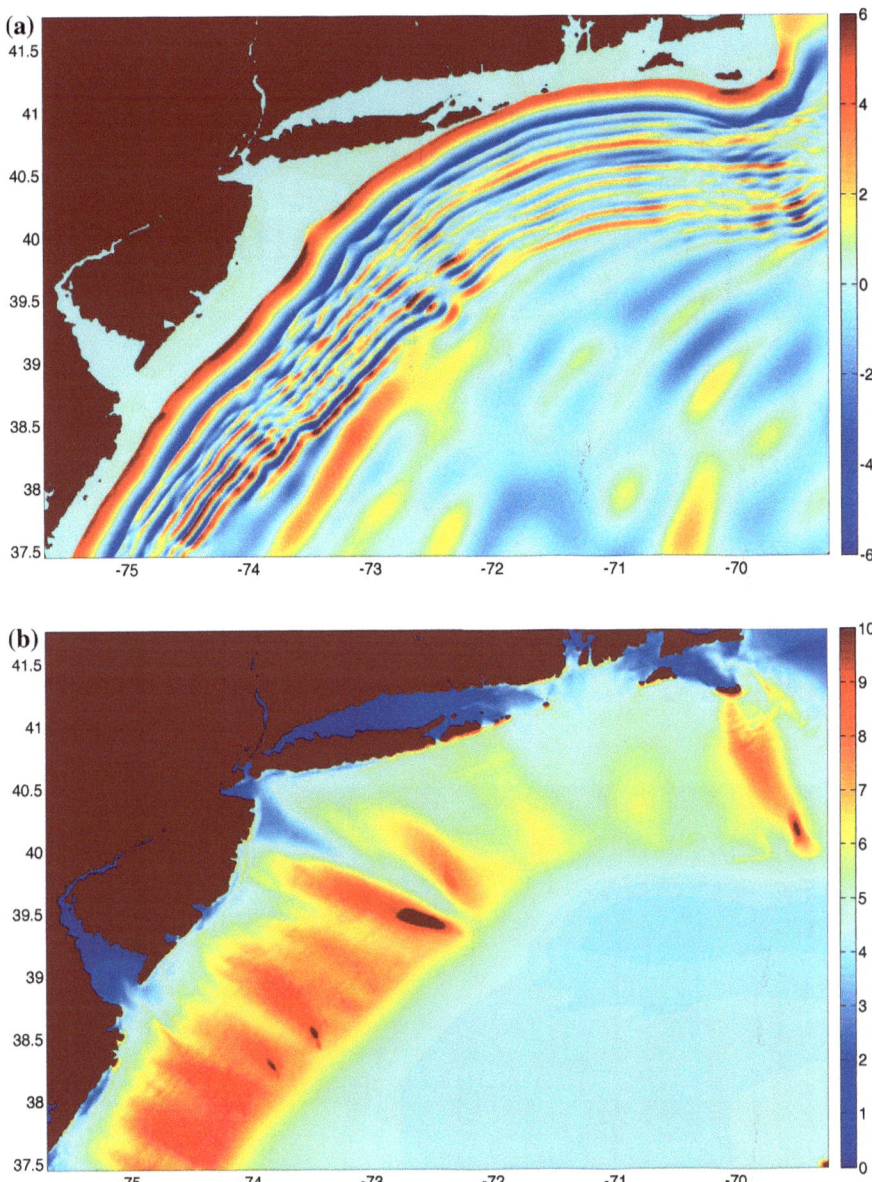

Figure 11
Surface elevation (*color scale* in meter) computed for the CVV 450 km³ flank collapse scenario, in 20 arc-sec N. US regional grid, off of New Jersey, New York, and Massachusetts (see Fig. 16 for locations of US states): **a** instantaneous elevation at 8h10′; and **b** maximum surface envelope up to 9h30′, after the start of the event. *Axes* are Long. E. (deg.) and Lat. N. (deg.)

transect $h(x)$ (<0) (DEAN and DALRYMPLE 1991). This formula expresses the dissipation, ϵ_D, by bottom friction, of energy flux, $E_f = E\,c_g$ (with $E = (1/2)\rho g a^2$ the period-averaged wave energy and c_g the group velocity), as a function of the cross-shelf distance x, i.e.,

$$\frac{dE_f}{dx} = -\epsilon_D \quad \text{with} \quad \epsilon_D = \frac{\rho f}{6\pi} u_{bm}^3 = \frac{\rho f}{6\pi}\left(\frac{a\omega}{\sinh kh}\right)^3 \quad (1)$$

where ρ is water density, g is the gravitational acceleration, k is the wavenumber, ω the wave

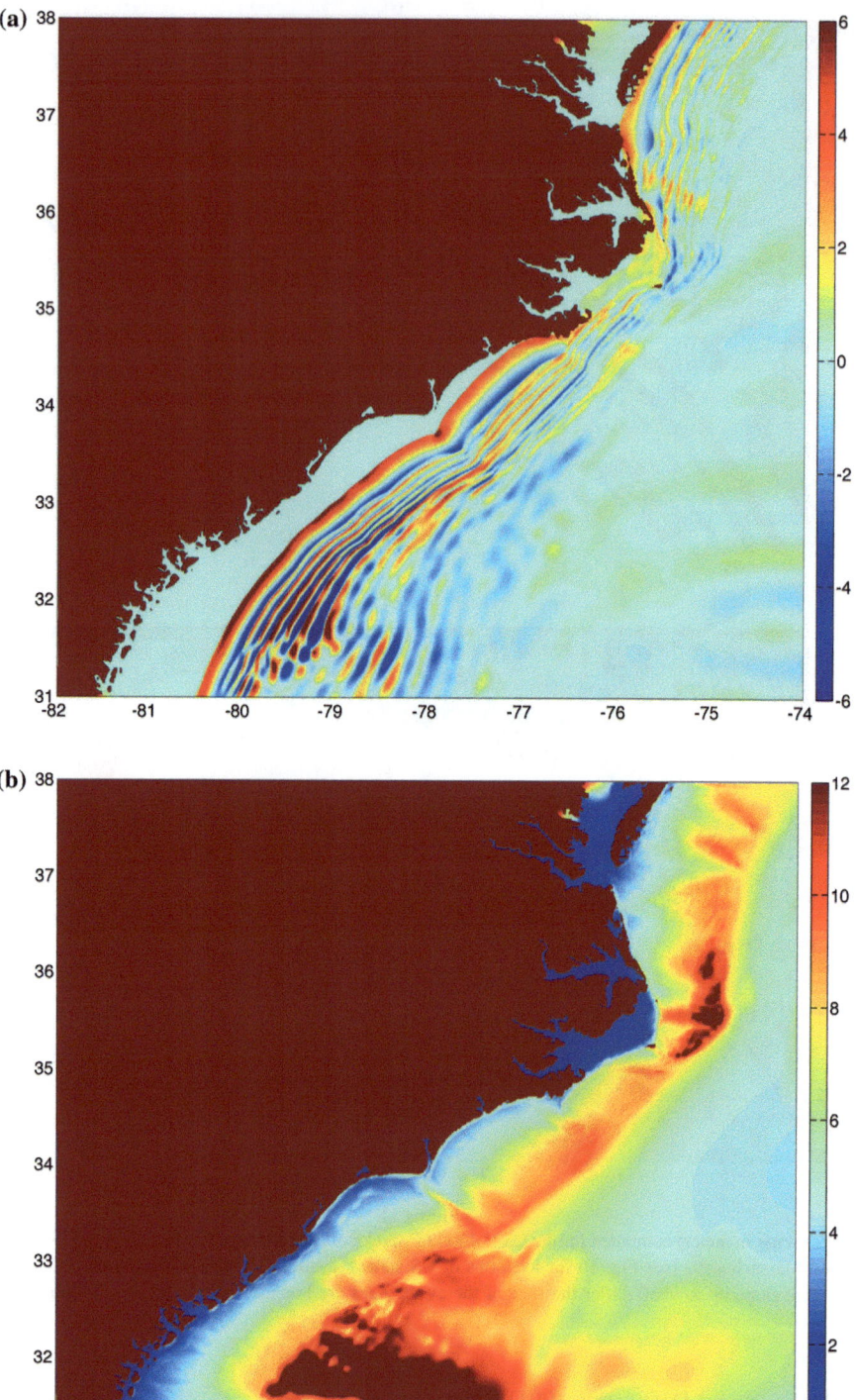

Figure 12
Surface elevation (*color scale* in meter) computed for the CVV 450 km³ flank collapse scenario, in 20 arc-sec Mid US regional grid, off of South/North Carolina, Virginia, Maryland and Delaware (see Fig. 16 for locations of US states): **a** instantaneous elevation at 8h55′; and **b** maximum surface envelope up to 10h40′, after the start of the event. *Axes* are Long. E. (deg.) and Lat. N. (deg.)

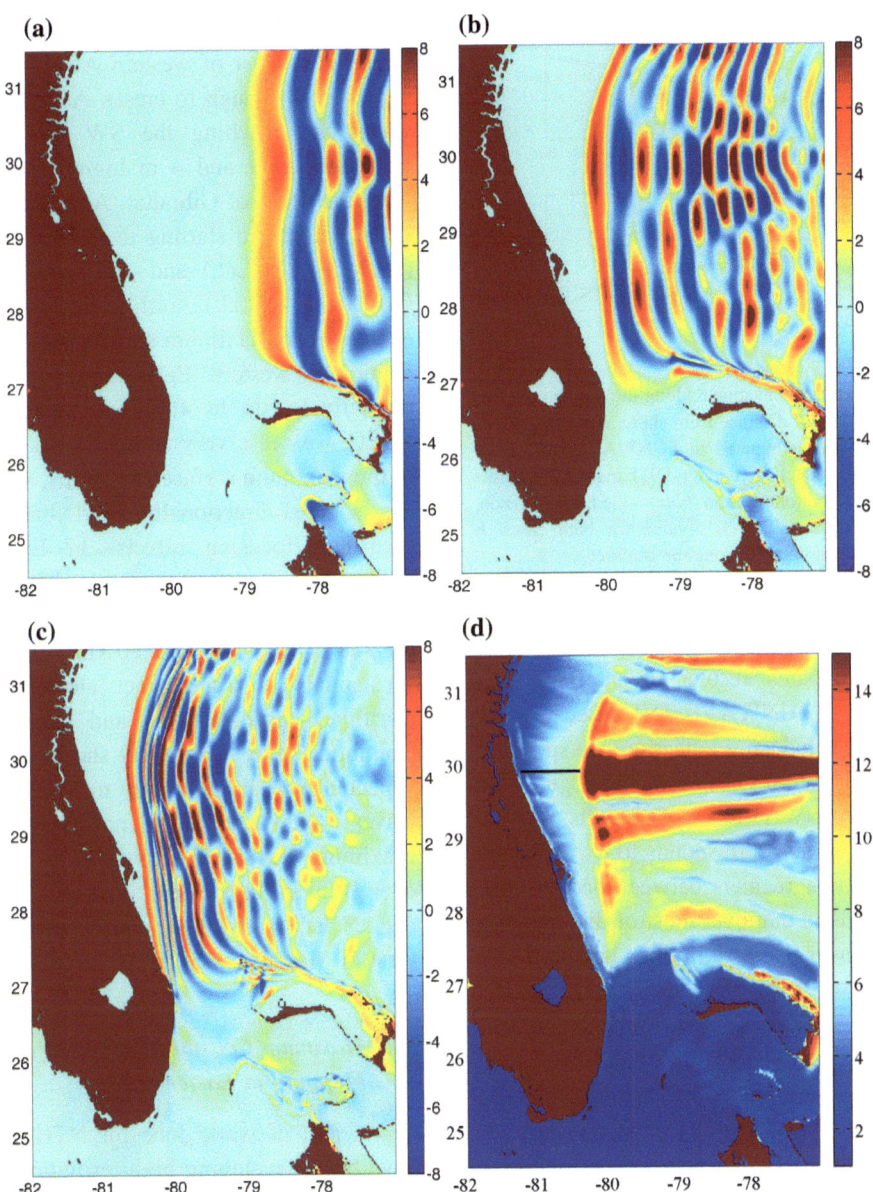

Figure 13
Surface elevation (*color scale* in meter) computed for the CVV 450 km³ flank collapse scenario, in 20 arc-sec S. US regional grid, off of Florida (see Fig. 16 for locations of US states): **a** instantaneous elevation at 7h55′; **b** instantaneous elevation at 8h25′; **c** instantaneous elevation at 8h55′; and **d** maximum surface envelope up to 10h40′, after the start of the event; the *black line* marks a cross-shore transect where dissipation is analyzed in more detail in Fig. 14. *Axes* are Long. E. (deg.) and Lat. N. (deg.)

angular frequency, and u_{bm} denotes the maximum horizontal velocity on the bottom. Assuming linear long waves (i.e., $c = \omega/k \simeq \sqrt{gh}$ and $\sinh kh \simeq kh$), and the friction coefficient $f = 8C_d$, which corresponds to FUNWAVE-TVD's parameterization of bottom friction, we find,

$$\frac{da}{dx} = -\frac{4C_d}{3\pi}\left(\frac{a}{h}\right)^2 + \frac{a}{4\,|\,h\,|}\frac{dh}{dx} \quad (2)$$

The latter formula has two terms in its right-hand-side, the first one representing bottom friction dissipation effects and the second one predicting

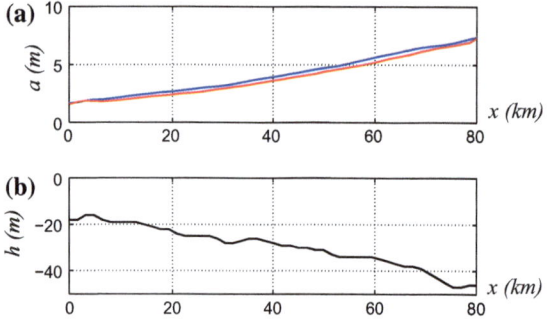

Figure 14
Analysis of decay of surface elevation a along the transect marked in Fig. 13d, as a function of distance from shore x: **a** Comparison between surface elevation computed in FUNWAVE-TVD (*red*) and predicted with DEAN and DALRYMPLE (1991) analytical formula including dissipation by bottom friction (*blue*)—the initial offshore wave elevation is FUNWAVE-TVD's value for both cases. **b** Variation of depth h along the transect

amplitude changes due to shoaling over a varying depth.

Figure 14 compares results of applying this formula to those of FUNWAVE-TVD. Because the formula does not include any reflection, we specified the initial wave elevation onshore of the edge of the shelf, based on FUNWAVE-TVD's results; for the transect of Fig. 13d, this corresponds to $a_0 = 7.5$ m. We observe a good agreement between both results. Moreover, we verified that over most of this transect, the wave height over depth ratio does not reach the breaking limit (set here to $H/h = 0.8$). Hence, bottom friction is the only source of energy dissipation in this cross-section. Because of the agreement between numerical and analytical results, we conclude that the model is behaving properly over the continental shelf, and that the large decay in wave elevation is caused by bottom friction rather than numerical dissipation.

Results in regional grid in western Europe and Africa Figure 15 shows the tsunami impact caused by the 80 km³ scenario in western Europe and western Africa, computed in the 15 arc-sec regional grid (W. Europe and W. Africa grid; Fig. 4; Table 1). Figure 15a–c show instantaneous surface elevations computed at 1h20′, 2h20′ and 2h50′ into the event. The first two snapshots are for the same times as in Fig. 6a, b for the 450 km³ scenario in the 1 arc-min grid; we see that wave patterns and phases appear to be similar, although wave amplitudes are smaller by a factor of 2.5 to 3. After 1h20′, the tsunami is about to impact the coast of western Africa with over 10 m high waves (trough to crest). After 2h20′, 6 m high waves are reaching the SW tip of the western European coast and 4 m high waves are about to enter the Strait of Gibraltar. After 2h50′ waves of 8–10 m height are starting to impact the Lisbon area (38.7 deg. N Lat.) and are approaching Coimbra (40.15 deg. N Lat.), north of it. Figure 15d, e show envelopes of maximum surface elevations computed along the western European and African coasts, respectively, up to 4h of simulations. As for the USEC, we see a very strong alongshore modulation of the maximum surface elevations, again due both to the source directionality and bathymetric wave focusing/defocusing effects. In Europe, although much of the tsunami energy is directed away from the continent (Figs. 9, 15b), Fig. 15d shows that, even for the more moderate 80 km³ scenario, there is substantial tsunami impact along the Portuguese coastline, particularly in and around Lisbon and Coimbra. Figure 15e finally shows that, as could be expected from its proximity to La Palma, very large waves would impact the northwest African coast. Maximum surface elevations along the coast reach over 10 m in the western Sahara in Morocco (25–27 deg. N lat.) and further north between Agadir and west of Marrakech (30.5–31.5 deg. N Lat.).

5.2. *Maximum Coastal Impact and Inundation Mapping in Eastern US*

As part of work done for NTHMP, the authors have been developing high-resolution tsunami inundation maps for the most critical or exposed areas of the USEC, by way of numerical simulations. These maps represent envelopes of maximum inundation caused by extreme near- and far-field tsunami sources in the Atlantic Ocean basin. The CVV 80 and 450 km³ flank collapse scenarios presented here are two of the sources considered in this NTHMP work. Initial inundation mapping efforts were based on the latter, most extreme CVV scenario, but more recently, in view of its likely very long return period as compared to other sources, it was decided in coordination with NTHMP leadership to base the USEC inundation mapping on the former ECWCS.

Far-Field Tsunami Impact in the North Atlantic Basin

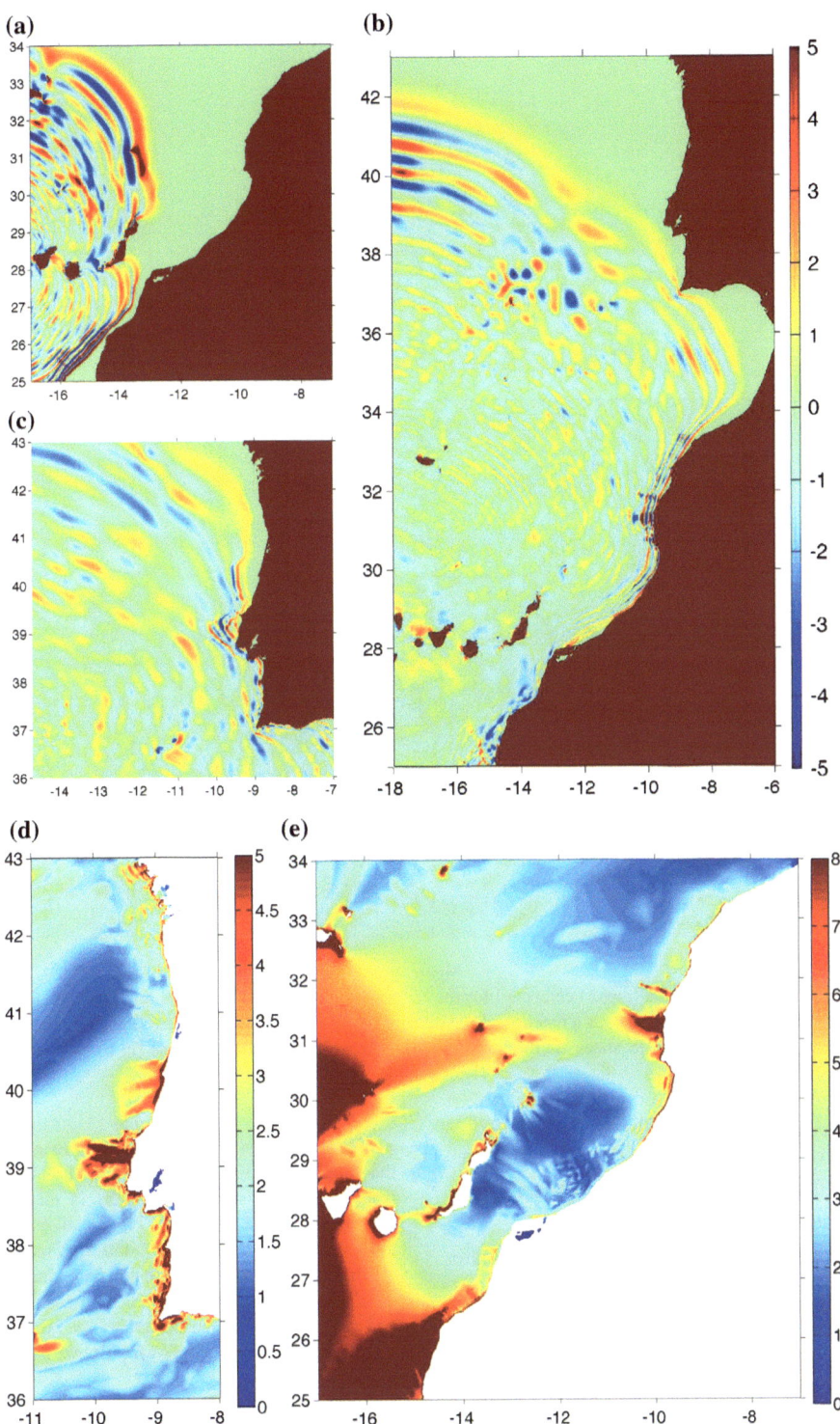

Figure 15
Surface elevation (*color scale* in meter) computed for the CVV 80 km³ flank collapse scenario, in 15 arc-sec regional grid, off of western Africa and western Europe (Fig. 4, Table 1). Instantaneous elevation at: **a** 1h20′; **b** 2h20′; and **d** 2h50′. Maximum surface envelope, up to 4h (**d**, **e**). Axes are Long. E. (deg.) and Lat. N. (deg.)

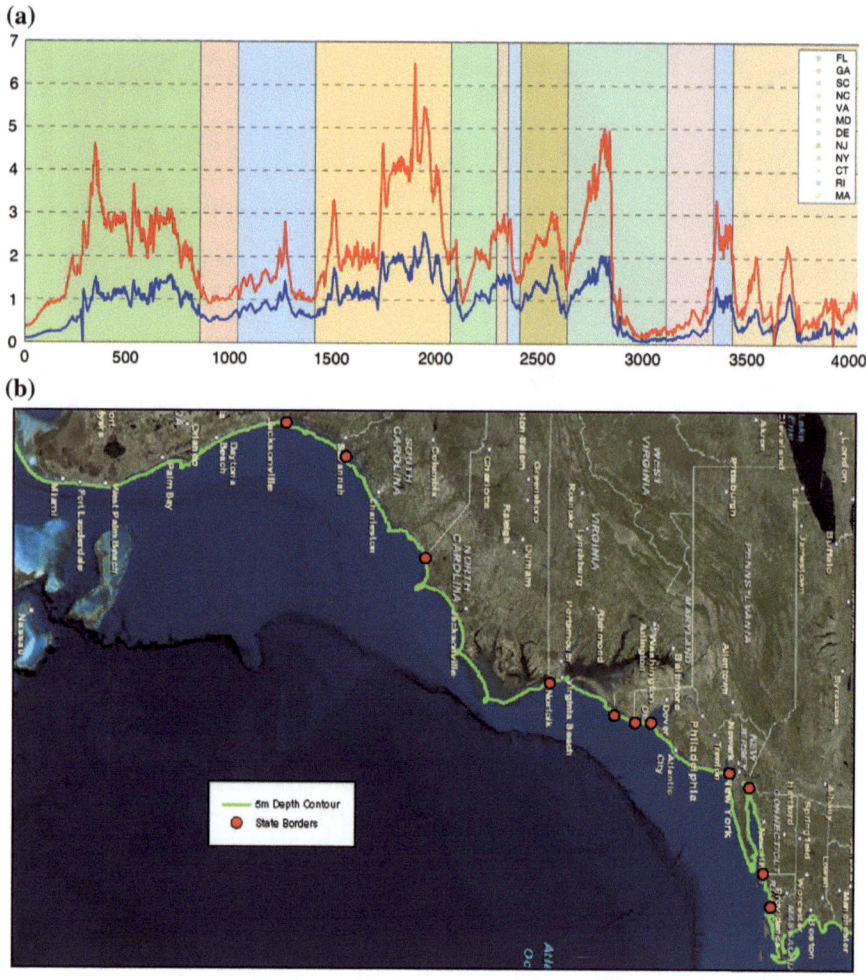

Figure 16
a Transects of maximum surface elevation (meter) computed for the 80 km³ (*blue*) and 450 km³ (*red*) scenarios, along a 5 m depth contour parallel to the US eastern coastline (show in **b**), as a function of the distance calculated along the contour (km), from south to north. Limits of the various US states are marked on both the transect **a** and the coastline (in **b**)

Other tsunami hazard assessment work, however, performed for critical coastal infrastructures such as power plants still considered the very conservative 450 km³ scenario.

Results of simulations discussed above indicate that a tsunami from the 450 km³ source would significantly impact the entire USEC and, particularly, the mid-Atlantic states and northern Florida. However, impact of a tsunami from the smaller 80 km³ scenario would also be quite large in many locations.

Surface elevation along the coast for two CVV scenarios Figure 16 compares maximum tsunami elevation simulated for the two CVV scenarios, along a 5 m depth contour parallel to the USEC (plotted as a function of the distance calculated along the contour, from south to north); limits of the various US states are marked on both the contour and the surface elevation plot. In general, surface elevations

Figure 17 ▶
a–c Wave rays (*solid red*) computed for long waves propagating in the Atlantic Ocean, from the CVV to the USEC (*color scale and contours are bathymetry/topography in meter*); the three panels correspond to grid areas from Fig. 10. **d, e** (*solid red*) Maximum surface elevation for the CVV 450 km³ scenario along the 5 m depth contour from Fig. 16 (see this figure for the definition of each state's *color code*): **d** normalized by maximum value, **e** in meter; and (*solid black*) ray concentration, (*dashed black*) wave height obtained from rays, following BOUWS and BATTJES (1982)

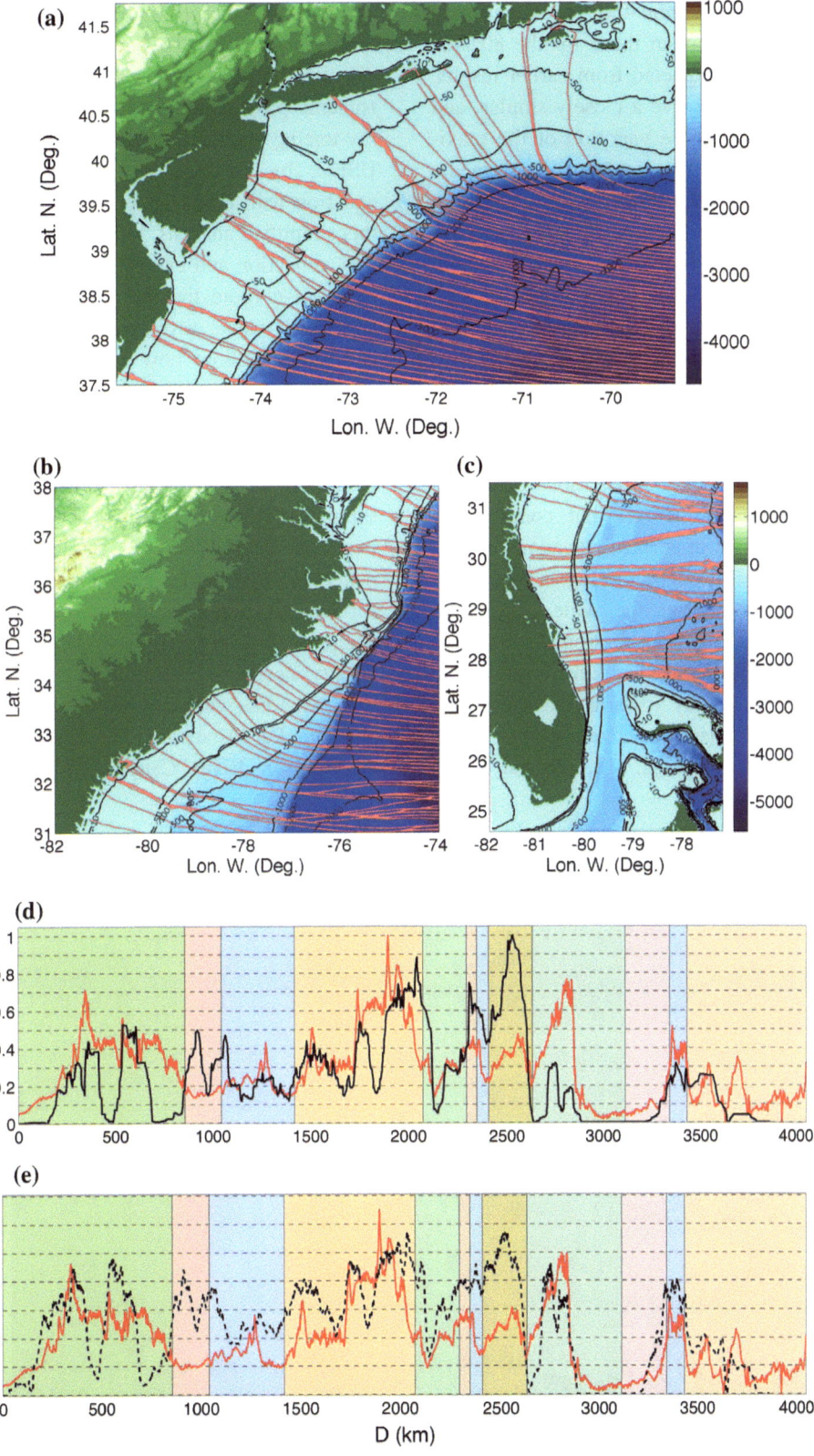

computed for the most extreme 450 km³ scenario are about 2.5–3 times larger than those for the 80 km³ ECWCS. However, as expected from earlier results, for both scenario, we observe a closely similar, and fairly significant, alongshore variation of the maximum surface elevation. This is due in part to the directionality of the CVV tsunami sources (Figs. 3, 9), but mostly to focusing and de-focusing of wave energy flux caused by refraction over the nearshore bathymetry (e.g., succession of canyons and ridges). This important feature of wave propagation over the wide USEC shelf is further discussed below. As also seen in previous results, maximum tsunami impact occurs around the mid-Atlantic states, particularly in North Carolina, but also in New York and Florida. Looking at maximum surface elevations computed for the 450 km³ scenario, we see that, compared to the incident time series of surface elevations plotted at the 200 m bathymetric contour in Fig. 8, these have significantly decreased over the wide continental shelf, due essentially to bottom friction (see Fig. 14) and some local effects of wave breaking, dissipations. This wave elevation decay towards the shore was apparent in plots of maximum envelopes of surface elevations detailed above (Figs. 11b, 12b and 13d).

Analysis of bathymetric control on nearshore wave propagation To better assess the controlling effect of the wide shelf bathymetry on coastal tsunami propagation and impact, we performed a ray-tracing analysis from the CVV source to the USEC, by solving the geometric optic eikonal equation with a fast-marching algorithm. Results of this simplified analysis, which neglects wave diffraction, reflection and energy dissipation, are plotted in Fig. 17a–c (note that for clarity, only 1 % of the computed wave rays were plotted). These figures clearly show how incident long waves start refracting in great depth and gradually bend over the shelf in a manner that the rays eventually become nearly orthogonal to the nearshore bathymetry. In doing so, areas of ray convergence occur over submarine ridges (or equivalent) and ray divergence over submarine canyons, that closely match the patterns of low and high values of coastal tsunami surface elevation observed in Fig. 16. This is because where ray convergence occurs, the wave energy flux density increases, leading to increased surface elevations, whereas where ray divergence occurs, it is the opposite. For instance, in Fig. 17a we see that, due to the Hudson River Canyon V-shaped bathymetry,

Figure 18
Aerial view of Ocean City, MD, illustrating the vulnerability to inundation of the heavily developed barrier island

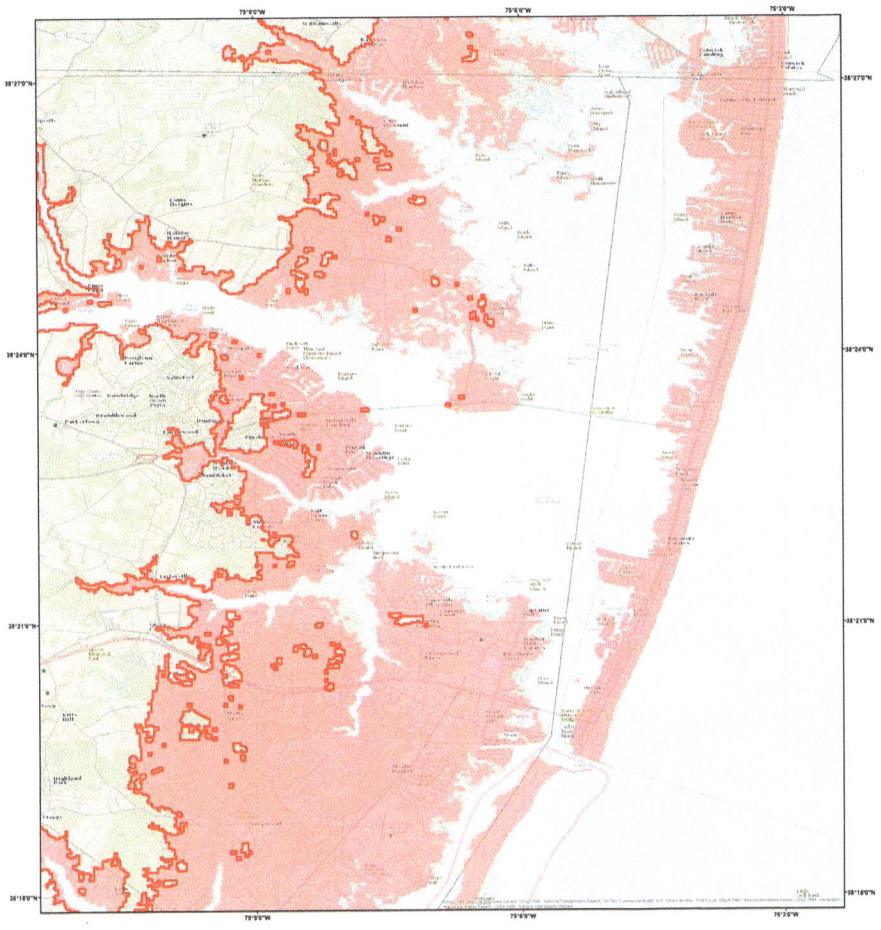

Figure 19
Tsunami inundation map (10 m resolution grid) for the 450 km³ CVV flank collapse case (ABADIE *et al.* 2012): inundation limit (*thick red line*); inundated area (*pink area*)

tsunami wave rays refract towards the northern parts of New Jersey and the western half of Long Island, NY. Thus, this bathymetric feature mitigates tsunami impact on New York harbor. A similar behavior is observed in Fig. 17a over the Delaware Bay Canyon, where tsunami wave rays refract towards Atlantic City, NJ in the north and Ocean City, MD in the south. Figure 17b and c show similar results for areas further south along the USEC.

We expect these wave ray patterns to be closely related to the alongshore variation of maximum surface elevation shown in Fig. 16, particularly in areas where energy dissipation over the shelf is not too large. This was verified (and quantified) two ways for the CVV 450 km³ scenario, along the 5 m depth contour: (i) in Fig. 17d, we compare the normalized number of wave rays that intersect a circle of given radius (here we used a 0.1° radius or about 10 km), at many equally spaced locations, versus the normalized surface elevation computed from Fig. 16; and (ii) in Fig. 17e, we compare the surface elevation computed based on the wave rays with the method of BOUWS and BATTJES (1982), to the surface elevation computed from Fig. 16; this method is based on the conservation of energy flux between the rays' initial and end points. The alongshore variation of the two different metrics based on wave rays is in general in reasonable agreement with the surface elevations computed with FUNWAVE-TVD, but more particularly so north of New Jersey, in south Florida and the Corolinas (except around Cape Hatteras, NC), where we see a higher correlation. In other areas, although

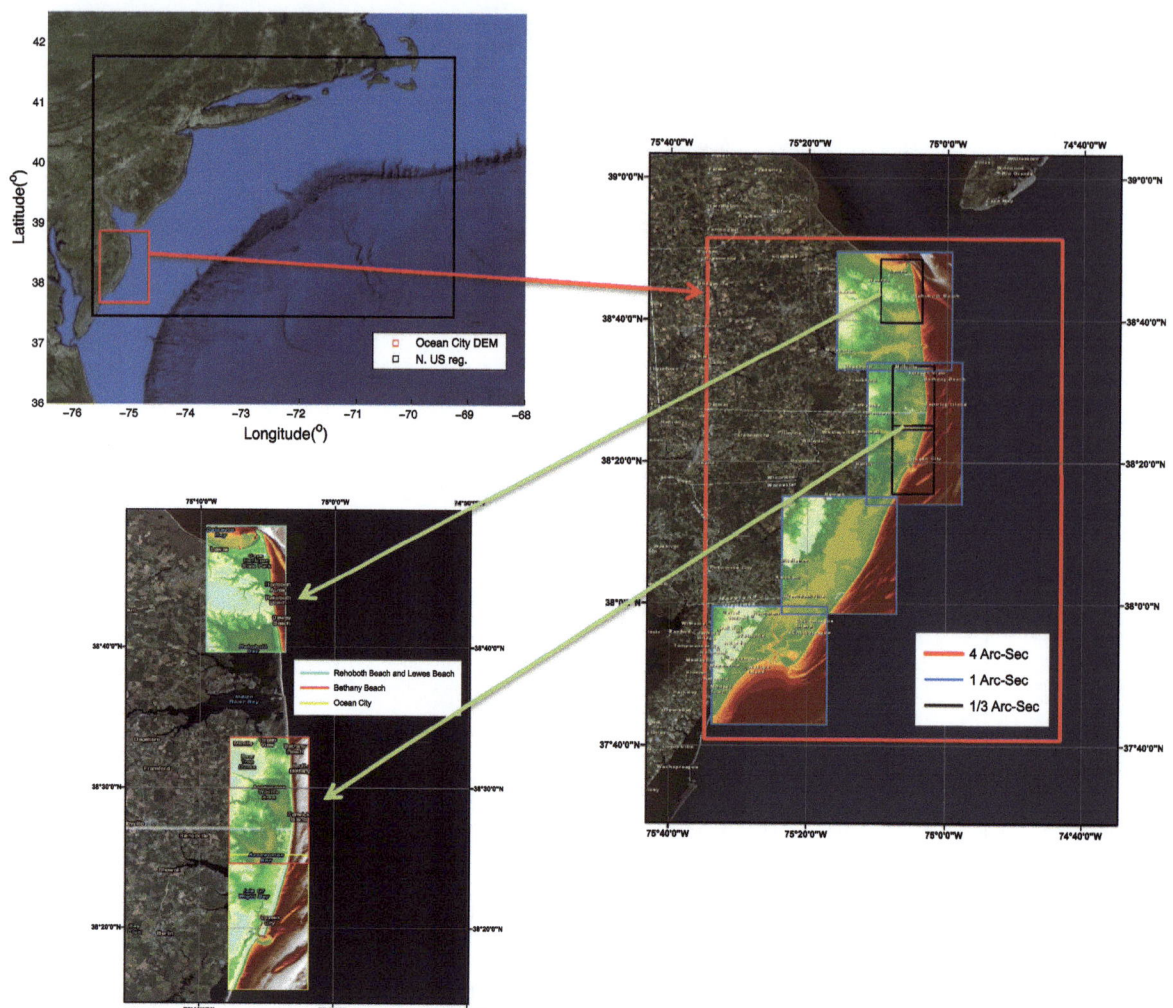

Figure 20
Nested FUNWAVE-TVD grids used for developing high-resolution inundation maps for the Ocean City, MD area. N. US reg. grid (20 arc-sec resolution) is forced along its boundary by the far-field tsunami sources simulated in the 1 arc-min ocean basin grid (Fig. 4). The *red box* marks the boundary of the "Ocean City DEM" 4 arc-sec (about 125 m) resolution grid. OC-1 to OC-4 are 1 arc-sec (about 30 m) resolution grids nested within this grid, and a few finer 1/3 arc-sec (about 10 m) resolution grids are nested within those, in areas of greatest interest

the trend is reasonably well predicted by the ray analysis, wave heights are typically overestimated due to neglecting dissipation from bottom friction and breaking. Nevertheless, these results confirm the controlling effect of a wide continental shelf, and of the refraction it induces on incoming tsunamis, for surface elevations at the coast.

Moreover, because long wave refraction patterns are only bathymetry (and not frequency) dependent, any incoming tsunami should refract in a similar manner, once close to shore, in shallow enough water. Hence tsunamis caused by different CVV flank collapse mechanisms or volumes should have similar nearshore areas of ray convergence or divergence, and thus enhanced or reduced wave elevation, whatever their incidence angle. This is supported by the similar alongshore variations of the maximum surface elevations shown in Fig. 16 for the CVV 80 and 450 km^3 scenarios.

Detailed inundation mapping in Ocean City MD
In the mid-Atlantic region, as is apparent on Fig. 18, the resort town of Ocean City, MD (at the southern limit of DE on Fig. 16), is particularly vulnerable to coastal flooding, from both hurricanes and potentially

tsunamis, because it is made of heavily developed low-elevation barrier beaches and islands, which also pose significant evacuation problems. Although this aspect of our work will not be presented exhaustively here, Fig. 19 shows an example of a high-resolution tsunami inundation map prepared for Ocean City based on the CVV 450 km^3 extreme flank collapse scenario, which could be used for emergency management purpose. This map shows the extent of the flooded area and the maximum penetration of the tsunami, computed in a 10 m resolution grid, based on results of the 20 arc-sec Mid US regional grid (Fig. 12), by one-way coupling in a series of additional nested grids with 4 arc-sec (125 m), 30 m and 10 m resolution, shown in Fig. 20. Other flow parameters such as maximum velocity and vorticity, and momentum flux based on flow depth, can be similarly plotted and used for estimating tsunami effects on navigation, coastal erosion, and forces on structures. Examples of maps generated in the same area for such metrics, for hazard resulting from near-field underwater landslide sources, can be found in GRILLI et al. (2015b).

6. Conclusions

We simulated tsunami generation, propagation, and far-field impact in the Atlantic Ocean Basin, for two extreme flank collapse scenarios of the Cumbre Vieja Volcano (CVV) in La Palma (Canary Islands), defined by ABADIE et al. (2012): (i) the ECWCS (based on a slope stability analysis), with a 80 km^3 volume; and (ii) the most extreme scenario with a 450 km^3 volume. Both of these scenarios can be qualified as being very conservative and although their return period is unknown, earlier geological studies indicate that massive CVV flank collapses are associated with return periods on the order of 100,000 years or more.

For each scenario, slide motion and the resulting initial tsunami generation were computed by ABADIE et al. (2012) using THETIS, a 3D multi-fluid Navier-Stokes model with a VOF interface tracking; the motion of the subaerial slide material was modeled as a heavy, nearly inviscid, fluid with a density similar to that of basalt. At 5 min into the event, when the energy from the slide has been transferred to the water motion, the wave elevation and velocity for the generated tsunami are used to initialize FUNWAVE-TVD, a 2D (horizontal) fully nonlinear Boussinesq long wave model with extended dispersion properties, to continue simulating their propagation in the near-field, around the Canary islands, in a 500 m resolution Cartesian grid. Then, at 20 min into the event, surface elevation and velocity computed in this grid are again specified as initial condition for a 1 arc-min Atlantic Ocean Basin grid, to compute the transoceanic tsunami propagation and far-field impact. Coastal tsunami impact in the far-field is finally computed by one-way coupling in a series of finer regional nested grids, with 15–20 arc-sec resolution and, in some areas of the USEC, in additional Cartesian nested grids down to a 10 m resolution. Details of the selection of slide scenarios, modeling of 3D slide motion and tsunami source generation with THETIS, and analysis of results around La Palma and in the near-field can be found in ABADIE et al. (2012).

Our overall findings regarding near- and far-field wave generation are qualitatively consistent with earlier results (GISLER et al. 2006; LØVHOLT et al. 2008), but near-field waves computed in this study appear to be notably higher than in these earlier works and attenuation rates during their transoceanic propagation smaller. In the far-field, the incoming tsunamis appear to be made of wave trains of 3–5 significantly large (long-crested) waves of 9- to 12-min period. This pattern confirms the importance of using a dispersive long wave model such as FUNWAVE-TVD to simulate landslide generated waves, which have relatively shorter wavelengths than coseismic tsunamis; this conclusion was also reached in a number of earlier landslide tsunami studies by these and other authors (see earlier discussions in the paper).

Along the US east coast, for the most extreme CVV scenario, while wave heights at the 200 m isobath are in the 10–20 m range (trough to crest), because of significant decay in wave height due to energy dissipation over the wide shelf, the maximum nearshore surface elevations along the coast (at the 5 m isobath) become significantly less, in the 1–6 m range, and less than 3 m at most locations, except in a few areas, such as off of West Palm Beach (FL),

Cape Hatteras and the outer Banks (NC), and Ocean City (MD). For the 80 km³ CVV ECWCS, maximum surface elevations are less than 2 m along the coast. A more detailed analysis of energy dissipation over the wide shelf indicates that wave elevation decay is essentially due to bottom friction, with dissipation from wave breaking being only locally important. Hence the computed tsunami coastal inundation and impact would be sensitive to the assumed value of the bottom friction coefficient; here, we used a uniformly low value, corresponding to coarse sand, which should ensure conservative results as far as coastal flooding. More analyses regarding this aspect will be left out for future work.

For both CVV scenarios, we observe a significant alongshore modulation of the maximum computed flow depth near the shore, following a closely similar pattern. This results in part from the source directionality, but more importantly from bathymetric wave focusing and defocusing effects over the wide continental shelf, which become most important nearshore, in shallower water. The controlling effect of the continental shelf on the alongshore pattern of surface elevations was confirmed by performing a simplified ray-tracing analysis. As a corollary, this bathymetric control implies that the long waves of any tsunami will essentially refract in a similar way to shore, whatever their initial angle of incidence. Here, this means that tsunamis generated from different CVV flank collapse mechanisms and volumes should focus their impact on the USEC in a similar manner, with areas of enhanced coastal tsunami impact and flow depth being in large part independent from details of the initial tsunami source.

Additional simulations in regional grids show that, along the western European and African coasts, the impact from both CVV scenarios would be much larger than on the USEC, because of the proximity to La Palma. For the smaller 80 km³ source, results indicate that, after 1h30′, western Sahara and NW Morocco would be impacted by over 10 m surface elevations and, 2h30′ to 3h into the event, the areas of Lisbon and Coimbra in Portugal would face over 5 m surface elevations. Similar to the USEC, the coastal impact from the most extreme CVV source would be even more dire.

Tsunami coastal impacts mentioned above, however, were computed in still rather coarse regional grids, (with 15–20 arc-sec resolution), and more accurate and detailed inundation simulations must be conducted in finer nested grids to accurately assess site specific coastal tsunami threat for the most impacted, exposed, or vulnerable areas. Such detailed inundation mapping is underway for most of the USEC, under the auspices of the US NTHMP and, in this paper, we show one example of a high-resolution tsunami inundation map computed for Ocean City (MD), based on the most extreme CVV flank collapse scenario; in this low laying area, we find that the extent of inundation and tsunami penetration would be very large. Other maps are currently being prepared for similarly highly impacted areas, based on these CVV flank collapse scenarios and other relevant extreme tsunami sources in the Atlantic Ocean Basin. Such maps can be used to both assess and mitigate the impact of future large tsunamis, through appropriate coastal development, and education and training of the populations regarding evacuation routes and sites.

Acknowledgments

Partial funding for this work was provided by grant #NA10NMS4670010 of the National Tsunami Hazards Mitigation Program (NTHMP), grant #EAR-09-11499 of the US National Science Foundation, and grant #037058 of the European Commission.

REFERENCES

ABADIE S., CALTAGIRONE J.P. and P. WATREMEZ, *Splash-up generation in a plunging breaker*. Comptes Rendus de l'Académie des Sciences, Ser. IIB, *326*:553559, 1998.

ABADIE S., MORICHON D., GRILLI S.T. and S. GLOCKNER, *Splash-up generation in a plunging breaker*. La Houille Blanche, *1* (Feb. 2008):21–26, doi:10.1051/lhb:2008001, 2008.

ABADIE S., MORICHON D., GRILLI S.T. and S. GLOCKNER, *Numerical simulation of waves generated by landslides using a multiple-fluid Navier-Stokes model*. Coastal Engineering, 57:779–794, doi:10.1016/j.coastaleng.2010.03.003, 2010.

ABADIE S., HARRIS J.C., GRILLI S.T. and R. FABRE, *Numerical modeling of tsunami waves generated by the flank collapse of the*

Cumbre Vieja Volcano (La Palma, Canary Islands) : tsunami source and near field effects. Journal of Geophysical Research, 117:C05030, doi:10.1029/2011JC007646, 2012.

BARKAN, R., TEN BRICK, U.S. and LIN, J. *Far field tsunami simulations of the 1755 Lisbon earthquake:Implication for tsunami hazard to the U.S. East Coast and the Caribbean.* Marine Geology, 264:109–122, 2009.

BOUWS, E., and J. A. BATTJES, *A Monte Carlo approach to the computation of refraction of water waves.* Journal of Geophysical Research: Oceans, 87(C8): 5718–5722, 1982.

CARRACEDO J., DAY S., GUILLO, H. and P. GRAVESTOCK, *Later stage of volcanic evolution of La Palma, Canary Islands: rift evolution, giant landslides, and the genesis of the Caldera de Taburiente.* Bulletin of the Geological Society of America, 111:755–768, 1999.

COCHONAT P., LENAT J. F., BACHELERY P., BOIVIN P., CORNIGLIA B., DENIEL C., LABAZUY P., LIPMAN P., OILIER G., SAVOYE B., VINCENT P. and M. VOISSET, *Importance des dépôts gravitaires dans la mise en place d'un système volcano-sédimentaire sous-marin (Volcan de la Fournaise, Ile de la Réunion).* Comptes Rendus de l' Académie des Sciences, Ser. IIB., 311:679–686, 1990.

DAYS S.J., WATTS P., GRILLI S.T. and J.T. KIRBY, *Mechanical models of the 1975 Kalapana, Hawaii earthquake and tsunami.* Marine Geology, 215:59–92, doi:10.1016/j.margeo.2004.11.008, 2005.

DEAN, R. G., and DALRYMPLE, R. A., *Water wave mechanics for engineers and scientists.* World Scientific, Advanced Series on Ocean Engineering, Prentice-Hall, 1991.

FRITZ, H.M. and J.C. BORRERO, *Somalia field survey of the 2004 Indian Ocean Tsunami.* Earthquake Spectra 22(S3):S219–S233, 2006.

GEIST E., P. LYNETT, and J. CHAYTOR, *Hydrodynamic modeling of tsunamis from the Currituck landslide.* Marine Geology, 264:41–52, doi:10.1016/j.margeo.2008.09.005, 2009.

GISLER G., WEAVER R. and M. GITTINGS, *SAGE calculations of the tsunami threat from La Palma.* Science of Tsunami Hazards, 24:288–301, 2006.

GLIMSDAL S., PEDERSEN G.K., HARBITZ C.B., and LØVHOLT F., *Dispersion of tsunamis: does it really matter ?* Nat. Hazards Earth Syst. Sci., 13:1507–1526, doi:10.5194/nhess-13-1507-2013, 2013.

GRILLI, A.R., GRILLI S.T., DAVID, E. and C. COULET, Modeling of tsunami propagation in the Atlantic Ocean Basin for tsunami hazard assessment along the North Shore of Hispaniola. In Proc. 25th Offshore and Polar Engng. Conf. (ISOPE15, Kona, HI, USA. June 21–26, 2015). Intl. Society of Offshore and Polar Engng., pps. 733–740, 2015a.

GRILLI S.T., DUBOSQ S., POPHET N., PÉRIGNON Y., KIRBY J.T., and F. SHI, *Numerical simulation and first-order hazard analysis of large co-seismic tsunamis generated in the Puerto Rico trench: near-field impact on the North shore of Puerto Rico and far-field impact on the US East Coast.* Natural Hazards and Earth System Sciences, 10:2109–2125, doi:10.5194/nhess-2109-2010, 2010.

GRILLI, S.T., HARRIS, J., F. SHI, J.T. KIRBY, T.S. TAJALLI BAKHSH, E. ESTIBALS and B. TEHRANIRAD, Numerical modeling of coastal tsunami dissipation and impact. In *Proc. 33rd Intl. Coastal Engng. Conf.* (P. Lynett and J. Mc Kee Smith, eds.) (ICCE12, Santander, Spain, July, 2012), 12 pps. World Scientific Publishing Co. Pte, 2012.

GRILLI, S.T., J.C. HARRIS, T. TAJALI-BAKHSH, T.L. MASTERLARK, C. KYRIAKOPOULOS, J.T. KIRBY and F. SHI, *Numerical simulation of the 2011 Tohoku tsunami based on a new transient FEM co-seismic source: Comparison to far- and near-field observations.*

Pure and Applied Geophysics, 170:1333–1359, doi:10.1007/s00024-012-0528-y, 2013.

GRILLI S.T., O'REILLY C., HARRIS J.C., TAJALLI-BAKHSH T., TEHRANIRAD B., BANIHASHEMI S., KIRBY J.T., BAXTER C.D.P., EGGELING T., MA G. and F. SHI, *Modeling of SMF tsunami hazard along the upper US East Coast: Detailed impact around Ocean City, MD.* Natural Hazards, 76(2):705–746, doi:10.1007/s11069-014-1522-8, 2015b.

GRILLI, S.T., TAYLOR, O.-D. S., BAXTER, D.P. and S. MARETZKI, *Probabilistic approach for determining submarine landslide tsunami hazard along the upper East Coast of the United States.* Marine Geology, 264(1–2):74–97, doi:10.1016/j.margeo.2009.02.010, 2009.

GRILLI S.T., IOUALALEN M., ASAVANANT J., SHI F., KIRBY J.T., and P. WATTS, *Source constraints and model simulation of the December 26, 2004 Indian Ocean tsunami.* Journal of Waterway, Port, Coastal, and Ocean Engineering, 33:414–428, doi: 10.1061/(ASCE)0733-950X(2007)133:6(414), 2007.

GRILLI, S.T. and P. WATTS. *Tsunami generation by submarine mass failure Part I : Modeling, experimental validation, and sensitivity analysis.* J. Waterway Port Coastal and Ocean Engng., 131(6):283–297, doi:10.1061/(ASCE)0733-950X(2005)131:6(283), 2005.

HIRT C.W. and B.D. NICHOLS, *Volume of fluid (VOF) method for the dynamics of free boundaries.* Journal of Computational Physics, 39:201–225, 1981.

HOLCOMB R. T. and R.C. SEARLE, *Large landslides from oceanic volcanoes.* Marine Geotechnology, 10:19–32, 1991.

HUNT J.E., WYNN R.B., MASSON D.G., TALLING P.J., and D.A.H. TEAGLE, *Sedimentological and geochemical evidence for multistage failure of volcanic island landslides: A case study from Icod landslide on north Tenerife, Canary Islands.* Geochem. Geophys. Geosyst., 12(12), 2011.

HUNT J.E., WYNN R.B., TALLING P.J. and D.G. MASSON, *Multistage collapse of eight western Canary Island landslides in the last 1.5 Ma: Sedimentological and geochemical evidence from subunits in submarine flow deposits.* Geochem. Geophys. Geosyst., 14(7):1525–2027, 2013.

INOUE K., *Shimabara-Shigatsusaku Earthquake and topographic changes by Shimabara Catastrophe in 1792.* Geographical Reports Tokyo Metropolitan University, 35:59–69, 2000.

IOUALALEN M., ASAVANANT J., KAEWBANJAK N., GRILLI S.T., KIRBY J.T. and P. WATTS, *Modeling the 26th December 2004 Indian Ocean tsunami: Case study of impact in Thailand.* Journal of Geophysical Research, 112:C07024, doi:10.1029/2006JC003850, 2007.

KAISER G., SCHEELE L., KORTENHAUS A., LVHOLT F., RMER H., and LESCHKA S., *The influence of land cover roughness on the results of high resolution tsunami inundation modeling*, Nat. Hazards Earth Syst. Sci., 11:2521–2540, doi:10.5194/nhess-11-2521-2011, 2011.

KARLSSON J.M., SKELTON A., SANDEN M., IOUALALEN M., KAEWBANJAK N., POPHET N., ASAVANANT, J. and A. VON MATERN, *Reconstructions of the coastal impact of the 2004 Indian Ocean tsunami in the Khao Lak area, Thailand.* Journal of Geophysical Research, 114:C10023, 2009.

KIRBY J.T., SHI F., TEHRANIRAD B., HARRIS J.C. and S.T. GRILLI, *Dispersive tsunami waves in the ocean: Model equations and sensitivity to dispersion and Coriolis effects.* Ocean Modeling, 62:39–55, doi:10.1016/j.ocemod.2012.11.009, 2013.

LEGROS, F., *The mobility of long-runout landslides.* Engineering Geology, *63*:301–331, 2002.

LØVHOLT F., PEDERSEN G. and G. GISLER, *Oceanic propagation of a potential tsunami from the La Palma Island.* Journal of Geophysical Research, *113*:C09026, doi:10.1029/2007JC004603, 2008.

LUBIN P., VINCENT S., ABADIE S. and J.P. CALTAGIRONE, *Three-dimensional large eddy simulation of air entrainment under plunging breaking waves.* Coastal Engineering, *53*:631–655, 2006.

MADER C.L., *Modeling the La Palma landslide tsunami.* Science of Tsunami Hazards., *19*:150–170, 2001.

MADSEN P.A., D.R. FUHRMAN and H. A. SCHAFFER, *On the solitary wave paradigm for tsunamis.* J. Geophys. Res., *113*:C12012, doi:10.1029/2008JC004932, 2008.

MASSON D., WATTS A., GEE M., URGELES R., MITCHELL N., BAS T.L. and M. CANALS, *Slope failures on the flanks of the western Canary Islands.* Earth-Science Review, *57*:1–35, 2002.

MOHAMMED, F. and FRITZ, H.M., *Physical modeling of tsunamis generated by three-dimensional deformable granular landslides.* J. Geophys. Res. Oceans, *117*:C11015, doi:10.1029/2011JC007850, 2012.

MOORE J.G., CLAGUE D.A., HOLCOMB R.T., LIPMAN P.W., NORMARK W.R. and M.E. TORRESAN, *Prodigious submarine landslides on the Hawaiian Ridge.* Journal of Geophysical Research, *94*:17465–17484, 1989.

MORICHON D. and S. ABADIE, *Vague générée par un glissement de terrain, influence de la forme initiale et de la loi de déformabilité du glissement.* La Houille Blanche, *1*:111–117, 2010.

OEHLER J.F., LABAZUY P. and J.F. LÉNAT, *Recurrence of major flank landslides during the last 2 Ma history of Réunion Island.* Bulletin Volcanology, *66*:585–595, 2004.

PARARAS-CARAYANNIS G., *Evaluation of the threat of mega tsunamis generation from postulated massive slope failures of island stratovolcanoes on La Palma, Canary Islands, and on the island of Hawaii.* Science of Tsunami Hazards, *20*:251, 2002.

PÉRIGNON Y., Tsunami hazard modeling. Master's thesis, University of Rhode Island and Ecole Centrale de Nantes, 2006.

RISS J., TRIC E., FABRE R., LEBOURG T. and S. ABADIE, *Potential collapse of the Cumbre Vieja volcanic edifice (Canary Island, Spain).* Geophys. Res. Abstr., 12, EGU2010-4843, EGU General Assembly, 2010.

ROBINSON J.E. and B.W. EAKINS, *Calculated volumes of individual shield volcanos at the young end of the Hawaiian Ridge.* Volcanologic Geothermal Research, *151*:309–317, 2006.

RYAN W.B.F., CARBOTTE S.M., COPLAN J.O., O'HARA S., MELKONIAN A., ARKO R., WEISSEL A., FERRINI V., GOODWILLIE A., NITSCHE F., BONCZKOWSKI J. and R. ZEMSKY, *Global Multi-Resolution Topography synthesis.* Geochemistry Geophysics Geosystems, *10*:Q03014, 2009.

SHI F., KIRBY J.T., HARRIS J.C., GEIMAN J.D., and S.T. GRILLI, *A high-order adaptive time-stepping TVD solver for Boussinesq modeling of breaking waves and coastal inundation.* Ocean Modelling, *43–44*:36–51, doi:10.1016/j.ocemod.2011.12.004, 2012.

TAPPIN D.R., WATTS P. and S.T. GRILLI, *The Papua New Guinea tsunami of 1998: anatomy of a catastrophic event.* Natural Hazards and Earth System Sciences, *8*:243–266, www.nat-hazards-earth-syst-sci.net/8/243/2008/, 2008.

TAPPIN D.R., GRILLI S.T., HARRIS J.C., GELLER R.J., MASTERLARK T., KIRBY J.T., F. SHI, G. Ma, K.K.S. THINGBAIJAMG, and P.M. MAIG, *Did a submarine landslide contribute to the 2011 Tohoku tsunami?*, Marine Geology, *357*:344–361 doi:10.1016/j.margeo.2014.09.043, 2014.

TEHRANIRAD B., SHI F., KIRBY, J.T., HARRIS J.C. and S.T. GRILLI, Tsunami benchmark results for fully nonlinear Boussinesq wave model FUNWAVE-TVD, Version 1.0. Technical report, No. CACR-11-02, Center for Applied Coastal Research, University of Delaware, 2011.

TEN BRINK U.S., CHAYTOR J.D., GEIST E.L., BROTHERS D.S. and B.D. ANDREWS, *Assessment of tsunami hazard to the U.S. Atlantic margin.* Marine Geology, *353*:31–54, doi:10.1016/j.margeo.2014.02.011, 2014.

TINTI S., MANUCCI A., PAGNONI G., ARMIGLIATO A. and F. ZANIBONI, *The 30th December 2002 landslide-induced tsunami in Stromboli: sequence of the events reconstructed from eyewitness accounts.* Natural Hazards Earth System Science, *5*:763–775, 2005.

WARD S. N. and S. DAY, *Cumbre Vieja Volcano potential collapse at La Palma, Canary Islands.* Geophysical Research Letter, *28*:397–400, 2001.

WATTS P., GRILLI S.T., KIRBY J.T., FRYER G.J. and D.R. TAPPIN, *Landslide tsunami case studies using a Boussinesq model and a fully nonlinear tsunami generation model.* Natural Hazards and Earth System Sciences, *3*:391–402, 2003.

WEI G., KIRBY J.T., GRILLI S.T. and R. SUBRAMANYA, *A fully nonlinear Boussinesq model for free surface waves. Part I: Highly nonlinear unsteady waves.* Journal of Fluid Mechanics, *294*:71–92, 1995.

WYNN R. and D. MASSON, Canary Islands landslides and tsunami generation: Can we use turbidite deposits to interpret landslide processes. In: Locat J, Mienert J (eds) *Submarine Mass Movements and Their Consequences*, 325–332. Kluwer Academic Publishers Dordrecht Netherlands, 2003.

ZHOU H., MOORE C.W., WEI Y. and V.V. TITOV, *A nested-grid Boussinesq type approach to modelling dispersive propagation and runup of landslide generated tsunamis.* Natural Hazards and Earth System Sciences, *11*:2677–2697, doi:10.5194/nhess-11-2677-2011, 2011.

(Received April 30, 2014, revised June 21, 2015, accepted June 29, 2015, Published online July 21, 2015)

Pure Appl. Geophys. 172 (2015), 3617–3638
© 2015 Springer Basel
DOI 10.1007/s00024-015-1069-y

Pure and Applied Geophysics

Earthquake Scenario-Based Tsunami Wave Heights in the Eastern Mediterranean and Connected Seas

OCAL NECMIOGLU[1] and NURCAN MERAL ÖZEL[1]

Abstract—We identified a set of tsunami scenario input parameters in a 0.5° × 0.5° uniformly gridded area in the Eastern Mediterranean, Aegean (both for shallow- and intermediate-depth earthquakes) and Black Seas (only shallow earthquakes) and calculated tsunami scenarios using the SWAN-Joint Research Centre (SWAN-JRC) code (MADER 2004; ANNUNZIATO 2007) with 2-arcmin resolution bathymetry data for the range of 6.5—Mw_{max} with an Mw increment of 0.1 at each grid in order to realize a comprehensive analysis of tsunami wave heights from earthquakes originating in the region. We defined characteristic earthquake source parameters from a compiled set of sources such as existing moment tensor catalogues and various reference studies, together with the Mw_{max} assigned in the literature, where possible. Results from 2,415 scenarios show that in the Eastern Mediterranean and its connected seas (Aegean and Black Sea), shallow earthquakes with Mw ≥ 6.5 may result in coastal wave heights of 0.5 m, whereas the same wave height would be expected only from intermediate-depth earthquakes with Mw ≥ 7.0 . The distribution of maximum wave heights calculated indicate that tsunami wave heights up to 1 m could be expected in the northern Aegean, whereas in the Black Sea, Cyprus, Levantine coasts, northern Libya, eastern Sicily, southern Italy, and western Greece, up to 3-m wave height could be possible. Crete, the southern Aegean, and the area between northeast Libya and Alexandria (Egypt) is prone to maximum tsunami wave heights of >3 m. Considering that calculations are performed at a minimum bathymetry depth of 20 m, these wave heights may, according to Green's Law, be amplified by a factor of 2 at the coastline. The study can provide a basis for detailed tsunami hazard studies in the region.

Key words: Tsunami hazard, Eastern mediterranean, Aegean sea, Black sea.

1. Introduction

Tsunamis, as infrequent events, have the potential to cause massive loss of life and destruction of the infrastructure, including critical facilities, resulting in large economic losses that may require long recovery periods (LØVHOLT *et al.* 2012). The Indonesian tsunami led to the death of approximately 250,000 people, caused property and business damage totalling more than $4.4 billion, and left approximately 700,000 people homeless, leaving an unprecedented level of damage to the economy and infrastructure of the region (OZEL *et al.* 2011). The Tohoku event resulted in ~20,000 casualties and ~300,000 refugees with an economic cost of ~$122 billion. The critical damage at the Fukushima Daiichi Nuclear Power Plant resulted in severe releases of radioactivity, which resulted in serious concerns regarding long-term health and environmental hazards. According to the "Sendai Report—Managing Disaster Risks for a resilient future" published by the World Bank (2012), economic losses from disasters over the past 30 years are estimated at $3.5 trillion. The report "Lessons Learned From the Fukushima Nuclear Accident for Improving Safety of US Nuclear Plants" by a committee of the US National Research Council calls for nuclear plant licensees and regulators to continually look for new scientific information about nuclear plant hazards and methodologies for estimating their magnitudes, frequencies, and potential impacts and to incorporate new findings and methodologies as they become available. The report also emphasizes taking timely actions to implement countermeasures when new information results in substantial changes to the risk profiles at nuclear plants and also focuses on "beyond-design-basis events," which include low-frequency but high-magnitude "extreme" events—such as the earthquake and tsunami that damaged the Fukushima Daiichi plant on 11 March 2011 (SHOW-STACK 2014). Moreover, the Tohoku disaster showed that a long-term forecast should be based on only prehistoric paleoseismological data, and tsunami

---
[1] Kandilli Observatory and Earthquake Research Institute, Boğaziçi University, Istanbul, Turkey. E-mail: ocal.necmioglu@boun.edu.tr; ozeln@boun.edu.tr

hazard maps may need to be prepared for infrequent gigantic earthquakes as well as more frequent smaller-sized earthquakes (SATAKE 2011).

According to the available tsunami catalogues, tsunamis are most frequently generated by large submarine earthquakes ($\sim$75 % of the cases according to tsunami catalogues), and less frequently by volcanic activity and landslides ($\sim$5 and 10 %, respectively) and 10–15 % of the known tsunamis occurred in the Mediterranean (CIESM 2011). Despite the fact that 65 % of the referenced events in the historical catalogues should be considered as doubtful (SALAMON et al. 2007) and a critical evaluation of these catalogues is a necessity (AMBRASEYS and SYNOLAKIS 2010), reliable sources of information, including catalog information with full confidence, paleotsunami and paleoseismology evidence still confirms the bitter truth: the Eastern Mediterranean and its connected seas (the Aegean, Marmara and Black Seas) are prone to tsunami hazard. Since the basins and sub-basins in the Eastern Mediterranean and its connected seas are small in comparison with large oceans, the challenges are greater than other parts of the world due to the short arrival times (less than 5–10 min) of the leading tsunami waves. With a fully developed infrastructure along the coast and with millions of tourists in the area, it is imperative that tsunami hazard needs are quantified. Tsunami scenarios are useful means for defining and evaluating tsunami hazards and constitute the primary step in tsunami risk mitigation and preparedness, potentially leading to defining the risk and, hence, contributing to sustainable coastal zone development (TINTI et al. 2005). As a preliminary step for possible tsunami hazard analysis in the the Eastern Mediterranean and its connected seas, we have attempted in our study to evaluate the distribution of tsunami wave heights from a set of earthquake sources for each source point of consideration in the range of Mw 6.5—Mw$_{max}$ defined.

2. Seismotectonic Setting of the Study Area

The Black Sea Basin is composed of two main geological parts, namely the Western and Eastern, separated by the Mid Black Sea Ridge (Andrusov Ridge) that is formed from continental crust and overlain by sedimentary cover 5–6 km thick (TARI et al. 2000; NIKISHIN et al. 2003). The Western Black Sea Basin is underlain by oceanic to suboceanic crust and contains a sedimentary cover up to 19 km thick, whereas the Eastern Black Sea Basin is underlain by thinned continental crust approximately 10 km thick and up to 12 km of sediments (NIKISHIN et al. 2003). The rifting age has been suggested as the Aptian-Albian period (125–100.5 million years-Myr) for the entire Black Sea (GORUR 1997; TARI et al. 2000). The Black Sea began to close in the Eocene–Oligocene (56–23 Ma) time period following the closure of the Neotethys. Wheras the Eastern Black Sea has continued its closure from the Miocene up to the present (20–5 Myr), the central and western parts of the Black Sea have gone and undergo more complex neotectonics controlled by the by the northward motion of the African Plate, the western escape of the Anatolian block and Aegean extension, but with low-moderate seismic activity in the present times. Earthquakes took place on the borders of the deep-water basin, in general, with an average hypocentral depth of 5–20 km and a magnitude up to 6–8 mainly as a result of compressional stresses (VOLVOVSKY 1989; NIKISHIN et al. 2003). This dominant compressional nature, as a result of the collision between the Eurasian and Arabian plates, has been verified by the stress fields obtained from the structural data such as thrusts in the Pontides and the Crimean part of the Black Sea, earthquake data such as BARKA and REILINGER (1997) and REILINGER et al. (1997), stress field measurements in the Crimean and the Caucasus regions (RASTSVETAEV 1987) and global positioning system (GPS) data (BARKA and REILINGER 1997; REILINGER et al. 1997; NIKISHIN et al. 2003). GPS velocities indicate that there is a slight north–south shortening of the eastern half of the southern Black Sea coast, whereas a westward movement can be seen on the southwestern coast. The north–south motions in the Black Sea region are in the range of a few millimetres per year (mm/yr), in contrast to the velocities in the Anatolian region of $\sim$10–20 mm/yr. Geological and geophysical studies including offshore seismic reflection profiles, offshore morphology and recent seismicity indicates an active compressional tectonic regime in the Eastern Black

Sea region, which is also supported with the fault plane solution of studied events that took place in the last century (TARI et al. 2000). ALPTEKIN et al. (1986) provided the first evidence of active thrust faulting at the southern margin of the Black Sea in their analysis of the strongest instrumentally recorded 3 September 1968 Bartin earthquake to occur along the Black Sea margin in northwestern Turkey.

The history of the Eastern Mediterranean begins with the early Mesozoic breakup of Gondwana and is controlled by the collision of the African and Eurasian plates, the Arabian Eurasian convergence and current displacement of the Anatolian-Aegean microplates. The general agreement is that plate convergence takes place in this area (MCKENZIE 1970). A good summary of the development of tectonic research for the region together with main references on the tectonic processes and tectonic setting of the Eastern Mediterranean is given in ROBERTSON and MOUNTRAKIS 2006. The Eastern Mediterranean and the Levantine Basin are a remnant of the Mesozoic Neo-Tethys Ocean (ROBERTSON and DIXON 1984; STAMPFLI and BOREL 2002; GARFUNKEL 2004), where its northern arm is the Alpine chain between the African-Arabian and Eurasian plates as a result of convergence (BIJU-DUVAL et al. 1978). It accommodates important geomorphological features such as the Hellenic-Pliny-Strabo trenches, the Eastern Mediterranean Ridge, the Herodotus Basin, the Florence Rise and the Levantine Basin. A detailed discussion on the origin of the Eastern Mediterranean Basin is provided in various publications (GARFUNKEL 2004; AKSU et al. 2005). The Eastern Mediterranean region accommodates the Anatolia, Arabia, and Nubia plates and the Aegean microplate, which moves 30 ±1 mm/yr to the southwest relative to Eurasia (EBELING et al. 2012; MCCLUSKY et al. 2000, 2003) and at a slightly lower velocity to the southwest relative to Nubia (EBELING et al. 2012; KREEMER and CHAMOT-ROOKE 2004; REILINGER et al. 2006). Aegean tectonics are characterized by 35–40 mm/yr north–south extension in central and southern Aegea (EBELING et al. 2012; KIRATZI and LOUVARI 2003; KIRATZI and PAPAZACHOS 1995; LE PICHON and ANGELIER 1979; MCCLUSKY et al. 2000; MCKENZIE 1978); east–west extension in the inner Hellenic Arc (EBELING et al. 2012; KREEMER and CHAMOT-ROOKE 2004; MCCLUSKY et al. 2003); and thrust faulting in the outer Hellenic Arc (EBELING et al. 2012; BENETATOS et al. 2004; MCKENZIE 1972, 1978). The western end of the Pliny and Strabo trenches lies southeast of Crete, however, their eastern ends are not well defined, but the broad deformation of the Pliny-Strabo zone has been correlated with the Burdur-Fethiye Fault Zone in southwestern Turkey (EBELING et al. 2012; HALL et al. 2009). GPS-derived velocity field data for the interaction zone of the Arabian, African (Nubian, Somalian), and Eurasian plates showes a counter-clockwise rotation of the Arabian plate, adjacent parts of the Zagros and central Iran, Turkey, and the Aegean/Peloponnesus at rates of 20–30 mm/yr occurring within the framework of the relatively slow moving (5 mm/yr) Eurasian, Nubian, and Somalian plates (REILINGER et al. 2006). On the basis of the observed kinematics, the deformation in the Africa-Arabia-Eurasia collision zone is driven in large part by rollback of the subducting African lithosphere beneath the Hellenic and Cyprus trenches aided by slab pull on the southeastern side of the subducting Arabian plate along the Makran subduction zone, and the separation of Arabia from Africa is a response to plate motions induced by active subduction (REILINGER et al. 2006). A fundamental discussion on the active tectonics of the Aegean Sea and surrounding regions is given in MCKENZIE (1978). The Aegean and its surroundings form the most active part of the Africa–Eurasia collision zone, responsible for the high level of seismicity in this region. It constitutes more than 60 % of the expected seismicity in Europe up to Mw = 8.2 (MORATTO et al. 2007; PAPAZACHOS 1990) as a result of the compressional motion between Europe and Africa and the resulting tectonic processes such as subduction of the eastern Mediterranean lithosphere under the Aegean along the Hellenic Arc and the westward motion of the Anatolian Block along the North Anatolia Fault (MORATTO et al. 2007; MCKENZIE 1970.) The whole Aegean Back Arc is mainly represented by normal faults (MORATTO et al. 2007; MCKENZIE 1978), whereas the dense shallow seismicity with low-angle thrust faults occurs along the Hellenic Arc (MORATTO et al. 2007; PAPAZACHOS 1990).

The tectonic setting of the Aegean and Eastern Mediterranean is complex; so is the associated discussion. Several other sources of information used in this study are related to studies focusing on fault classification and seismic zonation. A zonation for seismogenic sources of intermediate-depth earthquakes in the southern Aegean area is given by PAPAZACHOS and PAPAIOANNOU (1993). The maximum expected ground motion in Greece using a deterministic seismic hazard analysis based on an homogeneous earthquake catalog for the period of 426 BC–2003 applied on a seismogenic source model with representative focal mechanisms and a set of velocity models after applying a smoothing algorithm to the main shocks in the catalogue have been estimated by MORATTO et al. (2007). PAPAIOANNOU and PAPAZACHOS (2000) provide a map of shallow earthquake sources based on previous work on seismic zonation in the Aegean and surrounding area, active tectonics and geological and geomorphological information. Maximum credible earthquake magnitudes (in the range 6.7–7.6 Mw) in the Aegean area constrained by tectonic moment release rates based on zonation corresponding to recent determinations of deformation rates from satellite data through the use of a merged historical and an instrumental earthquake catalog for the Aegean is provided by KORAVOS et al. (2003). A fault classification for the Aegean region is provided in SBORAS et al. (2011). CAPUTO et al. (2012) provides a repository of geological, tectonic and active-fault data for Northern Greece and the Northern Aegean Sea based on a collection of all available published and unpublished historical and instrumental seismicity data supported by seismogenic sources recognized on the basis of geological, structural, morphotectonic, paleo-seismological and geophysical investigations, which was followed by critical examination of all collected data with the aim of identifiying as many as possible seismogenic source as possible, as well as the parameters and the characateristics associated. Both SBORAS et al. (2011) and CAPUTO et al. (2012) are part of the SHARE Project (BASILI et al. 2013). MITSAKAKI et al. (2013) provide geometric characteristics of selected tectonic segments in the Hellenic Arc and Northern Aegean. A full description of the complexity of the seismotectonic setting of the study area is beyond the scope of this study and it has been addressed in many aspects by many studies, such as BAYRAK and BAYRAK 2011; BEN AVRAHAM et al. 2008; BOHNHOFF et al. 2005; DELIBASIS et al. 1999; DEWEY et al. 1973; EYIDOĞAN and JACKSON 1985; GANAS and PARSONS 2009; HATZFELD et al. 1993; HEURET et al. 2011; HOWE and BIRD 2010; HYNDMAN et al. 1997; JOLIVET et al. 2013; JOST et al. 2002; KARABULUT et al. 2006; MEIER et al. 2007; PAPADIMITRIOU and KARAKOSTAS 2008; PAPAZACHOS et al. 1991; PAPAZACHOS 1996; PAPAZACHOS and PAPAIOANNOU, 1999; PILIDOU et al. 2004; REILINGER et al. 2010; SALEH 2013; SAATÇILAR et al. 1999; SALAMON et al. 2003; SCHOLZ 1998; SHAW and JACKSON 2010; SNOPEK et al. 2007; STERN 2002; ŞENGÖR et al. 1985; STROBL et al. 2014; TAYMAZ et al. 1990, 1991; VANNUCCI and GASPERINI 2004; YOLSAL-ÇEVIKBILEN and TAYMAZ 2012.

3. An Overview of Previous Tsunami Hazard Research in the Study Area

In addition to the earthquake and moment tensor catalogues and various individual references on earthquakes and seismic hazard analysis in the study region, information from studies specifically targeting tectonic tsunami hazard has also been added to the compiled database.

3.1. Black Sea

DOTSENKO and INGEROV (2007) revised the quantitative characteristics of four historical events in the Black Sea based on spectral analysis of the digitized mareograms. They concluded that while the maximum heights of the recorded tsunami wave tide-gauge locations do not exceed 52 cm, an increase in the wave height with an increase in the magnitude is observed, as expected. They also reported that the typical periods of tsunami waves lie within the interval of 8–39 min (DOTSENKO and INGEROV 2007).

In their tsunami modeling study for the Black Sea, DOTSENKO and INGEROV (2010) concluded that the highest waves are formed at the coastal sites closest to the seismic source. They simulated mareogram records at 27 locations in the northern part of the

Black Sea based on 24 possible source zones and concluded that the maximum wave height should not exceed 3 m at any site while, for most of the sites, it is reasonable to expect a maximum of 1.5 m for $M = 7$.

3.2. Eastern Mediterranean and Aegean

A review of the terrestrial geologic records of the Aegean region showing that little geologic evidence has been identified for a many tsunamis reported in the catalogues was provided in DOMINEY-HOWES (2006). In an attempt to develop some simple scenarios of earthquake-generated tsunamis in the Mediterranean, TINTI et al. (2005) identified four different seismogenic areas in the western, central and eastern sections of the basin and defined a seismic fault capable of generating an earthquake with a magnitude equal or larger than the highest magnitude registered throughout history.

ALTINOK et al. (2005) studied the 1881 Chios-Cesme (Mw 6.5) and 1949 Chios-Karaburun (Mw 6.7) earthquakes where associated earthquakes and co-seismic underwater failure-originated tsunamis affected Chios Island and Cesme. SALAMON et al. (2007) collected and investigated ancient and modern tsunami reports in the Eastern Mediterranean to understand and model the typical tsunamigenic sources with possible characterization of the tsunami hazard along the Levant coasts. Their analysis indicated that only 35 % of the tsunami reports could be traced back to primary sources. In addition to critical compilation and assessment of the tsunami reports, they also modeled three typical scenarios and examined the more likely severe magnitudes, which leads toward the upper range of expected run-ups. They concluded that an offshore slump produced by a strong Dead Sea transform system earthquake may lead to 4- to 6-m run-up flooding the Syrian, Lebanese, and Israeli coasts, whereas tsunamis from remote earthquakes produce only 1- to 3-m run-ups in these coasts, but are more regional in extent. YOLSAL et al. (2007) attempted to synthesize historical tsunamis and tsunami propagation in the Eastern Mediterranean Sea region, focusing specifically to the Hellenic and the Cyprus arcs and the Levantine Basin, and provided focal mechanism parameters for the 11 May 1222 Paphos and 8 August 1303 Crete earthquakes while at the same time providing their interpretation of tsunami source zones in the Eastern Mediterranean and Aegean Seas.

In their study, LORITO et al. (2008) selected three potential source zones located at short, intermediate and large distances from Southern Italy to study the impact of a large set of tsunamis resulting from earthquakes generated by major fault zones of the Mediterranean Sea. They argued that only a systematic identification of all possible sources along with their correlative tsunami scenarios is needed to deal with the uncertain source-impact zones and there is a need for considering both distant and local sources.

OKAL et al. (2009) conducted a study on the 1956 Amorgos earthquake based on a normal mechanism fault with a rupture area of 75×40 km derived from the systematic relocation of the main shock and 34 associated events expressing extensional tectonics in the Hellenic subduction zone back-arc. Also supported with eye-witness reports, they argued that the observed tsunami is incompatible with a seismic dislocation source only and demonstrated this through hydrodynamic simulations using both the dislocation and landslide source models. This conclusion has been somehow further supported by BEISEL et al. (2009) based on the spectral analysis of the tide-gauge record in the port of Yafo (Israel), as the coseismic tsunami source failed to capture significant spectral energy components while simulations resulted in tsunami waves with their heights close to that obtained from the record measured at Yafo. In fact, BEISEL et al. (2009) succeeded in obtaining harmonics with frequencies very close to those measured at the tide gauge station when landslide movement, triggered by the main shock and/or by the largest aftershock, is suggested as a source of these tsunami waves.

BASILI et al. (2013) states that while tsunamis generated by Mw = 8 earthquakes affect the entire basin, the impact of tsunamis generated by Mw = 7 earthquakes should be expected to be strong at many localities and conclude that a set of scenarios with regard to the epistemic uncertainties in the parametric fault characterization in terms of geometry, kinematics, and assessment of activity rates should be considered in tsunami hazard analysis.

EBELING et al. (2012) reassessed four large (M ~ 7) historical earthquakes occurring in various regions of the Hellenic Arc (on 6 October 1947, in the Peloponnesus; 9 February 1948, near Karpathos; and a couplet east of Rhodos on 24 and 25 April 1957), where the first two are associated with the damaging near-field tsunamis involving submarine slumping. A systematic assessment of the earthquake-generated tsunami hazard for Rhodes Island in the SE Aegean Sea based on several hypothetical, credible, near-field 'worst case' scenarios associated with seismic events of magnitude 8.0–8.4 has been presented in MITSOUIDIS et al. (2012).

SØRENSEN et al. (2012) provided the first probabilistic estimate of earthquake-generated tsunami hazard for the entire Mediterranean Sea based on deterministic tsunami wave propagation scenarios corresponding to earthquake activity rates estimated from the observed seismicity. Their results indicated that while the highest hazard in the eastern Mediterranean is due to the earthquakes along the Hellenic Arc, in fact most of the Mediterranean coastline is prone to tsunami impact, the probability of a tsunami wave exceeding 1 m somewhere in the Mediterranean in the next 30 years is close to 100 % (SØRENSEN et al. 2012). Their source model consists of 21 homogeneous zones chosen to be small enough to represent regions of relatively homogeneous earthquake activity while at the same time sufficiently large enough allowing for a stable statistical analysis of the source zone characteristics. However, the sizes of the zone areas they determined would not allow capturing the complexity of the seismo-tectonic setting presented in Fig. 1 in our view.

YOLSAL-ÇEVIKBILEN and TAYMAZ (2012) numerically simulated the major and well-known earthquake-induced Eastern Mediterranean tsunamis of 365, 1222, 1303, 1481, 1494, 1822 and 1948 and proposed several hypothetical tsunami scenarios to demonstrate the characteristics of tsunami waves, propagations and effects of coastal topography. While they argue that the Cyprus Island acts as a natural barrier for tsunami waves in case of earthquakes along the Hellenic Arc, the Cyprus Arc and/or the Levantine Basin, they also emphasized the need for better geophysical, seismological and geological observations in the future for the improvement of tsunami simulations.

In their PTHA study, LØVHOLT et al. (2012) argued that in the Hellenic Arc, at the southern parts of Peloponnese and Crete, the run-up height for a return period of 500 years might well exceed 10 m.

A map of tsunamigenic zones in the Mediterranean region and its connected seas, including the Marmara Sea, the Black Sea and the SW Iberian Margin in the NE Atlantic Ocean, is provided in PAPADOPOULOS et al. (2014) based on combined analysis of various sources such as historical documents, onshore and offshore geological signatures, geomorphological imprints, observations from selected coastal archeological sites, as well as instrumental records, eyewitnesses accounts and pictorial material concerning both seismic and non-seismic origins with a variable tsunamigenic potential.

Various other studies such as PAPAZACHOS et al. (1986), SOLOVIEV (1990), RANGUELOV and GOSPODINOV (1994), DOTSENKO and KONOVALOV (1996), YALÇINER et al. (2004), PAPADOPOULOS et al. (2011a, b), PARARAS-CARAYANNIS (2011), MITSOUDIS et al. (2012) provide information on the tsunamis observed and/or analysed in the Black Sea, the Aegean Sea and the Eastern Mediterranean. All of these studies, whether presented here or not (except SØRENSEN et al. 2012), provide conclusions based on one or several selected earthquake-triggered tsunami scenarios. In this study, we have attempted to go beyond all of theses previous studies by not only considering all of the possible tsunamigenic earthquake sources but also considering their possible magnitude ranges.

4. Parameterization of Tsunamigenic Earthquake Sources and Creation of the Tsunami Scenario Database

Uncertainties in earthquake source parameters, such as strike, dip, rake and depth, and the effect of their variability on tsunami wave propagation and calculated wave heights at offshore coastal locations was discussed in NECMIOĞLU and ÖZEL (2014). Among the various sources used to assign characteristics of

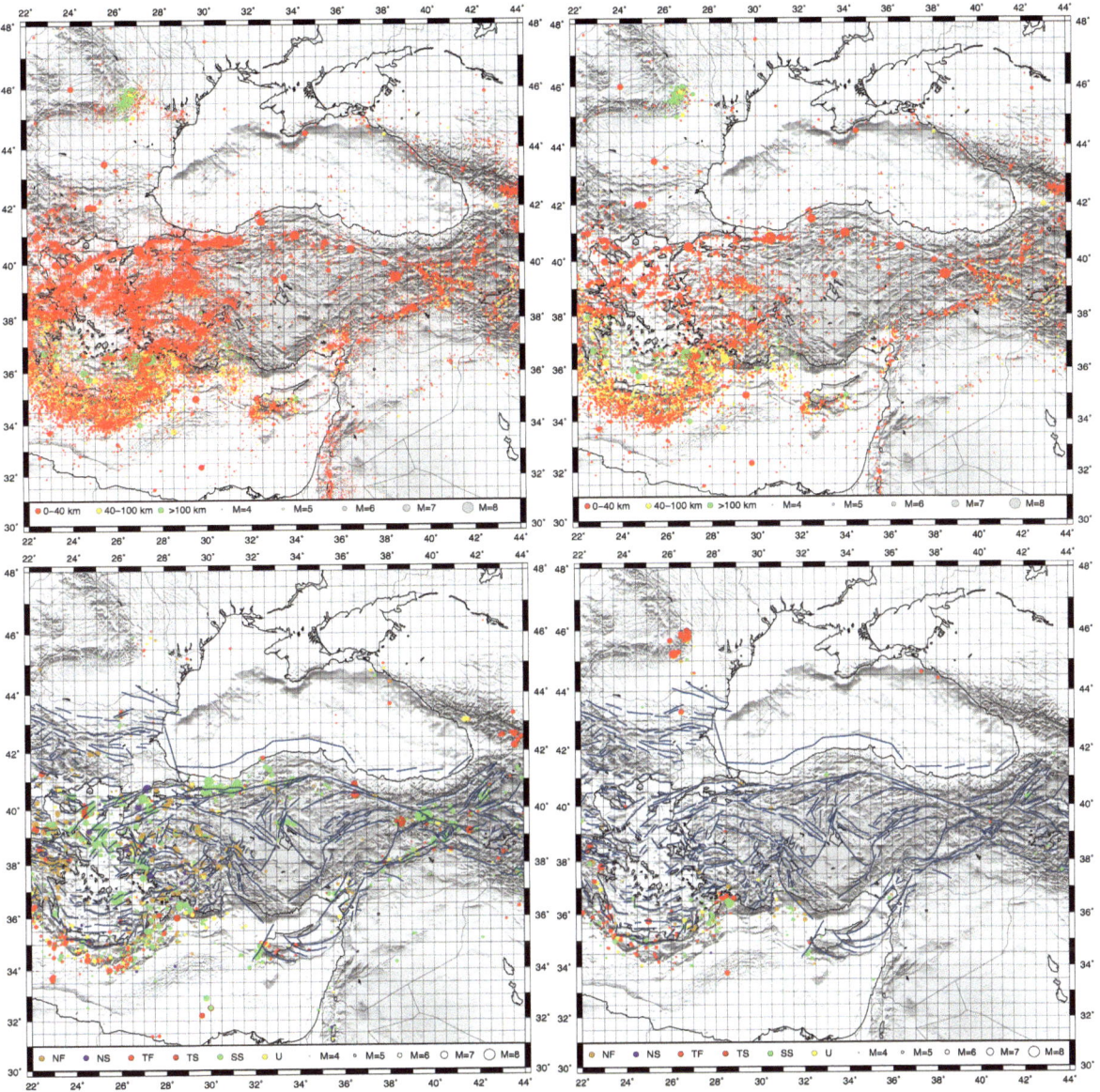

Figure 1
Consolidated seismicity maps from the International Seismological Centre (ISC), the Earthquake Mechanisms of the Mediterranean Area (EMMA; VANNUCCI and GASPERINI 2004) and KALAFAT et al. (2009) catalogs with no magnitude restriction (*upper left*) and with M > 4 (*upper right*). Fault types in the Eastern Mediterranean, Aegean and Black Seas according to Zoback (1992) derived from ISC, EMMA and KALAFAT et al. (2009) moment tensor catalogs for the depth range of 0-40 km (*bottom left*) and 40–100 km (bottom right). *NF* normal faulting, *NS* normal faulting combined with considerable strike-slip component, *TF* thrust faulting, *TS* thrust faulting combined with considerable strike-slip component, *SS* strike-slip faulting, *U* unclassified type of faulting. Fault lines are taken from BASILI et al. (2013). Figure modified from NECMIOĞLU and ÖZEL (2014)

earthquake sources at each source point in this study, the first and main source was the available moment tensor catalogues (ISC Focal Mechanism Database; VANNUCCI and GASPERINI 2004; KALAFAT et al. 2009). The seismicity maps from ISC, EMMA and KALAFAT et al. (2009) catalogues are given in Fig. 1. The events selected from the Reviewed ISC Bulletin focal mechanism database, which also includes important datasets such as the Global Centroid Moment Tensor (GCMT) Project, has a total of 466 events with

M ≥ 4.0 defined strike/dip/rake values out of 4,233 events for the range of 1 January 1911–1 January 2011 in the study area (30°–48°N/22°–44°E). The second release of the EMMA database (VANNUCCI and GASPERINI 2004) is a very detailed and comprehensive compilation of earthquake source parameters. A total of 735 events from the EMMA catalog were considered in this study; 827 events with M ≥ 4.0 can be found in KALAFAT et al. (2009), where earthquake source parameters of 46 % of the events included are originally calculated for that study; the rest were complied from other sources. Interpretation of the fault mechanism based on different moment tensor catalogues may lead to different conclusions. On the other hand, when combined, the catalogues may be used to identify zones with similar properties, as shown in Fig. 1. Seismicity maps shown in Fig. 1 helped to identify the locations of 0.5° × 0.5° bins at 0–40 and 40–100 km depth layers to be used in the source characterization and corresponding tsunami scenario creation (Fig. 2). The reason for considering two depth layers was to study the generation of tsunamis by intermediate-depth sources as suggested by NECMIOĞLU and ÖZEL (2014). The variation of strike, dip and rake values through polar plots have been evaluated subjectively at 312 0.5° × 0.5° uniformly gridded bins for a 0–40 km shallow depth layer and 92 bins for a 40–100 km intermediate-depth layer in order to determine the dominant strike/dip/rake parameters in each bin. An additional database was the European Database of Seismogenic Faults (EDSF) compiled in the framework of the EU Seismic Hazard Harmonization in Europe (SHARE) Project (BASILI et al. 2013; SBORAS et al. 2011), which includes only faults that are deemed to be capable of generating earthquakes of magnitude equal to or larger than 5.5 with the aims at ensuring an homogenous input for use in ground-shaking hazard assessment in the Euro-Mediterranean area. Faults considered from the EDSF are shown in Figs. 3 and 4. Last but not least, a total of 2,733 events compiled and provided within the Tsunami Risk and Strategies For the European Region (TRANSFER) Project (http://www.transferproject.eu/) have also been considered in this study.

The characteristic earthquake source database derived in this study for 0–40 and 40–100 km depth layers is a result of evaluating the compiled source databases referenced above supported by the strike/dip/rake distribution analysis and interpretation of the tectonic setting of the study region. Corresponding maps with representative focal mechanisms are given in Fig. 2. The derived database includes strike, dip, rake, depth, Mw_{max} and corresponding fault length (L), fault width (W) and slip (D) values for each bin together with associated parameters. Several parameters within the characteristic database are referenced directly from the ISC Moment Tensor

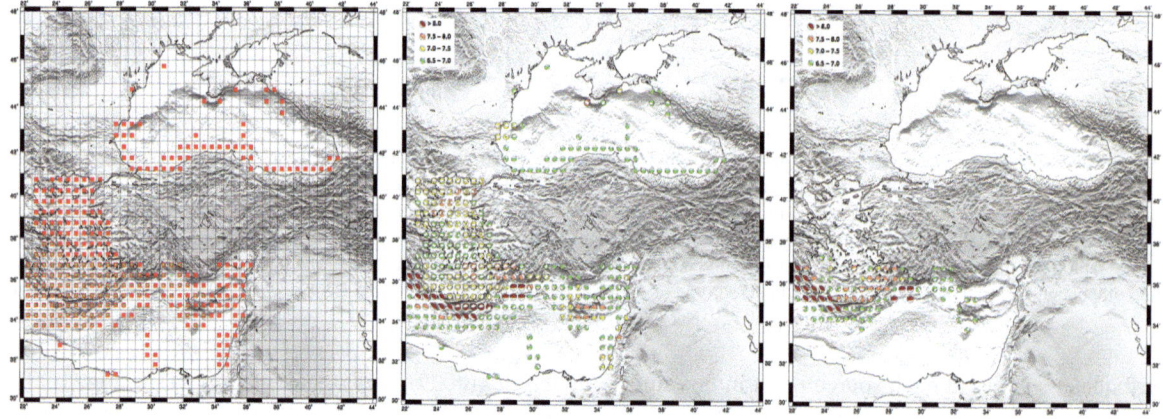

Figure 2
(*Left*) Locations of 0.5° × 0.5° bins at 0–40 km (*red squares*) and 40–100 km depth (*green squares*) layers to be used in the source characterization and corresponding tsunami scenario creation. The total number of bins is 404 (312 shallow–92 deep). *Fault lines* are taken from BASILI et al. (2013). Representation of characteristic earthquake focal mechanism parameters for 0–40 km (*center*) and 40–100 km (*right*) depth layer provided. Beach-ball sizes are proportional to the Mw_{max} assigned to each bin

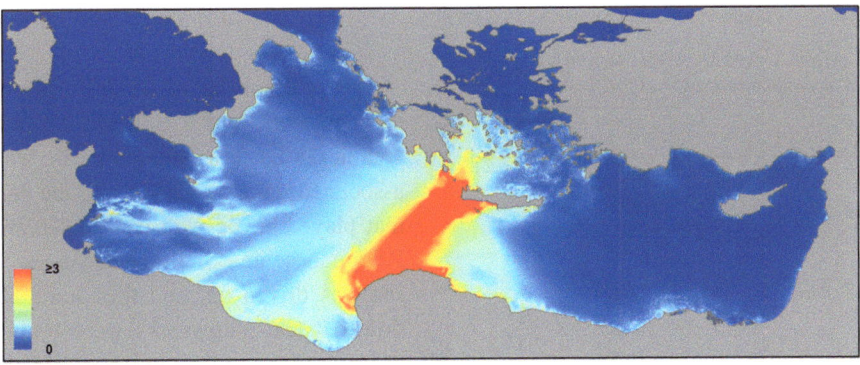

Figure 3
Maximum wave height distribution for the 365 AD event. Strong focusing between Banghazi and Darna provinces in the eastern part of the Libya is evident and a minimum offshore wave height of 1 m can be observed throughout the Eastern Mediterranean from Alexandria (Egypt) to Sicily and southern coasts of Italy

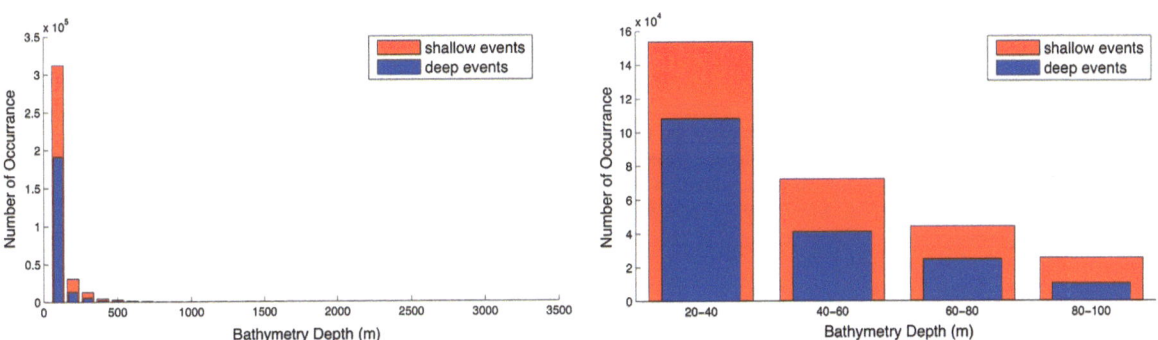

Figure 4
Distribution of calculated bathymetry depths for all depths (*left*) and for the depth range of 20–100 m (*right*). A vast majority of the calculations were performed within the 20–100 m bathymetry depth range and a high portion within this range corresponds to calculations at depths of 20–40 m. No calculations were performed in less than 20-m bathymetry depths during the coarse resolution modeling

Catalog, the EMMA Moment Tensor Catalog (VANNUCCI and GASPERINI 2004); KALAFAT et al. (2009), the SHARE database (BASILI et al. 2013; CAPUTO et al. 2012), the TRANSFER database (CONSTANTINESCU et al. 1966; DINEVA 1993; DZIEWONSKI et al. 1981; Swiss Federal Institute of Technology in Zurich (ETHZ) Moment Tensor Catalogue, RADULIAN et al. 2002; UDIAS et al. 1989) and the literature (ALPTEKIN et al. 1986; BEISEL et al. 2009; BERNARDI 2004; BOHNHOFF et al. 2005; EBELING et al. 2012; EVA et al. 1988; JOST et al. 2002; KORAVOS et al. 2003; LORITO et al. 2008; MCKENZIE 1978; MITSAKAKI et al. 2013; MITSOUDIS et al. 2012; MORATTO et al. 2007; OKAL et al. 2009; PAPAIOANNOU and PAPAZACHOS 2000; PAPAZACHOS 1996; PARKE 2001; PILIDOU et al. 2004; SALAMON et al. 2003, 2007; SHAW et al. 2008; SHAW and JACKSON 2010; SØRENSEN et al. 2012; TAYMAZ et al. 1990; TINTI et al. 2005; YOLSAL et al. 2007). Further studies on moment tensor catalogues, such as KIRATZI and LOUVARI (2003), PONDRELLI et al. (2002, 2004, 2011), and BENETATOS et al. (2004) have also been considered. If not provided in the references, L, W and D values are mostly calculated from LEONARD (2010) for the Mw_{max} earthquakes with strike/dip/rake parameters slightly modified; in certain cases, copied from individual studies such as LORITO et al. (2008), MITSOUDIS et al. (2012) etc. or scaled from Mw-Mo relations provided in KANAMORI (1977) and HANKS and KANAMORI (1979). Within the range of 6.5-Mw_{max}, excluding the case corresponding to Mw_{max}, L and W parameters are scaled exponentially with respect to L and W parameters defined for an Mw 6.5 scenario in consideration of the thickness of the seismogenic layer. Slip parameters have been

Table 1

Calculation of L, W, D from a given Mw and L, D from a given Mw and W according to LEONARD (2010)

	a	b
Mw -> L, W, D		
Mw = a × log (RA) + b		
DS	1.00	4
SS	1.00	3.99
Mw = a × log (L) + b		
DS	1.67	4.24
SS	1.67	4.17
log(D_{AV}) = a × log (L) + b		
DS	0.833	−1.30
SS	0.833	−1.34
Mw, W -> L, D		
Mw = a × log (RA) + b		
DS	1.00	4
SS	1.00	3.99
log(D_{AV}) = a × log (L) + b		
DS	0.833	−1.30
SS	0.833	−1.34

In the latter, W correspond to the assumed thickness of the seismogenic layer for the Mw_{max} defined. Fault classification based on the rake angle is the following: *DS* dip slip: thrust faults where $15° < \lambda < 165°$) and normal faults where $195° < \lambda < 345°$; *SS* strike slip: strike-slip faults where $345° \leq \lambda \leq 15°$ and $165° \leq \lambda \leq 195°$

inversely derived than from the moment calculated with given Mw, L, and W parameters where $\mu = 3.25E + 11$ dyn/cm^2. Reference formulas used from LEONARD (2010) are given in Table 1. In cases where no Mw_{max} information was available from the literature or other sources used in this study, a hypothetical value was assigned in accordance with the Mw_{max} value defined in the neighbouring bin. Focal mechanism solutions for each bin are shown in Fig. 2 where beach-ball sizes are proportional to the Mw_{max} assigned to each bin. Using the earthquake source parameters as described above, we calculated tsunami scenarios with the SWAN-JRC code (ANNUNZIATO 2007) which solves the non-linear shallow water equations by the finite difference numerical scheme based on SWAN code by MADER (2004). Initial conditions for the tsunami modeling is obtained using an analytical solution for surface deformation in an elastic half-space (OKADA 1985) embedded in the SWAN-JRC code by estimating the distribution of coseismic uplift and subsidence using the earthquake source parameters (ZAMORA et al. 2014).

5. Case Study: 365 AD Event

The earthquake of 21 July 365 AD was a significant event in the Hellenic Arc affecting a large area in the Eastern Mediterranean, where most of the associated damage was due to the seismic sea wave that played havoc with coastal settlements in Egypt, Peloponnese and Sicily (AMBRASEYS 2009). First, the sea was driven out and then huge masses of water flowed back; shipwrecks were found 2 km off the coast on the southwestern shore of Peloponnesus near Methoni; a tsunami was observed in Asia Minor and the coast of Sicily was flooded (SOLOVIEV et al. 2000; ALTINOK et al. 2011). PAPADOPOULOS (2011a, b) provides a detailed discussion on this event. According to AMBRASEYS (2009), there is evidence that this rather shallow earthquake was produced by thrust faulting off the southwest coast of Crete, which extended for about 100 km to the northwest, striking at 320°. The uplifting of the west coast of the island of Crete by 4–9 m may be associated with this or other earthquakes during that period. The seismic sea wave was caused by an offshore fault rupture, by (1) large scale landslide(s) from the bathymetric escarpments, or both. In their interpretation of the Eastern Mediterranean tectonics and tsunami hazard based on detailed investigation of the 365 AD earthquake, SHAW et al. (2008) presented evidence from field observations and radiocarbon data that western Crete was lifted by up to 10 m above sea level during the earthquake and suggested that the earthquake occurred on a fault dipping at around 30° within the overriding plate and not on the subduction interface. Their tsunami modelling provided open-ocean tsunami waves heights that are comparable to that of observed and modelled in the open ocean for the Sumatra 2004 tsunami. Furthermore, LORITO et al. (2008) argued that up to a 5-m tsunami wave could of been produced by the western Hellenic Arc source (365 AD event). The 365 AD event has been researched by many, such as TINTI et al. (2005); FOKAEFS and PAPADOPOULOS (2007); PAPADIMITRIOU and KARAKOSTAS (2008); SHAW and JACKSON (2010); and PARARAS-CARAYANNIS (2011).

Strike, dip, rake and depth parameters (strike 315°, dip 35°, rake 90°, depth 27.3 km) of the associated earthquake source model used in the tsunami modelling for the 365 AD event in this study were assigned from LORITO et al. (2008) with a slight modification, whereas we considered also a Mw 8.4 event with a smaller rupture area and displacement (L 120 km, W 77 km, and slip 16.7 m). The epicenter of the source is selected as 35.25°N 23.25°E. Our modelling indicates that the tsunami reaches northeastern Libya and western Peloponnese in half an hour, whereas the arrival time for eastern Sicily, Calabria and Rhodes is 1 h. Waves arrive in Antalya, western Cyprus, Alexandria, and Tripoli in 90 min and reach the Levantine coast in 2 h. Maximum wave height distribution for the 365 AD event, shown in Fig. 3, indicates a strong focusing between the Banghazi and Darna provinces in the eastern part of Libya and a minimum of 1 m offshore wave heights can be observed throughout the Eastern Mediterranean from Alexandria (Egypt) to Sicily and southern coasts of Italy. Extreme impact is evident in western Crete, eastern Peloponnese and Attica, and in the eastern part of Libya between the Banghazi and Darna provinces. Considerable impact is evident in the southern Aegean, in central and eastern Crete, in Rhodes and Turkish coasts opposite Rhodes, on the northwestern coasts of Egypt, in northern Libya between the Tripoli and Banghazi provinces, eastern Sicily, southern Italy and in western Peloppenese. The impact is relatively low in the Levantine and almost no impact is associated with the Northern Aegean, eastern Cyprus, Tunisia, western Sicily and beyond, and Albania. The results are consistent with the historical records, indicating the reliability of the model.

6. Evaluation of the Tsunami Scenario Database

We present varied statistical information on the created tsunami scenario database to reflect the overall characteristics of the data, issues related to the resolution of the bathymetry used in the modeling, followed by a qualitative and quantitative assessment of the results. There are 1,394 forecast points in the scenario database where tsunami wave heights have been calculated. No calculations were performed in less than 20-m bathymetry depths in this study (Fig. 4). The distribution of the calculated wave heights with respect to the bathymetry depth of the calculation points are given in Fig. 5. The effect of the source depth is clearly visible, with a ratio close to 2 between the wave heights calculated for shallow events (top of the fault at a 5-km depth) and intermediate deep events (top of the fault at a 40-km depth). The distribution of the calculated wave heights with respect to the distance between the calculation point and associated earthquake epicentre can also be found in Fig. 5. The effect of the source depth is clearly visible where the calculated wave heights are substantially higher for shallow events qwith a source-calculation point distance in the range of 0–200 km. The relative increase observed in the wave heights between 250 and 500 km is mostly associated with the earthquakes in the Hellenic Arc and calculations points in northern Africa, especially between Banghazi (Libya) and Alexandria (Egypt). The distribution of the calculated wave heights with respect to Mw is given in Fig. 6, indicating a ratio of 2 for calculated wave heights for shallow and deep events for the Mw range of 6.5–6.6, around 1.6 for the Mw range of 6.7–7.8 and 1.2 for the Mw range of 7.9–8.5 for both sets of data.

Maximum wave heights calculated in the Black Sea for the shallow earthquakes defined in Fig. 2 are provided in Fig. 7., which shows that <3-m tsunami wave heights could be expected in locations in the southern coasts of Crimea, the northwestern coast of Turkey, the Bulgarian coast and along the southeastern coasts of Romania, whereas along the eastern Black Sea coasts, the expected maximum tsunami wave height is <1 m. The results of the modeling are in accordance with historical tsunami events in the region.

Maximum wave heights calculated in the Central and Eastern Mediterranean for shallow earthquakes defined in Fig. 2 are provided in Fig. 8 (top), which shows that the expected maximum tsunami wave height is >3 m in locations in, around and orthogonal to the Hellenic Arc, whereas for the rest of the Eastern Mediterranean, Southern Aegean, Tripoli (Libya), eastern Sicily, Calabria and western coasts Greece, the expected maximum tsunami wave height is <3 m. In the northern Aegean, Tunisia, western Sardinia, southwest coasts of Italy, and western and northern coasts of Sicily, the expected maximum

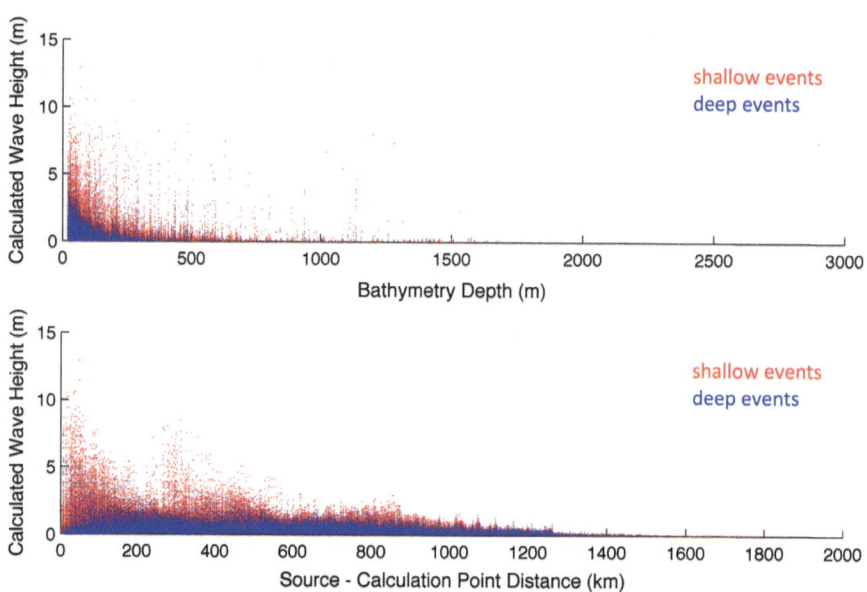

Figure 5
Top Distribution of the calculated wave heights with respect to the bathymetry depth of the calculation point. The effect of the source depth is clearly visible with a ratio close to 2 between the wave heights calculated for shallow events (*top* of the fault at a 5-km depth) and deep events (*top* of the fault at a 40-km depth). *Bottom* Distribution of the calculated wave heights with respect to the distance between the calculation point and associated earthquake epicenter. The effect of the source depth is clearly visible where the calculated wave heights are substantially higher for shallow events with a source-calculation point distance in the range of 0–200 km. The relative increase observed in the wave heights between 250 and 500 km is mostly associated with earthquakes in the Hellenic Arc and calculation points in northern Africa, especially between Banghazi (Libya) and Alexandria (Egypt)

tsunami wave height is <1 m. The results of the modeling are in accordance with the historical tsunami events in the region. It should be emphasized that the conclusions given above for both the Black Sea and Eastern Mediterranean are valid only for the shallow earthquake sources given in Fig. 2 and excluding any possibility of an associated submarine landslide. Descriptions and conclusions of tsunamis generated from shallow earthquake sources in the Eastern Mediterranean are also valid for tsunamis generated from intermediate-depth earthquake sources defined in Fig. 2 (Fig. 8 bottom).

In order to support our conclusions and avoid possible discussion associated with the definition of the maximum magnitude values considered in the study, we have decided to identify the minimum earthquake magnitude for each forecast point that would lead to 50-cm wave height at the coast line. Using Green's Law, defined as where $Aw_{1m} = (Aw_{CP})^{1/4} \times BD_{CP}$, where Aw_{1m} is the wave height at 1 m bathymetry depth, Aw_{CP} is the wave height calculated and BD_{CP} is the bathymetry depth of the calculation point, we considered all calculations where the wave height calculated offshore is higher than 20 cm; this was based on the consideration that the majority of the calculations took place in the 20–40 m bathymetry depth range and a 20 cm wave height calculated at 40 m bathymetry depth would correspond to a 50-cm wave height at the coastline according to Green's Law. On the other hand, it should be emphasized that while Green's Law provides an empirical approximation, the non-linear dynamics of the tsunami at the coastal zones may lead to deviations in actual wave heights.

Minimum values of earthquake magnitudes leading to a 50-cm wave height along the Black Sea coasts for the shallow earthquakes defined in Fig. 2 and associated earthquake magnitudes is shown in Fig. 9. While a higher magnitude Mw range of 7.5–7.9 is needed for the locations in the eastern Black Sea, an Mw range of 7.0–7.4 is sufficient enough to have 50-cm waves in most parts of the western Black Sea. Especially in the southern coasts

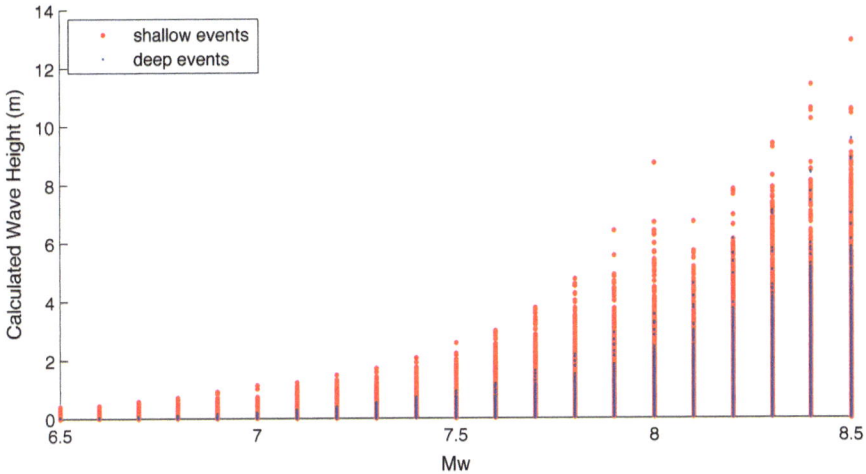

Figure 6
Distribution of the calculated wave heights with respect to Mw. the calculated mean varies around 2 for the Mw range of 6.5–6.6, around 1.6 for the Mw range of 6.7–7.8 and 1.2 for the Mw range of 7.9–8.5 for both sets of data

of Crimea, southern coasts of Romania, Bulgarian coasts and several locations in the north-western part of the Black sea region of Turkey, a Mw range of 6.5–6.9 is enough to result in 50-cm wave height at the coastline. Figure 10 shows minimum values of earthquake magnitudes leading to a 50-cm wave height on the Eastern Mediterranean and Aegean coasts for the shallow earthquakes defined in Fig. 2. Locations in an around the Hellenic Arc, the Aegean Sea and south- and eastern Cyprus can be subject to 50-cm coastal wave height for earthquakes ranging from Mw 6.5–6.9, whereas for southern coasts of Turkey, the Levantine coasts, northern Egypt and north-eastern Libya, a higher Mw range of 7.0–7.4 is needed. Earthquakes in the Mw range of 7.5–7.9 are capable of generating 50-cm coastal wave heights in northern Libya, in southern and eastern Sicily, in southern Italy, Albania and along western coasts of Greece. Only earthquakes in the Mw range of 8.0–8.5 are capable of generating a 50-cm coastal wave height in Tunisia, along western and northern Sicily, on the western and Adriatic coasts of Italy, and in the Gulf of Corinth. Minimum earthquake magnitudes leading to a 50-cm wave height at the Eastern Mediterranean and Aegean coasts for the deep earthquakes defined in Fig. 2 show that locations in an around the Hellenic Arc, the southern Aegean Sea and north-eastern Libya can be subject to 50-cm coastal wave heights for earthquakes ranging from

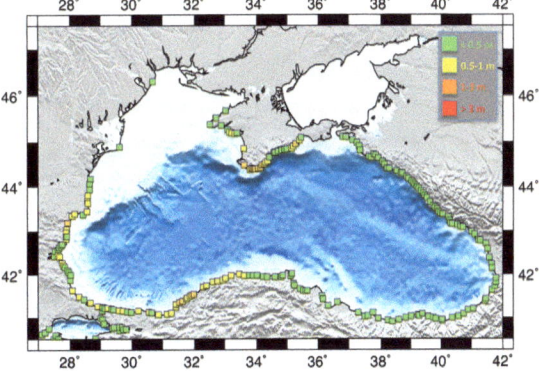

Figure 7
Maximum wave heights calculated in the Black Sea for the shallow earthquakes defined in Fig. 2 indicate that <3-m tsunami wave heights could be expected in locations along the southern coasts of Crimea, the northwestern coast of Turkey, the Bulgarian coast and the southeastern coasts of Romania, whereas along the eastern Black Sea coastline, the expected maximum tsunami wave height is <1 m. The results of the modeling are in accordance with historical tsunami events in the region. It should be emphasized that these conclusions are valid only for the earthquake sources given in Fig. 2 and excluding any possibility of an associated submarine landslide

7.0–7.4, whereas for the southern coasts of Turkey, western Cyprus, southern Levantine, northern Egypt, Tripoli (northern Libya), eastern Sicily, southern Italy, and western Greece, a higher Mw range of 7.5–7.9 is needed (Fig. 10). Only earthquakes in the Mw range of 8.0–8.5 are capable of generating 50-cm

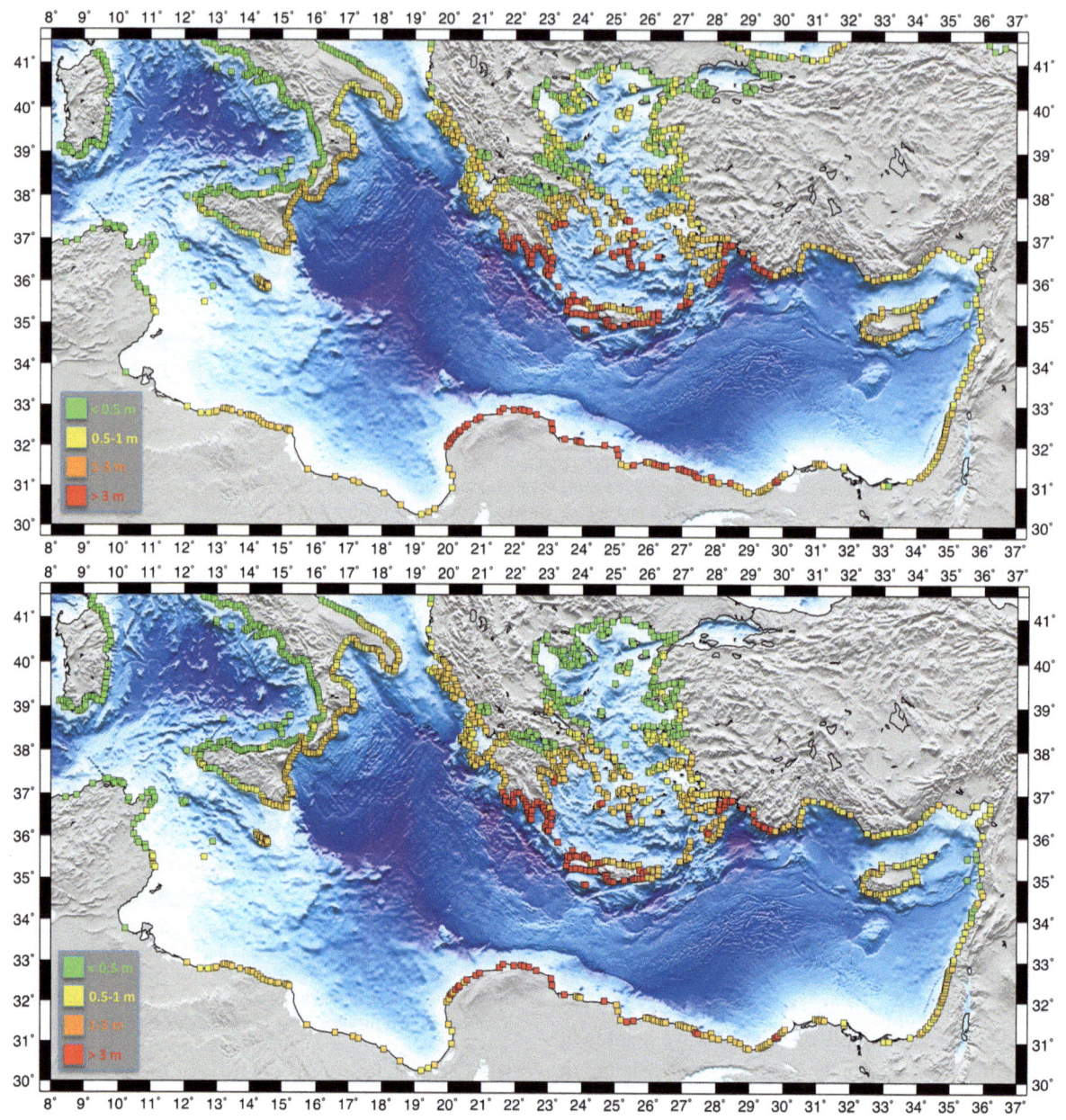

Figure 8
Maximum wave heights calculated in the Central and Eastern Mediterranean for the shallow- (*top*) and intermediate-depth (*bottom*) earthquakes defined in Fig. 2 indicate that the expected maximum tsunami wave height is >3 m in locations in, around and orthogonal to the Hellenic Arc, whereas for the rest of the Eastern Mediterranean, Southern Aegean, Tripoli (Libya), eastern Sicily, Calabria and western coasts Greece, a maximum tsunami wave height of <3 m is expected. In the northern Aegean, Tunisia, western Sardinia, southwest coasts of Italy, and western and northern coasts of Sicily, the expected maximum tsunami wave height is <1 m. The results of the modeling are in accordance with the historical tsunami events in the region. It should be emphasized that these conclusions are valid only for the earthquake sources given in Fig. 2 and excluding any possibility of an associated submarine landslide

coastal wave heights in Tunisia, western and northern Sicily, along the western and Adriatic coasts of Italy, in the Gulf of Corinth, in the central and northern Aegean, in eastern Cyprus and along the northern Levantine coasts. It should be emphasized that these conclusions are valid only for the shallow earthquake

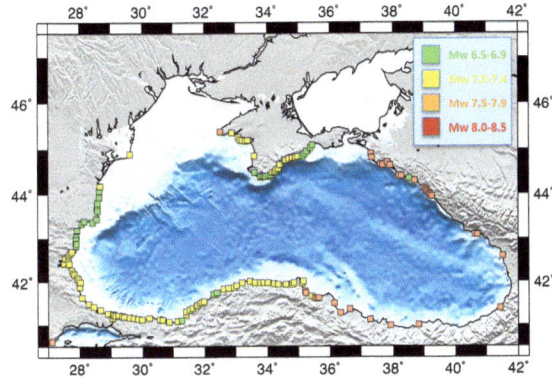

Figure 9
Minimum earthquake magnitude values leading to 50-cm coastal wave heights along the Black Sea coast for the shallow earthquakes defined in Fig. 2, and associated earthquake magnitudes. While a higher magnitude Mw range of 7.5–7.9 is needed for the locations in the eastern Black Sea, an Mw range of 7.0–7.4 is sufficient enough to have 50-cm waves in most parts of the western Black Sea. Especially in the southern coasts of Crimea, southern coasts of Romania, Bulgarian coasts and several locations in the northwestern part of the Black sea region of Turkey, a Mw range of 6.5–6.9 is enough to result in 50-cm wave heights at the coastline. It should be emphasized that these conclusions are valid only for the shallow earthquake sources given in Fig. 2 and excluding any possibility of an associated submarine landslide

sources given in Fig. 2 and excluding any possibility of an associated submarine landslide.

7. Discussion and Conclusions

There are various strong tsunami events in the Eastern Mediterranean with an apparent recurrence period of about 150–200 years associated to both seismic and non-seismic origins (YOLSAL et al. 2007). But more importantly, some extreme events, such as those occurring in 365 and 1303 AD, have resulted in basin-wide impacts, as also confirmed by the tsunami simulations performed in this study. The low recurrence of strong events in the Black Sea makes it difficult to conduct a preliminary tsunami hazard study in the region since there is only a certain amount of descriptive information concerning the historical tsunami events in the Black Sea. The most important fact is that the available historical data indicates that some historical tsunami waves were equal to 2–3 m and, thus, destructive (DOTSENKO and INGEROV 2007). In this study, we have tried to consider all possible locations of tectonic origin tsunamis with a possible and meaningful range of magnitudes. However, submarine landslide-generated tsunamis in the Black Sea constitute a major element of the tsunami hazard in the region and further comprehensive studies on this should be initiated. The importance of landslides in terms of tsunami hazard following a large earthquake should be an important element of tsunami hazard analysis for the Eastern Mediterranean, as in the case of the 1956 Amorgos earthquake and tsunami (OKAL et al. 2009).

Maximum wave heights calculated in the Black Sea for the shallow earthquakes defined in Fig. 2 indicate that <3-m tsunami wave heights could be expected along the southern coasts of Crimea, the northwestern coast of Turkey, the Bulgarian coast and the southeastern coasts of Romania, whereas along the eastern Black Sea coast, the expected maximum tsunami wave height is <1 m. A corresponding simplified maximum tsunami wave height zonation map is given Fig. 11. Minimum Mw values that may lead to 50-cm coastal wave height is in the range of 7.0–7.4 for the most part of the western Black Sea, whereas a higher magnitude Mw range of 7.5–7.9 is needed for the locations in the eastern Black Sea. Maximum wave height calculated for the Central and Eastern Mediterranean for the shallow earthquakes defined in Fig. 2 indicate that the expected maximum tsunami wave height is >3 m in locations in, around and orthogonal to the Hellenic Arc. In the rest of the Eastern Mediterranean, Southern Aegean, Tripoli (Libya), eastern Sicily, Calabria and western coasts Greece, the expected maximum tsunami wave height is <3 m. In the northern Aegean, Tunisia, western Sardinia, southwest coasts of Italy, and western and northern coasts of Sicily, the expected maximum tsunami wave height is <1 m. A corresponding simplified maximum tsunami wave height zonation map is given Fig. 12. Minimum Mw values that may possibly lead to 50-cm wave coastal wave height is in the range of 6.5–6.9 in locations in an around the Hellenic Arc, the Aegean Sea and south- and eastern Cyprus, whereas for southern coasts of Turkey, the Levantine coasts, northern Egypt and north-eastern Libya, a higher Mw range of 7.0–7.4 is needed. Earthquakes in the Mw range of 7.5–7.9 are capable of generating

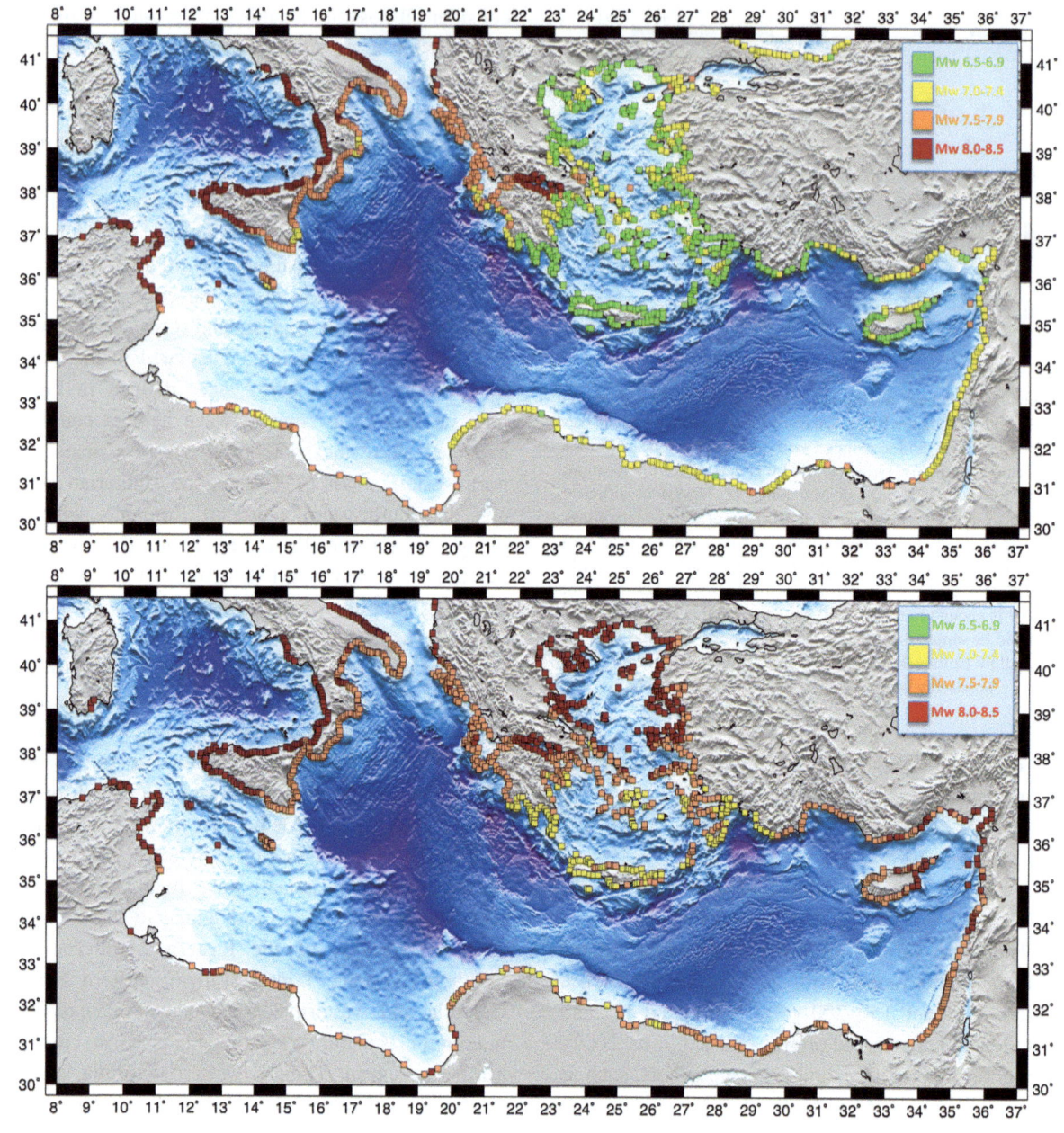

Figure 10
Minimum earthquake magnitude values leading to 50-cm coastal wave heights along the Eastern Mediterranean and Aegean coasts for the shallow- (*top*) and intermediate-depth (*bottom*) earthquakes defined in Fig. 2, and associated earthquake magnitudes. Locations in an around the Hellenic Arc, the Aegean Sea and south- and eastern Cyprus can be subject to 50-cm coastal wave heights for shallow earthquakes ranging from 6.5–6.9 and from 7.0–7.4 for intermediate-depth earthquakes. For the southern coasts of Turkey, ther Levantine coasts, northern Egypt and north-eastern Libya, a higher Mw range of 7.0–7.4 for shallow and 7.5–7.9 for the intermediate-depth earthquakes is needed. Shallow earthquakes in the Mw range of 7.5–7.9 are capable of generating a 50-cm coastal wave height in northern Libya, southern and eastern Sicily, southern Italy, Albania and along the western coasts of Greece. Only earthquakes in the Mw range of 8.0–8.5 are capable of generating 50-cm coastal wave height in Tunisia, in western and northern Sicily, at the western and Adriatic coasts of Italy, and in the Gulf of Corinth. It should be emphasized that these conclusions are valid only for the shallow earthquake sources given in Fig. 2 and excluding any possibility of an associated submarine landslide

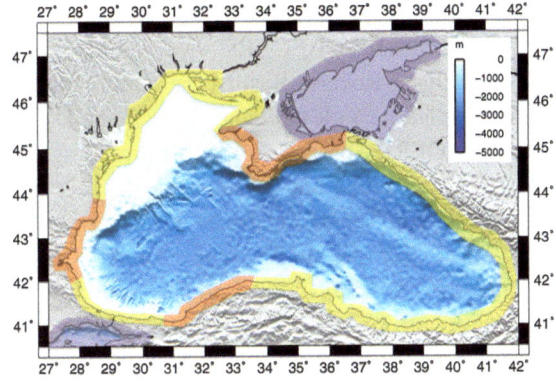

Figure 11
Simplified maximum tsunami wave height zonation map for the Black Sea derived from this study. *Colors* indicate the level of maximum tsunami wave height at the coastline of the respective zone (*orange* 1–3 m; *yellow* <1 m) due to earthquake sources given in Fig. 2 and excluding any possibly associated submarine landslide; source regions indicated by magenta on the map

50-cm coastal wave height in northern Libya, in southern and eastern Sicily, in southern Italy, Albania and along the western coast of Greece. Only earthquakes in the Mw range of 8.0–8.5 are capable of generating 50-cm coastal wave height in Tunisia, in western and northern Sicily, along the western and Adriatic coasts of Italy, and in the Gulf of Corinth. The results of the modeling are in accordance with the historical tsunami events in the study area. It should be emphasized that these conclusions should not be considered as complete for the areas shown in the maps since they are valid only for the earthquake sources given in Fig. 2 and excluding any possibility of an associated submarine landslide. On the other hand, minimum values of earthquake magnitudes leading to a 50-cm coastal wave height in the study area helps us to conclude that, when excluding secondary phenomenon such as a triggered submarine landslide, an Mw value of 6.5 could be accepted as a lower threshold for a tsunami early warning system in the region.

The first element of disaster risk reduction is assessment of the hazard. The main purpose of this study is to provide a basis for detailed tsunami hazard studies in Eastern Mediterranean, Aegean and Black Seas. Historical and instrumental studies reveal the complex nature of plate interactions and crustal deformation in the region, interactions that are mainly

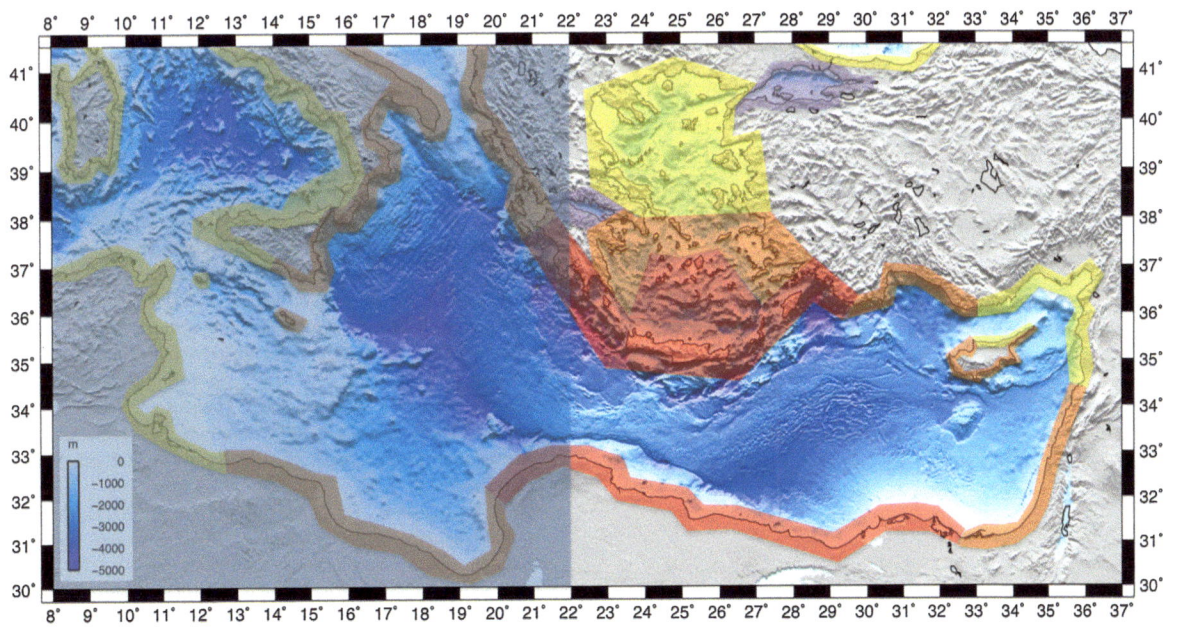

Figure 12
Simplified maximum tsunami wave height zonation map for the Central-Eastern Mediterranean and Aegean Sea derived from this study. *Colors* indicate the level of the maximum tsunami wave height at the coastlines of the respective zones (*red* >3 m; *orange* 1–3 m; *yellow* <1 m) due to earthquake sources given in Fig. 2 and excluding any possibly associated submarine landslide; earthquake source regions indicated by magenta on the map. Only offshore earthquake sources within 22°E–37°E are considered in this study, as shown in Fig. 2; thus, conclusions for the western part of the study area could be considered less reliable

evidenced by devastating earthquakes sometimes accompanied by catastrophic tsunamis. In addition to earthquakes as tsunami sources, massive land movements, such as in the case of the Santorini event around 1600 BC or the Fatsa Tsunami triggered by the Erzincan (Turkey) earthquake in 1939, gives a clear indication that the entire Eastern Mediterranean and its connected seas region is prone to tsunami events. A 365 AD-type earthquake could be expected approximately every 800 years and the fact that there has been only one other such event (in 1303) in the past 1,650 years, the modern-day tsunami hazard in the Eastern Mediterranean requires more attention (SHAW et al. 2008). Therefore, it is evident that a comprehensive tsunami hazard analysis is needed covering the Eastern Mediterranean, Aegean and Black Seas, since future tsunamis in the region could be even more damaging than past events, when considering increased population density and economic activity in the coastal zones, such as ports, shipyards, marinas, nuclear and thermoelectric power plants, oil refineries and coastal airports. In addition, the increased technological and organizational complexities of modern societies accompanied by a limitation of resources require both short- and long-term hazard assessment as an integrated element of sustainable development. ALPAR (2009) showed that the relative rise in sea-level calculated for Antalya is in the range of 2.6–4.3 mm/yr, which should not put the terrain bordering the Mediterranean in danger; however, the margins of deltaic plains in the region may be prone to inundation and local submergence, and storm surges will accelerate the vulnerability. This estimated sea-level rise could also mean that a 25-cm tsunami wave height would equate to twice the impact 50 years from now on. Therefore, long-term tsunami risk assessment studies should definitely consider the added effect of sea-level rise.

Acknowledgments

We would like to thank the European Commission's Joint Research Centre (EC-JRC), in particular Alessandro Annunziato, for his support during the realization of the modeling database, which was possible through a collaborative agreement between the Kandilli Observatory and Earthquake Research Institute (KOERI) and EC-JRC. We would like to especially thank Roberto Basili [The Istituto Nazionale di Geofisica e Vulcanologia (INGV)] and Doğan Kalafat (KOERI) for their supports providing the SHARE active faults and seismogenic sources and KOERI Moment Tensor databases, respectively. We would also like to thank Prof. Ahmet Cevdet Yalciner for providing the earthquake source database compiled in TRANSFER Project (http://www.transferproject.eu/) and for his feedback throughout the study. We would like to thank Dr. Ceren Özer Sözdinler for her feedback and support in the quality control of the tsunami scenario database. We also would like thank Mustafa Comoglu from KOERI for his assisstance in IT-related issues concerning the tsunami scenario database and to Dr. Mehmet Yılmazer for his support in the creation of the TTT maps. Last but not least, we would like to express our sincere gratitude to two anonymous reviewers who considerably helped us in improving the manuscript. Maps are produced with Generic Mapping Tools (GMT; WESSEL et al. 2013) and ArcMap 10 by Esri.

REFERENCES

AKSU, A. E., HALL, J., YALTIRAK, C. 2005, *Miocene to Recent tectonic evolution of the eastern Mediterranean: New pieces of the old Mediterranean puzzle*, Marine Geology *221*, 1–13.

ALPAR, B., 2009, "*Vulnerability of Turkish Coasts to Accelerated Sea-Level Rise*", Geomorphology, *107*(2009) 58–63.

ALPTEKIN, Ö., J. L. NÁBALEK, and N. TOKSÖZ, 1986, "*Source mechanism of the Bartin earthquake of September 3, 1968 in northwestern Turkey: Evidence for active thrust faulting at the southern Black Sea margin*", Tectonophysics, Vol. *122*, Issues 1–2, Pages 73–88.

ALTINOK, Y., B. ALPAR, N. OZER, and C. GAZIOGLU, 2005, "*1881 and 1949 earthquakes at the Chios-Cesme Strait (Aegean Sea) and their relation to tsunamis*", Natural Hazards and Earth System Sciences, *5*, 717–725.

ALTINOK, Y., ALPAR, B., ÖZER, N., and AYKURT, H. (2011), *Revision of the tsunami catalogue affecting Turkish coasts and surrounding regions*, Nat. hazards Earth Syst. Sci., *11*, 273–293.

AMBRASEYS, N. 2009. Earthquakes in the Mediterranean and Middle East, Cambridge University Press, ISBN 978-0-521-87292-8.

AMBRASEYS, N. N. and SYNOLAKIS C., 2010, "*Tsunami Catalogues for the Eastern Mediterranean*", Revisited, Journal of Earthquake Engineering, *14*, 3, 309–330.

ANNUNZIATO, A. (2007). *The Tsunami Assessment Modelling System by the Joint Research Centre*. Science of Tsunami Hazards *26*:2, 70–92.

BARKA, A., REILINGER, R., 1997. *Active Tectonics of the eastern Mediterranean region: deduced from GPS, neotectonic and seismicity data.* Annali di Geofisica XL (3), 586–608.

BASILI, R. TIBERTI, M. M, KASTELIC, V., ROMANO, F., PIATANESI, A., SELVA, J. and LORITO, S. (2013). *Integrating geologic fault data into tsunami hazard studies*; Nat. Hazards Earth Syst. Sci., *13*, 1025–1050.

BAYRAK, Y. and BAYRAK, E. (2011), *An Evaluation of Earthquake Hazard Potential for Different Regions in Western Anatolia Using the Historical and Instrumental Earthquake Data*, Pure Appl. Geophys., Volume *169*, Issue 10, pp 1859–1873.

BEISEL S., L. CHUBAROV, I. DIDENKULOVA, E. KIT, A. LEVIN, E. PELINOVSKY, Y. SHOKIN, and M. SLADKEVICH, 2009, "*The 1956 Greek tsunami recorded at Yafo, Israel, and its numerical modeling*", Journal of Geophysical Research, vol. *114*.

BEN-AVRAHAM Z, GARFUNKEL Z, LAZAR M (2008) *Geology and evolution of the southern Dead Sea fault with emphasis on subsurface structure.* Annu Rev Earth Planet Sci 36:357–387.

BENETATOS C., A. KIRATZI, C. PAPAZACHOS, G. KARAKAISIS, 2004, "*Focal mechanisms of shallow and intermediate depth earthquakes along the Hellenic arc*", Journal of Geodynamics, *37*, 253–296.

BERNARDI F. (2004) Earthquake source parameters in the Alpine-Mediterranean region from surface wave analysis, Diss. ETH Nr. 15652, Ph.D. Thesis.

BIJU-DUVAL, B.; LETOUZEY, J. and MONTADERT, L., 1978. Structure and evolution of the Mediterranean basins. In: HSUE, K. and MONTADERT et al. (Editors), Initial Report of the Deep Sea Drillin project, Vol. *42*, Part 1, pp. 951–984.

BOHNHOFF, M. HARJES, H-P and MEIER, T. (2005) *Deformation and stress regimes in the Hellenic subduction zone from focal Mechanisms* Journal of Seismology (2005) 9: 341–366.

CAPUTO R., C. ALEXANDROS, P. SPYROS, S. SOTIRIS, (2012), "*The Greek Database of Seismogenic Sources (GreDaSS): state-of-the-art for northern Greece*", Annals of Geophysics, 55, 5.

CONSTANTINESCU, L., RUPRECHTOVA, L. and ENESO, D., 1966. *Mediterranean-Alpine earthquake mechanisms and their seismotectonic implications.* Geophys. J.R. Astron. Soc., *10*: 347–368.

CIESM (2011). Marine geo-hazards in the Mediterranean. N° 42 in CIESMWorkshop Monographs [F. BRIAND Ed.], 192 pages, Monaco.

DELIBASIS, N., ZIAZIA, M., VOULGARIS, N., PAPADOPOULOS, T., STAVRAKAKIS, G., PAPANASTASSIOU, D., DRAKATOS, G. (1999), *Microseismic activity and seismotectonics of Heraklion Area (central Crete Island, Greece)*, Tectonophysics 308, 237–248.

DEWEY, J. F., PITMAN, W. C., RYAN, W. B. F. & BONNIN, J., 1973. "*Plate tectonics and the evolution of the Alpine system*", Geological Society of America Bulletin, 84:3137–3180.

DINEVA S. (1993) *Catalogue of Earthquakes in Bulgaria, 1981–1990*, Bulg. Acad. Sci., Geophys. Inst., Seismol. Dep., 39 pp.

DOMINEY-HOWES, D. T. M., G. S. HUMPHREYS, and P. P. HESSE, 2006, "*Tsunami and palaeotsunami depositional signatures and their potential value in understanding the late-Holocene tsunami record*", The Holocene, vol. *16*, 1095. doi:10.1177/0959683606069400.

DOTSENKO, S. F. and KONOVALOV, A. V. (1996). *Tsunami waves in the Black Sea in 1927: observations and numerical modeling (in Thermohydrodynamics of the Ocean)* Phys. Oceanogr.,Vol. 7, No. 6, pp. 389–401 (1996).

DOTSENKO, S. F. and A. V. INGEROV, 2007, "*Characteristics of Tsunami Waves In The Black Sea According to the Data of Measurements*", Physical Oceanography, Vol. *17*, No. 1.

DOTSENKO, S. F. and A. V. INGEROV, 2010, "*Numerical Modeling Of The Propagation And Strengthening Of Tsunami Waves Near The Crimean Peninsula And The Northeast Coast Of The Black Sea*", Physical Oceanography, Vol. *20*, No. 1, 2010.

DZIEWONSKI, A. M., CHOU T. A. and J. H. WOODHOUSE. (1981) *Determination of earthquake source parameters from waveform data for studies of global and regional seismicity.* J. Geophys. Res., *86*, 2825–2852 and subsequent quarterly papers on Phys. Earth Planet. Int.

EBELING, C. W., OKAL, E., KALLIGERIS, N., SYNOLAKIS, C. E. (2012) *Modern seismological reassessment and tsunami simulation of historical Hellenic Arc earthquakes*; Tectonophysics *530–531*, 225–239.

EVA C., RIUSCETTI M. and D. SLEJKO. (1988) *Seismicity of the Black Sea Region*, Boll. Geof. Teor. Appl., XXX, 117–118, 53–66.

EYIDOĞAN, H., J. A. JACKSON. (1985), "*A seismological study of normal faulting in the Demirci, Alaşehir and Gediz earthquakes of 1969–70 in western Turkey: implications for the nature and geometry of deformation in the continental crust*", Geophysical Journal of the Royal Astronomical Society, *81*, 569–607.

FOKAEFS, A. and PAPADOPOULOS, G. A. (2007). *Tsunami hazard in the Eastern Mediterranean: strong earthquakes and tsunamis in Cyprus and the Levantine Sea.* Natural Hazards 40:503–526.

GANAS, A. and T. PARSONS, 2009, "*Three-dimensional model of Hellenic Arc deformation and origin of the Cretan uplift*", Journal of Geophysical Research, vol. *114*.

GARFUNKEL, Z., 2004. *Origin of the Eastern Mediterranean basin: a reevaluation.* Tectonophysics *391*, 11–34.

Gorur, N. (1997) Crateceous syn- to postrift sedimantation on the Southern Continental Margin of the Western Black Sea Basin, in Regional and Petroleum Geology of the Black Sea and Surrounding Region, edited by A. G. ROBIN-SON, American Association of Petroleum Geologists (AAPG), AAPG Memoir 68, pp. 227–240.

HALL, J., AKSU, A., YALTIRAK, C., WINSOR, J., 2009. *Structural architecture of the Rhodes Basin: a deep depocentre that evolved since the Pliocene at the junction of the Hellenic and Cyprus arcs, eastern Mediterranean.* Marine Geology 258, 1–23.

HANKS, T. C. and KANAMORI, H. (1979). *A moment magnitude scale* Journal of Geophysical Research: Solid Earth, Volume *84*, Issue B5, pages 2348–2350.

HATZFELD, D., BESNARD, M., MAKROPOULOS, K. and HATZIDIMITRIOU, P. (1993), *Microearthquake seismicity and fault-plane solutions in the southern Aegean and its geodynamic implications*, Geophys. J. Int. *115*, 799–818.

HEURET, A., S. LALLEMAND, F. FUNICIELLO, C. PIROMALLO, and C. FACCENNA (2011), *Physical characteristics of subduction interface type seismogenic zones revisited*, Geochem. Geophys. Geosyst., Volume *12*, Issue 1, January 2011.

HOWE, T. M. and P. BIRD, 2010, "*Exploratory models of long-term crustal flow and resulting seismicity across the Alpine-Aegean orogen*", Tectonics, vol. *29*, 4.

HYNDMAN, R. D., YAMANO, M. & OLESKEVICH, D. A., 1997. *The seismogenic zone of subduction thrust faults*, Island Arcs, *6*, 244–260.

JOLIVET, L. FACCENNA, C., HUET, B., LABROUSSE, L., LE POURHIET, L., LACOMBE, O., LECOMTE, E., BUROV, E., DENÈLE, Y., BRUN, J-P., PHILIPPON, M., PAUL, A., SALAÜN, G., KARABULUT, H., PIROMALLO,

C., Monié, P., Gueydan, F., Okay, A., Oberhänsli, R., Pourteau, A., Augier, R., Gadenne, L., Driussiü, O. (2013), *Aegean tectonics: Strain localisation, slab tearing and trench retreat*, Tectonophysics 597–598, 1–33.

Jost, M. L., O. Knabenbauer, J. Cheng, H-P. Harjes, 2002, "*Fault plane solutions of microearthquakes and small events in the Hellenic arc*", Tectonophysics, *356*, 87–114.

Kalafat, K., Kekovali, K., Gunes, Y., Yilmazer, M., Kara, M., Deniz, P., Berberoglu, M. (2009). A catalogue of Source Parameters of Moderate and Strong Earthquakes for Turkey and its Surrounding Area (1938–2008), Bogazici University Press, ISBN 978-975-518-303-9.

Kanamori, H. (1977) *The energy release in great earthquakes*, Journal of Geophysical Research:Solid Earth and Planets, Volume 82, Issue 20, pages 2981–2987.

Karabulut, H., Z. Roumelioti, C. Benetatos, A.K. Mutlu, S. Özalaybey, M. Aktar and A. Kiratzi, 2006, "*A source study of the 6 July 2003 (Mw 5.7) earthquake sequence in the Gulf of Saros (northern Aegean Sea): Seismological evi- dence for the western continuation of the Ganos fault*", Tectonophysics, *412*, 195–216.

Kiratzi, A. A. and C. B. Papazachos. 1995. *Active Deformation of the Shallow Part of the Subducting Lithospheric Slab in the Southern Aegean*. J. Geodyn. *19*, 65–78.

Kiratzi, A., E. Louvari, 2003, "*Focal mechanisms of shallow earthquakes in the Aegean Sea and the surrounding lands determined by waveform modelling: a new database*", Journal of Geodynamics, *36*, 251–274.

Koravos, G. Ch., I. G. Main, T. M. Tsapanos and R. M. W. Musson, 2003, "*Maximum earthquake magnitudes in the Aegean area constrained by tectonic moment release rates*", Geophysical Journal International, *152*, 94–112.

Kreemer, C. and N. Chamot-Rooke, 2004, "*Contemporary kinematics of the southern Aegean and the Mediterranean Ridge*", Geophysical Journal International, *157*, 1377–1392.

Le Pichon, X. and J. Angelier, 1979. *The Hellenic Arc and Trench system: A Key to the Neotectonic Evolution of the Eastern Mediterranean Area*. Tectonophysics, *60*, 1–42.

Leonard, M., 2010, "*Earthquake Fault Scaling: Self-Consistent Relating of Rupture Length, Width, Average Displacement, and Moment Release*", Bulletin of the Seismological Society of America, Vol. *100*, No. 5A, pp. 1971–1988.

Lorito, S., Tiberti, M. M., Basili, R., Piatanesi, A. and Valensise, G. (2008). *Earthquake-generated tsunamis in the Mediterranean Sea: Scenarios of potential threats to Southern Italy*; Journal of Geophysical Research, Vol. *113*, B01301.

Løvholt, F., S. Glimsdal, C. B. Harbitz, N. Zamora, F. Nadim, P. Peduzzi, H. Dao, H. Smebye, 2012, "*Tsunami hazard and exposure on the global scale*", Earth-Science Reviews, *110*, 58–73.

Mader, C. L. (2004). Numerical Modeling of Water Waves, CRC Press; 2nd edition, ISBN 0-8493-2311-8.

McClusky, S., S. Balassanian, A. Barka, C. Demir, S. Ergintav, I. Georgiev, O. Gürkan, M. Hamburger, K. Hurst, H. Kahle, K. Kastens, G. Kekelidze, R. King, V. Kotzev, O. Lenk, S. Mahmoud, A. Mishin, M. Nadariya, A. Ouzounis, D. Paradissis, Y. Peter, M. Prilepin, R. Reilinger, I. Şanli, H. Seeger, A. Tealeb, M.N. Toksöz and G. Veis. 2000. *Global Positioning System Constraints on Plate Kinematics and Dynamics in the Eastern Mediterranean and Caucasus*. Journal of Geophysical Research, *105*, 5695–5719.

McClusky, S., R. Reilinger, S. Mahmoud, D. Ben-Sari and A. Tealeb, 2003, "*GPS constraints on Africa (Nubia) and Arabia plate motions*", Geophysical Journal International, *155*, 126–138.

McKenzie, D. P. (1970). *Plate Tectonics of the Mediterranean Region*. Nature *226*, 239–243.

McKenzie, D. P., (1972), "*Active tectonics of the Mediterranean region*", Geophys. J. R. Ast. Soc. *30*, 109–185.

McKenzie, D. P., (1978), "*Active tectonics of the Alpine–Himalayan belt: the Aegean Sea and surrounding regions*", Geophysical Journal of the Royal Astronomical Society, *55*, 217–254.

Meier, T., Becker, D., Endrun, B., Rische, M., Bohnhoff, M., Stöckhert, B. and Harjes, H.-P. (2007) *A model for the Hellenic subduction zone in the area of Crete based on seismological investigations*, Geological Society, London, Special Publications 2007, v. *291*; pp. 183–199.

Mitsakaki, C., M. G. Sakellariou, D. Tsinas, 2013, "*A study of the crust stress field for the Aegean region (Greece)*", Tectonophysics, Vol. *597–598*, 50–72.

Mitsoudis, D. A., E. T. Flour, N. Chrysoulakis, Y. Kamarianakis, E. A. Okal, C. E. Synolakis, 2012, "*Tsunami hazard in the southeast Aegean Sea*", Coastal Engineering, *60*, 136–148.

Moratto, L., B. Orlecka-Sikora, G. Costa, P. Suhadolc, Ch. Papaioannou, C. B. Papazachos, 2007, "*A deterministic seismic hazard analysis for shallow earthquakes in Greece*", Tectonophysics, *442*, 66–82.

Necmioğlu, Ö. and Özel, N. M. (2014). An Earthquake Source Sensitivity Analysis for Tsunami Propagation in the Eastern Mediterranean. Oceanography 27(2): 76–85.

Nikishin, A. M., Korotaev, M. V., Ershov, A. V., Brunet, M-F. (2003). *The Black Sea basin: tectonic history and Neogene–Quaternary rapid subsidence modelling*, Sedimentary Geology *156*,149–168.

Okada, Y., 1985, "*Surface Deformation due to Shear and Tensile Faults in a Half-Space*", Bulletin of the Seismological Society of America, *75*, 1135–1154.

Okal, E. A., C. E. Synolakis, B. Uslu, N. Kalligeris and E. Voukouvalas, 2009, "*The 1956 earthquake and tsunami in Amorgos, Greece*", Geophysical Journal International, *178*, 1533–1554.

Ozel, M. N., Necmioglu, O., Yalciner, A.C., Kalafat, D., Mustafa, E. (2011). *Tsunami hazard in the Eastern Mediterranean and its connected seas: Toward a Tsunami warning center in Turkey*, Soil Dynamics and Earthquake Engineering, Volume *31*, Issue 4, April 2011, Pages 598–610.

Papadimitriou, E. E., and Karakostas, V. G. (2008) *Rupture model of the great AD 365 Crete earthquake in the southwestern part of the Hellenic Arc*; Acta Geophysica, Volume 56, Issue 2, pp. 293–312.

Papadopoulos, G. (2011), The Seismic History of Crete, Ocelatos Publications, ISBN 978-960-9499-68-2.

Papadopoulos, G. A., G. Diakogianni, A. Fokaefs, and B. Ranguelov, 2011, "*Tsunami hazard in the Black Sea and the Azov Sea: a new tsunami catalogue*", Natural Hazards Earth System Science, *11*, 945–963.

Papadopoulos, G. A, Gràcia, E., Urgeles, R., Sallares, V., De Martini, P. M., Pantosti, D., M. González, Yalciner, A. C., Mascle, J., Sakellariou, D., Salamon, A., Tinti, S., Karastathis, V., Fokaefs, A., Camerlenghi, A., Novikova, T., Papageorgiou, A. (2014). *Historical and pre-historical tsunamis in the Mediterranean and its connected seas: Geological* signatures, generation mechanisms and coastal impacts. Marine Geology *354* (2014) 81–109.

PAPAIOANNOU, CH. A. and B. C. PAPAZACHOS, 2000, "*Time-Independent and Time-Dependent Seismic Hazard in Greece Based on Seismogenic Sources*, Bulletin of the Seismological Society of America, *90*, 1, 22–33.

PAPAZACHOS, B. C., 1990. *Seismicity of the Aegean and surrounding area*. Tectonophysics *178*, 287–308.

PAPAZACHOS, B. C., CH. KOUTITAS, P. M. HATZIDIMITRIOU, B. M., KARAKOSTAS AND C. A. PAPAIOANNOU, 1986, "*Tsunami hazard in Greece and the surrounding area*", Annales Geophysicae, *4*, 79–90.

PAPAZACHOS, B. C., 1996, "*Large Seismic Faults in the Hellenic Arc*", Annali di Geofisica, Vol. *XXXIX*, No.5.

PAPAZACHOS, B., A. KIRATZI, E. PARADIMITRIOU, 1991, "*Regional focal mechanism for earthquakes in the Aegean area*", Pure Applied Geophysics, *4*, 405–419.

PAPAZACHOS, B. C. and PAPAIOANNOU, CH. A. 1993. *Long-term earthquake prediction in the Aegean area based on a time and magnitude predictable model*. Pageoph, *140*, 593–612, 1993.

PAPAZACHOS, B. C. and CH. A. PAPAIOANNOU, 1999, "*Lithospheric boundaries and plate motions in the Cyprus area*", Tectonophysics, *308*, 193–204.

PARARAS-CARAYANNIS, G. (2011). *The earthquake and tsunami of July 21, 365 AD in the Eastern Mediterranean Sea – A review of Impact on the Ancient World – Assessment of Recurrence and Future Impact*. Science of Tsunami Hazards, Vol. *30*, No. 4, page 253.

PARKE, G., (2001), Active Tectonics and Sedimentary Process in Western Turkey, Ph.D. Dissertation, University of Cambridge.

PILIDOU, S., PRIESTLEY, K., JACKSON, J., and MAGGI, A. (2004). *The 1996 Cyprus earthquake: a large, deep event in the Cyprean Arc*. Geophys. J. Int. *158*, 85–97.

PONDRELLI, S., MORELLI, A., EKSTRÖM, G. (2004). *European-Mediterranean regional centroid-moment tensor catalog: solutions for years 2001 and 2002*. Physics of the Earth and Planetary Interiors *145*, 127–147.

PONDRELLI, S., SALIMBENIA, S., MORELLI, A., EKSTRÖM, G., POSTPISCHL, L., VANNUCCI, G., BOSCHI, E. (2011) *European–Mediterranean Regional Centroid Moment Tensor catalog: Solutions for 2005–2008*, Physics of the Earth and Planetary Interiors *185* (2011) 74–81.

PONDRELLI, S., MORELLI, A., EKSTRÖM, G., MAZZA, S., BOSCHI, E., DZIEWONSKI, A.M. (2002) *European–Mediterranean regional centroid-moment tensors: 1997–2000*, Physics of the Earth and Planetary Interiors *130*, 71–101.

RADULIAN M., POPESCU E., BALA A. and A. (2002) UTALE *Catalog of fault plane solutions for the earthquakes occurred on the Romanian territory*, Rom. Journ. Phys., *47*, 663–685.

RANGUELOV, B. and D. GOSPODINOV, 1994, "*Seismic activity after the earthquake of 31 March, 1901 in the Shabla-Kaliakra zone (in Bulgarian)*", Bulgarian Geophysical Journal, *20*, 44–49.

RASTSVETAEV, L. M., 1987. Tectono-dynamical environments of the Great Caucasus Alpine structure origin. In: Milanovsky, E. E., Koronovsky, N. V. (Eds.), Geology and Mineral Resources of the Great Caucasus. Nauka, Moscow, pp. 69–96. in Russian.

REILINGER, R. E., MCCLUSKY, S. C., ORAL, M. B., KING, R. W., TOKSOZ, M. N., BARKA, A. A., KINIK, I., LENK, O., SANLI, I. (1997). *Global positioning system measurements of present-day crustal move- ments in the Arabia – Africa – Eurasia plate collision zone*. Journal of Geophysical Research *102* (B5), 9983–9999.

REILINGER, R., MCCLUSKY, S., VERNANT, P., LAWRENCE, S., ERGINTAV, S., CAKMAK, R., OZENER, H., KADIROV, F. GULIEV, I., STEPANYAN, R., NADARIYA, M., HAHUBIA, G., MAHMOUD,S., SAKR,K., ARRAJEHI, A., PARADISSIS, D., AL-AYDRUS, A., PRILEPIN, M., GUSEVA, T., EVREN,E., DMITROTSA, A., FILIKOV, S. V., GOMEZ, F., AL-GHAZZI,R., and KARAM, G. (2006), *GPS constraints on continental deformation in the Africa-Arabia- Eurasia continental collision zone and implications for the dynamics of plate interactions*, Journal of Geophysical Research, Vol. *111*, B05411.

REILINGER, R., MCCLUSKY, S., PARADISSIS, D., ERGINTAV, S., VERNANT, P. (2010), *Geodetic constraints on the tectonic evolution of the Aegean region and strain accumulation along the Hellenic subduction zone*, Tectonophysics *488*, 22–30.

ROBERTSON, A. H. F., DIXON, J. E., 1984. *Introduction: aspects of the geological evolution of the Eastern Mediterranean*, Geological Evolution of the Eastern Mediterranean. Geol. Soc. Spec. Publ. London, vol. *17*, pp. 1–74.

ROBERTSON, A. H. F. AND MOUNTRAKIS, D. (2006) *Tectonic development of the Eastern Mediterranean region: an introduction*, Geological Society, London, Special Publications 2006; v. *260*; p. 1–9.

SAATÇILAR, R., ERGINTAV, S., DEMIRBAĞ, E., INAN, S. (1999), *Character of active faulting in the North Aegean Sea*, Marine Geology *160*, 339–353.

SALAMON, A., A. HOFSTETTER, Z. GARFUNKEL, and H. RON, 2003, "*Seismotectonics of the Sinai subplate: the Eastern Mediterranean region*", Geophysical Journal International, *155*, 149–173.

SALAMON, A., ROCKWELL, T., WARD, S. N., GUIDOBONI, E., and COMASTRI, A., 2007, *Tsunami hazard evaluation of the Eastern Mediterranean: Historical analysis and selected modeling*: Seismological Society of America Bulletin, v. *97*, p. 705–724.

SALEH, S. (2013), *3D Crustal and Lithospheric Structures in the Southeastern Mediterranean and Northeastern Egypt*, Pure Appl. Geophys. *170*, 2037–2074.

SATAKE, K. 2011. *Unforecasted Earthquake and Forgotten Tsunamis: Lessons from 2011 Tohoku Event*. American Geophysical Union, Fall Meeting 2011, Abstract #NH12A-05.

SBORAS, S., PAVLIDES, S., CAPUTO, R., CHATZIPETROS, A., MICHAILIDOU, A., VALKANIOTIS., A. and PAPATHANASIOU, G. (2011). Improving resolution of seismic hazard estimates for critical facilities: The Database of Greek Crustal Seismogenic Sources in the Frame of the SHARE Project. Proceedings of the 30° Convegno Nazionale GNGTS, 14–17 November, 2011, Trieste, Extended Abstracts, 232–235.

SCHOLZ, C. H., 1998. *Earthquakes and friction laws*. Nature, *391*, 37–42.

SHAW, B., AMBRASEYS, N. N., ENGLAND, P. C., FLOYD, M. A., GORMAN, G. J, HIGHAM, T. F. G., JACKSON, J. A., NOCQUET, J.-M., PAIN, C. C., and PIGGOTT, M. D. (2008). *Eastern Mediterranean tectonics and tsunami hazard inferred from the AD 365 earthquake*; Nature Geoscience Vol. *1* APRIL 2008.

SHAW, B. and JACKSON, J. (2010). *Earthquake mechanisms and active tectonics of the Hellenic subduction zone*. Geophysical Journal International *181*, 966–984.

SHOWSTACK, R. (2014). *Fukushima Nuclear Accident Report Calls for More Focus on Threats From Extreme Events*. Eos, Vol. *95*, No. 31.

SNOPEK, K., MEIER, T., ENDRUN, B., BOHNHOFF, M., CASTEN, U. (2007), *Comparison of gravimetric and seismic constraints on the structure of the Aegean lithosphere in the forearc of the*

Hellenic subduction zone in the area of Crete, Journal of Geodynamics.

Soloviev, S. L., 1990, "Tsunamigenic zones in the Mediterranean sea", Natural Hazards, 3, 183–202.

Soloviev, S. L., Solovieva, O. N., Go, C. N., Kim, K. S., Shchetnikov, N. A. (2000). Tsunamis in the Mediterranean Sea – 2000 B.C.–2000 A.D., Kluwer Academic Publishers, 237 pp.

Sørensen, M.B., M. Spada, A. Babeyko, S. Wiemer and G. Grünthal, 2012, "Probabilistic tsunami hazard in the Mediterranean Sea", Journal of Geophysical Research, vol. 117, B01305.

Stampfli, G. M., Borel, G. D., 2002. A plate tectonic model for the Paleozoic and Mesozoic constrained by dynamic plate boundaries and restored synthetic ocean isochrons. Earth Planet. Sci. Lett. 196, 17–33.

Stein, S., Geller, R.J., Liu, M. 2012. Why Earthquake Hazard Maps Often Fail and What to do About it? Tectonophysics 562–563 (2012) 1–25.

Stern, R. J., 2002. Subduction zones, Rev. Geophys., 40, 3–1.

Strobl, M., R. Hetzela, C. Fassoulas and P. W. Kubik, 2014, A Long-Term Rock Uplift Rate for Eastern Crete and Geodynamic Implications for the Hellenic Subduction Zone, Journal of Geodynamics, 78 (2014) 21–31.

Şengör, A. M. C., N. Görür, F. Saroğlu, 1985, Strike–slip faulting and related basin formation in zones of tectonic escape: Turkey as a case study, In: K. T. Biddle, N. Christie-Blick (Eds.), Strike – slip Deformation, Basin Formation and Sedimentation, Society of Economic Paleontologists and Mineralogists Special Publication, 37, 227–264.

Tari, E. Sahin, M., Barka, A., Reilinger, R., King, R.W., McClusky, S., Prilepin, M. (2000) Active tectonics of the Black Sea with GPS, Earth Planets Space, 52, 747–751.

Taymaz, T., J. A. Jackson and R. Westaway, 1990, "Earthquake mechanisms in the Hellenic trench near Create", Geophysical Journal International, 102, 695–731.

Taymaz, T., J. A. Jackson and D. McKenzie, 1991, "Active tectonics of the north and central Aegean Sea", Geophysical Journal International, 106, 433–490.

Tinti, S., A. Armigliato, G. Pagnoni, and F. Zaniboni. (2005). Scenarios of giant tsunamis of tectonic origin in the Mediterranean, ISET J. Earthquake Technol., 42, 171–188.

Tinti, S., Armigliato, A., Pagnoni, G., Zaniboni, F. and Tonini, R. (2011). Tsunamis in the Euro-Mediterranean region: emergency and long term counter measures. In CIESM, 2011. Marine geohazards in the Mediterranean. N° 42 in CIESMWorkshop Monographs [F. Briand Ed.], 192 pages, Monaco.

Udias A., Buforn E. and J. Ruiz De Gauna (1989). Catalogue of Focal Mechanisms of European Earthquakes. Department of Geophysics, Universidad Complutense, Madrid.

Vannucci, G. and Gasperini, P. (2004). The new release of the database of Earthquake Mechanisms of the Mediterranean Area (EMMA Version 2), Annals of Geophysics, Supplement to V. 47, N.1, 307–334.

Volvovsky, B. S., 1989. Seismicity. In: Beloussov, V. V., Volvovsky, B. S. (Eds.), Structure and Evolution of the Earth's Crust and Upper Mantle of the Black Sea. Nauka, Moscow, pp. 95–97. in Russian.

Wessel, P., W. H. F. Smith, R. Scharroo, J. F. Luis, and F. Wobbe, Generic Mapping Tools: Improved version released, EOS Trans. AGU, 94, 409–410, 2013.

Yalçıner, A. C., Pelinovsky, E., Talipova, T., Kurkin, A., Kozelkov, A. & Zaitsev, A. 2004. Tsunami in the Black Sea: comparison of the historical, instrumental and numerical data. Journal of Geophysical Research, 109, C12023, doi:10.1029/2003JC002113.

Yolsal, S., T. Taymaz and A. C. Yalçıner, 2007, "Understanding tsunamis, potential source regions and tsunami-prone mechanisms in the Eastern Mediterranean", Geological Society,.

Yolsal-Çevikbilen, S. and T. Taymaz, 2012, "Earthquake Source Parameters Along the Hellenic Subduction Zone and Numerical Simulations of Historical Tsunamis in the Eastern Mediterranean", Tectonophysics, 536–537, 61–100.

London, Special Publications, 291, 201–230.

Zamora, N., Franchello, G., and Anunziato, A. (2014). Validation of the JRC Tsunami Propagation and Inundation Codes; Science of Tsunami Hazards, Volume 33, Number 2, Pages 112–132.

(Received September 5, 2014, revised March 7, 2015, accepted March 12, 2015, Published online March 25, 2015)

Pure and Applied Geophysics

Tsunami Squares Approach to Landslide-Generated Waves: Application to Gongjiafang Landslide, Three Gorges Reservoir, China

LILI XIAO,[1] STEVEN N. WARD,[2] and JIAJIA WANG[1]

Abstract—We have developed a new method, named "Tsunami Squares", for modeling of landslides and landslide-generated waves. The approach has the advantages of the previous "Tsunami Ball" method, for example, separate, special treatment for dry and wet cells is not needed, but obviates the use of millions of individual particles. Simulations now can be expanded to spatial scales not previously possible. The new method accelerates and transports "squares" of material that are fractured into new squares in such a way as to conserve volume and linear momentum. The simulation first generates landslide motion as constrained by direct observation. It then computes induced water waves, given assumptions about energy and momentum transfer. We demonstrated and validated the Tsunami Squares method by modeling the 2008 Three Gorges Reservoir Gongjiafang landslide and river tsunami. The landslide's progressive failure, the wave generated, and its subsequent propagation and run-up are well reproduced. On a laptop computer Tsunami Square simulations flexibly handle a wide variety of waves and flows, and are excellent techniques for risk estimation.

Key words: Tsunami squares, numerical modeling, landslide, waves in reservoirs.

1. Introduction

Large, high-velocity landslides entering confined water bodies (reservoirs, lakes, fiords, and rivers) may cause potentially extreme tsunami run-up that increases the variety, range, and severity of effects attributable to the landslide alone. Alaska's 1958 Lituya Bay tsunami caused by a subaerial landslide has attracted the attention of scholars, and a series of experimental and numerical modeling research projects has been reported (BASU *et al.* 2010; FRITZ *et al.* 2009; WEISS *et al.* 2009). Two-dimensional cross-sections of the landslide and the proximal wave were modeled

---

[1] Engineering Faculty, China University of Geosciences (Wuhan), Wuhan, China. E-mail: xiao749@126.com
[2] Institute of Geophysics and Planetary Physics, University of California, Santa Cruz, USA.

numerically by MADER and GITTINGS (2002) and QUECEDO *et al.* (2004) with full Navier–Stokes hydrodynamic codes. WARD and DAY (2010) developed a novel "Tsunami Ball" method, which they demonstrated and validated by simulating the landslide and the propagation of the tsunami to the seaward end of Lituya Bay. WARD (2014) (https://www.youtube.com/watch?v=6COeNRToYqU) revisited this case by using the Tsunami Squares method described herein.

Other examples of well-documented rockslide tsunami include the 1934 Tafjord event in Norway, for which HARBITZ *et al.* (2014a, b) recently applied numerical models (denoted GloBouss and DpWaves) based on Boussinesq equations. A similar Boussinesq model (Geowave) has also been applied to several examples of landslide-generated waves, mainly for open-ocean submarine landslides (WATTS *et al.* 2003; POISSON and PEDREROS 2010; WATTS and TAPPIN 2012).

We have developed a new method, named "Tsunami Squares", for modeling of landslides and landslide-generated waves. Tsunami Squares can simulate landslide evolution, wave propagation, the moving mass of impacting water, and the triggering wave process. Tsunami Squares is suited both to long-term hazard assessment in regions where many thousands of possible landslides might need to be modeled, for example China's Three Gorges Reservoir, and to emergencies in which accelerating creep of an unstable slope has been detected and emergency managers require a rapid assessment of the tsunami hazard.

2. Overview of the Gongjiafang Landslide

On the afternoon of November 23, 2008, 120 km up-river of China's Three Gorges Dam and 4 km downstream of Wushan County (Figs. 1, 2), the Gongjiafang landslide slipped from the north shore of the Yangtze and generated a wave that crossed the river and washed the far bank to a height of 13 m. The tsunami damaged the Wuxia town docks 3.5 km upstream, some navigation aids, roads, orange trees, and a few small boats. Luckily, no ships were passing through the busy river corridor, and no one was hurt. The direct economic loss was limited to 800,000 USD (HUANG et al. 2012).

The Gongjiafang landslide happened after a test impoundment brought the Three Gorges Reservoir to a new high water level of 172.8 m. The landslide was located on a scarp slope on the west of Hengshixi anticline and the east of Wushan syncline (WU et al. 2010). The slide mass consisted of thinly-bedded soft marl stones intersected with thick limestone and dolomite limestone. Because of strong weathering and joint damage, the mass had been fractured into small rocks and vegetated soils. The high shale content, large developed fissures, and strong weathering resulted in numerous discontinuities in the rock. In November, the lower part of the slide mass was submerged as the reservoir level rose to its new high level. The clays in the marlstone and shale softened easily, aggravating a decrease of rock strength. After being submerged for a long period, the shallow rock structure at the slope toe began to fail along tension joints and the upper dry part soon followed.

3. Tsunami Squares Theory

The two-dimensional Tsunami Squares approach evolved from the "Tsunami Ball" method which proved to be effective for simulating flow and flow-like movements (WARD and DAY 2005, 2008, 2010). Tsunami Squares inherits advantages of the

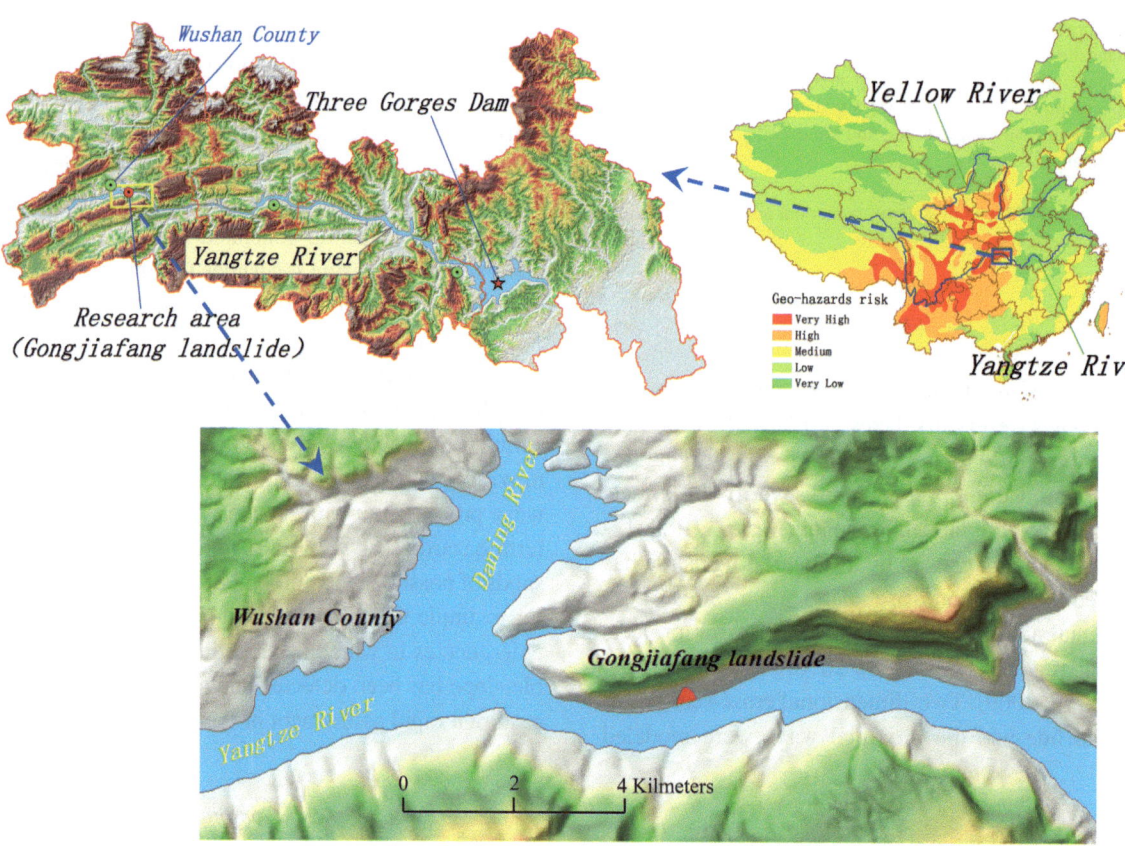

Figure 1
Location of Gongjifang Landslide. The lowest map shows the area 6 km up and down river from the landslide in which the tsunami was modeled. Populated Wushan County, on the left, is the area affected

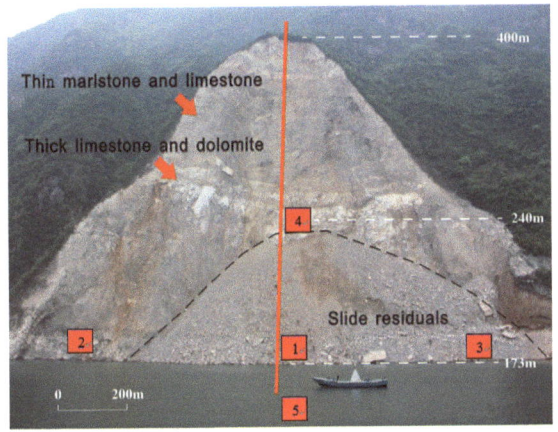

Figure 2
Landforms at Guongjiafang after the slide (Yichang Center of China Geological Survey). The slide stopped funnel-shaped on the riverbed, mainly distributed below 200 m but with thin deposits above 220 m. The numbers in the red boxes show the positions where slide velocities were tracked during modeling

Tsunami Ball method, for example not needing special treatment for dry versus wet cells. Moving materials in the Tsunami Squares method are, however, made from divisible squares, which obviates the need for millions of individual particles. Simulations can be extended to spatial scales not previously possible.

3.1. Mass Flow and Wave Propagation

Typical tsunami calculations for earthquake and landslide-generated tsunamis (SATAKE and TANIOKA 1995; LIU et al. 2003; TITOV and GONZALEZ 1997) solve non-linear, long wave continuity, and momentum equations for variations in water column thickness $H(\mathbf{r},t)$ and depth averaged horizontal water velocity $\mathbf{v}(\mathbf{r},t)$ at points $\mathbf{r} = (x,y)$

$$\frac{\partial H(\mathbf{r},t)}{\partial t} = -\nabla_h \bullet [\mathbf{v}(\mathbf{r},t) H(\mathbf{r},t)] \quad (1)$$

and

$$\frac{\partial H(\mathbf{r},t)\mathbf{v}(\mathbf{r},t)}{\partial t} = -\nabla_h \bullet [\mathbf{v}(\mathbf{r},t)\mathbf{v}(\mathbf{r},t) H(\mathbf{r},t)] \\ - gH(\mathbf{r},t)\nabla_h \zeta(\mathbf{r},t) \quad (2)$$

or

$$\frac{\partial \mathbf{v}(\mathbf{r},t)}{\partial t} = -\mathbf{v}(\mathbf{r},t) \bullet \nabla_h \mathbf{v}(\mathbf{r},t) - g\nabla_h \zeta(\mathbf{r},t)$$

where g is the acceleration of gravity, ∇_h is the horizontal gradient, $\zeta(\mathbf{r},t)$ is the elevation of the water, and t is time. (To relax the long wave assumption and include wave dispersion, replace Eq. (2) by Eq. (22) in Appendix 1. For our purposes here, water wave dispersion is unnecessary.)

For small steps dt, Eqs. (1) and (2) become:

$$H(\mathbf{r},t+dt) = H(\mathbf{r},t) - \nabla_h \bullet [\mathbf{v}(\mathbf{r},t)H(\mathbf{r},t)]dt \quad (3)$$

and

$$H(\mathbf{r},t+dt)v(\mathbf{r},t+dt) = H(\mathbf{r},t)\mathbf{v}(\mathbf{r},t) \\ - \nabla_h \bullet [\mathbf{v}(\mathbf{r},t)\mathbf{v}(\mathbf{r},t)H(\mathbf{r},t)]dt \\ - gH(\mathbf{r},t)\nabla_h \zeta(\mathbf{r},t)dt \quad (4)$$

Tsunami Squares solves equations that are equivalent to these, but by use of a new approach. It considers a regular set of N square cells with dimension D_c (the length of the uniform squares) and center points $\mathbf{r}_i = (x_i, y_i)$. At time t, each cell holds material (water or landslide mass) of thickness $H_i(t) = H(\mathbf{r}_i,t)$, with mean horizontal velocity $\mathbf{v}_i(t) = \mathbf{v}(\mathbf{r}_i,t)$ and mean horizontal acceleration $\mathbf{a}_i(t) = \mathbf{a}(\mathbf{r}_i,t)$ (Fig. 3, top left). For locations outside the flow, $H_i(t)$ would be zero. The entire concept of wave propagation or mass flow involves updating those conditions to time $t + dt$, where dt is some small time interval.

Pick one cell, for example $i = 10$ (red square, Fig. 3, top right). With its known velocity and acceleration, displace the cell material to a new center point:

$$\tilde{\mathbf{r}}_i = \mathbf{r}_i(t) + \mathbf{v}_i(t)dt + 0.5\mathbf{a}_i(t)dt^2 \\ = \mathbf{r}_i(t) + 0.5[\mathbf{v}_i(t) + \tilde{\mathbf{v}}_i]dt \\ = [\tilde{x}_i(t), \tilde{y}_i(t)] \quad (5)$$

and give it a new mean velocity:

$$\tilde{\mathbf{v}}_i = \mathbf{v}_i(t) + \mathbf{a}_i(t)dt \\ = [\tilde{v}_{xi}(t), \tilde{v}_{yi}(t)] \quad (6)$$

we wish to partition the volume and linear momentum of the material in the displaced cell among the N original cells. It is logical that the partitioned volume of the displaced ith cell into the jth original cell is:

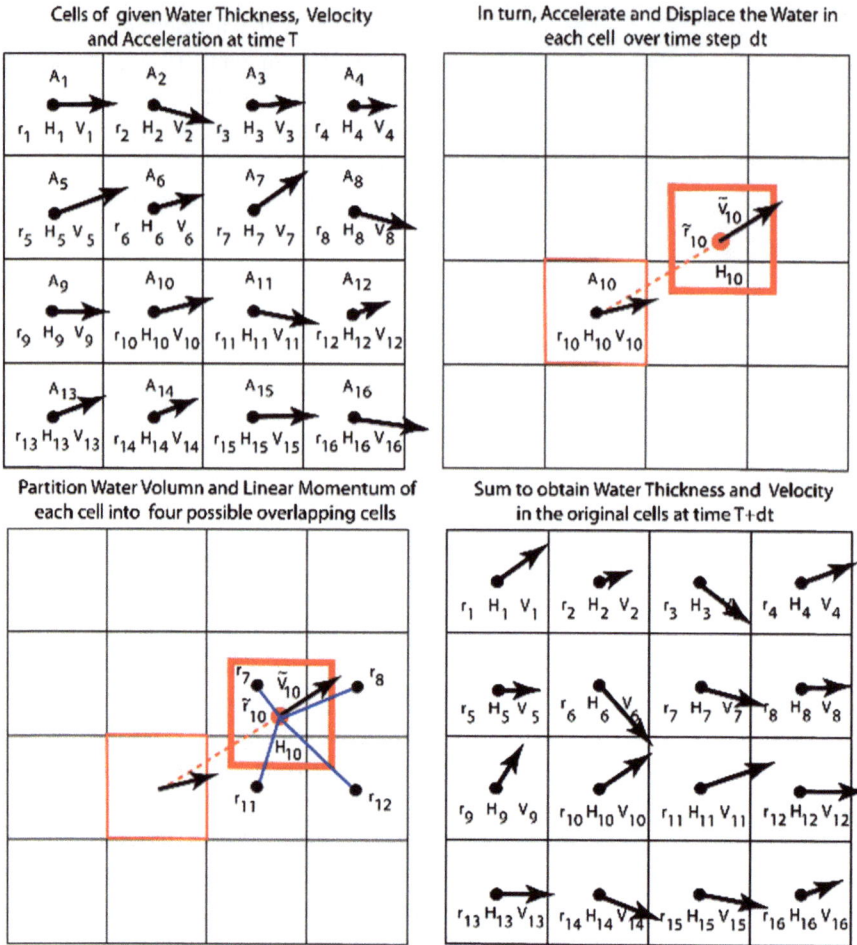

Figure 3
Tsunami Squares wave propagation and/or flow simulation concept

$$\delta V_{ji} = (H_i D_c^2)\left(1 - \frac{|\tilde{x}_i - x_j|}{D_c}\right)\left(1 - \frac{|\tilde{y}_i - y_j|}{D_c}\right)$$

if $\frac{|\tilde{x}_i - x_j|}{D_c} < 1$ and $\frac{|\tilde{y}_i - y_j|}{D_c} < 1$; otherwise $\delta V_{ji} = 0$

(7)

$$\delta \mathbf{M}_{ji} = (\rho_w H_i D_c^2 \tilde{\mathbf{v}}_i)\left(1 - \frac{|\tilde{x}_i - x_j|}{D_c}\right)\left(1 - \frac{|\tilde{y}_i - y_j|}{D_c}\right)$$

if $\frac{|\tilde{x}_i - x_j|}{D_c} < 1$ and $\frac{|\tilde{y}_i - y_j|}{D_c} < 1$; otherwise $\delta \mathbf{M}_{ji} = 0$

(8)

The product of the two terms on the right of Eq. (7) is simply a statement of the fractional area overlap of the ith displaced cell with the jth fixed cell. Clearly there is no need to run the partitioning through all $j = N$ cells because, at most, only four fixed cells overlap the displaced cell (Fig. 3, bottom left). Moreover because $\tilde{r}_i$ is known and the cells are square, it is simple to determine which four cells overlap.

Partitioning of vector linear momentum follows in the same way:

The N updated thickness $H_j(t + dt)$ and velocity $\mathbf{v}_j(t + dt)$ values in the jth fixed cell comes from summing and normalizing the volume (Eq. 7) and momentum (Eq. 8) contributions from all i displaced cells:

$$H_j(t + dt) = \frac{\sum_{i=1}^{N} \delta V_{ji}}{D_c^2} \quad (9)$$

$$\mathbf{v}_j(t + dt) = \frac{\sum_{i=1}^{N} \delta \mathbf{M}_{ji}}{\rho_w D_c^2 H_j(t + dt)} \quad (10)$$

Because only four of δV_{ji} and $\delta \mathbf{M}_{ji}$ are non-zero for each i, the sums in Eqs. (9) and (10) involve $4N$ terms (not N^2). Actually, there may be fewer than $4N$ terms because there is no need to displace and partition cells that are dry.

This process has time-stepped a wave propagation and/or flow simulation on a fixed set of cells while:

1. Conserving material volume exactly. It is possible to verify that the sum of the four non-zero partitioned volumes δV_{ji} in Eq. (7) is equal to $(H_i D_c^2)$, the volume of the displaced cell. Equation (9) simply replaces the "continuity equation" common to most numerical approaches (SATAKE AND TANIOKA 1995; LIU et al. 2003; HEINRICH 1992; WALDER et al. 2003) by tracking material from one cell to another.
2. Conserving linear momentum exactly. $\delta \mathbf{M}_{ji} = \rho_w \delta V_{ji} \tilde{\mathbf{v}}_i$ sums to the final momentum of the material in the ith cell. By tracking momentum from one cell to another, Eq. (10) replaces the advected component $[-\nabla_h \bullet \mathbf{v}(\mathbf{r},t)\mathbf{v}(\mathbf{r},t)H(\mathbf{r},t)]$ in the momentum equation (Eq. 2). For high-speed landslides and flows, conservation of momentum is critical for constructing realistic simulations. For deep water wave propagation, however, fluid velocities are small and advected momentum is negligible. This aspect of Tsunami Squares can be linearized by replacing Eq. (10) by $\mathbf{v}_j(t+dt) = \tilde{\mathbf{v}}_j$ where $\tilde{\mathbf{v}}_j$ is determined by use of Eq. (6)
3. Requiring no special treatment of dry cells or any mention of topography.
4. Reducing a N^2 summation to a $4N$ summation.
5. Obviating the need for a single numerical derivative.

To show that Eqs. (9) and (10) are equivalent to Eqs. (3) and (4), Eq. (9) is evaluated with dt very small ($dt \ll 1$) so that $\mathbf{v}_x(\mathbf{r},t)dt \ll D_c$ and $\mathbf{v}_y(\mathbf{r},t)dt \ll D_c$ for all cells, and terms with dt^2 are ~ 0

$$H(\mathbf{r}_{j,t}+dt) = H(\mathbf{r}_{j,t}) - H(\mathbf{r}_{j,t})\left(\frac{|\mathbf{v}_x(\mathbf{r}_{j,t})|}{D_c} + \frac{|\mathbf{v}_y(\mathbf{r}_{j,t})|}{D_c}\right)dt$$
$$+ \sum_{i=1}^{N=j} H(\mathbf{r}_j,t)\left(\frac{|\mathbf{v}_x(\mathbf{r}_i,t)|}{D_c}\right)dt + \sum_{i=1}^{N=j} H(\mathbf{r}_j,t)\left(\frac{|\mathbf{v}_y(\mathbf{r}_i,t)|}{D_c}\right)dt$$
(11)

The first RHS term in Eq. (11) is the original material in cell j. The second RHS term in Eq. (11) is the material originally in cell j that has moved into adjacent cells. The sums in Eq. (11) only include those cells in the x and y directions that overlap the jth cell after their displacements (Eq. 5). They account for the material originally in adjacent cells that has moved into cell j. Another statement of (Eq. 11) is:

$$H(\mathbf{r}_j, t+dt) = H(\mathbf{r}_j, t) - \nabla \bullet [\mathbf{v}(\mathbf{r}_j,t)H(\mathbf{r}_j,t)]dt$$
(12)

Hence, Eq. (11) is exactly equivalent to Eq. (3) for vanishingly small dt. We can evaluate Eq. (10) in the same way:

$$H(\mathbf{r}_{j,t}+dt)\mathbf{v}(\mathbf{r}_{j,t}) = H(\mathbf{r}_{j,t})\tilde{\mathbf{v}}(\mathbf{r}_{j,t}) - H(\mathbf{r}_{j,t})\mathbf{v}(\mathbf{r}_{j,t})$$
$$\times \left(\frac{|\mathbf{v}_x(\mathbf{r}_{j,t})|}{D_c} + \frac{|\mathbf{v}_y(\mathbf{r}_{j,t})|}{D_c}\right)dt$$
$$+ \sum_{i=1}^{N \neq j} H(\mathbf{r}_j,t)\mathbf{v}(\mathbf{r}_j,t)\left(\frac{|\mathbf{v}_x(\mathbf{r}_i,t)|}{D_c}\right)dt$$
$$+ \sum_{i=1}^{N \neq j} H(\mathbf{r}_j,t)\mathbf{v}(\mathbf{r}_j,t)\left(\frac{|\mathbf{v}_y(\mathbf{r}_i,t)|}{D_c}\right)dt$$
(13)

The first RHS term in Eq. (13) is the original momentum of cell j. The second RHS term in Eq. (13) is momentum originally of cell j that was transferred to adjacent cells. The sums in Eq. (13) account for momentum originally in adjacent cells transferred to cell j. Equation (13) is equivalent to the first three terms in Eq. (4) for vanishingly small dt. Tsunami Squares updates flow velocities through the slope of the surface (third RHS term in Eq. (4)) as a separate step.

Tsunami Squares satisfies the same non-linear continuity and momentum equations used in traditional methods, but does so in a way that is more intuitive (by just moving cell mass and momentum by appropriate partitioning).

3.2. Gravitational Acceleration

To complete the time step, it is necessary to update mean cell accelerations $\mathbf{a}_i(t)$. As is customary in "long wave" theory, the mean acceleration of material in the cell is proportional to the slope of upper surface $\zeta(\mathbf{r}_i,t)$:

$$\begin{aligned}\mathbf{a}_i(t) = \mathbf{a}_g(\mathbf{r}_i,t) &= -g\nabla_h\zeta(\mathbf{r}_i,t)\\ &= -g\nabla_h[T(\mathbf{r}_i) + H(\mathbf{r}_i,t)]\end{aligned} \quad (14)$$

$H(\mathbf{r}_i,t) = H_i(t)$ is material thickness found above, $T(\mathbf{r}_i)$ is the fixed topography over which it moves, g is the acceleration of gravity, and ∇_h is the horizontal gradient (Fig. 4).

The $\nabla_h\zeta(\mathbf{r}_i,t)$ in Eq. (14) is the only step in which a numerical derivative must be evaluated. Even this sole differentiation can, however, be avoided by fitting a plane to $\zeta(\mathbf{r}_i,t)$ and its eight adjacent neighbors, then fixing the horizontal gradient from the slope of that surface. This plane-fitting approach helps stabilize the calculation by estimating the gradient across a two-dimensional region as a function of adjacent points alone. Another advantage of the plane-fitting approach is the ability to "punch out" specific locations near $\mathbf{r}_i$ by excluding them from the fit. Where wet cells are near dry ones, the dry sites would normally be "punched out" during calculation of the slope of the surface $\zeta(\mathbf{r}_i,t)$: For example, if a steep dry cliff was adjacent to a wet area, the cliff surface would not be included when computing the slope of the fluid.

In most cases the computation grid is made sufficiently large that flows or waves do not reach the ends of the domain in the time period of interest. If need be, an "absorbing buffer zone" can be included near domain walls to minimize reflections.

4. Landslide Simulation

A difficulty in simulating landslides is that they behave partly like a solid, because of a cohesive fraction, and partly like a fluid, because they flow into valleys and channels (COUSSOT and MEUNIER 1996). Landslides have solid characteristics when they initially fail and when they near a stop (HUNGR and MCDOUGALL 2009). In between, after a few seconds of acceleration and before the final seconds of de-acceleration, slide material loses cohesion and takes on fluid characteristics. This is why Tsunami Squares does not differentiate fluids from moving slide material. For landslide material considered solid during the first few seconds, the driving force on the square depends on the slope of the bottom surface. For landslide material regarded as fluid-like, the driving force on the square depends upon the slope of the top surface (Eq. 14).

4.1. Frictional Acceleration

Two types of frictional acceleration are added to Eq. (14) to oppose sliding, one due to basal friction ($\mathbf{a}_b$) and one due to dynamic friction ($\mathbf{a}_d$):

$$\mathbf{a}_i(t) = \mathbf{a}_g(\mathbf{r}_i,t) + \mathbf{a}_d(\mathbf{r}_i,t) + \mathbf{a}_b(\mathbf{r}_i,t) \quad (15)$$

Basal friction is the simple static resistance of the sliding surface because of interactions between moving materials and the rough bed. It depends on material type, solid fraction, bed roughness, and normal stress. The acceleration of a cell of material as a result of basal friction is:

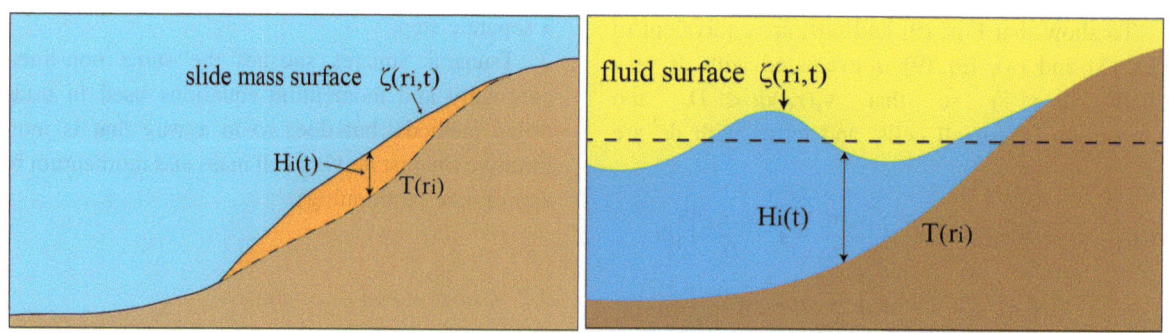

Figure 4
Geometry of slide mass and water flow. Tsunami Squares considers moving masses more fluid-like than solid, so the upper surface horizontal gradient drives the flow for both the slide and the water

$$\mathbf{a}_b(\mathbf{r}_i, t) = -\mu_b g \hat{\mathbf{v}}_{\text{slide}}(\mathbf{r}_i, t) \quad (16)$$

Here, $\hat{\mathbf{v}}_{\text{slide}}$ is the unit velocity vector and μ_b is the basal friction coefficient. Basal friction is treated as a solid-like moving resistance.

Dynamic friction originates from resistance encountered by the moving bulk material through air or water. Dynamic resistance is intrinsically velocity-dependent. "V-square" friction, as it is called, originates from pressure or viscous-like forces acting on the top and bottom surfaces of moving slides of thickness $H(\mathbf{r}_{i,t})$:

$$\mathbf{a}_d(\mathbf{r}_i, t) = -\mu_d \mathbf{v}(\mathbf{r}_i, t) |\mathbf{v}(\mathbf{r}_i, t)| / H(\mathbf{r}_i, t) \quad (17)$$

Here, μ_d is the dynamic friction coefficient that expresses all velocity-dependent properties of particle interaction. Because dynamic friction increases as $|v^2|$, it imparts a terminal velocity to motion, with thicker slides attaining a higher speed. It is weak during landslide initiation and stoppage, but is dominant for high-speed moving masses.

Both basal and dynamic friction act to slow the slide, but are not allowed to reverse the sliding direction. μ_d and μ_b can be functions of time and space and made as complicated as necessary. For example, μ_d of subaerial slides might be less than μ_d for subaqueous slides; μ_b might change from a low value to a high value as the velocity of the slide falls below a critical value (WARD and DAY 2006). Alternatively, the transition could be determined by an "angle of repose θ_{repose}" of the solidifying material. When the slope of the top surface drops below a specific angle, basal friction grows.

Transition between two types of friction enables Tsunami Squares to embrace the dual behavior of landslides. When slides move at high speeds, fluid behavior and dynamic friction dominate and the slope of the upper surface drives the motion. When slow or stopped, solid behavior and basal friction dominate.

4.2. Observational Constraints on the Landslide

The landslide model incorporates several observational constraints: the topography of the bed, the initial and final landslide shape, and the initial and final landslide thickness. The Yangtze River, at an elevation of 173 m in this region, has a generally parabolic cross-section 470 m wide and 122 m deep. Freshly exposed slip surfaces on the north bank reveal the isosceles triangular outline of the landslide (Fig. 2). The slide was 40 m wide at the top and 160 m wide at the water surface. It formed an upslope from 120 to 400 m above sea level (350 m above the deepest part of the river bed) with a horizontal distance of 290 m from the shoreline (Fig. 5), and dipped steeply with upper and lower slope angles of 63° and 44°. From measured centerline values, we interpolated landslide thicknesses across the slide face, arriving at a total volume of 380,000 m³ and an average thickness of 17 m. Slide deposits were distributed at elevations from 210 to 50 m above sea level, with the thickest part at approximately 90 m; the toe of the landslide deposit is at the floor of the river bed.

We are also fortunate to have an eyewitness video of the landslide. This shows that the lower part of the slide slipped first, followed by the upper part three seconds later, and that the whole sliding event lasted approximately 31 s. According to field investigation the landslide is divided into two parts, below and above 240 m. The program imposes sequential failure that keeps the upper part fixed for the first 3 s and then releases it.

4.3. Landslide Simulation Results

4.3.1 Model Setup

We ran the simulations on 10 m digital topography that defines the dimension of the cells, $D_c = 10$. In general, to maintain reasonable resolution in Eqs. (5)–(10), it is best to select a time step, dt, such that $\mathbf{v}_{\text{rep}} \times dt < D_c$ where $\mathbf{v}_{\text{rep}}$ is some representative peak velocity of the flow. Because $\mathbf{v}_{\text{rep}}$ is unknown before the fact, experimentation is needed to select dt. In these calculations we took $dt = 0.5$ s. For values of friction we set dynamic friction μ_d equal to 0.05 and 0.2 for landslide material moving on dry land and under water, respectively. For basal friction μ_b gradually changes from 0.2 to 6.0 when the slope angle becomes less than the 31° angle of repose measured from the final landslide shape. These values of friction were derived after several runs to best fit the observed field data.

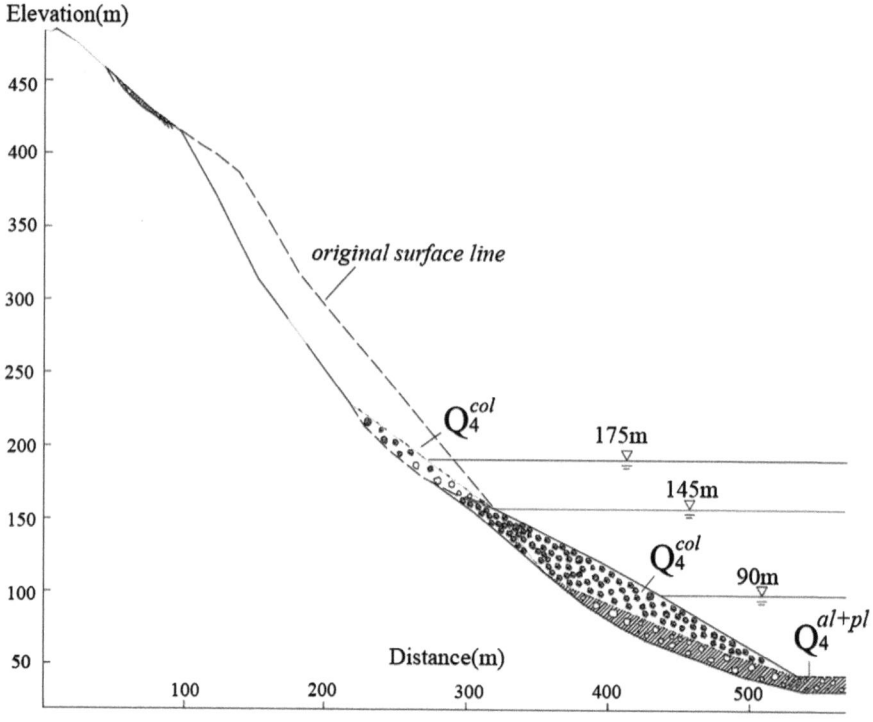

Figure 5
Engineering geological profile of the Guongjiafang landslide. The initial and final surfaces provide the thickness of the central section and residuals

4.3.2 Model Results

Figure 6 shows instantaneous profiles of the middle cross-section and plane view of the slide at four times from start to finish (a Quicktime animation of the landslide is available at http://es.ucsc.edu/~ward/Yangtze1.mov). Rainbow colors indicate sliding velocity. At $t = 4$ s, the speed of the middle part exceeded that of the upper part, a consequence of the sequential failure assumption. At $t = 10$ s, the upper part, now moving faster than the lower parts, helps to push the whole mass deeper into the river. The model slide moved downward for 26 s, stopping on the riverbed with a uniform angle of repose and only a thin layer depositing above water (Fig. 6, lower right and left), consistent with the behavior of the real landslide (Fig. 5).

Impact velocity and landslide shape are critical variables for calculation of wave heights in the laboratory (FRITZ et al. 2003). We tracked slide-impact velocities near the river surface and on the centerline above and below water (five squares in

Figure 6 ▶
Moving landslide profiles at $t = 0$, 4, 10 and 30 s. For the plane views (left), background colors brown and red indicate land above and underwater. The blue to white shading indicates the thickness of the landslide, thick to thin. The arrows indicate slide direction and speed. Arrow colors follow the rainbow legend. The red spots mark the positions where velocities are being tracked in Fig. 7. Points 1, 2, and 3 are at the water surface (173 m). Points 4 and 5 are above and under water, at 240 m and 140 m. For the cross sections (right), the rainbow colors represent landslide speeds from 0 to greater than 15 m/s. The lower part of the slide has a higher speed during first few seconds. The upper part moves fastest later because of higher gravitational potential energy. A Quicktime animation of the landslide is available at http://es.ucsc.edu/~ward/Yangtze1.mov

Fig. 2 (right) and five circles in Fig. 6 (left)). Figure 7 shows the velocity curves of the tracked spots. The orange line that follows the above-water portion at 240 m elevation (#4) has the highest peak speed of 21 m/s. For the initial four seconds, however, speeds there are actually slower than in other places before increasing sharply to peak at approximately nine seconds. The initial delay reflects the applied sequential bottom-to-top failure (Fig. 8).

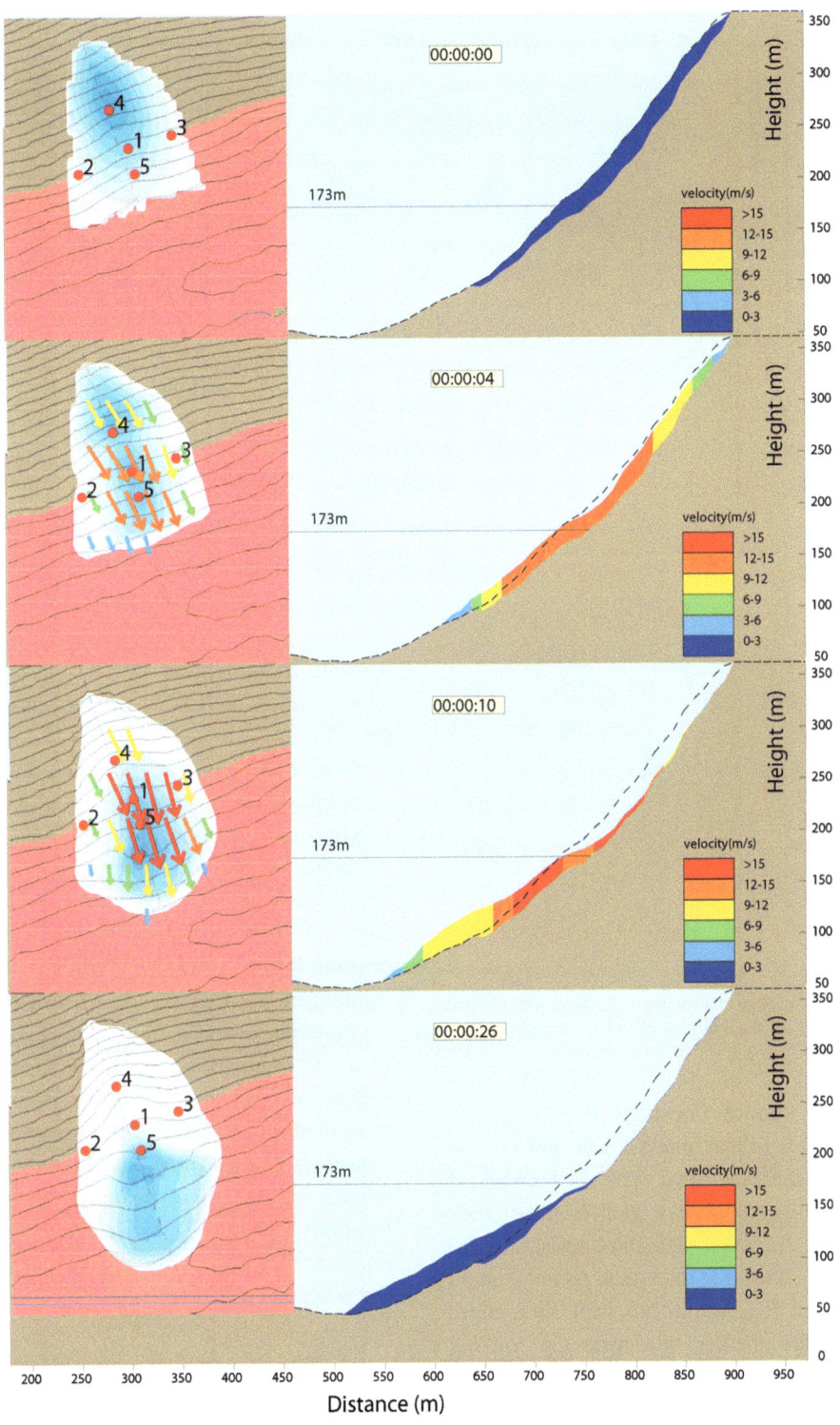

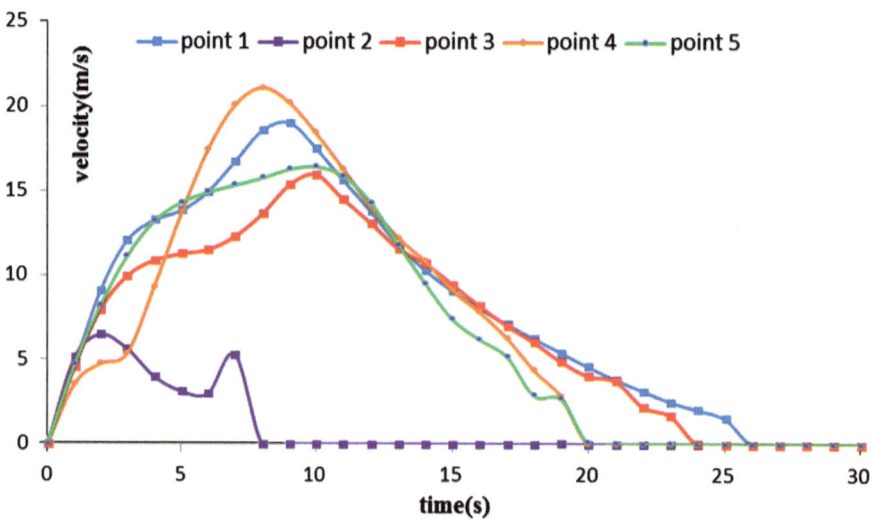

Figure 7
Slide velocities at five positions above, below, and at the water level. The *blue, purple, red, orange*, and *green lines* indicate velocities at positions 1–5 in Fig. 2

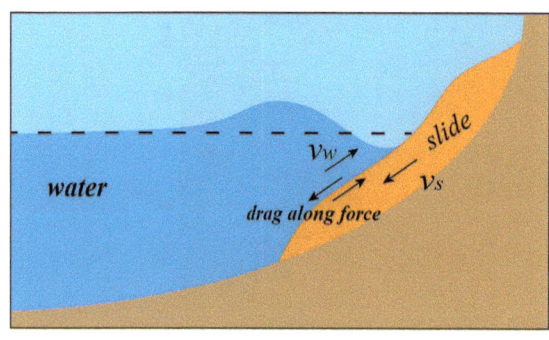

Figure 8
Landslide tsunami-generating mechanisms. "Lift-up" (NMT) and "drag-along" (DA) are the main sources of wave generation

At the water surface (locations #1, 2, and 3), velocity on the left (#2) is much lower and falls to zero faster than the other two. This is because of differences in slide thickness and upslope slide extent. V-square friction (18) offers resistance inversely proportional to slide thickness. As apparent in the left column of Fig. 6, thin bits near slide edges move slower than thicker bits near the middle. Upslope extent also affects slide speed and duration. The more material upslope, the longer it takes to pass by a given place. The higher on the slope it originated the higher velocity it acquires in transit. After the slide mass has completely passed by at the left, material continues to pass at the middle and right. As a result, more deposits pile on the middle and right sides, which agrees with the observed residual (Fig. 2).

With regard to peak speeds on the centerline (#1, 4, and 5), the above-water position (#4) ranks highest and the underwater position (#5) ranks lowest. This is expected, because greater frictional resistance was applied underwater. At $t = 20$ s, velocities at positions #4 and #5 vanish, but masses still pass by position #1. The last few moving masses slip into the water there but they stop just below the surface.

Previous analysis of the witness video (HUANG et al. 2012) inferred a peak slide speed of 11.65 m/s, somewhat lower than our value. HUANG et al. (2012) quantified the velocity of the uppermost part of the landslide, but we know the landslide deformed as it moved with different velocity profiles at different locations (e.g. Fig. 7). Measured velocity at the slide top does not necessarily reflect the slide velocity at the water level, where waves are produced.

5. Wave Generation, Propagation, and Inundation

5.1. Wave Sources

Given an initial distribution of still water $H(\mathbf{r}_i, t)$ and the landslide model computed above, there are several ways to introduce waves. The classical approach simply lifts or drops the water by an amount equal to the passing slide thickness. Gravitational energy is imparted to the water in this way, but there is no direct transfer of momentum from the slide to the water. We call this approach "no momentum transfer (NMT)". NMT may be adequate for some tsunami sources, for example submarine earthquakes, but for high speed slides into water it does not suitably describe wave generation.

Here, in addition to NMT, we introduce "drag-along (DA)". DA assumes that extra forces exist at the water–landslide interface. These forces act to slow the slide but also accelerate the water, much like anti-friction. For a submarine landside moving at velocity $\mathbf{v}_s(\mathbf{r}_i,t)$, drag-along acceleration of an overlying water layer of thickness $H(\mathbf{r}_i,t)$ would be:

$$\mathbf{a}_{da}(\mathbf{r}_i,t) = c_{da}\mathbf{v}_s(\mathbf{r}_i,t)|\mathbf{v}_s(\mathbf{r}_i,t)|/H(\mathbf{r}_i,t) \qquad (18)$$

The drag-along coefficient, c_{da}, may or may not equal the dynamic friction coefficient μ_d. Unlike friction (Eq. 17) that acts only in the direction opposite to fluid flow, DA (Eq. 18) can accelerate the flow in any direction that the slide is moving. DA transfers momentum from the slide to the water and enhances wave production beyond that of NMT alone.

5.2. Observational Constraints on the Tsunami

Field workers record tsunami heights at the shore in two ways: measurement of the wave trail on trees, docks, and structures, or measurement of the trace of the dry land–wet land transition. The former records wave height as it reaches those objects (this is denoted "flow depth"). The latter reveals the highest elevation the wave reaches on shore (run-up height). The two types of measurement can lead to significantly different results, especially where tsunamis inundate complex terrain, for example that around the Gongjiafang site.

A field investigation of wave run-up was conducted by two groups soon after the landslide (DAI et al. 2010; HUANG et al. 2012). They did not reveal the methods used for measurement nor did they indicate precision or local variability. We note the ten surveyed values from the two groups in Figs. 11 and 12. Wave run-up decayed both upstream and downstream from the landslide, and the further from the landslide, the slower was the rate of decrease. Waves ran up highest on the north shore where the landslide occurred. Three-hundred and 4,000 m upstream on the north shore the impulse wave ran up 13.1 and 1.1 m, respectively.

5.3. Tsunami Simulation Results

5.3.1 Model Setup

Using the same 10-m spaced digital topography as for the landslide simulation, the Yangtze reservoir was filled to 173 m elevation extending 6 km up and downstream and the simulated Gongjiafang landslide was allowed to fall into the River. For water flows, both dynamic and basal friction, μ_d and μ_b, are equal to 0, except near the shore where, μ_d increases to 0.02, because of the higher friction there. For this study we set the drag-along coefficient, c_{da}, at 0.2.

5.3.2 Model Results

Within seconds, the landslide "pushes up" and "drags along" the water to form a tsunami that spreads out over the river (Figs. 9, 10). The wave reaches the opposite bank in approximately 15 s, runs up on the land, then flows back as a reflected wave. The return wave peaks over the landslide bank at $t = 61$ s. The orange curve in each part of Fig. 9 shows the maximum flow depth up to that time along the river cross-section.

Figure 10 shows a plot of the tsunami waveforms at three positions (#1, #2, and #3 in Fig. 9). In general, the first wave crest is higher than the later ones. The 17-s interval between the first crests of the blue and green curves corresponds to the cross-river propagation time from point #1 to point #3. For long waves, the cross-river transit time Δt can be expressed as:

$$\Delta t = \int_{\#1}^{\#3} ds/\sqrt{gh(s)} \qquad (19)$$

where $h(s)$ is the water depth, g is the acceleration of gravity, and s is the along-path position. For known

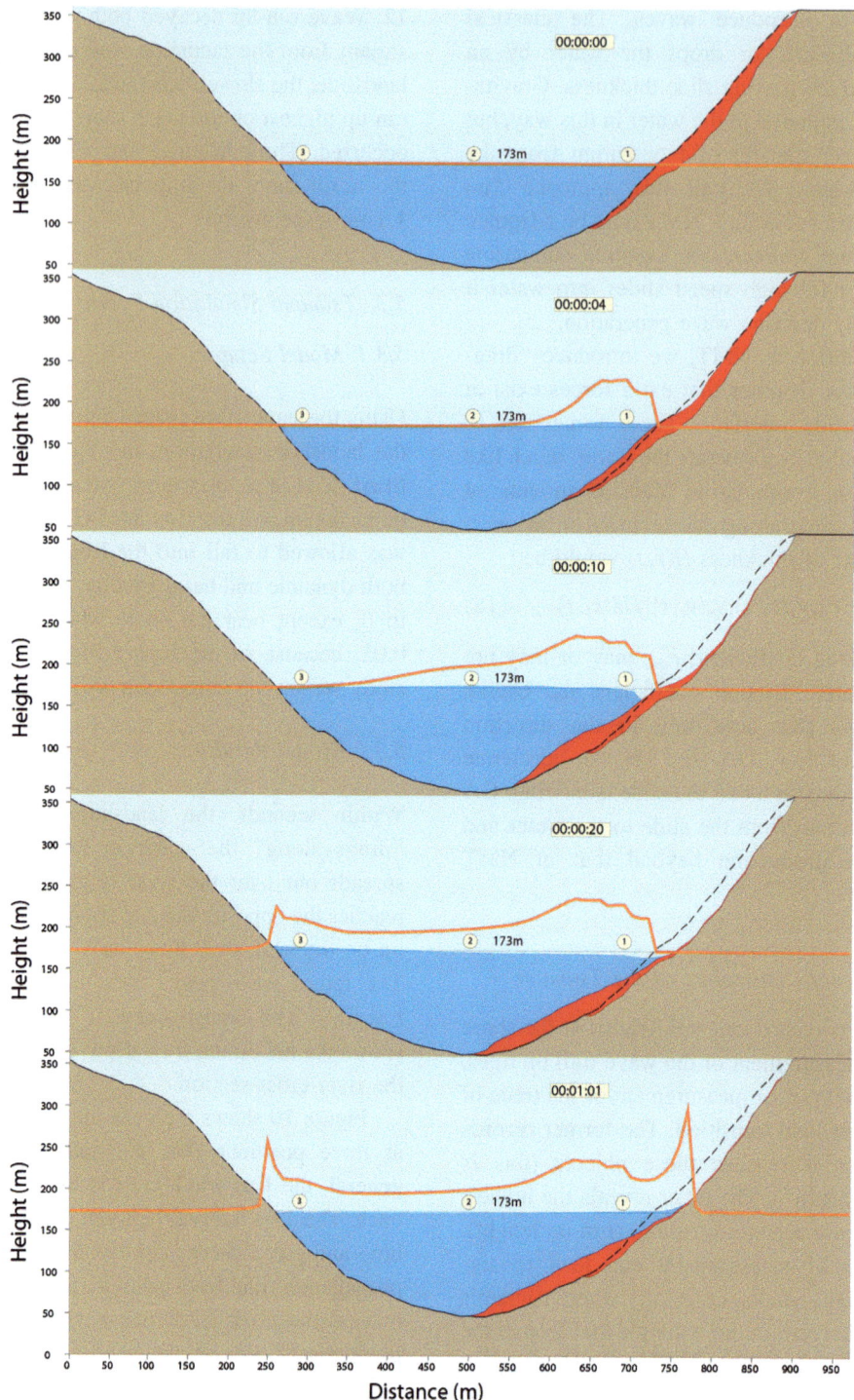

Figure 9
Waves propagate and reflect in river cross section at the 4th, 10th, 20th, and 61st seconds. The *orange lines* depict the highest flow height and have been exaggerated tenfold for better visualization. The *yellow spots 1* (*right*), *2* (*middle*), and *3* (*left*) on the water surface mark positions where flow height is tracked in Fig. 10. A Quicktime animation is available at http://es.ucsc.edu/~ward/Yangtze2.mov

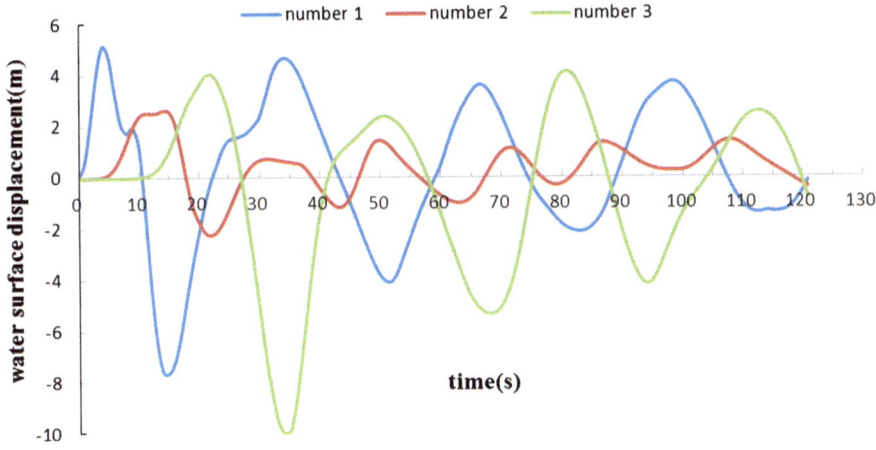

Figure 10
Wave trains at three positions on a cross-river line. The locations are shown in Fig. 9

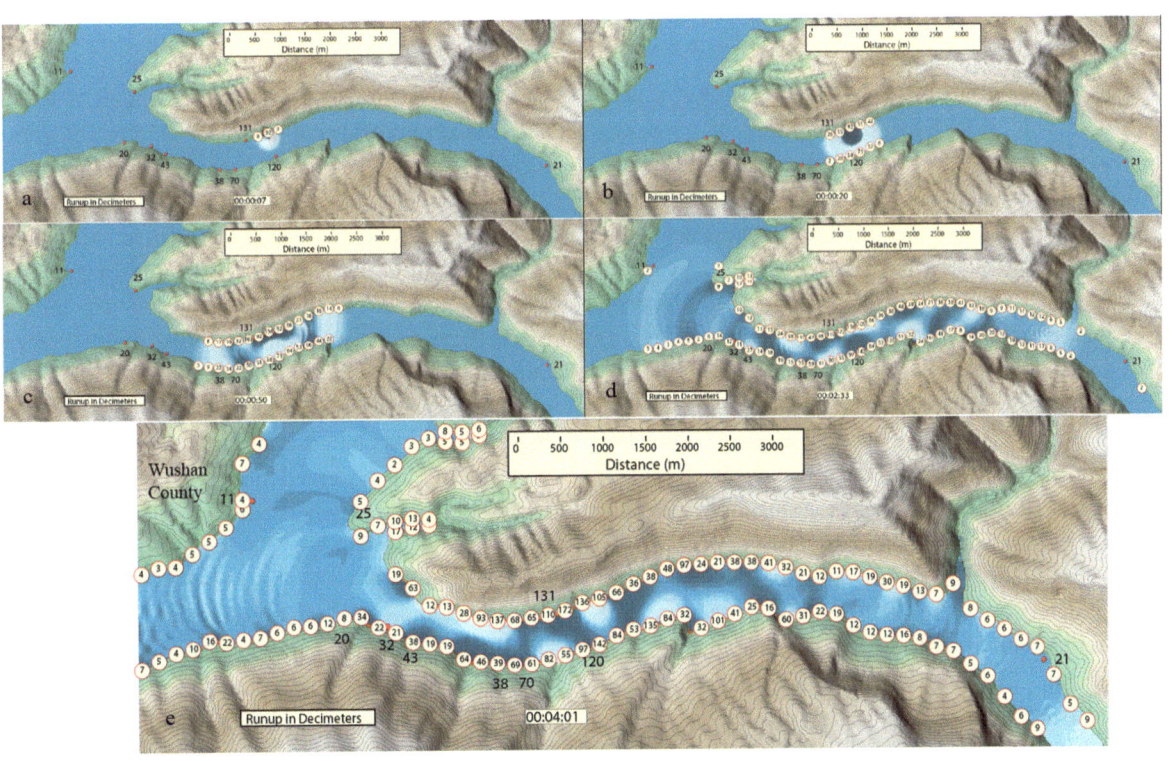

Figure 11
Extent of the river tsunami after $T = 7, 20$, and 50 s, and 2:33 and 4:01 min. *Numbers in circles* show run-up heights in decimeters. *Numbers beside red spots* are field data, also in decimeters. A Quicktime animation is available at http://es.ucsc.edu/~ward/Yangtze3.mov

water depth, Eq. (19) gives a cross-river transit time of 17 s, which perfectly matches the modeling result.

The dominant wave period at sites #1 and #3 is 32 s. Wave crests at #1 correlate with troughs at #3, and vice versa. The dominant period at mid-river site #2 is 16 s, much less than at the river edge sites.

These features fit the character of standing wave modes. The period of the nth river mode is:

$$T_n = 2\Delta t/n \qquad (20)$$

with Δt calculated by use of Eq. (19). $n/2$ corresponds to the number of wavelengths that fit, bank to bank.

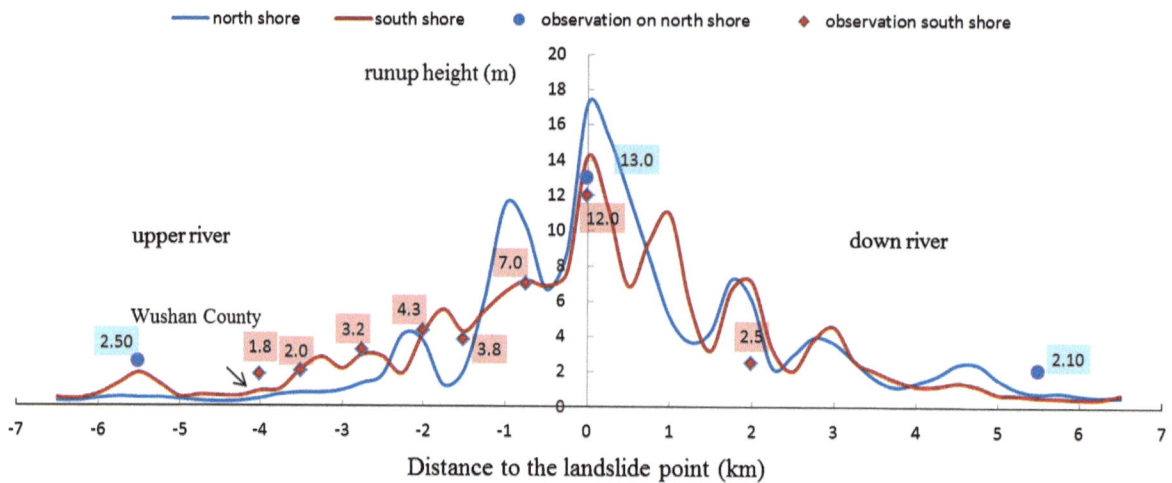

Figure 12
Run-up height decay as a function of distance along the north (*blue*) and south (*red*) shores. The landslide occurred on the north shore. The scattered dots are the observed values on the north (*blue*) and south (*red*) shores

The fundamental ("slosh") mode ($n = 1$) has a period of $T = 32$ s with peak amplitudes of opposite sign at the river banks, as is apparent from the blue and green curves in Fig. 9. The $n = 1$ mode has a node at the river center, so this oscillation is not seen in the red curve. The $n = 2$ overtone has a peak in mid-river and a period of $T = 16$ s. It is the largest contributor to the red curve.

Landslide-generated tsunami transit rapidly and negatively affects Yangtze shores for many kilometers in our model, as in the real event. Figure 11 shows a map of wave propagation at five typical moments. Within 7 s, slide masses hit the river and water waves radiate outward in an arc (Fig. 11a). They cross the river and wash the far bank in approximately 20 s (Fig. 11b). Within 50 s the waves have propagated a distance of 2 km (Fig. 11c). In 2.5 min the first wave reaches Wushan County (Fig. 11d). For several minutes afterwards, the signal echoes between shores before dying out (Fig. 11e).

Collection of simulated run-up heights along shore line enables preparation of a wave-decay curve (Fig. 12). This includes broad variations with distance especially within the first 4 km up and down river. These variations probably result from interference of many positive and negative waves reflected from the curved river shores (Fig. 11a, for example). Some sites are naturally prone to higher or lower run-up because of local geography. Likewise, waves run-up higher in such narrowing water channels as valley and branch stream outlets than at straight shore locations.

The highest wave run-up in the simulation, 17.2 m, was located near the landslide. Maximum run-up height on the opposite bank was 14.2 m. The wave height had dropped to 0.4 m by the time it reached the docks at Wushan County. Field observations measured 13 m at the landslide and 12 m at the opposite bank. For the south shore, the observed data (red dots) are scattered around the simulation result (red line). The observed north shore values (blue spots) are slightly higher than the values calculated. Considering that the specific run-up quantities measured in the field are not completely clear nor were any uncertainties assigned, we cannot make a formal statement of goodness of fit, but the run-up heights on both the north and south shores are reasonable fits with the observations, with correlation coefficient $R^2 = 0.88$.

6. Conclusions

This paper introduces Tsunami Squares, a new approach for modeling of landslide-generated waves. Tsunami Squares has the advantages of the previous Tsunami Ball method, for example, special separate treatment for dry and wet cells is not needed, but it obviates the use of millions of individual particles. Simulations can be expanded to spatial scales not possible previously. The new method accelerates and transports squares of material that are fractured into new squares in such a way as to conserve volume and linear momentum.

The simulation first computes landslide motions on dry land. A novel aspect of Tsunami Squares is that it considers landslides as part solid and part fluid. Landslides are more fluid-like within a few seconds of initiation after the material loses cohesion. Solid landslides are driven by the slope of the bottom surface. Fluid-like landslides are driven by the slope of the top surface.

The second step in the simulation introduces water. The falling landslide generates waves by two means. One is the simple uplift of the fluid, known as "no momentum transfer". The second mechanism is drag-along, a force representing the interaction between the slide and water. Velocity-dependent drag-along contributes substantially to tsunami wave generation from high-speed landslides.

We have demonstrated and validated Tsunami Squares by modeling the 2008 Three Gorges Reservoir Gongjiafang landslide and river tsunami. The landslide's progressive failure, generated wave, and subsequent propagation and run-up are well reproduced (WARD and XIAO 2013).

On a laptop computer, Tsunami Square simulations flexibly handle a wide variety of waves and flows, and are excellent techniques for risk estimation, hazard assessment, and emergency management applications.

Acknowledgments

L. Xiao was supported by the National Natural Science Foundation of China (no. 41202247). We thank Professor K. Yin, from the China University of Geosciences (Wuhan), for supporting the geology background research and field investigation. We thank Dr Simon Day, from University College London, for careful revision. We also thank Chongqing Three Gorges Reservoir Geological Hazards Prevention and Control Office for supplying the 10-meter digital elevation map.

Appendix 1: Inclusion of wave dispersion

Long wave theory assumes that the depth-averaged horizontal acceleration of a water column is proportional to the gradient of the fluid surface as stated in Eq. (21):

$$\frac{\partial \mathbf{v}(\mathbf{r},t)}{\partial t} = -\mathbf{v}(\mathbf{r},t) \bullet \nabla_h \mathbf{v}(\mathbf{r},t) - g\nabla_h \zeta(\mathbf{r},t) \quad (21)$$

According to linear dispersive wave theory (WARD AND DAY 2010), the depth-averaged horizontal acceleration of a water column is proportional to the gradient of the fluid surface smoothed over a dimension comparable with the water depth. Linear dispersion can be accommodated in tsunami squares simply by replacing Eq. (21) by Eq. (22)

$$\frac{\partial \mathbf{v}(\mathbf{r},t)}{\partial t} = -\mathbf{v}(\mathbf{r},t) \bullet \nabla_h \mathbf{v}(\mathbf{r},t) - g\nabla_h \zeta_{\text{smooth}}(\mathbf{r},t)$$
$$(22)$$

where

$$\zeta_{\text{smooth}}(\mathbf{r},t) = \zeta(\mathbf{r},t) \times S(\mathbf{r})$$
$$= \int \zeta(\mathbf{r}',t) S(\mathbf{r}-\mathbf{r}') d\mathbf{r}' \quad (23)$$

and

$$S(\mathbf{r}) = \text{Re} \int_k \frac{e^{i\mathbf{k}\cdot\mathbf{r}}}{4\pi^2} \frac{\tanh(kh)}{kh} d\mathbf{k} \quad (24)$$

This indicates that short wave (kh ≫ 1) contributions to the surface gradient impart less depth-averaged acceleration to the water column than do longer wave (kh ≪ 1) contributions. As a result, short waves fall behind long waves, as dictated by linear dispersive theory. For the applications in this paper, water wave dispersion is not important.

REFERENCES

BASU D, DAS K, GREEN S, JANETZKE R, STAMATAKOS J. (2010), *Numerical Simulation of Surface Waves Generated by a Subaerial Landslide at Lituya Bay Alaska*. Journal of Offshore Mechanics and Arctic Engineering, *132*, p 41101.

COUSSOT P, MEUNIER M. (1996), *Recognition, classification and mechanical description of debris flows*. Earth-Sci Rev, *40*(1996), 209–227.

Dai Y, Wang Y, Yin K, Chen L, Liu B. (2010), *Surge Survey and Calculation Analysis of a Landslide in Wushan County in the Three Gorges Reservoir*. Journal of Wuhan University of Technology, 32(19), 14–71 (In Chinese).

Fritz HM, Hager WH, Minor HE. (2003), *Landslide generated impulse waves. 1. Instantaneous flow fields*. Exp Fluids, 35, 505–519.

Fritz HM, Mohammed F, Yoo J. (2009), *Lituya Bay Landslide Impact Generated Mega-Tsunami 50th Anniversary*. Pure Appl Geophys (166), 153–175.

Harbitz CB, Glimsdal S, Lovholt F, Kveldsvik V, Pedersen GK, A Jensen. (2014), *Rockslide tsunamis in complex fjords: from an unstable rock slope at Akerneset to tsunami risk in western Norway*. Coast Eng, 88(2014), 101–122.

Harbitz CB, Vholt FL, Bungum H. (2014), *Submarine landslide tsunamis: how extreme and how likely?* Nat Hazards, 72(3), 1341–1374. doi:10.1007/s11069-013-0681-3.

Heinrich P. (1992), *Nonlinear water waves generated by submarine and aerial landslides*. Journal of Waterway, Port, Coastal and Ocean Engineering, 118, 249.

Huang B, Yin Y, Liu GN, Wang SC. (2012), *Analysis of waves generated by Gongjiafang landslide in Wu Gorge, Three Gorges Reservoir, on November 23, 2008*. Landslides, 9(3), 395–405.

Huang B, Yin Y, Wang S, Chen X, Liu G, Jiang Z, Liu J. (2013), *A physical similarity model of an impulsive wave generated by Gongjiafang landslide in Three Gorges Reservoir, China*. Landslides, 1–13.

Hungr O. (1995), *A model for the runout analysis of rapid flow slides, debris flows, and avalanches*. Can. Geotech. J., 32(1995), 610–623.

Hungr O, Mcdougall S. (2009), *Two numerical models for landslide dynamic analysis*. Computer and Geosciences, 35(2009), 978–992.

Liu PL, Lynett P, Synolakis CE. (2003), *Analytical solutions for forced long waves on a sloping beach*. J Fluid Mech, 478, 101–109.

Mader CL, Gittings ML. (2002), *Modeling the 1958 Lituya Bay mega-tsunami*. Science of Tsunami Hazards, 20(5), 241–250.

Poisson B, Pedreros R. (2010), *Numerical modeling of historical landslide-generated tsunamis in the French Lesser Antilles*. Natural Hazards and Earth System Sciences (10), 1281–1292.

Quecedo M, Pastor M, Herreros MI. (2004), *Numerical modelling of impulse wave generated by fast landslides*. Int J Numer Meth Eng, 59, 1633–1656.

Satake K, Tanioka Y. (1995), *Tsunami generation of the 1993 Hokkaido Nansei-Oki earthquake*. Pure Appl Geophys, 144(3–4), 803–821.

Titov VV, Gonzalez FI. (1997), *Implementation and testing of the method of splitting tsunami (MOST) model*: US Department of Commerce, National Oceanic and Atmospheric Administration, Environmental Research Laboratories, Pacific Marine Environmental Laboratory.

Walder JS, Watts P, Sorensen OE, Janssen K. (2003), *Tsunami generated by subaerial mass flows*. Journal of Geophysical Research, 108(B5), 2236–2254.

Ward SN. (2014), *Lituya Bay Tsunami*. https://www.youtube.com/watch?v=6COeNRToYqU.

Ward SN, Day S. (2010), *The 1958 Lituya bay landslide and tsunami–a tsunami ball approach*. Journal of Earthquake and Tsunami, 4(4), 285–319.

Ward SN, Day S. (2005), *Tsunami thoughts*. CSEG RECORDER.

Ward SN, Day S. (2006), *Particulate kinematic simulations of debris avalanches: interpretation of deposits and landslide seismic signals of Mount Saint Helens, 1980 May 18*. Geophys. J. Int, 167, 991–1004.

Ward SN, Day S. (2008), *Tsunami Balls: A Granular appproach to Tsunami runup and inundation*. Communications in computational Physics, 3(1), 222–249.

Ward SN, Xiao L. (2013), *Yangtze Tsunami*. YOUTUBE MOVIE https://www.youtube.com/watch?v=JBa8z9oPgLI.

Watts P, Tappin DR. (2012), *Geowave Validation with Case Studies: Accurate Geology Reproduces Observations*. In Y Yamada, K Kawamura, K Ikehara et al. (Eds.), Submarine Mass Movements and Their Consequences (31, pp. 517): Springer Netherlands.

Watts P, Grilli ST, Kirby JT, Fryer GJ, Tappin DR. (2003), *Landslide tsunami case studies using a Boussinesq model and a fully nonlinear tsunami generation model*. Natural Hazards and Earth System Sciences, 3, 391–402.

Weiss R, Fritz HM, Wunnemann K. (2009), *Hybrid modeling of the mega-tsunami runup in Lituya Bay after half a century*. Geophys Res Lett, 36(9), L9609.

Wu B, Li H, Yao M. (2010), *Deformation and Failure Mechanism of Slope in Area from Gongjiafang to Dulong of Wuxia County, Chongqing*. Chinese Journal of Underground Space and Engineering, 6(2010), 1656–1659 (in Chinese).

Yin K, Liu Y, Wang Y, Jiang Z. (2012), *Physical Model Experiments of landslide-induceds surge in Three Gorges reservoir*. Earth Science–Journal of China University of Geosicences, 37(5), 1067–1074 (in Chinese).

(Received August 8, 2014, revised January 15, 2015, accepted January 16, 2015, Published online February 14, 2015)

Pure Appl. Geophys. 172 (2015), 3655–3670
© 2015 Springer Basel
DOI 10.1007/s00024-015-1113-y

Pure and Applied Geophysics

Quantification of Tsunami Bathymetry Effect on Finite Fault Slip Inversion

QUENTIN BLETERY,[1] ANTHONY SLADEN,[1] BERTRAND DELOUIS,[1] and LIONEL MATTÉO[1]

Abstract—The strong development of tsunami instrumentation in the past decade now provides observations of tsunami wave propagation in most ocean basins. This evolution has led to the wide use of tsunami data to image the complexity of earthquake sources. In particular, the 2011 M_w9.0 Tohoku-Oki earthquake is the first mega-event for which such a tsunami instrumentation network was available with an almost complete azimuthal coverage. Source inversion studies have taken advantage of these observations which add a lot of constrain on the solutions, especially in the shallow part of the fault models where other standard data sets tend to lack resolution: while on-land data are quite insensitive to slip on the often-distant shallow part of a subduction fault interface, tsunami observations are directly sensitive to the shallowest slip. And it is in this shallow portion that steep bathymetry combined with horizontal motion, the so-called bathymetry effect, can contribute to the tsunami excitation, in addition to the direct vertical sea-bottom deformation. In this study, we carefully investigate the different steps involved in the calculation of this bathymetry effect, from the initial sea-floor deformation to the prediction of the tsunami records, and evaluate its contribution across the main subduction zones of the world. We find that the bathymetry effect locally exceeds 10 % of the tsunami excitation in all subduction zones and 25 % in those known to produce the largest tsunami, either from mega- or tsunami- earthquakes. We then show how the bathymetry effect can modify the tsunami wave predictions, with time shifts of the wavefront and amplitudes sometimes varying by a factor of two. If the bathymetry effect can have a strong impact on the simulated tsunami, it will also affect the solution of the finite-fault slip inversion. We illustrate this later aspect in the case of the Tohoku-Oki earthquake. We find that not accounting for the bathymetry effect will not necessarily cause strong variations in the spatial extent of the inferred coseismic rupture but can severely distort the solution. We also find that the bathymetry effect improves the consistency of the slip model inverted from tsunami data with seafloor geodesy observations, implying that taking the bathymetry effect into account reduces the epistemic uncertainties on tsunami modeling. Implementing this easily quantifiable effect in the tsunami early warning system could thus lead to improved estimates of the tsunami impact across ocean basins.

1. Introduction

Tsunami excitation caused by an earthquake is traditionally modeled assuming incompressible water and that the deformation is rapid compared to the time it takes for the gravity wave to propagate across the rupture area. These arguments allow copying the vertical displacement of the sea bottom to the surface of the ocean. TANIOKA and SATAKE (1996) showed that horizontal motion of the bottom of the sea, where the gradient of bathymetry is non-negligible could also contribute to vertical water displacement. This so-called bathymetry effect (BE) depends on the dip angle of the slab interface (which conditions the ratio of horizontal to vertical surface displacement), the amount of shallow slip at the surface, and the gradient of the bathymetry. The bathymetry effect is sometimes called horizontal effect, but we prefer the former term as horizontal effect might convey the idea that the effect is related to a transfer of horizontal momentum. Tsunami earthquakes, defined as events generating anomalously large tsunami compared to their seismic radiation (KANAMORI 1972), are for the large majority shallow megathrust earthquakes. The shallow part of megathrusts often combines low dip angle and steep bathymetry (due to the proximity of the trench). Therefore, BE is expected to be particularly strong for these events and could partly explain the anomalously large size of the generated tsunamis. More generally, large earthquakes preferentially occur on low dipping interfaces (SCHELLART and RAWLINSON 2013) and one may reasonably suspect that mega-earthquakes have enough energy to rupture up to the free surface. Thus, mega-earthquakes are expected to be particularly influenced

---

[1] CNRS, IRD, Observatoire de la Côte d'Azur, Géoazur UMR 7329, Univ. Nice Sophia Antipolis, 250 rue Albert Einstein, Sophia Antipolis, 06560 Valbonne, France. E-mail: bletery@geoazur.unice.fr; sladen@geoazur.unice.fr; delouis@geoazur.unice.fr; lionelmatteo93@gmail.com

by the BE. Among them, the 2011 M_w9.0 Tohoku-Oki earthquake (TO) combines a shallow dipping slab interface with very large shallow slip (e.g. FUJIWARA et al. 2011) and a wide bathymetry gradient towards the trench. Moreover, TO is the first megaevent for which tsunami instrumentation [and even other types of observations (e.g. FRITZ et al. 2012)] recorded the wave close to the source and with an almost complete azimuthal coverage. This makes TO an ideal case for studying the BE.

Numerous studies have taken advantage of tsunami observations to characterize the TO co-seismic rupture: either by performing tsunami-only inversions (SAITO 2011; SATAKE et al. 2013; MELGAR and BOCK 2013; MAEDA et al. 2011; KOKETSU et al. 2011), joint inversions including tsunami (SIMONS et al. 2011; YOKOTA et al. 2011; GUSMAN et al. 2012; ROMANO et al. 2012, 2014; HOOPER et al. 2013; MINSON et al. 2014; BLETERY et al. 2014), or tsunami forward predictions from various source models (LAY et al. 2011; YAMAZAKI et al. 2011a, 2013; WEI et al. 2012). In these studies, BE was often neglected, but HOOPER et al. (2013) and SATAKE et al. (2013), followed by ROMANO et al. (2014), then demonstrated its major impact on the finite-fault source imaging of the TO event. TANIOKA and SATAKE (1996) originally predicted that this effect could amplify the tsunami amplitude by 30 % in the case of the Java subduction. But for the specific case of TO, HOOPER et al. (2013) found that the additional coseismic contribution due to the horizontal translation of bathymmetry could locally exceed 100 % of the contribution of vertical deformation. And in terms of tsunami amplitude, SATAKE et al. (2013) found that it amplified the signals observed at Ocean Bottom Pressure sensors and GPS wave gauge stations by up to 60 %.

In this study, we detail the steps involved in the calculation of the BE, how it affects the tsunami generation, and how it varies across different subduction zones of the world. We then illustrate the impact of BE on the recovered slip distribution when the inversion is controlled by tsunami data.

2. Forward Modeling

Tsunami response to finite-fault dislocation is typically modeled in three steps: (1) the computation of the displacement field generated by a given dislocation at the bottom of the sea, (2) the tsunami excitation transferred through the water column, and (3) the tsunami propagation. We detail below the calculation of tsunami synthetic data, and the contribution of the BE, at each of those three steps.

2.1. Displacement Field Generated by Shallow Finite Fault Dislocation

In source inversion problems, the surface displacement field generated by finite-fault dislocation is commonly modeled using OKADA (1985)'s implementation of the equations given by MANSINHA and SMYLIE (1971), which are derived from the formulation of STEKETEE (1958). This theoretical formulation is linear and computationally cheap, two great advantages to perform finite-fault source inversions. To highlight the sensitivity of finite-fault source inversion to fault geometry (e.g. MORENO et al. 2009), we use OKADA (1985)'s formulation to compute the displacement field generated by a shallow dislocation along a 15-km fault plane in four different cases: a fault plane reaching the free surface with a dip angle of 5° (case 1) or with a dip angle of 10° (case 2). Cases 3 and 4 are similar to 1 and 2 except that the fault planes are 4 km deeper, respectively, and thus do not reach the surface. In each case, we imposed 1 m of pure dip-slip (rake angle $\lambda = 90°$) homogeneously distributed along the fault. We compute the response at the surface on a regular grid with 1 point every km².

We represent the induced surface deformation across a profile perpendicular to the trench in the four different cases (Fig. 1). Vertical displacement is, to the first order, the main source of tsunami excitation. We observe that for a fault plane reaching the trench (Fig. 1b), the uplift amplitude doubles when considering a dip angle of 10° instead of 5° [this is because $\sin(10°) \approx 2\sin(5°)$]. The vertical displacement field is also extremely sensitive to depth variations: peak amplitudes are two (for a dip of 10°) or three (for a dip of 5°) times larger when the fault plane is placed at 4 km depth than at the surface. Indeed, in the case of an embedded fault (top edge at 4 km) the horizontal motion is resisted elastically by the surrounding medium and is translated into vertical

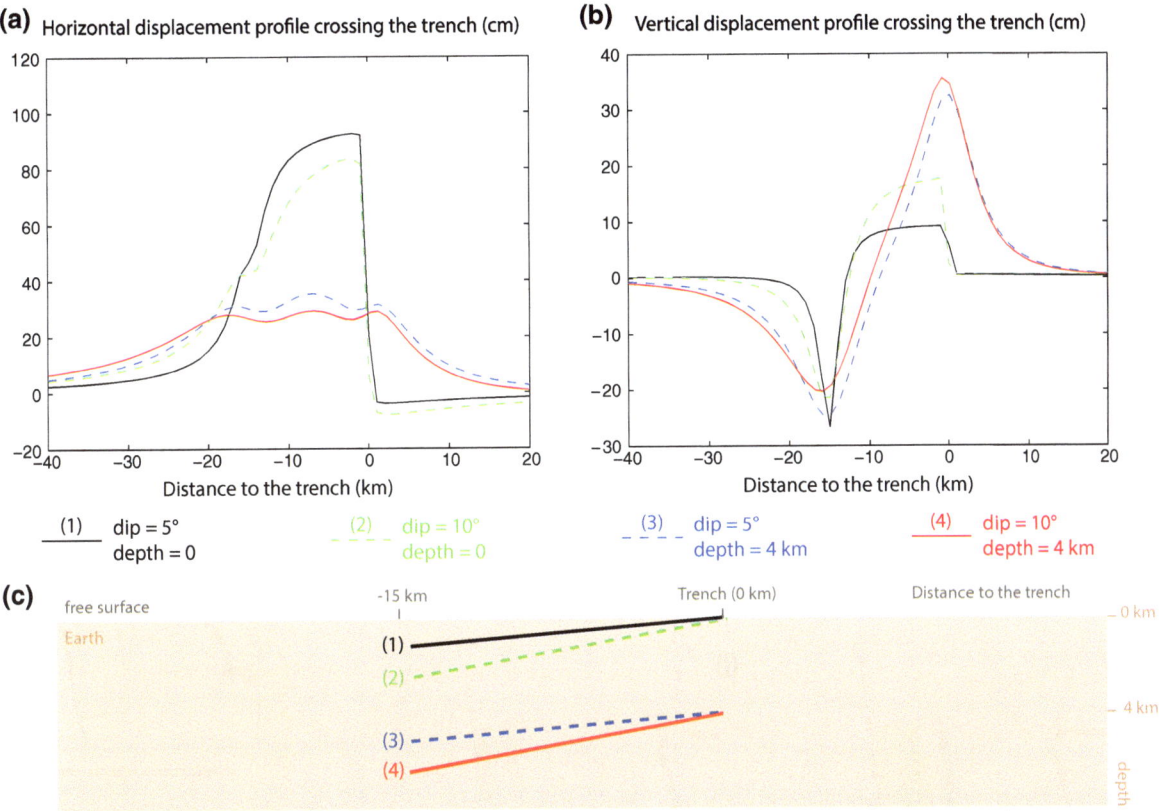

Figure 1
Effect of dip angle and depth on surface displacement generated by shallow fault dislocation across a profile perpendicular to the trench. 1 m of homogeneous pure dip slip is applied on a 15 km long fault in 4 cases (subfigure c): (*1*) fault dipping is 5° and fault reaches the free surface (*black lines*), (*2*) same as (*1*) but dipping is 10° (*green dashed lines*), (*3*) same as (*1*) but fault is 4 km deeper (*blue dashed lines*), (*4*) dipping angle is 10° and depth is 4 km (*red lines*). Subfigure **a** is horizontal deformation (taken positive in the eastern direction), **b** is vertical

motion. On the other hand, when the top of the fault is at the surface, the horizontal motion is not limited by the free surface so that no, or much less, horizontal motion translates into vertical motion. The effect of the dip on the horizontal component (Fig. 1a) is much lower (because $\cos(10°) \approx \cos(5°)$), but horizontal displacement field appears to be extremely depth dependent: peak amplitudes are three to four times larger for ruptures at the surface.

These simple tests highlight that the vertical deformation field generated by shallow dislocation is very sensitive to both depth and dip angle while the horizontal is extremely sensitive to depth. They also indicate that for subduction zones presenting a shallow dipping slab interface, such as the Tohoku slab region, the relative contribution of horizontal deformation to the total ocean elevation might become much larger because a wide region will combine low dip angle and shallow depth, resulting in both a lower contribution of the vertical deformation and a larger contribution of the bathymetry effect.

2.2. Calculation of the Bathymetry Effect

The contribution of horizontal motion combined with bathymetry to water elevation induced by an offshore earthquake is given by Eq. (1) (TANIOKA and SATAKE 1996):

$$\text{BE} = -\frac{\partial \beta}{\partial x} u_x - \frac{\partial \beta}{\partial y} u_y \quad (1)$$

where β is the bathymetry, and u_x and u_y are the displacements generated by the earthquake on the east and north directions, respectively.

As discussed in the previous section, large discrepancies exist in the displacement fields

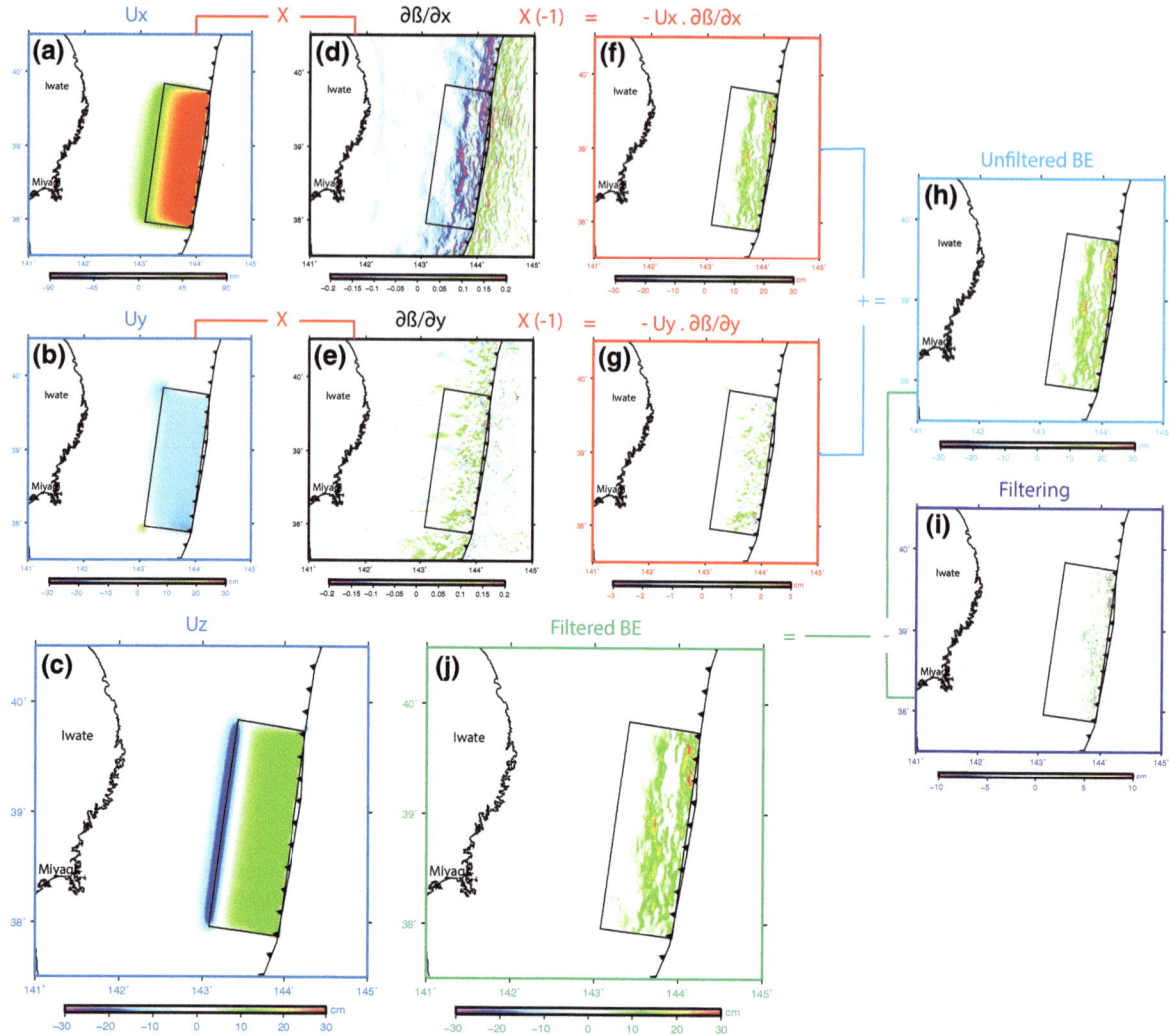

Figure 2
Calculation of the BE induced by a 1-m homogeneous slip along a 210×75 km^2 fault in the shallow megathrust part of the Meiji-Sanriku subduction zone. Fault parameters are: $\varphi = 188°$, $\delta = 6°$, $\lambda = 90°$ (pure dip-slip). Fault location is shown in all subfigures (*long black squares*). **a** Along east sea-floor displacement field generated by the fault dislocation: u_x. **b** Along north sea-floor displacement field: u_y. **c** Vertical sea-floor displacement field: u_z. **d** Bathymetry gradient along east (bathymetry grid resolution is 500 m): $\partial\beta/\partial x$. **e** Bathymetry gradient along north: $\partial\beta/\partial y$. **f** Unfiltered bathymetry effect along east: $-u_x \partial\beta/\partial x$. **g** Unfiltered bathymetry effect along north: $-u_y \partial\beta/\partial y$. **h** Unfiltered total bathymetry effect: $-u_x \partial\beta/\partial x$, $-u_y \partial\beta/\partial y$. **i** Residual from the 1/cosh(kh) spatial filter. **j** Filtered bathymetry effect

generated by different earthquakes such that u_x, u_y, and the vertical displacement u_z will take very different values depending on the fault geometry. We detail here the calculation of the BE for a synthetic rectangular fault source offshore Miyagi and Iwate prefectures (Fig. 2). 1 m of co-seismic slip is imposed homogeneously along a 210×75 km^2 rectangular fault. The fault contours are drawn in black in Fig. 2. The fault parameters φ, δ, λ (strike, dip, rake angles, respectively) are imposed at $\varphi = 188°$, $\delta = 6°$ [to mimic the SLAB1.0 model (HAYES *et al.* 2012) in the region] and $\lambda = 90°$ to model pure dip-slip (as strike-slip motion is generally low in subduction context and does not generate strong tsunami excitation anyway).

Considering the above parameters (low δ and $\varphi \sim 180°$), the displacement field generated by this synthetic source is mainly oriented in the east direction, resulting in amplitudes in u_x several times larger than both in u_y and u_z (Fig. 2a–c, be aware of the different color scale used for u_x). The bathymetry gradient—that we derived from the JTOPO30v2 500 m resolution topography data of the Marine information Research Center of the Japan Hydrographic Association (http://www.mirc.jha.jp/products/JTOPO30v2/)—also presents larger amplitudes in the east direction (Fig. 2d, e, note that the color scale is saturated at −0.2 but the gradient reaches −0.5 locally) because the trench is mainly oriented north–south. As a result, the calculation of the BE is dominated by the first term in Eq. (1) (Fig. 2f), the second one being negligible (Fig. 2e, plotted with a different color scale). The addition of the two terms reveals a short wavelength pattern (Fig. 2h) related to the high resolution of the bathymetry. But the smallest scale features of this field are then filtered with 1/cosh(kh) spatial filter to take into account the attenuation of the high wave number perturbations by the water column (k is the wave number and h the thickness of the column) (KAJIURA 1963) (Fig. 2i). The final BE distribution obtained (Fig. 2j) reveals amplitudes which are in general on the order of vertical deformation (Fig. 2c), but locally even larger. In the particular region of Miyagi-Iwate, the BE is especially large and induces much larger tsunami excitation. This region is an extreme example because it combines shallow slab interface and steep bathymetry but it is interesting to note that the reported inundations at the coast of Miyagi and Iwate were significantly larger than what is predicted by co-seismic slip models of the M8.5 1896 Meiji-Sanriku (KANAMORI 1972) and for the M_w9.0 2011 Tohoku-Oki earthquakes (SATAKE et al. 2013; MELGAR and BOCK 2013). But it is likely that some of these anomalously large inundations may be related to propagation effects due to the very complex bathymetry and topography of the coast in this part of the Japanese Pacific coast (SHIMOZONO et al. 2012, 2014; MELGAR and BOCK 2013).

Because the dip angle of subduction megathrust faults and their bathymetry gradients vary significantly from one subduction to another, we evaluate the contribution of the BE for the main subduction zones of the world (Figure S1). We apply the method described above, except that we rely on a dense distribution of cross-section profiles (following SLAB1.0 models (HAYES et al. 2012)) instead of finite sub-faults, to adjust the sometimes-rough variations in strike angles. BE is calculated for a homogeneous 1 m slip source on the different megathrusts. We then interpolate the bathymetry effect obtained on these cross-section profiles. The contribution of the BE is then evaluated as a percentage of the total tsunami excitation: BE/(u_z + BE). The results (Fig. 3) show that BE is a non-negligible part of tsunami excitation in all the main subduction zones (contribution exceeds 10 % locally in all the zones) and exceeds 25 % of the total tsunami excitation over wide areas of the Honshu, Kamchatka, Alaska, Nankai, Ryukyu, Scotia, Peru, Sumatra and Java subduction zones (Fig. 3). Except for the case of the Scotia arc, these regions are all known to experience mega- and/or tsunami-earthquakes. Thus, BE is likely important for all shallow earthquakes, but it is particularly critical for the simulation of tsunami induced in the aforementioned regions where it exceeds 25 %. We also observe large along-strike variations of the contribution of BE in tsunami excitation (Fig. 3), which in addition to changing the amplitude, might distort the distribution of the slip imaged if the BE is not included in the inversion.

2.3. Importance of BE in the Calculation of Tsunami Green's Functions

We showed that BE could be significant on a megathrust combining a shallow dipping interface and a steep bathymetry. We now evaluate how it affects the prediction of tsunami time series.

We consider a set of five square sub-faults at different depths (Fig. 4). The sub-faults parameters are the same as in the previous section: $\varphi = 188°$, $\delta = 6°$, $\lambda = 90°$ (pure dip-slip), and their size is 15×15 km^2. We compute both the tsunami excitations considering in the first case only the vertical displacement and in the second case the vertical displacement with the BE contribution. We then apply the 1/cosh(kh) filter mentioned above and propagate the obtained tsunami excitation patterns using the simulation code NEOWAVE (YAMAZAKI et al 2009, 2011b). This code is based on the shallow water equations and describes dispersive waves through the non-hydrostatic pressure and vertical velocity (dispersion becomes non-negligible at distant stations such as DARTs). We compare the obtained water

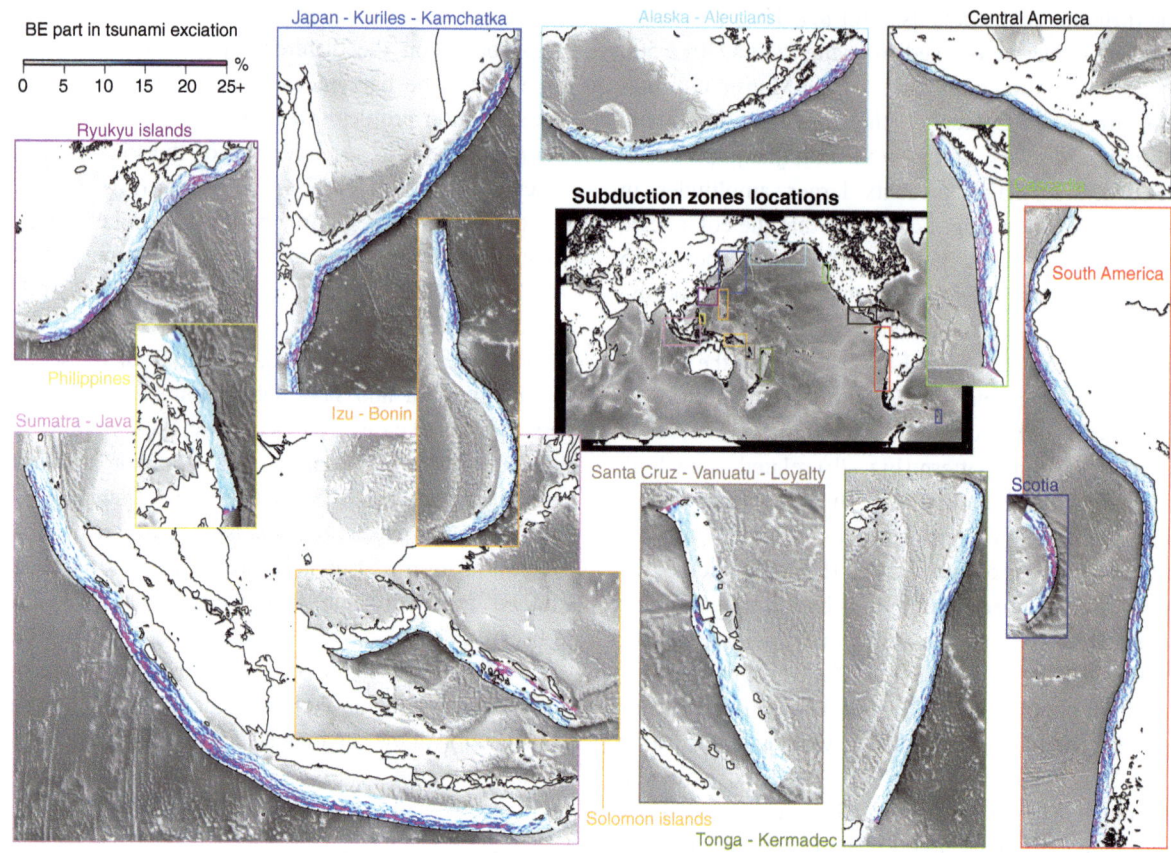

Figure 3
Part of BE in total tsunami excitation (BE/(u_z + BE)) in the main subduction zones

level time series at the pressure gauge station TM1, the tsunami record the closest to the sources, for the five different sub-faults. We see in Fig. 4, that for the shallowest sub-faults, the addition of BE can double the amplitude signal at station TM1. This order of magnitude is larger than what was found by TANIOKA and SATAKE (1996) and SATAKE et al. (2013), but can be explained by (1) the difference in sub-fault sizes and (2) the choice of dip angle of the shallow slab interface made in the different studies: Green's functions (GF) shown in Fig. 4 are calculated for a shallow sub-fault with a dip angle of 6° while the lower dip angle considered by SATAKE et al. (2013) is 8°. As shown in Fig. 1, dip angle variations of this order lead to large changes in the generated horizontal motion and consequently in the BE calculation.

We calculate the spectra of the different GF to investigate the impact of the BE on the frequency content of the time series (Fig. 4). We observe that the spectra are not much affected by BE. This comes from the spatial filter we apply to mimic the water column attenuation (see Sect. 2.2). This filter removes wavelengths shorter than three times the water depth (h) (KAJIURA 1963). For 15 × 15 km² sub-faults, both the vertical deformation field and the BE contain wavelengths shorter than this cut-off (we clearly see in Fig. 2 that the vertical deformation field contains shorter wavelengths than the sub-fault width). Thus, the minimum period we can expect in tsunami GF can be expressed in function of h and v the wave velocity:

$$T_{\min} = \frac{3h}{v} \quad (2)$$

Under the shallow water approximation (which is valid at these depths and distances)

$$v = \sqrt{gh} \quad \text{(with } g \text{ the gravitational acceleration)} \quad (3)$$

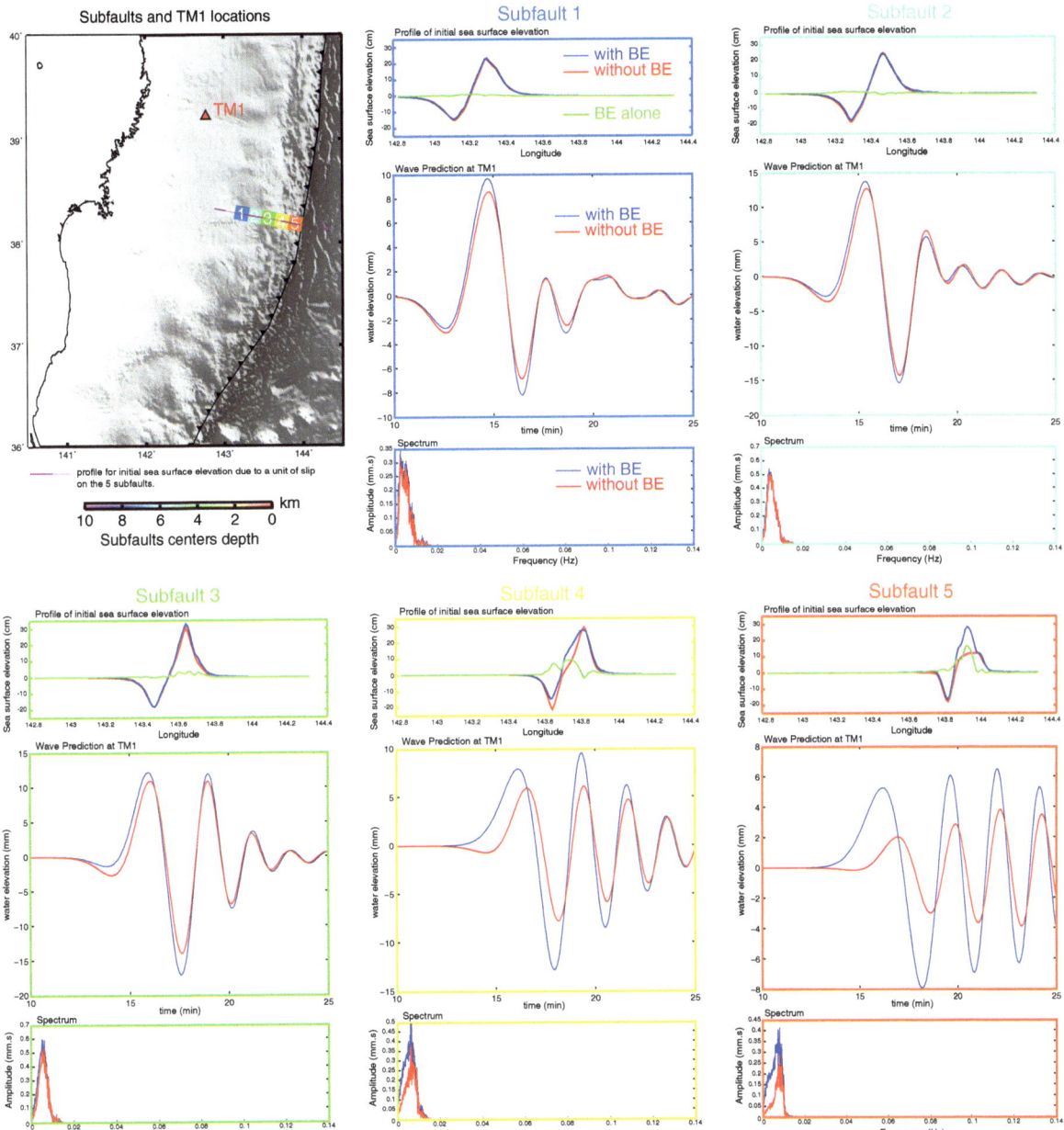

Figure 4
Importance of the BE in the GF for a set of 15 km long square sub-faults at station TM1 (sub-faults and TM1 locations are shown in *top left* subfigure). Colors correspond to sub-fault depths. For each sub-fault, top sub-figures show sea surface deformation profiles across the *dashed purple line* shown in the map with (*blue*), without (*red*) BE and their difference (*green*); center sub-figures show associated GF time series calculated taking BE into account (*blue curves*) and not taking it into account (*red ones*). Bottom subfigures show the respective spectrum of each sub-fault GF

Hence

$$T_{\min} = 3\sqrt{\frac{h}{g}} \quad (4)$$

For near-trench sub-faults, where the BE is the largest, the water depth is over 7 km, meaning that the minimum period we can model in GF is about 80 s and the maximum frequency 0.0125 Hz, which explains the spectra cut-off and waveforms similarity in Fig. 4. Even though the frequency content is very similar, for sub-faults 4 and 5, the first wave arrival is significantly delayed when neglecting BE (the delay for all shallow sub-faults in Figure S2 varies from 20 s to 1 min). This is because horizontal deformation extends farther (west in our case) than the vertical deformation (see Fig. 2a compared to c): for the sub-faults the closest to the trench, a large positive contribution of the BE is added to the negative u_z (in the western subsiding part of the subfaults) (see Fig. 2c and top sub-figures in Fig. 4). This causes the wave front to arrive earlier. As absolute tsunami arrival times are a critical constraint on the inverted slip models, the BE has the potential to distort the solution of finite-fault inversions.

The potentially very strong amplification of the tsunami by the BE has several important implications. First, flat slab regions should not be considered as associated with lower tsunami hazard because the BE is particularly large in these regions and compensates for the lower contribution of vertical sea-floor deformation. Then, neglecting the BE in forward simulations, e.g. in the critical case of tsunami-early warning, can lead to important errors. Moreover, the BE has likely a major impact on source inversions as it affects tsunami GF both in amplitude and phase. We now focus on this last implication, taking as a case study the 2011, M_w9.0 Tohoku-Oki earthquake, which is currently the best-recorded large shallow slip event.

3. Importance of BE on Slip Inversions

HOOPER et al. (2013), SATAKE et al. (2013), and ROMANO et al. (2014) illustrated the importance of BE by directly comparing tsunami data predictions with and without BE. But slip inversion is an ill-conditioned problem meaning that small errors in the data or in the GF can have a large impact on the solution. Here, we propose to test the influence of the BE on the inversion process. To do so, we use the TO earthquake as a test case and perform two separate slip inversions of tsunami data: in the first case using GF accounting for BE and in the second case not accounting for BE. In both cases, we use the fault discretization of BLETERY et al. (2014): a 3D fault geometry mimicking the SLAB1.0 model, which allows each sub-fault to match the free surface at the trench while matching the changes in dip angle of the megathrust interface. Sub-faults are square fault planes of 15×15 km^2 in the top 10 km, and of 30×30 km^2 in the deeper part of the slab interface (Figure S2). We choose to refine the grid in the shallow part because the timing and frequency content of the different sub-faults GF can clearly be resolved in the data (for instance looking at the wave arrival times for neighboring sub-faults in Fig. 4). We calculate the tsunami excitation for unit dislocations of pure dip-slip and pure strike-slip for 187 sub-faults using the formulation of OKADA (1985). We then apply a 1/cosh(kh) space filter (KAJIURA 1963) and calculate the predicted tsunami time series for each sub-fault at each tsunami station using the NEOWAVE simulation code described above.

3.1. Data

We use 15 time series of the tsunami wave height at different points of measurement from 4 DART (Deep-ocean Assessment and Reporting of Tsunamis) buoys (21418, 21401, 21413, 21419), 6 GPS buoys (GPS801, GPS802, GPS803, GPS804, GPS806, GPS807), 2 pressure gauges (TM1, TM2), and 3 cables (KPG1, KPG2, HPG). DART records are provided by the NOAA National Geophysical Data Center (http://ngdc.noaa.gov/hazard/dart/2011 honshu_dart.html) and have a sampling rate of 1 min. GPS buoys are given by the NOWPHAS system (http://nowphas.mlit.go.jp/info_eng.html) and have a sampling rate of 5 s. Pressure gauge records are described by MAEDA et al. (2011). Cable data are downloaded from the JAMSTEC cabled observatories web site (http://www.jamstec.go.jp/scdc/top_e.html), their sampling rate is very high frequency (1 Hz), but we band-pass filter them between 2 and 50 min to eliminate the effect of wind waves and tides. Stations locations are shown in Figs. 5 and 6. Their azimuthal coverage is very good (see rose diagrams in Fig. 5). In particular, DART stations provide the only robust information east of the source. Cable stations bound the propagation to the North and pressure gauges and GPS buoys provide records west and close to the source.

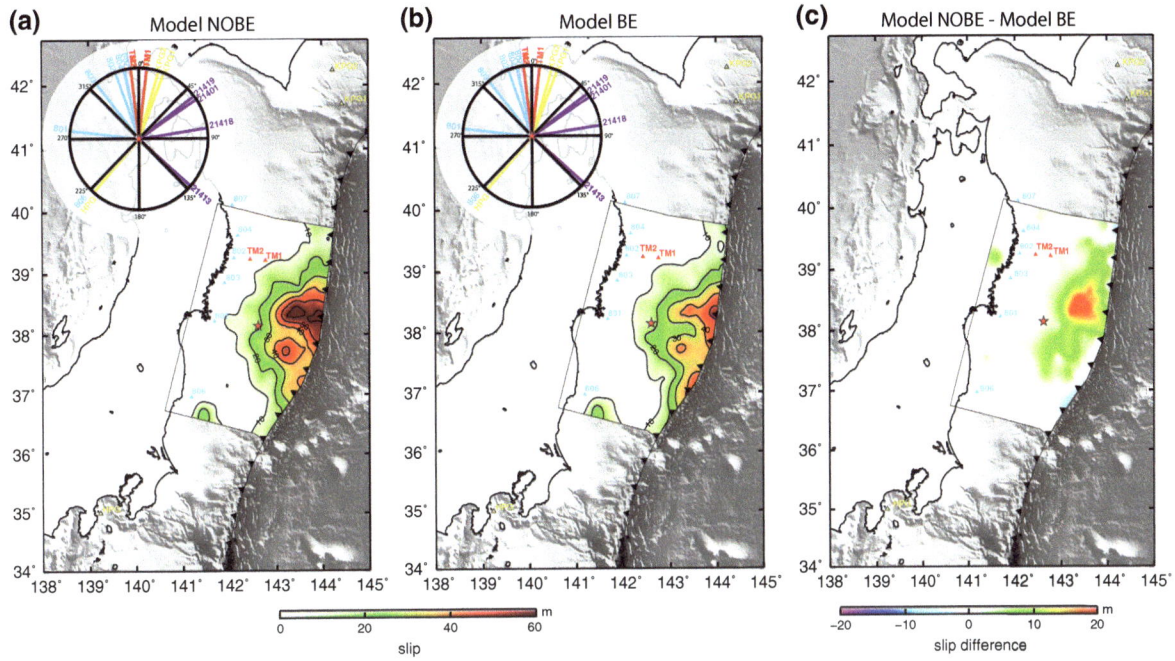

Figure 5

a Slip model obtained by inversion of tsunami data using Green's functions ignoring BE (model NOBE) (after removing the artifact discussed in the main text). Rose diagram at the *top left* of the figure shows station azimuthal coverage. **b** Slip model obtained by inversion of tsunami data using Green's functions accounting for BE (model BE) (after removing the artifact). **c** Difference: model NOBE—model BE

3.2. Inversion Procedure

We invert the aforementioned data set to estimate the slip distribution best explaining these observations. We choose to treat tsunami observations as kinematic data since, as shown by SATAKE et al. (2013), the static approximation can lead to significant bias in the inverted slip distribution when based on relatively near-field tsunami data. Our kinematic modeling follows the approach described by DELOUIS et al. (2002). The model hypocenter is—based on a seismic waveform and GPS inversion—imposed at 38.15 N, 142.61E and at a depth of 16.5 km (the depth imposed by seismic analysis is 24.5 but we translate the whole fault geometry so that it matches the free surface at the trench depth).

For each sub-fault, a local source time function is defined, corresponding to the rate of seismic moment locally released. It is represented by seven mutually overlapping isosceles triangular functions of 12 s duration, allowing the local source time function to last for a maximum of 48 s. For each of the 187 sub-faults, the parameters inverted for are the slip onset time, the rake angle, and the amplitudes of the seven triangular functions. Rupture onset times are bounded according to a minimum and a maximum rupture velocity of 1.1 and 3.1 km/s, respectively. The rake angle can vary between 60° and 120° to smoothly compensate the significant strike variations along the fault.

A non-linear inversion of the aforementioned data is performed using a simulated annealing optimization algorithm. The convergence criterion is based on the simultaneous minimization of the root mean square (RMS) data misfit and of the total seismic moment. Data are weighted to compensate for their relative amplitudes (see Table 1).

3.3. Neglecting the Bathymetry Effect can Induce Large Over-Prediction of the Inverted Slip Distribution

Both inversions, performed using GF accounting for BE (model BE) and using GF not accounting for BE (model NOBE), reveal slip distributions

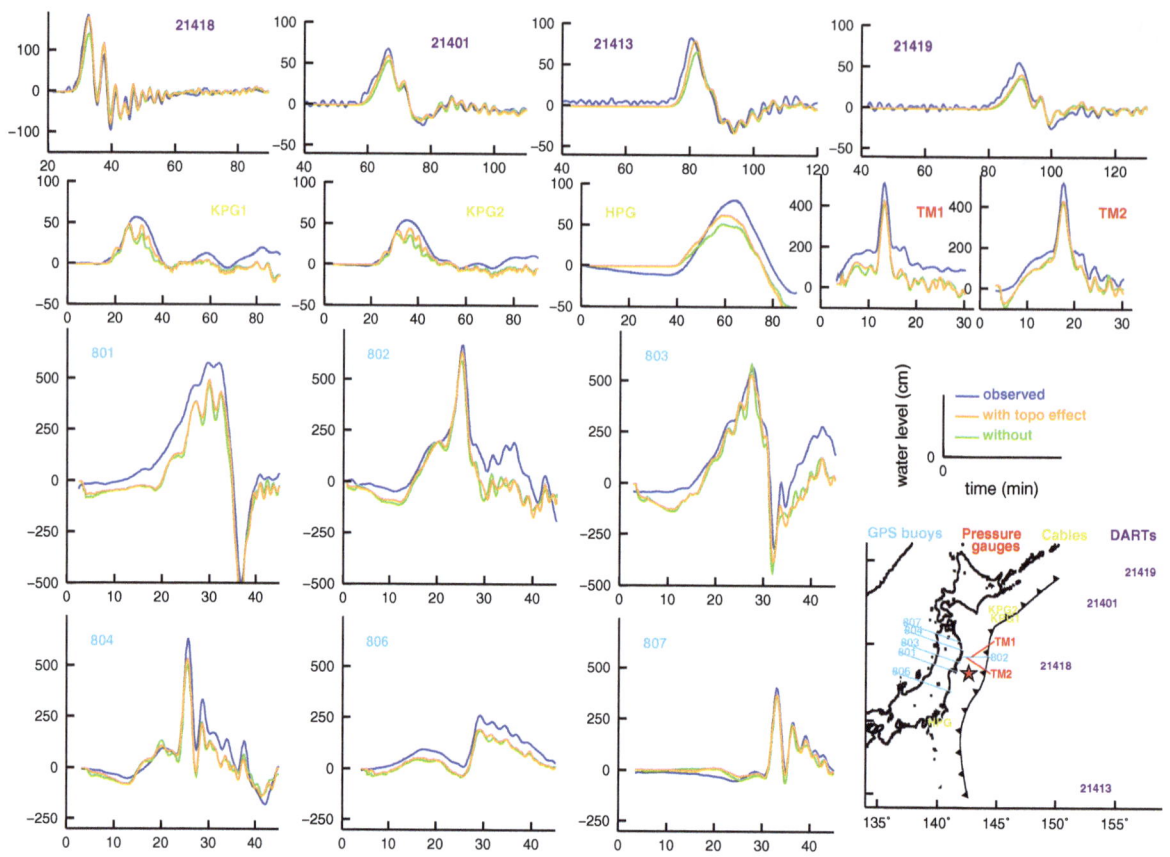

Figure 6
Comparison between observed (*blue*) and predicted tsunami data obtained by inversions accounting for BE (*orange*) and not (*green*). In both cases predicted data are calculated after removing the artifact as discussed in the main text. Stations locations are shown in the *bottom right* figure. *Colors* correspond to data type

dominated by shallow slip with very large amplitudes (60 m) up to the free surface (Figure S3). An isolated patch is present in both slip models (Figure S3), on the southwestern edge of the fault plane. As shown in Figure S4, because of the lack of observations between stations 801 and 806, tsunami response to sub-fault dislocation in this area is eight times smaller in terms of amplitude (GF norms) than in the best-constrained fault area, resulting in a low sensitivity relative to other areas on the interface. Additionally, because this part of the fault is deep, it generates a very low frequency tsunami signal (frequency is not taken into account in the sensitivity calculation of Figure S4), which is much less constraining than the higher-frequency GF generated by shallow sub-fault dislocation. This implies that the inherent resolution of this deep patch is even lower than what the sensitivity map suggests (Figure S4). We check the effect of this patch on the predicted waveforms and found that it only adds a very low frequency component to the nearby gps801 and gps806 stations and marginally improves the fit: deep source dislocations do not generate large tsunami (OKAL and SYNOLAKIS 2004); thus resolution of tsunami data on source models dramatically decreases with depth. For this reason, we consider this patch as an artifact of the inversion and remove it from both slip models for the sake of clarity (Fig. 5a, b). While the contribution of this artifact to the solution is negligible, at least in terms of contribution to the total seismic moment, its presence illustrates the inability of tsunami data to resolve long wavelength deformation patterns due to deep sources close to the coast.

Assuming a shear modulus of 45 GPa, the seismic moment associated with the slip model accounting for BE (model BE) is $M_0 = 5.1 \times 10^{22}$ N m ($M_w = 9.1$),

Table 1

Data weighting imposed in inversions to compensate the relative amplitudes of the observations

Data	Weight
DART21418	4
DART21401	10
DART21413	10
DART21419	10
TM1	1
TM2	1
GPS801	1
GPS802	1
GPS803	1
GPS804	1
GPS806	2
GPS807	1
KPG1	1
KPG2	1
HPG	1

and the one associated with the model ignoring BE (model NOBE) is $M_0 = 6.3 \times 10^{22}$ N m ($M_w = 9.2$) (for both models the artifact was removed from the calculation). The seismic moment associated with model BE and model NOBE are 21 and 50 % larger than the one found by SATAKE *et al.* (2013) ($M_0 = 4.2 \times 10^{22}$ N m) respectively. Model BE (Fig. 5b) is very similar to the slip model obtained by BLETERY *et al.* (2014) performing a joint inversion of a large number of different data (GPS, high-rate GPS, strong-motion, teleseismic and the same tsunami records), which indicates that tsunami data are critical to constrain offshore ruptures with shallow slip.

A detailed discussion on the TO physical rupture process is not in the scope of this study, but we show on an indicative basis the rupture history obtained from the slip inversion including BE (Figure S5). We find a rupture duration of 140 s, with the downdip spurious patch interpreted as an artifact appearing late in the sequence. Another patch along the southern edge of the fault geometry, because it is small and isolated from the rest of the slip distribution, might also be an artifact. But although co-seismic slip in this southern part of the fault is in debate (KATO and IGARASHI 2012; BLETERY *et al.* 2014), tsunami sensitivity is too low in this area (because of the lack of observation south of the source area) (Figure S4) to make any conclusive statement. The delayed large shallow slip found north of the fault in the kinematic tsunami inversion performed by SATAKE *et al.* (2013) is absent of both slip distributions obtained accounting or not for BE (Fig. 5). We can point out here that because the most constraining information brought by tsunami data is the timing of the first phase, the resolution decreases with time as waves arriving later can be explained by slip in many other parts of the megathrust. Thus, the tsunami data available might not be able to bring a definitive argument on the debate of a possible delayed source.

The difference between Fig. 5a and b illustrates the impact of the BE on the slip distribution inferred using the tsunami observations of TO (Fig. 5c). The slip difference is concentrated in the shallowest 10 km of the megathrust where it reaches 20 m locally. Interestingly, the larger discrepancy is not strictly observed at the trench, where BE is the largest, but it is found in an area where there are both large slip amplitudes (Fig. 5a, b) and high sensitivity (Figure S4). Because finite-fault source inversion is an ill-conditioned problem, changes in GF on one sub-fault can also affect slip at other distant sub-faults depending on their respective sensitivity. The correlation between the model's discrepancy and the sensitivity map reflects that in an inverse problem the adjustment is made preferentially on well-resolved parameters: sub-faults with high sensitivity have more impact on the fit of the data. We think this is the main explanation but the time shift of the GF caused by the BE might also distort the solution.

The large discrepancy between model NOBE and model BE is way too large to neglect the BE in the simulation of tsunami waves initiated by shallow ruptures. And as slip inversions in subduction zones are almost always only constrained in their shallowest part by tsunami observations (e.g. YOKOTA *et al.* 2011; WEI *et al.* 2012; HILL *et al.* 2012; BLETERY *et al.* 2014), neglecting BE in joint inversion of offshore earthquakes is likely to affect the inverted slip distributions in a similar manner as in this tsunami-only example.

3.4. Data Fit and Resolution Tests

We find that the two slip models obtained by inversion with/without BE produce very similar fit to the data (Fig. 6) (Fig. 6 shows the tsunami fit after removing the marginal effect of the coastal artifact patch). Indeed, model BE only slightly improves the fit compared to model NOBE. This means that the difference in the GF is mainly accommodated in terms of shallow slip distribution (Fig. 5c). The slight difference in the fit comes from changes in arrival time of some GF (illustrated in Fig. 4), which modify the solution space. The sea surface deformations associated with these predictions are shown in figure S6. It confirms, as suggested by the larger amplitudes in the waveforms predicted by model BE, that the BE more than compensates the smaller moment associated with model BE compared to model NOBE: model NOBE predicts more slip but less sea surface deformation because of the importance of the BE close to the trench.

To support this result we perform a resolution test (Figure S7). The target pattern is made of 60×60 km^2 patches (Figure S7.a) and the synthetic data are computed considering BE. Then, we invert these synthetic data using GF accounting for BE (Figure S7.b), or not accounting for it (Figure S7.c). We find that when we do not introduce prediction errors in the GF (accounting for BE), the pattern is very well recovered (Figure S7.b). When we introduce prediction errors by removing the BE, the slip amplitudes in the shallowest part of the slab interface are 50–100 % larger. This level of amplification confirms the large potential bias introduced in the recovered slip distribution.

The aforementioned similarity of the waveforms predicted by models BE and NOBE comes from the absence of independent observations. Indeed, when inverted jointly with a large number of other observations, tsunami data are better adjusted when accounting for the BE (BLETERY et al. 2014). This suggests that taking the BE into account improves the consistency of tsunami modeling with independent observations.

3.5. Ability of the Tsunami-Based Model to Predict the Geodetic Measurements

In order to evaluate the consistency of the bathymetry effect, we compute the coseismic displacements predicted by our two inverted slip models—model NOBE (Fig. 5a) and model BE (Fig. 5b) (after removing the artifact)—and compare them to GPS and seafloor geodesy measurements. We use the same simple rheology as in our tsunami GF calculations to simplify the comparison of the model predictions. GPS offsets are provided by the ARIA (Advanced Rapid Imaging and Analysis) team at the Jet Propulsion Laboratory/California Institute of Technology and seafloor geodesy measurements by SATO et al. (2011) and KIDO et al. (2011).

The slip model NOBE leads to important over prediction of the seafloor geodesy records while the slip model BE succeeds in explaining the amplitudes of these records (Fig. 7). This improvement in the fit of the geodetic records—not included in the inversion—is a strong evidence of the improvement brought by the BE in the computation of the tsunami GF and in the resolution of the slip model. And the better fit of the geodetic data comes in addition to the improvement of the fit to the tsunami waveforms as discussed above.

For both models, however, the orientation of the seafloor geodetic displacement vectors is not well recovered. This is probably due to a poor estimation of the slip rake angles on the different sub-faults (Figure S3) as strike-slip motion does not produce as much tsunami excitation as dip-slip motion, and is thus not well resolved. Another explanation for the misfit of the seafloor geodetic measurements is the contamination of the data by pre- and post-seismic signals: the seafloor geodetic measurements include a strong foreshocks sequence—including a $M_w7.4$ event—localized very close to the stations (NETTLES et al. 2011)—and between 16 and 31 days of post-seismic signal (SATO et al. 2011; KIDO et al. 2011). The predicted orientation of the horizontal motion is much better for on-land GPS stations for both slip models probably due to their greater distance to the source which forces the GPS vector azimuths to be compatible with the average orientation, and not the details, of the slip model. Model BE slightly underpredicts the on-land observations compared to model NOBE. This discrepancy is probably due to the inability of tsunami records to precisely resolve deep slip, with the model NOBE artificially compensating by larger slip amplitudes near the surface because it doesn't include the BE.

This interpretation is also coherent with the fact that the predicted vertical offsets are in relatively good agreement with seafloor geodetic measurements for model BE, but significantly and systematically larger for slip model NOBE. On-land GPS vertical prediction is poor for both models as expected from the low sensitivity of tsunami data to deep slip: vertical displacement is directly sensitive to slip beneath the stations while horizontal deformation tends to affect a much wider area (see Fig. 4a, c).

If we try to quantify the misfit by calculating the normalized RMS (Root Mean Square) of the error on the prediction of geodetic measurements Eq. (5),

$$RMS = \sqrt{\sum \frac{(observed - predicted)^2}{observed^2}}, \quad (5)$$

we obtain a much better fit (RMS = 0.23) for model BE than for model NOBE (RMS = 0.38). This pleads again for a significant improvement brought by the bathymetry effect in the estimation of the solid Earth-ocean coupling and the inference of the coseismic slip model using tsunami data.

4. Conclusion

The development of tsunami instrumentation has brought critical constraints on offshore co-seismic deformation and dramatically improved the resolution of subduction earthquake slip models close to the trench. However, this gain in resolution comes with the need to more accurately model the associated GF, in particular in the shallow portion of the

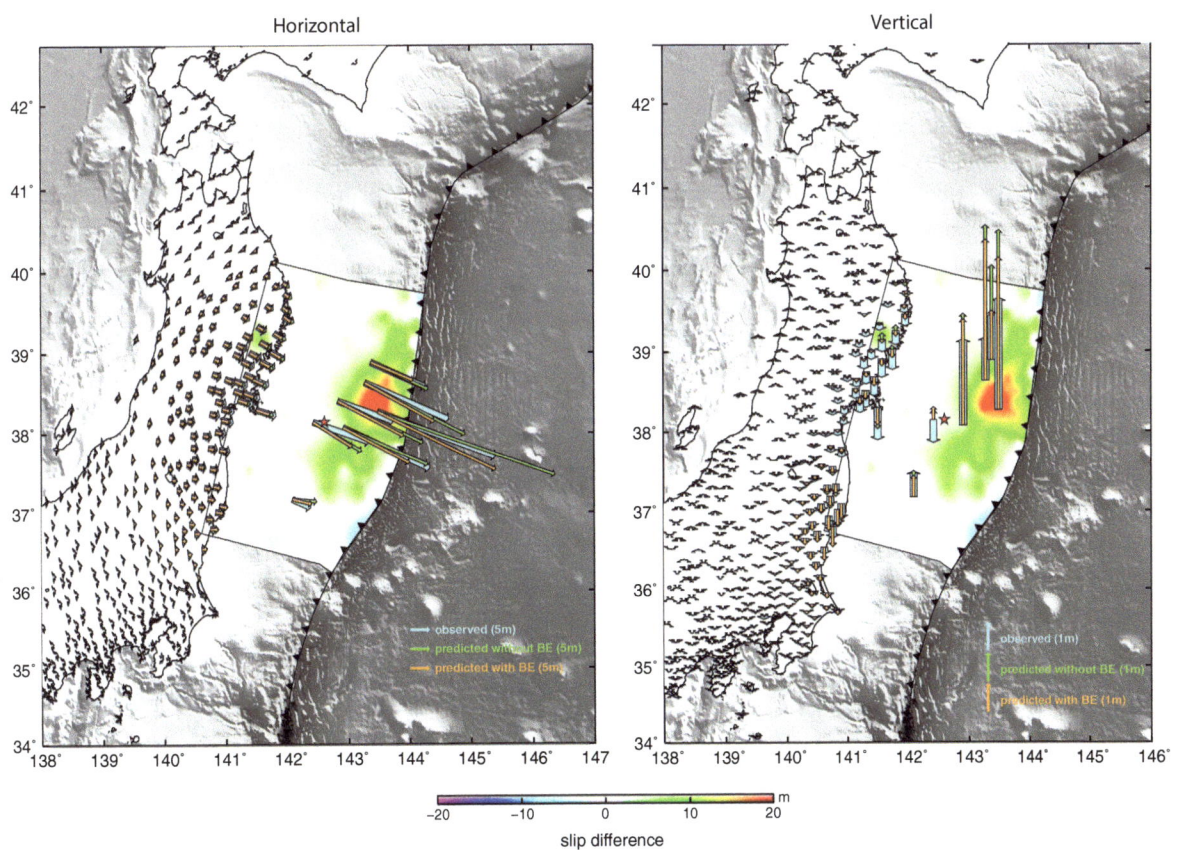

Figure 7
Comparison between geodesy measurements (*blue*) and synthetic data predicted by model NOBE (*green*) and model BE (*orange*). *Left* is horizontal, *right* is vertical. The colored map shows the difference between the two slip models. We notice a better prediction of seafloor geodesy data for model BE

megathrusts, where the tsunami sensitivity is the highest. TANIOKA and SATAKE (1996) showed the contribution of horizontal motion to the tsunami generation could be significant in this particular shallow portion. A series of studies suggested that the effect was particularly large for the $M_w 9.0$ Tohoku-Oki earthquake (HOOPER et al. 2013; SATAKE et al. 2013; ROMANO et al. 2014). Yet, the computation of BE has been overlooked in many other studies. Here, we detailed the steps involved in the calculation of BE and show that this effect is non-negligible—locally exceeds 10 % of the total tsunami excitation—in all the main subduction zones and significant—locally exceeds 25 % of total tsunami excitation—in many of them. Areas exceeding 25 % are particularly wide in the Sumatra, Java, Honshu, Kamchatka and Alaska subductions, all known to generate giant and/or tsunami earthquakes. The addition of an effect of this order is likely to induce strong change in the amplitude of the predicted tsunami waveforms, further implying that the BE should be included in forward simulations, including in the case of early-warning, and in the inversions relying on tsunami data. This later statement is valid for both tsunami-only and joint inversions (i.e. based on several data sets) as there is still no network capable of constraining the shallow slip equivalent to the existing tsunami open-ocean pressure sensor network.

To support our conclusion, we computed the perturbation due to the BE in tsunami prediction for different synthetic sources offshore Honshu and showed that for the shallowest sources: (1) BE is on the order of the contribution of the vertical deformation and (2) BE can even shift the timing of the wavefront. The time shifts tend to be relatively small in the case of forward simulations, but large enough to induce differences in finite-fault inversions. We took advantage of the unprecedented quality and coverage of tsunami records for the $M_w 9.0$ Tohoku-Oki earthquake to show that, in the context of this event, neglecting BE can introduce large slip over-prediction (>30 % locally) in source inversion over a wide zone ($\sim 80 \times 80$ km^2 in this case) (Fig. 5). Additionally, we find that accounting for BE significantly improves the consistency between geodetic and tsunami measurements, which is a strong evidence that the BE improves tsunami GF accuracy.

Acknowledgments

We thank the National Oceanic and Atmospheric Administration (NOAA), National Ocean Wave Information network for Ports and HArbourS (NOWPHAS), Agency for Marine-Earth Science and TEChnology (JAMSTEC) for their valuable data. This work was granted access to the HPC and visualization resources of "Centre de Calcul Interactif" hosted by "Université Nice Sophia Antipolis" and partly supported by the French National Research Agency (ANR) TO-EOS project ANR-11-JAPN-008, the French ministry of research and higher education, the University Nice Sophia Antipolis and the Centre National de la Recherche Scientifique (CNRS). We are grateful to Dr Y. Yamazaki for providing his tsunami simulation code NEOWAVE and for his valuable comments.

REFERENCES

BLETERY, Q., A. SLADEN, B. DELOUIS, M. VALLÉE, J.-M. NOCQUET, L. ROLLAND, J. JIANG (2014), *A detailed source model for the Mw9.0 Tohoku-Oki earthquake reconciling geodesy, seismology and tsunami records*. Journal of Geophysical Research: Solid Earth, doi: 10.1002/2014JB011261.

DELOUIS, B., D. GIARDINI, P. Lundgren, and J. SALICHON (2002), *Joint inversion of InSAR, GPS, teleseismic, and strong-motion data for the spatial and temporal distribution of earthquake slip: Application to the 1999 Izmit mainshock*, Bulletin of the Seismological Society of America, 92 (1), 278–299, doi:10.1785/0120000806.

FRITZ, H.M., PHILLIPS, D.A., OKAYASU, A., SHIMOZONO, T., LIU, H., MOHAMMED, F., SKANAVIS, V., SYNOLAKIS, C.E., TAKAHASHI, T. (2012), *The 2011 Japan tsunami current velocity measurements from survivor videos at Kesennuma Bay using LiDAR*, Geophys. Res. Lett., 39, L00G23, doi:10.1029/2011GL050686.

FUJIWARA, T., S. KODAIRA, T. No, Y. KAIHO, N. TAKAHASHI, Y. KANEDA (2011), *The 2011 Tohoku-Oki Earthquake: Displacement Reaching the Trench Axis*, Science 334, 6060, 10.1126/science.1211554.

GUSMAN, A. R., Y. TANIOKA, S. SAKAI, H. TSUSHIMA (2012), *Source model of the great 2011 Tohoku earthquake estimated from tsunami waveforms and crustal deformation data*, Earth and Planetary Science Letters 341–344, 234–242, 10.1016/j.epsl.2012.06.006.

HAYES, G. P., D. J. WALD, and R. L. JOHNSON (2012), *Slab1. 0: A three-dimensional model of global subduction zone geometries*, Journal of Geophysical Research: Solid Earth *(19782012)*, 117 (B1).

HILL, E.M., J.C. BORRERO, Z. HUANG, Q. QIU, P. BANERJEE, D.H. NATAWIDJAJA, P. ELOSEGUI, H.M. FRITZ, B.W. SUWARGADI, I.R. PRANANTYO, L. LIN, K.A. MACPHERSON, V. SKANAVIS, C.E.

SYNOLAKIS, and K. SIEH (2012), *The 2010 Mw 7.8 Mentawai earthquake: Very shallow source of a rare tsunami earthquake determined from tsunami field survey and near-field GPS*, Journal of Geophysical. Reseearch: Solid Earth, *117, B06402*, doi:10.1029/2012JB009159.

HOOPER, A., et al. (2013), *Importance of horizontal seafloor motion on tsunami height for the 2011 Mw = 9.0 tohoku-oki earthquake*, Earth and Planetary Science Letters, *361*, 469–479, doi:10.1016/j.epsl.2012.11.013.

KAJIURA, K. (1963), *The leading wave of a tsunami*, Bulletin of the Earthquake Research Institute, *41* (33), 535–571.

KANAMORI H. (1972), *Mechanism of tsunami earthquakes*, Phys. Earth Planet. Interiors, *6*, 346–359.

KATO, A., and T. IGARASHI (2012), *Regional extent of the large coseismic slip zone of the 2011 Mw 9.0 Tohoku-Oki earthquake delineated by on-fault aftershocks*, Geophysical Research Letters, *39* (15), n/an/a, doi:10.1029/2012GL052220.

KOKETSU, K., Y. YOKOTA, N. NISHIMURA, Y. YAGI, S. I. MIYAZAKI, K. SATAKE, and T. OKADA (2011), *A unified source model for the 2011 Tohoku earthquake*, Earth and Planetary Science Letters, *310* (34), 480–487, doi:10.1016/j.epsl.2011.09.009.

KIDO, M., Y. OSADA, H. FUJIMOTO, R. HINO, Y. ITO (2011), *Trench-normal variation in observed seafloor displacements associated with the 2011 Tohoku-Oki earthquake*, Geophysical Research Letters 38, 24, doi:10.1029/2011GL050057.

LAY, T., Y. YAMAZAKI, C. J. AMMON, K. F. CHEUNG, and H. KANAMORI (2011), *The 2011 Mw 9.0 off the pacific coast of Tohoku earthquake: Comparison of deep-water tsunami signals with finite-fault rupture model predictions*, Earth, planets and space, *63* (7), 797801.

MAEDA, T., T. FURUMURA, S. SAKAI, and M. SHINOHARA (2011), *Significant tsunami observed at ocean-bottom pressure gauges during the 2011 off the pacific coast of Tohoku earthquake*, Earth, Planets and Space, *63* (7), 803808.

MANSINHA L. and D. E. SMYLIE (1971), *The displacement fields of inclined faults*, Bulletin of the Seismological Society of America, *61* (5), 1433–1440.

MELGAR, D., and Y. BOCK (2013), *Near-field tsunami models with rapid earthquake source inversions from land- and ocean-based observations: The potential for forecast and warning*, Journal of Geophysical Research: Solid Earth, *118* (11), 2013JB010,506, doi:10.1002/2013JB010506.

MINSON, S. E., M. SIMONS, J. L. BECK, F. ORTEGA, J. JIANG, S. E. OWEN, A. W. MOORE, A. INBAL, and A. SLADEN (2014), *Bayesian inversion for finite fault earthquake source models II the 2011 great Tohoku-Oki, japan earthquake*, Geophys. J. Int., 198(2), 922–944, doi: 10.1093/gji/ggu170.

MORENO M. S., J. BOLTE, J. KLOTZ and D. MELNICK (2009), *Impact of megathrust geometry on inversion of coseismic slip from geodetic data: Application to the 1960 Chile earthquake*, GRL, doi:10.1029/2009GL039276).

NETTLES, M., G. EKSTROM, and H. C. KOSS (2011), *Centroid-moment-tensor analysis of the 2011 off the Pacific coast of Tohoku earthquake and its larger foreshocks and aftershocks*, Earth Planets Space, *63*(7), 519–523.

OKADA, Y. (1985), *Surface deformation due to shear and tensile faults in a half-space*, Bulletin of the Seismological Society of America, *75* (4), 1135–1154.

OKAL, E. A., and C. E. SYNOLAKIS (2004), *Source discriminants for near-field tsunamis*, Geophysical Journal International, *158* (3), 899{912, doi:10.1111/j.1365-246X.2004.02347.x.

ROMANO, F., A. PIATANESI, S. LORITO, N. D'AGOSTINO, K. HIRATA, S. ATZORI, Y. YAMAZAKI, and M. COCCO (2012), *Clues from joint inversion of tsunami and geodetic data of the 2011 Tohoku-Oki earthquake*, Scientific Reports, *2*, doi: 10.1038/srep00385.

ROMANO, F., E., TRASATTI, S. LORITO, C. PIROMALLO, A. PIATANESI, Y. ITO, D. ZHAO, K. HIRATA, P. LANUSARA, M. COCCO (2014), *Structural control on the Tohoku earthquake rupture process investigated by 3D FEM, tsunami and geodetic data*, Scientific Reports, *4*, 10.1038/srep05631.

SAITO, T., Y. ITO, D. INAZU, and R. HINO (2011), *Tsunami source of the 2011 Tohoku-Oki earthquake, Japan: Inversion analysis based on dispersive tsunami simulations*, Geophysical Research Letters, *38* (7), n/an/a, doi:10.1029/2011GL049089.

SATAKE, K., Y. FUJII, T. HARADA, and Y. NAMEGAYA (2013), *Time and space distribution of coseismic slip of the 2011 Tohoku earthquake as inferred from tsunami waveform data*, Bulletin of the Seismological Society of America, *103* (2B), 1473-1492, doi:10.1785/0120120122.

SCHELLART, W. P. and N. RAWLINSON (2013), *Global correlations between maximum magnitudes of subduction zone interface thrust earthquakes and physical parameters of subduction zones*, Physics of the Earth and Planetary Interiors, *225*, 41–67, doi:10.1016/j.pepi.2013.10.001.

SATO, M., ISHIKAWA, T., UJIHARA, N., YOSHIDA, S., FUJITA, M., MOCHIZUKI, M., and A. ASADA (2011), *Displacement above the hypocenter of the 2011 Tohoku-Oki earthquake*. Science, *332*(6036), 1395

SHIMOZONO, T., SATO, S., OKAYASU, A., TAJIMA, Y., FRITZ, H.M., LIU, H., TAKAGAWA, T. (2012). *Propagation and Inundation Characteristics of the 2011 Tohoku Tsunami on the Central Sanriku Coast*, Coastal Eng. J., *54*(1):1250004, doi:10.1142/S057856 3412500040.

SHIMOZONO, T., CUI, H., PIETRZAK, J.D., FRITZ, H.M., OKAYASU, A., HOOPER, A.J. (2014). *ShortWave Amplification and Extreme Runup by the 2011 Tohoku Tsunami*. Pure Appl. Geophys., *171*(12):3217–3228, doi:10.1007/s00024-014-0803-1.

SIMONS, M., S. E. MINSON, A. SLADEN, F. ORTEGA, J. JIANG, S. E. OWEN, L. MENG, J.-P. AMPUERO, S. WEI, R. CHU, D. V. HELMBERGER, H. KANAMORI, E. HETLAND, A. W. MOORE and F. H. WEBB (2011), *The 2011 magnitude 9.0 Tohoku-Oki earthquake: Mosaicking the megathrust from seconds to centuries*, Science, *332* (6036), 1421–1425, doi: 10.1126/science.1206731.

STEKETEE, J. A. (1958). *On Volterra's dislocations in a semi-infinite elastic medium*. Canadian Journal of Physics, 36(2), 192–205.

TANIOKA, Y., and K. SATAKE (1996), *Tsunami generation by horizontal displacement of ocean bottom*, Geophysical Research Letters, *23* (8), 861864.

WEI, S., R. GRAVES, D. HELMBERGER, J.-P. AVOUAC, and J. JIANG (2012), *Sources of shaking and flooding during the Tohoku-Oki earthquake: A mixture of rupture styles*, Earth and Planetary Science Letters, *333*, 91100.

YAMAZAKI, Y., Z. KOWALIK, and K. FAI CHEUNG (2009), *Depth-integrated, non-hydrostatic model for wave breaking and run-up*, International journal for numerical methods in fluids, *61* (5), 473497.

YAMAZAKI, Y., T. LAY, K. F. CHEUNG, H. YUE, and H. KANAMORI (2011a), *Modeling near-field tsunami observations to improve finite-fault slip models for the 11 March 2011 Tohoku earthquake*, Geophysical Research Letters, *38* (7), n/an/a, doi:10.1029/2011GL049130.

YAMAZAKI, Y., K. F. CHEUNG, and Z. KOWALIK (2011b), *Depth-integrated, non-hydrostatic model with grid nesting for tsunami generation, propagation, and run-up*, International Journal for Numerical Methods in Fluids, *67*, 20812107.

YAMAZAKI, Y., K. F. CHEUNG, and T. LAY (2013), *Modeling of the 2011 Tohoku near-field tsunami from finite-fault inversion of seismic waves*, Bulletin of the Seismological Society of America, *103* (2B), 14441455.

YOKOTA, Y., K. KOKETSU, Y. FUJII, K. SATAKE, S. SAKAI, M. SHINOHARA, and T. KANAZAWA (2011), *Joint inversion of strong motion, teleseismic, geodetic, and tsunami datasetsfor the rupture process of the 2011 Tohoku earthquake*, Geophysical Research Letters, *38* (7), n/an/a, doi:10.1029/2011GL050098.

(Received May 3, 2014, revised May 11, 2015, accepted May 27, 2015, Published online June 12, 2015)

If you have any concerns about our products,
you can contact us on
ProductSafety@springernature.com

In case Publisher is established outside the EU,
the EU authorized representative is:
**Springer Nature Customer Service Center GmbH
Europaplatz 3, 69115 Heidelberg, Germany**

Printed by Libri Plureos GmbH
in Hamburg, Germany